Biochar Ecotechnology for Sustainable Agriculture and Environment

Biochar Ecotechnology for Sustainable Agriculture and Environment

Edited by

Abhay Kumar

Department for Innovation in Biological, Agro-food and Forest Systems, University of Tuscia, Viterbo, Italy

Majeti Narasimha Vara Prasad

Retired Professor, School of Life Sciences, University of Hyderabad, Hyderabad, Telangana, India

Pallavi Kumari

Centre of Polymer and Carbon Materials, Polish Academy of Sciences, Zabrze, Poland

Manoj Kumar Solanki

Department of Life Sciences and Biological Sciences, IES University, Bhopal, Madhya Pradesh, India

ELSEVIER

Elsevier
Radarweg 29, PO Box 211, 1000 AE Amsterdam, Netherlands
125 London Wall, London EC2Y 5AS, United Kingdom
50 Hampshire Street, 5th Floor, Cambridge, MA 02139, United States

ISBN: 978-0-443-29855-4

For Information on all Elsevier publications
visit our website at https://www.elsevier.com/books-and-journals

Publisher: Candice Janco
Acquisitions Editor: Jessica Mack
Editorial Project Manager: Teddy Lewis
Production Project Manager: Kumar Anbazhagan
Cover Designer: Miles Hitchen

Typeset by MPS Limited, Chennai, India

Contents

List of contributors

Milkias Ahmed
Department of Horticulture and Plant Science, College of Agriculture and Vetrnary Medicine, Jimma University, Jimma, Ethiopia

Beauty Akter
Department of Environmental Sciences, Jahangirnagar University, Savar, Dhaka, Bangladesh

Gisele Amaral-Labat
Coordination of Applied Research and Technological Development, National Institute for Space Research (INPE), São José dos Campos- São Paulo, Brazil

Joao Antonangelo
Department of Crop and Soil Sciences, Washington State University, Pullman, WA, United States

Nataliya Apanovich
School of Landscape Architecture and Planning, University of Arizona, Tucson, AZ, United States

Mohd Asmadi
Chemical Reaction Engineering Group (CREG), Faculty of Chemical and Energy Engineering, Universiti Teknologi Malaysia, Johor Bahru, Malaysia

Nur Aina Najwa Mohd Nor Azman
Chemical Reaction Engineering Group (CREG), Faculty of Chemical and Energy Engineering, Universiti Teknologi Malaysia, Johor Bahru, Malaysia

Benginur Baştabak
Ege University, Solar Energy Institute, Izmir, Turkey

Jhony Benavides
School of Natural Resources and Environmental Engineering—EIDENAR, College of Engineering, Universidad del Valle, Cali, Colombia

Kushalindu Biswas
Academy of Scientific and Innovative Research (AcSIR), Ghaziabad, Uttar Pradesh, India; CSIR—Institute of Minerals and Materials Technology, Bhubaneswar, Odisha, India

Jutishna Bora
Amity Institute of Biotechnology, Amity University, Ranchi, Jharkhand, India

Lucélia Borgo
Department of Soil Science, Federal University of Lavras, Lavras, Minas Gerais, Brazil

Wassila Bouta
Laboratory of Agro-Industrial and Medical Biotechnologies, Faculty of Science and Techniques, University of Sultan Moulay Slimane, Beni Mellal, Morocco; Office of Professional Training and Employment Promotion, Agricultural Center, City of Trades and Skills (CMC-BMK), Beni Mellal, Morocco

Flavia Lega Braghiroli
Research Forest Institute (Institut de recherche sur les forêts - IRF), University of Québec at Abitibi-Témiscamingue (UQAT), Rouyn-Noranda, QC, Canada

Deivisson Ferreira da Silva
Federal Institute Catarinense, Araquari, Santa Catarina, Brazil; Department of Soil Science, Federal University of Lavras, Lavras, Minas Gerais, Brazil

Paurabi Das
Crop Production and Protection Division, CSIR—Central Institute of Medicinal and Aromatic Plants, Lucknow, Uttar Pradesh, India

Alisson de Carli Souza
Department of Soil Science, Federal University of Lavras, Lavras, Minas Gerais, Brazil

Sara de Jesus Duarte
Latin America Institute of Biochar, Ciudad de Mexico, Mexico

Olívia Bibiana Souza Dias
Department of Soil Science, Federal University of Lavras, Lavras, Minas Gerais, Brazil

Chandrama Dipa
Academy of Scientific and Innovative Research (AcSIR), Ghaziabad, Uttar Pradesh, India; CSIR—Institute of Minerals and Materials Technology, Bhubaneswar, Odisha, India

Bayu Dume
Department of Natural Resource managment, College of Agriculture and Vetrnary Medicine, Jimma University, Jimma, Ethiopia

Ahmed El Moukhtari
Laboratory of Ecology and Environment, Faculty of Sciences Ben Msik, Hassan II University, Casablanca, Morocco

Taoufik El Rasafi
Laboratory of Health and Environment, Faculty of Sciences Ain Chock, Hassan II University, Casablanca, Morocco

Samer Fawzy
Trinity Biocarbon, London, United Kingdom; School of Chemistry and Chemical Engineering, Queen's University Belfast, Belfast, United Kingdom

Alejandro Fernandez
Department of Earth and Environmental Sciences, University of Minnesota, Minneapolis, MN, United States

Gertrudis Pentón Fernández
Indio Hatuey Pastures and Forages Experimental Station, University of Matanzas, Matanzas, Cuba; Latin American Institute of Biochar, Mexico City, Mexico

Assima Firdoos
Department of Botany, School of Bioengineering and Biosciences, Lovely Professional University, Phagwara, Punjab, India

Karen Ghazaryan
Laboratory of Applied Biology and Ecology, Faulty of Biology, Yerevan State University, Yerevan, Armenia

Elif Gödekmerdan
Ege University, Solar Energy Institute, Izmir, Turkey

Daniel Guarín
Department of Ecosystem Science and Management, Pennsylvania State University, University Park, PA, United States

Praveen Guleria
Department of Biotechnology, Faculty of Life Sciences, DAV University, Jalandhar, Punjab, India

Abdelmajid Haddioui
Laboratory of Agro-Industrial and Medical Biotechnologies, Faculty of Science and Techniques, University of Sultan Moulay Slimane, Beni Mellal, Morocco

Riti Thapar Kapoor
Centre for Plant and Environmental Biotechnology, Amity Institute of Biotechnology, Amity University Uttar Pradesh, Noida, Uttar Pradesh, India

Vashista Kotra
Amity Institute of Organic Agriculture, Amity University, Noida, Uttar Pradesh, India

Abhay Kumar
Department for Innovation in Biological, Agro-food and Forest Systems, University of Tuscia, Viterbo, Italy

Chitranjan Kumar
Amity Institute of Organic Agriculture, Amity University, Noida, Uttar Pradesh, India

Manish Kumar
Academy of Scientific and Innovative Research (AcSIR), Ghaziabad, Uttar Pradesh, India; CSIR—Institute of Minerals and Materials Technology, Bhubaneswar, Odisha, India

Vineet Kumar
Department of Biotechnology, School of Bioengineering and Biosciences, Lovely Professional University, Phagwara, Punjab, India

Pallavi Kumari
Centre of Polymer and Carbon Materials, Polish Academy of Sciences, Zabrze, Poland

Smitha John Kundukulam
Department of Soil Science & Agricultural Chemistry, College of Agriculture, Kerala Agricultural University, Vellanikkara, Kerala, India

Sumira Malik
Amity Institute of Biotechnology, Amity University, Ranchi, Jharkhand, India; Department of Biotechnology, University Center for Research & Development (UCRD) Chandigarh University, Mohali, Punjab, India

Melkamu Mamuye
Department of Natural Resource managment, College of Agriculture and Vetrnary Medicine, Jimma University, Jimma, Ethiopia

Saglara Mandzhieva
Academy of Biology and Biotechnology, Southern Federal University, Rostov-on-Don, Russia

Sara Marinari
Department for Innovation in Biological Agricultural Food and Forestry Systems, University of Tuscia, Viterbo, Italy

Luisa Massaccesi
National Research Council of Italy, Institute for Agriculture and Forestry Systems in the Mediterranean (ISAFOM-CNR), Perugia, Italy

Odelin Brea Maure
Agricultural Faculties, University of Matanzas, Matanzas, Cuba

V. Mazzurco Miritana
National Research Council of Italy, Research Institute on Terrestrial Ecosystems (IRET-CNR), Rome, Italy; Department of Energy Technologies and Renewable Sources TERIN-ENEA - Italian National Agency for New Technologies, Energy and Sustainable Economic Development, Roma, Italy

Hugh McLaughlin
Trinity Biocarbon, London, United Kingdom

Leônidas Carrijo Azevedo Melo
Department of Soil Science, Federal University of Lavras, Lavras, Minas Gerais, Brazil

Tatiana Minkina
Academy of Biology and Biotechnology, Southern Federal University, Rostov-on-Don, Russia

Sibadutta Mohanty
Academy of Scientific and Innovative Research (AcSIR), Ghaziabad, Uttar Pradesh, India; CSIR—Institute of Minerals and Materials Technology, Bhubaneswar, Odisha, India

Sagar Mondal
Amity Institute of Biotechnology, Amity University, Ranchi, Jharkhand, India

Abebe Nigussie
Department of Natural Resource managment, College of Agriculture and Vetrnary Medicine, Jimma University, Jimma, Ethiopia

I. Nogués
National Research Council of Italy, Research Institute on Terrestrial Ecosystems (IRET-CNR), Rome, Italy

Abdallah Oukarroum
AgroBioSciences Department (AgBS), University Mohammed VI Polytechnic (UM6P), Ben Guerir, Morocco

Mahesh Pallikonda
Department of Engineering Technology, Austin Peay State University, Clarksville, TN, United States

Sangeeta Pandey
Amity Institute of Organic Agriculture, Amity University, Noida, Uttar Pradesh, India

Mannat Parbhakar
Department of Biotechnology, School of Bioengineering and Biosciences, Lovely Professional University, Phagwara, Punjab, India

Amir Parnian
National Salinity Research Center (NSRC), Agricultural Research, Education and Extension Organization (AREEO), Yazd, Yazd Province, Iran

L. Passatore
National Research Council of Italy, Research Institute on Terrestrial Ecosystems (IRET-CNR), Rome, Italy

Akshyurna Pattnaik
Academy of Scientific and Innovative Research (AcSIR), Ghaziabad, Uttar Pradesh, India; CSIR—Institute of Minerals and Materials Technology, Bhubaneswar, Odisha, India

Majeti Narasimha Vara Prasad
Retired Professor, School of Life Sciences, University of Hyderabad, Hyderabad, Telangana, India

Swati Priya
Amity Institute of Biotechnology, Amity University, Ranchi, Jharkhand, India

Flávio Henrique Silveira Rabêlo
Luiz de Queiroz College of Agriculture, University of São Paulo, Piracicaba, São Paulo, Brazil

Mohd Rafatullah
Division of Environmental Technology, School of Industrial Technology, Universiti Sains Malaysia, Penang, Malaysia

Shafiqur Rahman
Agricultural Research and Development Program, Central State University, Wilberforce, OH, United States

Priyadarshani Rajput
Academy of Biology and Biotechnology, Southern Federal University, Rostov-on-Don, Russia

Vishnu D. Rajput
Academy of Biology and Biotechnology, Southern Federal University, Rostov-on-Don, Russia

Nicolás Rangel
Geothermal Energy and Geofluids, Institute of Geophysics, ETH Zürich, Zurich, Switzerland

Arati Ray
Academy of Scientific and Innovative Research (AcSIR), Ghaziabad, Uttar Pradesh, India; CSIR—Institute of Minerals and Materials Technology, Bhubaneswar, Odisha, India

Prosanta Saha
Department of Botany, Durgapur Government College, Durgapur, West Bengal, India

Ayesha Sarker
Agricultural and Environmental Research Station, West Virginia State University, Institute, WV, United States

Md. Rayhan Shaheb
Department of Agricultural and Life Sciences, Central State University, Wilberforce, OH, United States

Mashura Shammi
Department of Environmental Sciences, Jahangirnagar University, Savar, Dhaka, Bangladesh

Abhishek Singh
Laboratory of Applied Biology and Ecology, Faulty of Biology, Yerevan State University, Yerevan, Armenia

Naveen Kumar Singh
Department of Life Science, Central University of South Bihar, Gaya, Bihar, India

Manoj Kumar Solanki
Department of Life Sciences and Biological Sciences, IES University, Bhopal, Madhya Pradesh, India

Tarah Sullivan
Department of Crop and Soil Sciences, Washington State University, Pullman, WA, United States

Ajay Tomar
Amity Institute of Organic Agriculture, Amity University, Noida, Uttar Pradesh, India

M. Zacchini
National Research Council of Italy, Research Institute on Terrestrial Ecosystems (IRET-CNR), Rome, Italy

Muzakkir Mohammad Zainol
School of Chemical Engineering, College of Engineering, Universiti Teknologi MARA, Shah Alam, Selangor, Malaysia

Aziz Ziouti
Laboratory of Health and Environment, Faculty of Sciences Ain Chock, Hassan II University, Casablanca, Morocco

Preface

Biochar agro-ecotechnologies involve applying biochar based on agronomic principles to enhance environmental protection, exploring technological methods to create carbon-negative products from renewable biological waste resources, and advancing the circular bioeconomy. In recent years, the interaction of agriculture and environmental management has become a focus of global discourse. As the challenges of climate change, soil degradation, and sustainable food production intensify, innovative solutions are being sought to bridge these pressing concerns. The pressures of reduced arable land, rapid population growth, and economic development have driven agricultural producers to rely heavily on synthetic chemical fertilizers to meet global food and forage demands. This extensive use of synthetic chemical fertilizers has made agriculture a major source of anthropogenic nitrous oxide, a potent greenhouse gas. Consequently, there is an urgent need for a well-defined management strategy to enhance fertilizer use efficiency and minimize chemical losses.

"Biochar Ecotechnology for Sustainable Agriculture and Environment" delves into this remarkable material, exploring its role as a transformative agent in modern ecological practices. Biochar, a stable form of carbon produced through the pyrolysis of organic materials, has gained significant attention for its ability to enhance soil health, sequester carbon, and reduce greenhouse gas emissions. Its applications extend beyond agriculture, touching on aspects of waste management, energy production, water purification, pollutant removal, and animal feed system. Additionally, biochar-based fertilizers have received a lot of interest as an agronomic management method in precision agriculture. Although biochar application offers significant agronomic benefits, advancing low-cost, local-scale biochar production technology is crucial for enhancing its acceptance and widespread adoption. This progress is essential for boosting circular and rural bioeconomies, ultimately leading to socioeconomic benefits.

Thus this book offers a comprehensive overview of biochar's potential, combining scientific research with practical insights. It summarizes current achievements in biochar ecotechnologies for enhancing agricultural production, promoting sustainable waste management, and advancing a circular bioeconomy through innovations in the pyrolysis process, both at small and large scales. Key topics include how biochar contributes to decarbonization, soil carbon management, and mitigation of global warming. The book also addresses the enhancement of the circular bioeconomy through effective biowaste management and the use of biochar or biochar-based fertilizers to improve soil fertility and food production.

Recent breakthroughs in biochar agro-ecotechnologies are highlighted, focusing on advances in the pyrolysis process and biochar–fertilizer interactions, including soil nutrient biogeochemical cycles, nutrient retention, soil-microbial responses, greenhouse gas emissions, and plant responses. Additionally, the book explores nanotechnological advancements in biochar-based slow-release fertilizers, emphasizing their nano characteristics that enhance fertilizer efficiency and minimize chemical losses for sustainable agriculture and environmental management.

It is intended for a diverse audience, including researchers, practitioners, policymakers, and anyone interested in sustainable practices. Throughout its chapters, readers will find discussions on the principles of biochar production, its impact on soil fertility, its role in mitigating climate change, and case studies demonstrating its real-world applications. By examining both successes and challenges, this book offers a balanced perspective on how biochar can be effectively integrated into existing systems to foster resilience and sustainability. As audiences delve into the study of biochar ecotechnology, we aim to inspire a deeper appreciation for how innovative approaches can foster a more sustainable and balanced relationship with our environment. Addressing our environmental challenges is a continuous journey, and biochar offers one of many promising pathways toward a more resilient and sustainable future.

We extend our gratitude to the many contributors and researchers who have shared their insights and expertise in this field. Their dedication and passion are reflected in the pages that follow, making this book a valuable resource for advancing our understanding and application of biochar in both agriculture and environmental management.

Welcome to "Biochar Ecotechnology for Sustainable Agriculture and Environment." We invite you to embark on a journey with us to explore this transformative technology and discover its potential to drive significant positive change in our world.

Acknowledgments

We extend our heartfelt thanks to all contributors for their invaluable research and dedication to this book. Our sincere appreciation goes to Theodore (Teddy) Lewis and the Elsevier publishing team for their enthusiastic support, cooperation, and swift resolution of submission and publishing issues. We also acknowledge the exceptional work of the Elsevier production team, whose efforts were instrumental in bringing this book to its final form.

PART

1

Biochar for waste management, circular economy, & carbon neutrality

CHAPTER

1 An overview of biochar production and its multifaceted applications for sustainable agriculture and environmental benefits

Abhay Kumar[1], Pallavi Kumari[2], Manoj Kumar Solanki[3] and Majeti Narasimha Vara Prasad[4]

[1]*Department for Innovation in Biological, Agro-food and Forest Systems, University of Tuscia, Viterbo, Italy*

[2]*Centre of Polymer and Carbon Materials, Polish Academy of Sciences, Zabrze, Poland*

[3]*Department of Life Sciences and Biological Sciences, IES University, Bhopal, Madhya Pradesh, India*

[4]*Retired Professor, School of Life Sciences, University of Hyderabad, Hyderabad, Telangana, India*

1.1 Introduction

Global warming poses a threat to food security by reducing agricultural productivity, limiting food availability, and diminishing nutritional quality (FAO, 2018). Climate change exacerbates these issues through erratic weather patterns, increased extreme events, and heightened occurrences of pests and diseases, all of which detrimentally impact agricultural output (Intergovernmental Panel on Climate Change, 2021). Simultaneously, population growth increases food demand, strains agricultural systems, and aggravating resource scarcity, thereby intensifying the pressure on agricultural productivity and necessitating increased food production from limited arable land and water resources (Kopittke et al., 2019). This increased demand accelerates environmental degradation, including soil erosion, water depletion, and biodiversity loss, which collectively threaten the sustainability of agricultural practices (Bateman & Muñoz-Rojas, 2019). Additionally, the rapid increase in global population and urbanization have led to significant increases in solid waste production, creating further management and environmental challenges (Voukkali et al., 2024). These stresses contribute to socio-economic problems such as rising food prices, increased poverty and inequality, and potential social instability (Parsons et al., 2024). Addressing these complex issues requires integrated climate adaptation and mitigation strategies to sustain agricultural productivity and secure food resources for future generations in the face of advancing global warming.

Biochar Ecotechnology for Sustainable Agriculture and Environment. DOI: https://doi.org/10.1016/B978-0-443-29855-4.00002-3

A variety of soil husbandry or management techniques are designed to sustainably manage soil health and fertility, aiming to enhance agricultural productivity while mitigating the impacts of global warming, and sustainable environment (Kumar, Antoniella, et al., 2024). Among many, biochar production and their application for sustainable agriculture and environment have gained much scientific attention. Biochar, a stable form of carbon (C) produced through the pyrolysis of organic biomass in low-oxygen conditions, is emerging as a pivotal tool in sustainable agriculture (Joseph et al., 2021). The pyrolysis process decomposes organic material at high temperatures (300°C–800°C) in an oxygen-limited environment, yielding biochar, syngas, and bio-oil (Ippolito et al., 2020; Zheng, Zhu, et al., 2023). Biochar production involves the pyrolysis of diverse organic materials, known as feedstocks, each imparting distinct characteristics to the resultant biochar (Amalina et al., 2022). Key feedstocks include agricultural residues (e.g., crop stalks, husks, and manure), forestry waste (e.g., wood chips, sawdust, bark, and branches), and organic waste from municipal sources (e.g., green waste and food scraps) (Ghodake et al., 2021). Extensive research efforts have rigorously addressed the problem, demonstrating various solutions, with the conversion of various solid biowaste into value-added materials proving particularly attractive.

Historically utilized in the Amazon Basin's Terra Preta soils, biochar is now extensively researched for its multifaceted benefits, including enhancing soil health, increasing crop yields, and mitigating climate change (Garcia-Nunez et al., 2017; Heitkötter & Marschner, 2015). Its unique properties like high C content, ion exchange capacity, and large surface area make biochar a promising solution for various agronomic and environmental challenges (Ippolito et al., 2020). The choice of feedstock significantly influences the physical and chemical properties of biochar, including its C content, porosity, and nutrient profile (Kumar, Sharma, et al., 2024).

In recent years, the challenges of achieving sustainable agriculture have grown, particularly in managing of biodegradable waste, including municipal solid waste and sewage sludge (Hazrati et al., 2021). Despite these challenges, there are successful examples of energy recovery from these biowastes, which enhance soil properties, reduce the reliance on mineral fertilizers and plant protection (Kumar, Joseph, et al., 2021), mitigate greenhouse gas (GHG) emissions (Sri Shalini et al., 2021), and promote C sequestration in soils (Mona et al., 2021). During pyrolysis, C in organic matter is converted into a stable form that resists decomposition for several years to decades (Ling et al., 2022). These characteristic position biochar as a powerful tool for mitigating climate change by reducing atmospheric CO_2 levels (Abagandura et al., 2019). Biochar also shows promise in remediating contaminated soils and water due to its high surface area and adsorption capacity, enabling it to bind toxic heavy metals (HMs) and organic pollutants, reducing their bioavailability and mitigating their impact on ecosystems (Kumar et al., 2016). This capability makes biochar an effective material for environmental cleanup efforts, particularly in areas affected by industrial pollution (Huang, Song, et al., 2019). To address these issues globally, innovative solutions are needed to reconcile these practices with sustainable development principles and to find effective, scalable methods for managing waste and improving agricultural sustainability.

Despite the growing research on biochar, significant knowledge gaps persist. Thus dedicated reviews are needed to synthesize the latest scientific findings, case studies, and practical applications, offering a comprehensive understanding of biochar's advantages and limitations. This resource can inform and inspire researchers, policymakers, farmers, and environmentalists to adopt biochar practices that enhance soil health, sequester C, and reduce GHG emissions. Biochar production promotes a circular bioeconomy by converting waste biomass into a valuable resource. Additionally, this comprehensive review also highlights the economic and environmental benefits related to biochar production and application, encouraging industries and municipalities to adopt biochar production technologies. Through this effort, the aim of this chapter is to discuss the biochar production and its multifaceted applications for sustainable agriculture and environmental benefits.

1.2 Biochar production technology

Biochar production technology involves the thermal decomposition (pyrolysis) of biomass materials such as wood, agricultural residues, or organic waste at temperatures typically ranging from 350°C to 800°C in the absence of oxygen (Joseph et al., 2021; Fig. 1.1). A database search conducted across Web of Science, Elsevier, SpringerLink, and Wiley Online Library using the keyword "Biochar production technologies," focusing exclusively on research articles, reviews, and book chapters in English, reveals a total of 39,996 publications from

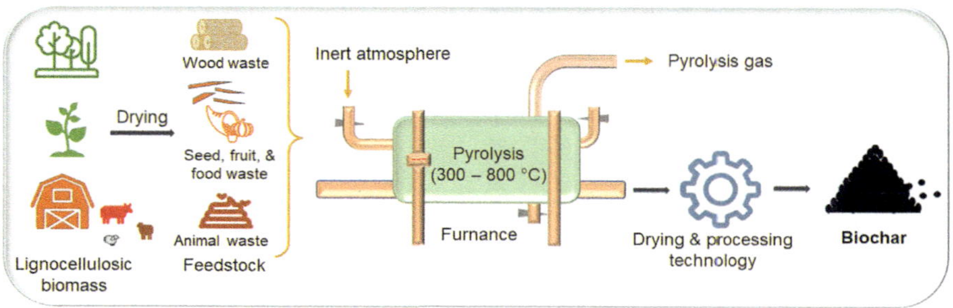

FIGURE 1.1

Biochar production through pyrolysis process. Schematic representation of biochar production from various biomass feedstocks using a pyrolysis reactor. This figure illustrates the different types of biomass materials (such as agricultural residues, forestry by-products, and organic waste) that can be processed to produce biochar.

2016 to July 2024 (Fig. 1.2). A thorough data search revealed that the primary production methods for biochar include slow pyrolysis, fast pyrolysis, torrefaction, hydrothermal carbonization, and gasification. Each of these methods affects the biochar's properties, yield, and potential applications (Table 1.1; Seow et al., 2022; Sharma et al., 2024). Slow pyrolysis, characterized by lower temperatures and longer residence times, maximizes biochar yield, making it ideal for soil amendment and C sequestration (Russell et al., 2017). In contrast, fast and flash pyrolysis, operating at higher temperatures and shorter residence times, favor the production of bio-oil and syngas, which are useful for bioenergy applications (Safarian, 2023). Optimal pyrolysis conditions carefully controlled temperature, heating rate, and residence time are crucial for tailoring biochar properties for specific applications. Key parameters, such as pyrolysis temperature and heating rate, determine the biochar's surface area, pore structure, and chemical composition, critical for its effectiveness in environmental and agricultural uses (Namaswa et al., 2023). Alternatively, biochar production can occur via gasification, a process where biomass undergoes initial pyrolysis followed by further oxidation to produce hydrogen, methane, and carbon monoxide-rich gases (Puglia et al., 2023). Compared to gasification, pyrolysis typically yields 20%–30% more biochar from equivalent biomass inputs (Afshar & Mofatteh, 2024). Research into biochar production from various thermochemical processes and diverse biomass feedstock for targeted applications continues to receive immense attention. Advanced technologies aim to optimize yield, quality, and energy efficiency while minimizing emissions, making biochar a versatile tool for soil improvement, C sequestration, and sustainable waste management.

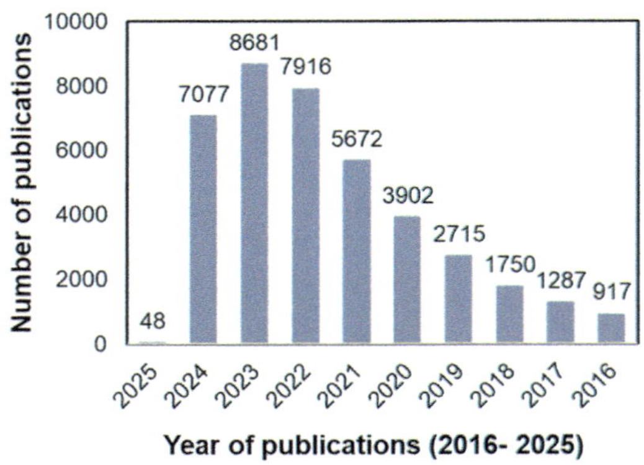

FIGURE 1.2

Publications on biochar production technologies (2016–July 2024). Total number of publications identified from 2016 to July 2024 through a search across Elsevier, SpringerLink, and Wiley Online Library using the keyword "Biochar production technologies." The search focused exclusively on research articles, reviews, and book chapters in English, revealing a total of 39,996 publications.

Table 1.1 Biomass thermochemical conversion technologies, process conditions, and product yield.

Technology	Pyrolysis conditions			Biochar yield (%)
	Temperature (°C)	Residence time (min)	Heating rate (°C/min)	
Hydrothermal carbonization	100–300	Long (1–16 h)	Slow	45–75 (hydrochar)
Torrefaction	200–300	Relatively long: 10–120 min	Slow: <20	60–80 (charcoal)
Slow pyrolysis	300–700	Long (from minutes to days)	Slow: <30 (generally, 5–10)	30–50
Fast pyrolysis	300–1000	Very short: <2 s	Very fast: 1000°C/s	12–25
Intermediate pyrolysis	300–500	Moderate: 1–15 min	Moderate: 1–10°C/s	30–40
Flash carbonization	300–600	Moderate: <30 min	Very fast	35–50 (charcoal)
Gasification	600–1500	Short: 10–20 s	Moderate-very fast	5–15
Microwave	350–650 (400–2700 W)	Moderate	Fast: 25–50	15%–80%

Source: *Reproduced and modified from Sharma, T., Hakeem, I. G., Gupta, A. B., Joshi, J., Shah, K., Vuppaladadiyam, A. K., & Sharma, A. (2024). Parametric influence of process conditions on thermochemical techniques for biochar production: A state-of-the-art review. Journal of the Energy Institute, 113. https://doi.org/10.1016/j.joei.2024.101559.*

1.2.1 Biochar-pyrolysis methods and technology

Biochar production has sparked significant interest, leading to a proliferation of diverse reactor designs accessible online or through commercial vendors (Garcia-Nunez et al., 2017). Biochar reactors, also known as pyrolysers, encompass a range of designs from small stoves and ovens to larger kilns and retorts, aligning with conventional definitions (Namaswa et al., 2023). However, these reactors are purpose-built for biochar production and require meticulous design to meet specific criteria. Various types of reactors optimize the pyrolysis process to produce biochar efficiently, each offering distinct advantages in terms of yield, quality, energy efficiency, and scalability (Table 1.2). The selection of a reactor type depends on factors such as the type of feedstock, desired biochar characteristics, process scale, and economic considerations within biochar production systems.

1.2.2 Biomass feedstock for biochar production

Biomass, which includes living or once-living materials, serves as the feedstock for biochar production. This category encompasses a wide range of reasonably dry

Table 1.2 Overview of biochar production technologies and their operational mechanisms.

Biochar production technology	Specific operation mechanisms	References
Batch pyrolysis reactors	Batch pyrolysis reactors process biomass in discrete batches within a controlled environment to produce biochar. This method offers flexibility and precision in controlling pyrolysis conditions, but typically involves intermittent operation and longer processing times	Reynolds et al. (2018)
Traditional kilns	Basic, low-tech devices that burn biomass in a low-oxygen environment to produce biochar. traditional kilns operate in batch mode, where biomass is loaded, pyrolyzed, and then cooled before biochar removal. They typically lack the efficiency and control of modern systems, leading to inconsistent quality and yield	Sparrevik et al. (2015)
Retorts	These are closed vessels that allow for controlled pyrolysis under reduced oxygen conditions, ensuring higher biochar quality and yield compared to open systems. They offer improved quality and yield consistency compared to traditional kilns	Ighalo et al. (2024)
Continuous pyrolysis reactors	Continuous pyrolysis reactors produce biochar by continuously feeding biomass into a high-temperature, low-oxygen chamber, allowing for constant and efficient conversion. These systems offer consistent biochar quality and high production rates	Veiga et al. (2021)
Rotary kilns	Rotary kilns produce biochar by rotating a cylindrical chamber filled with biomass, ensuring even heating and carbonization. This method provides consistent biochar quality and efficient heat transfer and processing, especially for large-scale operations	Moser et al. (2023)
Auger pyrolyzers	In this system, biomass is continuously fed into a screw conveyor system, where it undergoes pyrolysis as it moves through a heated chamber. This method ensures efficient, uninterrupted processing of biomass into biochar, delivering consistent quality and minimizing operational downtime	Brassard et al. (2017)
Fluidized bed reactors	In fluidized bed reactors, biomass particles are suspended in a hot gas stream, ensuring rapid heat transfer and efficient pyrolysis. This versatile method promotes uniform biochar and bio-oil production, offering high efficiency and consistent quality across a range of feedstocks	Liu et al. (2020)

(*Continued*)

Table 1.2 Overview of biochar production technologies and their operational mechanisms. *Continued*

Biochar production technology	Specific operation mechanisms	References
Microwave pyrolysis reactors	Microwave pyrolysis reactors use microwave radiation to rapidly and uniformly heat biomass, inducing pyrolysis in a thermally insulated chamber. This method allows for precise control over temperature, enhances energy efficiency, and produces consistent biochar quality while potentially reducing processing time	Nhuchhen et al. (2018)
Gasification reactors		
Fixed bed gasifiers	Fixed bed gasifiers convert biomass into synthesis gas (syngas) through partial oxidation, followed by the progressive pyrolysis of the remaining material into biochar. This method ensures controlled pyrolysis conditions and delivers consistent biochar quality across various biomass types	Akgül et al. (2024)
Fluidized bed gasifiers	Fluidized bed gasifiers convert biomass into biochar and syngas by suspending it in a hot, turbulent gas stream, which enhances gas-solid contact and heat transfer. This method ensures efficient pyrolysis, uniform biochar quality, and high processing efficiency	Marrugo et al. (2017)
Hybrid reactors	Hybrid reactors use integrated heating methods like microwave and conventional heating, or combine pyrolysis with gasification to optimize biochar and syngas production. They blend features of fixed and fluidized bed systems for better control, continuous feed, and efficient heat transfer, enhancing biochar quality and processing efficiency	Ryšavý et al. (2024)
A stove	A stove gasifier burns biomass in a controlled environment to produce biochar and syngas efficiently. It heats the stove and surrounding space, making it ideal for small-scale and residential use	Gitau et al. (2019)

organic materials such as wood, bark, nutshells, coffee waste crop residues, invasive species, and various animal by-products like manures, animal waste, and biosolids (Fig. 1.3; Mlonka-Mędrala et al., 2021; Silva et al., 2022; Xia et al., 2025). Optimal biochar for agricultural enhancement often comes from high-nutrient-density biomass and fine-grained residues from agriculture, animal husbandry, and other clean biosolids waste streams (Amalina et al., 2022). Lignocellulosic biomass, comprising cellulose, hemicellulose, and lignin, is critical for biochar production (Kumar, Sharma, et al., 2024). Cellulose provides C-rich char with good structural integrity, while

FIGURE 1.3

Feedstock sources for biochar production. Various sources of feedstock and biowaste utilized in biochar production processes.

lignin enhances stability and C content. Furthermore, effective biochar production also requires managing the balance of these components, as they influence the pyrolysis process and the quality of the biochar. For effective CO_2 removal through pyrolysis and biochar production, the biomass should ideally be sourced from fast-growing plants (Ding et al., 2024) or from waste and residues that would otherwise emit CO_2 through combustion or natural decomposition. Managing the moisture content of biomass is critical; biomass with over 20% moisture typically needs drying to ensure efficient production of high-quality biochar (Gholizadeh et al., 2024). The properties of biochar are significantly influenced by both the type of biomass and the specific pyrolysis reactor used. Understanding the fundamental composition, characteristics, and requirements of biomass feedstock is essential for optimizing biochar production processes and its application (refer chapter 22 in this book for details) (Gholizadeh et al., 2024). Therefore sustainable sourcing of biomass is imperative to maintain biochar's environmental benefits, ensuring the feedstock is free from toxic organic and inorganic contaminants for use in agriculture and animal production systems.

1.3 Biochar characteristics

1.3.1 Physical characteristics

Biochar possesses various physical characteristics that influence its effectiveness and suitability for different applications (Downie et al., 2012). Key physical properties include bulk density, porosity, specific surface area, hydrophobicity, and particle density, all of which actively contribute to the overall characteristics of biochar. Bulk density, defined as mass per unit volume, significantly affects biochar's distribution, soil structure improvement, and its effectiveness in various applications, particularly in agriculture and environmental management (Kuryntseva et al., 2003). Typically, biochar has a low bulk density around 0.5 g/cm^3; however, depending on the feedstock types and production methods, some biochars can exhibit higher bulk densities ranging from 1.5 to 2 g/cm^3 (Brewer et al., 2009). Particle size distribution impacts surface area, porosity, and uniform mixing with soil, which are crucial for nutrient retention and water management (Khaledi et al., 2023). Porosity, encompassing both macro and micro levels, determines biochar's potential for water retention, air exchange capacity, and support for soil microbial communities (Liu et al., 2017). The surface area directly correlates with biochar's adsorption capacity for water, nutrients, and pollutants, which is essential for enhancing soil fertility and environmental remediation (Kumar, Joseph, et al., 2018). Optimizing these properties through appropriate feedstock selection and production methods is crucial for maximizing biochar's positive impacts on soil health, agricultural productivity, and environmental sustainability.

1.3.2 Chemical properties

Similar to its physical properties, biochar exhibits diverse chemical properties essential for tailoring its applications in agriculture, environmental management, and soil remediation strategies (Adhikari et al., 2022; Amalina et al., 2022). Biochar is primarily composed of organic elements such as C, hydrogen (H), oxygen (O), and nitrogen (N) (Cao et al., 2022). The C content in biochar varies from 43% to 75%, influenced significantly by the type of feedstock and pyrolysis temperature (Ippolito et al., 2020; Mlonka-Mędrala et al., 2021). This C exists in forms of stable organic C or carbonate salts, contributing to biochar's longevity in soil, potentially lasting hundreds to thousands of years. The hydrogen to organic C (H/Corganic) ratio is a crucial indicator of biochar's persistence in soil environments (Lehmann & Joseph, 2015). Moreover, a rectangular model was adopted to predict the aromatic cluster sizes of biochars based on their H/C ratios, with the resulting robust structure-sorption relationship demonstrating that the H/C atomic ratio serves as a universal indicator for predicting pyrolytic temperatures, aromatic cluster sizes, and sorption characteristics (Xiao et al., 2016).

The inorganic fraction of biochar, often referred to as residual noncarbonaceous materials or ash, includes both macro and micronutrients along with HMs (Ippolito et al., 2020). The composition and concentration of these constituents are vital for assessing biochar's suitability for agricultural applications and ensuring compliance with environmental standards. These inorganic components, primarily minerals and trace elements inherent to the biomass feedstock, include calcium (Ca), potassium, magnesium (Mg), phosphorus, silicon, sodium, iron (Fe), zinc (Zn), copper (Cu), and manganese (Mn). The composition and quantity of ash vary depending on biomass source, growth conditions, and pyrolysis parameters (Kumar, Joseph, et al., 2021).

1.3.3 Heavy metals in biochar

Biochar derived from the pyrolysis of organic biomass contains both beneficial and potentially toxic HMs or pollutants. Essential micronutrients for plants, such as Cu, Fe, Mn, nickel (Ni), and Zn, are present in biochar in trace amounts (Hawryluk-Sidoruk et al., 2024). These elements are crucial for various biochemical processes in plants and are beneficial at low concentrations. However, biochar can also contain toxic HMs like mercury (Hg), arsenic (As), cadmium (Cd), chromium (Cr), and lead (Pb), which are strictly regulated due to their potential risks to plants, animals, and humans (Gao et al., 2020; Hawryluk-Sidoruk et al., 2024). Biochars intended for soil applications generally exhibit low levels of HMs such as Cd, Pb, Cr, Hg, As, Cu, and Zn. However, biochars derived from sewage sludge, urban wood, and contaminated residues may have higher concentrations of HM. HMs in biochar originate from the biomass feedstock such as sewage sludge, naturally present in varying concentrations (Gao et al., 2020). During pyrolysis, metals can become concentrated, reduced, or volatilized, depending on their properties and pyrolysis conditions (Fan et al., 2023). Additionally, most biochars contain trace amounts of polycyclic aromatic hydrocarbons and dioxins (Kończak et al., 2019). Therefore it is crucial to assess the bioavailability of these toxins and determine safe application rates, particularly if concentrations are high. Nevertheless, monitoring and controlling HM concentrations in biochar is crucial to meet regulatory standards and avoid risks to soil health, crop productivity, and environmental quality.

1.4 Biochar application for sustainable agriculture and environment

The application of biochar in agriculture is a sustainable practice with multifaceted benefits, enhancing soil health, reducing dependency on chemical inputs, mitigating climate change, and aiding in environmental remediation. The integration of biochar into agricultural systems represents a viable pathway toward achieving ecological balance (Prodana et al., 2019) and promoting long-term agricultural productivity (Kumar, Elad, et al., 2018). Biochar's application offers significant agronomic advantages, including improved soil fertility (Higashikawa et al., 2023), enhanced water retention (Fascella et al., 2020), and the promotion of beneficial microbial activity (Xu et al., 2018). Its porous structure enables the retention of nutrients and moisture, making them available to plants over extended periods and thereby reducing the need for chemical fertilizers (Rezaei Rashti et al., 2019). From an environmental perspective, biochar contributes to by trapping C that would otherwise be released into the atmosphere as CO_2, thereby playing a role in climate change mitigation (Papageorgiou et al., 2021). Additionally, biochar helps reduce GHG emissions from soils, such as N_2O, by creating conditions that inhibit the activities of denitrifying bacteria (Wen et al., 2020). Furthermore, biochar has the capacity to remediate polluted soils by adsorbing HMs and organic contaminants, thus preventing their uptake by plants and leaching into groundwater (Kumar et al., 2016). These properties make biochar an invaluable tool for restoring degraded lands, rangeland, and forest soils as well as abandoned mines and other severely degraded areas.

1.4.1 Biochar for sustainable agriculture

The efficacy of biochar in enhancing crop yields, plant disease resistance, food quality, and soil health evolves as the biochar ages, influenced by its type, soil properties, agronomic practices, and environmental factors, as presented in Table 1.3. Biochar can also act as a carrier for plant nutrients, producing efficient long-term mineral and organic fertilizers (Gwenzi et al., 2018). These fertilizers prevent nutrient leaching, ensuring nutrient availability as needed by plants (Rafael et al., 2019). By stimulating microbial symbiosis, plants can uptake nutrients from the porous C matrix of biochar (He et al., 2021). Mixing biochar with organic waste like manure and slurry produces organic C-based fertilizers that are as efficient as conventional fertilizers without their adverse environmental effects (Mardiansyah et al., 2020; Yu et al., 2019). Pyrolysis also generates various by-products that adhere to biochar's pores and surfaces, potentially enhancing plant immune systems and resistance to pathogens (Kumar, Elad, et al., 2018). Biochar organomineral mixes are generally more effective than fresh wood biochar for improving crop yields, disease resistance, and the abundance of beneficial microbes (refer chapter 22 in this book) (Joseph et al., 2018). The response of plants, soil properties, and profitability to specific biochars depends on application rates, where over-application can promote disease or negatively impact growth (Xiang et al., 2021). In fact, the high adsorption of plant nutrients by biochar can initially hinder plant growth. Improved soil health and higher crop yields contribute to local and regional food security, making biochar an essential component in advancing the circular bioeconomy and supporting sustainable agricultural practices. Kumar, Joseph, et al. (2021) observed that low-dose foliar application of aqueous extracts from organomineral-activated biochars derived from pine wood/clay/sand and wheat straw/bird-manure enhances lettuce growth. C-coated mineral and biochar nanoparticles penetrate stomata, increasing photosynthetic pigments thereby inducing plant growth and strengthens resilience to stressors with effective results at lower biochar application rates compared to conventional soil applications (Kumar, Joseph, et al., 2021). By improving soil fertility, enhancing water retention, and reducing the need for chemical fertilizers, biochar has the potential to revolutionize agricultural practices, contributing to food security and agricultural sustainability (Fig. 1.4).

Table 1.3 A list of recently published review articles on the role of biochar in sustainable agriculture.

References	Title	Key focus
Rahim et al. (2024)	Biochar aged or combined with humic substances: Fabrication and implications for sustainable agriculture and environment-a review	C-based materials; Humic substances; Nutrient availability; Soil microbial activities; Soil
Pandian et al. (2024)	Biochar—a sustainable soil conditioner for improving soil health, crop production and environment under changing climate: A review	Soil conditioner; Soil health; HMs; Climate change; GHGs; Crop yield
Mashamaite et al. (2024)	Assessing the potential of biochar as a viable alternative to synthetic fertilizers in sub-Saharan Africa smallholder farming: A review	Soil physicochemical and biological properties; Crop growth and yield; Smallholder farming
Lavagi et al. (2024)	Recycling agricultural waste to enhance sustainable greenhouse agriculture: Analyzing the cost-effectiveness and agronomic benefits of bokashi and biochar byproducts as soil amendments in citrus nursery production	Soil amendments; Resource use efficiency; Economic sustainability; C sequestration; Nutrient management
Khan et al. (2024)	Biochar production and characteristics, its impacts on soil health, crop production, and yield enhancement: A review	Crop productivity; Organic amendments; Microbial activity; Soil fertility; Sustainable agriculture
Gao et al. (2024)	The potential and prospects of modified biochar for comprehensive management of salt-affected soils and plants: A critical review	Health assessment; Soils and plants; Salt-affected soils; Salt stress
Brar et al. (2024)	Nanobiochar: Soil and plant interactions and their implications for sustainable agriculture	C nanotubes; Nano-composites; Organic biomass; Plant growth; Soil health
Arshad et al. (2024)	Biochar: Black gold for sustainable agriculture and fortification against plant pathogens—A review	Disease control; Plant pathogens; Induce resistance; Foliar pathogens; Plant defense mechanism
Zheng, Ouyang, et al. (2023)	Synthesis, characterization, safety design, and application of NPs@BC for contaminated soil remediation and sustainable agriculture	Nanoparticle-decorated biochar; Soil remediation; Sustainable agriculture; Soil remediation
Utkarsh et al. (2023)	Biochar: A feasible and visible solution for agricultural sustainability	Organic fertilizer; Biodegradation; Bioremediation; Sustainability; Microbiome; Nano-agri formulations

(Continued)

Table 1.3 A list of recently published review articles on the role of biochar in sustainable agriculture. *Continued*

References	Title	Key focus
Safarian (2023)	Performance analysis of sustainable technologies for biochar production: A comprehensive review	Biochar production; Biochar yield; Process parameters; Thermochemical technologies
Rex et al. (2023)	Agricultural biomass waste to biochar: A review on biochar applications using machine learning approach and circular economy	Microwave pyrolysis; Agricultural biomass waste; Circular bio-economy; Machine learning; Technology readiness
Hamidzadeh et al. (2023)	Biomass-derived biochar and its application in agriculture	Pyrolysis; Fertilizer; Soil properties; Microorganisms
Chen et al. (2023)	Sustainable application for agriculture using biochar-based slow-release fertilizers: A review	Slow-release fertilizers; Slow-release mechanisms; Agricultural waste recycling; High efficiency
Brichi et al. (2023)	Organic residues and their impact on soil health, crop production and sustainable agriculture: A review including bibliographic analysis	Crop yield; Organic residue; Organic waste; Soil quality; Sustainability
Ajeng et al. (2023)	Biochar-*Bacillus* consortium for a sustainable agriculture: physicochemical and soil stability analyses	Process optimization; Plant growth promoting bacteria; Adhesion kinetics; Isotherms
Zhu et al. (2022)	Life-cycle assessment of pyrolysis processes for sustainable production of biochar from agro-residues	Circular economy; C neutrality; Renewable energy; Engineered biochar; Activated C; Sustainable waste management
Wang et al. (2022)	Biochar-based slow-release of fertilizers for sustainable agriculture: A mini review	Slow-release fertilizer; Sustainability; Soil amendment; Agricultural production
Seow et al. (2022)	A review on biochar production from different biomass wastes by recent carbonization technologies and its sustainable applications	Feedstock; Thermal treatment; Technologies
Rombel et al. (2022)	Sustainable biochar-based soil fertilizers and amendments as a new trend in biochar research	Fertilization; Soil amendment; Slow-release; Sustainable agriculture; Circular economy
Rasul et al. (2022)	Biochar-induced priming effects in soil via modifying the status of soil organic matter and microflora: A review	Soil organic matter; Priming effect; Microbial community; Mineralization
Malik et al. (2022)	Unlocking the potential of co-applied biochar and plant growth-promoting rhizobacteria (PGPR) for sustainable agriculture under stress conditions	Soil amendments; Soil quality; Crop growth; PGPR; Drought stress; Salinity stress

(*Continued*)

Table 1.3 A list of recently published review articles on the role of biochar in sustainable agriculture. *Continued*

References	Title	Key focus
Sim et al. (2021)	Encapsulated biochar-based sustained release fertilizer for precision agriculture: A review	Slow/controlled release; Nutrient release mechanism; Encapsulating technology; Soil amendment; Plant growth
Nguyen (2021)	Potential use of silica-rich biochar for the formulation of adaptively controlled release fertilizers: A mini review	Biomass; Silica; Controlled release fertilizer
Mona et al. (2021)	Toward sustainable agriculture with carbon sequestration, and greenhouse gas mitigation using algal biochar	Algal biomass; C sequestration; Potential fertilizer; GHG
Pathy et al. (2020)	Biochar amendments and its impact on soil biota for sustainable agriculture	Soil biota; Microbes; Plant growth; Microbial interaction
Ippolito et al. (2020)	Feedstock choice, pyrolysis temperature and type influence biochar characteristics: A comprehensive meta-data analysis review	Total elemental analysis; Plant-available elemental; Physico-chemical characteristics;
Ajeng et al. (2020)	Bioformulation of biochar as a potential inoculant carrier for sustainable agriculture	Biofertilizer; Cells immobilization; Plant growth-promoting rhizobacteria; Pyrolysis; Sustainability

C, *carbon;* GHG, *greenhouse gases;* HMs, *heavy metals.*

1.4.2 Biochar-impact on soil microbes

Biochar-mediated soil health improvement has garnered significant attention for its potential to enhance soil fertility, structure, and sustainability. Its application in agricultural systems has been shown to increase crop yields, enhance soil pH, and reduce GHG emissions. Biochar serves as a beneficial habitat and C source for soil microbes, fostering microbial diversity and activity essential for nutrient cycling, disease suppression, and overall soil health (Paz-Ferreiro et al., 2015). Its unique physicochemical properties such as high surface area, porosity, and significant cation exchange capacity create an environment conducive to microbial colonization, enhancing soil fertility and structure (Han et al., 2017). Biochar's ability to modify soil pH and retain nutrients directly affects microbial metabolism and community composition (Song et al., 2019), promoting the growth of beneficial microbes such as N-fixing bacteria (Chen et al., 2015). This interaction improves nutrient availability, soil aeration, and water retention, while also reducing soil compaction and mitigating the bioavailability of contaminants, thus promoting a healthier soil ecosystem (Chung et al., 2021). The impact of biochar on microbial communities is complex and influenced by factors such as biochar type, application rate, soil properties,

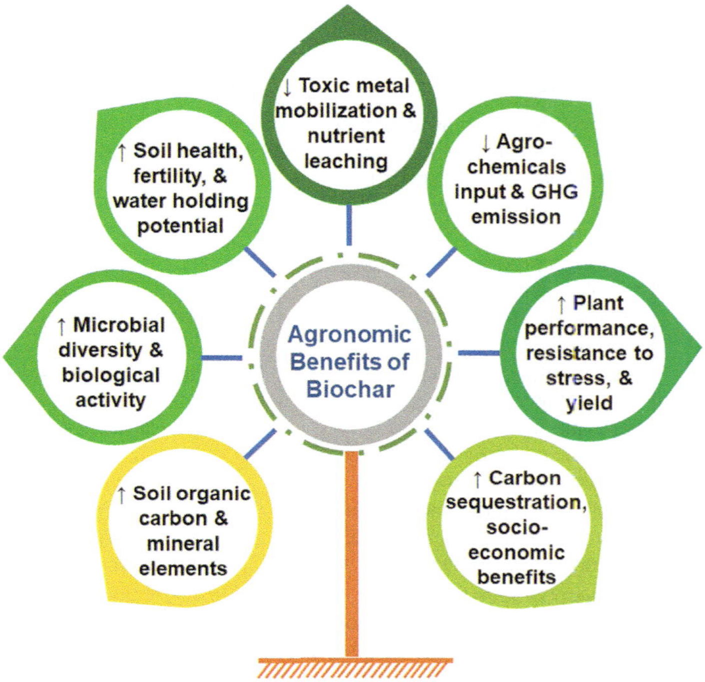

FIGURE 1.4

Agronomic benefits of biochar. Schematic representation of the agronomic benefits of biochar for promoting sustainable agriculture.

and crop types (Kolton et al., 2017). These factors affect microbial abundance and the bacteria-to-fungal ratio, which in turn influences soil ecosystem functions. The interaction between biochar and soil organic matter stimulates microbial decomposition processes, accelerating nutrient cycling and further enhancing soil health (Tian et al., 2016).

1.4.3 Biochar for sustainable environment

Biochar's versatility across various soil types makes it a cost-effective and sustainable soil conditioner (Kumar et al., 2016; Kumar, Tsechansky, et al., 2018). Biochar's chemical and biological resilience allows it to persist in soil for extended periods, aiding in C sequestration and enhancing soil quality and C stocks over the long term (Lehmann & Joseph, 2015). The advantages of biochar include promoting a circular bioeconomy and enhancing agronomic outcomes (Oni et al., 2019). It enhances soil organic C (SOC) mineralization, nutrient levels (Ling et al., 2022), water retention (Adhikari et al., 2022), while also alleviating environmental stress and reducing GHG emissions. Through these mechanisms, biochar production and their utilization can support sustainable and clean environment as discussed in various review articles listed in Table 1.4.

Table 1.4 A list of recently published review articles on the role of biochar for sustainable environment.

References	Work title	Keywords
Yu et al. (2024)	A review on recent advances of biochar from agricultural and forestry wastes: Preparation, modification and applications in wastewater treatment	Agriculture and forestry waste; Biochar modification; Wastewater treatment
Lan et al. (2024)	Research progress of biochar modification technology and its application in environmental remediation	Biomass; Soil environmental remediation; Preparation methods; Modification technology
Fu et al. (2024)	A comprehensive review on the preparation of biochar from digestate sources and its application in environmental pollution remediation	Characteristics of digestate; Modification method; Anaerobic digestion; Resistance genes
Afshar and Mofatteh (2024)	Biochar for a sustainable future: Environmentally friendly production and diverse applications	Sustainable production; Environmental applications; C sequestration
Lin et al. (2023)	Low-temperature biochar production from torrefaction for wastewater treatment: A review	Torrefaction; Low-temperature biochar; Wastewater treatment; Pollutant adsorbent; HMs
Gusiatin and Rouhani (2023)	Application of selected methods to modify pyrolyzed biochar for the immobilization of metals in soil: A review	Modified biochar; Soil remediation; HMs immobilization
Gęca et al. (2023)	Surface treatment of biochar-methods, surface analysis and potential applications: A comprehensive review	Biomass conversion; Thermochemical treatment; Surface chemical modification; Coupling agents; Biochar implementation
Elkhlifi et al. (2023)	Potential role of biochar on capturing soil nutrients, carbon sequestration and managing environmental challenges: A review	C sequestration; Feedstock; Soil pollution; Pyrolysis condition; Soil remediation
Danesh et al. (2023)	Biochar production: Recent developments, applications, and challenges	Conversion techniques; Pollutants removal; Soil fertility
Blenis et al. (2023)	Biochar production, modification, and its uses in soil remediation: A review	Pyrolysis; Cation exchange capacity; Specific surface area; Soil remediation
Amalina et al. (2023)	Recent advancement and applications of biochar technology as a multifunctional component toward sustainable environment	Contaminants; Biochar hazard; Environmental sustainability
Bolan et al. (2023)	The potential of biochar as a microbial carrier for agricultural and environmental applications	Microbial carrier; Agricultural application; Environmental implications

(*Continued*)

Table 1.4 A list of recently published review articles on the role of biochar for sustainable environment. *Continued*

References	Work title	Keywords
Zheng, Xu, et al. (2022)	The effects of biochar and its applications in the microbial remediation of contaminated soil: A review	Soil bioremediation; Microbial growth; Electron mediator
Zeghioud et al. (2022)	A comprehensive review of biochar in removal of organic pollutants from wastewater: Characterization, toxicity, activation/functionalization and influencing treatment factors	Characterization; Toxicity; Contaminant removal; Functionalization; Magnetization
You et al. (2022)	Energy, economic, and environmental impacts of sustainable biochar systems in rural China	Sustainable development; goals; Green and sustainable energy, waste management
Rajput et al. (2022)	Nano-biochar: A novel solution for sustainable agriculture and environmental remediation	Nano-BC; Remediation; Microbial community; Risk assessment; Toxic metals; Antibiotics
Quispe et al. (2022)	Use of biochar-based column filtration systems for greywater treatment: A systematic literature review	Column filtration system; Greywater; Water reuse; Water-scarce regions
Qiu et al. (2022)	Biochar for the removal of contaminants from soil and water: A review	HMs ions; Organic pollutants; Water; Soil
Wang, Shi, et al. (2021)	Analysis of the long-term effectiveness of biochar immobilization remediation on heavy metal contaminated soil and the potential environmental factors weakening the remediation effect: A review	Immobilization; Long-term effectiveness; Potential factors; Weakening
Luo et al. (2021)	A critical review on the application of biochar in environmental pollution remediation: Role of persistent free radicals (PFRs)	Organic and inorganic contaminants; Persistent free radicals; Degradation; Reactive oxygen species
Liang, Xi, et al. (2021)	Review of organic and inorganic pollutants removal by biochar and biochar-based composites	Biochar magnetic composites; Nanometallic oxide/hydroxide biochar composites; Biochar based 2D membrane; 3D biochar-based macrostructures; Biological toxicity
Amusat et al. (2021)	Ball-milling synthesis of biochar and biochar-based nanocomposites and prospects for removal of emerging contaminants: A review	Ball-milled biochar; Biochar; nanoparticle composite; Nanomaterials; Modification of biochar-nanoparticles

C, carbon; GHG, *greenhouse gases;* HMs, *heavy metals.*

1.4.3.1 Pollution remediation

Biochar is increasingly recognized for its efficacy in remediating polluted soils and water bodies, offering significant environmental benefits (Fan et al., 2024). Due to its high surface area and adsorptive properties, biochar effectively captures and immobilizes contaminants such as HMs, pesticides, and organic pollutants (Kumar et al., 2020; Schreiter et al., 2020). This capability reduces the bioavailability of harmful substances, thereby mitigating their impact on ecosystems and human health. Moreover, biochar can be utilized in water filtration systems to purify drinking water and treat industrial effluents (Palansooriya et al., 2020). The remediation potential of biochar extends to diverse applications, including addressing polluted soils at Brownfields and other contaminated sites (Kumar, Tsechansky, et al., 2018). These sites often harbor elevated concentrations of pollutants or toxic HMs and lack vegetation cover, posing environmental risks. Biochar application not only immobilizes contaminants but also enhances soil fertility and promotes ecological restoration (Manikandan et al., 2023). By improving soil physical, chemical, and biological properties, biochar mitigates soil toxicity and supports processes like phytoremediation and phytostabilization as represented in Fig. 1.5.

Biochar's effectiveness in adsorbing contaminants depends on its surface area and the presence of oxygenated surface acidic groups, which vary with production conditions. As biochar ages in soil, its sorption capacity for metals tends to increase, enhancing long-term remediation efficacy (Liu et al., 2021). Furthermore, alkaline biochars can precipitate insoluble metal compounds, thereby reducing dissolved metal concentrations in soil environment (Xiao et al., 2019). HMs and metalloids are particularly concerning due to their toxicity and potential bioaccumulation in the food chain (Sytar et al., 2013). Biochar's ability to immobilize these contaminants through surface adsorption, precipitation, and complexation reactions significantly reduces their bioavailability in soils (Kumar, Joseph, et al., 2018). This immobilization is influenced by soil pH, with biochar amendments typically increasing soil alkalinity and decreasing metal solubility. In practical applications, biochar amendments have demonstrated substantial reductions in bioavailable metal concentrations in contaminated soils, supporting plant growth and facilitating phytoremediation efforts (Kumar, Tsechansky, et al., 2018). For instance, studies have shown significant decreases in bioavailable cadmium, zinc, and lead concentrations in soils treated with biochar, enhancing plant tolerance to metal toxicity (Azadi & Raiesi, 2021). Overall, biochar represents a versatile and sustainable approach to pollution remediation, offering simultaneous benefits of environmental protection and soil improvement.

1.4.3.2 Biochar for environmental stress alleviation

Biochar has emerged as a potent agent for mitigating both biotic and abiotic stresses in plants through its impact on systemic defense mechanisms and soil properties. Its role in enhancing plant defense against pathogens is well-documented;

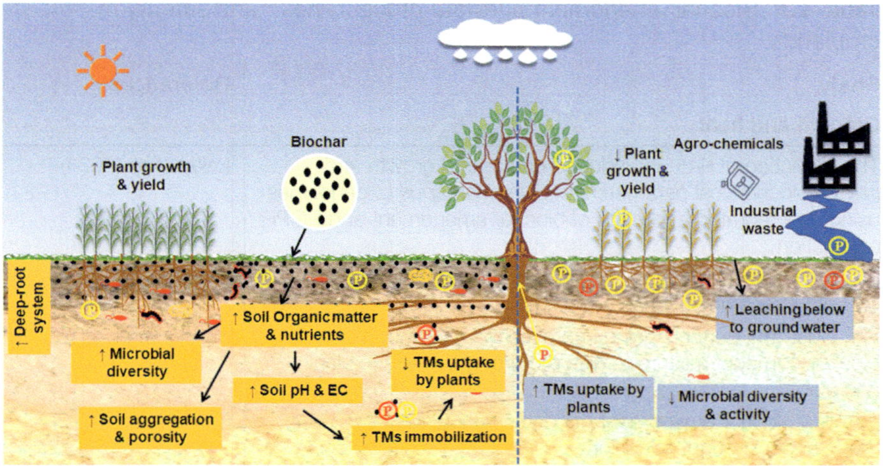

FIGURE 1.5

Interaction between biochar, pollutants, and plant. Functions of biochar in mitigating pollutants (P) and enhancing soil health. This figure illustrates the emission of toxic pollutants from industrial waste and biochar's role in pollutant removal and environmental cleanup. The dashed line indicates the comparative effectiveness of biochar, with the right side showing conditions without biochar and the left side with biochar amendments. Mechanisms depicted include biochar's influence on soil physicochemical properties, pollutant immobilization, plant uptake, and pollutant degradation through increased microbial activity, contributing to sustainable agriculture. The letter "P" within the circle denotes pollutants. Mechanisms by which biochar modifies soil properties, lowers pollutant bioavailability, enhances pollutant adsorption/absorption, and stimulates degradation through increased microbial activity, thereby supporting sustainable agriculture.

biochar stimulates systemic acquired resistance mediated by salicylic acid and induced systemic resistance mediated by jasmonic acid and ethylene, thus bolstering plant immunity (Jaiswal et al., 2024). In addition to ameliorating biotic stress and HMs in-soil immobilzation and reducing-induced phytotoxicity, biochar significantly influences plant responses to other abiotic stresses such as salt, heat, and drought (Zhang et al., 2023; Zulfiqar et al., 2022). Over the past decade, several researchers have made significant studies in demonstrating how biochar helps alleviate stress from salt, drought, and heat, as detailed in Table 1.5. Studies suggested that high surface area and porous structure of biochar involved in improving soil water retention and aeration, essential for alleviating drought stress (Védère et al., 2023). By enhancing soil moisture availability, biochar supports plant hydration during arid periods, thereby reducing water stress and increasing plant resilience. In saline conditions, biochar acts as a buffer, reducing soil salinity and promoting plant growth. Its elevated cation exchange capacity facilitates

Table 1.5 Biochar amendment alleviate drought, heat, and salinity in croplands.

Study	References
Drought and heat	
The alleviation of drought-induced damage to growth and physio-biochemical parameters of *Brassica napus* L. genotypes using an integrated approach of biochar amendment and PGPR application	Lalay et al. (2024)
Biochar derived from hazelnut shells mitigates the impact of drought stress on soybean seedlings	Gullap et al. (2024)
The older, the better: Ageing improves the efficiency of biochar-compost mixture to alleviate drought stress in plant and soil	Védère et al. (2023)
Ash and biochar amendment of coarse sandy soil for growing crops under drought conditions	Bruun et al. (2022)
Distinctive in-planta acclimation responses to basal growth and acute heat stress were induced in *Arabidopsis* by cattle manure biochar	Kumar, Friedman, et al. (2021)
Biochar application rate does not improve plant water availability in soybean under drought	Mannan et al. (2021)
Does biochar mitigate the adverse effects of drought on the agronomic traits and yield components of soybean	Gavili et al. (2019)
In-situ aged and fresh biochar on soil hydraulic conditions and microbial C use under drought	Paetsch et al. (2018)
Biochar soil amendment for alleviation of drought and salt stress in plants—a critical review	Ali et al. (2017)
Biochar Application for drought—stress physiological characteristics of *Silybum marianum*	Keshavarz Afshar et al. (2016)
Salinity	
Enhancing tomato plants' tolerance to combined heat and salt stress—The role of arbuscular mycorrhizae and biochar	Sousa et al. (2024)
Biochar amendments changed soil properties and improved cereal crop growth under salt stress	Ababsa et al. (2023)
Effects of two biochar types on mitigating drought and salt stress in tomato seedlings	Zhang et al. (2023)
Biochar for improving soil biological properties and mitigating salt stress in plants on salt-affected soils	Egamberdieva et al. (2022)
Biochar application alleviated rice salt stress via modifying soil properties and regulating soil bacterial abundance and community structure	Huang et al. (2022)
Irrigation and biochar effects on pearl millet and kinetics of ammonia volatilization from saline sandy soils	Ding et al. (2022)
Optimizing biochar application to improve soil physical and hydraulic properties in saline-alkali soils	Liang, Li, et al. (2021)

(Continued)

Table 1.5 Biochar amendment alleviate drought, heat, and salinity in croplands. *Continued*

Study	References
Apply biochar to ameliorate soda saline-alkali land, improve soil function and increase corn nutrient availability in the Songnen plain	Zhao et al. (2020)
Biochar addition alleviate the negative effects of drought and salinity stress on soybean productivity and water use efficiency	Zhang et al. (2020)
Biochar impacts on NH_3-volatilization kinetics and growth of sweet basil (*Ocimum basilicum* L.) under saline conditions	Ding et al. (2020)
Biochar alleviated the salt stress of induced saline paddy soil and improved the biochemical characteristics of rice seedlings differing in salt tolerance	Zhang et al. (2019)
Effect of biochar on sweet corn and soil salinity under conjunctive irrigation with brackish water in coastal saline soil	Huang, Zhang, Zhu, et al. (2019)
Effect of straw biochar on soil properties and wheat production under salinewater irrigation	Huang, Zhang, Zhai, et al. (2019)
Residual effect of biochar on growth, antioxidant defence and cadmium (Cd) accumulation in rice in a Cd contaminated saline soil	Hafeez et al. (2019)
Benefits of soil biochar amendments to tomato growth under saline water irrigation	She et al. (2018)
Application of biochar in mitigation of negative effects of salinity stress in wheat (*Triticum aestivum* L.)	Kanwal et al. (2018)

Source: *From Prasad, M. N. V., & Kumar, C. (2023). Biochar applications for sustainable agriculture and environmental management. In* Agroecological approaches for sustainab e soil management *(pp. 165–199). Wiley. https://doi.org/10.1002/9781119911999.ch7.*

better retention of essential nutrients while decreasing toxic ion concentrations, mitigating salt-induced stress (Mehdizadeh et al., 2020). Despite these promising benefits, research into biochar's effects on abiotic stressors is still limited. Multiyear field trials, such as those with pepper plants, have demonstrated improved productivity under extreme environmental conditions (Kumar, Elad, et al., 2018), suggesting that biochar induces acclimation responses that enhance plant resilience. Studies have shown that biochar enhances Arabidopsis growth under both basal and heat stress conditions by modulating oxidative stress, metabolic pathways, and hormone levels, leading to greater biomass and inflorescence height (Kumar, Friedman, et al., 2021). These findings underscore biochar's potential in facilitating plant acclimation to various stressors, thereby improving growth and resilience.

1.4.3.3 Waste management and resource utilization

Effective waste management and resource utilization are critical for sustainable development. Advanced methods, such as recycling, composting, and waste-to-energy technologies, reduce the environmental footprint of waste. Converting agricultural waste into biochar through pyrolysis offers a sustainable solution to managing organic residues generated by agricultural activities (Haque et al., 2023). Traditionally, these residues, such as crop residues, prunings, and livestock manure, are either burned or left to decompose, releasing GHG like methane and CO_2 into the atmosphere (Bhagawati et al., 2020). Thus biochar production transforms these wastes into a stable form of C, reducing waste volumes and preventing emission of these GHG (Alonso-Gómez et al., 2024). This waste-to-resource approach not only mitigates environmental pollution but also creates a valuable soil amendment that enhances soil health and fertility (Fan et al., 2023). This sustainable waste management strategy aligns with the circular economy by recycling organic materials, promoting environmental sustainability and resource efficiency. Producing biochar from agricultural waste effectively manages farming residues and supports sustainable agricultural practices. Advanced research and policy support are crucial to scaling biochar technology for widespread environmental and economic benefits.

1.4.3.4 Supporting a circular economy

Integrating biochar production and application into agricultural practices promotes the sustainability of a circular economy by converting waste into valuable resources, reducing reliance on external inputs, and enhancing the natural resource base on which agriculture depends (Yrjälä et al., 2022). This exemplifies the principles of circular economy, aiming for circular formation in consumption and production systems. Similar to biochar production technology, a database search conducted across Elsevier, SpringerLink, and Wiley Online Library using the keyword "Biochar for circular bioeconomy," focusing exclusively on research articles, reviews, and book chapters in English, reveals a total of 1578 publications from 2016 to July 2024 (Fig. 1.6). The database search indicates a growing interest in research studies on this topic, with a 52-fold increase from 2016 to July 2024. A circular bioeconomy strategy not only reduces waste production across specific supply chains but also utilizes food waste and related organic waste products such as farm residues, animal excreta, food processing waste, and wastewater within the chain to recycle nutrients (Sani et al., 2023; Silva et al., 2022; Wu et al., 2023). This practice supports sustainable land management, ensuring agricultural production can meet ongoing needs while enhancing farming system resilience and productivity, contributing to broader environmental goals. Biochar annually sequesters between 1 and 35 gigatons of CO_2, with a cumulative potential of 78−477 gigatons over the century (Shoudho et al., 2024). Additionally, biochar's by-products can be harnessed for renewable energy. Its capacity to generate C-credits and diminish emissions aligns with the Sustainable Development Goals (SDGs). The United Nations SDGs strongly support food waste management to achieve SDG 2 (zero hunger), SDG 3 (good health and well-being), SDG 6 (clean water and sanitation), SDG 7 (affordable and clean energy), and SDG 12 (responsible consumption and production) (Zabaniotou et al., 2015). For example, (Kumar, Kimta, et al., 2024) provide a comprehensive

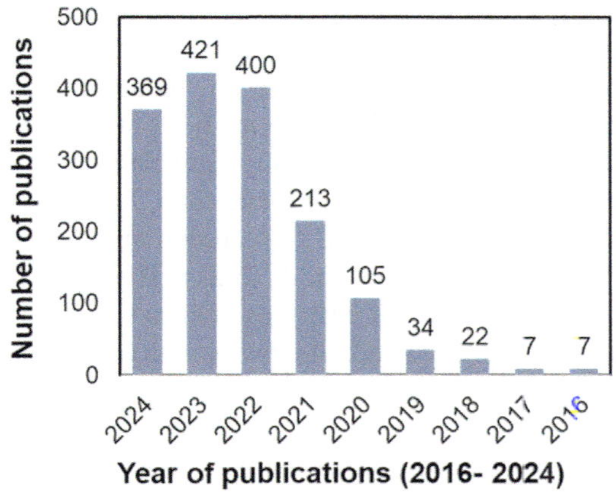

FIGURE 1.6

Publications on biochar for circular bioeconomy (2016—July 2024). Total number of publications identified from 2016 to July 2024 through a search across Elsevier, SpringerLink, and Wiley Online Library using the keyword "biochar for circular bioeconomy." The search focused exclusively on research articles, reviews, and book chapters in English, revealing a total of 1578 publications.

review on using nonedible fruit seeds to synthesize metallic nanoparticles, C dots, biochar, bio-sorbent, and biodiesel. This review also explores the potential of these seeds as functional foods and feed ingredients and addresses safety aspects related to their utilization in a circular bioeconomy (Haque et al., 2023). Additionally, Jukka et al. (2022) evaluate three circular bioeconomy models in southern Finland: willow cultivation on marginal lands, wood-based biochar production, and wood chip-heated GHG. Using a multimethod approach, they assess financial sustainability (net present value, internal rate of return) and environmental sustainability (such as C footprint). Their results indicate positive correlations between financial and environmental indicators, suggesting win—win outcomes. Potential closed loops are identified, enhancing raw material flexibility and supply security, though imbalanced revenue allocation poses a financial risk. Overall, returning agricultural waste to the field through biochar production is a holistic approach addressing waste management, soil health, C sequestration, and crop productivity (Chen et al., 2024).

1.4.3.5 Carbon sequestration and climate change mitigation

In addition to boost soil fertility, biochar also plays a crucial role in C sequestration, offering significant environmental benefits. Composed primarily of stable C, biochar resists decomposition and remains in the soil for centuries to millennia (Lehmann & Joseph, 2015). The ability of biochar to sequester C is central to its environmental impact. Its stable C content locks C in the soil for long periods, preventing its release back into the atmosphere and stabilization of C within natural systems. Furthermore, applying biochar to soils, substantial amounts of CO_2 are effectively removed from the atmosphere and stored in a stable form. This process helps mitigate climate change by reducing GHG concentrations (Kumar, Antoniella, et al., 2024). Life-cycle analyses shows that biochar reduces GHG emissions by about 0.4–1.2 tons of CO_2 equivalents (CO_2eq) per ton of dry feedstock. In a long-term study conducted in Sardinia, Italy, researchers compared the C footprint and economic value of C for cardoon crops grown under five different N fertilization methods (Solinas et al., 2021). The study found that the low N with cover crop treatment had the best GHG balance, with 0.14 tons of CO_2eq per production unit. On the other hand, the low N with biochar treatment achieved the highest C storage in biomass, at 61.1 tons CO_2eq per ton (Solinas et al., 2021).

Biochar enhances soil C sequestration through its porous structure, promoting microbial activity, soil stabilization, and plant growth, thereby facilitating the accumulation of SOC (Zheng, Liu, et al., 2022). This process involves enhancing rhizospheric microbial biomass, nutrient dynamics, and mineralization processes (Ling et al., 2022). Application of biochar can increase SOC levels by up to 30%, depending on biochar type and application rate (Wang et al., 2023). Long-term field studies with maize indicated substantial increases in SOC, particulate organic C, and microbial biomass C with biochar application, demonstrating enhanced soil quality and C sequestration potential over time (Yang et al., 2018). Research done by Ling et al. (2022) on maize-derived biochar pyrolyzed at different temperatures revealed shifts in labile and persistent SOC pools, impacting dissolved organic matter diversity and potential SOC loss due to mineralization. Biochar-amended soils exhibited higher levels of lipids, carbohydrates, and proteins/amino sugars, with CO_2-C emissions primarily from persistent organic matter decomposition. In addition to biochar application, implementing a no-tillage system has been observed to further enhance the C sequestration efficacy of biochar (Virk et al., 2021). This is achieved by minimizing soil disturbance and thereby extending the residence time of biochar within the soil. Additionally, biochar application can foster the formation of soil inorganic C in high pH calcareous soils, further contributing to long-term soil C storage (Liu et al., 2023).

Recent interest in biochar as a slow-release nutrient-based fertilizer has grown, driven by its potential to enhance crop growth while mitigating chemical losses, such as reactive N compounds (Ramesh & Raghavan, 2024). Field studies over 3 years evaluated various biochar types at 10 Mg/ha and observed reductions in CO_2 and N_2O emissions in sandy loam soils, but limited effects in clay loam soils. Biochar combined with manure also showed promise in reducing N_2O emissions in sandy soils (Abagandura et al., 2019). However, widespread adoption of biochar in agriculture necessitates careful consideration of production methods, transport logistics, soil pH adjustments, and nutrient interactions. Integration of biochar into soils also reduces reliance on synthetic chemical fertilizers, thereby cutting input costs (Wang, Nie, et al., 2021).

Additionally, incorporating biochar production and utilization into agricultural practices not only enhances crop yields but also supports environmental management. Biochar's capacity for C sequestration opens opportunities in the C-credit market (Amonette et al., 2021). Producers of biochar, including farmers and communities, may qualify for C-credits, which offer financial incentives for sustainable practices and contribute to achieving climate action goals. Capitalizing on CO_2 sequestration through C markets can improve the economic feasibility of biochar production systems. This financial approach has the potential to make biochar production profitable and reduce its costs to a level that supports widespread adoption across various sectors. These benefits underscore biochar's role in advancing soil sustainability while also fostering economic opportunities through ecological responsibility.

1.4.3.6 Renewable energy production

Biochar production extends far beyond its environmental benefits, significantly impacting local economies and contributing to energy sustainability (Senthil & Lee, 2021). The process involves multiple stages, beginning with biomass collection, followed by pyrolysis technology for biochar production, and the distribution of both biochar and its by-products (Fig. 1.7). Among various biomass feedstocks,

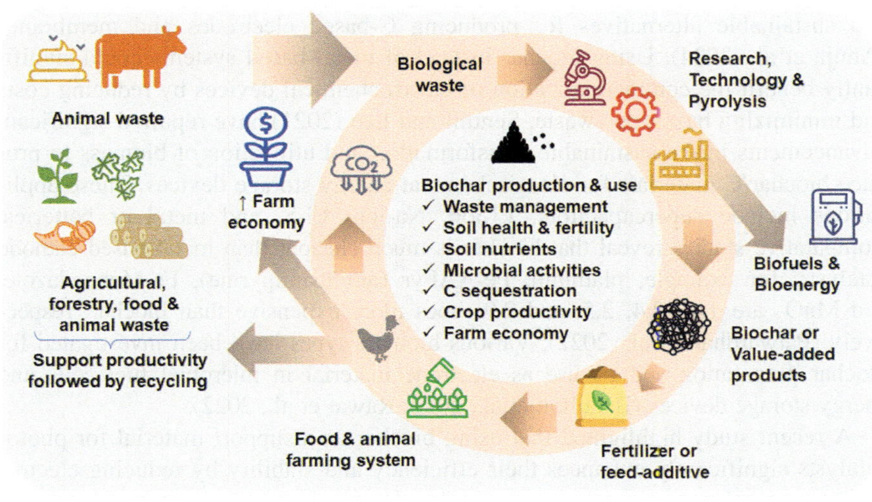

FIGURE 1.7

Biochar production, associated benefits, and its broader applications. Technological advancements in biochar production aimed at its integration into sustainable practices, reinforcing its value in the circular bioeconomy, socio-economic benefits and improving agricultural and environmental sustainability. Biochar production focusing on embedding biochar into sustainable practices, enhancing its role within the circular bioeconomy, and advancing agricultural and environmental sustainability.

livestock and poultry waste contains numerous valuable and recyclable components, including nutrients, organic matter, solids, energy, and fiber. These components can be utilized as fertilizer, soil amendments, energy sources, and even building materials. Manure is also a rich source of inexpensive cellulosic biomass, comprising lignocellulose and mineral nutrients. Generally, the C content in manure ranges from 38% to 54%, while the N content ranges from 1% to 5%. However, its composition varies depending on the type of animal and its diet. Therefore biochar produced from animal manure biomass could be used as a bioresource for developing renewable energy and value-added products (Hoyos-Sebá et al., 2024). Furthermore, the coproduction of syngas and bio-oil during biochar pyrolysis adds substantial economic value (Urrutia et al., 2022). Syngas, a mixture of CO, H, and methane, can be utilized for heat generation, electricity production through combustion, or further processed into synthetic fuels (Woolcock & Brown, 2013). Bio-oil, a liquid product rich in organic compounds, serves as a valuable feedstock for renewable fuels and chemicals production, enhancing the economic viability of biochar production systems (Maresca et al., 2024; Zhang et al., 2024; Zheng, Zhu, et al., 2023).

In addition, biomass composed of carbohydrates, proteins, and lipids, contains essential elements such as C, N, H, and O, together with chlorine, sulfur, and trace metals (Kumar, Elad, et al., 2018). The recent advancements highlight the potential of C-rich biomass residues and organic waste materials as cost-effective and sustainable alternatives for producing C-based electrodes and membranes (Ahuja et al., 2024). Using biochar instead of metal-based systems could significantly benefit the commercialization of electrochemical devices by reducing costs and minimizing hazardous waste. Senthil and Lee (2021) have reported significant advancements in the sustainable transformation and utilization of biomass to produce biochar/C materials for electrochemical energy storage devices. These applications include supercapacitors, Li-ion, Na-ion, Li-S, and metal-air batteries. Comparative studies reveal that biochar is much cheaper than metal-based cathode catalysts; for example, platinum, Fe-AAPyr (aminoantipyrine), Fe-Mebendazole, and MnO_2 are 115, 2.4, 2.5, and 2.9 times more expensive than biochar, respectively (Patwardhan et al., 2022). Various biomass types have been investigated for biochar preparation and its use as electrode material in microbial fuel cells and energy storage devices (Bataillou et al., 2022; Rawat et al., 2022).

A recent study highlighted that using biochar as a support material for photocatalysts significantly enhances their efficiency and stability by reducing electron–hole recombination rates, narrowing band gaps, improving charge separation, and enhancing electron transport (Mohtaram et al., 2024). In photocatalytic CO_2 reduction, biochar increases stability, recyclability, and reusability of the catalytic system. The inherent stability of biochar's carbonaceous material protects the photocatalyst from degradation and harsh reaction conditions, extending its operational life. Integrating biochar into these systems enhances photocatalytic activity, contributing to sustainable C management by reducing band gaps, enhancing electron transport and charge separation, and increasing surface area. Ongoing research promises advancements in solar fuel production and CO_2 conversion, guiding us toward a sustainable, low-C energy future (Mohtaram et al., 2024; Senthil & Lee, 2021).

1.5 Feed additive and aquaculture

Biochar, derived from the pyrolysis of organic biomass, is gaining traction as an innovative animal feed supplement (Fig. 1.7; Man et al., 2021). Its highly porous structure allows for the adsorption of toxins and mycotoxins, which enhances digestive health and nutrient uptake. This property not only boosts growth performance in ruminants (cattle and goats), pigs, poultry (chickens and ducks), and fish but also contributes to environmental sustainability by reducing methane emissions from ruminants (Lan et al., 2016; Schmidt et al., 2019; Toth & Dou, 2015), a significant GHG. Biochar supplementation has demonstrated various benefits, such as improved blood profiles, egg production, pathogen resistance, and better overall animal health (Islam et al., 2014; Prasai et al., 2018). Yıldızlı et al. (2021) reported that biochar directly reduced inflammatory cytokine levels in activated mammalian macrophages, indicating its potential to modulate immune responses. Furthermore, the mineral content in biochar, including Ca and Mg, has been shown to decrease the incidence of cracked eggs in poultry (Islam et al., 2014). It also supports the growth and survival of fish and improves lipid profiles in ducks by increasing high-density lipoproteins and decreasing low-density lipoproteins levels. Furthermore, biochar can act as an alternative to antibiotics, promoting a healthier gut microbiome and reducing pathogenic bacteria (Willson et al., 2019). Its application in animal farming also aids in nutrient recycling, reducing the need for chemical fertilizers (Osman et al., 2022). McAvoy et al. (2020) also demonstrated that adding biochar to grain-based diets improved diet digestibility and influenced some ruminal parameters in lambs, although these benefits did not translate into significant improvements in average daily gain or feed conversion efficiencies. Additionally, when given a choice, lambs showed a moderate preference for biochar-enriched rations, incorporating around 1.2% biochar into their diet without negatively affecting ration palatability (McAvoy et al., 2020). However, ensuring consistent biochar quality, effective proportion in animal diets, and adhering to regulatory standards, such as controlled pyrolysis temperatures, are crucial for its safe and effective use.

1.6 Economic and social benefits

Biochar technology provides substantial environmental, economic, and social benefits, making it a critical tool for sustainable development (Latawiec et al., 2021) (Fig. 1.7). By minimizing waste and optimizing resource efficiency, biochar supports closed-loop waste management systems and addresses various sustainability challenges (Rogers et al., 2022). Economically, biochar offers significant advantages by creating new revenue streams for farmers and local entrepreneurs, particularly through small-scale production units in rural areas (Joseph, 2012). Transforming agricultural waste into biochar improves soil fertility and water retention while reducing dependence on expensive chemical fertilizers and irrigation. This not only lowers production costs but also increases crop yields and financial stability for farmers, thereby stimulating local economic growth (Latawiec et al., 2017).

Additionally, community-based biochar initiatives foster social cohesion and collaboration, leading to the development of new markets and job opportunities. Furthermore, Afshar and Mofatteh (2024) have reported that the biochar industry, valued at approximately US$541.8 million in 2023, however, it is anticipated to experience significant growth, with a projected annual expansion rate of 13.9% from 2024 to 2030. Gwenzi et al. (2015) have demonstrated that biochar feedstocks in Zimbabwe are estimated at 9.9 Mt/year, primarily from manure (88%) and firewood (10%). This would produce 3.5 Mt/year of biochar, 1.7 Mt/year of bio-oil, and 3.1 Mt/year of synthetic gas. Applying 3.5 Mt/year of biochar, which contains approximately 63% C, could sequester about 2.2 Mt/year of soil C. This approach, as an integral component of the circular bioeconomy, would reduce environmental and public health risks, lower GHG emissions, and enhance socioeconomic benefits, thereby advancing SDGs through its emphasis on waste reduction and resource efficiency (Mohammed et al., 2024). A systems-oriented analysis using life-cycle assessment (LCA) and cost–benefit analysis (CBA) for biochar in a rural Indonesian village found that soil amendment for agricultural production was the most beneficial, with annual improvements of −26 ecopoints (LCA) and −173 USD (CBA) per household (Sparrevik et al., 2014). The C sequestration and increased agricultural yield outweighed the negative impacts of biochar production costs. Conversely, using biochar in briquettes for cooking fuel resulted in negative net effects (85 ecopoints and 176 USD), due to high production emissions and labor intensity. These findings highlight the need to prioritize C storage and agricultural benefits in biochar utilization for sustainable practices (Sparrevik et al., 2014). Therefore, affective policies advocacy for biochar research, production, and applications can further enhance its impact on building resilient and regenerative systems.

1.7 Several other uses of biochar

Biochar's potential extends beyond soil amendment, offering valuable applications across various industries as listed in Table 1.6. Schmidt (2013) highlighted that biochar can serve as a storage medium for volatile nutrients, and insulation in construction. Activated C, with its odor-neutralizing properties, is highly effective

Table 1.6 Additional applications of biochar or biochar-based products beyond agriculture and environmental uses.

Specific use	Biochar type	Specific response	References
Insulating material	Poultry litter waste	Biochar at 5% in cryogels increased its compressive strength by 30% and thermal conductivity	Lazzari et al. (2024)
Biochar-based bricks	–	Biochar in concrete insulation saves about 59–65 kg of CO_2/t while offering clear economic benefits	Osman et al. (2024)
Biodiesel production	–	Biochar from agricultural waste can be used for biodiesel production	Kusuma et al. (2024)
Supercapacitor	–	Metal oxide-agricultural waste biochar composites showed potential for sustainable high-energy supercapacitors	Khedulkar et al. (2024)
Air purification	Bark-derived biochar	Biochar adsorbs volatile organic compounds effectively, with up to 75% efficiency for formaldehyde and 69% for methanol. Smaller particles (<100 μm) offer superior adsorption and are ideal for indoor air remediation	Zouari et al. (2024)
Controlling CO_2 and particulate matter 2.5 ($PM_{2.5}$)	Raw bamboo	Activated bamboo biochar achieved 91.2% CO_2 and 89.2% $PM_{2.5}$ removal efficiency	Jagnade and Panwar (2024)
Protection against electromagnetic (EM) radiation	pinewood, coffee husk, sugarcane bagasse, and maize cob	Biochar loading improved biochar/HDPE composites: 30% pine wood biochar enhanced thermal conductivity to 2.219 W/mK, while 40% pine wood biochar maximized electrical conductivity to 4.67×10^{-7} S/cm and EM shielding effectiveness to 44.03 dB at 2.1 GHz	Fenta and Ali (2024)
Indoor plasters with excellent insulation	–	Biochar enhanced the strength, and durability of cement mortar through pozzolanic action	Habibur Rahman Sobuz et al. (2024)
Adsorbs smells and toxins	Tea plant stem	Metal-organic adsorbents on tea stems effectively removed ammonia, toluene, and butyric acid with high efficiency	Dai et al. (2024)

(Continued)

Table 1.6 Additional applications of biochar or biochar-based products beyond agriculture and environmental uses. *Continued*

Specific use	Biochar type	Specific response	References
biochar-mud plaster	–	5% biochar improved compressive strength and reduced GWP by 720 kg CO_2/t, achieving a 200% saving	Labianca et al. (2024)
Active carbon filter	–	Biochar-modified fillers and bioretention guide the application and optimization of biochar as a bioretention filler	Xiong et al. (2022)
Composting toilets	Bamboo	The biochar-based formulation effectively degrades human waste, reduces odors, lowers heavy metals, and improves compost quality, with hemicellulose and cellulose degradation rates of 73.9% and 62.4%, respectively	Borker et al. (2024)

in deodorants and oral care products. Consequently, biomass-derived activated C offers significant potential for improving human health and well-being through its diverse applications in cosmetics (Budiman et al., 2024). For example, the study assessed the impact of incorporating orange peel biochar particles into amino silane-grafted areca fiber on the mechanical, thermal conductivity, and dielectric properties of epoxy resin biocomposites, suggesting that the enhanced epoxy biocomposite could be suitable for applications in food storage containers and packaging materials (Alshahrani & Prakash, 2024). It also has uses in energy storage within batteries, water filtration in sewage plants, and as a silage agent or feed supplement (Man et al., 2021). These applications maximize its utility before eventual integration into agricultural or waste management systems. In the medium term, biochar could potentially replace oil as a primary industrial raw material, provided that sustainable practices are adopted. This shift is essential for ensuring long-term environmental and societal viability, demonstrating biochar's pivotal role in a sustainable industrial future.

1.8 Conclusion and future perspective

Biochar has emerged as a significant tool in sustainable agriculture and environmental management due to its ability to enhance soil health, increase crop yields, and mitigate climate change impacts. Produced through the pyrolysis of organic biomass, biochar offers numerous benefits, including C sequestration, soil remediation, and reduction of GHG emissions. The diverse feedstocks used in biochar production contribute to its varied physical and chemical properties, making it adaptable to different agronomic and environmental contexts. Despite its potential, substantial research gaps remain. This comprehensive reviews and interdisciplinary discussions can address these gaps, facilitating knowledge exchange and innovation in biochar technology. Such efforts will help integrate biochar into sustainable practices, promoting its use as a valuable resource in a circular bioeconomy and enhancing overall agricultural and environmental sustainability (Fig. 1.7). Additionally, scaling up biochar production and application processes is essential for its broader adoption. Future research should focus on optimizing feedstock selection and pyrolysis conditions to tailor biochar properties for specific applications. Long-term studies are needed to evaluate the sustainability and efficacy of biochar in various environmental settings and agricultural systems. Analyzing the economic factors and policy frameworks supporting biochar implementation is crucial. This involves evaluating cost–benefit dynamics and the role of financial incentives and regulations in promoting its adoption.

Acknowledgments

P.K. thankfully acknowledges financial support by the European Union's Horizon 2020 research and Innovation Program under the Marie Skłodowska-Curie grant agreement No 847639 and from the Ministry of Education and Science. A.K. is thankful to University of Tuscia, Viterbo, Italy for providing researcher grant from the Excellence Landscape 4.0.

Credit author statement

A.K.: conceptualization, formal analysis, investigation, methodology, resources, visualization, writing—original draft, writing—review and editing, supervision. P.K.: conceptualization, methodology, investigation, writing—original draft, writing—review and editing; M.K.S.: conceptualization, writing—original draft. M.N.V.P.: conceptualization, writing—review and editing.

Conflicts of interest

The authors declare that they have no known competing financial interests or personal relationships that could have appeared to influence the work reported in this paper.

References

Ababsa, N., Boudjabi, S., & Chenchouni, H. (2023). Biochar amendments changed soil properties and improved cereal crop growth under salt stress. *Journal of Soil Science and Plant Nutrition*, *23*(4), 4912–4925. Available from https://doi.org/10.1007/s42729-023-01453-7, https://www.springer.com/journal/42729.

Abagandura, G. O., Chintala, R., Sandhu, S. S., Kumar, S., & Schumacher, T. E. (2019). Effects of biochar and manure applications on soil carbon dioxide, methane, and nitrous oxide fluxes from two different soils. *Journal of Environmental Quality*, *48*(6), 1664–1674. Available from https://doi.org/10.2134/jeq2018.10.0374, https://www.agronomy.org/publications/jeq.

Adhikari, S., Timms, W., & Mahmud, M. A. P. (2022). Optimising water holding capacity and hydrophobicity of biochar for soil amendment – a review. *Science of the Total Environment*, *851*. Available from https://doi.org/10.1016/j.scitotenv.2022.158043, http://www.elsevier.com/locate/scitotenv.

Afshar, M., & Mofatteh, S. (2024). Biochar for a sustainable future: Environmentally friendly production and diverse applications. *Results in Engineering*, *23*, 102433. Available from https://doi.org/10.1016/j.rineng.2024.102433.

Ahuja, V., Palai, A. K., Kumar, A., Patel, A. K., Farooque, A. A., Yang, Y. H., & Bhatia, S. K. (2024). Biochar: Empowering the future of energy production and storage. *Journal of Analytical and Applied Pyrolysis*, *177*. Available from https://doi.org/10.1016/j.jaap.2024.106370, https://www.sciencedirect.com/science/journal/01652370.

Ajeng, A. A., Abdullah, R., & Ling, T. C. (2023). Biochar-Bacillus consortium for a sustainable agriculture: Physicochemical and soil stability analyses. *Biochar*, *5*(1). Available from https://doi.org/10.1007/s42773-023-00215-z, https://www.springer.com/journal/42773.

Ajeng, A. A., Abdullah, R., Ling, T. C., Ismail, S., Lau, B. F., Ong, H. C., Chew, K. W., Show, P. L., & Chang, J. S. (2020). Bioformulation of biochar as a potential inoculant carrier for sustainable agriculture. *Environmental Technology and Innovation*, *20*. Available from https://doi.org/10.1016/j.eti.2020.101168, http://www.journals.elsevier.com/environmental-technology-and-innovation/.

Akgül, A., Akın, S. Ş., Güleç, F., & Kazanç, F. (2024). Biochar gasification: Insights from pyrolysis atmospheres and gasification heating rates. *Fuel*, *360*, 130469. Available from https://doi.org/10.1016/j.fuel.2023.130469.

Ali, S., Rizwan, M., Qayyum, M. F., Ok, Y. S., Ibrahim, M., Riaz, M., Arif, M. S., Hafeez, F., Al-Wabel, M. I., & Shahzad, A. N. (2017). Biochar soil amendment on alleviation of drought and salt stress in plants: A critical review. *Environmental Science and Pollution Research*, *24*(14), 12700–12712. Available from https://doi.org/10.1007/s11356-017-8904-x, http://www.springerlink.com/content/0944-1344.

Alonso-Gómez, L. A., Celis-Carmona, D. D., Rodríguez-Sánchez, Y. F., Castro-Ladino, J. R., & Solarte-Toro, J. C. (2024). Biochar production from cassava waste biomass: A techno-economic development approach in the Colombian context. *Bioresource Technology Reports*, *26*, 101872. Available from https://doi.org/10.1016/j.biteb.2024.101872.

Alshahrani, H., & Prakash, V. R. A. (2024). Effect of silane-grafted orange peel biochar and areca fibre on mechanical, thermal conductivity and dielectric properties of epoxy resin composites. *Biomass Conversion and Biorefinery*, *14*(6), 8081–8089. Available from https://doi.org/10.1007/s13399-022-02801-w, https://www.springer.com/journal/13399.

Amalina, F., Krishnan, S., Zularisam, A. W., & Nasrullah, M. (2023). Recent advancement and applications of biochar technology as a multifunctional component towards sustainable environment. *Environmental Development*, *46*, 100819. Available from https://doi.org/10.1016/j.envdev.2023.100819.

Amalina, F., Syukor Abd Razak, A., Krishnan, S., Sulaiman, H., Zularisam, A. W., & Nasrullah, M. (2022). Advanced techniques in the production of biochar from lignocellulosic biomass and environmental applications. *Cleaner Materials*, *6*, 100137. Available from https://doi.org/10.1016/j.clema.2022.100137.

Amonette, J. E., Archuleta, J. G., Fuchs, M. R., Hills, K. M., Yorgey, G. G., Flora, G., et al. (2021). *Biomass to biochar: Maximizing the carbon value*. Center for Sustaining Agriculture and Natural Resources.

Amusat, S. O., Kebede, T. G., Dube, S., & Nindi, M. M. (2021). Ball-milling synthesis of biochar and biochar−based nanocomposites and prospects for removal of emerging contaminants: A review. *Journal of Water Process Engineering*, *41*. Available from https://doi.org/10.1016/j.jwpe.2021.101993, http://www.journals.elsevier.com/journal-of-water-process-engineering/.

Arshad, U., Altaf, M. T., Liaqat, W., Ali, M., Shah, M. N., Jabran, M., & Ali, M. A. (2024). Biochar: Black gold for sustainable agriculture and fortification against plant pathogens—a review. *Journal of Crop Health*, *76*(2), 385−396. Available from https://doi.org/10.1007/s10343-023-00952-y.

Azadi, N., & Raiesi, F. (2021). Biochar alleviates metal toxicity and improves microbial community functions in a soil co-contaminated with cadmium and lead. *Biochar*, *3*(4), 485−498. Available from https://doi.org/10.1007/s42773-021-00123-0, http://springer.com/journal/42773.

Bataillou, G., Lee, C., Monnier, V., Gerges, T., Sabac, A., Vollaire, C., & Haddour, N. (2022). Cedar wood-based biochar: Properties, characterization, and applications as anodes in microbial fuel cell. *Applied Biochemistry and Biotechnology*, *194*(9), 4169−4186. Available from https://doi.org/10.1007/s12010-022-03997-3, https://www.springer.com/journal/12010.

Bateman, A. M., & Muñoz-Rojas, M. (2019). To whom the burden of soil degradation and management concerns. *Advances in chemical pollution, environmental management and protection*, *4*, 1−22. Available from https://doi.org/10.1016/bs.apmp.2019.07.001, http://www.sciencedirect.com/bookseries/advances-in-chemical-pollution-environmental-management-and-protection.

Bhagawati, R., Paul, S., & Dambale, A. S. (2020). *Farm waste management Agronomy*.

Blenis, N., Hue, N., Maaz, T. M. C., & Kantar, M. (2023). Biochar production, modification, and its uses in soil remediation: A review. *Sustainability*, *15*(4), 3442. Available from https://doi.org/10.3390/su15043442.

Bolan, S., Hou, D., Wang, L., Hale, L., Egamberdieva, D., Tammeorg, P., Li, R., Wang, B., Xu, J., Wang, T., Sun, H., Padhye, L. P., Wang, H., Siddique, K. H. M., Rinklebe, J., Kirkham, M. B., & Bolan, N. (2023). The potential of biochar as a microbial carrier for agricultural and environmental applications. *Science of the Total Environment*, *886*, 163968. Available from https://doi.org/10.1016/j.scitotenv.2023.163968.

Borker, S. S., Thakur, A., Pandey, K. K., Sharma, P., Manyapu, V., Khatri, A., & Kumar, R. (2024). Nutrient recycling of source-separated human faeces using biochar immobilized indigenous psychrotrophic bacteria for sustaining the agroecosystems of northwestern Himalaya. *Applied Biological Chemistry*, *67*(1). Available from https://doi.org/10.1186/s13765-024-00887-6, https://www.springer.com/journal/13765.

Brar, B., Saharan, B. S., Seth, C. S., Kamboj, A., Surekha., Bala, K., Rajput, V. D., Minkina, T., Wong, M. H., Kumar, D., Sadh, P. K., & Duhan, J. S. (2024). Nanobiochar: Soil and plant interactions and their implications for sustainable agriculture. *Biocatalysis and Agricultural Biotechnology*, *57*. Available from https://doi.org/10.1016/j.bcab.2024.103077, https://www.sciencedirect.com/science/journal/18788181.

Brassard, P., Godbout, S., & Raghavan, V. (2017). Pyrolysis in auger reactors for biochar and bio-oil production: A review. *Biosystems Engineering*, *161*, 80–92. Available from https://doi.org/10.1016/j.biosystemseng.2017.06.020, http://www.elsevier.com/inca/publications/store/6/2/2/7/9/5/index.htt.

Brewer, C. E., Schmidt-Rohr, K., Satrio, J. A., & Brown, R. C. (2009). Characterization of biochar from fast pyrolysis and gasification systems. *Environmental Progress and Sustainable Energy*, *28*(3), 386–396. Available from https://doi.org/10.1002/ep.10378, http://www3.interscience.wiley.com/cgi-bin/fulltext/122544837/PDFSTART.

Brichi, L., Fernandes, J. V. M., Silva, B. M., Vizú, J. D. F., Junior, J. N. G., & Cherubin, M. R. (2023). Organic residues and their impact on soil health, crop production and sustainable agriculture: A review including bibliographic analysis. *Soil Use and Management*, *39*(2), 686–706. Available from https://doi.org/10.1111/sum.12892, http://onlinelibrary.wiley.com/journal/10.1111/(ISSN)1475-2743.

Bruun, E. W., Müller-Stöver, D., Pedersen, B. N., Hansen, L. V., & Petersen, C. T. (2022). Ash and biochar amendment of coarse sandy soil for growing crops under drought conditions. *Soil Use and Management*, *38*(2), 1280–1292. Available from https://doi.org/10.1111/sum.12783, http://onlinelibrary.wiley.com/journal/10.1111/(ISSN)1475-2743.

Budiman, I., Amrita, R., Yuliansyah, Widyaningrum, B. A., & Fatriasari, W. (2024). Activated carbon for cosmetics applications. In E. T. Arung, W. Fatriasari, I. W. Kusuma, H. Kuspradini, K. Shimizu, Y-u. Kim, N. I. W. Azelee, & Z. Edis (Eds.), *Biomass-based cosmetics: Research Trends and Future Outlook* (pp. 217–237). Singapore: Springer Nature.

Cao, Q., An, T., Xie, J., Liu, Y., Xing, L., Ling, X., & Chen, C. (2022). Insight to the physiochemical properties and DOM of biochar under different pyrolysis temperature and modification conditions. *Journal of Analytical and Applied Pyrolysis*, *166*, 105590. Available from https://doi.org/10.1016/j.jaap.2022.105590.

Chen, B., Zeng, H., Yang, F., Yang, Y., Qiao, Z., Zhao, X., Wang, L., & Wu, F. (2024). Functional biochar as sustainable precursors to boost the anaerobic digestion of waste activated sludge from a circular economy perspective: A review. *Biochar*, *6*(1). Available from https://doi.org/10.1007/s42773-024-00345-y, https://www.springer.com/journal/42773.

Chen, J., Liu, X., Li, L., Zheng, J., Qu, J., Zheng, J., Zhang, X., & Pan, G. (2015). Consistent increase in abundance and diversity but variable change in community composition of bacteria in topsoil of rice paddy under short term biochar treatment across three sites from South China. *Applied Soil Ecology*, *91*, 68–79. Available from https://doi.org/10.1016/j.apsoil.2015.02.012, http://www.elsevier.com/inca/publications/store/5/2/4/5/1/8/index.htt.

Chen, Z. S., Liu, T., Dong, J. F., Chen, G., Li, Z., Zhou, J. L., & Chen, Z. (2023). Sustainable application for agriculture using biochar-based slow-release fertilizers: A review. *ACS Sustainable Chemistry and Engineering*, *11*(1), 1–12. Available from https://doi.org/10.1021/acssuschemeng.2c05691, http://pubs.acs.org/journal/ascecg.

Chung, W. J., Chang, S. W., Chaudhary, D. K., Shin, J. D., Kim, H., Karmegam, N., Govarthanan, M., Chandrasekaran, M., & Ravindran, B. (2021). Effect of biochar amendment on compost quality, gaseous emissions and pathogen reduction during in-vessel composting of chicken manure. *Chemosphere*, *283*. Available from https://doi.org/10.1016/j.chemosphere.2021.131129, http://www.elsevier.com/locate/chemosphere.

Dai, X., Wang, L., Man, Z., He, Y., Wang, W., Lin, F., Zhu, J., Liu, D., Xiao, H., & Wang, K. (2024). In situ synthesis of porous metal-organic frameworks NH2-UiO-66 on tea stem biochar and application in odours adsorption. *Environmental Pollution*, *353*, 124168. Available from https://doi.org/10.1016/j.envpol.2024.124168.

Danesh, P., Niaparast, P., Ghorbannezhad, P., & Ali, I. (2023). Biochar production: Recent developments, applications, and challenges. *Fuel*, *337*, 126889. Available from https://doi.org/10.1016/j.fuel.2022.126889.

Ding, Z., Hamann, K. T., & Grundmann, P. (2024). Enhancing circular bioeconomy in Europe: Sustainable valorization of residual grassland biomass for emerging bio-based value chains. *Sustainable Production and Consumption*, *45*, 265–280. Available from https://doi.org/10.1016/j.spc.2024.01.008, https://www.sciencedirect.com/science/journal/23525509.

Ding, Z., Majrashi, M. A., Ghoneim, A. M., Ali, E. F., Eissa, M. A., & Shal, R. E. (2022). Irrigation and biochar effects on pearl millet and kinetics of ammonia volatilization from saline sandy soils. *Journal of Soil Science and Plant Nutrition*, *22*(2), 1546–1558. Available from https://doi.org/10.1007/s42729-021-00753-0, https://www.springer.com/life + sciences/plant + sciences/journal/42729.

Ding, Z., Zhou, Z., Lin, X., Zhao, F., Wang, B., Lin, F., Ge, Y., & Eissa, M. A. (2020). Biochar impacts on NH3-volatilization kinetics and growth of sweet basil (*Ocimum basilicum* L.) under saline conditions. *Industrial Crops and Products*, *157*. Available from https://doi.org/10.1016/j.indcrop.2020.112903, http://www.elsevier.com/inca/publications/store/5/2/2/8/2/5.

Downie, A., Crosky, A., & Munroe. (2012). *Physical properties of biochar. Biochar for environmental management* (pp. 45–64). Routledge.

Egamberdieva, D., Alaylar, B., Kistaubayeva, A., Wirth, S., & Bellingrath-Kimura, S. D. (2022). Biochar for improving soil biological properties and mitigating salt stress in plants on salt-affected soils. *Communications in Soil Science and Plant Analysis*, *53*(2), 140–152. Available from https://doi.org/10.1080/00103624.2021.1993884, http://www.tandfonline.com/loi/lcss20.

Elkhlifi, Z., Iftikhar, J., Sarraf, M., Ali, B., Saleem, M. H., Ibranshahib, I., Bispo, M. D., Meili, L., Ercisli, S., Torun Kayabasi, E., Alemzadeh Ansari, N., Hegedűsová, A., & Chen, Z. (2023). Potential role of biochar on capturing soil nutrients, carbon sequestration and managing environmental challenges: A review. *Sustainability*, *15*(3), 2527. Available from https://doi.org/10.3390/su15032527.

Fan, S., Wang, S., Zhang, W., Fan, X., Huang, Z., Zhou, N., & Xu, H. (2024). KOH activated biochar loaded with MnAl-layered double hydroxides for superior adsorption of tetracycline from the aquatic environment. *Journal of Industrial and Engineering Chemistry*, *134*, 474–483. Available from https://doi.org/10.1016/j.jiec.2024.01.010, https://www.sciencedirect.com/science/journal/1226086X.

Fan, Z., Zhou, X., Peng, Z., Wan, S., Fan Gao, Z., Deng, S., Tong, L., Han, W., & Chen, X. (2023). Co-pyrolysis technology for enhancing the functionality of sewage sludge biochar and immobilizing heavy metals. *Chemosphere*, *317*, 137929. Available from https://doi.org/10.1016/j.chemosphere.2023.137929.

FAO. (2018). *The future of food and agriculture: Alternative pathways to 2050. Summary version*. Food and Agriculture Organization (FAO). Available from https://openknowledge.fao.org/server/api/core/bitstreams/e51e0cf0-4ece-428c-8227-ff6c51b06b16/content.

Fascella, G., Mammano, M. M., D'Angiolillo, F., Pannico, A., & Rouphael, Y. (2020). Coniferous wood biochar as substrate component of two containerized Lavender species: Effects on morpho-physiological traits and nutrients partitioning. *Scientia Horticulturae*, *267*, 109356. Available from https://doi.org/10.1016/j.scienta.2020.109356.

Fenta, A. A., & Ali, A. N. (2024). Development of biochar/HDPE composites and characterization of the effects of carbon loadings on the electromagnetic shielding properties. *Heliyon*, *10*(2). Available from https://doi.org/10.1016/j.heliyon.2024.e24424, https://www.sciencedirect.com/science/journal/24058440.

Fu, Z., Zhao, J., Guan, D., Wang, Y., Xie, J., Zhang, H., Sun, Y., Zhu, J., & Guo, L. (2024). A comprehensive review on the preparation of biochar from digestate sources and its application in environmental pollution remediation. *Science of the Total Environment*, *912*, 168822. Available from https://doi.org/10.1016/j.scitotenv.2023.168822.

Gao, G., Yan, L., Tong, K., Yu, H., Lu, M., Wang, L., & Niu, Y. (2024). The potential and prospects of modified biochar for comprehensive management of salt-affected soils and plants: A critical review. *Science of the Total Environment*, *912*, 169618. Available from https://doi.org/10.1016/j.scitotenv.2023.169618.

Gao, N., Kamran, K., Quan, C., & Williams, P. T. (2020). Thermochemical conversion of sewage sludge: A critical review. *Progress in Energy and Combustion Science*, *79*. Available from https://doi.org/10.1016/j.pecs.2020.100843, https://www.journals.elsevier.com/progress-in-energy-and-combustion-science.

Garcia-Nunez, J. A., Pelaez-Samaniego, M. R., Garcia-Perez, M. E., Fonts, I., Abrego, J., Westerhof, R. J. M., & Garcia-Perez, M. (2017). Historical developments of pyrolysis reactors: A review. *Energy and Fuels*, *31*(6), 5751−5775. Available from https://doi.org/10.1021/acs.energyfuels.7b00641, http://pubs.acs.org/journal/enfuem.

Gavili, E., Moosavi, A. A., & Kamgar Haghighi, A. A. (2019). Does biochar mitigate the adverse effects of drought on the agronomic traits and yield components of soybean? *Industrial Crops and Products*, *128*, 445−454. Available from https://doi.org/10.1016/j.indcrop.2018.11.047, http://www.elsevier.com/inca/publications/store/5/2/2/8/2/5.

Ghodake, G. S., Shinde, S. K., Kadam, A. A., Saratale, R. G., Saratale, G. D., Kumar, M., Palem, R. R., AL-Shwaiman, H. A., Elgorban, A. M., Syed, A., & Kim, D. Y. (2021). Review on biomass feedstocks, pyrolysis mechanism and physicochemical properties of biochar: State-of-the-art framework to speed up vision of circular bioeconomy. *Journal of Cleaner Production*, *297*. Available from https://doi.org/10.1016/j.jclepro.2021.126645, https://www.journals.elsevier.com/journal-of-cleaner-production.

Gholizadeh, M., Meca, S., Zhang, S., Clarens, F., & Hu, X. (2024). Understanding the dependence of biochar properties on different types of biomass. *Waste Management*, *182*, 142−163. Available from https://doi.org/10.1016/j.wasman.2024.04.011, https://www.sciencedirect.com/science/journal/0956053X.

Gitau, J. K., Sundberg, C., Mendum, R., Mutune, J., & Njenga, M. (2019). Use of biochar-producing gasifier cookstove improves energy use efficiency and indoor air quality in rural households. *Energies*, *12*(22). Available from https://doi.org/10.3390/en12224285, https://www.mdpi.com/1996-1073/12/22.

Gullap, M. K., Severoglu, S., Karabacak, T., Yazici, A., Ekinci, M., Turan, M., & Yildirim, E. (2024). Biochar derived from hazelnut shells mitigates the impact of drought stress on soybean seedlings. *New Zealand Journal of Crop and Horticultural Science*, *52*(1), 19−37. Available from https://doi.org/10.1080/01140671.2022.2079680.

Gusiatin, M. Z., & Rouhani, A. (2023). Application of selected methods to modify pyrolyzed biochar for the immobilization of metals in soil: A review. *Materials*, *16*(23). Available from https://doi.org/10.3390/ma16237342, http://www.mdpi.com/journal/materials.

Gwenzi, W., Chaukura, N., Mukome, F. N. D., Machado, S., & Nyamasoka, B. (2015). Biochar production and applications in sub-Saharan Africa: Opportunities, constraints, risks and uncertainties. *Journal of Environmental Management*, *150*, 250–261. Available from https://doi.org/10.1016/j.jenvman.2014.11.027, https://www.sciencedirect.com/journal/journal-of-environmental-management.

Gwenzi, W., Nyambishi, T. J., Chaukura, N., & Mapope, N. (2018). Synthesis and nutrient release patterns of a biochar-based N–P–K slow-release fertilizer. *International Journal of Environmental Science and Technology*, *15*(2), 405–414. Available from https://doi.org/10.1007/s13762-017-1399-7, http://www.springerlink.com/content/1735-1472.

Gęca, M., Khalil, A. M., Tang, M., Bhakta, A. K., Snoussi, Y., Nowicki, P., Wiśniewska, M., & Chehimi, M. M. (2023). Surface treatment of biochar—methods, surface analysis and potential applications: A comprehensive review. *Surfaces*, *6*(2), 179–213. Available from https://doi.org/10.3390/surfaces6020013, http://www.mdpi.com/journal/surfaces.

Habibur Rahman Sobuz, M., Khan, M. H., Kawsarul Islam Kabbo, M., Alhamami, A. H., Aditto, F. S., Saziduzzaman Sajib, M., Johnson Alengaram, U., Mansour, W., Hasan, N. M. S., Datta, S. D., & Alam, A. (2024). Assessment of mechanical properties with machine learning modeling and durability, and microstructural characteristics of a biochar-cement mortar composite. *Construction and Building Materials*, *411*. Available from https://doi.org/10.1016/j.conbuildmat.2023.134281, https://www.sciencedirect.com/science/journal/09500618.

Hafeez, F., Rizwan, M., Saqib, M., Yasmeen, T., Ali, S., Abbas, T., Zia-Ur-Rehman, M., & Qayyum, M. F. (2019). Residual effect of biochar on growth, antioxidant defence and cadmium (Cd) accumulation in rice in a Cd contaminated saline soil. *Pakistan Journal of Agricultural Sciences*, *56*(1), 197–204. Available from https://doi.org/10.21162/PAKJAS/19.7546, https://pakjas.com.pk/papers/2945.pdf.

Hamidzadeh, Z., Ghorbannezhad, P., Ketabchi, M. R., & Yeganeh, B. (2023). Biomass-derived biochar and its application in agriculture. *Fuel*, *341*, 127701. Available from https://doi.org/10.1016/j.fuel.2023.127701.

Han, G., Lan, J., Chen, Q., Yu, C., & Bie, S. (2017). Response of soil microbial community to application of biochar in cotton soils with different continuous cropping years. *Scientific Reports*, *7*(1). Available from https://doi.org/10.1038/s41598-017-10427-6, http://www.nature.com/srep/index.html.

Haque, F., Fan, C., & Lee, Y. Y. (2023). From waste to value: Addressing the relevance of waste recovery to agricultural sector in line with circular economy. *Journal of Cleaner Production*, *415*, 137873. Available from https://doi.org/10.1016/j.jclepro.2023.137873.

Hawryluk-Sidoruk, M., Raczkiewicz, M., Krasucka, P., Duan, W., Mašek, O., Zarzycki, R., Kobyłecki, R., Pan, B., & Oleszczuk, P. (2024). Effect of biochar chemical modification (acid, base and hydrogen peroxide) on contaminants content depending on feedstock and pyrolysis conditions. *Chemical Engineering Journal*, *481*, 148329. Available from https://doi.org/10.1016/j.cej.2023.148329.

Hazrati, S., Farahbakhsh, M., Cerdà, A., & Heydarpoor, G. (2021). Functionalization of ultrasound enhanced sewage sludge-derived biochar: Physicochemical improvement and its effects on soil enzyme activities and heavy metals availability. *Chemosphere*, *269*, 128767. Available from https://doi.org/10.1016/j.chemosphere.2020.128767.

He, Q., Raheem, A., Ding, L., Xu, J., Cheng, C., & Yu, G. (2021). Combining wet torrefaction and pyrolysis for woody biochar upgradation and structural modification. *Energy Conversion and Management*, *243*, 114383. Available from https://doi.org/10.1016/j.enconman.2021.114383.

Heitkötter, J., & Marschner, B. (2015). Interactive effects of biochar ageing in soils related to feedstock, pyrolysis temperature, and historic charcoal production. *Geoderma*, *245-246*, 56–64. Available from https://doi.org/10.1016/j.geoderma.2015.01.012, http://www.elsevier.com/inca/publications/store/5/0/3/3/3/2.

Higashikawa, F. S., Silva, C. A., Carducci, C. E., Jindo, K., Kurtz, C., de Araújo, E. R., Sousa Resende, R., & Alves, D. P. (2023). Effects of the application of biochar on soil fertility status, and nutrition and yield of onion grown in a no-tillage system. *Archives of Agronomy and Soil Science*, *69*(2), 212–227. Available from https://doi.org/10.1080/03650340.2021.1978073, http://www.tandf.co.uk/journals/titles/03650340.asp.

Hoyos-Sebá, J. J., Arias, N. P., Salcedo-Mendoza, J., & Aristizábal-Marulanda, V. (2024). Animal manure in the context of renewable energy and value-added products: A review. *Chemical Engineering and Processing - Process Intensification*, *196*, 109660. Available from https://doi.org/10.1016/j.cep.2023.109660.

Huang, J., Zhu, C., Kong, Y., Cao, X., Zhu, L., Zhang, Y., Ning, Y., Tian, W., Zhang, H., Yu, Y., & Zhang, J. (2022). Biochar application alleviated rice salt stress via modifying soil properties and regulating soil bacterial abundance and community structure. *Agronomy*, *12*(2), 409. Available from https://doi.org/10.3390/agronomy12020409.

Huang, M., Zhang, Z., Zhai, Y., Lu, P., & Zhu, C. (2019). Effect of straw biochar on soil properties and wheat production under salinewater irrigation. *Agronomy*, *9*(8). Available from https://doi.org/10.3390/agronomy9080457, https://www.mdpi.com/2073-4395/9/8/457/pdf.

Huang, M., Zhang, Z., Zhu, C., Zhai, Y., & Lu, P. (2019). Effect of biochar on sweet corn and soil salinity under conjunctive irrigation with brackish water in coastal saline soil. *Scientia Horticulturae*, *250*, 405–413. Available from https://doi.org/10.1016/j.scienta.2019.02.077, http://www.elsevier.com/inca/publications/store/5/0/3/3/1/6.

Huang, Q., Song, S., Chen, Z., Hu, B., Chen, J., & Wang, X. (2019). Biochar-based materials and their applications in removal of organic contaminants from wastewater: State-of-the-art review. *Biochar*, *1*(1), 45–73. Available from https://doi.org/10.1007/s42773-019-00006-5.

Ighalo, J. O., Adeyanju, C. A., Iwuozor, K. O., Emenike, E. C., & Adeniyi, A. G. (2024). Retort co-carbonization of *Daniellia oliveri* leaves: Effect of cow dung co-feed on biochar properties. *Waste and Biomass Valorization*, *15*(7), 4235–4246. Available from https://doi.org/10.1007/s12649-024-02461-z, https://www.springer.com/journal/12649.

Intergovernmental Panel on Climate Change. (2021). *Unpublished content climate change 2021: The physical science basis*. Working group I contribution to the IPCC sixth assessment report. Cambridge University Press.

Ippolito, J. A., Cui, L., Kammann, C., Wrage-Mönnig, N., Estavillo, J. M., Fuertes-Mendizabal, T., Cayuela, M. L., Sigua, G., Novak, J., Spokas, K., & Borchard, N. (2020). Feedstock choice, pyrolysis temperature and type influence biochar characteristics: a comprehensive meta-data analysis review. *Biochar*, *2*(4), 421–438. Available from https://doi.org/10.1007/s42773-020-00067-x, http://springer.com/journal/42773.

Islam, M. M., Ahmed, S. T., Kim, Y. J., Mun, H. S., Kim, Y. J., & Yang, C. J. (2014). Effect of sea tangle (*Laminaria japonica*) and charcoal supplementation as alternatives to antibiotics on growth performance and meat quality of ducks. *Asian-Australasian*

Journal of Animal Sciences, *27*(2), 217–224. Available from https://doi.org/10.5713/ajas.2013.13314, http://www.ajas.info/.

Jagnade, P., & Panwar, N. L. (2024). Experiment investigation on the performance of activated bamboo biochar for the adsorption of CO_2 and PM 2.5. *Emergent Materials*, *7*(3), 1227–1238. Available from https://doi.org/10.1007/s42247-024-00624-1, https://www.springer.com/journal/42247.

Jaiswal, A. K., Frenkel, O., Debode, J., & Graber, E. R. (2024). *How does biochar influence plant biotic stress? Biochar for environmental management* (pp. 589–612). Informa UK Limited. Available from https://doi.org/10.4324/9781003297673-23.

Joseph, S. (2012). *Socio-economic assessment and implementation of small-scale biochar projects. Biochar for environmental management: Science and technology* (pp. 359–373). Taylor and Francis. Available from http://www.taylorandfrancis.com/books/details/9781849770552/, 10.4324/9781849770552.

Joseph, S., Cowie, A. L., Van Zwieten, L., Bolan, N., Budai, A., Buss, W., Cayuela, M. L., Graber, E. R., Ippolito, J. A., Kuzyakov, Y., Luo, Y., Ok, Y. S., Palansooriya, K. N., Shepherd, J., Stephens, S., Weng, Z., & Lehmann, J. (2021). How biochar works, and when it doesn't: A review of mechanisms controlling soil and plant responses to biochar. *GCB Bioenergy*, *13*(11), 1731–1764. Available from https://doi.org/10.1111/gcbb.12885, http://onlinelibrary.wiley.com/journal/10.1111/(ISSN)1757-1707.

Joseph, S., Kammann, C. I., Shepherd, J. G., Conte, P., Schmidt, H. P., Hagemann, N., Rich, A. M., Marjo, C. E., Allen, J., Munroe, P., Mitchell, D. R. G., Donne, S., Spokas, K., & Graber, E. R. (2018). Microstructural and associated chemical changes during the composting of a high temperature biochar: Mechanisms for nitrate, phosphate and other nutrient retention and release. *Science of the Total Environment*, *618*, 1210–1223. Available from https://doi.org/10.1016/j.scitotenv.2017.09.200, http://www.elsevier.com/locate/scitotenv.

Jukka, L., Miika, M., Lauri, L., Mirja, M., Ville, U., & Lassi, L. (2022). A financial and environmental sustainability of circular bioeconomy: A case study of short rotation coppice, biochar and greenhouse production in southern Finland. *Biomass and Bioenergy*, *163*, 106524. Available from https://doi.org/10.1016/j.biombioe.2022.106524.

Kanwal, S., Ilyas, N., Shabir, S., Saeed, M., Gul, R., Zahoor, M., Batool, N., & Mazhar, R. (2018). Application of biochar in mitigation of negative effects of salinity stress in wheat (*Triticum aestivum* L.). *Journal of Plant Nutrition*, *41*(4), 526–538. Available from https://doi.org/10.1080/01904167.2017.1392568.

Keshavarz Afshar, R., Hashemi, M., DaCosta, M., Spargo, J., & Sadeghpour, A. (2016). Biochar application and drought stress effects on physiological characteristics of *Silybum marianum*. *Communications in Soil Science and Plant Analysis*, *47*(6), 743–752. Available from https://doi.org/10.1080/00103624.2016.1146752.

Khaledi, S., Delbari, M., Galavi, H., Bagheri, H., & Chari, M. M. (2023). Effects of biochar particle size, biochar application rate, and moisture content on thermal properties of an unsaturated sandy loam soil. *Soil and Tillage Research*, *226*, 105579. Available from https://doi.org/10.1016/j.still.2022.105579.

Khan, S., Irshad, S., Mehmood, K., Hasnain, Z., Nawaz, M., Rais, A., Gul, S., Wahid, M. A., Hashem, A., Abd_Allah, E. F., & Ibrar, D. (2024). Biochar production and characteristics, its impacts on soil health, crop production, and yield enhancement: A review. *Plants*, *13*(2), 166. Available from https://doi.org/10.3390/plants13020166.

Khedulkar, A. P., Dang, V. D., Thamilselvan, A., Doong, Ra, & Pandit, B. (2024). Sustainable high-energy supercapacitors: Metal oxide-agricultural waste biochar composites paving the way for a greener future. *Journal of Energy Storage*, *77*. Available from https://doi.org/10.1016/j.est.2023.109723, https://www.sciencedirect.com/science/journal/2352152X.

Kolton, M., Graber, E. R., Tsehansky, L., Elad, Y., & Cytryn, E. (2017). Biochar-stimulated plant performance is strongly linked to microbial diversity and metabolic potential in the rhizosphere. *New Phytologist*, *213*(3), 1393–1404. Available from https://doi.org/10.1111/nph.14253, http://onlinelibrary.wiley.com/journal/10.1111/(ISSN)1469-8137.

Kopittke, P. M., Menzies, N. W., Wang, P., McKenna, B. A., & Lombi, E. (2019). Soil and the intensification of agriculture for global food security. *Environment International*, *132*. Available from https://doi.org/10.1016/j.envint.2019.105078, http://www.elsevier.com/locate/envint.

Kończak, M., Gao, Y., & Oleszczuk, P. (2019). Carbon dioxide as a carrier gas and biomass addition decrease the total and bioavailable polycyclic aromatic hydrocarbons in biochar produced from sewage sludge. *Chemosphere*, *228*, 26–34. Available from https://doi.org/10.1016/j.chemosphere.2019.04.029, http://www.elsevier.com/locate/chemosphere.

Kumar, A., Elad, Y., Tsechansky, L., Abrol, V., Lew, B., Offenbach, R., & Graber, E. R. (2018). Biochar potential in intensive cultivation of *Capsicum annuum* L. (sweet pepper): Crop yield and plant protection. *Journal of the Science of Food and Agriculture*, *98*(2), 495–503. Available from https://doi.org/10.1002/jsfa.8486, http://onlinelibrary.wiley.com/journal/10.1002/(ISSN)1097-0010.

Kumar, A., Joseph, S., Tsechansky, L., Privat, K., Schreiter, I. J., Schüth, C., & Graber, E. R. (2018). Biochar aging in contaminated soil promotes Zn immobilization due to changes in biochar surface structural and chemical properties. *Science of the Total Environment*, *626*, 953–961. Available from https://doi.org/10.1016/j.scitotenv.2018.01.157, http://www.elsevier.com/locate/scitotenv.

Kumar, A., Tsechansky, L., Lew, B., Raveh, E., Frenkel, O., & Graber, E. R. (2018). Biochar alleviates phytotoxicity in Ficus elastica grown in Zn-contaminated soil. *Science of the Total Environment*, *618*, 188–198. Available from https://doi.org/10.1016/j.scitotenv.2017.11.013, http://www.elsevier.com/locate/scitotenv.

Kumar, A., Friedman, H., Tsechansky, L., & Graber, E. R. (2021). Distinctive in-planta acclimation responses to basal growth and acute heat stress were induced in Arabidopsis by cattle manure biochar. *Scientific Reports*, *11*(1). Available from https://doi.org/10.1038/s41598-021-88856-7, http://www.nature.com/srep/index.html.

Kumar, A., Joseph, S., Graber, E. R., Taherysoosavi, S., Mitchell, D. R. G., Munroe, P., Tsechansky, L., Lerdahl, O., Aker, W., & Sæbø, M. (2021). Fertilizing behavior of extract of organomineral-activated biochar: Low-dose foliar application for promoting lettuce growth. *Chemical and Biological Technologies in Agriculture*, *8*(1). Available from https://doi.org/10.1186/s40538-021-00222-x, http://chembioagro.springeropen.com/.

Kumar, A., Schreiter, I. J., Wefer-Roehl, A., Tsechansky, L., Schüth, C., & Graber, E. R. (2016). *Production and utilization of biochar from organic wastes for pollutant control on contaminated sites. Environmental materials and waste: Resource recovery and pollution prevention* (pp. 91–116). Elsevier Inc. Available from http://www.sciencedirect.com/science/book/9780128038376, 10.1016/B978-0-12-803837-6.00005-6.

Kumar, A., Joseph, S., Tsechansky, L., Schreiter, I. J., Schüth, C., Taherysoosavi, S., Mitchell, D. R. G., & Graber, E. R. (2020). Mechanistic evaluation of biochar potential for plant growth promotion and alleviation of chromium-induced phytotoxicity in *Ficus elastica*. *Chemosphere*, *243*, 125332. Available from https://doi.org/10.1016/j.chemosphere.2019.125332.

Kumar, H., Kimta, N., Guleria, S., Cimler, R., Sethi, N., Singh Dhanjal, D., Singh, R., Duggal, S., Verma, R., Prerna, P., Kumar Pathera, A., Alomar, S. Y., & Kuca, K. (2024). Valorization of non-edible fruit seeds into valuable products: A sustainable approach towards circular bioeconomy. *Science of the Total Environment*, *922*, 171142. Available from https://doi.org/10.1016/j.scitotenv.2024.171142.

Kumar, V., Sharma, N., Panneerselvam, B., Kumar Dasarahally Huligowda, L., Umesh, M., Gupta, M., Muzammil, K., Zahrani, Y., & Malmutheibi, M. (2024). Lignocellulosic biomass for biochar production: A green initiative on biowaste conversion for pharmaceutical and other emerging pollutant removal. *Chemosphere*, *360*, 142312. Available from https://doi.org/10.1016/j.chemosphere.2024.142312.

Kumar, A., Antoniella, G., Blasi, E., & Chiti, T. (2024). *Agronomic practices for storing soil carbon and reducing greenhouse gas emission in the Mediterranean region* (pp. 445–480). Elsevier BV. Available from 10.1016/b978-0-443-13607-8.00024-9.

Kuryntseva, P., Karamova, K., Galitskaya, P., Selivanovskaya, S., & Evtugyn, G. (2003). Biochar functions in soil depending on feedstock and pyrolyzation properties with particular emphasis on biological properties. *Agriculture*, *13*.

Kusuma, H. S., Az-Zahra, K. D., Saputri, R. W., Utomo, M. D. P., Jaya, D. E. C., Amenaghawon, A. N., & Darmokoesoemo, H. (2024). Unlocking the potential of agricultural waste as biochar for sustainable biodiesel production: A comprehensive review. *Bioresource Technology Reports*, *26*. Available from https://doi.org/10.1016/j.biteb.2024.101848, https://www.sciencedirect.com/science/journal/2589014X.

Labianca, C., Zhu, X., Ferrara, C., Zhang, Y., De Feo, G., Hsu, S. C., & Tsang, D. C. W. (2024). A holistic framework of biochar-augmented cementitious products and general applications: Technical, environmental, and economic evaluation. *Environmental Research*, *245*. Available from https://doi.org/10.1016/j.envres.2023.118026, https://www.sciencedirect.com/science/journal/00139351.

Lalay, G., Ullah, A., Iqbal, N., Raza, A., Asghar, M. A., & Ullah, S. (2024). The alleviation of drought-induced damage to growth and physio-biochemical parameters of *Brassica napus* L. genotypes using an integrated approach of biochar amendment and PGPR application. *Environment, Development and Sustainability*, *26*(2), 3457–3480. Available from https://doi.org/10.1007/s10668-022-02841-2, https://www.springer.com/journal/10668.

Lan, T. T., Preston, T. R., & Leng, R. A. (2016). Feeding biochar or charcoal increased the growth rate of striped catfish (*Pangasius hypophthalmus*) and improved water quality. *Livestock Research for Rural Development*, *28*(5). Available from http://www.lrrd.org/lrrd28/5/lan28084.html.

Lan, W., Zhao, X., Wang, Y., Jin, X., Ji, J., Cheng, Z., Yang, G., Li, H., & Chen, G. (2024). Research progress of biochar modification technology and its application in environmental remediation. *Biomass and Bioenergy*, *184*, 107178. Available from https://doi.org/10.1016/j.biombioe.2024.107178.

Latawiec, A., Królczyk, J., Kuboń, M., Szwedziak, K., Drosik, A., Polańczyk, E., Grotkiewicz, K., & Strassburg, B. (2017). Willingness to adopt biochar in agriculture: The producer's perspective. *Sustainability*, *9*(4), 655. Available from https://doi.org/10.3390/su9040655.

Latawiec, A. E., Koryś, A., Koryś, K. A., Kuboń, M., Sadowska, U., Gliniak, M., Sikora, J., Drosik, A., Niemiec, M., Klimek-Kopyra, A., Sporysz, M., Usowicz, B., & Medeiros, B. (2021). Economic analysis of biochar use in soybean production in Poland. *Agronomy*, *11* (11), 2108. Available from https://doi.org/10.3390/agronomy11112108.

Lavagi, V., Kaplan, J., Vidalakis, G., Ortiz, M., Rodriguez, M. V., Amador, M., Hopkins, F., Ying, S., & Pagliaccia, D. (2024). Recycling agricultural waste to enhance sustainable greenhouse agriculture: Analyzing the cost-effectiveness and agronomic benefits of bokashi and biochar byproducts as soil amendments in citrus nursery production. *Sustainability*, *16*(14), 6070. Available from https://doi.org/10.3390/su16146070.

Lazzari, L. K., Ornaghi, H. L., Neves, R. M., Kerche, E. F., Zattera, A. J., & Santana, R. M. C. (2024). Bio-based cellulose-biochar-peg cryogels for thermal insulation. *Cellulose Chemistry and Technology*, *58*(1–2), 91–99. Available from https://doi.org/10.35812/CelluloseChemTechnol.2024.58.09, https://www.cellulosechemtechnol.ro/pdf/CCT1-2(2024)/p.91-99.pdf.

Lehmann, J., & Joseph, S. (2015). *Biochar for environmental management: Science, technology and implementation*. Routledge.

Liang, J., Li, Y., Si, B., Wang, Y., Chen, X., Wang, X., Chen, H., Wang, H., Zhang, F., Bai, Y., & Biswas, A. (2021). Optimizing biochar application to improve soil physical and hydraulic properties in saline-alkali soils. *Science of the Total Environment*, *771*. Available from https://doi.org/10.1016/j.scitotenv.2020.144802, http://www.elsevier.com/locate/scitotenv.

Liang, L., Xi, F., Tan, W., Meng, X., Hu, B., & Wang, X. (2021). Review of organic and inorganic pollutants removal by biochar and biochar-based composites. *Biochar*, *3*(3), 255–281. Available from https://doi.org/10.1007/s42773-021-00101-6, http://springer.com/journal/42773.

Lin, S. L., Zhang, H., Chen, W. H., Song, M., & Kwon, E. E. (2023). Low-temperature biochar production from torrefaction for wastewater treatment: A review. *Bioresource Technology*, *387*. Available from https://doi.org/10.1016/j.biortech.2023.129588, http://www.elsevier.com/locate/biortech.

Ling, L., Luo, Y., Jiang, B., Lv, J., Meng, C., Liao, Y., Reid, B. J., Ding, F., Lu, Z., Kuzyakov, Y., & Xu, J. (2022). Biochar induces mineralization of soil recalcitrant components by activation of biochar responsive bacteria groups. *Soil Biology and Biochemistry*, *172*, 108778. Available from https://doi.org/10.1016/j.soilbio.2022.108778.

Liu, Y., Luo, H., Tie, B., Li, D., Liu, S., Lei, M., & Du, H. (2021). The long-term effectiveness of ferromanganese biochar in soil Cd stabilization and reduction of Cd bioaccumulation in rice. *Biochar*, *3*(4), 499–509. Available from https://doi.org/10.1007/s42773-021-00113-2, http://springer.com/journal/42773.

Liu, Y., Song, Y., Ran, C., Siyal, A. A., Chtaeva, P., Dai, J., Jiang, Z., Deng, Z., Zhang, T., Ao, W., & Fu, J. (2020). Pyrolysis of furfural residue in a bubbling fluidized bed reactor: Biochar characterization and analysis. *Energy*, *211*, 118966. Available from https://doi.org/10.1016/j.energy.2020.118966.

Liu, Z., Dugan, B., Masiello, C. A., & Gonnermann, H. M. (2017). Biochar particle size, shape, and porosity act together to influence soil water properties. *PLoS One*, *12*(6). Available from https://doi.org/10.1371/journal.pone.0179079, http://journals.plos.org/plosone/article/file?id = 10.1371/journal.pone.0179079&type = printable.

Liu, Z., Jia, M., Li, Q., Lu, S., Zhou, D., Feng, L., Hou, Z., & Yu, J. (2023). Comparative analysis of the properties of biochars produced from different pecan feedstocks and pyrolysis temperatures. *Industrial Crops and Products*, *197*, 116638. Available from https://doi.org/10.1016/j.indcrop.2023.116638.

Luo, K., Pang, Y., Wang, D., Li, X., Wang, L., Lei, M., Huang, Q., & Yang, Q. (2021). A critical review on the application of biochar in environmental pollution remediation: Role of persistent free radicals (PFRs). *Journal of Environmental Sciences*, *108*, 201−216. Available from https://doi.org/10.1016/j.jes.2021.02.021.

Malik, L., Sanaullah, M., Mahmood, F., Hussain, S., Siddique, M. H., Anwar, F., & Shahzad, T. (2022). Unlocking the potential of co-applied biochar and plant growth-promoting rhizobacteria (PGPR) for sustainable agriculture under stress conditions. *Chemical and Biological Technologies in Agriculture*, *9*(1). Available from https://doi.org/10.1186/s40538-022-00327-x, http://chembioagro.springeropen.com/.

Man, K. Y., Chow, K. L., Man, Y. B., Mo, W. Y., & Wong, M. H. (2021). Use of biochar as feed supplements for animal farming. *Critical Reviews in Environmental Science and Technology*, *51*(2), 187−217. Available from https://doi.org/10.1080/10643389.2020.1721980, http://www.tandf.co.uk/journals/titles/10643389.asp.

Manikandan, S., Vickram, S., Subbaiya, R., Karmegam, N., Woong Chang, S., Ravindran, B., & Kumar Awasthi, M. (2023). Comprehensive review on recent production trends and applications of biochar for greener environment. *Bioresource Technology*, *388*. Available from https://doi.org/10.1016/j.biortech.2023.129725, http://www.elsevier.com/locate/biortech.

Mannan, M. A., Mia, S., Halder, E., & Dijkstra, F. A. (2021). Biochar application rate does not improve plant water availability in soybean under drought stress. *Agricultural Water Management*, *253*, 106940. Available from https://doi.org/10.1016/j.agwat.2021.106940.

Mardiansyah, K., Usup, A., & Asie, E. R. (2020). The effects of biochar and chicken manure application on red ginger (*Zingiber officinale* Rosc.) growth and yield of semi paludiculture in tropical peatlands. *Journal of Tropical Peatlands*, *10*(2), 9−16. Available from https://doi.org/10.52850/jtpupr.v10i2.2070.

Maresca, V., Fedeli, R., Vannini, A., Munzi, S., Corrêa, A., Cruz, C., & Loppi, S. (2024). Wood distillate enhances seed germination of chickpea, lettuce, and basil. *Applied Sciences*, *14*(2), 631. Available from https://doi.org/10.3390/app14020631.

Marrugo, G., Valdés, C. F., & Chejne, F. (2017). Biochar gasification: An experimental study on colombian agroindustrial biomass residues in a fluidized bed. *Energy and Fuels*, *31*(9), 9408−9421. Available from https://doi.org/10.1021/acs.energyfuels.7b00665, http://pubs.acs.org/journal/enfuem.

Mashamaite, C. V., Motsi, H., Manyevere, A., & Poswa, S. B. (2024). Assessing the potential of biochar as a viable alternative to synthetic fertilizers in Sub-Saharan Africa smallholder farming: A review. *Agronomy*, *14*(6). Available from https://doi.org/10.3390/agronomy14061215, http://www.mdpi.com/journal/agronomy/.

McAvoy, D. J., Burritt, B., & Villalba, J. J. (2020). Use of biochar by sheep: Impacts on diet selection, digestibility, and performance. *Journal of Animal Science*, *98*(12). Available from https://doi.org/10.1093/JAS/SKAA380, https://academic.oup.com/jas.

Mehdizadeh, L., Moghaddam, M., & Lakzian, A. (2020). Amelioration of soil properties, growth and leaf mineral elements of summer savory under salt stress and biochar application in alkaline soil. *Scientia Horticulturae*, *267*, 109319. Available from https://doi.org/10.1016/j.scienta.2020.109319.

Mlonka-Mędrala, A., Evangelopoulos, P., Sieradzka, M., Zajemska, M., & Magdziarz, A. (2021). Pyrolysis of agricultural waste biomass towards production of gas fuel and high-quality char: Experimental and numerical investigations. *Fuel*, *296*, 120611. Available from https://doi.org/10.1016/j.fuel.2021.120611.

Mohammed, S., Fatumah, N., Abasi, K., Olupot, M., Egesa, M., Rubhara, T., Augustyniak, A., O'Connor, T., Tsolakis, N., Gaffey, J., McMahon, H., & Anastasiadis, F. (2024). Co-designing sustainable biochar business models with sub-Saharan African communities for inclusive socio-economic transformation. *Scientific Reports*, *14*(1). Available from https://doi.org/10.1038/s41598-024-66120-y, https://www.nature.com/srep/.

Mohtaram, S., Sina Mohtaram, M., Sabbaghi, S., You, X., Wu, W., & Golsanami, N. (2024). Enhancement strategies in CO_2 conversion and management of biochar supported photocatalyst for effective generation of renewable and sustainable solar energy. *Energy Conversion and Management*, *300*, 117987. Available from https://doi.org/10.1016/j.enconman.2023.117987.

Mona, S., Malyan, S. K., Saini, N., Deepak, B., Pugazhendhi, A., & Kumar, S. S. (2021). Towards sustainable agriculture with carbon sequestration, and greenhouse gas mitigation using algal biochar. *Chemosphere*, *275*. Available from https://doi.org/10.1016/j.chemosphere.2021.129856, http://www.elsevier.com/locate/chemosphere.

Moser, K., Wopienka, E., Pfeifer, C., Schwarz, M., Sedlmayer, I., & Haslinger, W. (2023). Screw reactors and rotary kilns in biochar production – a comparative review. *Journal of Analytical and Applied Pyrolysis*, *174*, 106112. Available from https://doi.org/10.1016/j.jaap.2023.106112.

Namaswa, T., Burslem, D. F. R. P., & Smith, J. (2023). Emerging trends in appropriate kiln designs for small-scale biochar production in low to middle income countries. *Bioresource Technology Reports*, *24*. Available from https://doi.org/10.1016/j.biteb.2023.101641, https://www.journals.elsevier.com/bioresource-technology-reports.

Nguyen, M. N. (2021). Potential use of silica-rich biochar for the formulation of adaptively controlled release fertilizers: A mini review. *Journal of Cleaner Production*, *307*, 127188. Available from https://doi.org/10.1016/j.jclepro.2021.127188.

Nhuchhen, D. R., Afzal, M. T., Dreise, T., & Salema, A. A. (2018). Characteristics of biochar and bio-oil produced from wood pellets pyrolysis using a bench scale fixed bed, microwave reactor. *Biomass and Bioenergy*, *119*, 293–303. Available from https://doi.org/10.1016/j.biombioe.2018.09.035, http://www.journals.elsevier.com/biomass-and-bioenergy/.

Oni, B. A., Oziegbe, O., & Olawole, O. O. (2019). Significance of biochar application to the environment and economy. *Annals of Agricultural Sciences*, *64*(2), 222–236. Available from https://doi.org/10.1016/j.aoas.2019.12.006, http://www.elsevier.com/journals/annals-of-agricultural-sciences/0570-1783/.

Osman, A. I., Farghali, M., Dong, Y., Kong, J., Yousry, M., Rashwan, A. K., Chen, Z., Al-Fatesh, A., Rooney, D. W., & Yap, P. S. (2024). Reducing the carbon footprint of buildings using biochar-based bricks and insulating materials: a review. *Environmental Chemistry Letters*, *22*(1), 71–104. Available from https://doi.org/10.1007/s10311-023-01662-7, https://www.springer.com/journal/10311.

Osman, A. I., Fawzy, S., Farghali, M., El-Azazy, M., Elgarahy, A. M., Fahim, R. A., Maksoud, M. I. A. A., Ajlan, A. A., Yousry, M., Saleem, Y., & Rooney, D. W. (2022). Biochar for agronomy, animal farming, anaerobic digestion, composting, water treatment, soil remediation, construction, energy storage, and carbon sequestration: a review. *Environmental Chemistry Letters*, *20*(4), 2385–2485. Available from https://doi.org/10.1007/s10311-022-01424-x, http://springerlink.metapress.com/app/home/journal.asp?wasp = d86tgdwvtg0yvw9gvkwp&referrer = parent&backto = browsepublicationsresults.140.541.

Paetsch, L., Mueller, C. W., Kögel-Knabner, I., Von Lützow, M., Girardin, C., & Rumpel, C. (2018). Effect of in-situ aged and fresh biochar on soil hydraulic conditions and microbial C use under drought conditions. *Scientific Reports*, *8*(1). Available from https://doi.org/10.1038/s41598-018-25039-x, http://www.nature.com/srep/index.html.

Palansooriya, K. N., Yang, Y., Tsang, Y. F., Sarkar, B., Hou, D., Cao, X., Meers, E., Rinklebe, J., Kim, K. H., & Ok, Y. S. (2020). Occurrence of contaminants in drinking water sources and the potential of biochar for water quality improvement: A review. *Critical Reviews in Environmental Science and Technology*, *50*(6), 549–611. Available from https://doi.org/10.1080/10643389.2019.1629803, http://www.tandf.co.uk/journals/titles/10643389.asp.

Pandian, K., Vijayakumar, S., Mustaffa, M. R. A. F., Subramanian, P., & Chitraputhirapillai, S. (2024). Biochar – a sustainable soil conditioner for improving soil health, crop production and environment under changing climate: A review. *Frontiers in Soil Science*, *4*.

Papageorgiou, A., Azzi, E. S., Enell, A., & Sundberg, C. (2021). Biochar produced from wood waste for soil remediation in Sweden: Carbon sequestration and other environmental impacts. *Science of the Total Environment*, *776*, 145953. Available from https://doi.org/10.1016/j.scitotenv.2021.145953.

Parsons, E. S., Jowell, A., Veidis, E., Barry, M., & Israni, S. T. (2024). Climate change and inequality. *Pediatric Research*. Available from https://doi.org/10.1038/s41390-024-03153-z, https://www.nature.com/pr/.

Pathy, A., Ray, J., & Paramasivan, B. (2020). Biochar amendments and its impact on soil biota for sustainable agriculture. *Biochar*, *2*(3), 287–305. Available from https://doi.org/10.1007/s42773-020-00063-1, http://springer.com/journal/42773.

Patwardhan, S. B., Pandit, S., Kumar Gupta, P., Kumar Jha, N., Rawat, J., Joshi, H. C., Priya, K., Gupta, M., Lahiri, D., Nag, M., Kumar Thakur, V., & Kumar Kesari, K. (2022). Recent advances in the application of biochar in microbial electrochemical cells. *Fuel*, *311*. Available from https://doi.org/10.1016/j.fuel.2021.122501, http://www.journals.elsevier.com/fuel/.

Paz-Ferreiro, J., Méndez, A., & Gascó, G. (2015). *Application of biochar for soil biological improvement. Agricultural and environmental applications of biochar: Advances and barriers* (pp. 145–173). Wiley. Available from http://onlinelibrary.wiley.com/book/9780891189671, 10.2136/sssaspecpub63.2014.0041.5.

Prasai, T. P., Walsh, K. B., Midmore, D. J., & Bhattarai, S. P. (2018). Effect of biochar, zeolite and bentonite feed supplements on egg yield and excreta attributes. *Animal Production Science*, *58*(9), 1632–1641. Available from https://doi.org/10.1071/AN16290, http://www.publish.csiro.au/nid/72.htm?nid = 73&aid = 65.

Prodana, M., Silva, C., Gravato, C., Verheijen, F. G. A., Keizer, J. J., Soares, A. M. V. M., Loureiro, S., & Bastos, A. C. (2019). Influence of biochar particle size on biota responses. *Ecotoxicology and Environmental Safety*, *174*, 120–128. Available from https://doi.org/10.1016/j.ecoenv.2019.02.044, http://www.elsevier.com/inca/publications/store/6/2/2/8/1/9/index.htt.

Puglia, M., Morselli, N., Lumi, M., Santunione, G., Pedrazzi, S., & Allesina, G. (2023). Assessment of hemp hurd-derived biochar produced through different thermochemical processes and evaluation of its potential use as soil amendment. *Heliyon*, *9*(4), e14698. Available from https://doi.org/10.1016/j.heliyon.2023.e14698.

Qiu, M., Liu, L., Ling, Q., Cai, Y., Yu, S., Wang, S., Fu, D., Hu, B., & Wang, X. (2022). Biochar for the removal of contaminants from soil and water: A review. *Biochar*, *4*(1).

Available from https://doi.org/10.1007/s42773-022-00146-1, http://springer.com/journal/42773.

Quispe, J. I. B., Campos, L. C., Mašek, O., & Bogush, A. (2022). Use of biochar-based column filtration systems for greywater treatment: A systematic literature review. *Journal of Water Process Engineering*, *48*. Available from https://doi.org/10.1016/j.jwpe.2022.102908, http://www.journals.elsevier.com/journal-of-water-process-engineering/.

Rafael, R. B. A., Fernández-Marcos, M. L., Cocco, S., Ruello, M. L., Fornasier, F., & Corti, G. (2019). Benefits of biochars and NPK fertilizers for soil quality and growth of cowpea (*Vigna unguiculata* L. Walp.) in an acid arenosol. *Pedosphere*, *29*(3), 311−333. Available from https://doi.org/10.1016/S1002-0160(19)60805-2, http://pedosphere.issas.ac.cn.

Rahim, H. U., Allevato, E., Vaccari, F. P., & Stazi, S. R. (2024). Biochar aged or combined with humic substances: Fabrication and implications for sustainable agriculture and environment—a review. *Journal of Soils and Sediments*, *24*(1), 139−162. Available from https://doi.org/10.1007/s11368-023-03644-2, https://www.springer.com/journal/11368.

Rajput, V. D., Minkina, T., Ahmed, B., Singh, V. K., Mandzhieva, S., Sushkova, S., Bauer, T., Verma, K. K., Shan, S., van Hullebusch, E. D., & Wang, B. (2022). Nano-biochar: A novel solution for sustainable agriculture and environmental remediation. *Environmental Research*, *210*. Available from https://doi.org/10.1016/j.envres.2022.112891, http://www.elsevier.com/inca/publications/store/6/2/2/8/2/1/index.htt.

Ramesh, K., & Raghavan, V. (2024). Agricultural waste-derived biochar-based nitrogenous fertilizer for slow-release applications. *ACS Omega*, *9*(4), 4377−4385. Available from https://doi.org/10.1021/acsomega.3c06687, http://pubs.acs.org/journal/acsodf.

Rasul, M., Cho, J., Shin, H.-S., & Hur, J. (2022). Biochar-induced priming effects in soil via modifying the status of soil organic matter and microflora: A review. *Science of the Total Environment*, *805*, 150304. Available from https://doi.org/10.1016/j.scitotenv.2021.150304.

Rawat, S., Boobalan, T., Krishna, B. B., Sathish, M., Hotha, S., & Bhaskar, T. (2022). Biochar for supercapacitor application: A comparative study. *Chemistry – An Asian Journal*, *17*(24). Available from https://doi.org/10.1002/asia.202200982, http://onlinelibrary.wiley.com/journal/10.1002/(ISSN)1861-471X.

Rex, P., Mohammed Ismail, K. R., Meenakshisundaram, N., Barmavatu, P., & Sai Bharadwaj, A. V. S. L. (2023). Agricultural biomass waste to biochar: A review on biochar applications using machine learning approach and circular economy. *ChemEngineering*, *7*(3). Available from https://doi.org/10.3390/chemengineering7030050, http://www.mdpi.com/journal/ChemEngineering.

Reynolds, A., Joseph, S. D., Verheyen, T. V., Chinu, K., Taherymoosavi, S., Munroe, P. R., Donne, S., Pace, B., van Zwieten, L., Marjo, C. E., Thomas, T., Rawal, A., & Hook, J. (2018). Effect of clay and iron sulphate on volatile and water-extractable organic compounds in bamboo biochars. *Journal of Analytical and Applied Pyrolysis*, *133*, 22−29. Available from https://doi.org/10.1016/j.jaap.2018.05.007, https://www.journals.elsevier.com/journal-of-analytical-and-applied-pyrolysis.

Rezaei Rashti, M., Esfandbod, M., Phillips, I. R., & Chen, C. (2019). Rhizosphere management by biochar and leaching improved plant performance in fresh bauxite residue sand. *Journal of Cleaner Production*, *219*, 66−74. Available from https://doi.org/10.1016/j.jclepro.2019.02.013, https://www.journals.elsevier.com/journal-of-cleaner-production.

Rogers, P. M., Fridahl, M., Yanda, P., Hansson, A., Pauline, N., & Haikola, S. (2022). Socio-economic determinants for biochar deployment in the southern highlands of Tanzania. *Energies*, *15*(1). Available from https://doi.org/10.3390/en15010144, https://www.mdpi.com/1996-1073/15/1/144/pdf.

Rombel, A., Krasucka, P., & Oleszczuk, P. (2022). Sustainable biochar-based soil fertilizers and amendments as a new trend in biochar research. *Science of the Total Environment*, *816*, 151588. Available from https://doi.org/10.1016/j.scitotenv.2021.151588.

Russell, S. H., Turrion-Gomez, J. L., Meredith, W., Langston, P., & Snape, C. E. (2017). Increased charcoal yield and production of lighter oils from the slow pyrolysis of biomass. *Journal of Analytical and Applied Pyrolysis*, *124*, 536–541. Available from https://doi.org/10.1016/j.jaap.2017.01.028.

Ryšavý, J., Čespiva, J., Kuboňová, L., Dej, M., Szramowiat-Sala, K., Molchanov, O., Niedzwiecki, L., Yan, W.-M., & Thangavel, S. (2024). Co-gasification of pistachio shells with wood pellets in a semi-industrial hybrid cross/updraft reactor for producer gas and biochar production. *Fire*, *7*(3), 87. Available from https://doi.org/10.3390/fire7030087.

Safarian, S. (2023). Performance analysis of sustainable technologies for biochar production: A comprehensive review. *Energy Reports*, *9*, 4574–4593. Available from https://doi.org/10.1016/j.egyr.2023.03.111, http://www.journals.elsevier.com/energy-reports/.

Sani, M. N. H., Amin, M., Siddique, A. B., Nasif, S. O., Ghaley, B. B., Ge, L., Wang, F., & Yong, J. W. H. (2023). Waste-derived nanobiochar: A new avenue towards sustainable agriculture, environment, and circular bioeconomy. *Science of the Total Environment*, *905*. Available from https://doi.org/10.1016/j.scitotenv.2023.166881, http://www.elsevier.com/locate/scitotenv.

Schmidt, H.-P., Hagemann, N., Draper, K., & Kammann, C. (2019). The use of biochar in animal feeding. *PeerJ*, *7*(7), e7373. Available from https://doi.org/10.7717/peerj.7373.

Schmidt, H.-P. (2013). 55 uses of biochar. *Ithaka Journal*.

Schreiter, I. J., Schmidt, W., Kumar, A., Graber, E. R., & Schüth, C. (2020). Effect of water leaching on biochar properties and its impact on organic contaminant sorption. *Environmental Science and Pollution Research*, *27*(1), 691–703. Available from https://doi.org/10.1007/s11356-019-06904-2, https://link.springer.com/journal/11356.

Senthil, C., & Lee, C. W. (2021). Biomass-derived biochar materials as sustainable energy sources for electrochemical energy storage devices. *Renewable and Sustainable Energy Reviews*, *137*, 110464. Available from https://doi.org/10.1016/j.rser.2020.110464.

Seow, Y. X., Tan, Y. H., Mubarak, N. M., Kansedo, J., Khalid, M., Ibrahim, M. L., & Ghasemi, M. (2022). A review on biochar production from different biomass wastes by recent carbonization technologies and its sustainable applications. *Journal of Environmental Chemical Engineering*, *10*(1). Available from https://doi.org/10.1016/j.jece.2021.107017, http://www.journals.elsevier.com/journal-of-environmental-chemical-engineering/.

Sharma, T., Hakeem, I. G., Gupta, A. B., Joshi, J., Shah, K., Vuppaladadiyam, A. K., & Sharma, A. (2024). Parametric influence of process conditions on thermochemical techniques for biochar production: A state-of-the-art review. *Journal of the Energy Institute*, *113*. Available from https://doi.org/10.1016/j.joei.2024.101559, https://www.sciencedirect.com/science/journal/17439671.

She, D., Sun, X., Gamareldawla, A. H. D., Nazar, E. A., Hu, W., Edith, K., & Yu, S. (2018). Benefits of soil biochar amendments to tomato growth under saline water irrigation. *Scientific Reports*, *8*(1). Available from https://doi.org/10.1038/s41598-018-33040-7, http://www.nature.com/srep/index.html.

Shoudho, K. N., Khan, T. H., Ara, U. R., Khan, M. R., Shawon, Z. B. Z., & Hoque, M. E. (2024). Biochar in global carbon cycle: Towards sustainable development goals. *Current Research in Green and Sustainable Chemistry*, *8*. Available from https://doi.org/10.1016/j.crgsc.2024.100409, https://www.sciencedirect.com/science/journal/26660865.

Silva, W. R., Carregosa, J. D. C., Almeida-Couto, J. M. F., Cardozo-Filho, L., & Wisniewski, A. (2022). Management of de-oiled coffee beans biomass through pyrolysis process: Towards a circular bioeconomy. *Journal of Analytical and Applied Pyrolysis*, *168*. Available from https://doi.org/10.1016/j.jaap.2022.105763, https://www.journals.elsevier.com/journal-of-analytical-and-applied-pyrolysis.

Sim, D. H. H., Tan, I. A. W., Lim, L. L. P., & Hameed, B. H. (2021). Encapsulated biochar-based sustained release fertilizer for precision agriculture: A review. *Journal of Cleaner Production*, *303*, 127018. Available from https://doi.org/10.1016/j.jclepro.2021.127018.

Solinas, S., Tiloca, M. T., Deligios, P. A., Cossu, M., & Ledda, L. (2021). Carbon footprints and social carbon cost assessments in a perennial energy crop system: A comparison of fertilizer management practices in a Mediterranean area. *Agricultural Systems*, *186*. Available from https://doi.org/10.1016/j.agsy.2020.102989, http://www.elsevier.com/inca/publications/store/4/0/5/8/5/1.

Song, D., Xi, X., Zheng, Q., Liang, G., Zhou, W., & Wang, X. (2019). Soil nutrient and microbial activity responses to two years after maize straw biochar application in a calcareous soil. *Ecotoxicology and Environmental Safety*, *180*, 348–356. Available from https://doi.org/10.1016/j.ecoenv.2019.04.073, http://www.elsevier.com/inca/publications/store/6/2/2/8/1/9/index.htt.

Sousa, B., Soares, C., Sousa, F., Martins, M., Mateus, P., Rodrigues, F., Azenha, M., Moutinho-Pereira, J., Lino-Neto, T., & Fidalgo, F. (2024). Enhancing tomato plants' tolerance to combined heat and salt stress – the role of arbuscular mycorrhizae and biochar. *Science of the Total Environment*, *948*, 174860. Available from https://doi.org/10.1016/j.scitotenv.2024.174860.

Sparrevik, M., Lindhjem, H., Andria, V., Fet, A. M., & Cornelissen, G. (2014). Environmental and socioeconomic impacts of utilizing waste for biochar in rural areas in Indonesia—a systems perspective. *Environmental Science and Technology*, *48*(9), 4664–4671. Available from https://doi.org/10.1021/es405190q, http://pubs.acs.org/journal/esthag.

Sparrevik, M., Adam, C., Martinsen, V., Jubaedah., & Cornelissen, G. (2015). Emissions of gases and particles from charcoal/biochar production in rural areas using medium-sized traditional and improved "retort" kilns. *Biomass and Bioenergy*, *72*, 65–73. Available from https://doi.org/10.1016/j.biombioe.2014.11.016.

Sri Shalini, S., Palanivelu, K., Ramachandran, A., & Vijaya Raghavan. (2021). Biochar from biomass waste as a renewable carbon material for climate change mitigation in reducing greenhouse gas emissions—a review. *Biomass Conversion and Biorefinery*, *11*(5), 2247–2267. Available from https://doi.org/10.1007/s13399-020-00604-5, http://www.springer.com/engineering/energy + technology/journal/13399.

Sytar, O., Kumar, A., Latowski, D., Kuczynska, P., Strzałka, K., & Prasad, M. N. V. (2013). Heavy metal-induced oxidative damage, defense reactions, and detoxification mechanisms in plants. *Acta Physiologiae Plantarum*, *35*(4), 985–999. Available from https://doi.org/10.1007/s11738-012-1169-6, https://rd.springer.com/journal/11738.

Tian, J., Wang, J., Dippold, M., Gao, Y., Blagodatskaya, E., & Kuzyakov, Y. (2016). Biochar affects soil organic matter cycling and microbial functions but does not alter microbial community structure in a paddy soil. *Science of the Total Environment*, *556*, 89–97. Available from https://doi.org/10.1016/j.scitotenv.2016.03.010, http://www.elsevier.com/locate/scitotenv.

Toth, J. D., & Dou, Z. (2015). *Use and impact of biochar and charcoal in animal production systems. Agricultural and environmental applications of biochar: Advances and barriers* (pp. 199–224). Wiley. Available from http://onlinelibrary.wiley.com/book/9780891189671, 10.2136/sssaspecpub63.2014.0043.5.

Urrutia, R. I., Gutierrez, V. S., Stefanazzi, N., Volpe, M. A., & Werdin González, J. O. (2022). Pyrolysis liquids from lignocellulosic biomass as a potential tool for insect pest management: A comprehensive review. *Industrial Crops and Products*, *177*. Available from https://doi.org/10.1016/j.indcrop.2022.114533, http://www.elsevier.com/inca/publications/store/5/2/2/8/2/5.

Utkarsh, K., Thakur, N., & Shukla, S. K. (2023). Biochar: A feasible and visible solution for agricultural sustainability. *Arabian Journal of Geosciences*, *16*(4). Available from https://doi.org/10.1007/s12517-023-11320-5.

Veiga, P. A. D. S., Cerqueira, M. H., Gonçalves, M. G., Matos, T. T. D. S., Pantano, G., Schultz, J., Andrade, J. B. D., & Mangrich, A. S. (2021). Upgrading from batch to continuous flow process for the pyrolysis of sugarcane bagasse: Structural characterization of the biochars produced. *Journal of Environmental Management*, *285*. Available from https://doi.org/10.1016/j.jenvman.2021.112145, https://www.sciencedirect.com/journal/journal-of-environmental-management.

Virk, A. L., Kan, Z. R., Liu, B. Y., Qi, J. Y., He, C., Liu, Q. Y., Zhao, X., & Zhang, H. L. (2021). Impact of biochar water extract addition on soil organic carbon mineralization and C fractions in different tillage systems. *Environmental Technology and Innovation*, *21*. Available from https://doi.org/10.1016/j.eti.2020.101193, http://www.journals.elsevier.com/environmental-technology-and-innovation/.

Voukkali, I., Papamichael, I., Loizia, P., & Zorpas, A. A. (2024). Urbanization and solid waste production: prospects and challenges. *Environmental Science and Pollution Research*, *31*(12), 17678–17689. Available from https://doi.org/10.1007/s11356-023-27670-2, https://www.springer.com/journal/11356.

Védère, C., Lebrun, M., Biron, P., Planchais, S., Bordenave-Jacquemin, M., Honvault, N., Firmin, S., Savouré, A., Houben, D., & Rumpel, C. (2023). The older, the better: Ageing improves the efficiency of biochar-compost mixture to alleviate drought stress in plant and soil. *Science of the Total Environment*, *856*, 158920. Available from https://doi.org/10.1016/j.scitotenv.2022.158920.

Wang, C., Luo, D., Zhang, X., Huang, R., Cao, Y., Liu, G., Zhang, Y., & Wang, H. (2022). Biochar-based slow-release of fertilizers for sustainable agriculture: A mini review. *Environmental Science and Ecotechnology*, *10*, 100167. Available from https://doi.org/10.1016/j.ese.2022.100167.

Wang, J., Shi, L., Zhai, L., Zhang, H., Wang, S., Zou, J., Shen, Z., Lian, C., & Chen, Y. (2021). Analysis of the long-term effectiveness of biochar immobilization remediation on heavy metal contaminated soil and the potential environmental factors weakening the remediation effect: A review. *Ecotoxicology and Environmental Safety*, *207*, 111261. Available from https://doi.org/10.1016/j.ecoenv.2020.111261.

Wang, X., Nie, J., Wang, P., Zhao, J., Yang, Y., Wang, S., Zeng, Z., & Zang, H. (2021). Does the replacement of chemical fertilizer nitrogen by manure benefit water use efficiency of winter wheat – summer maize systems? *Agricultural Water Management*, *243*, 106428. Available from https://doi.org/10.1016/j.agwat.2020.106428.

Wang, Y., Pang, J., Zhang, M., Tian, Z., Wei, T., Jia, Z., Ren, X., & Zhang, P. (2023). Is adding biochar be better than crop straw for improving soil aggregates stability and organic carbon contents in film mulched fields in semiarid regions? Evidence of 5-year field experiment. *Journal of Environmental Management*, *338*, 117711. Available from https://doi.org/10.1016/j.jenvman.2023.117711.

Wen, T., Yang, L., Dang, C., Yang, M., Miki, T., Bai, H., & Nagasaka, T. (2020). Effect of modified basic oxygen furnace slag on the controlled release of nitrate nitrogen and the functional microbial community in soil. *Journal of Environmental Management*, *261*, 110191. Available from https://doi.org/10.1016/j.jenvman.2020.110191.

Willson, N. L., Van, T. T. H., Bhattarai, S. P., Courtice, J. M., McIntyre, J. R., Prasai, T. P., Moore, R. J., Walsh, K., & Stanley, D. (2019). Feed supplementation with biochar may reduce poultry pathogens, including Campylobacter hepaticus, the causative agent of Spotty Liver Disease. *PLoS One*, *14*(4). Available from https://doi.org/10.1371/journal.pone.0214471, https://journals.plos.org/plosone/article/file?id = 10.1371/journal.pone.0214471&type = printable.

Woolcock, P. J., & Brown, R. C. (2013). A review of cleaning technologies for biomass-derived syngas. *Biomass and Bioenergy*, *52*, 54–84. Available from https://doi.org/10.1016/j.biombioe.2013.02.036.

Wu, G., Tham, P. E., Chew, K. W., Munawaroh, H. S. H., Tan, I. S., Wan-Mohtar, W. A. A. Q. I., Sriariyanun, M., & Show, P. L. (2023). Net zero emission in circular bioeconomy from microalgae biochar production: A renewed possibility. *Bioresource Technology*, *388*. Available from https://doi.org/10.1016/j.biortech.2023.129748, http://www.elsevier.com/locate/biortech.

Xia, X., Zeng, S., Li, K., Zeng, L., & Miao, S. (2025). Unraveling the outstanding catalytic efficiency of unprocessed bone-derived biochar: A deep dive into the mechanisms of native organic encapsulation and defective nitrogen doping in boosting persulfate activation for tetracycline degradation. *Separation and Purification Technology*, *353*, 128571. Available from https://doi.org/10.1016/j.seppur.2024.128571.

Xiang, L., Liu, S., Ye, S., Yang, H., Song, B., Qin, F., Shen, M., Tan, C., Zeng, G., & Tan, X. (2021). Potential hazards of biochar: The negative environmental impacts of biochar applications. *Journal of Hazardous Materials*, *420*, 126611. Available from https://doi.org/10.1016/j.jhazmat.2021.126611.

Xiao, R., Wang, P., Mi, S., Ali, A., Liu, X., Li, Y., Guan, W., Li, R., & Zhang, Z. (2019). Effects of crop straw and its derived biochar on the mobility and bioavailability in Cd and Zn in two smelter-contaminated alkaline soils. *Ecotoxicology and Environmental Safety*, *181*, 155–163. Available from https://doi.org/10.1016/j.ecoenv.2019.06.005, http://www.elsevier.com/inca/publications/store/6/2/2/8/1/9/index.htt.

Xiao, X., Chen, Z., & Chen, B. (2016). H/C atomic ratio as a smart linkage between pyrolytic temperatures, aromatic clusters and sorption properties of biochars derived from diverse precursory materials. *Scientific Reports*, *6*. Available from https://doi.org/10.1038/srep22644, http://www.nature.com/srep/index.html.

Xiong, J., Liang, L., Shi, W., Li, Z., Zhang, Z., Li, X., & Liu, Y. (2022). Application of biochar in modification of fillers in bioretention cells: A review. *Ecological*

Engineering, *181*, 106689. Available from https://doi.org/10.1016/j.ecoleng.2022.106689.

Xu, W., Wang, G., Deng, F., Zou, X., Ruan, H., & Chen, H. Y. H. (2018). Responses of soil microbial biomass, diversity and metabolic activity to biochar applications in managed poplar plantations on reclaimed coastal saline soil. *Soil Use and Management*, *34*(4), 597–605. Available from https://doi.org/10.1111/sum.12460, http://onlinelibrary.wiley.com/journal/10.1111/(ISSN)1475-2743.

Yang, X., Wang, D., Lan, Y., Meng, J., Jiang, L., Sun, Q., Cao, D., Sun, Y., & Chen, W. (2018). Labile organic carbon fractions and carbon pool management index in a 3-year field study with biochar amendment. *Journal of Soils and Sediments*, *18*(4), 1569–1578. Available from https://doi.org/10.1007/s11368-017-1874-2, http://www.springerlink.com/content/1439-0108.

You, S., Li, W., Zhang, W., Lim, H., Kua, H. W., Park, Y. K., Igalavithana, A. D., & Ok, Y. S. (2022). Energy, economic, and environmental impacts of sustainable biochar systems in rural China. *Critical Reviews in Environmental Science and Technology*, *52*(7), 1063–1091. Available from https://doi.org/10.1080/10643389.2020.1848170, http://www.tandf.co.uk/journals/titles/10643389.asp.

Yu, J., Tang, L., Pang, Y., Zeng, G., Wang, J., Deng, Y., Liu, Y., Feng, H., Chen, S., & Ren, X. (2019). Magnetic nitrogen-doped sludge-derived biochar catalysts for persulfate activation: Internal electron transfer mechanism. *Chemical Engineering Journal*, *364*, 146–159. Available from https://doi.org/10.1016/j.cej.2019.01.163, http://www.elsevier.com/inca/publications/store/6/0/1/2/7/3/index.htt.

Yu, S., Zhang, W., Dong, X., Wang, F., Yang, W., Liu, C., & Chen, D. (2024). A review on recent advances of biochar from agricultural and forestry wastes: Preparation, modification and applications in wastewater treatment. *Journal of Environmental Chemical Engineering*, *12*(1), 111638. Available from https://doi.org/10.1016/j.jece.2023.111638.

Yıldızlı, G., Coral, G., & Ayaz, F. (2021). Biochar as a biocompatible mild anti-inflammatory supplement for animal feed and agricultural fields. *Chemistry and Biodiversity*, *18*(6). Available from https://doi.org/10.1002/cbdv.202001002, http://onlinelibrary.wiley.com/journal/10.1002/(ISSN)1612-1880.

Yrjälä, K., Ramakrishnan, M., & Salo, E. (2022). Agricultural waste streams as resource in circular economy for biochar production towards carbon neutrality. *Current Opinion in Environmental Science & Health*, *26*, 100339. Available from https://doi.org/10.1016/j.coesh.2022.100339.

Zabaniotou, A., Rovas, D., Libutti, A., & Monteleone, M. (2015). Boosting circular economy and closing the loop in agriculture: Case study of a small-scale pyrolysis-biochar based system integrated in an olive farm in symbiosis with an olive mill. *Environmental Development*, *14*, 22–36. Available from https://doi.org/10.1016/j.envdev.2014.12.002, http://www.sciencedirect.com/science/journal/22114645.

Zeghioud, H., Fryda, L., Djelal, H., Assadi, A., & Kane, A. (2022). A comprehensive review of biochar in removal of organic pollutants from wastewater: Characterization, toxicity, activation/functionalization and influencing treatment factors. *Journal of Water Process Engineering*, *47*, 102801. Available from https://doi.org/10.1016/j.jwpe.2022.102801.

Zhang, J., Bai, Z., Huang, J., Hussain, S., Zhao, F., Zhu, C., Zhu, L., Cao, X., & Jin, Q. (2019). Biochar alleviated the salt stress of induced saline paddy soil and improved the biochemical characteristics of rice seedlings differing in salt tolerance. *Soil and Tillage Research*, *195*, 104372. Available from https://doi.org/10.1016/j.still.2019.104372.

Zhang, W., Wei, J., Guo, L., Fang, H., Liu, X., Liang, K., Niu, W., Liu, F., & Siddique, K. H. M. (2023). Effects of two biochar types on mitigating drought and salt stress in tomato seedlings. *Agronomy*, *13*(4), 1039. Available from https://doi.org/10.3390/agronomy13041039.

Zhang, X., Yin, J., Ma, Y., Peng, Y., Fenton, O., Wang, W., Zhang, W., & Chen, Q. (2024). Unlocking the potential of biostimulants derived from organic waste and by-product sources: Improving plant growth and tolerance to abiotic stresses in agriculture. *Environmental Technology & Innovation*, *34*, 103571. Available from https://doi.org/10.1016/j.eti.2024.103571.

Zhang, Y., Ding, J., Wang, H., Su, L., & Zhao, C. (2020). Biochar addition alleviate the negative effects of drought and salinity stress on soybean productivity and water use efficiency. *BMC Plant Biology*, *20*(1). Available from https://doi.org/10.1186/s12870-020-02493-2, http://www.biomedcentral.com/bmcplantbiol/.

Zhao, W., Zhou, Q., Tian, Z., Cui, Y., Liang, Y., & Wang, H. (2020). Apply biochar to ameliorate soda saline-alkali land, improve soil function and increase corn nutrient availability in the Songnen Plain. *Science of the Total Environment*, *722*, 137428. Available from https://doi.org/10.1016/j.scitotenv.2020.137428.

Zheng, H., Liu, D., Liao, X., Miao, Y., Li, Y., Li, J., Yuan, J., Chen, Z., & Ding, W. (2022). Field-aged biochar enhances soil organic carbon by increasing recalcitrant organic carbon fractions and making microbial communities more conducive to carbon sequestration. *Agriculture, Ecosystems & Environment*, *340*, 108177. Available from https://doi.org/10.1016/j.agee.2022.108177.

Zheng, X., Xu, W., Dong, J., Yang, T., Shangguan, Z., Qu, J., Li, X., & Tan, X. (2022). The effects of biochar and its applications in the microbial remediation of contaminated soil: A review. *Journal of Hazardous Materials*, *438*, 129557. Available from https://doi.org/10.1016/j.jhazmat.2022.129557.

Zheng, J. L., Zhu, Y. H., Dong, Y. Y., & Zhu, M. Q. (2023). Life-cycle assessment and techno-economic analysis of the production of wood vinegar from Eucommia stem: a case study. *Frontiers of Chemical Science and Engineering*, *17*(8), 1109–1121. Available from https://doi.org/10.1007/s11705-022-2296-2, https://www.springer.com/journal/11705.

Zheng, T., Ouyang, S., & Zhou, Q. (2023). Synthesis, characterization, safety design, and application of NPs@BC for contaminated soil remediation and sustainable agriculture. *Biochar*, *5*(1). Available from https://doi.org/10.1007/s42773-022-00198-3, https://www.springer.com/journal/42773.

Zhu, X., Labianca, C., He, M., Luo, Z., Wu, C., You, S., & Tsang, D. C. W. (2022). Life-cycle assessment of pyrolysis processes for sustainable production of biochar from agro-residues. *Bioresource Technology*, *360*. Available from https://doi.org/10.1016/j.biortech.2022.127601, http://www.elsevier.com/locate/biortech.

Zouari, M., Hribernik, S., & Schwarzkopf, M. (2024). Indoor air remediation using biochar from bark: Impact of particle size and pollutant concentration. *Indoor Air*, *2024*. Available from https://doi.org/10.1155/2024/1537588, https://www.hindawi.com/journals/ina/.

Zulfiqar, B., Raza, M. A. S., Saleem, M. F., Aslam, M. U., Iqbal, R., Muhammad, F., Amin, J., Ibrahim, M. A., & Khan, I. H. (2022). Biochar enhances wheat crop productivity by mitigating the effects of drought: Insights into physiological and antioxidant defense mechanisms. *PLoS One*, *17*(4). Available from https://doi.org/10.1371/journal.pone.0267819, https://journals.plos.org/plosone/article/file?id = 10.1371/journal.pone.0267819&type = printable.

CHAPTER

2 Biochar production to support circular bioeconomy: from waste biomass to a valuable product

Luisa Massaccesi[1], I. Nogués[2], V. Mazzurco Miritana[2,3], L. Passatore[2], M. Zacchini[2] and Sara Marinari[4]

[1]*National Research Council of Italy, Institute for Agriculture and Forestry Systems in the Mediterranean (ISAFOM-CNR), Perugia, Italy*

[2]*National Research Council of Italy, Research Institute on Terrestrial Ecosystems (IRET-CNR), Rome, Italy*

[3]*Department of Energy Technologies and Renewable Sources TERIN-ENEA - Italian National Agency for New Technologies, Energy and Sustainable Economic Development, Roma, Italy*

[4]*Department for Innovation in Biological Agricultural Food and Forestry Systems, University of Tuscia, Viterbo, Italy*

2.1 Introduction

Biochar is a carbon-rich material in which the carbon is arranged in aromatic structures and occasionally in graphite-like layers (Zimmerman, 2010). Because of these properties, biochar is expected to remain stable for hundreds to thousands of years when applied to soil. In terms of carbon sequestration, biochar plays a critical role in mitigating climate change by sequestering carbon that would otherwise be released into the atmosphere as carbon dioxide (CO_2) during the decomposition of organic matter. By sequestering carbon in the soil, biochar helps to reduce net greenhouse gas (GHG) emissions. The properties of biochar, such as carbon yield from the pyrolysis process, thermal stability, and recalcitrance, are directly related to its suitability and ability to sequester carbon in the soil over long periods of time (Li & Tasnady, 2023).

The potential of biochar as a soil improver has been widely studied, particularly the ability to improve soil physical and biochemical properties and increase soil fertility and productivity, through increasing soil aggregation, water retention, pH, and microbial activities has been proved (Nogués et al., 2023). The extent of biochar's impact on soil physiochemical properties varies depending on biochar source, type, size, inherent soil characteristics, cropping system, etc. Moreover,

Biochar Ecotechnology for Sustainable Agriculture and Environment. DOI: https://doi.org/10.1016/B978-0-443-29855-4.00003-5

biochar has significant potential in soil and water remediation, especially through its unique adsorption and chemical properties capable of capturing and immobilizing pollutants such as metal(loid)s, organic pollutants, and hazardous emerging contaminants such as microplastics. However, a knowledge gap remains as to understanding the long-term persistence of biochar on agroecosystems, the optimal biochar application rate for the diversity of biochar-soil-crop-environmental conditions, specific mechanisms of biochar's effect on soil biotic properties, and regulating nutrients biogeochemical cycle. Moreover, synergy or potential antagonistic effects with other carbon sources such as compost, manure, and agricultural residues, over the long term, still need to be explored in depth.

This chapter aims to describe state-of-the-art biochar production from a circular bioeconomy perspective in agricultural farming systems. This chapter is divided into two subchapters: the first one describes the latest knowledge about the increasing soil carbon storage capacity using agricultural waste biomass for biochar production, and the second one aims to define the impact of biochar on regulating soil properties such as nutrients biogeochemical cycle and water retention. Finally, we describe BIOCHAR LATIUM case study as an example of strategy for the valuable exploitation of lignocellulosic wastes from the Lazio region through the production of biochar in a circular bioeconomy perspective.

2.2 State of the art about the biochar role in increasing soil carbon storage

2.2.1 Biochar and soil carbon sequestration

The carbon sequestration potential of biochar is influenced by several factors, including the choice of feedstock, pyrolysis conditions, and posttreatment processes. In particular, the carbon yield (C-yield) obtained during the pyrolysis process is a key determinant of the ability of biochar to sequester carbon. Research suggests that the type of feedstock plays a more significant role in influencing C-yield than the specific pyrolysis temperature (Mašek et al., 2013; Mukome et al., 2013). As the pyrolysis temperature increases, carbon undergoes aromatization and is converted into a more condensed and recalcitrant form (Almutairi et al., 2023). At the same time, nitrogen, hydrogen, and oxygen are lost through volatilization. Biochar with high thermal stability typically has low H/C and O/C atomic ratios, indicating a higher degree of carbonization and aromaticity (Almutairi et al., 2023). This structural composition confers high resistance to chemical reactions and microbial degradation in soils. The H/C and O/C atomic ratios are commonly used as indicators of biochar stability. According to the International Biochar Initiative (IBI) guidelines, biochar should not exceed atomic ratio thresholds of 0.7 for H/C and 0.4 for O/C. These limits help distinguish biochar from biomass that has undergone only minor thermochemical changes (Lehmann & Joseph, 2015). Following these guidelines ensures that biochar maintains its stability and can effectively contribute to long-term carbon sequestration when applied to soils.

When produced under the same pyrolysis conditions, the feedstock, or the type of biomass material used, plays a significant role in determining the chemical composition of the resulting biochar. A recent meta-analysis reported that wood biochar had a significantly higher average C content compared to corn stover biochar and switchgrass biochar (Li et al., 2023). These differences can be attributed to differences in cellulose, hemicellulose, and lignin content between the three feedstock types (Li et al., 2023). Indeed, woody species generally contain more lignin and more cellulose than herbaceous or grassy species and carbonization involves the breakdown of complex organic compounds, particularly lignocellulosic structures, leading to the formation of aromatic structures. Lignocellulosic materials are rich in carbon and have a relatively low proportion of volatile components. On the other hand, feedstocks with higher levels of volatile matter, such as some manures or green plant materials, may yield biochar with lower carbon content. The chemical composition of the feedstock influences the yield, stability, and nutrient-retention properties of the resulting biochar.

On the other hand, the physical and chemical properties of biochar can be improved by posttreatment (Conte et al., 2021). Posttreatment refers to additional processing steps or modifications applied after the initial pyrolysis to further refine the physical and chemical properties of biochar. Posttreatment of biochar for carbon sequestration purposes typically involves methods that increase its stability and effectiveness in storing carbon. These treatments include activation, chemical modification, mineralization, and surface coating. In this context, research by Cao et al. (2023) has shown that ultrasound-assisted acid washing improves the micropore structure of biochar and consequently its CO_2 adsorption capacity. Furthermore, Mašek et al. (2019) showed that the introduction of potassium as a low-concentration additive during biochar production can increase its carbon sequestration potential by up to 45%.

Several physicochemical properties and parameters of biochar, such as particle size, surface area, porosity, pore size distribution, pH, and cation exchange capacity (CEC), are crucial and indirectly impact its potential for carbon sequestration. Together, these properties contribute to the overall efficiency of biochar as a carbon sink (Li et al., 2023). Biochar consists of two distinct carbon pools: persistent aromatic carbon (PAC) and semipersistent carbon (SPC) (Schmidt et al., 2022). PAC is recognized by larger clusters of aromatic carbon rings, often exceeding seven rings. It is not susceptible to degradation, remaining unaffected by common environmental factors such as soil moisture, temperature swings, freeze-thaw cycles, and the effects of biological activity or agricultural practices like tillage. The PAC pool has a mean residence time (MRT) of more than 1000 years (Bowring et al., 2022; Howell et al., 2022). The SPC pool is composed of aliphatic, small aromatic, and heteroaromatic carbon species, and is more susceptible to degradation in the soil (Rombolà et al., 2016). Specific compounds in the

SPC pool may experience degradation during the first year of soil application, while others exhibit long-lasting durability, spanning decades or even centuries. The persistence of biochar is dependent on the chemistry of the aliphatic and small aromatic compounds and their physical placement within the porous structure of the biochar (Schmidt et al., 2022).

Biochar incubation research has shown that the degradation rate decelerates exponentially over time. Further, the decomposition process is mostly described with a double exponential decay model (Lehmann & Joseph, 2015; Ventura et al., 2019; Wang et al., 2016). In this context, the persistence of biochar in soils has been studied in many experiments (Maestrini et al., 2014; Ventura et al., 2015). In a meta-analysis study, MRTs estimations ranged from 3 to 891 years (Wang et al., 2016). The wide range of MRTs can be explained by the variety of experimental designs used in different trials. Some of the critical variables to consider include the length of the study, whether it was a laboratory or field experiment, the characteristics of the soil, the climatic conditions, and the properties of the biochar being tested (Wang et al., 2016). Another method to assess the biochar stability in the soil uses proximate analysis that determines moisture content and the labile (volatile matter) and stable carbon (fixed carbon) fractions (Rombolà et al., 2023). Chemical oxidation methods have also been used to predict biochar stability through the removal of the labile carbon fraction. Chemical oxidation using hydrogen peroxide with heat treatment (Cross & Sohi, 2013) and potassium dichromate (Liu et al., 2020) have been reported.

The Intergovernmental Panel on Climate Change (IPCC) put forward a methodology to assess carbon sequestration in agricultural soil through the application of biochar. Nonetheless, the approach only accounted for pyrolysis production conditions as a metric for measuring biochar longevity (Woolf et al., 2021). Woolf et al. (2021) have recommended a simple approach for estimating carbon sequestration with biochar application. This method estimates the fraction of biochar carbon that remains in soil over 100 years based on either H/C_{org} or pyrolysis temperature. It does not require detailed soil property or environmental condition information, except for soil temperature. However, Schmidt et al. (2022) have recently argued that assessing the degree of aromatization using the parameters H/C_{org} and pyrolysis temperature has several limitations. In most industrial pyrolyzers, accurately measuring the actual temperature inside the biomass particle during conversion is not possible, and other pyrolysis-related factors such as heating rate, residence time, particle size, and pressure may also affect PAC formation (Santín et al., 2017). Moreover, though the H/C ratio can be measured with precision, low H:C fractions of biochars may not necessarily be PAC and could only be considered semipersistent (Howell et al., 2022). Schmidt et al. (2022) suggested making an assumption that biochar with an H:C ratio below 0.4 can be best described by a two-pool-model. A PAC pool of 75% with an MRT of >1000 (1400–14,400 years) and a SPC pool with an MRT of at least 50 years.

2.2.2 Use of biochar as a carbon farming practice

Carbon farming encompasses a business model that aims to upscale climate mitigation by encouraging farmers to adopt climate-friendly farm management practices (European Commission, 2021). Carbon farming is the land management practice of transferring carbon from the atmosphere into terrestrial sinks such as soil and living biomass. These practices include peatland rewetting and restoration, agroforestry management, conservation and enhancement of total organic carbon in soil and livestock, and manure management (McDonald et al., 2021). Such measures have been proposed for mitigating GHG emissions in numerous countries (Parikh & Winfield, 2020; Winfield, 2020). Among the agronomic practices aimed at increasing the total organic carbon in the soil, organic amendments, cover cropping, and improved crop rotations offer promising approaches for both mitigating the impact of climate change and retaining organic carbon. In this context, soil amendment using biochar has garnered significant and growing interest over the past decade. Various recent literature reviews have delved into the effect of biochar on soil properties and nutrient cycles (Agegnehu et al., 2017; Gul & Whalen, 2016). As stated above, biochar has been extensively researched for its potential to reduce gas emissions and increase carbon sequestration (Kalakodio et al., 2018; Li et al., 2018; Majumder et al., 2019; Siedt et al., 2021). Studies by Vijay et al. (2021), Yang and Lu (2021), and others have also supported its potential benefits. It is estimated that by 2050, carbon removal via biochar will be in the range of 0.3−2 Gt CO_2/year (Fawzy et al., 2020).

The effect of biochar amendment on soil C stock change depends on biochar application rate (Mg biochar/year) and quality, including organic C content (Mg C/Mg biochar) and fraction (Mg C/Mg biochar C) which can sustain for 100 years. However, the effect of biochar addition in soil on GHG fluxes is usually evaluated only within land applications, without considering industrial processes. As highlighted by Canatoy et al. (2022), higher temperature of pyrolysis improves biochar stability but significantly increase GHG emissions. Only with the installation of a syngas recycling system, biochar application demonstrated to be effective in decreasing GHG fluxes, and higher pyrolysis temperature of the process seemed to be more effective in mitigating global warming (Canatoy et al., 2022).

Despite the promising results from the scientific community, the carbon credit market linked to biochar amendment still has difficulties in raising its status (Campbell et al., 2018). Notwithstanding the evidence of these potential benefits in total organic carbon (TOC) sequestration practices, farmers need urgent policies and incentives for their adoption (Jat et al., 2022). Although farmers are aware of the benefits of applying carbon farming, policy uncertainty is a barrier to farmers participating in carbon farming policy programs (Dumbrell et al., 2016). The need for economic incentives for a transition to low-carbon agriculture has been extensively highlighted in the scientific literature (Borychowski et al., 2022; Verde & Chiaramonti, 2021). In this context, the 2019 refinement to the 2006 IPCC Guidelines provided a basis for the utilization of biochar as an organic amendment in the international C markets (Intergovernmental Panel on Climate Change, 2019).

Thus far, the carbon economy has been built on regulatory mechanisms such as carbon taxation, as well as mandatory and voluntary market-based mechanisms like emission trading systems. Carbon credits are sold and traded on carbon markets and can broadly be divided between compliance such as the European Emissions Trading System and voluntary, for example, companies wanting to fulfill climate targets (Sörman, 2023). Small-scale biochar production potentially has better conditions for voluntary markets due to more variations in the demand for credits. In compliance markets, buying countries often want to buy large amounts of credits, and a majority of countries do not recognize biochar as a carbon sink for countries' own target fulfillment (Sörman, 2023). In last years, a number of voluntary carbon removal platforms pioneered by three different organizations have been created. Most of them are based on block-chain technology and offers a high degree of transparency. The current carbon removal credits linked to biochar application are offered at 119 Euro per ton CO_2, which includes 19% value added tax (Fawzy et al., 2021).

Specialized broker companies pay farmers to produce credits through specific agronomic practices. Credits are then sold at about $40 per ton of captured carbon to big companies such which aim to offset GHG emissions from their operations and bolster their green image. Millions of dollars of soil credits have already been sold, and companies are ramping up aggressively to claim a piece of an industry that could overall be worth 45 billion of euros by 2030, according to the consulting firm McKinsey & Company (Popkin, 2023). Anyway, some researcher raised its doubts on how soils store and release carbon, highlighting that the knowledge on these processes and the models to estimate the climate benefits of farming practices are too uncertain to support an industry claiming to be cooling the planet (Ogle et al., 2010). They accuse companies of exaggerating the benefits of their programs. One of the biggest impediments to the widespread adoption of climate-friendly farming is the lack of a practical way to quantify the soil carbon gained through a regenerative tactic. Precise measurements would require expensive instruments and soil coring campaigns (Popkin, 2023). Policies grounded in independent and relevant scientific research should be implemented to regulate the emerging carbon removal marketplace. This will ensure that these climate solutions are optimized for the benefit of the planet.

2.3 Impact of biochar on regulating soil properties, biogeochemical cycle, and water retention

The nutrient value of biochar is widely recognized and it is supplied either directly by providing a range of plant macro- and micronutrients or indirectly by improving the soil quality (Fig. 2.1). Several biochar properties interacting with

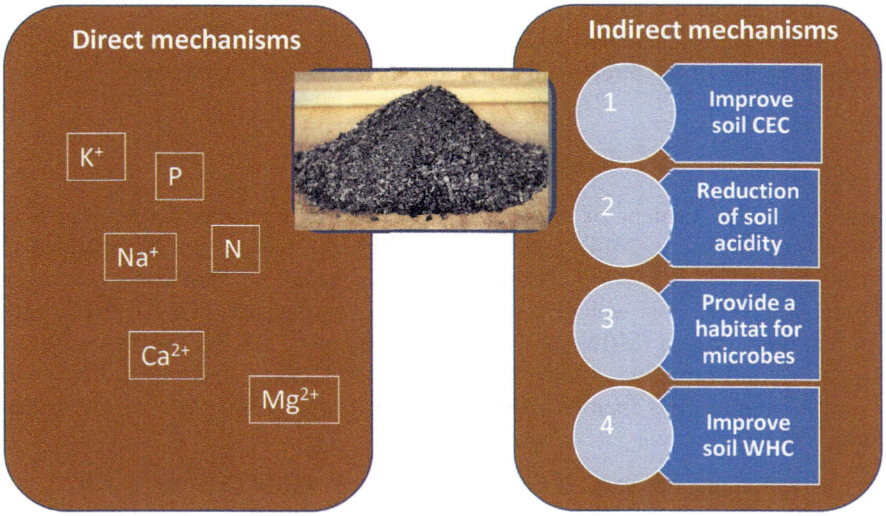

FIGURE 2.1 Direct and indirect mechanisms associated with biochar as a soil amendment for improving soil quality and increasing crop production.

K^+, Potassium cation; *Na^+*, sodium cation; *Ca^{2+}*, calcium cation; *Mg^{2+}*, manganese cation; *P*, phosphorus; *N*, nitrogen. Direct and indirect mechanisms involved in the biochar-amended soil.

soil components can contribute to increasing the availability of the main nutrients required by plants for growth such as nitrogen (N), phosphorous (P), and potassium (K). The nutrient content of biochar depends mainly on the nature of the feedstock materials and the pyrolytic conditions (i.e., pyrolytic temperature, residence time, and gaseous environment). The pyrolysis temperature has a significant effect, not only on the nutrient content but also on the nutrient availability of biochar. It was reported that biochars pyrolyzed at low temperatures (250°C–300°C) were a potential source of P and K, while higher temperatures (>300°C) reduced the fertilizing effect of the biochar (Gunes et al., 2015). This is because, at higher pyrolysis temperatures, nutrient elements are oxidized and become less soluble, as discussed in the former section. Moreover, biochars produced from nutrient-rich feedstocks, such as manure and crop residues, generally have higher N, P, and K values than those from lignocellulosic feedstocks and are therefore best suited to be used as a soil fertilizer (Hossain et al., 2020). The biochar nutrient availability was also influenced by its minerals content, such as KOH, NaOH, $MgCO_3$, and $CaCO_3$, which could elevate pH values of the soil (Ippolito et al., 2016). Yuan et al. (2011) found that the soil pH was increased by 0.59–1.05 in 2 months, depending on the biochar applied, thus improving the soil fertility. However, the information on the available nutrient content of biochars has not always been included in reports of experimental results, making it difficult to assess the agronomic values of biochars used (Xu et al., 2017). We will present below the role of biochar in the processes involved in the biogeochemical cycles of the three main soil nutrients (N, P, and K) and water retention.

2.3.1 Biochar impact on the N cycle

Despite the low content of available N, which was estimated to be less than 0.01% (Ippolito et al., 2014), and the recalcitrance of the organic N contained in the biochar, it can play an important role in influencing the processes involved in (1) soil active N pools which include dissolved organic nitrogen (DON), microbial biomass nitrogen (MBN), and dissolved inorganic N (DIN; NH_4-N and NO_3-N), (2) N transformations (mineralization, immobilization, and nitrification), (3) N input (biological N_2 fixation), and (4) N outputs (plant N uptake, NH_3 volatilization, N_2O emissions).

1. Based on a meta-analysis conducted by Liu et al. (2018), biochar has no significant effect on soil-DON but it significantly increases soil microbial biomass N by 12%. Dempster et al. (2012a) also reported that in two contrasting agroecosystems, biochar application did not influence DON turnover and did not reduce dissolved organic N leaching, since most of the DON carries a net negative charge, but it significantly increases soil MBN. Biochar has been shown to stimulate the abundance of a variety of soil microorganisms associated with soil N cycling (Prommer et al., 2014; Sánchez-García et al., 2014) due to its fine porous structure, high surface area, hydrophilicity, and mineral nutrients content (Nogués et al., 2023). Conversely, Adebajo et al. (2022) in a greenhouse experiment showed an increase in microbial biomass carbon (MBC) compared to a decrease in microbial biomass nitrogen (MBN) at the tomato harvesting stage indicating that rice husk biochar in soil acted as a carbon source rather than a nitrogen source for soil microbes. Moreover, several recent meta-analyses (Gao et al., 2019; Liu et al., 2018; Nguyen et al., 2017) showed that reductions in soil mineral N are a widespread response to biochar application. Across approximately 900 studies, (Nguyen et al., 2017) found biochar application decreased NH_4-N by an average of 11% and NO_3-N by 10%. Evidence that microbial N immobilization caused this decline in NH_4-N and NO_3-N was provided by reports of increased CO_2 emissions and increases in microbial abundances associated with N cycling, indicating an increased microbial N demand following biochar application (Deenik et al., 2010; Ducey et al., 2013; Zheng et al., 2012). Dempster et al. (2012b) also reported that biochar application significantly decreases NO_3-N leaching, through the NH_4^+-N adsorption in the biochar's surface, especially those produced at low temperatures.
2. Given that nitrogen mineralization and immobilization occur simultaneously in soil, it is of great importance to distinguish between gross and net

mineralization and immobilization due to the biochar additions. Based on a meta-analysis conducted by Liu et al. (2018), the overall effects of biochar on soil N transformations suggest significant increases in gross immobilization and of both gross and net nitrification. The enhanced abundance of microorganisms following biochar application can on one hand increase the mineralization process facilitating the degradation of soil organic N but also enhance the N immobilization through the synthesis of proteins and other nitrogen-containing organic compounds by the microorganisms themselves. So soil net N mineralization shows no significant response to biochar addition (Liu et al., 2018). Even if biochar may not alter the size of soil organic N pool in the short term, the accelerated N turnover by biochar addition may facilitate the transformation from a recalcitrant pool to a more labile pool of soil organic N (Nelissen et al., 2012).

3. The application of biochar may increase the biological N fixation in legumes, on average, by 63%, and this effect mainly occurs in soil with a pH less than 5 and depends on the rate of applications and on the row materials used for biochar production (Liu et al., 2018). Lai et al. (2022) reported that biochar addition increased the nodulation, biological nitrogen fixation, and yield of various legume species, including soybean, alfalfa, and red clover. Van Zwieten et al. (2015) reported an increase in biological N fixation of beans with the biochar application at the rate of 78 and 100 t/ha. Mia et al. (2014) observed that biochar addition increased the total biomass, nodule number, and biological N fixation. Possible processes driving the impacts of biochar addition on biological N fixation in legumes are (1) higher pH of soil improves the Mo availability, an essential nutrient needed in biological N fixation mechanism, (2) higher concentration of available N immobilization that is associated with biological N fixation enhancement, and (3) strong impact of biochar on nodule production via efficient use of Nod and flavonoids feature.
4. Biochar addition generally increases plant biomass and N uptake in acidic soils, and the impact was greater in soils with poor structure, and/or low CEC due to the increase in soil pH and water-holding capacity. Biochar also stimulates soil NH_3 volatilization to a larger extent in acidic soils with clay texture, while it shows little effect on neutral or alkaline soils. The increase of NH_3 volatilization is also streakily linked to soil organic carbon content. Conversely, biochar addition significantly decreases N_2O emission. This effect is maximized in loam soils and minimized in soils with low organic carbon ($\leq$5 g/kg) (Liu et al., 2018).

2.3.2 Biochar impact on the P cycle

The meta-analysis of 108 pairwise comparisons by Glaser and Lehr (2019) and the data from 124 peer-reviewed papers resumed by Gao et al. (2019) both showed that plant-available P in agricultural soils was significantly increased by the application of biochar produced from different materials and pyrolysis conditions, as well as for a variety of climate and soil types. The increase in soil pH due to the alkaline biochar application induces a change to P dynamics in acid soil through the solubilization of Al- or Fe-P minerals and/or Al or Fe-(hydr)oxides such as goethite. Moreover, the biochar inhibits the development of phosphate precipitates, and hereafter, the phosphorous availability rises in the plants. Xu et al. (2016) reported that biochar added to soil increases the available phosphorous and maize yield, in a field study in Indonesia. They found that biochar application increases the phosphorous for the plants even in less available soil phosphorous conditions. Finally, Li and Chan (2022) reported that sharp decline in P content in clay soil after biochar addition; it was due to the chemical adsorption of P on the surfaces of clay-mineral and temporary immobilization of P via soil microbes. Kelly et al. (2015) noticed a significant enhancement in the P content of clay soil after hardwood-derived biochar application with the rate of 5, 10, and 15 g/kg. An increase of 54% P extractability ($\sim$20 mg/kg) was achieved in the clay soil amended with biochars, whereas the increase was 42% (11 mg/kg) in the control. Soil P fixation increases in biochar-amended alkaline soils, through the precipitation of Ca–P minerals and P sorption onto calcite, due to additional input of Ca (Bornø et al., 2018). Higher contents of P were found in biochars produced from feedstocks of animal origins, sewage sludge, and broiler litter, compared to those from plants (e.g., wood).

2.3.3 Biochar impact on K cycle

Biochar typically contains a large amount of potassium (K), whose concentration usually ranges from 0.7 to 116 g/kg (Ippolito et al., 2014). Potassium-enriched biochar has been manufactured from K-rich biomass, such as animal manures (Subedi et al., 2016), banana peduncles (Karim et al., 2019), rice straw (Peng et al., 2011), and seaweeds (Eibisch et al., 2015). Availability of biochar-derived K nutrients to plants will depend on several physicochemical factors such as the solubility of K compounds in water and/or high ionic strength solution, the extent and rate of dissolution of these compounds from biochar to soils, and properties of the soils such as texture, exchange capacity, pH, and water content. A meta-analysis of 371 independent studies strongly suggested that biochar application resulted in increased soil K content and plant K tissue concentration. (Wang et al., 2018) conducted a pot experiment to examine the effect of biochar addition (0, 5, 10, and 25 g/kg soil) on the soil K dynamics in two types of soil (Alfisol and Entisol). In both soils, K increased in response to biochar addition, with a prominent impact in the Alfisol.

2.3.4 Impact of biochar on water retention

Biochar added to soil has been reported to not only affect its chemical and biological properties but also change its physical properties (Haider et al., 2022). In this regard, Blanco-Canqui (2017) has reported how biochar can generally improve the physical environment of the soil, for example by reducing bulk density, which gradually decreases as the amount of biochar applied increases. The reduction in bulk density in the long term may be due to the interaction of biochar with soil particles, thereby improving aggregation and porosity (Liu et al., 2016). Soil physical properties have a major impact on water movement and retention capacity. Thus the improvement in soil textural and structural properties generally reported after biochar addition (Rasa et al., 2018) can directly alter soil moisture properties. Biochar quantity, type, and particle size are reported to have a direct effect on soil water retention. In this regard, it is worth highlighting that soil water retention is a key factor for crop productivity as it affects water uptake and transport by plants. Amoakwah et al. (2017) highlighted that the application of maize cob biochar at 20 t/ha resulted in a significant increase in soil water retention compared to a lower biochar rate, by increasing soil microporosity. Willow biochar, characterized by the pore size of 50 and 10 μm, added to soil in a laboratory experiment (5% on a dry weight basis) was reported by Rasa et al. (2018) to increase soil porosity and the amount of water readily available to plants. Two different application rates (22 and 44 t/ha) of biochar derived from orchard pruning biomass to the soil of an Italian vineyard were observed to reduce soil bulk density while increasing soil water content by up to 9.2% compared to unamended soil (Baronti et al., 2014). In a laboratory experiment reported by Villagra-Mendoza and Horn (2018), the reduction in total soil porosity and air capacity, and the increase in soil available water content, paralleled the increase in biochar addition. Biochar produced from the mixture of paper fiber sludge and grain husks (in a 1:1 per weight ratio) and applied at 10 and 20 t/ha was observed to improve plant-available water, particularly at the highest application rate by reducing bulk density by 12% and increasing porosity by 12% (Toková et al., 2020).

In addition to the rate of application of biochar to the soil, the quality of biochar (i.e., derived from different biomasses) and particle size have also been highlighted as key factors in improving soil water retention capacity. Regarding quality, Ndede et al. (2022) showed that the addition of four different types of biochar (e.g., wood chips, *Ludwigia grandiflora* waterweed, poultry litter, and bagasse) to sandy soils on a laboratory scale at various concentrations resulted in a reduction in bulk density for all biochar types, but although the soil water-holding capacity increased with application rate, the readily available water reached its highest value at 5% (v/v) biochar content. Different types of biochar (produced from rice straw, maize straw, wheat straw, rice husk, and bamboo) were applied by Yang and Lu (2021) in a field trial on paddy soil at a rate of 11.25 and 22.5 t/ha, respectively, highlighting the increase in soil macroaggregate content and soil water-holding capacity. Accordingly, the increase in available water content in the soil with all biochar types added, especially at the highest rate, was observed. Biochar from rice husk, bean harvest residues, and corn cobs added to loam and sandy loam soils in a greenhouse experiment was reported by

Günal et al. (2018) to improve water-holding capacity at biochar application rates of 11.25 (0.5%), 22.5 (1.0%), 45 (2.0%), and 67.5 t/ha (3.0%).

The increase in available water content varied with biochar type (19.9% for rice husk, 6.0% for bean crop residues and maize) and application rate (maximum values at 2% and 3%). Different effects of the application of two types of biochar (from water hyacinth and mesquite plants) on soil water-holding capacity and water content were observed by Hussain et al. (2020), with both biochar types increasing these parameters while decreasing water infiltration rate, also in a positive relationship with the application rates (5%–10%). Similar results were reported by Huang et al. (2021) using water hyacinth, chicken manure, and wood biochar at 5% and 10% (w/w) rates, showing that the addition of three biochars (at 10%) significantly improved soil water retention, with water hyacinth biochar highlighting the best performance. Bamboo-derived biochar was found by Qian et al. (2020) to cause a greater improvement in soil physical properties related to water status, reducing bulk density while increasing saturated water content, field capacity and readily available water compared to rice straw and tobacco, with effects progressively increasing with increasing application rate (2%, 5%, and 10%, w/w). In the literature, an effect of biochar particle size on soil physical properties affecting water status has been widely reported. In this context, using biochar of three different sizes (<0.15, 0.15–2, and >2 mm), de Jesus Duarte et al. (2019) observed an increase in water retention in both soil types considered when a smaller biochar (<0.15 mm particle size) was applied. Consistent with this finding, Alghamdi et al. (2020) found the greatest increase in both field capacity and available water content when the smallest particle size biochar (<0.1 mm) was applied, due to the increased microporosity of the soil as a result of the larger internal surface area and porous structure of the biochar particle.

Biochar with two different particle size ranges (0.06–0.5 and 2–4 mm), both sieved or not, was applied to soil by Liao and Thomas (2019), who demonstrated how small particle size biochar increased water retention capacity, with sieved biochars showing 91%–258% greater water retention capacity than ground biochars of similar particle size, likely due to higher elongation of sieved particles than ground particles, thus increasing soil pore volume. A different effect of biochar particle size on plant-available water was reported by Zhang et al. (2021), depending on whether this soil parameter was expressed in gravimetric terms or volumetric terms, with little or increasing effects on plant-available water content with decreasing biochar particle size, respectively. The application of biochar with different characteristics (i.e., quality, quantity, particle size) has been reported to affect soil properties such as bulk density and maximum water-holding capacity differently depending on the soil type (Verheijen et al., 2019). In this regard, several authors have highlighted that soil type is a critical factor in assessing the effect of biochar on soil properties related to water infiltration and retention, although there are conflicting reports on this issue. In both sandy loam and loamy soils, the decrease in total porosity following biochar amendment was described by Günal et al. (2018), with the positive effect of biochar addition on available water content more evident in the loamy soils. Irrespective of other factors (i.e., biochar quality, quantity, particle size), the effect of biochar on water-holding capacity is reported to be more pronounced in sandy soils than in clayey soils (Adhikari et al., 2022), due to an increase in soil micropores and macropores in biochar-amended soils.

In tropical soils, Santos et al. (2022) observed an improvement in water-holding capacity after biochar application in coarse-textured soils, while in clayey soils biochar increased soil porosity and water transmission. In this context, Soinne et al. (2020) reported that biochar, even at high doses, did not alter the water-holding capacity of boreal agricultural clay soils. However, reports on the extent of biochar effects on soil properties related to water status for different soil types are quite contrasting. For example, Omondi et al. (2016) highlighted the more pronounced effect of biochar on bulk density in the fine-textured soils more than the medium- and coarse-textured soils. In contrast, Blanco-Canqui (2017) reported a decrease in bulk density in the coarse-textured soil more than in the fine-textured soils after biochar addition. Regarding such controversial results, Rasa et al. (2018) have argued that the main effect attributed to biochar, that is, the increase in soil porosity, is due to its internal porosity (direct mechanism), thus hypothesizing that this effect is not related to soil type. However, biochar is also known to modify soil properties that affect soil moisture properties (indirect mechanism), with this effect being particularly dependent on biochar and soil type, which may explain the contrasting observations on biochar application on soil water retention.

2.4 Biochar production for circular bioeconomy in agricultural farming systems: BIOCHAR LATIUM case study

A circular bioeconomy is a new economic model that combines bioeconomy and circular economy concepts emphasizing the use of renewable natural capital and focusing on minimizing waste, replacing the wide range of nonrenewable, fossil-based products currently in use. Based on the concept of circular bioeconomy, organic wastes and by-products can be reused as valuable resources for biochar and bioenergy production through pyrolysis technologies. This process not only transforms wastes into valuable products but can reduces disposal costs and energy consumption. Within this context, the project titled "Valorisation of the biochar produced from the recovery of woody wastes from Lazio supply chains" is a research project funded by Lazio Innova (Lazio Regional Development Agency) within the POR-FESR framework, which aims at designing a strategy for the valuable exploitation of lignocellulosic wastes from the Lazio region through the production of biochar. The project deals with important issues such as environmental sustainability, green economy, and carbon storage (climate change mitigation).

The project's specific goal, reached through an interdisciplinary approach was to test the efficacy of the production of biochar from woody wastes generated in the hazelnut and olive local supply chains and its application as a soil amendment. For this purpose, the project assessed:

- the characteristics and potentialities of the produced biochar;
- the effects and the most efficient application mode (alone or with compost) on nutrient-deficient soils;
- the effects and the most efficient application mode (alone or with microorganisms) on contaminated soils.

Two experiments were set up: (1) a field trial for soil fertility and *lettuce sativa* yield amelioration through the application of biochar, compost, and compost–biochar mixture; and (2) a mesocosm trial for multicontaminated soil remediation (hydrocarbons and copper), through the application of biochar and microorganisms. The main results of the first trial showed that the biomass of lettuce was increased in all amended soil with respect to the control soil; the compost-amended soil was the one that registered the higher effect on microbial biomass activity (increase in ß-glucosidase and butyrate esterase activities, and synthetic enzymatic index), and structure (shift through G-bacteria, total fungi, and arbuscular mycorrhizal fungi-AMF) indicating that the organic matter added with the compost is more readily mineralized than that one added with the biochar. Moreover, the biochar used, derived from woody wastes, registered the lowest metabolic quotient (qCO_2) and the highest microbial quotient (qMIC), indicating a C immobilization process into microbial biomass and a greater limitation for N with respect to C, offset by an increase in N-cycling enzyme activity (Fig. 2.2).

The main results of the second trial showed the effects of applying biochar alone or in combination with microorganisms (bio-activators), on hydrocarbons contaminated soils. Experiments conducted in soil mesocosms under greenhouse conditions provided interesting results. The addition of the biochar to the soil, other than increasing the cations exchange capacity, promoted the degradation of $C > 12$ hydrocarbons, probably enhancing the autochthonous soil microbial community and mitigating the soil toxicity linked to the presence of copper. In particular, the addition of biochar and bio-activators to the soil promotes higher total microbial abundance in terms of bacteria and fungi (Fig. 2.3). The project results contribute to demonstrating the suitability of biochar for application in agriculture and for the treatment of contaminated soils.

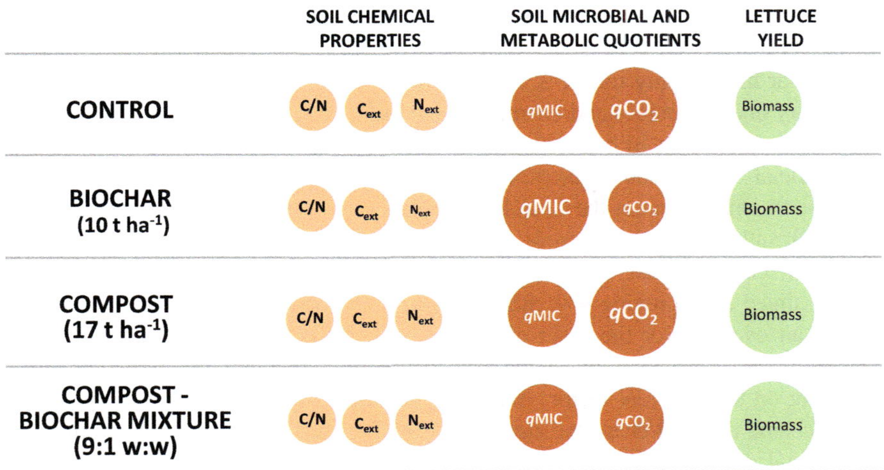

FIGURE 2.2 Main resultsobtained from the field trial of the Biochar Latium research project.

C/N, Total organic C total N ratio; C_{ext}, C extracted by 0.5 M K_2SO_4 solution; N_{ext}, N extracted by 0.5 M K_2SO_4 solution; q*MIC*, microbial quotient; qCO_2, metabolic quotient. Main results of the Biochar Latium research project about the field trial.

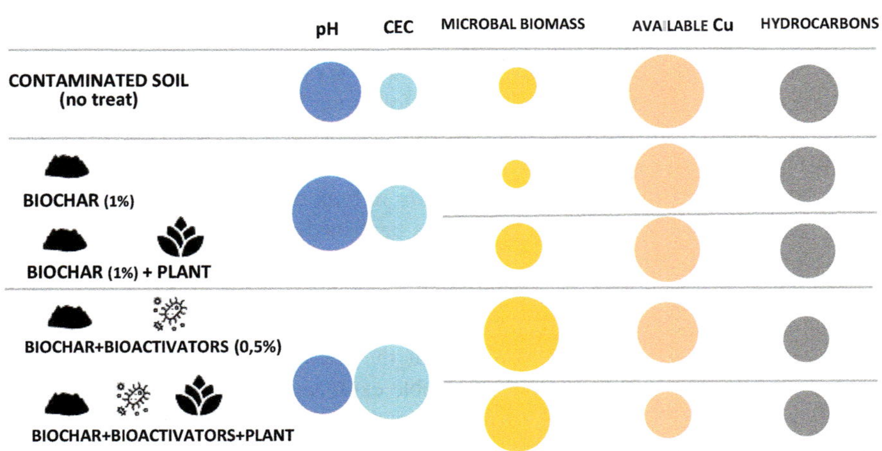

FIGURE 2.3 Main results obtained from the mesocosms trial of the Biochar Latium research project.

CEC, Cation exchange capacity. Main results of the Biochar Latium research project about the mesocosms trial.

2.5 Conclusion

The circular bioeconomy model is designed to preserve resources by reducing the use of auxiliary energy sources and chemicals, and utilizing waste products in the cycle of the process production. In this context, biochar obtained from agricultural residues can be beneficial in promoting bioeconomy at a local scale providing carbon- and nutrient-rich amendment at the farm level. Biochar is a soil amendment that can present significantly different properties depending on the characteristics of the row materials. Such characteristics are important to define to establish the best agricultural practice aimed at improving soil chemical, physical, and biological properties. In this chapter, an overview of the main effects associated with biochar soil amendment is reported. Positive, negative, or neutral effects of biochar addition to soil are described, depending on the different factors mainly associated with the biochar type and application scheme and soil characteristics. Changes of the soil environment can affect the fundamental services it provides, in particular, soil pH regulates nutrient availability affecting several biogeochemical cycles. So that the alkaline reaction of biochar may improve the environment of acidic soil, conversely may be worse for saline or alkaline soils. For this reason, it is also important to pay particular attention to the characteristics of the biomass used to produce biochar. In the case of biomasses particularly rich in lignin and cellulose. For example, since the C/N ratio of biochar is particularly high, attention is required when it is applied to agricultural soils because the biogeochemical cycle of nitrogen can be substantially modified enhancing N immobilization process into microbial biomass. In this context, it is worthy to underline that biochar application is claimed to reduce soil mineral N while increasing the abundance of a variety of soil microorganisms in relation to soil N cycling. Instead, soil P and K levels are commonly reported to be increased following biochar addition, with a positive impact on the availability of these elements to plants. Soil physical properties are also highlighted as positively affected by biochar, especially for that regards the increase of water-holding capacity, the decrease in bulk density, and the enhancement of soil porosity, albeit contrasting evidence is described about the extension of these effects depending on the soil type. Even though a relevant number of works on the addition of biochar to soil were published in the last few years, field studies taking into consideration the main environmental variables are still requested. In this chapter, the key outcomes of a local-scale project in the Mediterranean environment were reported, highlighting a strategy for the valuable exploitation of lignocellulosic wastes under a circular bioeconomy model to improve the fertility and water availability of agricultural soil with quite low organic matter content under Mediterranean climate.

Acknowledgment

The project "Biochar Latium" N. CUP J85F21000410002 funded by Lazio Region – Lazio Innova (Italy), within POR FESR Lazio 2014–2020 program.

Author contribution

L. Massaccesi: Conceptualization, writing—original draft, writing review and editing, visualization, luisa.massaccesi@cnr.it; I. Nogués: writing—original draft, writing review and editing, mariaisabel.noguesgonzalez@cnr.it; V. Mazzurco Miritana: writing—original draft, valentina.mazzurco@enea.it; L. Passatore: writing—original draft, laura.passatore@cnr.it; M. Zacchini: writing—original draft, massimo.zacchini@cnr.it; S. Marinari: conceptualization, writing—original draft, writing review and editing, marinari@unitus.it.

Conflict of interest

The authors declare that they have no known competing financial interests or personal relationships that could have appeared to influence the work reported in this chapter.

Statement on the use of AI tools for writing

AI and AI-assisted technologies should not be listed as an author or coauthor, or be cited as an author.

References

Adebajo, S. O., Oluwatobi, F., Akintokun, P. O., Ojo, A. E., Akintokun, A. K., & Gbodope, I. S. (2022). Impacts of rice-husk biochar on soil microbial biomass and agronomic performances of tomato (*Solanum lycopersicum* L.). *Scientific Reports*, *12* (1). Available from https://doi.org/10.1038/s41598-022-05757-z. Available from: http://www.nature.com/srep/index.html.

Adhikari, S., Timms, W., & Mahmud, M. A. P. (2022). Optimising water holding capacity and hydrophobicity of biochar for soil amendment – a review. *Science of the Total Environment*, *851*, 158043. Available from https://doi.org/10.1016/j.scitotenv.2022.158043.

Agegnehu, G., Srivastava, A. K., & Bird, M. I. (2017). The role of biochar and biochar-compost in improving soil quality and crop performance: A review. *Applied Soil Ecology*, *119*, 156–170. Available from https://doi.org/10.1016/j.apsoil.2017.06.008, http://www.elsevier.com/inca/publications/store/5/2/4/5/1/8/index.htt.

Alghamdi, A. G., Alkhasha, A., & Ibrahim, H. M. (2020). Effect of biochar particle size on water retention and availability in a sandy loam soil. *Journal of Saudi Chemical Society*, *24*(12), 1042–1050. Available from https://doi.org/10.1016/j.jscs.2020.11.003, http://www.sciencedirect.com/science/journal/13196103.

Almutairi, A. A., Ahmad, M., Rafique, M. I., & Al-Wabel, M. I. (2023). Variations in composition and stability of biochars derived from different feedstock types at varying pyrolysis temperature. *Journal of the Saudi Society of Agricultural Sciences*, *22*(1), 25–34. Available from https://doi.org/10.1016/j.jssas.2022.05.005, https://www.journals.elsevier.com/journal-of-the-saudi-society-of-agricultural-sciences/.

Amoakwah, E., Frimpong, K. A., Okae-Anti, D., & Arthur, E. (2017). Soil water retention, air flow and pore structure characteristics after corn cob biochar application to a tropical sandy loam. *Geoderma*, *307*, 189–197.

Baronti, S., Vaccari, F. P., Miglietta, F., Calzolari, C., Lugato, E., Orlandini, S., Pini, R., Zulian, C., & Genesio, L. (2014). Impact of biochar application on plant water relations in *Vitis vinifera* (L.). *European Journal of Agronomy*, *53*, 38–44. Available from https://doi.org/10.1016/j.eja.2013.11.003, http://www.elsevier.com/inca/publications/store/6/0/0/1/0/8.

Blanco-Canqui, H. (2017). Biochar and soil physical properties. *Soil Science Society of America Journal*, *81*(4), 687–711. Available from https://doi.org/10.2136/sssaj2017.01.0017, https://dl.sciencesocieties.org/publications/sssaj/pdfs/81/4/687.

Bornø, M. L., Müller-Stöver, D. S., & Liu, F. (2018). Contrasting effects of biochar on phosphorus dynamics and bioavailability in different soil types. *Science of the Total Environment*, *627*, 963–974. Available from https://doi.org/10.1016/j.scitotenv.2018.01.283, http://www.elsevier.com/locate/scitotenv.

Borychowski, M., Grzelak, A., & Popławski, Ł. (2022). What drives low-carbon agriculture? The experience of farms from the Wielkopolska region in Poland. *Environmental Science and Pollution Research*, *29*(13), 18641–18652. Available from https://doi.org/10.1007/s11356-021-17022-3. Available from, https://link.springer.com/journal/11356.

Bowring, S. P. K., Jones, M. W., Ciais, P., Guenet, B., & Abiven, S. (2022). Pyrogenic carbon decomposition critical to resolving fire's role in the Earth system. *Nature Geoscience*, *15*(2), 135–142. Available from https://doi.org/10.1038/s41561-021-00892-0, http://www.nature.com/ngeo/index.html.

Campbell, R. M., Anderson, N. M., Daugaard, D. E., & Naughton, H. T. (2018). Financial viability of biofuel and biochar production from forest biomass in the face of market price volatility and uncertainty. *Applied Energy*, *230*, 330–343. Available from https://doi.org/10.1016/j.apenergy.2018.08.085, http://www.elsevier.com/inca/publications/store/4/0/5/8/9/1/index.htt.

Canatoy, R. C., Jeong, S. T., Galgo, S. J. C., Kim, P. J., & Cho, S. R. (2022). Biochar as soil amendment: Syngas recycling system is essential to create positive carbon credit. *Science of the Total Environment*, *809*. Available from https://doi.org/10.1016/j.scitotenv.2021.151140, http://www.elsevier.com/locate/scitotenv.

Cao, W., Xu, H., Zhang, X., Xiang, W., Qi, G., Wan, L., & Gao, B. (2023). Novel post-treatment of ultrasound assisting with acid washing enhance lignin-based biochar for CO2 capture: Adsorption performance and mechanism. *Chemical Engineering Journal*, *471*, 144523. Available from https://doi.org/10.1016/j.cej.2023.144523.

Conte, P., Bertani, R., Sgarbossa, P., Bambina, P., Schmidt, H. P., Raga, R., Lo Papa, G., Chillura Martino, D. F., & Lo Meo, P. (2021). Recent developments in understanding biochar's physical–chemistry. *Agronomy*, *11*(4). Available from https://doi.org/10.3390/agronomy11040615, https://www.mdpi.com/2073-4395/11/4/614/pdf.

Cross, A., & Sohi, S. P. (2013). A method for screening the relative long-term stability of biochar. *GCB Bioenergy*, *5*(2), 215–220. Available from https://doi.org/10.1111/gcbb.12035, http://onlinelibrary.wiley.com/journal/10.1111/(ISSN)1757-1707.

Deenik, J. L., McClellan, T., Uehara, G., Antal, M. J., & Campbell, S. (2010). Charcoal volatile matter content influences plant growth and soil nitrogen transformations. *Soil Science Society of America Journal*, *74*(4), 1259–1270. Available from https://doi.org/10.2136/sssaj2009.0115.

Dempster, D. N., Jones, D. L., & Murphy, D. V. (2012a). Clay and biochar amendments decreased inorganic but not dissolved organic nitrogen leaching in soil. *Soil Research*, *50*(3), 216–221. Available from https://doi.org/10.1071/SR11316.

Dempster, D. N., Jones, D. L., & Murphy, D. V. (2012b). Organic nitrogen mineralisation in two contrasting agro-ecosystems is unchanged by biochar addition. *Soil Biology and Biochemistry*, *48*, 47–50. Available from https://doi.org/10.1016/j.soilbio.2012.01.013.

Ducey, T. F., Ippolito, J. A., Cantrell, K. B., Novak, J. M., & Lentz, R. D. (2013). Addition of activated switchgrass biochar to an aridic subsoil increases microbial nitrogen cycling gene abundances. *Applied Soil Ecology*, *65*, 65–72. Available from https://doi.org/10.1016/j.apsoil.2013.01.006.

Dumbrell, N. P., Kragt, M. E., & Gibson, F. L. (2016). What carbon farming activities are farmers likely to adopt? A best-worst scaling survey. *Land Use Policy*, *54*, 29–37. Available from https://doi.org/10.1016/j.landusepol.2016.02.002, http://www.elsevier.com/inca/publications/store/3/0/4/5/1/.

Eibisch, N., Durner, W., Bechtold, M., Fuß, R., Mikutta, R., Woche, S. K., & Helfrich, M. (2015). Does water repellency of pyrochars and hydrochars counter their positive effects on soil hydraulic properties? *Geoderma*, *245-246*, 31–39. Available from https://doi.org/10.1016/j.geoderma.2015.01.009, http://www.elsevier.com/inca/publications/store/5/0/3/3/3/2.

European Commission, Commission Implementing Regulation (EU) 2021/1165 of 15 July 2021. (2021). Authorising certain products and substances for use in organic production and establishing their lists. *Official Journal of European Communities*, *50*, 13-48.

Fawzy, S., Osman, A. I., Doran, J., & Rooney, D. W. (2020). Strategies for mitigation of climate change: A review. *Environmental Chemistry Letters*, *18*(6), 2069–2094. Available from https://doi.org/10.1007/s10311-020-01059-w, http://springerlink.metapress.com/app/home/journal.asp?wasp = d86tgdwvtg0yvw9gvkwp&referrer = parent&backto = browsepublicationsresults,140,541.

Fawzy, S., Osman, A. I., Yang, H., Doran, J., & Rooney, D. W. (2021). Industrial biochar systems for atmospheric carbon removal: A review. *Environmental Chemistry Letters*, *19*(4), 3023–3055. Available from https://doi.org/10.1007/s10311-021-01210-1. Available from: http://springerlink.metapress.com/app/home/journal.asp?wasp = d86tgdwvtg0yvw9gvkwp&referrer = parent&backto = browsepublicationsresults,140,541.

Gao, S., DeLuca, T. H., & Cleveland, C. C. (2019). Biochar additions alter phosphorus and nitrogen availability in agricultural ecosystems: A meta-analysis. *Science of the Total Environment*, *654*, 463–472. Available from https://doi.org/10.1016/j.scitotenv.2018.11.124, http://www.elsevier.com/locate/scitotenv.

Glaser, B., & Lehr, V. I. (2019). Biochar effects on phosphorus availability in agricultural soils: A meta-analysis. *Scientific Reports*, *9*(1). Available from https://doi.org/10.1038/s41598-019-45693-z, http://www.nature.com/srep/index.html.

Gul, S., & Whalen, J. K. (2016). Biochemical cycling of nitrogen and phosphorus in biochar-amended soils. *Soil Biology and Biochemistry*, *103*, 1–15. Available from https://doi.org/10.1016/j.soilbio.2016.08.001, http://www.elsevier.com/inca/publications/store/3/3/2.

Günal, E., Erdem, H., & Çelik, İ. (2018). Effects of three different biochars amendment on water retention of silty loam and loamy soils. *Agricultural Water Management*, *208*, 232–244. Available from https://doi.org/10.1016/j.agwat.2018.06.004. Available from, http://www.journals.elsevier.com/agricultural-water-management/.

Gunes, A., Inal, A., Sahin, O., Taskin, M. B., Atakol, O., & Yilmaz, N. (2015). Variations in mineral element concentrations of poultry manure biochar obtained at different pyrolysis temperatures, and their effects on crop growth and mineral nutrition. *Soil Use and Management*, *31*(4), 429–437. Available from https://doi.org/10.1111/sum.12205, http://www.wiley.com/bw/editors.asp?ref = 0266-0032&site = 1.

Haider, F. U., Coulter, J. A., Cai, L., Hussain, S., Cheema, S. A., Wu, J., & Zhang, R. (2022). An overview on biochar production, its implications, and mechanisms of biochar-induced amelioration of soil and plant characteristics. *Pedosphere*, *32*(1), 107–130. Available from https://doi.org/10.1016/S1002-0160(20)60094-7, http://pedosphere.issas.ac.cn.

Hossain, M. Z., Bahar, M. M., Sarkar, B., Donne, S. W., Ok, Y. S., Palansooriya, K. N., Kirkham, M. B., Chowdhury, S., & Bolan, N. (2020). Biochar and its importance on nutrient dynamics in soil and plant. *Biochar*, *2*(4), 379–420. Available from https://doi.org/10.1007/s42773-020-00065-z, http://springer.com/journal/42773.

Howell, A., Helmkamp, S., & Belmont, E. (2022). Stable polycyclic aromatic carbon (SPAC) formation in wildfire chars and engineered biochars. *Science of The Total Environment*, *849*, 157610. Available from https://doi.org/10.1016/j.scitotenv.2022.157610.

Huang, H., Reddy, N. G., Huang, X., Chen, P., Wang, P., Zhang, Y., Huang, Y., Lin, P., & Garg, A. (2021). Effects of pyrolysis temperature, feedstock type and compaction on water retention of biochar amended soil. *Scientific Reports*, *11*(1). Available from https://doi.org/10.1038/s41598-021-86701-5, http://www.nature.com/srep/index.html.

Hussain, R., Bordoloi, S., Gupta, P., Garg, A., Ravi, K., Sreedeep, S., & Sahoo, L. (2020). Effect of biochar type on infiltration, water retention and desiccation crack potential of a silty sand. *Biochar*, *2*(4), 465–478. Available from https://doi.org/10.1007/s42773-020-00064-0, http://springer.com/journal/42773.

Intergovernmental Panel on Climate Change. (2019). Special report.

Ippolito, J. A., Ducey, T. F., Cantrell, K. B., Novak, J. M., & Lentz, R. D. (2016). Designer, acidic biochar influences calcareous soil characteristics. *Chemosphere*, *142*, 184–191. Available from https://doi.org/10.1016/j.chemosphere.2015.05.092, http://www.elsevier.com/locate/chemosphere.

Ippolito, J. A., Stromberger, M. E., Lentz, R. D., & Dungan, R. S. (2014). Hardwood biochar influences calcareous soil physicochemical and microbiological status. *Journal of Environmental Quality*, *43*(2), 681–689. Available from https://doi.org/10.2134/jeq2013.08.0324, https://www.agronomy.org/publications/jeq/pdfs/43/2/681.

Jat, M. L., Chakraborty, D., Ladha, J. K., Parihar, C. M., Datta, A., Mandal, B., Nayak, H. S., Maity, P., Rana, D. S., Chaudhari, S. K., & Gerard, B. (2022). Carbon sequestration potential, challenges, and strategies towards climate action in smallholder agricultural systems of South Asia. *Crop and Environment*, *1*(1), 86–101. Available from https://doi.org/10.1016/j.crope.2022.03.005, http://www.elsevier.com/locate/issn/2773126X.

de Jesus Duarte, S., Glaser, B., & Pellegrino Cerri, C. (2019). Effect of biochar particle size on physical, hydrological and chemical properties of loamy and sandy tropical soils. *Agronomy*, *9*(4), 165. Available from https://doi.org/10.3390/agronomy9040165.

Kalakodio, L., Bakayoko, M., Kalagodio, A., Abo, B. O., Muhoza, J. P., & Ismaila, E. M. (2018). Progress on the amendment in biochars and its effects on the soil-plant-microorganism-biochar system. *De Reviews on Environmental Health*, *33*(3), 281–293. Available from https://doi.org/10.1515/reveh-2018-0007, http://www.degruyter.com/view/j/reveh.

Karim, A. A., Kumar, M., Mohapatra, S., & Singh, S. K. (2019). Nutrient rich biomass and effluent sludge wastes co-utilization for production of biochar fertilizer through different thermal treatments. *Journal of Cleaner Production*, *228*, 570–579. Available from https://doi.org/10.1016/j.jclepro.2019.04.330, https://www.journals.elsevier.com/journal-of-cleaner-production.

Kelly, C. N., Calderón, F. C., Acosta-Martínez, V., Mikha, M. M., Benjamin, J., Rutherford, D. W., & Rostad, C. E. (2015). Switchgrass biochar effects on plant biomass and microbial dynamics in two soils from different regions. *Pedosphere*, *25*(3), 329–342. Available from https://doi.org/10.1016/S1002-0160(15)30001-1, http://pedosphere.issas.ac.cn.

Lai, W., Wu, Y., Zhang, C., Dilinuer, Y., Pasang, L., Lu, Y., Wang, Y., Chen, H., & Li, Z. (2022). Combination of biochar and phosphorus solubilizing bacteria to improve the stable form of toxic metal minerals and microbial abundance in lead/cadmium-contaminated soil. *Agronomy*, *12*(5), 1003. Available from https://doi.org/10.3390/agronomy12051003.

Lehmann, J., & Joseph, S. (2015). *Biochar for environmental management: Science, technology and implementation*. Taylor & Francis Group.

Li, L., Long, A., Fossum, B., & Kaiser, M. (2023). Effects of pyrolysis temperature and feedstock type on biochar characteristics pertinent to soil carbon and soil health: A meta-analysis. *Soil Use and Management*, *39*(1), 43–52. Available from https://doi.org/10.1111/sum.12848, http://onlinelibrary.wiley.com/journal/10.1111/(ISSN)1475-2743.

Li, S., & Chan, C. Y. (2022). Will biochar suppress or stimulate greenhouse gas emissions in agricultural fields? Unveiling the dice game through data syntheses. *Soil Systems*, *6*(4), 73. Available from https://doi.org/10.3390/soilsystems6040073.

Li, S., & Tasnady, D. (2023). Biochar for soil carbon sequestration: Current knowledge, mechanisms, and future perspectives. *Journal of Carbon Research*, *9*(3), 67. Available from https://doi.org/10.3390/c9030067.

Li, Y., Hu, S., Chen, J., Müller, K., Li, Y., Fu, W., Lin, Z., & Wang, H. (2018). Effects of biochar application in forest ecosystems on soil properties and greenhouse gas emissions: A review. *Journal of Soils and Sediments*, *18*(2), 546–563. Available from https://doi.org/10.1007/s11368-017-1906-y, http://www.springerlink.com/content/1439-0108.

Liao, W., & Thomas, S. (2019). Biochar particle size and post-pyrolysis mechanical processing affect soil pH, water retention capacity, and plant performance. *Soil Systems*, *3*(1), 14. Available from https://doi.org/10.3390/soilsystems3010014.

Liu, B., Liu, Q., Wang, X., Bei, Q., Zhang, Y., Lin, Z., Liu, G., Zhu, J., Hu, T., Jin, H., Wang, H., Sun, X., Lin, X., & Xie, Z. (2020). A fast chemical oxidation method for predicting the long-term mineralization of biochar in soils. *Science of the Total Environment*, *718*, 137390. Available from https://doi.org/10.1016/j.scitotenv.2020.137390.

Liu, Q., Zhang, Y., Liu, B., Amonette, J. E., Lin, Z., Liu, G., Ambus, P., & Xie, Z. (2018). How does biochar influence soil N cycle? A meta-analysis. *Plant and Soil*, *426*(1-2), 211–225. Available from https://doi.org/10.1007/s11104-018-3619-4, http://www.wkap.nl/journalhome.htm/0032-079X.

Liu, Z., Dugan, B., Masiello, C. A., Barnes, R. T., Gallagher, M. E., & Gonnermann, H. (2016). Impacts of biochar concentration and particle size on hydraulic conductivity and DOC leaching of biochar-sand mixtures. *Journal of Hydrology*, *533*, 461–472. Available from https://doi.org/10.1016/j.jhydrol.2015.12.007, http://www.elsevier.com/inca/publications/store/5/0/3/3/4/3.

Maestrini, B., Abiven, S., Singh, N., Bird, J., Torn, M. S., & Schmidt, M. W. I. (2014). Carbon losses from pyrolysed and original wood in a forest soil under natural and increased N deposition. *Biogeosciences*, *11*(18), 5199–5213. Available from https://doi.org/10.5194/bg-11-5199-2014, http://www.biogeosciences.net/volumes_and_issues.html.

Majumder, S., Neogi, S., Dutta, T., Powel, M. A., & Banik, P. (2019). The impact of biochar on soil carbon sequestration: Meta-analytical approach to evaluating environmental and economic advantages. *Journal of Environmental Management*, *250*, 109466. Available from https://doi.org/10.1016/j.jenvman.2019.109466.

Mašek, O., Brownsort, P., Cross, A., & Sohi, S. (2013). United Kingdom influence of production conditions on the yield and environmental stability of biochar. *Fuel*, *103*, 151–155. Available from https://doi.org/10.1016/j.fuel.2011.08.044.

Mašek, O., Buss, W., Brownsort, P., Rovere, M., Tagliaferro, A., Zhao, L., Cao, X., & Xu, G. (2019). Potassium doping increases biochar carbon sequestration potential by 45%, facilitating decoupling of carbon sequestration from soil improvement. *Scientific reports*, *9*(1), 5514.

McDonald, H., Frelih-Larsen, A., Keenleyside, C., Lóránt, A., Duin, L., Andersen, S. P., Costa, G., Aubert, G., & Hiller, N. (2021). *Carbon farming: Making agriculture fit for 2030*. IEEP.

Mia, S., van Groenigen, J. W., van de Voorde, T. F. J., Oram, N. J., Bezemer, T. M., Mommer, L., & Jeffery, S. (2014). Biochar application rate affects biological nitrogen fixation in red clover conditional on potassium availability. *Agriculture, Ecosystems and Environment*, *191*, 83–91. Available from https://doi.org/10.1016/j.agee.2014.03.011, http://www.elsevier.com/inca/publications/store/5/0/3/2/9/8.

Mukome, F. N. D., Zhang, X., Silva, L. C. R., Six, J., & Parikh, S. J. (2013). Use of chemical and physical characteristics to investigate trends in biochar feedstocks. *Journal of Agricultural and Food Chemistry*, *61*(9), 2196–2204. Available from https://doi.org/10.1021/jf3049142.

Ndede, E. O., Kurebito, S., Idowu, O., Tokunari, T., & Jindo, K. (2022). The potential of biochar to enhance the water retention properties of sandy agricultural soils. *Agronomy*, *12*(2). Available from https://doi.org/10.3390/agronomy12020311, https://www.mdpi.com/2073-4395/12/2/311/pdf.

Nelissen, V., Rütting, T., Huygens, D., Staelens, J., Ruysschaert, G., & Boeckx, P. (2012). Maize biochars accelerate short-term soil nitrogen dynamics in a loamy sand soil. *Soil Biology and Biochemistry*, *55*, 20–27. Available from https://doi.org/10.1016/j.soilbio.2012.05.019.

Nguyen, T. T. N., Xu, C. Y., Tahmasbian, I., Che, R., Xu, Z., Zhou, X., Wallace, H. M., & Bai, S. H. (2017). Effects of biochar on soil available inorganic nitrogen: A review and meta-analysis. *Geoderma*, *288*, 79–96. Available from https://doi.org/10.1016/j.geoderma.2016.11.004, http://www.elsevier.com/inca/publications/store/5/0/3/3/3/2.

Nogués, I., Mazzurco Miritana, V., Passatore, L., Zacchini, M., Peruzzi, E., Carloni, S., Pietrini, F., Marabottini, R., Chiti, T., Massaccesi, L., & Marinari, S. (2023). Biochar soil amendment as carbon farming practice in a Mediterranean environment. *Geoderma Regional*, *33*, e00634. Available from https://doi.org/10.1016/j.geodrs.2023.e00634.

Ogle, S. M., Breidt, F. J., Easter, M., Williams, S., Killian, K., & Paustian, K. (2010). Scale and uncertainty in modeled soil organic carbon stock changes for US croplands using a process-based model. *Global Change Biology*, *16*(2), 810–822. Available from https://doi.org/10.1111/j.1365-2486.2009.01951.x.

Omondi, M. O., Xia, X., Nahayo, A., Liu, X., Korai, P. K., & Pan, G. (2016). Quantification of biochar effects on soil hydrological properties using meta-analysis of literature data. *Geoderma*, *274*, 28–34. Available from https://doi.org/10.1016/j.geoderma.2016.03.029, http://www.elsevier.com/inca/publications/store/5/0/3/3/3/2.

Parikh, S. J., & Winfield, E. (2020). *Climate-smart agriculture: Biochar amendments*. USDA.

Peng, X., Ye, L. L., Wang, C. H., Zhou, H., & Sun, B. (2011). Temperature- and duration-dependent rice straw-derived biochar: Characteristics and its effects on soil properties of an Ultisol in southern China. *Soil and Tillage Research*, *112*(2), 159–166. Available from https://doi.org/10.1016/j.still.2011.01.002.

Popkin, G. (2023). Shaky ground. *Science*, *381*(6656), 369–373. Available from https://doi.org/10.1126/science.adj9318, https://www.science.org/doi/10.1126/science.adj9318.

Prommer, J., Wanek, W., Hofhansl, F., Trojan, D., Offre, P., Urich, T., Schleper, C., Sassmann, S., Kitzler, B., Soja, G., Hood-Nowotny, R. C., & Jose, S. (2014). Biochar decelerates soil organic nitrogen cycling but stimulates soil nitrification in a temperate arable field trial. *PLoS ONE*, *9*(1), e86388. Available from https://doi.org/10.1371/journal.pone.0086388.

Qian, Z., Tang, L., Zhuang, S., Zou, Y., Fu, D., & Chen, X. (2020). Effects of biochar amendments on soil water retention characteristics of red soil at south China. *Biochar*, *2*(4), 479–488. Available from https://doi.org/10.1007/s42773-020-00068-w, http://springer.com/journal/42773.

Rasa, K., Heikkinen, J., Hannula, M., Arstila, K., Kulju, S., & Hyväluoma, J. (2018). How and why does willow biochar increase a clay soil water retention capacity? *Biomass and Bioenergy*, *119*, 346–353. Available from https://doi.org/10.1016/j.biombioe.2018.10.004, http://www.journals.elsevier.com/biomass-and-bioenergy/.

Rombolà, A. G., Fabbri, D., Meredith, W., Snape, C. E., & Dieguez-Alonso, A. (2016). Molecular characterization of the thermally labile fraction of biochar by hydropyrolysis and pyrolysis-GC/MS. *Journal of Analytical and Applied Pyrolysis*, *121*, 230–239. Available from https://doi.org/10.1016/j.jaap.2016.08.003.

Rombolà, A. G., Greggio, N., Fabbri, D., Facchin, A., Torri, C., Pulcher, R., Carlini, C., Balugani, E., Marazza, D., Zannoni, D., & Buscaroli, A. (2023). Changes of labile, stable and water-soluble fractions of biochar after two years in a vineyard soil. *Environmental Science: Advances*, *2*(11), 1587–1599. Available from https://doi.org/10.1039/d3va00197k, http://www.rsc.org/journals-books-databases/about-journals/environmental-science-advances.

Sánchez-García, M., Roig, A., Sánchez-Monedero, M. A., & Cayuela, M. L. (2014). Biochar increases soil N_2O emissions produced by nitrification-mediated pathways. *Frontiers in Environmental Science*, *2*. Available from https://doi.org/10.3389/fenvs.2014.00025. Available from, http://journal.frontiersin.org/article/10.3389/fenvs.2014.00025/pdf.

Santín, C., Doerr, S. H., Merino, A., Bucheli, T. D., Bryant, R., Ascough, P., Gao, X., & Masiello, C. A. (2017). Carbon sequestration potential and physicochemical properties differ between wildfire charcoals and slow-pyrolysis biochars. *Scientific Reports*, *7*(1). Available from https://doi.org/10.1038/s41598-017-10455-2, http://www.nature.com/srep/index.html.

Santos, J. A., Gonzaga, M. I. S., Santos, W. M., & Silva. (2022). Water retention and availability in tropical soils of different textures amended with biochar. *Catena*, *219*.

Schmidt., Abiven., Hagemann., & Meyer zu Drewer. (2022). Permanence of soil applied biochar. *The Biochar Journal.*, *2022*.

Siedt, M., Schäffer, A., Smith, K. E. C., Nabel, M., Roß-Nickoll, M., & van Dongen, J. T. (2021). Comparing straw, compost, and biochar regarding their suitability as agricultural soil amendments to affect soil structure, nutrient leaching, microbial communities, and the fate of pesticides. *Science of the Total Environment*, *751*, 141607. Available from https://doi.org/10.1016/j.scitotenv.2020.141607.

Soinne, H., Keskinen, R., Heikkinen, J., Hyväluoma, J., Uusitalo, R., Peltoniemi, K., Velmala, S., Pennanen, T., Fritze, H., Kaseva, J., Hannula, M., & Rasa, K. (2020). Are there environmental or agricultural benefits in using forest residue biochar in boreal agricultural clay soil? *Science of the Total Environment*, *731*, 138955. Available from https://doi.org/10.1016/j.scitotenv.2020.138955.

Sörman, L. (2023). *Enabling small-scale production of biochar in carbon markets: A multi-actor governance approach*. KTH.

Subedi, R., Taupe, N., Pelissetti, S., Petruzzelli, L., Bertora, C., Leahy, J. J., & Grignani, C. (2016). Greenhouse gas emissions and soil properties following amendment with manure-derived biochars: Influence of pyrolysis temperature and feedstock type. *Journal of Environmental Management*, *166*, 73–83. Available from https://doi.org/10.1016/j.jenvman.2015.10.007, https://www.sciencedirect.com/journal/journal-of-environmental-management.

Toková, L., Igaz, D., Horák, J., & Aydin, E. (2020). Effect of biochar application and re-application on soil bulk density, porosity, saturated hydraulic conductivity, water content and soil water availability in a silty loam haplic luvisol. *Agronomy*, *10*(7), 1005. Available from https://doi.org/10.3390/agronomy10071005.

Van Zwieten, L., Rose, T., Herridge, D., Kimber, S., Rust, J., Cowie, A., & Morris, S. (2015). Enhanced biological N_2 fixation and yield of faba bean (*Vicia faba* L.) in an acid soil following biochar addition: dissection of causal mechanisms. *Plant and Soil*, *395*(1-2), 7–20. Available from https://doi.org/10.1007/s11104-015-2427-3, http://www.wkap.nl/journalhome.htm/0032-079X.

Ventura, M., Alberti, G., Panzacchi, P., Vedove, G. D., Miglietta, F., & Tonon, G. (2019). Biochar mineralization and priming effect in a poplar short rotation coppice from a 3-year field experiment. *Biology and Fertility of Soils*, *55*(1), 67–78. Available from https://doi.org/10.1007/s00374-018-1329-y, http://link.springer.de/link/service/journals/00374/index.htm.

Ventura, M., Alberti, G., Viger, M., Jenkins, J. R., Girardin, C., Baronti, S., Zaldei, A., Taylor, G., Rumpel, C., Miglietta, F., & Tonon, G. (2015). Biochar mineralization and priming effect on SOM decomposition in two European short rotation coppices. *GCB Bioenergy*, *7*(5), 1150–1160. Available from https://doi.org/10.1111/gcbb.12219, http://onlinelibrary.wiley.com/journal/10.1111/(ISSN)1757-1707.

Verde, S. F., & Chiaramonti, D. (2021). *The biochar system in the EU: The pieces are falling into place, but key policy questions remain*. European University Institute.

Verheijen, F. G. A., Zhuravel, A., Silva, F. C., Amaro, A., Ben-Hur, M., & Keizer, J. J. (2019). The influence of biochar particle size and concentration on bulk density and maximum water holding capacity of sandy vs sandy loam soil in a column experiment. *Geoderma*, *347*, 194–202. Available from https://doi.org/10.1016/j.geoderma.2019.03.044, http://www.elsevier.com/inca/publications/store/5/0/3/3/3/2.

Vijay, V., Shreedhar, S., Adlak, K., Payyanad, S., Sreedharan, V., Gopi, G., Sophia van der Voort, T., Malarvizhi, P., Yi, S., Gebert, J., & Aravind, P. V. (2021). Review of large-scale biochar field-trials for soil amendment and the observed influences on crop yield variations. *Frontiers in Energy Research*, *9*. Available from https://doi.org/10.3389/fenrg.2021.710766, http://www.frontiersin.org/Energy_Research/.

Villagra-Mendoza, K., & Horn, R. (2018). Effect of biochar addition on hydraulic functions of two textural soils. *Geoderma*, *326*, 88–95. Available from https://doi.org/10.1016/j.geoderma.2018.03.021, http://www.elsevier.com/inca/publications/store/5/0/3/3/3/2.

Wang, J., Xiong, Z., & Kuzyakov, Y. (2016). Biochar stability in soil: Meta-analysis of decomposition and priming effects. *GCB Bioenergy*, *8*(3), 512–523. Available from https://doi.org/10.1111/gcbb.12266. Available from: http://onlinelibrary.wiley.com/journal/10.1111/(ISSN)1757-1707.

Wang, L., Xue, C., Nie, X., Liu, Y., & Chen, F. (2018). Effects of biochar application on soil potassium dynamics and crop uptake. *Journal of Plant Nutrition and Soil Science*, *181*(5), 635–643. Available from https://doi.org/10.1002/jpln.201700528, http://onlinelibrary.wiley.com/journal/10.1002/(ISSN)1522-2624.

Winfield, E. (2020). *Climate-smart agriculture: Soil health & carbon farming*. USDA.

Woolf, D., Lehmann, J., Ogle, S., Kishimoto-Mo, A. W., McConkey, B., & Baldock, J. (2021). Greenhouse gas inventory model for biochar additions to soil. *Environmental Science and Technology*, *55*(21), 14795–14805. Available from https://doi.org/10.1021/acs.est.1c02425, http://pubs.acs.org/journal/esthag.

Xu, G., Zhang, Y., Sun, J., & Shao, H. (2016). Negative interactive effects between biochar and phosphorus fertilization on phosphorus availability and plant yield in saline sodic soil. *Science of the Total Environment*, *568*, 910–915. Available from https://doi.org/10.1016/j.scitotenv.2016.06.079, http://www.elsevier.com/locate/scitotenv.

Xu, X., Zhao, Y., Sima, J., Zhao, L., Mašek, O., & Cao, X. (2017). Indispensable role of biochar-inherent mineral constituents in its environmental applications: A review. *Bioresource Technology*, *241*, 887–899. Available from https://doi.org/10.1016/j.biortech.2017.06.023, http://www.elsevier.com/locate/biortech.

Yang, C. D., & Lu, S. G. (2021). Effects of five different biochars on aggregation, water retention and mechanical properties of paddy soil: A field experiment of three-season crops. *Soil and Tillage Research*, *205*, 104798. Available from https://doi.org/10.1016/j.still.2020.104798.

Yuan, J. H., Xu, R. K., & Zhang, H. (2011). The forms of alkalis in the biochar produced from crop residues at different temperatures. *Bioresource Technology*, *102*(3), 3488–3497. Available from https://doi.org/10.1016/j.biortech.2010.11.018.

Zhang, J., Amonette, J. E., & Flury, M. (2021). Effect of biochar and biochar particle size on plant-available water of sand, silt loam, and clay soil. *Soil and Tillage Research*, *212*, 104992. Available from https://doi.org/10.1016/j.still.2021.104992.

Zheng, R. L., Cai, C., Liang, J. H., Huang, Q., Chen, Z., Huang, Y. Z., Arp, H. P. H., & Sun, G. X. (2012). The effects of biochars from rice residue on the formation of iron plaque and the accumulation of Cd, Zn, Pb, As in rice (*Oryza sativa* L.) seedlings. *Chemosphere*, *89*(7), 856–862. Available from https://doi.org/10.1016/j.chemosphere.2012.05.008, http://www.elsevier.com/locate/chemosphere.

Zimmerman, A. R. (2010). Abiotic and microbial oxidation of laboratory-produced black carbon (biochar). *Environmental Science & Technology*, *44*, 1295–1301. Available from https://doi.org/10.1021/es903140c.

CHAPTER

A practical approach of biochar production and application toward sustainable agricultural waste management and achieving carbon neutrality

3

Mashura Shammi and Beauty Akter

Department of Environmental Sciences, Jahangirnagar University, Savar, Dhaka, Bangladesh

3.1 Introduction

Agricultural waste poses a significant challenge worldwide due to its adverse environmental, economic, and social impacts. The accumulation of agricultural residues, such as crop residues, animal manure, and agro-industrial byproducts, presents multifaceted issues ranging from soil degradation to greenhouse gas (GHG) emissions. Agricultural waste and its processing are a global concern because the most majority of it is now burnt or buried in soil, polluting the air, water, and contributing to global warming (Koul et al., 2022). Despite efforts to address this issue, effective management strategies remain elusive. The problem of agricultural waste management persists, manifesting in several interconnected dimensions. The first and foremost issue is the improper disposal practices of agricultural wastes, which lead to the contamination of soil and water resources through soil erosion, nutrient runoff, and water pollution, thus impacting aquatic ecosystems and biodiversity (Liu, 2023). Secondly, inadequate waste management practices lead to the decomposition of agricultural waste and the release of GHGs such as methane (CH_4) and nitrous oxide (N_2O). Subsequently, such practices contribute to increased GHG emission levels, exacerbating climate change and further intensifying global warming (Liu et al., 2019; Smith et al., 2014; Smith, 2016). Thirdly, the inefficient handling of agricultural waste results in economic losses for farmers and stakeholders. Failure to utilize waste materials for productive purposes represents a missed opportunity for resource recovery and cost savings (Duque-Acevedo et al., 2020).

At the community level, improper disposal and burning of agricultural waste generate air pollutants and particulate matter (PM), posing health risks to nearby

Biochar Ecotechnology for Sustainable Agriculture and Environment. DOI: https://doi.org/10.1016/B978-0-443-29855-4.00004-7

communities. Exposure to these pollutants increases the incidence of respiratory diseases and other health issues among vulnerable populations (Raza et al., 2022). The lack of comprehensive policies and regulatory frameworks hinders effective agricultural waste management. Inconsistent enforcement and limited incentives for waste reduction and recycling impede progress toward sustainable solutions (Bhuvaneshwari et al., 2019). Crop residue burning, a common practice in many parts of India and subcontinents, including Bangladesh, Pakistan, and other countries, releases harmful pollutants into the atmosphere, exacerbating air pollution levels in cities like Delhi. The pollutants released from crop residue burning include PM, carbon monoxide (CO), volatile organic compounds (VOCs), and GHGs such as methane (CH_4) and nitrous oxide (N_2O) (Kulkarni et al., 2020). These pollutants contribute to air pollution and affect public health. In Delhi, where air pollution levels often reach hazardous levels during the postharvest season, crop residue burning in neighboring state is a significant contributor which mostly depend on the rice–wheat cropping patterns (Jain et al., 2014; Saxena et al., 2021; Shyamsundar et al., 2019). Seasonal smog incurs significant expenses, including extensive transportation interruptions and the shutdown of educational institutions during the dry season (Shyamsundar et al., 2019). Addressing the issue of crop residue burning requires a multifaceted approach that includes policy interventions, technological solutions, and behavioral change.

Sustainable agricultural waste management is, therefore, imperative for reducing GHG emissions and air pollutants such as VOCs, smogs, and PMs. The agricultural sector encounters diverse challenges and obstacles when shifting from a linear resource consumption model to embracing a circular economy (Haque et al., 2023). Crop wastes have traditionally been used for combustion, animal food, roof thatching, composting, soil mulching, matchstick manufacturing, and paper making. However, lignocellulosic biomass may also be used as a sustainable source of biofuel (biodiesel, bioethanol, biogas, and biohydrogen) and bioenergy, helping to alleviate the fossil fuel scarcity and climate change challenges. Thus valorization of lignocellulosic wastes has the potential to impact the bioeconomy by providing value-added goods such as biofertilizers, bio-bricks, biocoal, bio-plastics, paper, biofuels, industrial enzymes, organic acids, and so on (Koul et al., 2022). Effective agricultural waste and crop residue management practices mitigate environmental impacts and enhance resource efficiency and agricultural productivity, thereby contributing to food security and economic development (Sarkar et al., 2020).

Addressing the problem of agricultural waste demands a holistic approach that integrates scientific research, technological innovation, policy intervention, and stakeholder collaboration. By acknowledging the multifaceted nature of this challenge and adopting context-specific strategies, we can mitigate the adverse impacts of agricultural waste and move toward a more sustainable farming system. The escalating GHG emissions pose a formidable challenge to global sustainability, with agriculture being a significant contributor. In this context, sustainable agricultural waste management is a crucial strategy to mitigate GHG emissions and foster environmental resilience. Among various approaches, using biochar holds promise for effectively managing agricultural waste while providing additional benefits to soil health and fertility. Biochar is a material rich in carbon produced by thermal decomposition of biomass in an oxygen-depleted atmosphere. It is an environmentally friendly adsorbent made from waste (agricultural, forestry, etc.) as it can remove pollutants from wastewater (Collivignarelli et al., 2024). Carbon sequestration potential is enhanced by converting agricultural residues into biochar, thereby mitigating GHG emissions. Furthermore, biochar application improves soil health, nutrient retention, and water management, contributing to sustainable agriculture and climate resilience (Woolf et al., 2010).

Biochar, a carbon (C)-rich compound derived from the thermochemical conversion of biomass in oxygen-limited settings, has been identified as one of the most promising materials for carbon sequestration and climate mitigation in soil. Biochar's contribution to C sequestration is determined not only by its fused aromatic structure but also by its abiotic and biotic interactions with soil components throughout its life cycle in the environment. Minerals and microbes, for example, can play a significant role in the mineralization or complexation of labile (soluble and readily decomposable) and even resistant biochar fractions, influencing C cycling and sequestration in soil (Luo et al., 2023). Recently, it has been common in agriculture to use inorganic fertilizers in the soil, causing an imbalance of nutrients. Blending biochar as an organic additive and fertilizer would be an appropriate solution to counterbalance this situation. Biochar amendments to soil improve soil quality through sequestering carbon, reducing emissions related to carbon dioxide and ammonia, lowering the compactness of soil, enhancing soil porosity and soil cation exchange capacity, increasing the availability of nutrients and water-holding capacity and acting as a slow release fertilizer by releasing micronutrients slowly. The rhizosphere microorganisms and mycorrhizal fungi are also stimulated by biochar properties, thus promoting plant growth (Akter et al., 2023; Nair et al., 2017; Shammi & Akter, 2024; Solaiman & Anawar, 2015). Biochar has a beneficial negative emission potential (0.7 GtCequiv./year), as well as the potential for lesser land impact, water consumption, nutrient effect, albedo impact, energy need, and cost, making it less disadvantageous than many other negative emission solutions (Smith, 2016). Addressing the problem of GHG emissions from agricultural waste necessitates a comprehensive approach that integrates sustainable waste management practices, such as biochar application. Biochar from agricultural waste supports the concept of circular economy as it is produced from waste materials but can potentially improve soil and water quality, reduce GHG emissions, and manage waste.

3.2 Decarbonization and carbon sequestration by converting agricultural waste into biochar

Through pyrolysis, organic agricultural residues are thermally decomposed in the absence of oxygen, yielding biochar as a byproduct. The resulting biochar contains a high proportion of carbon and exhibits a porous structure, which enhances its ability to adsorb and retain nutrients and water in soils. Moreover, biochar's recalcitrant nature makes it resistant to microbial degradation, ensuring long-term carbon sequestration. Biochar enhances soil fertility by improving nutrient retention and availability, increasing water-holding capacity, and promoting microbial activity (Lehmann et al., 2011). Globally, biochar systems could deliver emission reductions of 3.4–6.3 $PgCO_2e$, half of which constitutes CO_2 removal (Lehmann et al., 2021). Furthermore, biochar-amended soils exhibit reduced GHG emissions, improved crop yields, and enhanced resilience to environmental stressors such as drought and soil erosion (Spokas et al., 2012). Despite its potential benefits, the widespread adoption of biochar faces several challenges, including technological constraints, cost considerations, and knowledge gaps regarding its long-term effects on soil and crop performance (Singh et al., 2014). Additionally, the scalability and feasibility of biochar production and application methods require further research and development to optimize its efficacy and economic viability. The impediments include governance difficulties such as legislation and permitting, economic challenges such as expenses, funding, and market stability, technological obstacles such as methodologies and equipment, and a need for further research and greater scientific dissemination (Pierson et al., 2024).

Waste to biochar is an increasing interest in building a "circular economy" with the primary objective of reducing agricultural waste related emissions via proper planning. Biochar has lately acquired favor in the environmental industry as a flexible substance for reducing waste and improving the circular economy's efficiency (Hu et al., 2021; Singh, Mishra, et al., 2022). Fig. 3.1 shows that the circular economy is supported by applying biochar to soil produced from agricultural waste.

3.2.1 Diverse approaches to biochar production and pyrolysis technologies

Pyrolysis processes encompass a range of techniques with distinct advantages and disadvantages, influenced by factors such as feedstock selection, preparation methods, operating conditions, and desired biochar properties. This chapter examines different approaches to biochar production, including their variations in pyrolysis processes and discusses the influence of temperature and residence time on biochar properties. Pyrolysis involves the thermal decomposition of biomass in the absence of oxygen, producing biochar, bio-oil, and syngas (Bridgwater, 2012). Several variations of pyrolysis exist, including slow, fast, and intermediate, each characterized by specific operating conditions and end products.

3.2.1.1 Slow pyrolysis

Slow pyrolysis operates at low temperatures (300°C–700°C) and long residence times, typically several hours to days. Temperature is the most significant parameter for the slow pyrolysis process (Vieira et al., 2020). Residence time is longer, which varies from minutes to days with slow heating rate of less than 30°C/min (Safarian, 2023). This method maximizes biochar yield and quality while

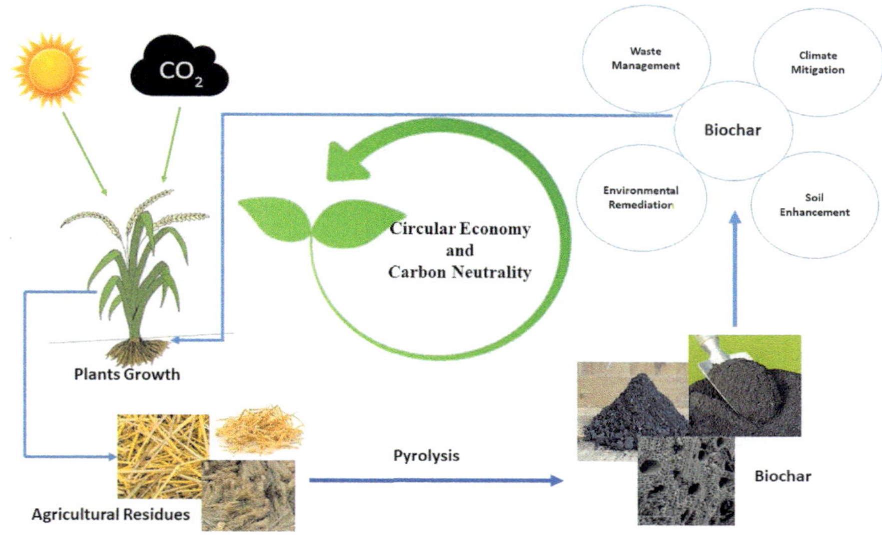

FIGURE 3.1 Agricultural waste to biochar promotes circular economy and waste management.

Pyrolysis of agricultural wastes can be converted into value-added biochar product for soil management and carbon sequestration while promoting circular economy. Agricultural crop residue is converted to biochar by pyrolysis. The application of biochar to soil promotes benefits, such as waste management, environmental remediation, soil enhancement, and climate mitigation. Carbon is sequestered in the soil, promoting a circular economy and carbon neutrality.

minimizing bio-oil and gas production (Lehmann et al., 2011). Slow pyrolysis is well-suited for producing biochar with high carbon content and stable properties, making it suitable for soil amendment and carbon sequestration applications. Advantages include high biochar yield, superior biochar quality, and potential for carbon sequestration (Brewer et al., 2009; Tsolis & Barouchas, 2023). However, disadvantages include longer processing times and higher energy requirements. Some of the recent examples of slow pyrolysis using agricultural biomass processes are given in Table 3.1.

Moreover, utilizing a pyrolysis environment consisting of CO_2 as a carrier instead of N_2 gas resulted in a notable increase in CO yield, a corresponding decrease in CO_2 production. These findings indicate that conducting biomass processing via pressurized slow pyrolysis under a CO_2 atmosphere holds promise for simultaneously generating two valuable outputs: biochar with suitable carbon sequestration potential and gas with a composition conducive to energy recovery (Manyà et al., 2018). Exploring the impact of CO_2 on biochar production through slow pyrolysis for industrial applications is valuable. Utilizing CO_2 from flue gas as the pyrolysis gas environment can enhance biochar production's economic

Table 3.1 Some recent examples of slow pyrolysis using agricultural biomass.

Biomass	Temperature	Remarks and references
Sugarcane bagasse, straw, and acid-treated biomass	Bench-scale fixed-bed reactor at 450°C, 550°C, and 650°C	Biochar had good stability and potential as an energy carrier (de Almeida et al., 2022)
Rice husk	300°C–500°C	Potential energy gains through pyrolysis (Vieira et al., 2020)
Coconut fiber	Bench-scale fixed-bed reactor at 350°C–600°C	Low-temperature biochars (350°C–450°C) are promising tools for solid fuels (Dhar et al., 2022)
Pigeon pea stalk and bamboo	Muffle furnace at 400°C–600°C	Potential adsorbent (Sahoo et al., 2021)
Co pyrolysis of rice straw and wheat straw	Muffle furnace at 600°C	Potential fuel and soil amendment additives (Bhatnagar et al., 2022)
Rice straw	Fixed-bed reactor at 300°C–600°C	Potential bio-oil source rich in phenolic compounds (Sakhiya et al., 2021)
Palm oil mill wastes	Fixed-bed reactor at 600°C	Potential bio-oil and biochar source (Chantanumat et al., 2023)
Banana biomass	Fixed-bed reactor at 400°C–600°C	Potential biofuel and soil additives (Abdullah et al., 2023)

feasibility, environmental sustainability, and properties. At elevated pyrolysis temperatures surpassing 700°C, the active properties of CO_2 strengthen the generation of pyrolytic gases (H_2, CO, and CH_4) by promoting the thermal cracking of tar, suppressing the formation of secondary char, and intensifying char gasification. Consequently, this decreases char yield (Premchand et al., 2023).

3.2.1.2 Fast pyrolysis

Fast pyrolysis operates at higher temperatures (300°C–100°C) and shorter residence times, typically a few seconds to minutes with very fast heating time of about 1000°C/s (Safarian, 2023). This rapid heating rate promotes bio-oil production as the primary product, with biochar generated as a byproduct. Fast pyrolysis offers advantages in process efficiency and bio-oil yield, making it suitable for bioenergy and biofuel production applications. Advantages include rapid processing, high bio-oil yield, and potential for bioenergy production (Bridgwater, 2012). Examples of fast pyrolysis and high-quality bio-oil yield from peanut shell at 650°C (Chang et al., 2021). Disadvantages include lower biochar yield and quality and challenges in biochar stabilization. The rapid depolymerization caused by the elevated heating rates during the fast pyrolysis trial led to reduced surface area and porosity in banana biomass biochar reported by Abdullah et al. (2023). It was determined that the formation mechanism of biochar in fast pyrolysis closely resembles that of slow pyrolysis. Biochar formation occurs in two stages: firstly, the production of free radicals resulting from covalent bond rupture at temperatures below 600°C, followed by the reduction of free radicals through condensation reactions occurring between 600°C and 800°C (Yuan et al., 2023).

3.2.1.3 Intermediate pyrolysis

Intermediate pyrolysis operates at moderate temperatures (400°C–600°C) and residence times between 1 and 15 minutes at heating time 1°C–10°C (Kazawadi et al., 2021; Safarian, 2023). This method balances biochar, bio-oil, and gas production, offering versatility in product distribution and application potential (Kazawadi et al., 2021). Advantages include moderate processing conditions, balanced product distribution, and flexibility in application (Hornung, 2013). Disadvantages may include intermediate biochar yield and quality compared to slow pyrolysis. In addition, torrefaction, microwave, gasification, flash carbonization, and hydrothermal carbonization are other sustainable technologies for biochar production (Safarian, 2023). The choice of feedstock significantly influences biochar properties and pyrolysis performance. Various biomass sources, including woody materials, agricultural residues, and organic waste, can serve as feedstocks for biochar production (Anand et al., 2023; Tomczyk et al., 2020). Moisture content, particle size, and chemical composition impact pyrolysis efficiency, biochar yield, and properties. Proper feedstock selection and preparation are essential to optimize biochar production and quality. Pyrolysis temperature and residence time are critical parameters influencing biochar properties, including carbon content, surface area, porosity, and chemical composition (Brewer et al., 2009; Suliman et al., 2016; Tomczyk et al., 2020;). In summary, higher pyrolysis temperatures typically result in biochars with higher carbon content and surface area but may also lead to increased ash content and reduced stability. Longer residence times enhance biomass conversion and biochar yield but may also increase energy consumption and emissions. Diverse approaches to biochar production offer opportunities to tailor biochar properties for various applications, from soil amendment to carbon sequestration and renewable energy. Understanding the advantages and disadvantages of different pyrolysis methods and the influence of feedstock selection, temperature, and residence time is essential for optimizing biochar production processes and maximizing its potential benefits in sustainable agricultural waste and environmental management.

3.2.2 Industrial pyrolysis of biochar

Industrial pyrolysis of biochar is a promising technology that converts biomass into a stable carbon-rich material with various applications. This essay explores the advantages and disadvantages of industrial biochar pyrolysis, presents case studies illustrating its practical implementation, discusses life cycle assessment (LCA), and evaluates its sustainability. LCA evaluates the environmental impacts of biochar production from the cradle to the grave, considering all stages, including feedstock cultivation, transportation, pyrolysis, and end-use. LCA methodologies quantify energy consumption, GHG emissions, and other environmental indicators to assess the sustainability of biochar production (Rogovska, 2019). Sustainable practices such as utilizing renewable biomass feedstock's, optimizing pyrolysis technologies, and implementing emission control measures can enhance the sustainability of biochar production (Jiang, 2021).

Biochar enhances soil fertility, water retention, and nutrient availability, promoting plant growth and crop productivity (Lehmann et al., 2011). Pyrolysis generates bio-oil and syngas as byproducts, which can be utilized for heat and power generation, contributing to renewable energy goals (Bridgwater, 2012). Industrial pyrolysis of biochar offers numerous advantages, including waste valorization, carbon sequestration, soil improvement, and renewable energy generation. However, energy intensity, emissions, and feedstock dependence must be addressed to enhance sustainability. Case studies demonstrate the practical application of industrial biochar pyrolysis, while LCA provides a holistic framework for evaluating its environmental impacts and sustainability. However, industrial biochar pyrolysis is not without disadvantages. Industrial-scale pyrolysis processes require significant energy inputs for heating biomass feedstock to high temperatures, leading to operational costs and environmental impacts (Brewer et al., 2009; Cusenza et al., 2021). Pyrolysis may emit pollutants such as VOCs and PM during biomass combustion and biochar production, contributing to air pollution (Conesa et al., 2009). The quality and properties of biochar depend on the type and composition of biomass feedstock used, limiting its applicability and market potential. Here, we give examples of two industrial biochar and biofuel production companies that produce from local cellulosic waste.

Example of Greenbelt Resource Corporation (USA) (https://www.greenbeltresources.com/): The Greenbelt production facility utilizes industrial-scale pyrolysis to convert agricultural residues and green waste into biofuel and biochar. The facility has demonstrated the viability of biochar and biofuel production as a waste management strategy and soil improvement tool. They focus on providing modular solutions that facilitate the localized processing of cellulosic waste generated in a specific area into products consumed within the same locality. Example of Pacific Biochar (USA) (https://pacificbiochar.com/): In California, Pacific Biochar transforms decommissioned biomass power plants into facilities for producing biochar. It gathers organic matter from forests prone to high fire risk, thereby mitigating the threat of wildfires. Subsequently, the biochar produced is distributed to agricultural suppliers and compost yards.

3.2.3 Small-scale pyrolysis of biochar

Small-scale pyrolysis of biochar offers a decentralized approach to biomass conversion, allowing individuals and communities to produce biochar locally for various applications. Small-scale pyrolysis enables local biochar production, reducing the need for long-distance biomass feedstock transportation and minimizing associated costs and emissions (Jeffery et al., 2011; Ronsse et al., 2013). It provides an effective solution for managing organic waste streams, including agricultural residues, forestry byproducts, and organic household waste, by converting them into a valuable resource (Jeffery et al., 2011). Small-scale biochar production empowers communities to take control of their waste management and agricultural practices, fostering local resilience and sustainability (Kloss et al., 2012; Tomczyk et al., 2020). By combining open-source tools, on-farm pyrolysis systems can be installed, which has potential for achieving net negative emissions through biochar production (Azzi et al., 2021). The community biochar initiatives using open-source tools can engage local communities in small-scale biochar production using simple, low-cost pyrolysis technologies even cookstoves (Shammi & Akter, 2024; Shammi et al., 2023). The initiative has demonstrated the feasibility of decentralized biochar production for soil improvement and carbon sequestration at the small farmer scale. However, to maximize biochar yield and carbon sequestration benefits, proper sizing of these systems is crucial, emphasizing the importance of sustainable biomass provision and efficient biochar utilization (Azzi et al., 2021).

As the scale is limited, small-scale pyrolysis systems have limited throughput and may not be suitable for efficiently processing large volumes of biomass, restricting their potential impact and scalability (Brewer et al., 2009). The technical complexity of small-scale pyrolysis technologies requires technical expertise for operation and maintenance, posing challenges for adoption and implementation in resource-constrained settings (Azzi et al., 2021). Between 2014 and 2019, a United Nations Industrial Development Organization (UNIDO) initiative developed an innovative small-pyrolysis system (the PPV300) in Switzerland. The PPV300 system was applied in Vietnam to reduce coffee husk waste. Pyrolysis of coffee husk using the PPV300 reduced smoke emissions and local air pollution while producing energy for coffee drying and biochar as a byproduct (Flammini et al., 2020). However, in most cases, the inadequate emission control measures in small-scale pyrolysis systems may result in air pollution and health hazards, particularly in densely populated areas (Conesa et al., 2009). The average abatement cost of biochar used in "lower-income nations" is $-$US\$58 Mg^{-1} CO_2e (financially possible), compared to $+$US\$93 Mg^{-1} CO_2e in "higher-income countries" (not financially feasible). Climate policy in low-income tropical countries should use biochar as an input for small-scale climate-smart agriculture to alleviate land degradation in tropical agricultural systems. Based on the recent research, biochar fertilizers, a value-added biochar product, may be an economically viable option for biochar value-chain growth in higher-income nations (Robb et al., 2020).

3.3 Understanding the persistence and stability of biochar in soil

Returning biomass-derived biochar to soil has been regarded one of the carbon sequestration approaches (Yin et al., 2022). Biochar is known for its high recalcitrance and stability in soil, making it an attractive option for carbon sequestration and long-term soil amendment. The persistence of biochar in soil is influenced by various factors, including its chemical composition, physical properties, soil characteristics, and environmental conditions. Recent studies have investigated the mechanisms underlying biochar persistence and stability, providing insights into its potential as a sustainable soil management tool. Chemical composition plays a crucial role in determining the persistence of biochar in soil. Biochar derived from different biomass feedstocks and pyrolysis conditions exhibit varying degrees of stability due to differences in their carbon content, surface functional groups, and aromaticity which increases with longer pyrolysis periods and lower ash content of the feedstocks (Lehmann et al., 2021). Biochar generated at lower pyrolytic temperatures (<500°C) had a stronger influence on microbial variety (both bacterial and fungal), with more varied bacterial populations in medium and coarse textured soils, and fungal diversity increasing in fine textured soils. Biochar with higher carbon content and aromaticity were more resistant to microbial degradation and exhibited longer persistence in soil. However, at higher rate, it can affect negatively (Singh, Northup, et al., 2022).

Furthermore, the physical properties of biochar, such as pore structure, surface area, and particle size, influence its stability and persistence in soil. Biochar with greater surface area and microporosity exhibit enhanced stability and carbon retention in soil, as they provide more sites for microbial colonization and organic matter sorption. Lehmann (2022) highlighted the importance of biochar pore structure in determining its persistence and carbon sequestration potential in soil. Biochar stability also improves soil health, moisture content, microbial activity, nutrient dynamics, and plant growth, which highlights the importance of biochar as soil amendment (Atkinson et al., 2010; Hossain et al., 2020; Jeffery et al., 2011; Wang et al., 2020). Overall, recent scientific research has advanced our understanding of the persistence and stability of biochar in soil, providing insights into its transformation processes, interactions with soil components, and implications for soil health and carbon cycling. Continued investigation into the factors influencing biochar stability, including feedstock type, pyrolysis conditions, soil properties, and environmental factors, will further enhance our ability to harness the potential of biochar for soil carbon sequestration and sustainable agriculture.

3.4 Role of biochar as a soil amendment for enhancing crop yield

Applying biochar as a soil amendment is considered an organic, sustainable nutrient management option (Ullah et al., 2021). Joseph et al. (2021) described biochar responses in soil in three stages: dissolution (1–3 weeks), reactive surface formation (1–6 months), and aging (>6 months). As biochar matures, it becomes mixed into soil aggregates, conserving the carbon content and supporting the stability of rhizodeposits and microbial products (Joseph et al., 2021). Biochar enhances nutrient uptake capacity and increases rice yield when applied to the soil with the combination of N fertilizer. The nitrogen concentration in plant leaves plays a crucial role in biochemical and physiological functions, and its availability determines crop yield. However, too much nitrogen generates several environmental issues, so adding anything organic that enhances N uptake capacity is essential. The authors stated significant differences among N fertilizer sources applied with or without biochar ($P < 0.05$). The biochar and N fertilizer mixture increases soil organic carbon by about 51.2% and nitrogen availability by 27.5% compared to N fertilizers alone. Applying biochar with synthetic N fertilizers has become a sustainable nutrient management strategy that can enhance N availability, improve soil properties, and increase rice production rate (Ullah et al., 2021). Biochar carbon remains in soil for hundreds to thousands of years. Biochars, by raising pH, porosity, and water availability, can promote root growth and microbial activity. Biochars can catalyze biotic and abiotic processes, particularly in the rhizosphere, increasing plant nutrient availability and absorption, lowering phytotoxins, stimulating plant growth, and increasing resistance to disease and environmental stresses (Joseph et al., 2021).

Biochar-based compound fertilizers and amendments have been verified to increase agricultural yield and alter soil properties positively (pH, organic matter contents, cation exchange capacity, soil porosity, water retention capacity, etc.) and have recently been commercially produced in several areas of the world (Chew et al., 2020). The information related to the effects of biochar on rice is minimal when applied continuously. Huang et al. (2019) studied a fixed field experiment conducted in early and late rice growing seasons from 2015 to 2017. They found a decreasing pattern of grain yield initially after adding biochar due to a decrease in grain weight. Gradually, they recorded an increasing pattern of rice yield. They noticed the positive effects of biochar depend on the duration of biochar application. Straw-derived biochar and its application to soil certainly is a good practice as it reduces the wastage of resources, increases soil fertility, and increases the capacity to retain soil phosphorus and, thus, crop yield. This study used a 40-day laboratory incubation experiment and a 3-year field to determine the impact of biochar on crop yield and P retention in alkaline soil. The study concluded that biochar addition enhances P retention and rice productivity in saline, alkaline soil (Wu et al., 2021).

Biochar amendment reduces soil acidity and improves soil quality and crop yield. A high dose of biochar (20 and 40 t/ha) combined with N fertilizer can significantly increase soil pH, available P content, and sweet potato yields in the first year. The integrated use of biochar and nitrogen fertilizer was a promising strategy to obtain higher and sustained productivity of the sweet potato cropping system in upland red soil (Jin et al., 2019). It can improve the soil water retention capacity in sandy loam soil after repeated apply of biochar for 2–3 years. Undisturbed soil samples were collected from 16 plots in two soil layers (5–10 and 15–20 cm). The impact of biochar on soil water retention capacity was most likely related to an effect on the overall porosity of the sandy loam soil, which was evident from an increase in saturated soil moisture (De Melo Carvalho et al., 2014).

In saline iodide soil, biochar helps enhance soil nutrients, soil enzyme activity, and rice yield. The results demonstrated that it significantly improved the availability of macronutrients, namely nitrogen, phosphorus, and potassium, as well as increased soil organic matter (SOM), C/N ratio, soil catalase activity, soil quality, and so on. It was also recorded that biochar amendment enhanced rice biomass, grain yield, and the harvest index. The management of saline-sodic soil using biochar will be a promising option for agricultural production (Yao et al., 2021). In silt loam soil biochar application enhanced soil quality and crop production. They studied the production rate of wheat and recorded that biochar amendment certainly increased SOM, N, P, K, and soil pH (Bista et al., 2019).

Guo et al. (2021) worked on the biochar application and its impact on tomato yield. The study was conducted in a greenhouse and applied four biochar levels (0, 30, 50, and 70 t/ha) with four N application rates (170, 190, 210, and 250 kg/ha). It was observed that adding too much biochar with less N fertilizer will not lead to satisfactory results on tomato yield. However, it significantly improved the production rate when it (50 t/ha) was applied to the soil with the combination of nitrogen fertilizer (24%). A review of crop residue biochar production and its responses to agricultural yield is documented in Table 3.2.

3.5 Benefits and challenges of using biochar in achieving carbon neutrality

The consumption of fossil fuel is increasing daily, which leads to increase atmospheric carbon dioxide from preindustrial value 280–415 ppm by 2021, resulting in several greenhouse effects. It is essential to reduce GHG emissions to reduce global greenhouse effects. Biochar from biomass is a promising option as it is highly recalcitrant organic carbon with a condensed aromatic structure with high carbon sequestration potential (Fu et al., 2022). Utilizing biochar produced from agricultural residues supports the concept of circular economy and promotes carbon neutrality. As biochar is made from biomass waste products, it is cost-effective. The porous structure of biochar improves soil quality. It can sequester atmospheric carbon (Yuan et al., 2023). To mitigate climate change, it is necessary to implement negative emission technologies to capture and store atmospheric carbon. Biochar has become a promising option for storing carbon in soils while serving several cobenefits. In a Norwegian research of LCA of biochar in Norwegian agriculture, all cases demonstrated harmful emissions ranging from -1.72 ± 0.45 to -7.18 ± 0.67 tons CO_2-equiv./(ha year) (Tisserant et al., 2022). It was evident that at the national level, biochar is able to offset 13%–40% GHG emission in the Norwegian agricultural sector.

Table 3.2 A review on plant responses to biochar produced from crop residues applied to agricultural soil.

Biochar type	Crop types	Experiment types	Crop responses and other benefits	Reasons	References
Cassava straw biochar	*Oryza sativa*	Pot experiment	Photosynthetic rate of leaves has increased 9.34% more and chlorophyll concentration was higher in the biochar treatment	Addition of biochar helps to increase soil organic carbon by 51.21% and the availability of N by 27.5%	Ullah et al. (2021)
Wheat straw biochar (400°C)	*O. sativa*	Grow bag experiment	There was an increase in rice yield and nutrients uptakes	Biochar helps to enhance N (40%), P (46%), Mg, K, and Na uptakes	Chew et al. (2020)
Rice husks biochar (600°C)	*O. sativa*	Field experiment	Rice yield increased with the addition of biochar but it depends on the duration of biochar application	Biochar helps to increase soil cation exchange capacity and enhance the rate of nutrient retention	Huang et al. (2019)
Rice straw biochar (450°C–500°C)	*O. sativa*	Field experiment	Rice biomass and yield were increased with rice straw and biochar application as compared to the control	Organic amendments application could improve soil structure and productivity	Wu et al. (2021)
Wheat straw biochar (450°C)	*Ipomoea batatas*	Field experiment	Reduce soil acidity and increase soil nutrients content and crop yields	Application of biochar helps to reduce soil acidity, enhance nutrients availability	Jin et al. (2019)
Eucalyptus wood biochar (450°C)	*O. sativa*	Field experiment	Increase the rice yield and plant-available water in a sandy loam soil	Biochar application to the soil increase soil porosity	De Melo Carvalho et al. (2014)
Peanut shell biochar	*O. sativa*	Field experiment	Increase rice yield and improved the soil catalase activity, soil alkaline phosphatase activity, soil urease activity, and soil sucrose activity	The addition of biochar significantly increased the availability of soil total N, available P, and available K	Yao et al. (2021)
Douglas wood biochar	Wheat*Triticum*	Field experiment	Biochar increased wheat shoot and root biomass and thus wheat growth	Increased soil organic matter, soil pH, P, K, and sulfur contents	Bista et al. (2019)
Rice straw biochar	*O. sativa*	Pot experiment	Combined application of biochar and N could elevate rice plant N content by 13%, plant height	Biochar can significantly compensate the toxicity of excessive nitrogen and deficiency of low N fertilizer on rice growth	Khan et al. (2021)
Tree branches biochar (450°C)	*Solanum lycopersicum*	Greenhouse experiment	Biochar addition at 50 t/ha and reducing N fertilizer by 24% achieved the greatest tomato yield	Significantly enhance microbial abundance, enzyme activity, and tomato growth compared	Guo et al. (2021)

Biochar used for soil amendment is considered a viable option for a bio-based circular economy, which offers several benefits and significantly reduces CO_2 emissions. In Malaysia, Palm oil empty bunches, fibers, and shells are the most abundant agricultural waste. The authors used cycle assessment and life cycle costing to determine the suitability of biochar production at different pyrolysis temperatures (300°C–600°C). The study's objectives were to produce and apply 1 ton of biochar and finally remove 1 ton of carbon dioxide from the air. The boundary of the cradle-to-grave system comprises all the stages, from the acquisition of feedstock to the use of biochar in soil. The potential of global warming potential is net negative in all scenarios, ranging from −436 to −2085 kg CO_2-equiv./t biochar and −660 to −933 kg CO_2−equiv./t CO_2 removed (Saharudin et al., 2024). Biochar helps reduce harmful GHG emissions by sequestering carbon through converting biomass into stable carbon content that has persistent nature to decomposition when applied to soil amendment, resulting in reduced net CO_2 in the atmosphere. In addition, applying biochar for soil amendment offers several benefits, including increasing crop production by improving soil conditions and fertility. Quantitatively, biochar-treated soils carbon footprint ranging from −1.43 to −2.799 kg CO_2-equiv./kg rice grain (Mohammadi, 2020).

During pyrolysis, energy is also produced in the form of syngas and bio-oil. The main challenges for biochar implementation at the farm level were associated with current biochar regulations and limited interest from local farmers. Most technologies require pretreatment, which requires some investment in instruments and labor for production. Farmers may not be willing to perform additional tasks, and the economy of such pretreatment on the farm level may also not be favorable. Another issue is posttreatment. After biochar production, its direct implementation is not recommended. Anaerobic digestion, cocomposting, and utilization as the bottom layer for farm bedding materials were beneficial. These posttreatment steps require additional efforts and requirements to gain significant benefits associated with biochar implementations (Heinrich et al., 2023). The profitability and attractiveness of biochar production and usage are very unclear and case-specific, dependent on factors such as location, feedstock, size, pyrolysis conditions, biochar pricing, grown crop, and the possible internalization of externalities, which impedes private investment. To improve biochar development and implementation, these variables must be carefully evaluated in each scenario (Campion et al., 2023). Fig. 3.2 shows the benefits, challenges, and uncertainties of biochar production and application in soil and crops.

Moreover, in many developing countries, rural locations have limited or no power supply to run the pyrolysis facilities. In addition, lack of investment, stakeholder participation, awareness, and research findings are also barriers to biochar production and its utilizations in developing countries. Examples of biochar production and agricultural usage in six developing and middle-income countries (Ethiopia, Indonesia, Kenya, Peru, Vietnam, and China) showed two types of production methods used in rural and urban regions: flame curtain kiln and gasifier. Net economic gains were only realized when low-cost basic kilns were employed

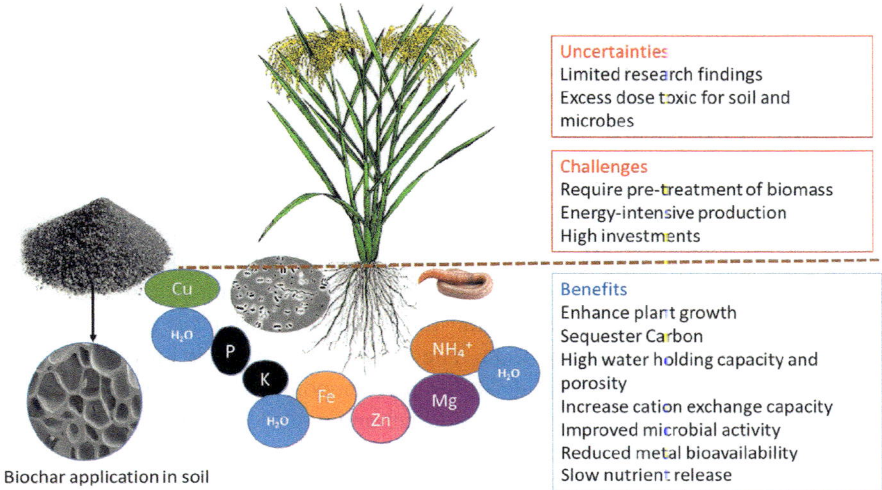

FIGURE 3.2 Benefits, challenges and uncertainties of biochar production and application.

Biochar production and application is not without advantages and disadvantages as well as uncertainties due to limited research findings. The benefits of biochar application include enhanced plant growth, carbon sequestration, high water-holding capacity and porosity, increased cation exchange capacity, improved microbial activity, reduced metal bioavailability, and slow nutrient release. Challenges of biochar include pretreatment of biomass, high energy and investments, limited research findings on disadvantages, and low interest and awareness.

in countries with low labor costs, such as Ethiopia, Kenya, and Vietnam (net profit ranged from 0.01 to 0.08 USD/kg of biowaste processed). Furthermore, significant investment and running expenses, as well as relatively low power return from replacing grid electricity, made gasifier scenarios economically untenable (net loss ranging from 0.29 to 1.58 USD/kg of biowaste processed) (Owsianiak et al., 2021). The role of the biochar pyrolysis equipment supply chain and the farmers or the processors culturally dependent operations heavily influences the choice of supply chain configuration. Farmers and processors may hesitate to engage in biochar production due to the significant investments required in equipment purchase (Flammini et al., 2020). In addition, technical person for pyrolysis equipment installing, repairing, and troubleshooting is still limited in many countries along with the product supply chain. Thus, when deciding whether to implement biochar technology in developing and middle-income countries, there are trade-offs to consider between positive environmental impacts for society and net market loss for the individual decision-maker (company or farmer) (Owsianiak et al., 2021). It is also evident that biochar is most likely to be used in areas with marginal land, high-value crops, and low-cost feedstock sources (Zilberman et al., 2023). However, to encourage the commercialization of biochar for large-scale applications, there must be a critical discussion of the benefits and drawbacks of biochar for various soil applications (e.g., soil amendment and/or decontamination) in terms of technical, environmental, economic, and social factors (Kamali et al., 2022).

Despite its numerous advantages, there are worries about the long-term safety and consequences of biochar usage, primarily since the processes influencing soil and organism health are poorly defined and understood (Brtnicky et al., 2021). High dosages of biochar in clay soils are expected to reduce available water content, whereas surface application of biochar in sandy soils is likely to increase erosion and PM emissions. Furthermore, biochar may increase the chance of excessive soil salinity and decreased soil fertility by increasing the pH of alkaline soils, triggering nutrient precipitation. The influence of biochar on (agro)chemicals and the role of biochar-borne hazardous compounds cannot be overlooked due to their obvious negative impacts on target and nontarget species, respectively (Brtnicky et al., 2021). It was reported that biochar produced under high pyrolysis temperatures caused agricultural yield to go down. Biochar also showed a selective nature for the adsorption of pollutants. For example, the application of biochar did not inhibit dichlorodiphenyltrichloroethane. It was also observed that biochar produced at high temperatures contained more ash content than at lower temperatures. Therefore lower crop production yields were expected if high-temperature-derived biochar was applied. Moreover, the feedstock type determines biochar's effectiveness for enhancing crop production. For example, biochar derived from coconut husk is highly effective for producing *Zea mays* biomass, while orange peel-derived biochar does not show similar results (Elkhlifi et al., 2023). The uncertainties of biochar application in different soil type should not be overlooked and require in-depth studies in the future to obtain a sustainable agricultural waste management and carbon sequestration.

3.6 Conclusion

The application of biochar derived from agricultural waste could vigorously promote the establishment of carbon neutrality and a circular economy. The multifunctional benefits of biochar, especially the porous structure, thermally stable nature, cost-effectiveness, and capacity to sequester carbon, contribute to a tremendous global impact. Biochar has made substantial breakthroughs in sequestering carbon, reducing GHG emissions and global warming, increasing soil productivity by increasing nutrients availability and water retention capacity, reducing contaminants' bioavailability, producing valuable products, and elevating closed-loop thinking. However, feedstock type and pyrolysis conditions determine the effectiveness of biochar and its sustainability in agricultural applications. Thus, before applying biochar, assessing some measures to determine the dose–response and its impacts on specific crop yield is essential, which may require some investments. Determining fast, slow, or intermediate pyrolysis technology is also crucial for

LCA of agricultural waste and biochar or biofuel production. However, biochar production from agricultural waste and its application to improve agrarian yield is an environmentally friendly approach that can inhibit climate-related issues, enhancing environmental and economic outputs. Despite that the regulatory issues and testing for the biomass feedstock and the biochar application in different soil and their adverse effects should not be overlooked.

Author contribution

Mashura Shammi and Beauty Akter wrote, edited, and reviewed the original manuscript.

AI declaration

No artificial intelligence (AI) tools have been used except for grammatical checks and proofreading. All the authors approved the manuscript for publication.

Declaration of competing interest

The authors declare that they have no known competing financial interests or personal relationships that could have appeared to influence the work reported in this chapter.

References

Abdullah, N., Mohd Taib, R., Mohamad Aziz, N. S., Omar, M. R., & Md Disa, N. (2023). Banana pseudo-stem biochar derived from slow and fast pyrolysis process. *Heliyon*, *9* (1). Available from https://doi.org/10.1016/j.heliyon.2023.e12940, http://www.journals.elsevier.com/heliyon/.

Akter, B., Shammi, M., Akbor, M. A., Yasmin, S., Nahar, A., Akhter, S., Nahar Jolly, Y., & Uddin, M. K. (2023). Preparation and characterization of biochar: A case study on textile and food industry sludge management. *Case Studies in Chemical and Environmental Engineering*, *7*, 100282. Available from https://doi.org/10.1016/j.cscee.2022.100282.

de Almeida, S. G. C., Tarelho, L. A. C., Hauschild, T., Costa, M. A. M., & Dussán, K. J. (2022). Biochar production from sugarcane biomass using slow pyrolysis: Characterization of the solid fraction. *Chemical Engineering and Processing—Process Intensification*, *179*. Available from https://doi.org/10.1016/j.cep.2022.109054, http://www.elsevier.com/inca/publications/store/5/0/4/0/8/1/index.htt.

Anand, A., Gautam, S., & Ram, L. C. (2023). Feedstock and pyrolysis conditions affect suitability of biochar for various sustainable energy and environmental applications.

Journal of Analytical and Applied Pyrolysis, *170*. Available from https://doi.org/10.1016/j.jaap.2023.105881, https://www.journals.elsevier.com/journal-of-analytical-and-applied-pyrolysis.

Atkinson, C. J., Fitzgerald, J. D., & Hipps, N. A. (2010). Potential mechanisms for achieving agricultural benefits from biochar application to temperate soils: A review. *Plant and Soil*, *337*(1), 1–18. Available from https://doi.org/10.1007/s11104-010-0464-5.

Azzi, E. S., Karltun, E., & Sundberg, C. (2021). Small-scale biochar production on Swedish farms: A model for estimating potential, variability, and environmental performance. *Journal of Cleaner Production*, *280*, 124873. Available from https://doi.org/10.1016/j.jclepro.2020.124873.

Bhatnagar, A., Singhal, A., Tolvanen, H., Valtonen, K., Joronen, T., & Konttinen, J. (2022). Effect of pretreatment and biomass blending on bio-oil and biochar quality from two-step slow pyrolysis of rice straw. *Waste Management*, *138*, 298–307. Available from https://doi.org/10.1016/j.wasman.2021.12.013, http://www.elsevier.com/locate/wasman.

Bhuvaneshwari, S., Hettiarachchi, H., & Meegoda, J. N. (2019). Crop residue burning in India: Policy challenges and potential solutions. *International Journal of Environmental Research and Public Health*, *16*(5). Available from https://doi.org/10.3390/ijerph16050832, https://www.mdpi.com/1660-4601/16/5/832/pdf.

Bista, P., Ghimire, R., Machado, S., & Pritchett, L. (2019). Biochar effects on soil properties and wheat biomass vary with fertility management. *Agronomy*, *9*(10), 623. Available from https://doi.org/10.3390/agronomy9100623.

Brewer, C. E., Schmidt-Rohr, K., Satrio, J. A., & Brown, R. C. (2009). Characterization of biochar from fast pyrolysis and gasification systems. *Environmental Progress and Sustainable Energy*, *28*(3), 386–396. Available from https://doi.org/10.1002/ep.10378, http://www3.interscience.wiley.com/cgi-bin/fulltext/122544837/PDFSTART.

Bridgwater, A. V. (2012). Review of fast pyrolysis of biomass and product upgrading. *Biomass and Bioenergy*, *38*, 68–94. Available from https://doi.org/10.1016/j.biombioe.2011.01.048.

Brtnicky, M., Datta, R., Holatko, J., Bielska, L., Gusiatin, Z. M., Kucerik, J., Hammerschmiedt, T., Danish, S., Radziemska, M., Mravcova, L., Fahad, S., Kintl, A., Sudoma, M., Ahmed, N., & Pecina, V. (2021). A critical review of the possible adverse effects of biochar in the soil environment. *Science of the Total Environment*, *796*, 148756. Available from https://doi.org/10.1016/j.scitotenv.2021.148756.

Campion, L., Bekchanova, M., Malina, R., & Kuppens, T. (2023). The costs and benefits of biochar production and use: A systematic review. *Journal of Cleaner Production*, *408*, 137138. Available from https://doi.org/10.1016/j.jclepro.2023.137138.

Chang, C., Liu, Z., Li, P., Wang, X., Song, J., Fang, S., & Pang, S. (2021). Study on products characteristics from catalytic fast pyrolysis of biomass based on the effects of modified biochars. *Energy*, *229*, 120818. Available from https://doi.org/10.1016/j.energy.2021.120818.

Chantanumat, Y., Phetwarotai, W., Sangthong, S., Palamanit, A., Abu Bakar, M. S., Cheirsilp, B., & Phusunti, N. (2023). Characterization of bio-oil and biochar from slow pyrolysis of oil palm plantation and palm oil mill wastes. *Biomass Conversion and Biorefinery*, *13*(15), 13813–13825. Available from https://doi.org/10.1007/s13399-021-02291-2, https://www.springer.com/journal/13399.

Chew, J., Zhu, L., Nielsen, S., Graber, E., Mitchell, D. R. G., Horvat, J., Mohammed, M., Liu, M., van Zwieten, L., Donne, S., Munroe, P., Taherymoosavi, S., Pace, B., Rawal,

A., Hook, J., Marjo, C., Thomas, D. S., Pan, G., Li, L., ... Fan, X. (2020). Biochar-based fertilizer: Supercharging root membrane potential and biomass yield of rice. *Science of the Total Environment*, *713*, 136431. Available from https://doi.org/10.1016/j.scitotenv.2019.136431.

Cristina Collivignarelli, M., Illankoon, W. A. M. A. N., Milanese, C., Calatroni, S., Maria Caccamo, F., Medina-Llamas, M., Girella, A., & Sorlini, S. (2024). Preparation and modification of biochar derived from agricultural waste for metal adsorption from urban wastewater. *Water*, *16*(5), 698. Available from https://doi.org/10.3390/w16050698.

Conesa, J. A., Font, R., Fullana, A., Martín-Gullón, I., Aracil, I., Gálvez, A., Moltó, J., & Gómez-Rico, M. F. (2009). Comparison between emissions from the pyrolysis and combustion of different wastes. *Journal of Analytical and Applied Pyrolysis*, *84*(1), 95–102. Available from https://doi.org/10.1016/j.jaap.2008.11.022.

Cusenza, M. A., Longo, S., Cellura, M., Guarino, F., Messineo, A., Mistretta, M., & Volpe, M. (2021). Environmental assessment of a waste-to-energy practice: The pyrolysis of agro-industrial biomass residues. *Sustainable Production and Consumption*, *28*, 866–876. Available from https://doi.org/10.1016/j.spc.2021.07.015, http://www.journals.elsevier.com/sustainable-production-and-consumption/.

De Melo Carvalho, M. T., De Holanda Nunes Maia, A., Madari, B. E., Bastiaans, L., Van Oort, P. A. J., Heinemann, A. B., Da Silva, M. A. S., Petter, F. A., Marimon, B. H., & Meinke, H. (2014). Biochar increases plant-available water in a sandy loam soil under an aerobic rice crop system. *Solid Earth*, *5*(2), 939–952. Available from https://doi.org/10.5194/se-5-939-2014.

Dhar, S. A., Sakib, T. U., & Hilary, L. N. (2022). Effects of pyrolysis temperature on production and physicochemical characterization of biochar derived from coconut fiber biomass through slow pyrolysis process. *Biomass Conversion and Biorefinery*, *12*(7), 2631–2647. Available from https://doi.org/10.1007/s13399-020-01116-y, http://www.springer.com/engineering/energy + technology/journal/13399.

Duque-Acevedo, M., Belmonte-Ureña, L. J., Yakovleva, N., & Camacho-Ferre, F. (2020). Analysis of the circular economic production models and their approach in agriculture and agricultural waste biomass management. *International Journal of Environmental Research and Public Health*, *17*(24), 9549. Available from https://doi.org/10.3390/ijerph17249549.

Elkhlifi, Z., Iftikhar, J., Sarraf, M., Ali, B., Saleem, M. H., Ibranshahib, I., Bispo, M. D., Meili, L., Ercisli, S., Torun Kayabasi, E., Alemzadeh Ansari, N., Hegedűsová, A., & Chen, Z. (2023). Potential role of biochar on capturing soil nutrients, carbon sequestration and managing environmental challenges: A review. *Sustainability*, *15*(3), 2527. Available from https://doi.org/10.3390/su15032527.

Flammini, A., Brundin, E., Grill, R., & Zellweger, H. (2020). Supply chain uncertainties of small-scale coffee husk-biochar production for activated carbon in Vietnam. *Sustainability*, *12*(19), 8069. Available from https://doi.org/10.3390/su12198069.

Fu, J., Li, P., Lin, Y., Du, H., Liu, H., Zhu, W., & Ren, H. (2022). Fight for carbon neutrality with state-of-the-art negative carbon emission technologies. *Eco-Environment & Health*, *1*(4), 259–279. Available from https://doi.org/10.1016/j.eehl.2022.11.005.

Guo, L., Yu, H., Kharbach, M., Zhang, W., Wang, J., & Niu, W. (2021). Biochar improves soil-tomato plant, tomato production, and economic benefits under reduced nitrogen application in Northwestern China. *Plants*, *10*(4), 759. Available from https://doi.org/10.3390/plants10040759.

Haque, F., Fan, C., & Lee, Y.-Y. (2023). From waste to value: Addressing the relevance of waste recovery to agricultural sector in line with circular economy. *Journal of Cleaner Production*, *415*, 137873. Available from https://doi.org/10.1016/j.jclepro.2023.137873.

Heinrich, T., Park, H., Orozco, R., Ding, Z., Álvarez-López, V., Mosquera-Losada, M. R., Steinbeis, L., & Hoffmann, T. (2023). Biochar production from late-harvest grass – Challenges and potential for farm-scale implementation. *Sustainable Production and Consumption*, *37*, 256–267. Available from https://doi.org/10.1016/j.spc.2023.02.019, http://www.journals.elsevier.com/sustainable-production-and-consumption/.

Hornung, A. (2013). *Intermediate pyrolysis of biomass. Biomass combustion science, technology and engineering* (pp. 172–186). Elsevier Inc. Available from http://www.sciencedirect.com/science/book/9780857091314, 10.1533/9780857097439.2.172.

Hossain, M. Z., Bahar, M. M., Sarkar, B., Donne, S. W., Ok, Y. S., Palansooriya, K. N., Kirkham, M. B., Chowdhury, S., & Bolan, N. (2020). Biochar and its importance on nutrient dynamics in soil and plant. *Biochar*, *2*(4), 379–420. Available from https://doi.org/10.1007/s42773-020-00065-z, http://springer.com/journal/42773.

Hu, Q., Jung, J., Chen, D., Leong, K., Song, S., Li, F., Mohan, B. C., Yao, Z., Prabhakar, A. K., Lin, X. H., Lim, E. Y., Zhang, L., Souradeep, G., Ok, Y. S., Kua, H. W., Li, S. F. Y., Tan, H. T. W., Dai, Y., Tong, Y. W., ... Wang, C. H. (2021). Biochar industry to circular economy. *Science of the Total Environment*, *757*. Available from https://doi.org/10.1016/j.scitotenv.2020.143820, http://www.elsevier.com/locate/scitotenv.

Huang, M., Fan, L., Jiang, L.-G., Yang, S.-Y., Zou, Y.-B., & Uphoff, N. (2019). Continuous applications of biochar to rice: Effects on grain yield and yield attributes. *Journal of Integrative Agriculture*, *18*(3), 563–570. Available from https://doi.org/10.1016/s2095-3119(18)61993-8.

Jain, N., Bhatia, A., & Pathak, H. (2014). Emission of air pollutants from crop residue burning in India. *Aerosol and Air Quality Research*, *14*(1), 422–430. Available from http://aaqr.org/VOL14_No1_February2014/40_AAQR-13-01-OA-0031_422-430.pdf, 10.4209/aaqr.2013.01.0031India.

Jeffery, S., Verheijen, F. G. A., van der Velde, M., & Bastos, A. C. (2011). A quantitative review of the effects of biochar application to soils on crop productivity using meta-analysis. *Agriculture, Ecosystems and Environment*, *144*(1), 175–187. Available from https://doi.org/10.1016/j.agee.2011.08.015, http://www.elsevier.com/inca/publications/store/5/0/3/2/9/8.

Jiang, Y. (2021). Sustainable production of biochar: A review of feedstocks, pyrolysis technologies, and applications. *Journal of Cleaner Production*, *281*.

Jin, Z., Chen, C., Chen, X., Jiang, F., Hopkins, I., Zhang, X., Han, Z., Billy, G., & Benavides, J. (2019). Soil acidity, available phosphorus content, and optimal biochar and nitrogen fertilizer application rates: A five-year field trial in upland red soil, China. *Field Crops Research*, *232*, 77–87. Available from https://doi.org/10.1016/j.fcr.2018.12.013.

Joseph, S., Cowie, A. L., Van Zwieten, L., Bolan, N., Budai, A., Buss, W., Cayuela, M. L., Graber, E. R., Ippolito, J. A., Kuzyakov, Y., Luo, Y., Ok, Y. S., Palansooriya, K. N., Shepherd, J., Stephens, S., Weng, Z., & Lehmann, J. (2021). How biochar works, and when it doesn't: A review of mechanisms controlling soil and plant responses to biochar. *GCB Bioenergy*, *13*(11), 1731–1764. Available from https://doi.org/10.1111/gcbb.12885, http://onlinelibrary.wiley.com/journal/10.1111/(ISSN)1757-1707.

Kamali, M., Sweygers, N., Al-Salem, S., Appels, L., Aminabhavi, T. M., & Dewil, R. (2022). Biochar for soil applications-sustainability aspects, challenges and future prospects. *Chemical Engineering Journal*, *428*, 131189. Available from https://doi.org/10.1016/j.cej.2021.131189.

Kazawadi, D., Ntalikwa, J., Kombe, G., & Saydut, A. (2021). A review of intermediate pyrolysis as a technology of biomass conversion for coproduction of biooil and adsorption biochar. *Journal of Renewable Energy*, *2021*, 1–10. Available from https://doi.org/10.1155/2021/5533780.

Khan, Z., Nauman Khan, M., Luo, T., Zhang, K., Zhu, K., Rana, M. S., Hu, L., & Jiang, Y. (2021). Compensation of high nitrogen toxicity and nitrogen deficiency with biochar amendment through enhancement of soil fertility and nitrogen use efficiency promoted rice growth and yield. *GCB Bioenergy*, *13*(11), 1765–1784. Available from https://doi.org/10.1111/gcbb.12884, http://onlinelibrary.wiley.com/journal/10.1111/(ISSN)1757-1707.

Kloss, S., Zehetner, F., Dellantonio, A., Hamid, R., Ottner, F., Liedtke, V., Schwanninger, M., Gerzabek, M. H., & Soja, G. (2012). Characterization of slow pyrolysis biochars: Eff ects of feedstocks and pyrolysis temperature on biochar properties. *Journal of Environmental Quality*, *41*(4), 990–1000. Available from https://www.agronomy.org/publications/jeq/pdfs/41/4/990, 10.2134/jeq2011.0070Austria.

Koul, B., Yakoob, M., & Shah, M. P. (2022). Agricultural waste management strategies for environmental sustainability. *Environmental Research*, *206*, 112285. Available from https://doi.org/10.1016/j.envres.2021.112285.

Kulkarni, S. H., Ghude, S. D., Jena, C., Karumuri, R. K., Sinha, B., Sinha, V., Kumar, R., Soni, V. K., & Khare, M. (2020). How much does large-scale crop residue burning affect the air quality in Delhi? *Environmental Science & Technology*, *54*(8), 4790–4799. Available from https://doi.org/10.1021/acs.est.0c00329.

Lehmann, J., Rillig, M. C., Thies, J., Masiello, C. A., Hockaday, W. C., & Crowley, D. (2011). Biochar effects on soil biota: A review. *Soil Biology and Biochemistry*, *43*(9), 1812–1836. Available from https://doi.org/10.1016/j.soilbio.2011.04.022.

Lehmann, J., Cowie, A., Masiello, C. A., Kammann, C., Woolf, D., Amonette, J. E., Cayuela, M. L., Camps-Arbestain, M., & Whitman, T. (2021). Biochar in climate change mitigation. *Nature Geoscience*, *14*(12), 883–892. Available from https://doi.org/10.1038/s41561-021-00852-8, http://www.nature.com/ngeo/index.html.

Lehmann, J. (2022). Role of biochar properties in its persistence and carbon sequestration in soil: A review. *Environmental Science & Technology (Elmsford, N.Y.)*, *56*(2), 759–771.

Liu, J. (2023). Environmental impacts of agricultural waste disposal: A global perspective. *Science of the Total Environment* (806).

Liu, Y., Tang, H., Muhammad, A., & Huang, G. (2019). Emission mechanism and reduction countermeasures of agricultural greenhouse gases – a review. *Greenhouse Gases: Science and Technology*, *9*(2), 160–174. Available from https://doi.org/10.1002/ghg.1848, https://onlinelibrary.wiley.com/page/journal/21523878/homepage/forauthors.html.

Luo, L., Wang, J., Lv, J., Liu, Z., Sun, T., Yang, Y., & Zhu, Y. G. (2023). Carbon sequestration strategies in soil using biochar: Advances, challenges, and opportunities. *Environmental Science and Technology*, *57*(31), 11357–11372. Available from https://doi.org/10.1021/acs.est.3c02620, http://pubs.acs.org/journal/esthag.

Manyà, J. J., Azuara, M., & Manso, J. A. (2018). Biochar production through slow pyrolysis of different biomass materials: Seeking the best operating conditions. *Biomass and Bioenergy*, *117*, 115–123. Available from https://doi.org/10.1016/j.biombioe.2018.07.019, http://www.journals.elsevier.com/biomass-and-bioenergy/.

Mohammadi, A. (2020). A critical review on advancement and challenges of biochar application in paddy fields: Environmental and life cycle cost analysis. *Processes*, *8* (10), 1275.

Nair, V. D., Nair, P. K. R., Dari, B., Freitas, A. M., Chatterjee, N., & Pinheiro, F. M. (2017). Biochar in the agroecosystem-climate-change-sustainability Nexus. *Frontiers in Plant Science*, *8*. Available from https://doi.org/10.3389/fpls.2017.02051, http://journal.frontiersin.org/article/10.3389/fpls.2017.02051/full.

Owsianiak, M., Lindhjem, H., Cornelissen, G., Hale, S. E., Sørmo, E., & Sparrevik, M. (2021). Environmental and economic impacts of biochar production and agricultural use in six developing and middle-income countries. *Science of the Total Environment*, *755*, 142455. Available from https://doi.org/10.1016/j.scitotenv.2020.142455.

Pierson, D., Anderson, N., Brewen, J., Clark, N., Hardy, M. C., McCollum, D., McCormick, F. H., Morisette, J., Nicosia, T., Page-Dumroese, D., Rodriguez-Franco, C., & Tirocke, J. (2024). Beyond the basics: a perspective on barriers and opportunities for scaling up biochar production from forest slash. *Biochar*, *6*(1). Available from https://doi.org/10.1007/s42773-023-00290-2, https://www.springer.com/journal/42773.

Premchand, P., Demichelis, F., Chiaramonti, D., Bensaid, S., & Fino, D. (2023). Biochar production from slow pyrolysis of biomass under CO_2 atmosphere: A review on the effect of CO_2 medium on biochar production, characterisation, and environmental applications. *Journal of Environmental Chemical Engineering*, *11*(3). Available from https://doi.org/10.1016/j.jece.2023.110009, http://www.journals.elsevier.com/journal-of-environmental-chemical-engineering/.

Raza, M. H., Abid, M., Faisal, M., Yan, T., Akhtar, S., & Mehedi Adnan, K. M. (2022). Environmental and health impacts of crop residue burning: Scope of sustainable crop residue management practices. *International Journal of Environmental Research and Public Health*, *19*(8), 4753. Available from https://doi.org/10.3390/ijerph19084753.

Robb, S., Joseph, S., Abdul Aziz, A., Dargusch, P., & Tisdell, C. (2020). Biochar's cost constraints are overcome in small-scale farming on tropical soils in lower-income countries. *Land Degradation and Development*, *31*(13), 1713–1726. Available from https://doi.org/10.1002/ldr.3541, http://onlinelibrary.wiley.com/journal/10.1002/(ISSN)1099-145X.

Rogovska, N. (2019). Environmental life cycle assessment of biochar production and use: A review. *Journal of Cleaner Production*, *222*, 228–244.

Ronsse, F., van Hecke, S., Dickinson, D., & Prins, W. (2013). Production and characterization of slow pyrolysis biochar: Influence of feedstock type and pyrolysis conditions. *GCB Bioenergy*, *5*(2), 104–115. Available from https://doi.org/10.1111/gcbb.12018, http://onlinelibrary.wiley.com/journal/10.1111/(ISSN)1757-1707.

Safarian, S. (2023). Performance analysis of sustainable technologies for biochar production: A comprehensive review. *Energy Reports*, *9*, 4574–4593. Available from https://doi.org/10.1016/j.egyr.2023.03.111, http://www.journals.elsevier.com/energy-reports/.

Saharudin, D. M., Jeswani, H. K., & Azapagic, A. (2024). Biochar from agricultural wastes: Environmental sustainability, economic viability and the potential as a negative emissions technology in Malaysia. *Science of the Total Environment*, *919*. Available from https://doi.org/10.1016/j.scitotenv.2024.170266, https://www.sciencedirect.com/science/journal/00489697.

Sahoo, S. S., Vijay, V. K., Chandra, R., & Kumar, H. (2021). Production and characterization of biochar produced from slow pyrolysis of pigeon pea stalk and bamboo. *Cleaner*

Engineering and Technology, *3*. Available from https://doi.org/10.1016/j.clet.2021.100101, https://www.journals.elsevier.com/cleaner-engineering-and-technology.

Sakhiya, A. K., Anand, A., Aier, I., Vijay, V. K., & Kaushal, P. (2021). Suitability of rice straw for biochar production through slow pyrolysis: Product characterization and thermodynamic analysis. *Bioresource Technology Reports*, *15*. Available from https://doi.org/10.1016/j.biteb.2021.100818, https://www.journals.elsevier.com/bioresource-technology-reports.

Sarkar, S., Skalicky, M., Hossain, A., Brestic, M., Saha, S., Garai, S., Ray, K., & Brahmachari, K. (2020). Management of crop residues for improving input use efficiency and agricultural sustainability. *Sustainability*, *12*(23), 9808. Available from https://doi.org/10.3390/su12239808.

Saxena, P., Sonwani, S., Srivastava, A., Jain, M., Srivastava, A., Bharti, A., Rangra, D., Mongia, N., Tejan, S., & Bhardwaj, S. (2021). Impact of crop residue burning in Haryana on the air quality of Delhi, India. *Heliyon*, *7*(5). Available from https://doi.org/10.1016/j.heliyon.2021.e06973, http://www.journals.elsevier.com/heliyon/.

Shammi, M., Winter, J., Islam, M. M., Akter, B., & Hasan, N. (2023). *Cookstoves for biochar production and carbon capture. Green sustainable process for chemical and environmental engineering and science: Carbon dioxide capture and utilization* (pp. 53–68). Elsevier. Available from https://www.sciencedirect.com/book/9780323994293, 10.1016/B978-0-323-99429-3.00025-4.

Shammi, M., & Akter, B. (2024). *CO_2 sequestration for conventional utilization and industrial application. Decarbonization strategies and drivers to achieve carbon neutrality for sustainability* (pp. 299–319). Elsevier BV. Available from 10.1016/b978-0-443-13607-8.00011-0.

Shyamsundar, P., Springer, N. P., Tallis, H., Polasky, S., Jat, M. L., Sidhu, H. S., Krishnapriya, P. P., Skiba, N., Ginn, W., Ahuja, V., Cummins, J., Datta, I., Dholakia, H. H., Dixon, J., Gerard, B., Gupta, R., Hellmann, J., Jadhav, A., Jat, H. S., ... Somanathan, R. (2019). Fields on fire: Alternatives to crop residue burning in India. *Science (New York, N.Y.)*, *365*(6453), 536–538. Available from https://doi.org/10.1126/science.aaw4085, https://science.sciencemag.org/content/365/6453/536/tab-pdf.

Singh, B., MacDonald, L. M., Kookana, R. S., Van Zwieten, L., Butler, G., Joseph, S., Weatherley, A., Kaudal, B. B., Regan, A., Cattle, J., Dijkstra, F., Boersma, M., Kimber, S., Keith, A., & Esfandbod, M. (2014). Opportunities and constraints for biochar technology in Australian agriculture: Looking beyond carbon sequestration. *Soil Research*, *52*(8), 739–750. Available from https://doi.org/10.1071/SR14112, http://www.publish.csiro.au/nid/84/aid/2406.htm.

Singh, E., Mishra, R., Kumar, A., Shukla, S. K., Lo, S. L., & Kumar, S. (2022). Circular economy-based environmental management using biochar: Driving towards sustainability. *Process Safety and Environmental Protection*, *163*, 585–600. Available from https://doi.org/10.1016/j.psep.2022.05.056, http://www.elsevier.com/wps/find/journaldescription.cws_home/713889/description#description.

Singh, H., Northup, B. K., Rice, C. W., & Prasad, P. V. V. (2022). Biochar applications influence soil physical and chemical properties, microbial diversity, and crop productivity: A meta-analysis. *Biochar*, *4*(1). Available from https://doi.org/10.1007/s42773-022-00138-1, http://springer.com/journal/42773.

Smith., Bustamante, M., Ahammad, H., Clark, H., Dong, H., Elsiddig, E. A., Haberl, H., Harper, R., House, J., Jafari, M., & Masera, O. (2014). *Climate change 2014: Mitigation of climate change. Contribution of working group* (pp. 811–922). Cambridge University Press.

Smith, P. (2016). Soil carbon sequestration and biochar as negative emission technologies. *Global Change Biology*, *22*(3), 1315–1324. Available from https://doi.org/10.1111/gcb.13178, http://www.blacksci.co.uk/~cgilib/jnlpage.bin?Journal = gcb&File = gcb&Page = aims.

Solaiman, Z. M., & Anawar, H. M. (2015). Application of biochars for soil constraints: Challenges and solutions. *Pedosphere*, *25*(5), 631–638. Available from https://doi.org/10.1016/S1002-0160(15)30044-8, http://pedosphere.issas.ac.cn.

Spokas, K. A., Cantrell, K. B., Novak, J. M., Archer, D. W., Ippolito, J. A., Collins, H. P., Boateng, A. A., Lima, I. M., Lamb, M. C., McAloon, A. J., Lentz, R. D., & Nichols, K. A. (2012). Biochar: A synthesis of its agronomic impact beyond carbon sequestration. *Journal of Environmental Quality*, *41*(4), 973–989. Available from https://www.agronomy.org/publications/jeq/pdfs/41/4/973, 10.2134/jeq2011.0069, United States.

Suliman, W., Harsh, J. B., Abu-Lail, N. I., Fortuna, A. M., Dallmeyer, I., & Garcia-Perez, M. (2016). Influence of feedstock source and pyrolysis temperature on biochar bulk and surface properties. *Biomass and Bioenergy*, *84*, 37–48. Available from https://doi.org/10.1016/j.biombioe.2015.11.010, http://www.journals.elsevier.com/biomass-and-bioenergy/.

Tisserant, A., Morales, M., Cavalett, O., O'Toole, A., Weldon, S., Rasse, D. P., & Cherubini, F. (2022). Life-cycle assessment to unravel co-benefits and trade-offs of large-scale biochar deployment in Norwegian agriculture. *Resources, Conservation and Recycling*, *179*, 106030. Available from https://doi.org/10.1016/j.resconrec.2021.106030.

Tomczyk, A., Sokołowska, Z., & Boguta, P. (2020). Biochar physicochemical properties: pyrolysis temperature and feedstock kind effects. *Reviews in Environmental Science and Biotechnology*, *19*(1), 191–215. Available from http://www.kluweronline.com/issn/1569-1705, 10.1007/s11157-020-09523-3.

Tsolis, V., & Barouchas, P. (2023). Biochar as soil amendment: The effect of biochar on soil properties using VIS-NIR diffuse reflectance spectroscopy, biochar aging and soil microbiology—a review. *Land*, *12*(8), 1580. Available from https://doi.org/10.3390/land12081580.

Ullah, S., Zhao, Q., Wu, K., Ali, I., Liang, H., Iqbal, A., Wei, S., Cheng, F., Ahmad, S., Jiang, L., Gillani, S. W., Amanullah., Anwar, S., & Khan, Z. (2021). Biochar application to rice with 15N-labelled fertilizers, enhanced leaf nitrogen concentration and assimilation by improving morpho-physiological traits and soil quality. *Saudi Journal of Biological Sciences*, *28*(6), 3399–3413. Available from https://doi.org/10.1016/j.sjbs.2021.03.003, http://www.ksu.edu.sa/sites/Colleges/CollegeofScinces/Botany Department/SJBS/Default.aspx.

Vieira, F. R., Romero Luna, C. M., Arce, G. L. A. F., & Ávila, I. (2020). Optimization of slow pyrolysis process parameters using a fixed bed reactor for biochar yield from rice husk. *Biomass and Bioenergy*, *132*, 105412. Available from https://doi.org/10.1016/j.biombioe.2019.105412.

Wang, L., O'Connor, D., Rinklebe, J., Ok, Y. S., Tsang, D. C. W., Shen, Z., & Hou, D. (2020). Biochar aging: Mechanisms, physicochemical changes, assessment, and implications for field applications. *Environmental Science and Technology*, *54*(23),

14797–14814. Available from https://doi.org/10.1021/acs.est.0c04033, http://pubs.acs.org/journal/esthag.

Woolf, D., Amonette, J. E., Street-Perrott, F. A., Lehmann, J., & Joseph, S. (2010). Sustainable biochar to mitigate global climate change. *Nature Communications*, *1*(5). Available from https://doi.org/10.1038/ncomms1053, http://www.nature.com/ncomms/index.html.

Wu, L., Zhang, S., Chen, M., Liu, J., & Ding, X. (2021). A sustainable option: Biochar addition can improve soil phosphorus retention and rice yield in a saline–alkaline soil. *Environmental Technology & Innovation*, *24*, 102070. Available from https://doi.org/10.1016/j.eti.2021.102070.

Yao, T., Zhang, W., Gulaqa, A., Cui, Y., Zhou, Y., Weng, W., Wang, X., Liu, Q., & Jin, F. (2021). Effects of peanut shell biochar on soil nutrients, soil enzyme activity, and rice yield in heavily saline-sodic paddy field. *Journal of Soil Science and Plant Nutrition*, *21*(1), 655–664. Available from https://doi.org/10.1007/s42729-020-00390-z, https://www.springer.com/life + sciences/plant + sciences/journal/42729.

Yin, J., Zhao, L., Xu, X., Li, D., Qiu, H., & Cao, X. (2022). Evaluation of long-term carbon sequestration of biochar in soil with biogeochemical field model. *Science of the Total Environment*, *822*, 153576. Available from https://doi.org/10.1016/j.scitotenv.2022. 153576.

Yuan, X., Cao, Y., Li, J., Patel, A. K., Dong, C.-D., Jin, X., Gu, C., Yip, A. C. K., Tsang, D. C. W., & Sik Ok, Y. (2023). Recent advancements and challenges in emerging applications of biochar-based catalysts. *Biotechnology Advances*, *67*, 108181. Available from https://doi.org/10.1016/j.biotechadv.2023.108181.

Zilberman, D., Laird, D., Rainey, C., Song, J., & Kahn, G. (2023). Biochar supply-chain and challenges to commercialization. *GCB Bioenergy*, *15*(1), 7–23. Available from https://doi.org/10.1111/gcbb.12952, http://onlinelibrary.wiley.com/journal/10.1111/(ISSN)1757-1707.

CHAPTER

4 Biochar production and use among smallholder farmers in rural Uganda: challenges and opportunities for circular economy

Nataliya Apanovich

School of Landscape Architecture and Planning, University of Arizona, Tucson, AZ, United States

4.1 Introduction

Biochar benefits are well documented and range from environmental to agronomic; however, many of those who could potentially benefit from biochar are often unaware of its nature and uses. This is especially apparent in Africa where limited access to knowledge on sustainable agricultural practices, such as biochar, is one of the barriers that prevents smallholder farmers from improving their livelihoods. A study in South Africa and Zimbabwe, Didarali and Gambiza (2019) found that, in addition to high labor inputs, lack of knowledge on permaculture practices limited their adoption. A study in Nigeria found that lack of knowledge on organic farming and its economic value hindered farmers from adopting it (Adebiyi & Olabisi, 2022). A study in Ghana, Zakaria et al. (2020) identified previous farming experience, access to mass media, and training as important factors for the adoption of climate-smart agricultural technologies. In Nigeria, farmers reported uncertainty, fear of failure, and cost implications as the biggest challenges to the adoption of agricultural innovations (Ikehi et al., 2022).

Understanding factors that explain farmers' willingness to adopt biochar is especially important now when smallholders are facing multiple challenges, including climate change, soil degradation, and food insecurity. By using biochar as a soil amendment, smallholder farmers can bypass synthetic fertilizers that are not only polluting but also expensive and hard to get. With farmers having limited resources and cash on hand, biochar can be a feasible option to improve their livelihoods in a quick and sustainable way. A study in Tanzania, Rogers et al. (2022) found that crop yield, food security, and family income were the main motivators behind biochar production and use. It also established that low

Biochar Ecotechnology for Sustainable Agriculture and Environment. DOI: https://doi.org/10.1016/B978-0-443-29855-4.00005-9

education and income contributed to low biochar adoption rates, especially among males ages 40–60. Participatory trials of on-farm biochar production in Ghana showed that, rather than the efficiency of the carbonization technology, the availability of feedstock and labor considerations determined the technology selected by farmers for biochar production (Steiner et al., 2018). A five-year long study on microgasifiers for heat and biochar production in Tanzania found that the households did not amend soils with biochar due to a combination of local practices and perceptions, and a lack of education and awareness programs (Eltigani et al., 2022). The study also recommended subsidies or microloans for making biochar stoves more affordable.

In addition to the socioeconomic barriers to a new technology adoption, some identified more complex factors that influence farmers' decision-making (Table 4.1). For example, Mutsvangwa-Sammie argues that in order for an agricultural innovation to become successfully adopted, power, beliefs, and values within a society need to be taken into account (Mutsvangwa-Sammie, 2020). A similar study by Branca et al. (2022) established that a sustainable agricultural technology needs to be adapted to the complexity and diversity of smallholder farming systems for its successful implementation.

Farmers' awareness levels, perceptions, and the technological potential of biochar need to be studied in a larger socioeconomic and cultural framework to address such an important development issue as food security. This is especially relevant in Uganda where soil fertility issues persist due to population growth and agricultural intensification. Currently, almost half of Uganda's population is experiencing one form of food insecurity. Biochar could be one way to address this issue in a cost-effective and sustainable way. This will lead not only to improved livelihoods and increased resilience but it will also contribute to the promotion of circular economy where, instead of being discarded, local materials can be recovered, reused, and recycled to create new products. The circular economy paradigm has been recognized as an appropriate mechanism to attain sustainable development goals through technological innovations (Kurniawan et al., 2023).

Thus, in addition to serving as a valuable product that can improve yields and food security, biochar can also bring environmental benefits, further contributing to circular economy through carbon neutrality. It is estimated that 12% of CO_2 emissions by land use can be offset annually via biochar for soil enrichment. Biochar can also sequester CO_2 in the long term and prevent the release of carbon back into the atmosphere after its decomposition. This practice could sequester 2.5 gigatons (Gt) of CO_2 annually (Kurniawan et al., 2023). While biochar's specific benefits depend on the feedstock and technology used, it is undeniable that even simple, homemade gasification systems are capable of producing biochar with positive impacts on the soil and environment. Preserving unused resources such as biomass is important not only to synergize between reducing material footprints and greenhouse gase emissions but also to ensure enough resources for present and future generations (Kapoor et al., 2022).

Table 4.1 Biochar adoption challenges as identified in the literature.

Challenge	Description
Lack of awareness and education	Many farmers may not be familiar with biochar and its potential benefits. The lack of awareness and education about biochar's applications, production methods, and positive impacts on soil heath can hinder its adoption (Fridahl et al., 2021; Latawiec et al., 2017).
Limited access to information	Farmers may face challenges in accessing reliable and relevant information about biochar, especially in regions with limited extension services. Access to knowledge about the proper use, application rates, and potential benefits of biochar is essential for successful adoption (Descheemaeker et al., 2016; Rogers et al., 2022).
Initial cost and investments	The upfront costs associated with acquiring biochar production equipment or purchasing biochar products can be a significant barrier. Farmers may be reluctant to invest in biochar technologies without a clear understanding of the potential long-term benefits and return on investment (Latawiec et al., 2017).
Inconsistent supply chain	The availability and accessibility of biochar products can be inconsistent in some regions. Farmers may hesitate to adopt biochar if there is uncertainty about a reliable and continuous supply chain for biochar products or feedstocks for on-farm production (Zilberman et al., 2023).
Perceived risks and uncertainties	Farmers may be cautious about adopting new practices, especially if they perceive risks or uncertainties related to the effects of biochar on their crops, soils, or overall farming practices. A lack of confidence in the efficacy and safety of biochar may deter adoption (Gwenzi et al., 2015).
Need for tailored research and demonstration	Local soil and climate conditions can vary widely, and farmers may be hesitant to adopt biochar without evidence of its effectiveness in their specific context. Conducting research trials and providing on-farm demonstrations tailored to local conditions can address this barrier (Rogers et al., 2022).
Policy and regulatory issues	Lack of supportive policies or unclear regulations related to biochar production and application can impede adoption. Farmers may be hesitant to adopt biochar if there is uncertainty about compliance with regulations or if policies do not incentivize or support its use (Heinrich et al., 2023).
Technology transfer challenges	The transfer of biochar technology from research institutions to farmers may face challenges. Effective extension services, training programs, and knowledge-sharing platforms are essential for bridging the gap between research and on-farm adoption (Pourhashem et al., 2019).
Limited infrastructure	In some regions, the lack of infrastructure for biochar production or distribution can hinder adoption. This includes challenges related to transportation, storage, and processing facilities for biochar products (Heinrich et al., 2023; Otte & Vik, 2017).

On-farm research using elements of participatory methodology is an appropriate way to investigate the potential of biochar and accelerate its adoption. Collaboration and joint learning between farmers and researchers can help to ensure that the new technology is relevant to the needs of the farmers, build trust between researchers and farmers, provide farmers with the opportunity to learn about the technology and to develop the skills needed to use it, create a sense of ownership of the technology among users. Overall, participatory approaches can play an important role in increasing the adoption of new agricultural technologies and improving the livelihoods of farmers and their communities (Jones et al., 2014). A semiparticipatory approach was used in Ghana to study biochar's challenges and opportunities. The farmers produced biochar using a simple top-lit updraft gasifier and then tested it on lettuce fields (Steiner et al., 2018). Another study found that participatory approaches proved to be effective in enabling small-scale farmers and local decision makers to identify, develop, and disseminate technologies (Rambim, 2022).

While most literature focuses on biochar production on an industrial scale, this chapter looks at biochar production on a household scale. In the countries of the Global South where farmers have limited resources, including access to credit, purchasing industrially produced biochar is simply unrealistic. Additionally, biochar remains to be a niche product with limited availability and distribution. For example, in Uganda, there are only a handful companies that produce biochar and with additional barriers of transportation, access to biochar remains elusive. With farmers' purchasing power being at near zero, any solutions proposed to improve rural livelihoods have to reflect this economic reality. Additionally, the focus is on the production of biochar as a soil amendment only compared to the dual production of energy and biochar. While many energy–biochar systems are used with great success, such systems are unrealistic in a country like Uganda where insufficient amounts of adequate feedstock prevent an efficient use of such systems. For example, a gasifier that was introduced to Tiribogo village in Mpigi District in central Uganda did not turn out to be a feasible solution as the farmers struggled with biomass provision (Okure et al., 2014). The system functioned exclusively on coffee husks and maize stover and their requirements turned out to be too burdensome for the village to procure to ensure an uninterrupted energy–biochar production. Using agricultural residues exclusively for energy and/or biochar production is not recommended as many residues already have established uses (i.e., livestock feed, mulch, and fuel for cooking). Using woody biomass for a similar system is also unrealistic given rapid deforestation, population growth, and general resource degradation. In fact, many studies found poor adoption of gasifier stoves designed to produce heat for cooking and biochar due to difficulty with fuel availability (Akolgo et al., 2018; Gitau et al., 2019; Shammi et al., 2023; Sundberg et al., 2020). Thus, in light of this information, the paper focuses on a small-scale biochar production only systems.

This chapter relies on participatory approach to analyze biochar production and use challenges and opportunities among smallholder farmers in southcentral Uganda.

Specifically, this work examines farmers' knowledge base on biochar, their willingness and capacity to produce and use it for soil amelioration, and their ability to continue the practice after the completion of the study. The chapter makes recommendations for policies and strategies that could incentivize the use of biochar as a pathway to bioeconomy. Overall, this study positions the role of biochar in advancing circular economy in the Global South.

4.2 Methodology

4.2.1 Sampling

Farmer interviews were conducted in 2015–2016 in Masaka District (Fig. 4.1) covering six subcounties (Bukakata, Mukungwe, Buwunga, Kabonera, Kyanamukaka, and Kyesiga) and one division (Katwe-Butego). In total, 150 smallholder farmers were interviewed representing 92 villages and 30 parishes. Farmers interviewed represented a range of locations and farming systems. District-level administrative units consist of subcounties, parishes, and villages. To choose 150 farmers, five villages were randomly chosen per parish. Then, one to six farmers were identified in each village with an average of two households per village. Farmers were identified from either farmer training records kept by local extension services or by village leaders. A list of all potential farmers (approximately 500 farmers) was created and then used to randomly select 150. To capture cross-village variation, the number of farmers surveyed per village was limited to a maximum of six. All interviews were conducted in the local language with the help of an extension agent and always in the presence of the lead researcher.

The interview questions covered demographics, socioeconomic conditions, farmers' awareness of and willingness to use biochar as a soil amendment, and availability of feedstocks suitable for biochar production and their present uses. To determine a relationship between biochar awareness and farmers' demographic and socioeconomic characteristics, the data were analyzed in R software using Chi-square test. Chi-square test allows to test whether two categorical variables are related to each other. Through literature review, the author identified the following variables that were found to be associated with biochar awareness and adoption: age (divided into groups of 0–24, 25–44, 45–64, and >64), gender (male or female), land size (divided into groups of 0–1.5, 1.6–3, 3.1–5, and >5), and education (divided into groups of 0–4, 5–7, 8–10, and >10). Excel spreadsheets were used for descriptive and inferential statistics.

The majority (66%) of those interviewed were women age of 30 or older. Most farming households had four to seven members. Most farm sizes were between one to two hectares, with the average size of 1.61 ha. Seven or eight years of primary schooling was the most common educational level achieved. Only 32% of farmers surveyed had access to electricity. Only 10% reported their soil quality as either excellent or good. Most didn't practice any soil management practices.

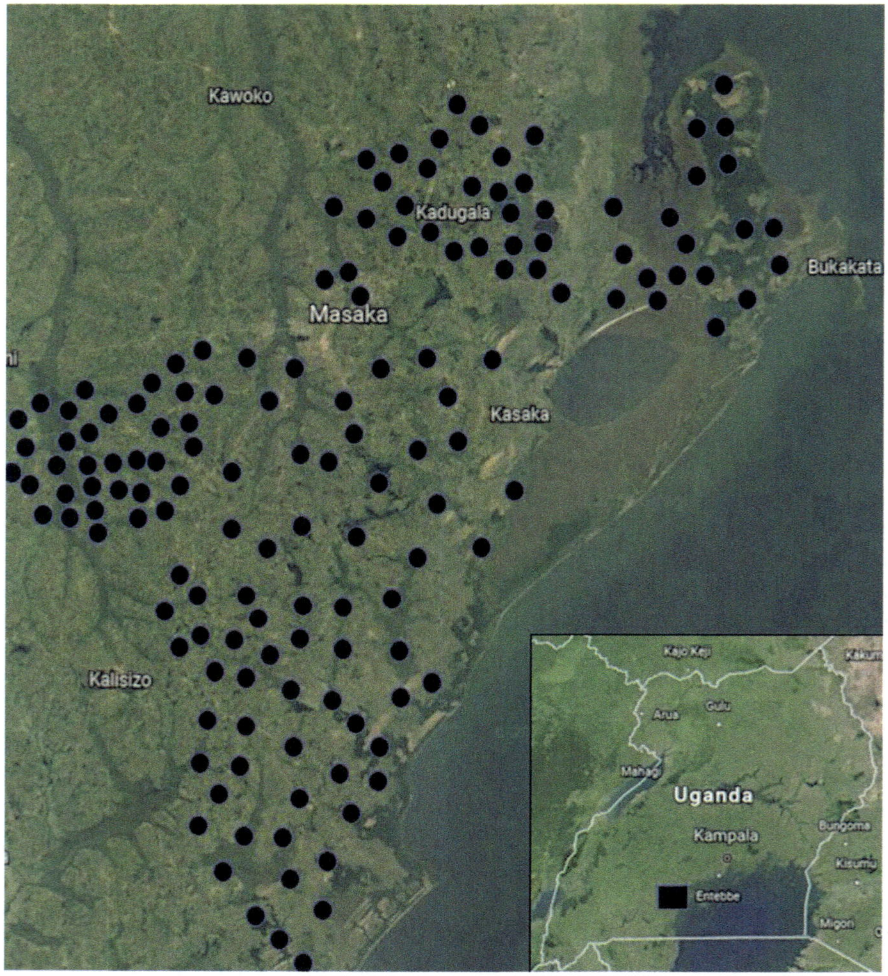

FIGURE 4.1

Map of Masaka District and sampled villages. Map of sampled villages in Masaka District, southcentral Uganda. Black dots indicate the location of each village sampled in Masaka District, southcentral Uganda.

After the interviews were completed, three farmers were selected to host a biochar demonstration site on their land. They were located in different villages and represented the average farmer in the area. Each demonstration plot was divided into two with biochar on one side and no biochar on the other side. The farmers helped the author prepare the land, incorporate biochar into the soil, and plant common beans (*Phaseolus vulgaris* L.; Fig. 4.2). All of this was done using the tools and practices that the farmers were familiar with. For example, one

FIGURE 4.2

Farmers incorporate biochar on demonstration plots in Mpungwe (A) and Lwanyi (B).

Two images of farmers demonstrating biochar application in (A) Mpugwe and (B) Lwanyi villages.

farmer decided to add chicken manure to his demonstration plot because this was how he had always grown his beans. The author procured biochar from Biochar Uganda Organics Ltd, a private company that produces and distributes its own proprietary blend of biochar. Biochar was outsourced because it was impossible to produce it on site in quantities sufficient for the land size of demonstration plots. Because the goal was not to test the agronomic efficacy of biochar but to visually demonstrate its potential to the farmers, neither the soil nor the biochar properties were analyzed. Germination rate and plant canopy size were the only measurements taken to assess whether the visual differences between beans grown with and without biochar were present. Visual efficacy of a new agricultural technology is important for the farmers and often impacts its adoption.

Additionally, the author demonstrated to the three farmers how to produce biochar at no cost using a simple top-lit updraft gasifier consisting of a repurposed oil barrel with a chimney (Fig. 4.3). The materials were sourced either from the farmer or within the village. Each carbonization system was put together in collaboration with the farmer, and each detail and step was carefully explained. The farmer then produced biochar and, after it was sufficiently chilled, applied it to the soil. All of this, in addition to demonstration plots, was intended to educate and encourage the farmers to continue using biochar as a soil amendment. The farmers were contacted again a year later to determine if they continued with producing and applying biochar on their land.

4.2.2 Study site

Masaka District is one of the most densely populated (245 people per square kilometer) in Uganda due to its favorable climate and proximity to Lake Victoria. Annual rainfall ranges from 1100 to 1200 mm during two rainy seasons, which allows for two growing seasons per year. Temperatures range between 10°C and

FIGURE 4.3

Farmers practicing biochar production using top-lit gasification system made from repurposed oil barrels.

People involved in biochar production using locally available carbonization technology.

30°C. Topography of the region is characterized by hills and ridges with an average altitude of 1150 m above the sea level (Masaka District, 2023). Masaka District has a total area of 1603 km^2, half of which is land, and the other half is wetlands. The dominant coffee–banana crop production system is supplemented with cassava (Manihot esculenta Crantz), sweet potatoes (*Ipomoea batatas* L. Lam), beans, groundnuts (*Arachis hypogaea* L.), and maize (Apanovich & Lenssen, 2018).

High population density and growth have led to agricultural land fragmentation, cropping intensification, deforestation, and decreased soil fertility. Despite the region's good climate and overall fertile soils, the quality of those soils has been deteriorating at a high rate. In Masaka District, soil degradation was estimated to affect 50% of the area (Rehm & Grashey-Jansen, 2016). The situation is exacerbated by insufficient soil conservation practices due to lack of knowledge, interest or incentives. This makes Masaka District susceptible to food insecurity and a good site to study biochar adoption. An estimated 73% of all households in the district depend on farming for their livelihoods.

4.3 Results

Most (85%) of the farmers reported not being aware of biochar. Those who reported being aware acquired their knowledge from friends, extension agents, or local radio (Fig. 4.4). No farmer had ever used biochar on their land. Almost all

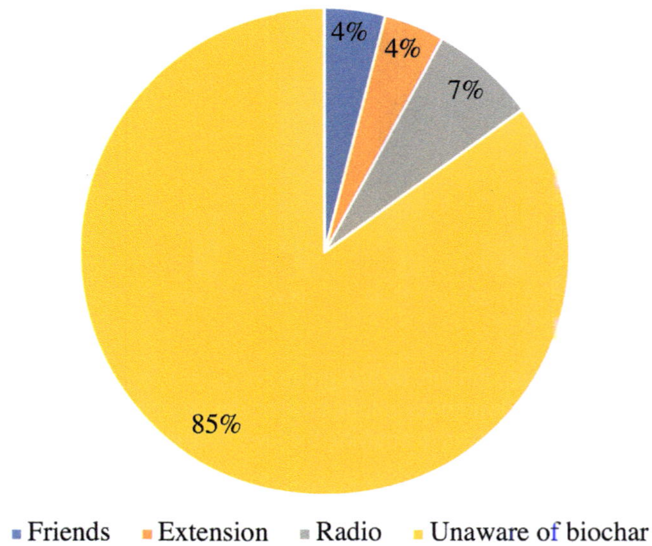

FIGURE 4.4

Sources of biochar awareness. Sources of biochar awareness among smallholder farming households. A graph with four segments showing the percentage distribution of sources of biochar awareness (friends, extension, radio, and unaware) among smallholder farming households.

farmers reported that they would be willing to learn about biochar production and application on their farm and host a demonstration plot. In general, the farmers expressed a level of confusion about the difference between biochar and ash. Many used ash from firewood as a soil amendment.

A Chi-square test showed that gender, education, and land size were not significantly associated with biochar awareness. Age, on the other hand, was significantly ($P > .001$) associated with biochar knowledge. The age group that had the greatest number of biochar aware farmers was 64 and older and the age group with the most biochar unaware farmers was 25–44 (Fig. 4.5).

After being informed on biochar, the majority (70%) perceived biochar to have a positive impact on the soil (Fig. 4.6). Only a small percentage (1%) of the farmers thought biochar would have no impact on the soil and almost 29% of the farmers remained uncertain about biochar's impacts.

The farmers reported a range of different feedstocks that they could use for biochar production on the farm. Their choices were based on what they were already wasting without value extraction. Most farmers reported that they could dedicate weeds, tree leaves, and some agricultural residues to biochar production. The farmers indicated leaves from mango, avocado, and jackfruit trees as potential feedstocks due to their thickness and slow decomposition. This type of biomass is normally swept away from the compound and burned without value

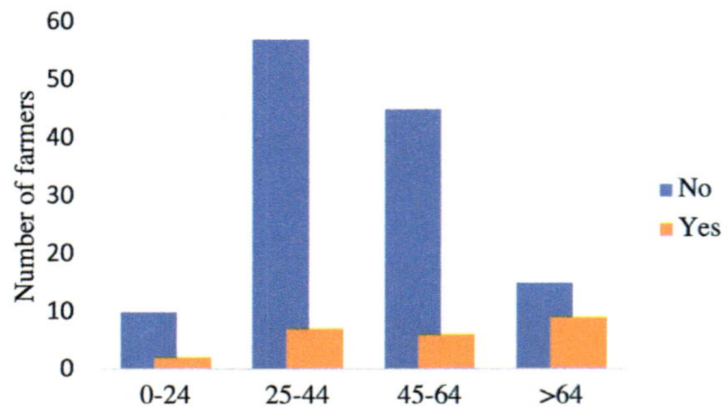

FIGURE 4.5

Distribution of biochar aware farmers by age groups $n_{0-24} = 1$, $n_{25-44} = 5$, $n_{45-64} = 7$, $n_{<64} = 9$. Level of biochar awareness among smallholder farming households. A graph with differently colored pastel bars quantifying the level of biochar awareness among smallholder farming households.

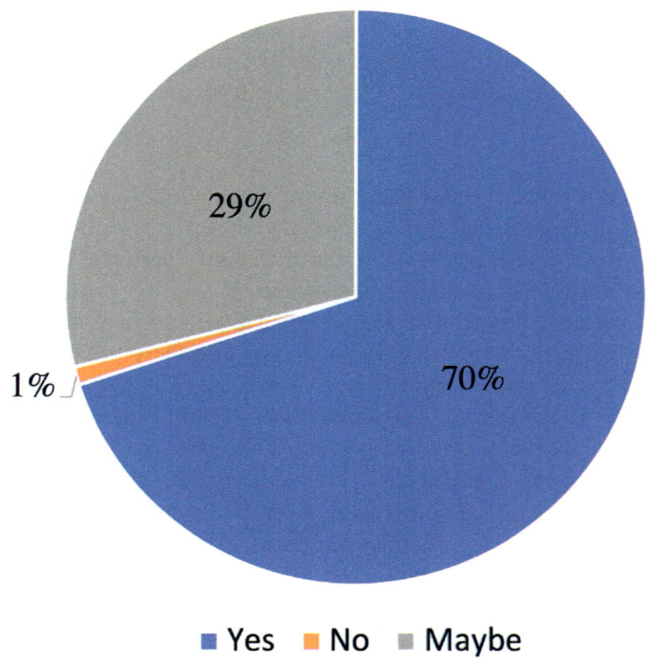

FIGURE 4.6

Farmers' perceptions of biochar's impact on soil. Farmers' perceptions of biochar's efficacy as a soil amendment. A pie chart showing the percentage distribution of farmers' perceptions of biochar efficacy as a soil amendment (yes, no, and maybe).

extraction. From agricultural residues, the farmers identified bean residues, maize stover, groundnut shells, banana leaves, and cassava and sweet potato peelings as potential biochar feedstocks. Some of these residues are also discarded without energy extraction. For example, farmers believed that groundnut residues were toxic to the soil and needed to be removed from the fields. They are either discarded or burned without value extraction. A small percentage proposed woody biomass, kitchen waste, and animal manure (Fig. 4.7). It is worth noting that because the farmers were not initially familiar with biochar, they hesitated to propose a feedstock. After being provided with sufficient information, the farmers proposed feedstocks for which they had no established uses.

The data from the demonstration plots indicate that there was a visible difference between biochar and control sides. Germination rate (Fig. 4.8) and plant canopy size (Fig. 4.9) at second and third weeks differed in all three locations indicating that the farmers were able to observe at least some biochar's impact on their crops. In Lwanyi, where the demonstration plot was conveniently located off the main road, the farmer reported receiving questions and comments from the village passersby about the rigor of his beans. The author hoped that such visual interaction with biochar would encourage the farmers to continue using it.

When contacted a year later, neither farmer still practiced biochar production or application. All farmers stopped the practice due to a loss of land access. The female farmer lost access to the land after her husband separated from her. The second farmer was evicted from the land by his family, who had legal rights to it, and the third was evicted by his landlord for being unable to pay rent.

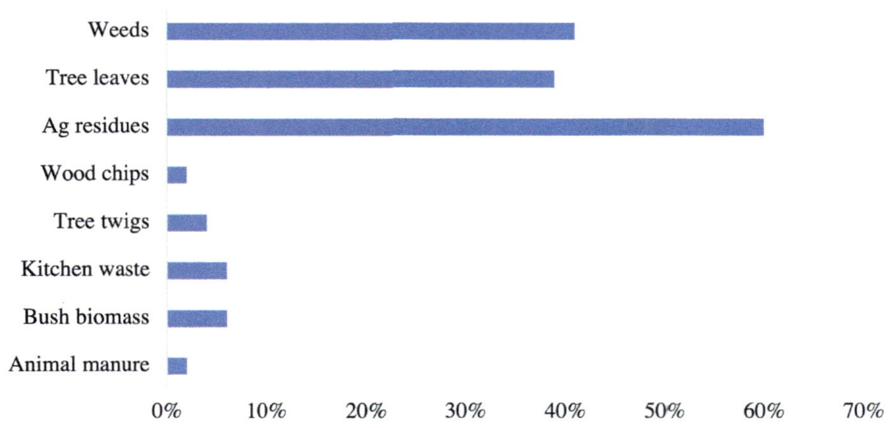

FIGURE 4.7

The farmer recommended feedstocks for at-home biochar production. Feedstock types available for biochar production as reported by the farmers. A bar graph showing distribution of feedstock types available for biochar production as reported by the farmers.

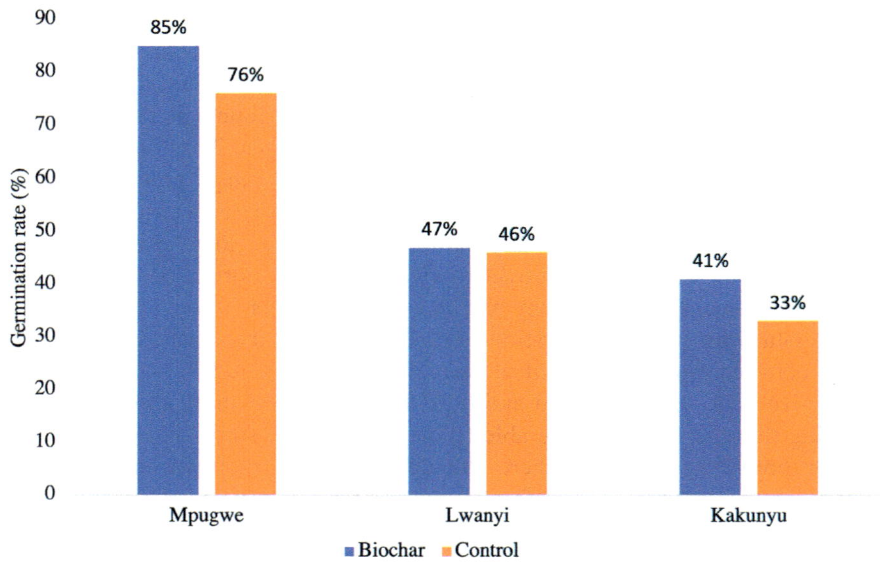

FIGURE 4.8

Germination rate (%) during the second week of growth at three demonstration plots. Bean germination rate on biochar and control fields in Mpugwe, Lwanyi, and Kakunyu villages. A bar graph showing bean germination rate on biochar and control fields in Mpugwe, Lwanyi, and Kakunyu villages.

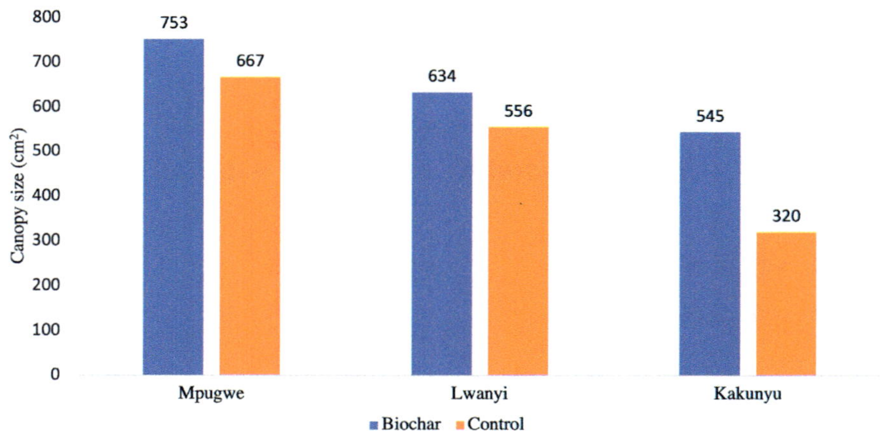

FIGURE 4.9

Plant canopy size (cm^2) during the third week of growth at three demonstration plots. Bean canopy size on biochar and control fields in Mpugwe, Lwanyi, and Kakunyu villages. A bar graph showing bean canopy size on biochar and control fields in Mpugwe, Lwanyi, and Kakunyu villages.

4.4 Discussion

The study discovered a low level of biochar awareness among smallholder farmers. This can be explained by poor agricultural extension services, low levels of education, and potentially other more farmer- and village-specific factors. Other studies confirm that more contact with extension officers and other sources of information are associated with higher probabilities of technology adoption (Ayenew et al., 2020; Kibue, 2018). A study in Ghana reported that farmers' access to extension services increased the likelihood of technology adoption by 12.3% (Adams et al., 2021). In Nepal, adoption decisions were mainly affected by extension-related variables—training, membership in a farmers' group, and off-farm employment (Suvedi et al., 2017). It is no surprise then that those who were aware of biochar learned about it from extension services, radio, and friends. This highlights the significance of social capital in new technology adoption among smallholder farmers. As such, initiatives needs to be taken to strengthen community associations and networks. Such novel approaches as farmer-to-farmer extension have already been proven to increase technology adoption, agricultural production, food availability, and create the multiplier effect in information sharing. In Uganda, where extension services are partially hindered by poor infrastructure which prevents agents from visiting isolated villages, having a farmer-to-farmer approach can be a valuable way to spread much needed agricultural information.

It was also observed that older farmers were more aware of biochar than the younger farmers. It can be explained by the fact that older farmers had more time and opportunities to forge networks and relationships that eventually led them to biochar knowledge. This assumption is supported by the fact that the age group with the greatest number of biochar unaware farmers was in the range of 25–44. Even though the younger farmers are more educated, it is the older ones that have a greater access to information. Other studies found that older farmers are the owners of key assets such as land, labor, and finances, which allows them to experiment with and invest in their resources, which is different compared with younger age groups (Rogers et al., 2022). Younger people are often discouraged from farming as they lack access to capital, which contributes to rural exodus to urban areas. This implies that the future of biochar may be threatened if young farmers are not supported in their access to resources and information.

Even though the farmers in this study were vastly unaware of biochar and its uses, they almost unanimously expressed an interest in learning about and even producing biochar at home. The farmers expressed their willingness for biochar adoption by identifying various feedstocks and converting them into biochar using cost-free homemade gasification systems. The farmers witnessed the benefits of biochar from their own demonstration plots. However, long-term biochar use was still unsuccessful. It was not the technical or even financial barriers that prevented the farmers from biochar adoption. It was land access issues that hindered a long-term biochar use. This serves contrary to what has been reported in the literature.

Most studies on biochar adoption in an African context cite, among other factors, feedstock availability and labor considerations as the biggest constrains (Karim et al., 2020; Rogers et al., 2022; Steiner et al., 2018; Sutradhar et al., 2021). This observation is consistent with the two previous ones that support the idea that older farmers tend to have more resources, including land, which creates a sense of stability and promotes experimental mindset. The three farmers who hosted demonstration plots were under the age of 40 with one having a university-level education. They all were motivated to implement a technology that had a potential to improve their livelihoods. This highlights the fact that limited land access in Uganda is a serious issue that cuts through different demographic and socioeconomic groups, and that it must be addressed for biochar to be widely adopted.

Land access issues in Uganda are complicated and common. With the fast-growing population that leads to land fragmentation and a colonial legacy that created a complex land tenure system, many Ugandans are impacted by land insecurity. Such a system leads to land conflicts, which have a negative impact on crop yields. For example, plots with conflicts have 20% lower yields than those without conflicts. Plots with eviction conflicts have 37% lower yield than their counterparts without conflicts (Mwesigye & Matsumoto, 2016). Land insecurity was also found to negatively impact the adoption of climate-smart technologies among smallholder farmers (Chavula & Turyasingura, 2022). A study conducted in Burkina Faso found that formal land rights lead to more technology adoption (Zahonogo & Séogo, 2019). A study in Benin established that the intensity of the adoption of agri-environmental practices was consistently higher on owned plots than borrowed, rented or sharecropped plots (Lawin & Tamini, 2019). A study in India established that land tenure enhances farm investment in the form of modern rice technology adoption (Paltasingh, 2018). A study in Ethiopia found that improvements in the land rental market could substantially reduce the gap between land demand and supply and improve household welfare through improved crop variety adoption (Zeng et al., 2018). A study in Vietnam established that in addition to improved seed adoption, land titles also encouraged households to take loans to expand agricultural production (Nguyen, 2020). It becomes evident that land security serves not only as an important prerequisite for a new technology or practice adoption but also functions as a gateway to financial capital.

Additionally, the misinterpretation of traditional law and a lack of enforcement of contemporary legal rights leaves many women landless (Naybor, 2014). With women in Uganda providing over 70% of labor for agriculture but owning only 7% of all agricultural land, creates limitations on what women can and cannot do on the land (Acidri, 2014). Secure land tenure empowers women to actively engage in implementing biochar on their own plots. Moreover, women with secure land tenure are more likely to invest time and effort into long term, regenerative practices like biochar application, as they have a vested interest in the health and productivity of their land. While many households report husbands and wives as joint owners of the land, women are less likely to be listed on ownership documents and may lose these claims if their marriage dissolves (Doss et al., 2014). This was exactly what happened in this study when one of the farmers' husband separated from her, effectively blocking her access to land. By allowing women to access land, biochar adoption can enhance soil fertility and water retention, thus alleviating some of the burdens traditionally placed on women in agriculture.

Currently, no research has been conducted on the link between women's land ownership and biochar adoption. More insight is necessary on this link if biochar as a soil amendment is to ever have a chance to be adopted in the Global South where women are the main stewards of land. The following recommendations have been proposed (Chavula & Turyasingura, 2022) to address women's land ownership issues: (1) support women's land rights upon divorce or death of husband, fathers, or brothers; (2) support initiatives that allow farmers to register and file land rights in a cost-effective, participatory manner; (3) advance efforts to proactively register and record the land rights of women and other vulnerable groups; and (4) allow for greater flexibility in the transfer of rights including those governing temporary (i.e., rental) and permanent (i.e., sale) transfers to increase opportunities for landholders to capture the full value of an investment.

The study also showed that biochar production on a household scale using locally available feedstocks was feasible. This points to a potential for a circular economy. By using what is otherwise considered waste creates an opportunity to generate value-added products in a sustainable and circular way. Unlike typical linear economy, in which resources are created, used, and then disposed eventually, a circular economy decouples growth from resource consumption by making the most of the limited natural resources (Das & Ghosh, 2023). The circular economy paradigm keeps resources in use as long as possible, minimizes waste and creates added value with technological benefits. By using efficient processes, waste prevention, reuse, and recycling, a circular economy practice reduces resources consumption by focusing on materials cycle after their consumption (Das & Ghosh, 2023). Additionally, the technology that the farmers used to produce biochar in this study was low-cost, simple to use, and could be easily put together by the farmer. By having full ownership of their homemade gasification systems, farmers can decide how and when to use it, eliminating the pressure of continued use or maintenance issues. This also allows the farmer to have control over what feedstocks to use and how much. With crop yields becoming increasingly unpredictable due to climate change, flexibility in feedstock choice can be an important deciding factor in biochar adoption. Finally, such small-scale, low-cost, homemade gasification systems can eliminate the need for costly and harmful to human health and environment fertilizers. Overall, biochar production and use can contribute to circular economy becoming a viable pathway to sustainable development in Uganda and potentially other countries with similar socioeconomic and agricultural production characteristics.

The feedstocks that the farmers reported could be available for biochar production are currently treated as nuisance even though generated in small amounts. For examples, the leaves from mango, avocado, and jackfruit trees are burned without value extraction due to their reported thickness and slow decomposition. Normally, these trees grow around farmers' houses and shed leaves on the compound. According to local custom, the compound must be clean and free of any organic material. To make this happen, every morning farmers sweep any fallen leaves into a pile and burn them. The farmers also reported burning groundnut shells, bean and soy residues, and weeds. While these residues are generally burned, many others have established uses. For example, many peelings are fed to livestock, coffee husks are used for animal bedding, and maize cobs are used as cooking fuel. It is imperative, therefore, not to divert these residues to biochar production as doing so can create additional problems for the farmers. By converting only the residues that are wasted, a waste product is transformed to a value-added product that can benefit the farmer not only socioeconomically through improved yields and food security but also environmentally by improving soil quality, promoting biomass growth, and storing carbon. The fact that biochar carries a cumulative effect can help farmers rip long-term benefits even from small amounts of biochar. In short, biochar's integration into smallholder systems exemplifies the circular economy's principles by transforming waste into a resource, enhancing resource efficiency, and contributing to the regeneration of natural systems. Its applications in soil health align with the goal of creating a more sustainable and regenerative economic model.

While it is apparent that Ugandan farmers are curious about and willing to adopt biochar as a soil amendment, the abovementioned land tenure insecurity issues combined with women's land ownership challenges must be addressed first before biochar adoption efforts are put forth. The government needs to acknowledge the role of biochar in sustainable development and create conditions necessary for its long-term adoption. Because the farmers reported learning about biochar from friends, extension agents, and radio, these channels of communication should be incorporated in educating rural areas about the uses and benefits of biochar as a soil amendment. Specifically, the government should (1) invest in its extension services, (2) provide a greater access to land and other forms of capital to young farmers, (3) address land tenure issues, especially among women, and (4) increase social capital by encouraging farmers form groups and organizations designed for information sharing. Only by addressing these issues can biochar have a fair chance at getting adopted and the farmers contributing to circular economy. With climate change wreaking havoc in the Global South, it seems that only carbon neutral solutions should be amplified and supported, especially in such an agrarian country like Uganda where most of the population depends on farming for survival.

4.5 Study limitations

Even though this study sheds new light on biochar adoption challenges and opportunities in Uganda, there are several limitations that need to be acknowledged. First, because only three farmers were targeted for long-term biochar adoption, their failure to do so cannot be representative of other farmers. A more representative sample would have provided a much more robust picture on biochar adoption challenges. Second, the study assumed that it would be possible to produce agronomic quality biochar from the farmer reported feedstocks such as weeds, tree leaves, and agricultural residues using a simple top-lit gasification technology. Because no analysis was performed on any of the produced biochar, the agronomic and environmental benefits of it remain speculative. Third, no data were collected on farmers' perceptions on biochar during the time the farmers hosted demonstrations on their land. Interviewing farmers during that time would have allowed for a more detailed understanding of their experiences with biochar. Fourth, because no soil analysis was performed before and after biochar use, it is impossible to determine its impact on the soil. Perhaps such analysis could have served as additional encouragement for the farmers to consider biochar for a long-term agronomic use. Finally, because of biochar's novel nature in southcentral Uganda, focus group discussions might have solicited more nuanced information about farmers' biochar perceptions and attitudes than structured interviews.

4.6 Conclusions

This study looked at biochar awareness and adoption in Masaka District, southcentral Uganda through 150 farmer interviews and three demonstration plots. Even though most of the farmers expressed an interest in learning about and producing their own biochar, the three farmers who were selected to host a demonstration plot failed to adopt it long-term due to land tenure insecurity. It was also established that older farmers were more aware about biochar than younger farmers. The study found that gender, education, and land size were not associated with biochar awareness. Knowing this, it is recommended that the government addresses the issue of land tenure and encourages the production and use of biochar on a household scale through educational programs and campaigns that rely on extension services, radio, and social capital. Future work should focus on the role of women's land ownership in biochar adoption. Additionally, a more gendered approach is needed to understand the role of social capital in uptake of sustainable agricultural practices. Overall, biochar has a potential to contribute to circular economy if produced on a small scale and with the feedstocks that do not have established uses. Biochar adoption even on a small scale has a potential to have a transformative impact on rural livelihoods and economies. As such, it is fundamental that the national government takes measures to incentivize a small-scale conversion of wasted organic materials into biochar.

Acknowledgment

The author would like to acknowledge all Ugandan farmers who participated in the study and Debora Katasi for interpreting farmer interviews.

Author contribution

The author conceptualized the study, collected and analyzed the data, and wrote the paper.

Conflict of interest

The author claims no conflict of interest.

Statement on the use of AI tools for writing

No AI tools were used to write this paper.

References

Acidri, D. (2014). Women's rights to land ownership in Uganda: Policy and practice. *Critical Social Thinking*, *6*, 184–203.

Adams, A., Jumpah, E. T., & Caesar, L. D. (2021). The nexuses between technology adoption and socioeconomic changes among farmers in Ghana. *Technological Forecasting and Social Change*, *173*. Available from https://doi.org/10.1016/j.techfore.2021.121133, https://www.journals.elsevier.com/technological-forecasting-and-social-change.

Adebiyi, J. A., & Olabisi, L. S. (2022). Participatory causal loop mapping of the adoption of organic farming in Nigeria. *Environmental Management*, *69*(2), 410–428. Available from https://doi.org/10.1007/s00267-021-01580-w, http://link.springer.de/link/service/journals/00267/index.htm.

Akolgo, G. A., Essandoh, E. O., Gyamfi, S., Atta-Darkwa, T., Kumi, E. N., & Maia, C. M. Bd. F. (2018). The potential of a dual purpose improved cookstove for low income earners in Ghana – improved cooking methods and biochar production. *Renewable and Sustainable Energy Reviews*, *82*, 369–379. Available from https://doi.org/10.1016/j.rser.2017.09.044, https://www.journals.elsevier.com/renewable-and-sustainable-energy-reviews.

Apanovich, N., & Lenssen, A. (2018). Cropping systems and soil quality and fertility in south-central Uganda. *African Journal of Agricultural Research*, *13*(15), 792–802.

Ayenew, W., Lakew, T., Kristos, E. H., Ayenew, W., Lakew, T., & Kristos, E. H. (2020). Agricultural technology adoption and its impact on smallholder farmers welfare in Ethiopia. *African Journal of Agricultural Research*, *15*(3), 431–445. (2020).

Branca, G., Cacchiarelli, L., Haug, R., & Sorrentino, A. (2022). Promoting sustainable change of smallholders' agriculture in Africa: Policy and institutional implications from a socio-economic cross-country comparative analysis. *Journal of Cleaner Production*, *358*, 131949. Available from https://doi.org/10.1016/j.jclepro.2022.131949.

Chavula, P., & Turyasingura, B. (2022). Land tenurial system influence among smallholder farmers' climate smart agriculture technologies adoption, Sub-Sahara Africa: A review paper. *International Journal of Food Science and Agriculture*, *6*(1), 8–16. Available from https://doi.org/10.26855/ijfsa.2022.03.003.

Das, A., & Ghosh, A. (2023). Vision net zero: A review of decarbonisation strategies to minimise climate risks of developing countries. *Environment, Development and Sustainability*. Available from https://doi.org/10.1007/s10668-023-03318-6, https://www.springer.com/journal/10668.

Descheemaeker, K., Oosting, S. J., Homann-Kee Tui, S., Masikati, P., Falconnier, G. N., & Giller, K. E. (2016). Climate change adaptation and mitigation in smallholder crop–livestock systems in sub-Saharan Africa: A call for integrated impact assessments. *Regional Environmental Change*, *16*(8), 2331–2343. Available from https://doi.org/10.1007/s10113-016-0957-8, http://springerlink.metapress.com/app/home/journal.asp?wasp = 64cr5a4mwldrxj984xaw&referrer = parent&backto = browsepublicationsresults,451,542;.

Didarali, Z., & Gambiza, J. (2019). Permaculture: Challenges and benefits in improving rural livelihoods in South Africa and Zimbabwe. *Sustainability*, *11*(8), 2219. Available from https://doi.org/10.3390/su11082219.

Doss, C., Meinzen-Dick, R., & Bomuhangi, A. (2014). Who owns the land? Perspectives from rural Ugandans and implications for large-scale land acquisitions. *Feminist Economics*, *20*(1), 76–100. Available from https://doi.org/10.1080/13545701.2013.855320, http://www.tandf.co.uk/journals/titles/13545701.asp.

Eltigani, A., Olsson, A., Krause, A., Ernest, B., Fridahl, M., Yanda, P., & Hansson, A. (2022). Exploring lessons from five years of biochar-producing cookstoves in the Kagera region, Tanzania. *Energy for Sustainable Development*, *71*, 141–150. Available from https://doi.org/10.1016/j.esd.2022.09.015, http://www.elsevier.com.

Fridahl, M., Haikola, S., Rogers, P. M., & Hansson, A. (2021). Biochar deployment drivers and barriers in least developed countries. *Handbook of climate change management: Research, leadership, transformation*, *1*, 119–148. Available from https://doi.org/10.1007/978-3-030-57281-5_324, https://link.springer.com/book/10.1007/978-3-030-57281-5.

Gitau, K. J., Mutune, J., Sundberg, C., Mendum, R., & Njenga, M. (2019). Factors influencing the adoption of biochar-producing gasifier cookstoves by households in rural Kenya. *Energy for Sustainable Development*, *52*, 63–71. Available from https://doi.org/10.1016/j.esd.2019.07.006, http://www.elsevier.com.

Gwenzi, W., Chaukura, N., Mukome, F. N. D., Machado, S., & Nyamasoka, B. (2015). Biochar production and applications in sub-Saharan Africa: Opportunities, constraints, risks and uncertainties. *Journal of Environmental Management*, *150*, 250–261. Available from https://doi.org/10.1016/j.jenvman.2014.11.027, https://www.sciencedirect.com/journal/journal-of-environmental-management.

Heinrich, T., Park, H., Orozco, R., Ding, Z., Álvarez-López, V., Mosquera-Losada, M. R., Steinbeis, L., & Hoffmann, T. (2023). Biochar production from late-harvest grass – challenges and potential for farm-scale implementation. *Sustainable Production and Consumption*, *37*, 256–267. Available from https://doi.org/10.1016/j.spc.2023.02.019, http://www.journals.elsevier.com/sustainable-production-and-consumption/.

Ikehi, M. E., Ejiofor, T. E., Ifeanyieze, F. O., Nwachukwu, C. U., & Ali, C. C. (2022). Adoption of agricultural innovations by farmers in Enugu State, Nigeria. *International Journal of Agricultural Technology*, *18*(1), 123–140. Available from http://www.ijat-aatsea.com/pdf/v18_n1_2022_January/9_IJAT_18(1)_2022_Ikehi,%20M.%20E.pdf.

Jones, K., Glenna, L. L., & Weltzien, E. (2014). Assessing participatory processes and outcomes in agricultural research for development from participants' perspectives. *Journal of Rural Studies*, *35*, 91–100. Available from https://doi.org/10.1016/j.jrurstud.2014.04.010, http://www.elsevier.com/inca/publications/store/3/4/8/.

Kapoor, R. T., Treichel, H., & Shah, M. P. (2022). *Biochar and its application in bioremediation. Biochar and its application in bioremediation* (pp. 1–481). Springer Nature. Available from https://link.springer.com/book/10.1007/978-981-16-4059-9, https://doi.org/10.1007/978-981-16-4059-9.

Karim, M. R., Fardaus, T., Khan, M. R., Noman, M. R. F., & Huda, S. (2020). Knowledge of tribal women on biochar promotion for homestead gardening. *Journal of Science and Technology* (25 ().

Kibue, G. W. (2018). Use of biochar for increased crop yields and reduced climate change impacts from agricultural ecosystems: Chinese farmers perception and adoption strategy. *African Journal of Agricultural Research, 13*(21).

Kurniawan, T. A., Othman, M. H. D., Liang, X., Goh, H. H., Gikas, P., Chong, K. K., & Chew, K. W. (2023). Challenges and opportunities for biochar to promote circular economy and carbon neutrality. *Journal of Environmental Management*, *332*. Available from https://doi.org/10.1016/j.jenvman.2023.117429, https://www.sciencedirect.com/journal/journal-of-environmental-management.

Latawiec, A., Królczyk, J., Kuboń, M., Szwedziak, K., Drosik, A., Polańczyk, E., Grotkiewicz, K., & Strassburg, B. (2017). Willingness to adopt biochar in agriculture: The producer's perspective. *Sustainability*, *9*(4), 655. Available from https://doi.org/10.3390/su9040655.

Lawin, K. G., & Tamini, L. D. (2019). Land tenure differences and adoption of agri-environmental practices: Evidence from Benin. *Journal of Development Studies*, *55*(2), 177–190. Available from https://doi.org/10.1080/00220388.2018.1443210, http://www.tandf.co.uk/journals/titles/00220388.asp.

Masaka District. (2023). *Higher Local Government Statistical Abstract. Masaka District local government*.

Mutsvangwa-Sammie, E. P. (2020). Impact narratives of agricultural innovations and their implication on rural livelihoods in south west Zimbabwe. *Outlook on Agriculture*, *49*(4), 330–339. Available from https://doi.org/10.1177/0030727020953534, https://journals.sagepub.com/home/OAG.

Mwesigye, F., & Matsumoto, T. (2016). The effect of population pressure and internal migration on land conflicts: Implications for agricultural productivity in Uganda. *World Development*, *79*, 25–39. Available from https://doi.org/10.1016/j.worlddev.2015.10.042, http://www.journals.elsevier.com/world-development/.

Naybor, D. (2014). Land as fictitious commodity: The continuing evolution of women's land rights in Uganda. *Gender, Place & Culture*, *22*(6), 884–900. Available from https://doi.org/10.1080/0966369x.2014.917275.

Nguyen, L. (2020). Land rights and technology adoption: Improved rice varieties in Vietnam. *The Journal of Development Studies*, *56*(8), 1489–1507. Available from https://doi.org/10.1080/00220388.2019.1677889.

Okure, M.A., Musinguzi, W., Wabwire, A., & Bagenda, S. (2014). Experience from rural electrification in Uganda: A case study of a husk powered system in Tiribogo Village. In *9th Sida Reg. conference*, Entebbe, Uganda.

Otte, P. P., & Vik, J. (2017). Biochar systems: Developing a socio-technical system framework for biochar production in Norway. *Technology in Society*, *51*, 34–45. Available from https://doi.org/10.1016/j.techsoc.2017.07.004, http://www.elsevier.com/inca/publications/store/3/8/4/.

Paltasingh, K. R. (2018). Land tenure security and adoption of modern rice technology in Odisha, Eastern India: Revisiting Besley's hypothesis. *Land Use Policy*, *78*, 236–244. Available from https://doi.org/10.1016/j.landusepol.2018.06.031, http://www.elsevier.com/inca/publications/store/3/0/4/5/1/.

Pourhashem, G., Hung, S. Y., Medlock, K. B., & Masiello, C. A. (2019). Policy support for biochar: Review and recommendations. *GCB Bioenergy*, *11*(2), 364–380. Available from https://doi.org/10.1111/gcbb.12582, http://onlinelibrary.wiley.com/journal/10.1111/(ISSN)1757-1707.

Rambim, D.A. (2022). *A participatory model for an enhanced adoption of ICT innovations by smallholder farmers* (Doctoral dissertation). JOOUST.

Rehm, R. P. M., & Grashey-Jansen, S. (2016). Soil sequence-studies on the tropical Buganda-catena. *African Journal of Soil Science*, *4*(1), 295–304.

Rogers, P. M., Fridahl, M., Yanda, P., Hansson, A., Pauline, N., & Haikola, S. (2022). Socio-economic determinants for biochar deployment in the southern highlands of Tanzania. *Energies*, *15*(1). Available from https://doi.org/10.3390/en15010144, https://www.mdpi.com/1996-1073/15/1/144/pdf.

Shammi, M., Winter, J., Islam, M. M., Akter, B., & Hasan, N. (2023). *Cookstoves for biochar production and carbon capture. Green sustainable process for chemical and environmental engineering and science: Carbon dioxide capture and utilization* (pp. 53–68). Elsevier. Available from https://www.sciencedirect.com/book/9780323994293, https://doi.org/10.1016/B978-0-323-99429-3.00025-4.

Steiner, C., Bellwood-Howard, I., Häring, V., Tonkudor, K., Addai, F., Atiah, K., Abubakari, A. H., Kranjac-Berisavljevic, G., Marschner, B., & Buerkert, A. (2018). Participatory trials of on-farm biochar production and use in Tamale, Ghana. *Agronomy for sustainable development*, *38*, 1–10.

Sundberg, C., Karltun, E., Gitau, J. K., Kätterer, T., Kimutai, G. M., Mahmoud, Y., Njenga, M., Nyberg, G., Roing de Nowina, K., Roobroeck, D., & Sieber, P. (2020). Biochar from cookstoves reduces greenhouse gas emissions from smallholder farms in Africa. *Mitigation and Adaptation Strategies for Global Change*, *25*(6), 953–967. Available from https://doi.org/10.1007/s11027-020-09920-7, http://www.wkap.nl/journalhome.htm/1381-2386.

Sutradhar, I., Jackson-deGraffenried, M., Akter, S., McMahon, S. A., Waid, J. L., Schmidt, H. P., Wendt, A. S., & Gabrysch, S. (2021). Introducing urine-enriched biochar-based fertilizer for vegetable production: Acceptability and results from rural Bangladesh. *Environment, Development and Sustainability*, *23*(9), 12954–12975. Available from https://doi.org/10.1007/s10668-020-01194-y, http://www.wkap.nl/journalhome.htm/1387-585X.

Suvedi, M., Ghimire, R., & Kaplowitz, M. (2017). Farmers' participation in extension programs and technology adoption in rural Nepal: A logistic regression analysis. *Journal of Agricultural Education and Extension*, *23*(4), 351–371. Available from https://doi.org/10.1080/1389224X.2017.1323653, http://www.tandfonline.com/toc/raee20/current.

Zahonogo, P., & Séogo, W. (2019). Land tenure system innovation and agricultural technology adoption in Burkina Faso: Comparing empirical evidence to the worsening situation of both rural people vulnerability and vulnerable groups' access to land. *African Journal of Science, Technology, Innovation and Development*, *11*(7), 833–842.

Zakaria, A., Alhassan, S. I., Kuwornu, J. K. M., Azumah, S. B., & Derkyi, M. A. A. (2020). Factors influencing the adoption of climate-smart agricultural technologies among rice farmers in Northern Ghana. *Earth Systems and Environment*, *4*(1), 257–271. Available from https://doi.org/10.1007/s41748-020-00146-w, https://link.springer.com/journal/41748.

Zeng, D., Alwang, J., Norton, G., Jaleta, M., Shiferaw, B., & Yirga, C. (2018). Land ownership and technology adoption revisited: Improved maize varieties in Ethiopia. *Land Use Policy*, *72*, 270–279. Available from https://doi.org/10.1016/j.landusepol.2017.12.047, http://www.elsevier.com/inca/publications/store/3/0/4/5/1/.

Zilberman, D., Laird, D., Rainey, C., Song, J., & Kahn, G. (2023). Biochar supply-chain and challenges to commercialization. *GCB Bioenergy*, *15*(1), 7–23. Available from https://doi.org/10.1111/gcbb.12952, http://onlinelibrary.wiley.com/journal/10.1111/(ISSN)1757-1707.

PART

2

Biochar production: techniques & process

CHAPTER 5

Transforming contaminated biomass from phytoremediation into biochar

Deivisson Ferreira da Silva[1,2], Olívia Bibiana Souza Dias[2], Alisson de Carli Souza[2], Amir Parnian[3], Lucélia Borgo[2], Leônidas Carrijo Azevedo Melo[2] and Flávio Henrique Silveira Rabêlo[4]

[1]*Federal Institute Catarinense, Araquari, Santa Catarina, Brazil*

[2]*Department of Soil Science, Federal University of Lavras, Lavras, Minas Gerais, Brazil*

[3]*National Salinity Research Center (NSRC), Agricultural Research, Education and Extension Organization (AREEO), Yazd, Yazd Province, Iran*

[4]*Luiz de Queiroz College of Agriculture, University of São Paulo, Piracicaba, São Paulo, Brazil*

5.1 Introduction

The world faces global warming and environmental contamination, largely due to anthropogenic activities such as the emission of greenhouse gases from the burning of fossil fuels and the disposal of industrial and urban waste, and the use of low-quality fertilizers. Other sources of environmental pollution include metals such as cadmium (Cd), nickel (Ni), zinc (Zn), and lead (Pb). However, depending on metal concentrations, one can use specific plants to cleaning-up the contaminated sites. Phytoremediation is an environmentally friendly technique that involves utilizing plants to remove contaminants from the soil while also enhancing carbon. Through phytoremediation, contaminated biomass containing metals from the soil is produced, necessitating specific disposal methods to properly handle the contaminated material. When it comes to contaminated biomass, proper disposal in a suitable location is necessary to prevent secondary pollution. Some methods used include incineration, compaction, leaching, and pyrolysis, among others. Pyrolysis is, from an environmental standpoint, an efficient method because it produces biochar, a beneficial product for the environment due to its carbon stability and other properties that improve soil health (Liu & Tran, 2021).

Several methods have been utilized for biochar production, including slow pyrolysis, fast pyrolysis, microwave-assisted pyrolysis, torrefaction, and hydrothermal carbonization (HTC; Amalina, Razak, Krishnan, Zularisam, & Nasrullah, 2022). The production process occurs in controlled environments and temperatures, with each method employing different levels of temperature and holding

Biochar Ecotechnology for Sustainable Agriculture and Environment. DOI: https://doi.org/10.1016/B978-0-443-29855-4.00006-0

time, resulting in a product more or less concentrated in nutrients and contaminants (Amalina, Razak, Krishnan, Sulaiman, et al., 2022). From this technology, biochar can be tailored for agronomic, industrial, and environmental purposes, serving mainly as a soil conditioner.

The level of contaminants in biochar must be lower than defined thresholds to avoid contamination risks. Pyrolysis of contaminated biomass generally concentrate the contaminants (e.g., potentially toxic elements [PTEs]), causing their concentration to be higher than the original feedstock, but their availability decreases especially with increasing pyrolysis temperature (Han et al., 2022). This limits the use of marginal biomasses, such as metal-rich sewage sludge, biomass from phytoremediation areas, demolition waste, or metal-treated wood for biochar production.

The mixture of biomass with varying levels of PTEs results in a process known as copyrolysis, which is an effective method for both waste management and the production of multifunctional biochar-based products with low levels of PTEs. Different biomass residues used in the copyrolysis process can have various positive synergistic effects, such as low energy consumption with a high calorific value of the synthesis gas, improved quality of pyrolysis products, and relatively low emissions of environmental pollutants (Li et al., 2021). Furthermore, copyrolysis has been found to enhance the economic and environmental value of biochar, including yield improvement, reduction of PTEs, and physicochemical properties. Copyrolysis of agricultural residues with manure or sludge can help immobilize pollutants and recover nutrients for safe application in agricultural soils (Kończak & Oleszczuk, 2020).

Another strategy to reduce PTEs in biochar is through temperature control during pyrolysis. High temperatures cause low-boiling-point PTEs to evaporate, potentially reducing their concentration in biochar derived from contaminated biomass (Yang et al., 2021). However, this approach requires the use of metallic filters in production systems to capture these metals. However, the production of biochar from contaminated biomass might be a green solution for proper disposal of these materials, but remains a topic in need of further study to refine existing methods, ensuring a product that can be safely used.

5.2 What to do with the contaminated biomass?

Phytoremediation is a technique in which soil contaminants are absorbed by plants, thereby reducing the environmental risks posed by these undesirable elements. Improper disposal of this biomass can create another environmental problem, causing secondary pollution, as the contaminants are not degraded by soil microorganisms. The main disposal options for contaminated biomass harvested during phytoremediation include composting, compaction, direct disposal, incineration, leaching, HTC, and pyrolysis (Vocciante et al., 2019). Some of these techniques are more efficient in environmental protection. In this section, we will discuss these options, highlighting their positive and negative aspects related to environmental threat mitigation.

5.2.1 Direct disposal

Among the techniques discussed in this section, direct disposal is one of the options that causes the most environmental damage due to its high levels of contaminants and slow reduction of biomass. If biomass is directly disposed in landfills, there is an additional emission of greenhouse gases of approximately 6 kg CO_2−equiv./m^2 to the operational value of the recovery, representing about 25% of the total impact on treatment cost (Vocciante et al., 2019). The main issue is the leaching of contaminants from biomass. Several leaching and solubility tests can be applied to assess and control the threat associated with direct disposal (Çoruh et al., 2013). Therefore direct disposal is prohibited due to associated environmental problems (Kovacs & Szemmelveisz, 2017; Šyc et al., 2012).

5.2.2 Composting

Composting with contaminated biomass has been used solely for volume reduction. The final compost may contain high levels of PTEs, preventing its use in agriculture. To produce compost for safe use, stabilization and inertization techniques have been tested, such as mixing with other materials or adding lime to reduce metal leaching (Kovacs & Szemmelveisz, 2017).

5.2.3 Compaction and leaching

The compaction process, also known as pelletization, involves a complex interaction among particles, where the compacted product is produced under pressure and leachate is collected separately. It is crucial to understand the basic compaction properties of different biomass species, such as sizes, shapes, chemical compositions, bulk densities, and particle densities, in order to optimize the process (Mani et al., 2006). The final products of this method are still considered hazardous waste, and unfortunately, there have been few studies conducted in this research area.

After compacting the contaminated biomass, it is exposed to different solvents for leaching of toxic metals, with the recovery rate dependent on time and pH value (Mulligan et al., 2001; Saeed et al., 2005). This method relies on the infiltration property of soluble metals through the transport medium, a characteristic that classifies heavy metal-contaminated biomass as hazardous (Kovacs et al., 2013). As a result, the residual biomass can be treated as nonhazardous material (Sas-Nowosielska et al., 2004), and the generated liquid treated as contaminated effluent. However, the available technologies are quite expensive.

5.2.4 Incineration

Incineration is the thermal degradation of contaminated biomass (combustion and gasification), which is similar to pyrolysis but includes oxygen in the process and mostly ashes remain. The fly ash (the fine fraction of ash) released by combustion causes pollution and requires the use of metallic filters to capture pollutants. The main advantages lie in reducing the volume of biomass by up to 99%, and reducing costs associated with transportation and treatment. However, despite being an approach with enormous potential, it needs to be improved, especially for large-scale applications, to become a viable option from both an economic and environmental standpoint (Sas-Nowosielska et al., 2004).

5.2.5 Hydrothermal carbonization

HTC involves converting contaminated biomass into hydrothermal biochar with higher energy density in a closed system that utilizes water, elevated temperature, and high pressure (Chai et al., 2022; Sharma et al., 2020). According to He et al. (2021), the effectiveness of this method in retaining contaminants within the biochar is achieved through the redistribution of metals between solid and liquid phases, which relies on dissolution, precipitation, and adsorption processes. The study also found that hydrothermal biochar possesses more oxygen-containing functional groups and active sites compared to biochar produced by pyrolysis, resulting in greater contaminant retention.

5.2.6 Pyrolysis

In this technique, biomass is heated to temperatures ranging from 350°C to 650°C under oxygen-free conditions, producing solid, liquid, and gas fractions (Wang et al., 2017). Pyrolysis is the most promising method for treating contaminated biomass, as it gradually transforms contaminants from highly toxic to less toxic and stable forms (Du et al., 2019). Table 5.1 summarizes the advantages and disadvantages of each disposal method for contaminated biomass.

The next section highlights the application of pyrolysis in the treatment of contaminated biomass and examines how the characteristics of pyrolysis influence the qualitative and quantitative features of the final product.

5.3 Production of biochar from pyrolysis of contaminated biomass

Biochar, charcoal-like material, is obtained through the thermal breakdown of biomass, resulting in a solid material that contains high carbon content, highly porous, and resistant to decomposition (Rangabhashiyam & Balasubramanian, 2019). The increased interest in its use in recent years is due to its versatility in

Table 5.1 Options of technologies for disposing the biomass contaminated by potentially toxic elements.

Techniques	Advantages	Disadvantages
Direct	Practicality	Higher environmental pollution; leachate monitoring
Composting	Volume reduction	Compost cannot be used in agriculture
Leaching	Uncontaminated waste	High-cost technology
Incineration	Volume reduction; power generation; metal recovery	Environmental pollution by fly ash
Hydrothermal carbonization	Retention of contaminants in biochar	Cautious use for soil conditioning
Pyrolysis	Retention of contaminants in biochar	Cautious use for soil conditioning

various agricultural and industrial applications. This usability of biochar is largely dependent on its final quality that depends on the concentrations of contaminants present, as well as its density, porosity, and pH. These characteristics are influenced by the type of biomass and pyrolysis temperature (Fig. 5.1; Amalina, Razak, Krishnan, Zularisam, & Nasrullah, 2022).

Pyrolysis methods can be classified into slow, fast, and microwave-assisted pyrolysis. The slow pyrolysis occurs at low temperatures, between 300°C and 550°C, with prolonged contact times ranging from minutes to days (El-Naggar et al., 2019). The high quality and yield of biochar produced by this process result from the combination of low heating rates, low temperature, and long residence times, which minimizes the production of liquid and gaseous products (Bertero & Sedran, 2015; Laird et al., 2009). Conversely, fast pyrolysis involves thermal decomposition in the absence of O_2 at temperatures ranging from 600°C to 650°C, primarily producing oil, vapors, and aerosols, along with small amounts of gas and biochar (Al-Rumaihi et al., 2022). Microwave-assisted biomass pyrolysis, on the other hand, involves energy transfer rather than heat, where electromagnetic energy is converted into thermal energy through conductive heat (Liu et al., 2018). Additionally, energy is released within the biomass volume, with temperatures ranging between 400°C and 800°C. Unlike conventional pyrolysis, the temperature at the center of the biomass is higher than at the surface (Zaker et al., 2019). The pyrolysis temperature is inversely proportional to the biochar yield (Al-Rumaihi et al., 2022).

Several factors influence the quality of biochar, with the most important being the type of biomass, residence time, pyrolysis temperature, and concentration of elements in the biomass. Table 5.2 illustrates how different types of biomass, combined with varying pyrolysis times and temperatures, affect the pH and specific surface area of biochar. Contaminated biomass typically contains 35%–50%

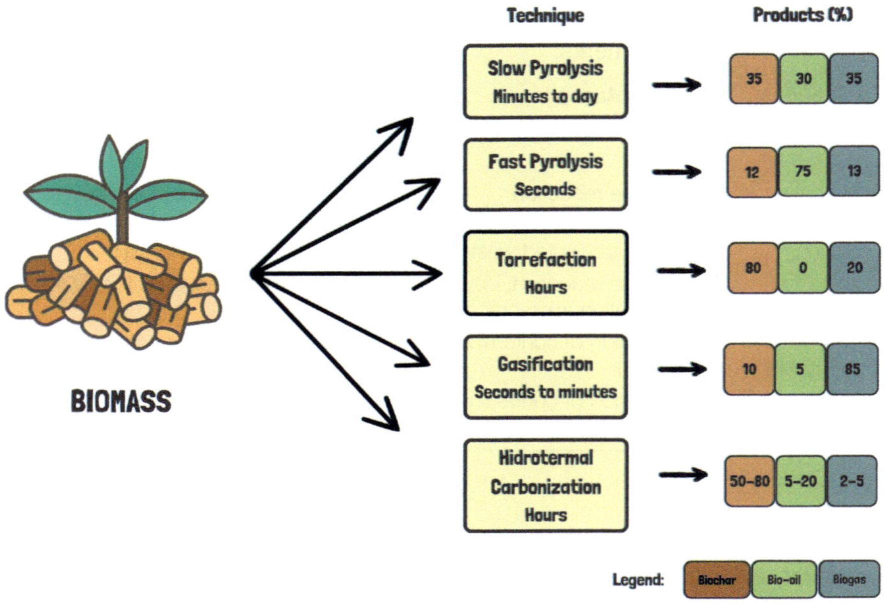

FIGURE 5.1

Techniques for biochar production from biomass. Figure shows a sequence of the main pyrolysis techniques for transforming biomass into biochar, bio-oil, and biogas, along with their respective yields.

Figure produced from data presented in the study of Amalina, Razak, Krishnan, Zularisam, and Nasrullah (2022).

cellulose, 20%–35% hemicellulose, and 10%–25% lignin on a dry basis. Thermal degradation of hemicellulose occurs between 220°C and 315°C, yielding more gaseous products compared to cellulose and lignin. Cellulose undergoes thermal degradation in the temperature range of 315°C–400°C, resulting in distinct product formations. Higher temperatures favor cellulose decomposition into gases, char, and secondary tar, whereas lower temperatures promote char formation through cellulose cross-linking and aromatization. In contrast, lignin is thermally more resistant than cellulose and hemicellulose, degrading in temperature ranging between 160°C and 900°C (Seah et al., 2023).

Biochar with higher surface area and high porosity enhance nutrient adsorption and water retention in soil (Elhessin Shareef & Zhao, 2017). As the temperature increases, there is an enrichment of basic cations in the ash, which may be associated with alkaline species such as carbonates, oxides, and hydroxides (Houben et al., 2013; Yuan et al., 2011), and a reduction in the concentration of surface acidic functional groups (Singh et al., 2010). Biochar with lower pH values has a narrower pore structure, which favors increased specific surface area and adsorption capacity (Truong et al., 2020). During pyrolysis, the high

Table 5.2 Different types of biomass and factors influencing biochar properties.

Feedstock	Temperature (°C)	Residence time	pH	Surface area (m^2/g)	References
Wood-pine	500	0.5 h	8.7	380	Leng et al. (2021)
Husk—rice	500	2 h	7.99	230.91	Kameyama et al. (2019)
Bamboo	600	2 h	7.9	470.4	Zhou et al. (2013)
Wheat straw	600	3 h	9.1	183.3	Alkurdi et al. (2019)
Herb-residue	600	3 h	10.1	51.3	Zhou et al. (2019)
Bamboo	450	–	8.7	10.2	Yao et al. (2012)
Gram	450	30 s	9.1	1.4	Oliveira et al. (2017)
Sugarcane bagasse	450	–	8.95	13.6	Yao et al. (2012)
Gram	800	30 s	11.2	17.2	Oliveira et al. (2017)
Sugarcane bagasse	600	2 h	7.5	557.4	Zhou et al. (2013)
Sugarcane straw	700	1 h	10.2	–	Puga et al. (2016)
Manure	500	4 h	10.5	13	Song et al. (2019)
Peanut shell	700	3 h	10.57	448.2	Thomas et al. (2019)

temperature causes some low-boiling point PTEs, such as arsenic (As), cadmium (Cd), and lead (Pb), to be vaporized, which can result in a reduction in the concentration of these metals in the biochar obtained from contaminated biomass. Yousaf et al. (2018) produced biochar from an invasive species that accumulates heavy metals and observed that the ecological risk associated with the use of biochar decreased with increasing pyrolysis temperature, heating rate, and smaller particle size of the raw material.

The data in Table 5.3 reveals that among the heavy metals analyzed, the concentration of Cd and Pb decreases in biochar as the pyrolysis temperature increases. The volatility rate of Pb and Cd was 75% and 74%, respectively, with the increase in pyrolysis temperature up to 900°C (Yang et al., 2021). Conversely, this trend was not observed for copper (Cu) and Zn, as higher temperatures led to increased concentrations of these metals in biochar. The

Table 5.3 Influence of pyrolysis temperature on the concentration of potentially toxic elements present in biochar produced from contaminated biomass.

Element	Biomass type	Biomass concentration (mg/kg DW)	Temperature (°C)	Time (h)	Concentration in biochar (mg/kg DW)	References
Cd	*Betula pendula*	2.2	450	4	0.70	Evangelou et al. (2014)
			700		b.q.c.	
Pb	*B. pendula*	9.8	450		7.6	
			700		6.6	
Zn	*B. pendula*	625	450		655	
			700		530	
Cu	Rice straw	12.69	300		18.61	Yang et al. (2021)
			700	1	32.66	
			900		41.92	
Zn		Rice straw	73.43	300	86.57	
				700	128.75	
				900	65.50	
Cd		Rice straw	0.73	300	1.09	
				700	0.62	
				900	0.19	
Pb		Rice straw	16.70	300	6.91	
				700	6.30	
				900	4.14	

b.q.c., *below the detection limit.*

accumulation of PTEs in biochar is primarily influenced by the thermal stability of the elements. Additionally, the addition of minerals, such as phosphate or calcium minerals, can also decrease the availability of PTEs (Liu et al., 2019; Shi et al., 2017). Therefore, by carefully controlling factors such as temperature, residence time, and addition of chemical additives, the risk of pollution might be mitigated by effectively retaining contaminants in biochar (Leng & Huang, 2018; Shi et al., 2017; Zong, Xiao, & Lu, 2021).

Temperature control and residence time represent effective strategies for producing biochar there is either free of contaminants or enriched with them, depending on the desired outcome. The removal of toxic elements during the pyrolysis process is beneficial for utilizing contaminated biomass to generate contaminant-free biochar, this also underscores the necessity of employing metal filters in biochar production facilities to ensure the purity and safety of the final product (Evangelou et al., 2014). The pyrolysis temperature is the key factor determining the yield and quality of biochar. Temperature is critical in determining the final characteristics of biochar, and the carbonization of vegetative materials begins at temperatures as low as 190°C (Song et al., 2012). The condensed aromatic structure of biochars can have varying forms, including amorphous C (which dominates at lower pyrolysis temperatures), turbostratic C (formed at higher temperatures), and graphite C (Keiluweit et al., 2010; Nguyen et al., 2010).

Process of biochar production had three stages: prepyrolysis; main-pyrolysis; and formation of carbonaceous soil products (Lee et al., 2017). The first stage (from ambient temperature to 200°C) is attributed to evaporation of moisture and light volatiles. The moisture evaporation causes breakage of bonds and formation hydroperoxide, −COOH and −CO groups (Cárdenas-Aguiar et al., 2017). The second stage (from 200°C to 500°C) was a devolatilized and decomposed of hemicelluloses and cellulose at a fast rate (Ding et al., 2014). The last stage (above 500°C) is degradation of lignin and other organic matter with stronger chemical bonds (Cárdenas-Aguiar et al., 2017). Cao et al. (2009) investigated the sorption capacities of dairy manure biochar produced at low temperatures (200°C and 350°C) and found that the biochar was six times more effective in removing Pb from wastewater than a commercial activated carbon. According to Chen and Yuan (2011), the biochar produced at high pyrolysis temperatures (400°C and 700°C) demonstrated higher efficiency in the sorption affinity of biochar-soil than biochar obtained at low pyrolysis temperatures (100°C and 300°C). During pyrolysis, most metals such as Pb, Ni, Cu, Zn, and Cr are retained in the biochar (solid fraction) because they have boiling points higher than the temperature typically used in pyrolysis (Lu et al., 2016; Van Wesenbeeck et al., 2014). Therefore pyrolysis concentrates metals in the biochar (Kameyama et al., 2019; Lu et al., 2016).

As the pyrolysis temperature increases, the concentration of metals in biochar also increases (Khanmohammadi et al., 2015; Lu et al., 2016; Yuan et al., 2015). However, when temperatures exceed 600°C, some PTEs such as Cd become volatile and exit the reactor along with the produced gases. Nevertheless, at temperatures below 600°C, the minor losses of PTEs observed are mainly due to the escape of fine particles laden with PTEs in the gas outlet stream (Kistler et al., 1987). Under these conditions, some metals leave the reactor during pyrolysis together with vapor and other gases because their boiling point is below 600°C (Nzihou & Stanmore, 2013).

Biochar particles exhibit a strong affinity for PTEs primarily due to their highly porous structure, large specific surface area, and abundance of functional groups on the surface (Kumar et al., 2018; Park et al., 2011). The amount of functional groups (carboxylic, hydroxyl, amino) decreases with increasing pyrolysis temperature due to higher degrees of carbonization (reduction in N/C, O/C, H/C ratios), resulting in a highly aromatic structure (Figueiredo et al., 2018). This aromatic structure contains a large number of delocalized π-electrons that can interact with cationic PTEs through cation-π interactions (Harvey et al., 2011; Uchimiya et al., 2010). Together, these changes determine the metal adsorption potential of biochars produced at different temperatures (Li et al., 2017).

Generally, increasing pyrolysis temperature reduces the amount of extractable PTEs due to partial fractionation and volatility of these metals at high temperatures, leading to reduced biochar toxicity (Gondek et al., 2014; Kończak & Oleszczuk, 2020). Pyrolyzing sewage sludge at 400°C is widely known to stabilize metals. According to the findings of He et al. (2010), pyrolysis temperatures higher than 350°C result in better stability of metals such as Cd, Pb, Zn, and Cu. The environmental damage and toxicity of metals in the copyrolyzed biochar decreased further, and copyrolysis at 600°C was found to be suitable for sewage sludge nutrient recycling (Jin et al., 2017; Yang et al., 2021). High pyrolysis temperatures promote the creation of more permanent aromatic bonds and the elimination of transitory aliphatic bonds (Gondek et al., 2014).

Pyrolysis also can stabilize PTEs by transforming active extractable and exchangeable and reducible fractions into stable fractions including oxidizable: metals bound to sulfides and organic matter, and residual fractions: nonsilicate bound metals, at low temperatures (300°C) (Wang et al., 2020). Multiple PTEs were evaluated in sediments, water, and soils, using the environmental risk index. High pyrolysis temperatures were found to reduce the potential environmental concern of PTEs, particularly Cu (Wang et al., 2020). Gondek et al. (2014) investigated the effects of low temperature (200°C) on the ecotoxicity of PTEs and metals mobile forms in biochars produced by copyrolyzing sewage sludge with various biomass materials such as bark, sawdust, rape straw, and wheat straw. The concentrations of PTEs in water-soluble form were substantially reduced in all of the tested cofeed materials as a result of thermal transformation (Gondek et al., 2014). The study also found low toxicity of PTEs extracts from biomass feedstocks for *Lepidium sativum* and *Vibrio fischeri*, irrespective of the transformation mechanism of PTEs. Kończak and Oleszczuk (2020) studied the effects of different temperatures (500°C–700°C) on the toxicity of biochars produced from sewage sludge and willow copyrolysis, and the increased temperature, particularly to 700°C, reduced the quantity of extractable metals for Al, Cu, Fe, Mn, and Zn. Similarly, Huang et al. (2017) studied the effects of different temperatures

(300°C − 700°C) on the leachable and total PTEs from biochars and the physicochemical properties of biochars from sewage sludge copyrolysis with rice straw or sawdust and reported that increased pyrolysis temperature concentrated the PTEs in the biochars.

The increased PTEs concentration in biochars might be related to the higher stability of metals than other sewage sludge compositions (Huang et al., 2017). Further, when temperature increases, metals in the biochar become more concentrated, due to the increased mass loss rate during copyrolysis (Huang et al., 2017; Kończak & Oleszczuk, 2020). Uchimiya et al. (2011) assessed the metal sorption efficacy of cottonseed hull–derived biochar prepared at five temperatures (200°C, 350°C, 500°C, 650°C, and 800°C), and concluded that the properties of the surface functional groups of biochars (volatile matter, oxygen contents, and pH_{PZC}) control the PTEs sequestration ability from the soil.

As already presented earlier, an alternative to reduce the concentration of PTEs in biochar is through copyrolysis, which involves mixing biomass with both high and low concentrations of PTEs under the same operating system of normal pyrolysis (Hassan et al., 2016). The synergistic effect of biochar and other substances can be achieved in a copyrolysis system, aiming to overcome the disadvantages of the biomass involved in the process, such as low yield, nutrient content, and high risk of PTE leaching. Therefore copyrolysis biochar is not simply a combination of two or more types of raw materials, but a product influenced by its synergistic effect (Chao Seah et al., 2023). The most common types of contaminated raw materials include phytoremediation residues, sewage sludge, and swine manure, which contain high levels of bioavailable and total PTEs, resulting in high concentrations of PTEs in their corresponding biochars.

Several studies indicate that the mixing ratio of different materials during copyrolysis is crucial for reducing PTEs concentrations and the toxicity of biochars produced from single-pyrolysis of contaminated biomass. Kończak and Oleszczuk (2020) reported that an appropriate ratio of willow and sewage sludge during copyrolysis is necessary to decrease the PTE concentration, as a smaller amount of biomass minimizes the "dilution effect" of PTEs in the produced biochars. Jin et al. (2017) observed that after copyrolysis of sewage sludge and low-PTE biomass such as sawdust and rice straw, the resulting product had lower concentrations compared to biochar produced solely from sewage sludge. Similar results were also obtained by Meng et al. (2018), where the copyrolysis of pig manure with rice straw resulted in lower levels of Cu and Zn compared to biochar produced solely from pig manure. Li et al. (2023) studied the pyrolysis and copyrolysis of industrial sludge and phytoremediation residue rich in Cd, Pb, and Zn at temperatures ranging from 350°C to 750°C. The results showed that compared to separate pyrolysis, introducing industrial sludge into the pyrolysis system of phytoremediation residue improved the immobilization performance of Cd, Pb, and Zn in the resulting biochar obtained at 750°C. Additionally, it was discovered that complexation of PTEs with functional groups, the interaction of PTEs with electrons, and the formation of crystalline compounds of metals could be the key reasons for enhancing the immobilization of PTEs during copyrolysis.

In addition to reducing PTE concentrations, copyrolysis influences other properties of biochars. Li et al. (2021) systematically compared the physicochemical properties of pure biochar and biochar derived from copyrolysis of various feedstocks. They found that most copyrolysis processes enhance yield, specific surface area, carbon content, and decrease the pH of biochar. Meanwhile, Wang et al. (2019) discovered that metal leaching from biochar was influenced by the amount of biomass used. Kończak et al. (2019) investigated the impacts of sewage sludge and copyrolysis with willow in different mixing ratios of 8:2 and 6:4 (w:w), under different inert gases N_2 and CO_2, and various temperatures ranging from 500°C to 700°C, on the physical and chemical properties of biochar. They observed that replacing N_2 with CO_2 and using different mixing ratios affected the physicochemical properties of the produced biochars. Kończak and Oleszczuk (2020) reported that the reduction in PTE leaching rates was further enhanced with increasing proportions of willow in sewage sludge. Copyrolysis of contaminated biomass leads to more effective transformations of mobile forms of metals into more stable fractions, indicating potential reductions in environmental risks associated with high concentrations of PTEs in biochars.

Copyrolysis of sewage sludge with hazelnut shell has been suggested as an effective and feasible method to reduce environmental toxicity and produce porous construction materials (Zhao et al., 2017). Yang et al. (2020) observed that copyrolysis of rice husk and sewage sludge reduced the content of Pb and Cd in the produced biochar, thereby lowering environmental risk levels by 1−2 grades. Additionally, the addition of cotton stalk to sewage sludge facilitated the transformation and migration of PTEs from bioavailable components to stable forms, thereby reducing potential toxic environmental impacts of PTEs and enabling significant recycling of sewage sludge (Wang et al., 2019).

Several metals such as Cu, Cr, Zn, Pb, Mn, and Ni are transformed into more stable states after copyrolysis, resulting in significant reductions in immediate phytotoxicity and bioavailability of biochar (Mohamed et al., 2023). It is notable how certain materials improve the physical and chemical properties of biochar resulting from copyrolysis. According to Ahmed and Hameed (2020), the properties of copyrolysis biochar are mainly attributed to the synergistic effect of two feedstocks during the copyrolysis process, which is essentially determined by the type of feedstocks, mixing ratio, temperature, and pyrolysis duration. Mohamed et al. (2023) reported that cotton stalk, hazelnut shell, and coconut shell have been identified as feedstocks that have positive impacts on reducing PTEs toxicity and improving the characteristics of biochars produced from copyrolysis. Therefore future studies may explore optimizing the copyrolysis process of contaminated biomass using different carrier gases, mixing ratios, residence times, and temperatures along with the use of chemical additives.

5.4 Fate of contaminants in biochar from contaminated biomass

Proper treatment of contaminated biomass effectively addresses the contaminants present in it. This process can transform contaminated biomass into organic fertilizer or soil conditioner, providing environmentally friendly disposal alternatives. By turning a former contaminated waste into an economic asset for commercialization, this approach offers both profitability and sustainability (Liu & Tran, 2021). The pyrolysis of contaminated biomass is a method capable of producing a product with potential applications in agriculture and industry. This process aims to concentrate the highest content of toxic metals in the solid fraction, with minimal content in the liquid and gas fractions. The biochar resulting from pyrolysis can be reused in agriculture; however, the contaminants present in the biomass are transferred to biochar. To ensure the safe use of biochar in agriculture and the environment, it is crucial to advocate for the removal of contaminants so that the stabilized metals do not leach or transfer to plants causing future contaminations (Shen et al., 2022).

5.4.1 Extraction and stabilization of contaminants from biochar

Metal extraction can occur either before or after pyrolysis. When extraction occurs before, the biomass is pressed (compacted) so that the liquid containing the contaminants (leachate) can be collected and treated for metal extraction. During pyrolysis, metals are primarily concentrated in the biochar, while the gases and bio-oil contain low or negligible percentages of metals due to the relatively low temperatures (300°C–400°C) at which the process occurs (Liu & Tran, 2021).

The removal of contaminants from biomass is achieved using extraction agents such as water, ammonium acetate, sulfuric acid, ammonium oxalate, potassium citrate, ethylenediaminetetraacetic acid, ammonium nitrate, and potassium tartrate. The extraction temperature ranges from 20°C to 90°C, with agitation apparatus operating at 100–700 rotations per minute, and extraction times ranging from 30 minutes to 16 hours. In apparatuses without an agitation function, the temperature ranges from 16°C to 60°C, with extraction times varying from 1 to 2 hours. This latter method, which uses only water for contaminant removal, is less efficient (Liu & Tran, 2021).

Despite being an option, the removal of contaminants from biomass before pyrolysis is challenging and costly to operate at scale. Therefore an alternative approach is to stabilize the contaminants, primarily PTEs, through pyrolysis. Wang et al. (2022) evaluated the stabilization of chromium (Cr), Ni, arsenic (As), antimony (Sb), mercury (Hg), and Pb in postphytoremediation plant residues from *Conyza canadensis*, *Gahnia tristis*, and *Betula luminifera* by preparing biochar at different pyrolysis temperatures (300°C, 450°C, and 600°C). The results of their environmental risk assessment showed that pyrolyzing contaminated biomass at the highest temperature (600°C) was effective at mitigating the environmental impact of As, Sb, and Hg. They also suggested further use of the contaminated biochar as an adsorbent for PTEs.

Increasing the pyrolysis temperature concentrates the PTEs in biochar while reducing their availability due to transformations that stabilizes the PTEs during pyrolysis. However, the total concentration of PTEs in biochar might limit its use in agriculture. To address this, Han et al. (2022) suggested copyrolyzing high-PTE biomass with low-PTE biomass (e.g., wood sawdust, rice straw, coffee husk) to reduce the PTE burden in the resulting biochar. The authors consider this the only reliable way to reduce the total PTE load of biochar. Additionally, copyrolyzing contaminated biomass with inorganic materials such as phosphate or calcium minerals can reduce both the availability and total concentration of PTEs in biochar. To comply with legislation threshold values, which are typically based on total PTE concentrations, copyrolysis aiming at reducing the total content is necessary.

5.4.2 Destination and reuse of metals from biochar

The reuse of extracted contaminants from biochar varies depending on the metal. Metals such as iron (Fe), manganese (Mn), Zn, Cu, and Ni can be repurposed for agricultural products, soil and foliar fertilizers (micronutrients), and phytosanitary products (fungicides). Lead can be reused to produce batteries and paints. PTEs such as As, barium (Ba), Cd, and Cr can be immobilized in building foundations using concrete, incorporated into mortar formulations in civil construction applications, and used in paving blocks for constructing streets, sidewalks, and patios, wherein these forms, the PTEs are immobilized without the risk of leaching, preventing further contamination (Altoé et al., 2021; Barros et al., 2020; Kumar et al., 2021).

The synthesis of metallic nanoparticles and metal oxide nanoparticles of Cu, Pb, Ni, and Zn is a promising technique for reusing contaminants. These nanoparticles can be used in various fields, such as electrochemistry, aviation industry, catalysis in organic synthesis, and pollutant degradation. However, this method remains costly and presents significant implementation challenges (Cui et al., 2021; Liu & Tran, 2021). Metals such as Fe, Cr, Mn, and Zn from solid waste can be safely incorporated into the production of glass and glass ceramics without risks of contamination to humans. These metals can effectively replace silicon (Si) and aluminum (Al) in the glass structure, forming stable chemical bonds such as Cr–O–Al and (Fe, Mn, and Zn)–O–Si. By substituting cations of the same valence, these metals can become solidified within the crystalline structure (Ker et al., 2015; Liu et al., 2023).

The process of extracting metals from biomass to produce reagents is known as phytomining. Hyperaccumulator plants are usually chosen for this purpose because they can extract significant quantities of metals from the soil without harming their metabolism. The extracted metal is concentrated through pyrolysis and then removed using leaching agents, as discussed earlier, where this residue is returned to the industry for extraction and purification. This process yields metallic salts with high analytical purity (Aravind Kumar et al., 2021; Chai et al., 2022).

Some mechanisms for extracting metals are well-established, yet the development of new methods for extracting contaminants from biomass is imperative in the environmental field. These innovations are necessary to ensure that treatments like pyrolysis can effectively remove contaminants from their byproducts, thereby allowing biochar to be utilized without concern. While there are existing techniques for immobilizing contaminants to facilitate their removal from circulation, reutilization could serve as an alternative to new production. However, the feasibility of reusing contaminants depends on their potential for reutilization, which can be better understood through new research on extraction and reuse (Aravind Kumar et al., 2021; Cui et al., 2021; Liu & Tran, 2021).

5.5 Future perspectives

The phytoremediation process currently demonstrates beneficial effects on the environment, effectively extracting contaminants without causing secondary damage. While hyperaccumulators are commonly used in phytoextraction, these plants cannot typically regrow and die after harvesting. An alternative approach involves utilizing plants of high biomass production in phytoremediation areas. Although their extraction capabilities may be lower, some of these plants can regrow multiple times, minimizing the need for repeated interventions in contaminated areas and thus enhancing process efficiency. Furthermore, such plants can provide a suitable and standardized source of biomass for producing biochar. Exploring new genotypes and plant species that contribute to both phytoremediation process and biochar production is a worthwhile endeavor (Rabêlo et al., 2021).

Through proper nutrition, plants of high biomass production have the potential to produce nutrient-rich biochar. Furthermore, their robust root system enable them to cover larger areas of soil both laterally and horizontally, conditioning the plants to tolerate potential stresses that may arise during their phenological cycle, including metal stress (Shen et al., 2022). Pyrolysis is a suitable treatment for biomasses containing PTEs with a medium to high boiling point. However, the fate of these contaminants still presents challenges for proper disposal and reuse. Despite the solutions discussed throughout the chapter for treating contaminants, metals like Hg can be lost through volatilization during pyrolysis. Therefore methods for capturing volatilized Hg still require further improvements and development (Cui et al., 2021). The production of biochar, alongside other thermal methods, has provided opportunities for biomass disposal. However, it still presents challenges as an expensive process that requires sophisticated facilities and trained technicians for optimal efficiency. Further studies are necessary to enhance the quality of biochar from contaminated biomass, thereby increasing its economic value and improving the efficiency of PTEs disposal (Liu & Tran, 2021).

5.6 Conclusion

The copyrolysis of contaminated biomass resulting from phytoremediation minimizes the risk of secondary pollution. Moreover, increasing the temperature for biochar production from contaminated biomass is an effective strategy for reducing As, Cd, and Pb. Considering biochar as the primary outcome of the process, the most efficient methods for its production include slow pyrolysis, torrefaction, and HTC. While biochar production systems can be economically viable, comprehensive cost analyses are still under development. To provide a thorough assessment, it is necessary to consider the biochar cycle for energy production, as well as its economic, sociocultural implications, and benefits for agricultural production. Biochar boasts multiple applications, spanning from urban and industrial waste management to soil improvement, water treatment, and green energy production. Recognized as an effective tool in climate change mitigation, biochar acts as a carbon sink, reducing greenhouse gas emissions, offering promising environmental benefits for the future. Hence, biochar produced through the pyrolysis of contaminated biomass can be repurposed for metal adsorption in aquatic systems and soil. To utilize biochar in agriculture effectively, the removal or immobilization of these metals is necessary. Further research will be necessary to ensure that the leaching process selectively removes only PTEs, leaving the nutrients intact in the final product, thereby enhancing the value of biochar.

Acknowledgments

The authors would like to thank the Brazilian Coordination for Improvement of Higher Level Personnel (CAPES), the Brazilian Council of Scientific and Technological Development (CNPq), and the Minas Gerais State Research Support Foundation (FAPEMIG) for financial support and scholarships provided to the authors, as well Elsevier for the opportunity.

Author contribution

Deivisson Ferreira da Silva, Olívia Bibiana Souza Dias, and Alisson de Carli Souza: writing—original draft preparation. Amir Parnian, Lucélia Borgo, and

Leônidas Carrijo Azevedo Melo: reviewing and supervision. Flávio Henrique Silveira Rabêlo: reviewing and supervision—original draft preparation.

Statement for conflict of interest

The authors declare that they have no known competing financial interests or personal relationships that could have appeared to influence the work reported in this chapter.

Statement on the use of AI tools for writing

The authors declare that they did not use any AI or AI-assisted technologies in the writing or the creation of images in this chapter.

References

Ahmed, M. J., & Hameed, B. H. (2020). Insight into the co-pyrolysis of different blended feedstocks to biochar for the adsorption of organic and inorganic pollutants: A review. *Journal of Cleaner Production*, *265*. Available from https://doi.org/10.1016/j.jclepro.2020.121762.

Alkurdi, S. S. A., Herath, I., Bundschuh, J., Al-Juboori, R. A., Vithanage, M., & Mohan, D. (2019). Biochar versus bone char for a sustainable inorganic arsenic mitigation in water: What needs to be done in future research? *Environment International*, *127*, 52–69. Available from https://doi.org/10.1016/j.envint.2019.03.012, http://www.elsevier.com/locate/envint.

Al-Rumaihi, A., Shahbaz, M., Mckay, G., Mackey, H., & Al-Ansari, T. (2022). A review of pyrolysis technologies and feedstock: A blending approach for plastic and biomass towards optimum biochar yield. *Renewable and Sustainable Energy Reviews*, *167*, 112715. Available from https://doi.org/10.1016/j.rser.2022.112715.

Altoé, S. P. S., Goés, I., Venancio Pinheiro Rotta, J., & Roberto Borin, M. (2021). Avaliação dos parâmetros sustentáveis para pavers confeccionados com resíduos industriais. *Atena Editora*, 468–475. Available from https://doi.org/10.22533/at.ed.81921081160.

Amalina, F., Razak, A. S. A., Krishnan, S., Sulaiman, H., Zularisam, A. W., & Nasrullah, M. (2022). Biochar production techniques utilizing biomass waste-derived materials and environmental applications – a review. *Journal of Hazardous Materials Advances*, *7*. Available from https://doi.org/10.1016/j.hazadv.2022.100134, https://www.journals.elsevier.com/journal-of-hazardous-materials-advances.

Amalina, F., Razak, A. S. A., Krishnan, S., Zularisam, A. W., & Nasrullah, M. (2022). A comprehensive assessment of the method for producing biochar, its characterization, stability, and potential applications in regenerative economic sustainability – a review. *Cleaner Materials*, *3*, 100045. Available from https://doi.org/10.1016/j.clema.2022.100045.

Aravind Kumar, J., Krithiga, T., Manigandan, S., Sathish, S., Annam Renita, A., Prakash, P., Naveen Prasad, B. S., Praveen Kumar, T. R., Rajasimman, M., Hosseini-Bandegharaei, A., Prabu, D., & Crispin, S. (2021). A focus to green synthesis of metal/metal based oxide nanoparticles: Various mechanisms and applications towards ecological approach. *Journal of Cleaner Production*, *324*, 129198. Available from https://doi.org/10.1016/j.jclepro.2021.129198.

Barros, S. V. A., Dantas, G. C. B., Neves, Gd. A., & Menezes, R. R. (2020). Immobilization of heavy metals present in quartzite residues through incorporation in mortars with total substitution of the natural aggregate. *Engenharia Sanitaria e Ambiental*, *25*(6), 833–845. Available from https://doi.org/10.1590/s1413-4152202020180063, http://www.scielo.br/pdf/esa/v25n6/1809-4457-esa-S1413-4152202020180063.pdf.

Bertero, M., & Sedran, U. (2015). *Coprocessing of bio-oil in fluid catalytic cracking. Recent advances in thermochemical conversion of biomass* (pp. 355–381). Elsevier Inc. Available from http://www.sciencedirect.com/science/book/97804, 10.1016/B978-0-444-63289-0.00013-2.

Cao, X., Ma, L., Gao, B., & Harris, W. (2009). Dairy-manure derived biochar effectively sorbs lead and atrazine. *Environmental Science and Technology*, *43*(9), 3285–3291. Available from https://doi.org/10.1021/es803092k.

Chai, Y., Chen, A., Bai, M., Peng, L., Shao, J., Yuan, J., Shang, C., Zhang, J., Huang, H., & Peng, C. (2022). Valorization of heavy metal contaminated biomass: Recycling and expanding to functional materials. *Journal of Cleaner Production*, *366*, 0959–6526. Available from https://doi.org/10.1016/j.jclepro.2022.132771.

Chao Seah, C., Hong Tan, C., Arifin., Hafriz., Salmiaton., Nomanbhay, S., & Shamsuddin. (2023). Co-pyrolysis of biomass and plastic: Circularity of wastes and comprehensive review of synergistic mechanism. *Results in Engineering*, *17*. Available from https://doi.org/10.1016/j.rineng.2023.100989.

Chen, B., & Yuan, M. (2011). Enhanced sorption of polycyclic aromatic hydrocarbons by soil amended with biochar. *Journal of Soils and Sediments*, *11*(1), 62–71. Available from https://doi.org/10.1007/s11368-010-0266-7.

Cui, X., Zhang, J., Wang, X., Pan, M., Lin, Q., Khan, K. Y., Yan, B., Li, T., He, Z., Yang, X., & Chen, G. (2021). A review on the thermal treatment of heavy metal hyperaccumulator: Fates of heavy metals and generation of products. *Journal of Hazardous Materials*, *405*, 123832. Available from https://doi.org/10.1016/j.jhazmat.2020.123832.

Cárdenas- Aguiar, E., Gascó, G., Paz-Ferreiro, J., & Méndez, A. (2017). The effect of biochar and compost from urban organic waste on plant biomass and properties of an artificially copper polluted soil. *International Biodeterioration and Biodegradation*, *124*, 223–232. Available from https://doi.org/10.1016/j.ibiod.2017.05.014.

Çoruh, S., Elevli, S., Ergun, O. N., & Demir, G. (2013). Assessment of leaching characteristics of heavy metals from industrial leach waste. *International Journal of Mineral Processing*, *123*, 165–171. Available from https://doi.org/10.1016/j.minpro.2013.06.005.

Ding, W., Dong, X., Ime, I. M., Gao, B., & Ma, L. Q. (2014). Pyrolytic temperatures impact lead sorption mechanisms by bagasse biochars. *Chemosphere*, *105*, 68–74. Available from https://doi.org/10.1016/j.chemosphere.2013.12.042.

Du, J., Zhang, L., Ali, A., Li, R., Xiao, R., Guo, D., Liu, X., Zhang, Z., Ren, C., & Zhang, Z. (2019). Research on thermal disposal of phytoremediation plant waste: Stability of potentially toxic metals (PTMs) and oxidation resistance of biochars. Institution of Chemical Engineers, China. *Process Safety and Environmental Protection*, *125*,

260–268. Available from https://doi.org/10.1016/j.psep.2019.03.035, http://www.elsevier.com/wps/find/journaldescription.cws_home/713889/description#description.

El-Naggar, A., Lee, S. S., Rinklebe, J., Farooq, M., Song, H., Sarmah, A. K., Zimmerman, A. R., Ahmad, M., Shaheen, S. M., & Ok, Y. S. (2019). Biochar application to low fertility soils: A review of current status, and future prospects. *Geoderma*, *337*, 536–554. Available from https://doi.org/10.1016/j.geoderma.2018.09.034, http://www.elsevier.com/inca/publications/store/5/0/3/3/3/2.

Elhessin Shareef, T. M., & Zhao, B. (2017). The fundamentals of biochar as a soil amendment tool and management in agriculture scope: An overview for farmers and gardeners. *Journal of Agricultural Chemistry and Environment*, *06*(01), 38–61. Available from https://doi.org/10.4236/jacen.2017.61003.

Evangelou, M. W. H., Brem, A., Ugolini, F., Abiven, S., & Schulin, R. (2014). Soil application of biochar produced from biomass grown on trace element contaminated land. *Journal of Environmental Management*, *146*, 100–106. Available from https://doi.org/10.1016/j.jenvman.2014.07.046, https://www.sciencedirect.com/journal/journal-of-environmental-management.

Figueiredo, C., Lopes, H., Coser, T., Vale, A., Busato, J., Aguiar, N., Novotny, E., & Canellas, L. (2018). Influence of pyrolysis temperature on chemical and physical properties of biochar from sewage sludge. *Archives of Agronomy and Soil Science*, *64*(6), 881–889. Available from https://doi.org/10.1080/03650340.2017.1407870.

Gondek, K., Baran, A., & Kopeć, M. (2014). The effect of low-temperature transformation of mixtures of sewage sludge and plant materials on content, leachability and toxicity of heavy metals. *Chemosphere*, *117*(1), 33–39. Available from https://doi.org/10.1016/j.chemosphere.2014.05.032.

Han, H., Buss, W., Zheng, Y., Song, P., Khalid Rafiq, M., Liu, P., Mašek, O., & Li, X. (2022). Contaminants in biochar and suggested mitigation measures – a review. *Chemical Engineering Journal*, *429*. Available from https://doi.org/10.1016/j.cej.2021.132287, http://www.elsevier.com/inca/publications/store/6/0/1/2/7/3/index.htt.

Harvey, O. R., Herbert, B. E., Rhue, R. D., & Kuo, L.-J. (2011). Metal interactions at the biochar-water interface: Energetics and structure-sorption relationships elucidated by flow adsorption microcalorimetry. *Environmental Science and Technology*, *45*(13), 5550–5556. Available from https://doi.org/10.1021/es104401h.

Hassan., Lim., & Hameed. (2016). Recent progress on biomass co-pyrolysis conversion into high-quality bio-oil. *Bioresource Technology*, *221*, 645–655. Available from https://doi.org/10.1016/j.biortech.2016.09.026.

He, Z., Li, Y., Chen, F., & Peng, F. (2010). The fate of Cu, Zn, Pb and Cd during the pyrolysis of sewage sludge at different temperatures. *Environmental Technology*, *31*(5), 567–574. Available from 10.1080/09593330903514466.

He, C., Zhang, Z., Xie, C., Giannis, A., Chen, Z., Tang, Y., & Qiu, R. (2021). Transformation behaviors and environmental risk assessment of heavy metals during resource recovery from Sedum plumbizincicola via hydrothermal liquefaction. *Journal of Hazardous Materials*, *410*, 124588. Available from https://doi.org/10.1016/j.jhazmat.2020.124588.

Houben, D., Evrard, L., & Sonnet, P. (2013). Mobility, bioavailability and pH-dependent leaching of cadmium, zinc and lead in a contaminated soil amended with biochar. *Chemosphere*, *92*(11), 1450–1457. Available from https://doi.org/10.1016/j.chemosphere.2013.03.055, http://www.elsevier.com/locate/chemosphere.

Huang, H.-J., Yang, T., Lai, F.-Y., & Wu, G.-Q. (2017). Co-pyrolysis of sewage sludge and sawdust/rice straw for the production of biochar. *Journal of Analytical and Applied Pyrolysis*, *125*, 61–68. Available from https://doi.org/10.1016/j.jaap.2017.04.018.

Jin, J., Wang, M., Cao, Y., Wu, S., Liang, P., Li, Y., Zhang, J., Zhang, J., Wong, M. H., Shan, S., & Christie, P. (2017). Cumulative effects of bamboo sawdust addition on pyrolysis of sewage sludge: Biochar properties and environmental risk from metals. *Bioresource Technology*, *228*, 218–226. Available from https://doi.org/10.1016/j.biortech.2016.12.103, http://www.elsevier.com/locate/biortech.

Kameyama, K., Miyamoto, T., & Iwata, Y. (2019). The preliminary study of water-retention related properties of biochar produced from various feedstock at different pyrolysis temperatures. *Materials*, *12*(11), 1732. Available from https://doi.org/10.3390/ma12111732.

Keiluweit, M., Nico, P. S., Johnson, M., & Kleber, M. (2010). Dynamic molecular structure of plant biomass-derived black carbon (biochar). *Environmental Science and Technology*, *44*(4), 1247–1253. Available from https://doi.org/10.1021/es9031419.

Ker, J. C., Nilton, C., & Schaefer, C. E. G. (2015). *Pedologia: Fundamentos*. Sociedade Brasileira de Ciência do Solo.

Khanmohammadi, Z., Afyuni, M., & Mosaddeghi, M. R. (2015). Effect of pyrolysis temperature on chemical and physical properties of sewage sludge biochar. *Waste Management and Research*, *33*(3), 275–283. Available from https://doi.org/10.1177/0734242X14565210.

Kistler, R. C., Brunner, P. H., & Widmer, F. (1987). Behavior of chromium, nickel, copper, zinc, cadmium, mercury, and lead during the pyrolysis of sewage sludge. *Environmental Science and Technology*, *21*(7), 704–708. Available from https://doi.org/10.1021/es00161a012.

Kovacs, H., & Szemmelveisz, K. (2017). Disposal options for polluted plants grown on heavy metal contaminated brownfield lands – a review. *Chemosphere*, *166*, 8–20. Available from https://doi.org/10.1016/j.chemosphere.2016.09.076, http://www.elsevier.com/locate/chemosphere.

Kovacs, H., Szemmelveisz, K., & Palotas, A. B. (2013). Solubility analysis and disposal options of combustion residues from plants grown on contaminated mining area. *Environmental Science and Pollution Research*, *20*(11), 7917–7925. Available from https://doi.org/10.1007/s11356-013-1673-2.

Kończak, M., & Oleszczuk, P. (2020). Co-pyrolysis of sewage sludge and biomass in carbon dioxide as a carrier gas affects the total and leachable metals in biochars. *Journal of Hazardous Materials*, *400*. Available from https://doi.org/10.1016/j.jhazmat.2020.123144.

Kończak, M., Oleszczuk, P., & Różyło, K. (2019). Application of different carrying gases and ratio between sewage sludge and willow for engineered (smart) biochar production. *Journal of CO_2 Utilization*, *29*, 20–28. Available from https://doi.org/10.1016/j.jcou.2018.10.019.

Kumar, A., Joseph, S., Tsechansky, L., Privat, K., Schreiter, I. J., Schüth, C., & Graber, E. R. (2018). Biochar aging in contaminated soil promotes Zn immobilization due to changes in biochar surface structural and chemical properties. *Science of the Total Environment*, *626*, 953–961. Available from https://doi.org/10.1016/j.scitotenv.2018.01.157.

Laird, D. A., Brown, R. C., Amonette, J. E., & Lehmann, J. (2009). Review of the pyrolysis platform for coproducing bio-oil and biochar. *Biofuels, Bioproducts and Biorefining. 1932104X.*, *3*(5), 547–562. Available from https://doi.org/10.1002/bbb.169United, http://www3.interscience.wiley.com/cgi-bin/fulltext/122589511/PDFSTART.

Lee, X. J., Lee, L. Y., Gan, S., Thangalazhy-Gopakumar, S., & Ng, H. K. (2017). Biochar potential evaluation of palm oil wastes through slow pyrolysis: Thermochemical characterization and pyrolytic kinetic studies. *Bioresource Technology*, *236*, 155–163. Available from https://doi.org/10.1016/j.biortech.2017.03.105.

Leng, L., & Huang, H. (2018). An overview of the effect of pyrolysis process parameters on biochar stability. *Bioresource Technology*, *270*, 627–642. Available from https://doi.org/10.1016/j.biortech.2018.09.030, http://www.elsevier.com/locate/biortech.

Leng, L., Xiong, Q., Yang, L., Li, H., Zhou, Y., Zhang, W., Jiang, S., Li, H., & Huang, H. (2021). An overview on engineering the surface area and porosity of biochar. *Science of the Total Environment*, *763*, 144204. Available from https://doi.org/10.1016/j.scitotenv.2020.144204.

Liu, Q., Fang, Z., Liu, Y., Liu, Y., Xu, Y., Ruan, X., Zhang, X., & Cao, W. (2019). Phosphorus speciation and bioavailability of sewage sludge derived biochar amended with CaO. *Waste Management*, *87*, 71–77. Available from https://doi.org/10.1016/j.wasman.2019.01.045, http://www.elsevier.com/locate/wasman.

Liu, Z., Singer, S., Tong, Y., Kimbell, L., Anderson, E., Hughes, M., Zitomer, D., & McNamara, P. (2018). Characteristics and applications of biochars derived from wastewater solids. *Renewable and Sustainable Energy Reviews*, *90*, 650–664. Available from https://doi.org/10.1016/j.rser.2018.02.040, https://www.journals.elsevier.com/renewable-and-sustainable-energy-reviews.

Liu, Z., & Tran, K. Q. (2021). A review on disposal and utilization of phytoremediation plants containing heavy metals. *Ecotoxicology and Environmental Safety*, *226*. Available from https://doi.org/10.1016/j.ecoenv.2021.112821, http://www.elsevier.com/inca/publications/store/6/2/2/8/1/9/index.htt.

Liu, J., Yu, Z., Zheng, J., Peng, H., Zhang, Z., Ye, H., Peng, S., Lin, X., Duan, W., & Wang, Q. (2023). Red mud/silica sand tailing-derived glasses and glass-ceramics for architectural decoration: Crystallization, structure, properties and immobilization of heavy metals. *Construction and Building Materials*, *400*, 132873. Available from https://doi.org/10.1016/j.conbuildmat.2023.132873.

Li, H., Dong, X., da Silva, E. B., de Oliveira, L. M., Chen, Y., & Ma, L. Q. (2017). Mechanisms of metal sorption by biochars: Biochar characteristics and modifications. *Chemosphere*, *178*, 466–478. Available from https://doi.org/10.1016/j.chemosphere.2017.03.072.

Li, Z., Huang, Y., Zhu, Z., Yu, M., Cheng, H., Shi, H., Zuo, W., Zhou, H., & Wang, S. (2023). Co-pyrolysis of industrial sludge with phytoremediation residue: Improving immobilization of heavy metals at high temperature. *Fuel*, *351*. Available from https://doi.org/10.1016/j.fuel.2023.128942.

Li, Y., Yu, H., Liu, L., & Yu, H. (2021). Application of co-pyrolysis biochar for the adsorption and immobilization of heavy metals in contaminated environmental substrates. *Journal of Hazardous Materials*, *420*. Available from https://doi.org/10.1016/j.jhazmat.2021.126655.

Lu, T., Yuan, H., Wang, Y., Huang, H., & Chen, Y. (2016). Characteristic of heavy metals in biochar derived from sewage sludge. *Journal of Material Cycles and Waste Management*, *18*(4), 725–733. Available from https://doi.org/10.1007/s10163-015-0366-y.

Mani, S., Tabil, L. G., & Sokhansanj, S. (2006). Effects of compressive force, particle size and moisture content on mechanical properties of biomass pellets from grasses.

Biomass and Bioenergy, *30*(7), 648–654. Available from https://doi.org/10.1016/j.biombioe.2005.01.004.

Meng, J., Liang, S., Tao, M., Liu, X., Brookes, P. C., & Xu, J. (2018). Chemical speciation and risk assessment of Cu and Zn in biochars derived from co-pyrolysis of pig manure with rice straw. *Chemosphere*, *200*, 344–350. Available from https://doi.org/10.1016/j.chemosphere.2018.02.138, http://www.elsevier.com/locate/chemosphere.

Mohamed, B. A., Ruan, R., Bilal, M., Khan, N. A., Kumar Awasthi, M., Amer, M. A., Leng, L., Hamouda, M. A., Nguyen Vo, D.-V., & Li, J. (2023). Co-pyrolysis of sewage sludge and biomass for stabilizing heavy metals and reducing biochar toxicity: A review. *Environmental Chemistry Letters*, *21*(2), 1231–1250. Available from https://doi.org/10.1007/s10311-022-01542-6.

Mulligan, C. N., Yong, R. N., & Gibbs, B. F. (2001). Remediation technologies for metal-contaminated soils and groundwater: An evaluation. *Engineering Geology*, *60*(1–4), 193–207. Available from https://doi.org/10.1016/S0013-7952(00)00101-0.

Nguyen, B. T., Lehmann, J., Hockaday, W. C., Joseph, S., & Masiello, C. A. (2010). Temperature sensitivity of black carbon decomposition and oxidation. *Environmental Science and Technology*, *44*(9), 3324–3331. Available from https://doi.org/10.1021/es903016y.

Nzihou, A., & Stanmore, B. (2013). The fate of heavy metals during combustion and gasification of contaminated biomass: A brief review. *Journal of Hazardous Materials*, *256–257*, 56–66. Available from https://doi.org/10.1016/j.jhazmat.2013.02.050.

Oliveira, F. R., Patel, A. K., Jaisi, D. P., Adhikari, S., Lu, H., & Khanal, S. K. (2017). Environmental application of biochar: Current status and perspectives. *Bioresource Technology*, *246*, 110–122. Available from https://doi.org/10.1016/j.biortech.2017.08.122, http://www.elsevier.com/locate/biortech.

Park, J. H., Choppala, G. K., Bolan, N. S., Chung, J. W., & Chuasavathi, T. (2011). Biochar reduces the bioavailability and phytotoxicity of heavy metals. *Plant and Soil*, *348*(1–2), 439–451. Available from https://doi.org/10.1007/s11104-011-0948-y.

Puga, A. P., Melo, L. C. A., de Abreu, C. A., Coscione, A. R., & Paz-Ferreiro, J. (2016). Leaching and fractionation of heavy metals in mining soils amended with biochar. *Soil and Tillage Research*, *164*, 25–33. Available from https://doi.org/10.1016/j.still.2016.01.008, http://www.elsevier.com/inca/publications/store/5/0/3/3/1/8.

Rabêlo, F. H. S., Vangronsveld, J., Baker, A. J. M., van der Ent, A., & Alleoni, L. R. F. (2021). Are grasses really useful for the phytoremediation of potentially toxic trace elements? A review. *Frontiers in Plant Science*, *12*. Available from https://doi.org/10.3389/fpls.2021.778275, https://www.frontiersin.org/journals/plant-science.

Rangabhashiyam, S., & Balasubramanian, P. (2019). The potential of lignocellulosic biomass precursors for biochar production: Performance, mechanism and wastewater application—a review. *Industrial Crops and Products*, *128*, 405–423. Available from https://doi.org/10.1016/j.indcrop.2018.11.041.

Saeed, A., Akhter, M., & Iqbal, M. (2005). Removal and recovery of heavy metals from aqueous solution using papaya wood as a new biosorbent. *Separation and Purification Technology*, *45*(1), 25–31. Available from https://doi.org/10.1016/j.seppur.2005.02.004.

Sas-Nowosielska, A., Kucharski, R., Małkowski, E., Pogrzeba, M., Kuperberg, J. M., & Kryński, K. (2004). Phytoextraction crop disposal – an unsolved problem. *Environmental Pollution*, *128*(3), 373–379. Available from https://doi.org/10.1016/j.envpol.2003.09.012, http://www.elsevier.com/inca/publications/store/4/0/5/8/5/6.

Sharma, H. B., Sarmah, A. K., & Dubey, B. (2020). Hydrothermal carbonization of renewable waste biomass for solid biofuel production: A discussion on process mechanism, the influence of process parameters, environmental performance and fuel properties of hydrochar. *Renewable and Sustainable Energy Reviews*, *123*. Available from https://doi.org/10.1016/j.rser.2020.109761, https://www.journals.elsevier.com/renewable-and-sustainable-energy-reviews.

Shen, X., Dai, M., Yang, J., Sun, L., Tan, X., Peng, C., Ali, I., & Naz, I. (2022). A critical review on the phytoremediation of heavy metals from environment: Performance and challenges. *Chemosphere*, *291*, 132979. Available from https://doi.org/10.1016/j.chemosphere.2021.132979.

Shi, L., Wang, L., Zhang, T., Li, J., Huang, X., Cai, J., Lü, J., & Wang, Y. (2017). Reducing the bioavailability and leaching potential of lead in contaminated water hyacinth biomass by phosphate-assisted pyrolysis. *Bioresource Technology*, *241*, 908–914. Available from https://doi.org/10.1016/j.biortech.2017.06.025, http://www.elsevier.com/locate/biortech.

Singh, B., Singh, B. P., & Cowie, A. L. (2010). Characterisation and evaluation of biochars for their application as a soil amendment. *Australian Journal of Soil Research*, *48*(7), 516–525. Available from https://doi.org/10.1071/SR10058.

Song, H., Wang, J., Garg, A., Lin, X., Zheng, Q., & Sharma, S. (2012). Quality variations of poultry litter biochar generated at different pyrolysis temperatures. *Journal of Analytical and Applied Pyrolysis*, *94*, 138–145. Available from https://doi.org/10.1016/j.jaap.2011.11.018.

Song, H., Wang, J., Garg, A., Lin, X., Zheng, Q., & Sharma, S. (2019). Potential of novel biochars produced from invasive aquatic species outside food chain in removing ammonium nitrogen: Comparison with conventional biochars and clinoptilolite. *Sustainability (Switzerland)*, *11*(24). Available from https://doi.org/10.3390/su11247136, https://res.mdpi.com/d_attachment/sustainability/sustainability-11-07136/article_deploy/sustainability-11-07136-v2.pdf.

Šyc, M., Pohořelý, M., Kameníková, P., Habart, J., Svoboda, K., & Punčochář, M. (2012). Willow trees from heavy metals phytoextraction as energy crops. *Biomass and Bioenergy*, *37*, 106–113. Available from https://doi.org/10.1016/j.biombioe.2011.12.025.

Thomas, P., Lai, C. W., & Bin Johan, M. R. (2019). Recent developments in biomass-derived carbon as a potential sustainable material for super-capacitor-based energy storage and environmental applications. *Journal of Analytical and Applied Pyrolysis*, *140*, 54–85. Available from https://doi.org/10.1016/j.jaap.2019.03.021.

Truong, H. B., Ike, I. A., Ok, Y. S., & Hur, J. (2020). Polyethyleneimine modification of activated fly ash and biochar for enhanced removal of natural organic matter from water via adsorption. *Chemosphere*, *243*. Available from https://doi.org/10.1016/j.chemosphere.2019.125454, http://www.elsevier.com/locate/chemosphere.

Uchimiya, M., Lima, I. M., Klasson, T., Chang, S., Wartelle, L. H., & Rodgers, J. E. (2010). Immobilization of heavy metal ions (CuII, CdII, NiII, and PbII) by broiler litter-derived biochars in water and soil. *Journal of Agricultural and Food Chemistry*, *58*(9), 5538–5544. Available from https://doi.org/10.1021/jf9044217.

Uchimiya, M., Wartelle, L. H., Klasson, K. T., Fortier, C. A., & Lima, I. M. (2011). Influence of pyrolysis temperature on biochar property and function as a heavy metal sorbent in soil. *Journal of Agricultural and Food Chemistry*, *59*(6), 2501–2510. Available from https://doi.org/10.1021/jf104206c.

Vocciante, M., Caretta, A., Bua, L., Bagatin, R., Franchi, E., Petruzzelli, G., & Ferro, S. (2019). Enhancements in phytoremediation technology: Environmental assessment including different options of biomass disposal and comparison with a consolidated approach. *Journal of Environmental Management*, *237*, 560–568. Available from https://doi.org/10.1016/j.jenvman.2019.02.104, https://www.sciencedirect.com/journal/journal-of-environmental-management.

Wang, S., Gao, B., Li, Y., Ok, Y. S., Shen, C., & Xue, S. (2017). Biochar provides a safe and value-added solution for hyperaccumulating plant disposal: A case study of *Phytolacca acinosa* Roxb. (Phytolaccaceae). *Chemosphere*, *178*, 59–64. Available from https://doi.org/10.1016/j.chemosphere.2017.02.121, http://www.elsevier.com/locate/chemosphere.

Wang, Z., Shu, X., Zhu, H., Xie, L., Cheng, S., & Zhang, Y. (2020). Characteristics of biochars prepared by co-pyrolysis of sewage sludge and cotton stalk intended for use as soil amendments. *Environmental Technology (United Kingdom)*, *41*(11), 1347–1357. Available from https://doi.org/10.1080/09593330.2018.1534891.

Wang, Q., Wang, B., Ma, Y., Zhang, X., Lyu, W., & Chen, M. (2022). Stabilization of heavy metals in biochar derived from plants in antimony mining area and its environmental implications. *Environmental Pollution*, *300*, 118902. Available from https://doi.org/10.1016/j.envpol.2022.118902.

Wang, Z., Xie, L., Liu, K., Wang, J., Zhu, H., Song, Q., & Shu, X. (2019). Co-pyrolysis of sewage sludge and cotton stalks. *Waste Management*, *89*, 430–438. Available from https://doi.org/10.1016/j.wasman.2019.04.033.

Van Wesenbeeck, S., Prins, W., Ronsse, F., & Antal, M. J. (2014). Sewage sludge carbonization for biochar applications. Fate of heavy metals. *Energy and Fuels*, *28*(8), 5318–5326. Available from https://doi.org/10.1021/ef500875c.

Yang, Y.-Q., Cui, M.-H., Ren, Y.-G., Guo, J.-C., Zheng, Z.-Y., & Liu, H. (2020). Towards understanding the mechanism of heavy metals immobilization in biochar derived from co-pyrolysis of sawdust and sewage sludge. *Bulletin of Environmental Contamination and Toxicology*, *104*(4), 489–496. Available from https://doi.org/10.1007/s00128-020-02801-4.

Yang, T., Meng, J., Jeyakumar, P., Cao, T., Liu, Z., He, T., Cao, X., Chen, W., & Wang, H. (2021). Effect of pyrolysis temperature on the bioavailability of heavy metals in rice straw-derived biochar. *Environmental Science and Pollution Research*, *28*(2), 2198–2208. Available from https://doi.org/10.1007/s11356-020-10193-5, https://link.springer.com/journal/11356.

Yao, Y., Gao, B., Zhang, M., Inyang, M., & Zimmerman, A. R. (2012). Effect of biochar amendment on sorption and leaching of nitrate, ammonium, and phosphate in a sandy soil. *Chemosphere*, *89*(11), 1467–1471. Available from https://doi.org/10.1016/j.chemosphere.2012.06.002, http://www.elsevier.com/locate/chemosphere.

Yousaf, B., Liu, G., Abbas, Q., Ali, M. U., Wang, R., Ahmed, R., Wang, C., Al-Wabel, M. I., & Usman, A. R. A. (2018). Operational control on environmental safety of potentially toxic elements during thermal conversion of metal-accumulator invasive ragweed to biochar. *Journal of Cleaner Production*, *195*, 458–469. Available from https://doi.org/10.1016/j.jclepro.2018.05.246, https://www.journals.elsevier.com/journal-of-cleaner-production.

Yuan, H., Lu, T., Huang, H., Zhao, D., Kobayashi, N., & Chen, Y. (2015). Influence of pyrolysis temperature on physical and chemical properties of biochar made from sewage sludge. *Journal of Analytical and Applied Pyrolysis*, *112*, 284–289. Available from https://doi.org/10.1016/j.jaap.2015.01.010.

Yuan, J. H., Xu, R. K., & Zhang, H. (2011). The forms of alkalis in the biochar produced from crop residues at different temperatures. *Bioresource Technology*, *102*(3), 3488–3497. Available from https://doi.org/10.1016/j.biortech.2010.11.018.

Zaker, A., Chen, Z., Wang, X., & Zhang, Q. (2019). Microwave-assisted pyrolysis of sewage sludge: A review. *Fuel Processing Technology*, *187*, 84–104. Available from https://doi.org/10.1016/j.fuproc.2018.12.011, https://www.journals.elsevier.com/fuel-processing-technology.

Zhao, B., Xu, X., Xu, S., Chen, X., Li, H., & Zeng, F. (2017). Surface characteristics and potential ecological risk evaluation of heavy metals in the bio-char produced by co-pyrolysis from municipal sewage sludge and hazelnut shell with zinc chloride. *Bioresource Technology*, *243*, 375–383. Available from https://doi.org/10.1016/j.biortech.2017.06.032.

Zhou, Y., Gao, B., Zimmerman, A. R., Fang, J., Sun, Y., & Cao, X. (2013). Sorption of heavy metals on chitosan-modified biochars and its biological effects. *Chemical Engineering Journal*, *231*, 512–518. Available from https://doi.org/10.1016/j.cej.2013.07.036.

Zhou, R., Zhang, M., Zhou, J., & Wang, J. (2019). Optimization of biochar preparation from the stem of Eichhornia crassipes using response surface methodology on adsorption of Cd2 + . *Scientific Reports*, *9*(1). Available from https://doi.org/10.1038/s41598-019-54105-1, http://www.nature.com/srep/index.html.

Zong, Y., Xiao, Q., & Lu, S. (2021). Biochar derived from cadmium-contaminated rice straw at various pyrolysis temperatures: Cadmium immobilization mechanisms and environmental implication. *Bioresource Technology*, *321*, 124459. Available from https://doi.org/10.1016/j.biortech.2020.124459.

CHAPTER

6 Biochar production methods and their diverse applications in improving plant production system to achieve agricultural sustainability

Elif Gödekmerdan and Benginur Baştabak

Ege University, Solar Energy Institute, Izmir, Turkey

6.1 Introduction

Addressing the escalating demands for food security driven by the increasing human population and consumer awareness necessitates a focus on sustainable agricultural practices (Sashidhar et al., 2020). In accordance with this purpose, various of materials, including activated carbon, straw, compost, zeolite, peat, vermiculite, perlite, and biochar, have been extensively studied for their potential to improve agricultural sustainability and soil management (Siedt et al., 2021). Among these materials, biochar has gained attention due to its cost-effectiveness and versatile agronomic attributes (Kim et al., 2020).

Biochar, a multifaceted substance, results from the thermal decomposition of organic feedstock through thermochemical conversion methods such as torrefaction, hydrothermal carbonization (HTC), pyrolysis, and gasification within the temperature range of 300°C–900°C (Kim et al., 2020; Leng et al., 2021). The physicochemical properties of biochar predominantly stem from feedstock types and process conditions, encompassing factors like temperatures, heating rates, and residence times, which also dictate the diverse fields of biochar application. Biochar obtained through pyrolysis boasts a highly porous structure and abundant surface functional groups, rendering it suitable for soil fertility improvement, carbon sequestration, and adsorption purposes (Brassard et al., 2016; Tripathi et al., 2016). Biochar's alkaline nature contributes to enhanced soil pH, playing a pivotal role in the immobilization of toxic metals via adsorption or precipitation of insoluble salts in soils, thereby impeding the transition of heavy metals to plants (Abdin et al., 2020). Beyond these benefits, empirical studies have underscored biochar's positive impact on diverse physicochemical properties of soil,

Biochar Ecotechnology for Sustainable Agriculture and Environment. DOI: https://doi.org/10.1016/B978-0-443-29855-4.00007-2

demonstrating improvements in seed germination rates, reduction of salt stress, and enhancement of plant growth (Kavitha et al., 2018; Shi et al., 2021). For instance, Yao et al. (2013) reported an increase in seed germination rates from 53% to 85% following biochar application, while Farhangi-Abriz and Torabian (2018) observed a significant reduction in salt stress on bean seedlings through the addition of 20% biochar, leading to improved growth hormones and reduced sodium concentrations.

However, alongside its potential benefits to soil and plant growth, biochar's composition may contain toxic compounds such as nonbioavailable heavy metals, volatile organic compounds, and polycyclic aromatic hydrocarbons (PAHs), posing potential risks to plants and soil microorganisms (Gezahegn et al., 2021). To address biochar-induced toxicity, researchers have explored various treatment methods, including washing, thermal drying, and pretreatment with ferric salts, with thermal treatment emerging as a more effective strategy for reducing biochar phytotoxicity (Intani et al., 2018; Schmidt et al., 2021). Moreover, organizations such as the European Biochar Certificate (EBC) and the International Biochar Initiative (IBI) have established guideline values to assess biochar quality during production, helping to avoid or mitigate potential adverse effects (Shi et al., 2021).

This chapter aims to provide a comprehensive exploration of various biochar production methods and their diverse applications in optimizing plant production systems for agricultural sustainability. The study encompasses aspects of an in-depth analysis of different biochar production techniques, ranging from torrefaction to modern high-tech pyrolysis processes. On the other hand, the multifaceted applications of biochar in improving soil health, nutrient management, water retention, and overall plant growth was investigated. Furthermore, this chapter explores the role of biochar in sequestering carbon, thereby contributing to climate change mitigation and carbon offset initiatives. The study draws on a wide range of relevant scientific data collected from reputable databases, including the Web of Science, Scopus, and Science Direct, covering the period from 2013 to 2023, as well as information obtained from the official websites of the EBC and IBI.

6.2 Production of biochar

6.2.1 Torrefaction

Torrefaction is a thermal process in that biomass is thermally degraded under an inert atmosphere in the absence of oxygen at relatively low temperatures (200°C–300°C). During the torrefaction process, the organic fraction of biomass is destroyed through devolatilization, deoxygenation, depolymerization, and carbonization stages, resulting in the production of biochar or torrefied product, condensable and noncondensable gases (Barskov et al., 2019). Torrefaction can enhance biomass properties and increase the quality of biomass by modifying surface functional groups through decomposition stages. Thanks to its a few characteristic features such as high energy density, carbon content, hydrophobicity, and grind ability, biochar or torrefied product can use as solid biofuel (Mamvura & Danha, 2020). On the other hand, some researchers reported that biochar can be used as feedstock in the gasification process due to low oxygen content. Thus the thermodynamic losses resulting from overoxidation during gasification are prevented, increasing the gasification efficiency and the syngas quality (Bach & Skreiberg, 2016; Mamvura & Danha, 2020).

6.2.2 Hydrothermal carbonization

On the contrary, wet biomass such as municipal waste and sewage sludge can be converted to biochar through HTC without mechanical dewatering (Kumar et al., 2020). In HTC, biomass is converted into value-added solid residue called hydrochar in the presence of subcritical liquid water at temperatures ranging from 180°C to 280°C and 2–10 MPa pressure (Table 6.1; Krysanova et al., 2019).

6.2.3 Pyrolysis

Another effective method is pyrolysis, which converts biomass into useful products such as high-quality biochar, bio-oil, and syngas through thermochemical decomposition of organic materials at temperatures ranging from 400°C to 1300°C (Table 6.1; Tripathi et al., 2016). In addition, pyrolysis technologies are categorized as slow, intermediate, fast, and microwave pyrolysis depending on temperature and heating rate. By this means, product characteristics can be adjusted suitable for various applications (Brassard et al., 2016). Slow pyrolysis occurs at relatively low temperatures (400°C–500°C) with low heating rates (5°C–7°C/min), whereas fast pyrolysis takes place at a high-temperature range between 800°C and 1300°C with the heating rate in the range of 10°C–200°C/min (Elkhalifa et al., 2019). Contrary to fast pyrolysis, biochar is obtained as the main product during slow pyrolysis (Kim et al., 2020). In addition to conventional pyrolysis, microwave-assisted pyrolysis involves microwave dielectric heating which provides uniform temperature distribution throughout biomass particles and thereby increases yield and quality of biochar. Owing to the removal of volatile matters uniformly from biomass with internal heating, the pore volume and the surface area of biochar enhance during microwave-assisted pyrolysis.

6.2.4 Gasification

Besides pyrolysis, gasification is also another option to produce biochar. The gasification process occurs at temperatures between 700°C and 1600°C using gasification agents such as air, steam, oxygen, or carbon dioxide (CO_2), which converts biomass into syngas as the main product and biochar and bio-oil (Table 6.1;

Table 6.1 Process, temperature, heating rate, pressure, residence time, biochar yield, usage area.

Process	Temperature (°C)	Heating rate (°C/min)	Pressure (MPa)	Microwave power level (W)	Residence time (min)	Biochar yield (%)	Usage area	References
Hydrothermal carbonization	180–280	<1	2–10	–	5–240	35–60	Solid fuel, soil amendment	Bach and Skreiberg (2016), Elkhalifa et al. (2019), Hussain et al. (2020), Kumar et al. (2020), Mamvura and Danha (2020), Tripathi et al. (2016)
Torrefaction	200–300	<50	–	–	≤ 60	60–80	Solid fuel, soil amendment	
Slow pyrolysis	400–500	5–7	–	–	≥60	30–60	Soil amendment, adsorbent	
Intermediate pyrolysis	~500	25–100	–	–	0.5–25	20–40	Soil amendment, adsorbent	
Fast pyrolysis	800–1300	10–200	–	–	Several seconds	15–20	Soil amendment, adsorbent	
Microwave pyrolysis	–	–	–	400–1200	10–120	27–60	Adsorbent	
Gasification	700–1600	–	–	–	–	5–10	Catalysts for tar removal and biodiesel production, adsorbent	

Elkhalifa et al., 2019). In terms of the biochar yield, the gasification process (5%–10% of the feedstock) produces less biochar than other thermochemical processes such as pyrolysis (15%–60% of the feedstock) and torrefaction (60%–80% of the feedstock) (Kumar et al., 2020; Oni et al., 2019). The choice of processes for biochar production depends on the properties of biomass material and intended use. Biomass materials with low moisture content (up to 10 wt.%) are feasible for pyrolysis, torrefaction, and gasification (Fakayode et al., 2020).

6.3 Diverse biochar applications in improving plant production

More recently, biochar has received attention for both agricultural and environmental applications due to its being cost-effective, high-efficiency, and environmentally friendly (Shan et al., 2020). The ability of biochar to be used in agricultural production depends mainly on the process temperature and raw material types, and its physicochemical properties, including specific surface area (SSA), surface functional groups, water retention capacity (WRC), pH, electrical conductivity, and cation exchange capacity (CEC) (Deng et al., 2021; Kuppusamy et al., 2016; Wang et al., 2021).

Through its versatile physicochemical properties, high quality biochar, which meets certain criteria (IBI and EBC biochar standards) for use in agricultural production, can be used for various applications (Fig. 6.1) such as soil fertility improvement, remediation of contaminated soils, carbon sequestration and mitigation of greenhouse gas emissions, soilless agricultural production, regenerative agriculture/agroecology/nature-based solutions, compost amendment, and pest and disease management (Cha et al., 2016).

6.4 Soil fertility improvement

6.4.1 Soil aggregation

Biochar, a carbon-rich material produced through the pyrolysis of organic matter, has gained considerable attention for its potential to enhance soil structure and promote agricultural sustainability. One of the key benefits of biochar application is its ability to improve soil aggregation, leading to better aeration and root penetration. Soil aggregation, a fundamental aspect of soil structure, is critical for various soil functions, including water retention, aeration, and root penetration.

The physicochemical properties of biochar can be modified by adjusting pyrolysis temperatures. Biochar produced at temperatures below 400°C typically exhibits lower surface areas and a less porous structure. This type of biochar contains a relatively higher proportion of oxygen-containing functional groups, such

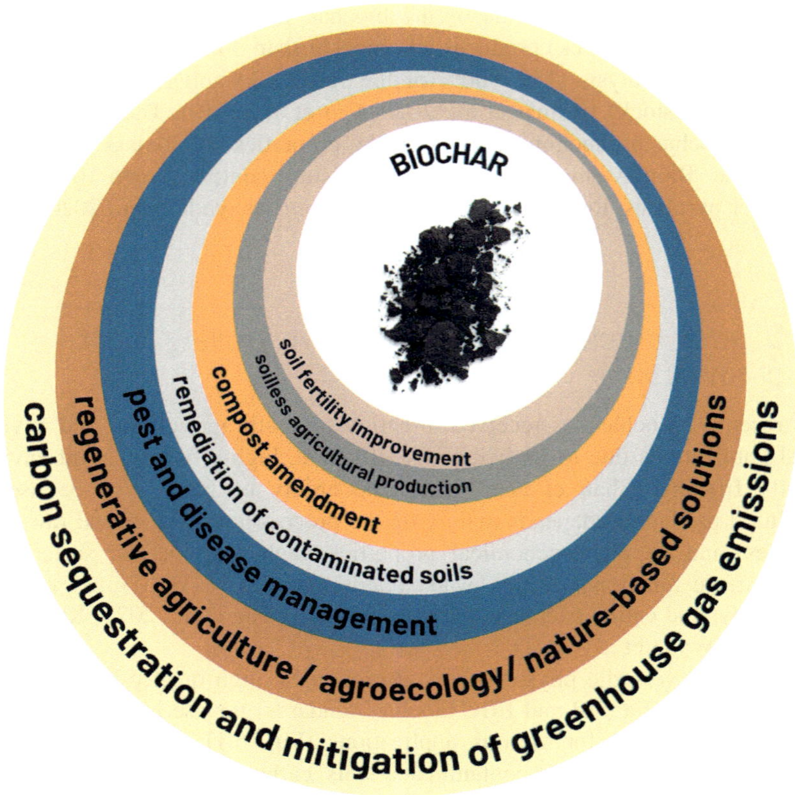

FIGURE 6.1

Diverse biochar applications in improving plant production.

as hydroxyl (−OH) and carboxyl (−COOH) groups (Hopkins & Hawboldt, 2020; Kuppusamy et al., 2016; Nuagah et al., 2020). In contrast, high-temperature pyrolysis (>500°C) yields biochar with a more porous structure and a significantly larger surface area, which enhances its water retention and aeration properties. High-temperature biochar has a reduced presence of oxygen-containing groups (Inyang et al., 2016). Islam et al. (2021) stated that soil structure improved by around 16% with woody biochar at high temperatures.

Due to its extremely porous structure, biochar acts as a physical scaffold within the soil matrix. Because biochar is porous, it offers a sink for helpful soil bacteria. Aggregation is enhanced by these microorganisms because they aid in binding soil particles together. For soil microorganisms, especially fungus and bacteria, these pores function as microhabitats. These microorganisms, through their metabolic activities and the production of extracellular polymeric substances (EPS), play a pivotal role in aggregating soil particles. The EPS that bacteria release binds soil particles together naturally. The presence of biochar provides a conducive environment for these microbial communities to thrive, augmenting their role in fostering soil aggregation. Furthermore, biochar's porous nature offers protection to these microorganisms from adverse environmental conditions, sustaining their aggregate-stabilizing functions over time (Zhou et al., 2020).

6.4.2 Water retention capacity

The capacity of the soil to hold water is crucial for agricultural systems because it affects crop growth, drought resistance, and overall agricultural sustainability. Biochar plays a pivotal role in improving soil aggregation, thereby influencing WRC and CEC in agricultural systems (Das et al., 2021; Kuppusamy et al., 2016). Biochar possesses a highly porous structure characterized by an intricate network of macro-, meso-, and micropores. When incorporated into soil, biochar contributes to the modification of the soil's pore structure, transforming it into a more conducive environment for water retention. Biochar's surface chemistry plays a pivotal role in its WRC. The surface of biochar contains functional groups, such as hydroxyl (−OH) and carboxyl (−COOH) groups, that exhibit strong affinity for water molecules. These functional groups create hydrogen bonds with water, effectively absorbing and holding moisture within the biochar matrix. The modification of pore structure, hydrophilic surface chemistry, and reduction in water loss via runoff and evaporation collectively result in more resilient and water-efficient soils. Some studies show that as the process temperature increases, its effect on WHC is positive. However, the positive effect on the soil is seen at different rates depending on the structure of the soil (Das et al., 2021; Wei et al., 2023).

6.4.3 Cation exchange capacity

Biochar's high surface area and porous structure also contribute to an increase in soil CEC (Das et al., 2021). At the same time, it was observed that the CEC decreased as the ash content in biochar increased (Domingues et al., 2020). CEC refers to the soil's ability to adsorb and exchange positively charged ions (cations) such as calcium (Ca^{2+}), magnesium (Mg^{2+}), and potassium (K^{+}) (Hussain et al., 2020; Kazemi Shariat Panahi et al., 2020). Munera-Echeverri et al. (2018) found that the CEC values of biochars ranged from 5 to 50 cmol/kg in the literature. The high surface area of biochar provides ample binding sites for cations. This increases the total CEC of the soil, making essential nutrients more available to plant roots. Additionally, the improved CEC ensures a more stable supply of essential nutrients to plants, potentially reducing the requirement for synthetic fertilizers. Furthermore, biochar's role in retaining and slowly releasing nutrients can mitigate nutrient runoff into water bodies, thus contributing to the reduction of water pollution and eutrophication. In conclusion, biochar's impact on soil aggregation extends to the enhancement of WHC and CEC, both of which are critical for sustainable agriculture. Through modification of soil pore structure, hydrophilic surface chemistry, and nutrient retention and release, biochar aids in conserving water resources and improving nutrient availability to plants. This multifaceted role underscores the significance of biochar as a valuable soil amendment for enhancing soil quality and agricultural productivity, while also promoting sustainable land and water management practices.

6.4.4 Organic carbon and humic fractions

Biochar application is associated with an increase in the production of humic substances in soils. Humic substances, humic acid (HA), fulvic acid (FA), and humin, are organic molecules renowned for their ability to act as natural binders. These substances aid in the creation of soil aggregates by promoting attractive forces between soil particles. The interaction between humic substances and soil particles is primarily electrostatic and van der Waals forces, facilitating the formation of stable aggregates. Biochar acts as a catalyst for this process by providing a platform for the adsorption and concentration of these humic substances, thereby intensifying their influence on soil aggregation. Zhao et al. (2018) stated that dissolved organic carbon (DOC), HA, and FA contents decreased in biochar produced at 500°C and 600°C increased in biochar produced at 300°C and 400°C).

6.5 Remediation of contaminated soils

Biochar exhibits substantial potential for the remediation of contaminated soils. Its porous structure, high surface area, and adsorption capacity make it an effective soil amendment for sequestering, immobilizing, and degrading a wide range of contaminants (Fig. 6.2). The interaction between biochar and the remediation of contaminated soils is characterized by several key mechanisms such as adsorption/immobilization and chemical transformation (Yaashikaa et al., 2020).

6.5.1 Adsorption and immobilization

Biochar's high surface area, porous structure, and chemical functional groups enable it to adsorb and immobilize a wide range of contaminants, reducing their bioavailability and potential for harm (El-Naggar et al., 2019). SSA and total pore volume of biochar range from 8 to 520 m^2/g and from 0.016 to 0.083 cm^3/g, respectively (Leng et al., 2021). Biochar derived from lignocellulosic biomass contains a more SSA and total pore volume than animal waste and sludge biochar due to the existence of different minerals in their biomass matrix (Das et al., 2021). This surface area allows for a significant adsorption capacity, enabling the binding of contaminants. Biochar's surface contains various chemical functional groups, including oxygen-containing moieties (e.g., hydroxyl, carboxyl, and

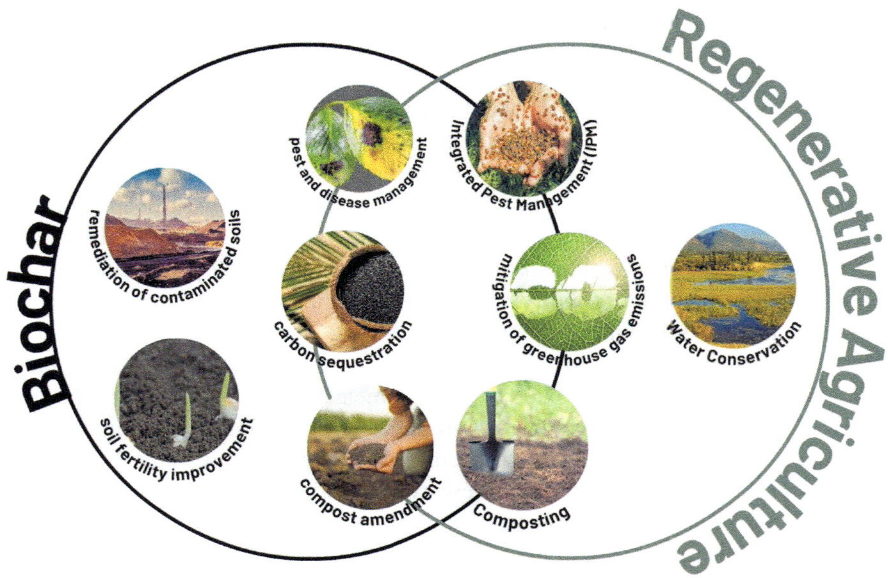

FIGURE 6.2

Biochar for regenerative agriculture.

phenolic groups). These groups can participate in chemical bonding with contaminants. For instance, metal ions may form bonds with oxygen groups through coordination chemistry. On the other hand, biochar can physically adsorb organic compounds, such as hydrophobic organic pollutants, through hydrophobic interactions between the nonpolar regions of the contaminants and the biochar surface (Mansoor et al., 2021).

Immobilization refers to the confinement or trapping of contaminants within the biochar matrix, rendering them less mobile and bioavailable. This dynamic interaction plays a pivotal role in the effectiveness of biochar in remediating contaminated soils. Biochar is particularly effective in immobilizing heavy metals. It can bind metals through ion exchange, precipitation, complexation, or surface adsorption, effectively reducing metal toxicity and preventing their leaching into groundwater. Biochar can adsorb a wide range of organic pollutants, including pesticides, PAHs, and pharmaceutical residues. The physical and chemical interactions between biochar and these pollutants can lead to their sequestration within the biochar matrix (Yaashikaa et al., 2020).

6.5.2 Chemical transformation

Chemical transformation refers to the alteration of contaminants' chemical structure or properties when they contact with biochar. This interaction is a key mechanism in the remediation of contaminated soils using biochar. Biochar can serve as an electron shuttle, facilitating redox (reduction–oxidation) reactions in the soil. It can act as an electron donor or acceptor, depending on the conditions, promoting the transformation of contaminants. For example, in reducing environments, biochar can reduce toxic elements such as chromium (Cr) from a more toxic, mobile form (Cr(VI)) to a less toxic, and immobile form (Cr(III)) (Kim et al., 2020). Biochar can influence soil pH, which can have significant implications for contaminant transformation. Changes in pH can affect the solubility and speciation of contaminants. For instance, biochar can raise soil pH in acidic soils, which may influence the mobility of certain contaminants, like heavy metals. In summary, biochar's chemical properties, including its redox potential, catalytic activity, and pH-adjusting capabilities, enable it to transform contaminants, making them less mobile, less toxic, or more amenable to degradation (Mansoor et al., 2021). In conclusion, biochar's multifaceted interactions with contaminated soils make it a promising tool for soil remediation while simultaneously supporting plant production. By addressing these considerations, biochar can play a crucial role in achieving agricultural sustainability by remediating contaminated soils and facilitating healthy plant growth.

6.6 Carbon sequestration and mitigation of greenhouse gas emissions

Biochar is a versatile agricultural amendment known for its potential to enhance plant production and mitigate climate change by sequestering carbon. Improving biochar for diverse applications in terms of carbon sequestration requires attention to various factors and their interaction with soil and plants (Shin et al., 2021). The carbon recalcitrance or stability of biochar is a key characteristic that determines its carbon sequestration potential. Many studies have been carried out to show the effects of the choice of pyrolysis conditions and feedstock during biochar production significantly influences its carbon sequestration capacity (Leng & Huang, 2018). Higher pyrolysis temperatures and longer residence times generally result in biochar with greater carbon stability, making it more effective in long-term carbon sequestration (Fig. 6.2). Different feedstocks (wood, crop residues, and manure) yield biochar with varying properties. Some feedstocks are more recalcitrant and resistant to decomposition, enhancing carbon retention in the soil (Lataf et al., 2022). Nzediegwu et al. (2021) reported that the high carbon stability of biochar derived from lignocellulosic feedstocks (canola straw, sawdust, and wheat straw) were significantly higher than that of nonlignocellulosic feedstock (manure pellet) biochar. The existence of DOC in biochar prepared from manure and crop is higher than that of biochar obtained from wood. This may be related to the ash contents of manure and crop, which is relatively high compared to that of wood (Han et al., 2020). According to Wei et al. (2019), total DOC contents of biochar prepared from rice straw, pinewood, pig manure, and sewage sludge were 15.9, 2.23, 18.3, and 12.2 g/kg, respectively. The main reason could be ascribed to differences in the chemical characteristics of the original feedstock.

Biochar can store carbon in the soil for an extended period, helping to mitigate climate change by reducing atmospheric CO_2 levels (Zhang, Tang, et al., 2023). Biochar, due to its high carbon content and recalcitrant nature, resists decomposition and can retain carbon for centuries to millennia. Biochar consists of different carbon fractions including a labile carbon pool like DOC and a stable carbon pool such as aliphatic carbon, aromatic carbon, and functional groups (Xing et al., 2021). During the pyrolysis process that transforms organic matter into biochar, a portion of the initial feedstock's carbon content may be converted into volatile compounds, contributing to the formation of DOC in the biochar matrix. A stable carbon pool provides long-term carbon sequestration because it is recalcitrant and has high ability to resist mineralization, on the other hand, a labile carbon pool changes the physicochemical properties of soil (Das et al., 2021; Wei et al., 2019). Although labile organic carbon in biochar rapidly mineralizes, the remaining recalcitrant organic carbon is mineralized very slowly by microorganisms (Han et al., 2020). The stability of biochar carbon is attributed to its unique physicochemical properties, including its high aromaticity, surface area, and structural integrity. Biochar's surface area and porosity influence its capacity to adsorb and stabilize organic matter and carbon compounds (Chung et al., 2023). Higher surface area biochar typically exhibits better carbon sequestration potential, as they provide more binding sites for carbon molecules. Biochar's chemical composition, including the presence of functional groups like hydroxyl (−OH) and carboxyl (−COOH), can affect its ability to interact with soil organic matter and enhance carbon sequestration (Leng & Huang, 2018).

Simultaneously, biochar serves as a potent tool for carbon sequestration by storing carbon in the soil, reducing greenhouse gas emissions such as CO_2, methane, and nitrous oxide (N_2O) from the soil, and promoting sustainable agricultural practices. The combined impact of these benefits positions' biochar as a valuable component of strategies aimed at achieving agricultural sustainability and mitigating climate change on a global scale. This enriched soil organic carbon content improves soil fertility by providing a continuous source of nutrients as it slowly decomposes. This carbon sequestration potential not only mitigates climate change by reducing greenhouse gas but also has profound effects on plant production (Wang et al., 2023).

N_2O is a greenhouse gas emitted into the atmosphere when fertilizers are excessive or misused in agricultural fields. Reducing the excessive use of nitrogen fertilizers is an effective way to decrease N_2O emissions. According to Zhang, Miao, et al. (2023), when traditional fertilization is reduced by 50% and combined with biochar applications at rates of 15 and 60 t/ha, cumulative N_2O emissions increased by 35.3% and 216%, respectively, but showed increases of up to 75% in rapeseed yield. Biochar combined with different nitrogen fertilization rates increased crop yield and greenhouse gas emissions in a rapeseed-soybean rotation system (Yang et al., 2019) indicated that 20 and 40 t/ha biochar application reduced global warming potential of methane and N_2O emissions by an average of 35.7% and 21.5%. The results showed up to 24.3% and 30.8% increases in rice yield and irrigation water productivity, respectively.

6.7 Soilless agricultural production

In recent years, research on soilless agriculture technologies has been increasing to meet the food needs of the increasing population. Some of this research is on materials that can be used instead of soil. Biochar has been the subject of research in soilless agricultural technologies due to its physical and chemical properties. Two hypotheses propose different mechanisms by which biochar can have a positive impact on plant growth. Biochar enhances plant growth by increasing the population of beneficial bacteria and fungi. This enhanced microbial activity can induce plant growth and help in the development of systemic resistance to plant diseases. Biochar contains low amounts of specific chemicals, including alkanoic acid, hydroxy and acetoxy acids, thiols, diol, benzoic acid, phenol, octadecane, and 2-phenoxy-ethanol. These chemical compounds available in biochar may stimulate plant growth when they interact with the plants' root systems or soil, leading to growth effects (Karakaş et al., 2017).

In a more recent study by Simiele et al. (2022), the addition biochar to a soilless substrate greatly enhanced the morphological and chemical characteristics of *Solanum lycopersicum* var. cerasiforme plants during the vegetative and fruit stages. They found that, *S. lycopersicum* var. cerasiforme plants treated with biochar had more fruits and a greater level of antioxidants compared to the untreated plants. The enhanced nutrient availability in the biochar-based substrate coincided with the enhancement of these plant traits. This study highlighted that biochar amendment has significant potential to increase tomato output and fruit quality by enhancing the physical and chemical properties of the growing medium. Zabaleta et al. (2023) studied the effects of the addition of almond shell biochar and perlite on Eruca vesicaria growth in soilless medium. The study indicated that the soilless Eruca vesicaria growing was enhanced by the addition of almond shell biochar compared to using only perlite. They showed that biochar produced from almond shell through pyrolysis process has lower negative environmental effects according to the circular economy idea.

6.8 Regenerative agriculture: agroecology and nature-based solutions

Soil is a living, breathing organism; traditional agricultural methods have the potential to degrade it. One of the most crucial components of agricultural sustainability is healthy soil since it regulates microbial activity, nutrient recycling, and decomposition in addition to providing several ecosystem services. Restoring soil quality can be facilitated by using natural based solutions, which will increase agricultural sustainability and productivity such as compost, animal manure and biochar (Mrunalini et al., 2022). Gopal et al. (2020) converted lignin-rich biomass residues remained by coconut plantations into biochar and investigated effects of coconut husk biochar on chilli growth. They found coconut husk biochar can be applied as an amendment that at a rate of 2 t/ha with coconut leaf vermicompost to support regenerative agriculture by boosting crop yield and improving soil health.

It is inevitable to holistic transition into regenerative agricultural practices immediately due to the negative impacts of conventional farming practices. In this context, carbon-negative technologies—such as the usage of biochar—as well as other practices can be used to mitigate greenhouse gas emissions dramatically and facilitate the shift to carbon-neutral economies and circular food chains. In addition, the usability of biochar in agricultural production from the perspective of nature-based solutions has also been emphasized by FAO and The Nature Conservancy (Miralles-Wilhelm, 2021).

6.8.1 Compost amendment

Compost amendment refers to the incorporation of biochar into composting processes, enhancing the quality and functionality of the resulting compost. This synergistic approach combines the benefits of biochar and compost, offering a potent soil amendment for improving plant production. To optimize biochar for diverse applications in compost amendment, optimal feedstock selection, biochar production conditions, particle size and surface area, biochar activation, composting process integration can be modified (Wang et al., 2018). Pyrolysis conditions can be designed to retain biochar's structural integrity and beneficial properties, ensuring compatibility with the composting matrix. Activating biochar through methods such as steam treatment or chemical activation can enhance its adsorption capacity, allowing it to capture and retain nutrients during composting. This activated biochar can act as a reservoir for slowly releasing essential elements. The incorporation of biochar into the composting process requires careful consideration of timing. Strategic addition during specific phases, such as the thermophilic phase, can maximize interactions between biochar, composting microbes, and organic matter decomposition (Siedt et al., 2021).

Biochar acts as a nutrient reservoir within the compost matrix. During composting, biochar adsorbs and retains nutrients, preventing leaching and providing a slow-release mechanism. This aids in nutrient availability for plants over an extended period. The porous structure of biochar provides a habitat for diverse microbial communities during composting. This enhanced microbial diversity contributes to the breakdown of organic matter and the transformation of nutrients into plant-available forms. Biochar's inherent alkalinity can buffer pH fluctuations during composting. This buffering effect helps maintain an optimal pH range for microbial activities and nutrient availability, supporting the overall quality of the compost. Biochar's hydrophilic nature improves water retention in the compost. This property aids in moisture management during composting and, when incorporated into the soil, contributes to improved water retention and soil structure (Tovihoudji et al., 2022).

Biochar contributes to the long-term stability of organic carbon in the compost. This stability translates into enhanced carbon sequestration when the compost–biochar blend is applied to the soil, promoting sustainable practices. The interaction between compost amendment and biochar results in a synergistic soil amendment that combines the nutrient-rich qualities of compost with the structural and adsorptive benefits of biochar. This blend contributes to improved plant growth, increased yields, and overall agricultural sustainability (Zhang, Tang, et al., 2023). The interaction between compost amendment and biochar underscores their combined potential for sustainable agricultural practices and plant production system improvement.

6.8.2 Pest and disease management

It is known that soil-borne pathogens reduce productivity in agricultural production and cause crop loss. The main mechanisms of action for plant-parasitic nematodes include modifications to the variety of the soil microbial population, the release of nematicidal chemicals, and the induction of plant defences (Poveda et al., 2021). Soil-borne pathogens, such as bacteria, fungus, nematodes, and certain viruses, can be minimized with appropriate treatments. According to Iacomino et al. (2022), biochar has shown suppressive effects when applied to pathogens. They stated that 86% of fungus, 100% of oomycetes, 100% of viruses, 96% of bacteria, and 50% of nematodes were suppressed by biochar.

6.9 Case studies and examples

Several authors have also observed biochar are essential for improving soil quality in the long term. For example, Tovihoudji et al. (2022) investigated that the effects of the compost and biochar on cotton growth and yield through the on-farm trials in Northern Benin. They used mineral fertilizer, cow dung–based compost, household waste–based compost, a combination of cow dung–based compost and biochar, and a combination of household waste–based compost and biochar. The results indicated that compared to the control, cotton yields enhanced 96% and 73% for combination of cow dung–based compost and biochar (Terra preta of Banikoara) and combination of household waste–based compost and biochar (Terra preta of Toffo), respectively. They found that the highest plant growth parameters and yield were observed with mineral fertilizer and combination cow dung–based compost and 15% biochar designated Terra peta at 300 kg/ha. They showed that the biochar–compost-based amendments could enhance growth and yields of cotton and farmers' livelihoods (Tovihoudji et al., 2022).

A similar result has been reported by Chiaramonti and Panoutsou (2019) who modeled a concrete case study evaluating economic support measures. In this study, biochar and compost applications were carried out within the context of improving sunflower cultivation in dry marginal lands in Italy. A mixture of 20% (w/w) biochar + 80% (w/w) compost was applied to both conventional and marginal land, at 0, 5, and 10 t biochar/ha. The results indicated that compared to the application of biochar, a combination of biochar and compost further increased crop yield by a range of 10%–15% on marginal land. They suggested that the application of biochar could be economically beneficial by increasing crop yield and boosting farmers' income. These results suggest that application of biochar sequestered carbon at low cost regarding the long-term effects of biochar when applied in large-scale farming systems.

Šimanský et al. (2022) studied the impact of two different biochar substrates such as biochar-farmyard manure and biochar-farmyard manure-digestate at two rates (10 and 20 t/ha) combined alone and with additional fertilization on two contrasting soils (a sandy Arenosol and a loamy Chernozem) in the field for 3 years. They stated that most of the soil's physical properties such as capillary porosity, capillary rise, maximum capillary water capacity, WRC, wilting point, available soil water content, soil water content, and soil structure were more favorably affected by the application of biochar blended with farmyard manure. Furthermore, they dedicated that applying biochar blended with farmyard manure at a rate of 20 t/ha was more effective than at a rate of 10 t/ha.

6.10 Future research and directions

The multifaceted applications of biochar contribute to the adoption of sustainable agricultural practices. The reduction in synthetic fertilizer dependence, improved nutrient management, and minimized environmental impacts position biochar as a key player in fostering eco-friendly and resilient agricultural systems. Previous studies have shown that biochar application to soil: (1) affects microbial activity and soil enzyme activity, (2) provides shelter for soil microbiota due to its large surface area and pore structures, (3) enhances soil nutrients and serves as a nutrient source for soil microbes, (4) helps to reduce the risk of nutrient losses, (5) increases the soil's CEC, therby decreasing nutrient losses and improving nutrient transfer to plants roots, and (6) increases carbon sequestration. To obtain the desired elemental composition and surface properties, biochar can be modified by adjusting production conditions, selecting various engineering techniques, or using different feedstocks. Many studies highlighting the advantages of biochar for sustainable agricultural production and climate change suggest that it is likely to become widespread in the future.

Case studies and research trials highlight biochar's ability to improve crop yields. However, there are important issues that need to be investigated to produce the right strategies regarding biochar usage in agricultural production. These strategies include environmental and agricultural socioeconomics impact. To understand the environmental impact of biochar, large-scale field trials should be carried out at multiple locations. In biochar production and application, applying certain quality criteria (IBI and EBC biochar standards) is critical to mitigate environmental impact of biochar. This not only addresses the environmental impact of biochar but also has significant implications for the livelihoods of farmers, ensuring increased productivity and economic stability. Application of biochar could be economically beneficial farmers' income by increasing crop yield and boosting farmers' income. Following this, biochar socioeconomics in agricultural systems merits considerable consideration. The integration of biochar technology into agricultural systems can stimulate local economies and promote sustainable development. Therefore the environmental and agricultural socioeconomics impact of biochar is substantial and far-reaching. This chapter explores biochar production methods and applications as a transformative technology that holds promise for ushering in a new era of sustainable agriculture. As we look forward, continued research, informed policymaking, and widespread adoption of biochar practices can contribute to a more resilient, productive, and environmentally conscious agricultural landscape.

Acknowledgments

The authors would like to acknowledge the support of Ege University Solar Energy Institute. They would like to thank the editor and the anonymous reviewers for their valuable feedback and insightful comments.

CRediT authorship contribution statement

Elif Gödekmerdan: conceptualization, methodology, writing—reviewing and editing, supervision. Benginur Baştabak: conceptualization, methodology, writing—reviewing and editing, visualization.

Declaration of competing interest

The authors declare that they have no known competing financial interests or personal relationships that could have appeared to influence the work reported in this paper.

References

Abdin, Y., Usman, A., Ok, Y. S., Tsang, Y. F., & Al-Wabel, M. (2020). Competitive sorption and availability of coexisting heavy metals in mining-contaminated soil: Contrasting effects of mesquite and fishbone biochars. *Environmental Research*, *181*. Available from http://www.elsevier.com/inca/publications/store/6/2/2/8/2/1/index.htt, https://doi.org/10.1016/j.envres.2019.108846.

Bach, Q. V., & Skreiberg, O. (2016). Upgrading biomass fuels via wet torrefaction: A review and comparison with dry torrefaction. *Renewable and Sustainable Energy Reviews*, *54*, 665–677. Available from https://www.journals.elsevier.com/renewable-and-sustainable-energy-reviews, https://doi.org/10.1016/j.rser.2015.10.014.

Barskov, S., Zappi, M., Buchireddy, P., Dufreche, S., Guillory, J., Gang, D., Hernandez, R., Bajpai, R., Baudier, J., Cooper, R., & Sharp, R. (2019). Torrefaction of biomass: A review of production methods for biocoal from cultured and waste lignocellulosic feedstocks. *Renewable Energy*, *142*, 624–642. Available from http://www.journals.elsevier.com/renewable-and-sustainable-energy-reviews/, https://doi.org/10.1016/j.renene.2019.04.068.

Brassard, P., Godbout, S., & Raghavan, V. (2016). Soil biochar amendment as a climate change mitigation tool: Key parameters and mechanisms involved. *Journal of Environmental Management*, *181*, 484–497. Available from https://www.sciencedirect.com/journal/journal-of-environmental-management, https://doi.org/10.1016/j.jenvman.2016.06.063.

Cha, J. S., Park, S. H., Jung, S. C., Ryu, C., Jeon, J. K., Shin, M. C., & Park, Y. K. (2016). Production and utilization of biochar: A review. *Journal of Industrial and Engineering Chemistry*, *40*, 1–15. Available from http://www.sciencedirect.com/science/journal/1226086X, https://doi.org/10.1016/j.jiec.2016.06.002.

Chiaramonti, D., & Panoutsou, C. (2019). Policy measures for sustainable sunflower cropping in EU-MED marginal lands amended by biochar: Case study in Tuscany, Italy. *Biomass and Bioenergy*, *126*, 199–210. Available from http://www.journals.elsevier.com/biomass-and-bioenergy/, https://doi.org/10.1016/j.biombioe.2019.04.021.

Chung, B. Y. H., Ang, J. C., Tang, J. Y., Chong, J. W., Tan, R. R., Aviso, K. B., Chemmangattuvalappil, N. G., & Thangalazhy-Gopakumar, S. (2023). Rough set approach to predict biochar stability and pH from pyrolysis conditions and feedstock characteristics. *Chemical Engineering Research and Design*, *198*, 221–233. Available from http://www.elsevier.com/wps/find/journaldescription.cws_home/713871/description#description, https://doi.org/10.1016/j.cherd.2023.09.003.

Das, S. K., Ghosh, G. K., & Avasthe, R. (2021). Applications of biomass derived biochar in modern science and technology. *Environmental Technology and Innovation*, *21*. Available from https://doi.org/10.1016/j.eti.2020.101306, http://www.journals.elsevier.com/environmental-technology-and-innovation/.

Deng, S., Chen, J., & Chang, J. (2021). Application of biochar as an innovative substrate in constructed wetlands/biofilters for wastewater treatment: Performance and ecological benefits. *Journal of Cleaner Production*, *293*. Available from https://www.journals.elsevier.com/journal-of-cleaner-production, https://doi.org/10.1016/j.jclepro.2021.126156.

Domingues, R. R., Sánchez-Monedero, M. A., Spokas, K. A., Melo, L. C. A., Trugilho, P. F., Valenciano, M. N., & Silva, C. A. (2020). Enhancing cation exchange capacity ofweathered soils using biochar: Feedstock, pyrolysis conditions and addition rate. *Agronomy*, *10*(6). Available from https://doi.org/10.3390/agronomy10060824, https://www.mdpi.com/2073-4395/10/6/824/pdf.

El-Naggar, A., Lee, S. S., Rinklebe, J., Farooq, M., Song, H., Sarmah, A. K., Zimmerman, A. R., Ahmad, M., Shaheen, S. M., & Ok, Y. S. (2019). Biochar application to low fertility soils: A review of current status, and future prospects. *Geoderma*, *337*, 536–554. Available from http://www.elsevier.com/inca/publications/store/5/0/3/3/3/2, https://doi.org/10.1016/j.geoderma.2018.09.034.

Elkhalifa, S., Al-Ansari, T., Mackey, H. R., & McKay, G. (2019). Food waste to biochars through pyrolysis: A review. *Resources, Conservation and Recycling*, *144*, 310–320. Available from https://doi.org/10.1016/j.resconrec.2019.01.024, http://www.elsevier.com/locate/resconrec.

Fakayode, O. A., Aboagarib, E. A. A., Zhou, C., & Ma, H. (2020). Co-pyrolysis of lignocellulosic and macroalgae biomasses for the production of biochar – a review. *Bioresource Technology*, *297*. Available from http://www.elsevier.com/locate/biortech, https://doi.org/10.1016/j.biortech.2019.122408.

Farhangi-Abriz, S., & Torabian, S. (2018). Biochar increased plant growth-promoting hormones and helped to alleviates salt stress in common bean seedlings. *Journal of Plant Growth Regulation*, *37*(2), 591–601. Available from https://doi.org/10.1007/s00344-017-9756-9, http://www.springer.com/life + sci/plant + sciences/journal/344.

Gezahegn, S., Sain, M., & Thomas, S. C. (2021). Phytotoxic condensed organic compounds are common in fast but not slow pyrolysis biochars. *Bioresource Technology Reports*, *13*. Available from https://www.journals.elsevier.com/bioresource-technology-reports, https://doi.org/10.1016/j.biteb.2020.100613.

Gopal, M., Gupta, A., Shahul Hameed, K., Sathyaseelan, N., Khadeejath Rajeela, T. H., & Thomas, G. V. (2020). Biochars produced from coconut palm biomass residues can aid regenerative agriculture by improving soil properties and plant yield in humid tropics. *Biochar*, *2*(2), 211–226. Available from https://doi.org/10.1007/s42773-020-00043-5, http://springer.com/journal/42773.

Han, L., Sun, K., Yang, Y., Xia, X., Li, F., Yang, Z., & Xing, B. (2020). Biochar's stability and effect on the content, composition and turnover of soil organic carbon. *Geoderma*, *364*. Available from http://www.elsevier.com/inca/publications/store/5/0/3/3/3/2, https://doi.org/10.1016/j.geoderma.2020.114184.

Hopkins, D., & Hawboldt, K. (2020). Biochar for the removal of metals from solution: A review of lignocellulosic and novel marine feedstocks. *Journal of Environmental Chemical Engineering*, *8*(4). Available from https://doi.org/10.1016/j.jece.2020.103975, http://www.journals.elsevier.com/journal-of-environmental-chemical-engineering/.

Hussain, R., Ravi, K., & Garg, A. (2020). Influence of biochar on the soil water retention characteristics (SWRC): Potential application in geotechnical engineering structures. *Soil and Tillage Research*, *204*. Available from http://www.elsevier.com/inca/publications/store/5/0/3/3/1/8, https://doi.org/10.1016/j.still.2020.104713.

Iacomino, G., Idbella, M., Laudonia, S., Vinale, F., & Bonanomi, G. (2022). The suppressive effects of biochar on above- and belowground plant pathogens and pests: A review. *Plants*, *11*(22). Available from https://doi.org/10.3390/plants11223144, http://www.mdpi.com/journal/plants.

Intani, K., Latif, S., Cao, Z., & Müller, J. (2018). Characterisation of biochar from maize residues produced in a self-purging pyrolysis reactor. *Bioresource Technology*, *265*, 224–235. Available from https://doi.org/10.1016/j.biortech.2018.05.103, http://www.elsevier.com/locate/biortech.

Inyang, M. I., Gao, B., Yao, Y., Xue, Y., Zimmerman, A., Mosa, A., Pullammanappallil, P., Ok, Y. S., & Cao, X. (2016). A review of biochar as a low-cost adsorbent for aqueous heavy metal removal. *Critical Reviews in Environmental Science and Technology*, *46*(4), 406–433. Available from https://doi.org/10.1080/10643389.2015.1096880, http://www.tandf.co.uk/journals/titles/10643389.asp.

Islam, M. U., Jiang, F., Guo, Z., & Peng, X. (2021). Does biochar application improve soil aggregation? A meta-analysis. *Soil and Tillage Research*, *209*. Available from http://www.elsevier.com/inca/publications/store/5/0/3/3/1/8, https://doi.org/10.1016/j.still.2020.104926.

Karakaş, C., Özçimen, D., & İnan, B. (2017). Potential use of olive stone biochar as a hydroponic growing medium. *Journal of Analytical and Applied Pyrolysis*, *125*, 17–23. Available from https://doi.org/10.1016/j.jaap.2017.05.005.

Kavitha, B., Reddy, P. V. L., Kim, B., Lee, S. S., Pandey, S. K., & Kim, K. H. (2018). Benefits and limitations of biochar amendment in agricultural soils: A review. *Journal of Environmental Management*, *227*, 146–154. Available from https://www.sciencedirect.com/journal/journal-of-environmental-management, https://doi.org/10.1016/j.jenvman.2018.08.082.

Kazemi Shariat Panahi, H., Dehhaghi, M., Ok, Y. S., Nizami, A. S., Khoshnevisan, B., Mussatto, S. I., Aghbashlo, M., Tabatabaei, M., & Lam, S. S. (2020). A comprehensive review of engineered biochar: Production, characteristics, and environmental applications. *Journal of Cleaner Production*, *270*. Available from https://www.journals.elsevier.com/journal-of-cleaner-production, https://doi.org/10.1016/j.jclepro.2020.122462.

Kim, J. Y., Oh, S., & Park, Y. K. (2020). Overview of biochar production from preservative-treated wood with detailed analysis of biochar characteristics, heavy metals behaviors, and their ecotoxicity. *Journal of Hazardous Materials*, *384*. Available from http://www.elsevier.com/locate/jhazmat, https://doi.org/10.1016/j.jhazmat.2019.121356.

Krysanova, K., Krylova, A., & Zaichenko, V. (2019). Properties of biochar obtained by hydrothermal carbonization and torrefaction of peat. *Fuel*, *256*. Available from http://www.journals.elsevier.com/fuel/, https://doi.org/10.1016/j.fuel.2019.115929.

Kumar, A., Saini, K., & Bhaskar, T. (2020). Hydochar and biochar: Production, physicochemical properties and techno-economic analysis. *Bioresource Technology*, *310*. Available from http://www.elsevier.com/locate/biortech, https://doi.org/10.1016/j.biortech.2020.123442.

Kuppusamy, S., Thavamani, P., Megharaj, M., Venkateswarlu, K., & Naidu, R. (2016). *Agronomic and remedial benefits and risks of applying biochar to soil: Current knowledge and future research directions*, *Environment International* (87, pp. 1–12). Available from http://www.elsevier.com/locate/envint, https://doi.org/10.1016/j.envint.2015.10.018.

Lataf, A., Jozefczak, M., Vandecasteele, B., Viaene, J., Schreurs, S., Carleer, R., Yperman, J., Marchal, W., Cuypers, A., & Vandamme, D. (2022). The effect of pyrolysis temperature and feedstock on biochar agronomic properties. *Journal of Analytical and Applied Pyrolysis*, *168*. Available from https://www.journals.elsevier.com/journal-of-analytical-and-applied-pyrolysis, https://doi.org/10.1016/j.jaap.2022.105728.

Leng, L., & Huang, H. (2018). An overview of the effect of pyrolysis process parameters on biochar stability. *Bioresource Technology*, *270*, 627–642. Available from http://www.elsevier.com/locate/biortech, https://doi.org/10.1016/j.biortech.2018.09.030.

Leng, L., Xiong, Q., Yang, L., Li, H., Zhou, Y., Zhang, W., Jiang, S., Li, H., & Huang, H. (2021). An overview on engineering the surface area and porosity of biochar. *Science of the Total Environment*, *763*, 144204. Available from https://doi.org/10.1016/j.scitotenv.2020.144204.

Mamvura, T. A., & Danha, G. (2020). Biomass torrefaction as an emerging technology to aid in energy production. *Heliyon*, *6*(3). Available from https://doi.org/10.1016/j.heliyon.2020.e03531, http://www.journals.elsevier.com/heliyon/.

Mansoor, S., Kour, N., Manhas, S., Zahid, S., Wani, O. A., Sharma, V., Wijaya, L., Alyemeni, M. N., Alsahli, A. A., El-Serehy, H. A., Paray, B. A., & Ahmad, P. (2021). Biochar as a tool for effective management of drought and heavy metal toxicity. *Chemosphere*, *271*. Available from http://www.elsevier.com/locate/chemosphere, https://doi.org/10.1016/j.chemosphere.2020.129458.

Miralles-Wilhelm, F. (2021). *Nature-based solutions in agriculture: Sustainable management and conservation of land, water and biodiversity*. Food and Agriculture Organization of the United Nations and The Nature Conservancy Virginia. Available from https://doi.org/10.4060/cb3140en.

Mrunalini, K., Behera, B., Jayaraman, S., Abhilash, P. C., Dubey, P. K., Swamy, G. N., Prasad, J. V. N. S., Rao, K. V., Krishnan, P., Pratibha, G., & Srinivasa Rao, C. (2022). Nature-based solutions in soil restoration for improving agricultural productivity. *Land Degradation and Development*, *33*(8), 1269–1289. Available from https://doi.org/10.1002/ldr.4207, http://onlinelibrary.wiley.com/journal/10.1002/(ISSN)1099-145X.

Munera-Echeverri, J. L., Martinsen, V., Strand, L. T., Zivanovic, V., Cornelissen, G., & Mulder, J. (2018). Cation exchange capacity of biochar: An urgent method modification. *Science of the Total Environment*, *642*, 190–197. Available from http://www.elsevier.com/locate/scitotenv, https://doi.org/10.1016/j.scitotenv.2018.06.017.

Nuagah, M. B., Boakye, P., Oduro-Kwarteng, S., & Sokama-Neuyam, Y. A. (2020). Valorization of faecal and sewage sludge via pyrolysis for application as crop organic fertilizer. *Journal of Analytical and Applied Pyrolysis*, *151*. Available from https://www.journals.elsevier.com/journal-of-analytical-and-applied-pyrolysis, https://doi.org/10.1016/j.jaap.2020.104903.

Nzediegwu, C., Arshad, M., Ulah, A., Naeth, M. A., & Chang, S. X. (2021). Fuel, thermal and surface properties of microwave-pyrolyzed biochars depend on feedstock type and pyrolysis temperature. *Bioresource Technology*, *320*. Available from http://www.elsevier.com/locate/biortech, https://doi.org/10.1016/j.biortech.2020.124282.

Oni, B. A., Oziegbe, O., & Olawole, O. O. (2019). Significance of biochar application to the environment and economy. *Annals of Agricultural Sciences*, *64*(2), 222–236. Available from https://doi.org/10.1016/j.aoas.2019.12.006, http://www.elsevier.com/journals/annals-of-agricultural-sciences/0570-1783/.

Poveda, J., Martínez-Gómez, Á., Fenoll, C., & Escobar, C. (2021). The use of biochar for plant pathogen control. *Phytopathology*, *111*(9), 1490–1499. Available from https://doi.org/10.1094/PHYTO-06-20-0248-RVW, https://apsjournals.apsnet.org/doi/10.1094/PHYTO-06-20-0248-RVW.

Sashidhar, P., Kochar, M., Singh, B., Gupta, M., Cahill, D., Adholeya, A., & Dubey, M. (2020). Biochar for delivery of agri-inputs: Current status and future perspectives. *Science of the Total Environment*, *703*. Available from http://www.elsevier.com/locate/scitotenv, https://doi.org/10.1016/j.scitotenv.2019.134892.

Schmidt, H. P., Kammann, C., Hagemann, N., Leifeld, J., Bucheli, T. D., Sánchez Monedero, M. A., & Cayuela, M. L. (2021). Biochar in agriculture – a systematic review of 26 global meta-analyses. *GCB Bioenergy*, *13*(11), 1708–1730. Available from https://doi.org/10.1111/gcbb.12889, http://onlinelibrary.wiley.com/journal/10.1111/(ISSN)1757-1707.

Shan, R., Shi, Y., Gu, J., Wang, Y., & Yuan, H. (2020). Single and competitive adsorption affinity of heavy metals toward peanut shell-derived biochar and its mechanisms in aqueous systems. *Chinese Journal of Chemical Engineering*, *28*(5), 1375–1383. Available from https://doi.org/10.1016/j.cjche.2020.02.012, https://www.journals.elsevier.com/chinese-journal-of-chemical-engineering.

Shi, Z., Yan, J., Ren, X., Wen, M., Zhao, Y., & Wang, C. (2021). Effects of biochar and thermally treated biochar on Eisenia fetida survival, growth, lysosomal membrane stability and oxidative stress. *Science of the Total Environment*, *770*. Available from http://www.elsevier.com/locate/scitotenv, https://doi.org/10.1016/j.scitotenv.2020.144778.

Shin, J. D., Park, D. G., Hong, S. G., Jeong, C., Kim, H., & Chung, W. (2021). Influence of activated biochar pellet fertilizer application on greenhouse gas emissions and carbon sequestration in rice (*Oryza sativa* L.) production. *Environmental Pollution*, *285*. Available from https://www.journals.elsevier.com/environmental-pollution, https://doi.org/10.1016/j.envpol.2021.117457.

Siedt, M., Schäffer, A., Smith, K. E. C., Nabel, M., Roß-Nickoll, M., & van Dongen, J. T. (2021). Comparing straw, compost, and biochar regarding their suitability as agricultural soil amendments to affect soil structure, nutrient leaching, microbial communities, and the fate of pesticides. *Science of the Total Environment*, *751*. Available from http://www.elsevier.com/locate/scitotenv, https://doi.org/10.1016/j.scitotenv.2020.141607.

Šimanský, V., Horák, J., & Bordoloi, S. (2022). Improving the soil physical properties and relationships between soil properties in arable soils of contrasting texture enhancement using biochar substrates: Case study in Slovakia. *Geoderma Regional*, *28*. Available from http://www.journals.elsevier.com/geoderma-regional/, https://doi.org/10.1016/j.geodrs.2021.e00443.

Simiele, M., Argentino, O., Baronti, S., Scippa, G. S., Chiatante, D., Terzaghi, M., & Montagnoli, A. (2022). Biochar enhances plant growth, fruit yield, and antioxidant content of cherry tomato (*Solanum lycopersicum* L.) in a soilless substrate. *Agriculture (Switzerland)*, *12*(8). Available from https://doi.org/10.3390/agriculture12081135, http://www.mdpi.com/journal/agriculture.

Tovihoudji, G. P., Diogo, R. V. C., Abiola, W. A., Akoha, F. B. R. M., & Godau, T. (2022). Profitability and agronomic potential of cotton (*Gossypium hirsutum* L.) under biochar-compost-based amendments in three agroecological zones of northern Benin. *Frontiers in Sustainable Food Systems*, *6*. Available from http://www.frontiersin.org/journals/sustainable-food-systems#, https://doi.org/10.3389/fsufs.2022.1036133.

Tripathi, M., Sahu, J. N., & Ganesan, P. (2016). Effect of process parameters on production of biochar from biomass waste through pyrolysis: A review. *Renewable and Sustainable Energy Reviews*, *55*, 467–481. Available from https://www.journals.elsevier.com/renewable-and-sustainable-energy-reviews, https://doi.org/10.1016/j.rser.2015.10.122.

Wang, Q., Awasthi, M. K., Ren, X., Zhao, J., Li, R., Wang, Z., Wang, M., Chen, H., & Zhang, Z. (2018). Combining biochar, zeolite and wood vinegar for composting of pig manure: The effect on greenhouse gas emission and nitrogen conservation. *Waste*

Management, *74*, 221–230. Available from http://www.elsevier.com/locate/wasman, https://doi.org/10.1016/j.wasman.2018.01.015.

Wang, X., Chang, V. W. C., Li, Z., Chen, Z., & Wang, Y. (2021). Co-pyrolysis of sewage sludge and organic fractions of municipal solid waste: Synergistic effects on biochar properties and the environmental risk of heavy metals. *Journal of Hazardous Materials*, *412*. Available from http://www.elsevier.com/locate/jhazmat, https://doi.org/10.1016/j.jhazmat.2021.125200.

Wang, Y., Gu, J., & Ni, J. (2023). Influence of biochar on soil air permeability and greenhouse gas emissions in vegetated soil: A review. *Biogeotechnics*, *1*(4), 100040. Available from https://doi.org/10.1016/j.bgtech.2023.100040.

Wei, B., Peng, Y., Lin, L., Zhang, D., Ma, L., Jiang, L., Li, Y., He, T., & Wang, Z. (2023). Drivers of biochar-mediated improvement of soil water retention capacity based on soil texture: A meta-analysis. *Geoderma*, *437*. Available from http://www.elsevier.com/inca/publications/store/5/0/3/3/3/2, https://doi.org/10.1016/j.geoderma.2023.116591.

Wei, S., Zhu, M., Fan, X., Song, J., Peng, P., Li, K., Jia, W., & Song, H. (2019). Influence of pyrolysis temperature and feedstock on carbon fractions of biochar produced from pyrolysis of rice straw, pine wood, pig manure and sewage sludge. *Chemosphere*, *218*, 624–631. Available from http://www.elsevier.com/locate/chemosphere, https://doi.org/10.1016/j.chemosphere.2018.11.177.

Xing, J., Xu, G., & Li, G. (2021). Comparison of pyrolysis process, various fractions and potential soil applications between sewage sludge-based biochars and lignocellulose-based biochars. *Ecotoxicology and Environmental Safety*, *208*. Available from http://www.elsevier.com/inca/publications/store/6/2/2/8/1/9/index.htt, https://doi.org/10.1016/j.ecoenv.2020.111756.

Yaashikaa, P. R., Kumar, P. S., Varjani, S., & Saravanan, A. (2020). A critical review on the biochar production techniques, characterization, stability and applications for circular bioeconomy. *Biotechnology Reports*, *28*. Available from http://www.journals.elsevier.com/biotechnology-reports/, https://doi.org/10.1016/j.btre.2020.e00570.

Yang, shihong, Xiao, Y. 'nan, Sun, X., Ding, J., Jiang, Z., & Xu, J. (2019). Biochar improved rice yield and mitigated CH4 and N2O emissions from paddy field under controlled irrigation in the Taihu Lake Region of China. *Atmospheric Environment*, *200*, 69–77. Available from https://doi.org/10.1016/j.atmosenv.2018.12.003.

Yao, Y., Gao, B., Chen, J., & Yang, L. (2013). Engineered biochar reclaiming phosphate from aqueous solutions: Mechanisms and potential application as a slow-release fertilizer. *Environmental Science and Technology*, *47*(15), 8700–8708. Available from https://doi.org/10.1021/es4012977.

Zabaleta, R., Sánchez, E., Fabani, P., Mazza, G., & Rodriguez, R. (2023). Almond shell biochar: characterization and application in soilless cultivation of Eruca sativa. *Biomass Conversion and Biorefinery*. Available from https://www.springer.com/journal/13399, https://doi.org/10.1007/s13399-023-04002-5.

Zhang, C., Miao, Y., Malghani, S., Liu, G., & Liao, X. (2023). Biochar combined with different nitrogen fertilization rates increased crop yield and greenhouse gas emissions in a rapeseed-soybean rotation system. *Journal of Environmental Management*, *345*. Available from https://www.sciencedirect.com/journal/journal-of-environmental-management, https://doi.org/10.1016/j.jenvman.2023.118915.

Zhang, T., Tang, Y., Li, H., Hu, W., Cheng, J., & Lee, X. (2023). A bibliometric review of biochar for soil carbon sequestration and mitigation from 2001 to 2020. *Ecotoxicology and Environmental Safety*, *264*. Available from http://www.elsevier.com/inca/publications/store/6/2/2/8/1/9/index.htt, https://doi.org/10.1016/j.ecoenv.2023.115438.

Zhao, S., Ta, N., Li, Z., Yang, Y., Zhang, X., Liu, D., Zhang, A., & Wang, X. (2018). Varying pyrolysis temperature impacts application effects of biochar on soil labile organic carbon and humic fractions. *Applied Soil Ecology*, *123*, 484–493. Available from http://www.elsevier.com/inca/publications/store/5/2/4/5/1/8/index.htt, https://doi.org/10.1016/j.apsoil.2017.09.007.

Zhou, X., Chen, Z., Li, Z., & Wu, H. (2020). Impacts of aeration and biochar addition on extracellular polymeric substances and microbial communities in constructed wetlands for low C/N wastewater treatment: Implications for clogging. *Chemical Engineering Journal*, *396*. Available from http://www.elsevier.com/inca/publications/store/6/0/1/2/7/3/index.htt, https://doi.org/10.1016/j.cej.2020.125349.

CHAPTER

Factors influencing biochar properties and relationships with soil applications

7

Hugh McLaughlin[1] and Samer Fawzy[1,2]

[1]*Trinity Biocarbon, London, United Kingdom*

[2]*School of Chemistry and Chemical Engineering, Queen's University Belfast, Belfast, United Kingdom*

7.1 Introduction

Naturally occurring biochar has been around for 600 million years, beginning when vegetation migrated out of the oceans. Land-based forests ignited due to lightning strikes, leaving 1%–2% of the biomass carbon inventory before the fire as "char," the intermediate form between biomass and ash. This small fraction of residual biochar in the soil evolved with life on earth to create the myriad of soil–plant–microbe synergies that formed the basis for agriculture before the age of fossil fuels, synthetic nitrogen fertilizers, and biocides. A succinct video introduction to biochar is available at: https://www.youtube.com/watch?v = AUh1vwDjaVo.

Biochar has returned as a commodity form of carbon due to the urgency to decarbonize the current fossil carbon-based economies and societies. Plants remove carbon dioxide from the atmosphere as they grow, and conversion to biochar yields fossil-free energy and stable carbon-rich materials that have applications of value and do not return the carbon to the skies as carbon dioxide (Fawzy et al., 2021). Biochar generally requires much more effort to produce than harvesting fossil carbon reserves but has become a necessary step as the atmosphere drifts away from the carbon dioxide range that promotes environments and climates compatible with current civilizations and lifestyles (Fawzy et al., 2020).

Since the biochar is destined for a century or more of utilization as a carbon-rich solid, the production processes strive for properties that will create value in subsequent uses. End uses range from soil amendments to mechanical materials such as inert fillers in plastics, concrete, and asphalt, to components in batteries and alternatives to activated carbon adsorbents—with additional value-added applications under development (Osman et al., 2022). This discussion will focus on the unique properties of biochars and how those properties are both created and lost during biochar production.

Biochar Ecotechnology for Sustainable Agriculture and Environment. DOI: https://doi.org/10.1016/B978-0-443-29855-4.00009-6

7.2 Influential properties of pre-pyrolysis biomass on biochar characteristics

Biomass feedstocks cover a wide range of biogenic materials, including biomass that has served one purpose and can be subsequently converted into biochar. Typically, except for moisture, all the nonorganic components in the incoming biomass are retained in the biochar. Inorganic minerals intrinsic to plant species as they grow, along with extraneous dirt and soil collected when the biomass is aggregated for pyrolysis processing, are the main contributors to inorganic "ash" or minerals in the finished biochar. In the case of animal-derived feedstocks, bones can be a major source of ash. If the biomass has been intentionally mixed with nonbiogenic materials, such as heavy metals for inhibiting decay in pressure-treated lumber, those materials may not disappear during pyrolysis and will generally significantly reduce the utility of the biochar except for applications such as inert fillers or nonagricultural soil applications.

Generally, it takes about four units (by weight) of dry organic biomass to make one unit of finished carbon-rich biochar. Since most biochar starting materials are mostly organic, the ash in the biomass is going to concentrate by a factor of about four in the corresponding biochar. If the biochar is going back into the soil, perhaps after an intermediate composting or blending step, the ash derived from the biomass is generally compatible with subsequent plant growth. Higher purity applications may require cleaning the incoming biomass or sourcing materials without unacceptable impurities.

Other typical pretreatment steps are to dry excess moisture from the biomass, to grind it into smaller pieces, and to densify it into pellets or briquettes. All these pretreatment steps are focused on increasing the throughput rate in the subsequent thermal conversion to biochar in the pyrolyzer. Since pyrolysis changes do not commence until the biomass is heated above 100°C, any excess moisture will need to be removed or evaporated. In freshly cut woody biomass, up to one-half of the incoming weight is moisture. The heat requirement to dry the biomass can be a major heat input requirement in any thermal conversion process, often more than the energy requirement to heat the dry biomass from 200°C to the final carbonization temperature, called the heat or highest treatment temperature (HTT).

7.3 Post-pyrolysis processing of biochar

In addition to weight loss, all biomass properties are altered by the conversion to biochar. Woody biomass condenses to about one-half the volume of the starting material. The local extent of consolidation depends on the HTT, the specific species, what part of the biomass is pyrolyzed (heartwood, sapwood, bark, leafy matter, etc.) and even direction of the grain. The friability (definition: the tendency of a solid substance to break into smaller pieces under duress or contact) of the biochar is much greater than the original biomass. Therefore any supplemental reduction of the particle size requirements, such as making a fine powder for aqueous suspension, is done after pyrolysis since it requires less mechanical energy.

Additional post-pyrolysis modifications of biochar are associated with modifying its properties to improve handling and/or subsequent performance in the end use application. Biochar exits the pyrolysis process as a hot and bone-dry material, which will react with air, ignite and convert into a red-hot glowing mass. Water addition is a common first step to both cool and reduce the tendency of the biochar to release dust, which is both a nuisance and inhalation risk. In addition, dry dust is an explosion concern if dispersed in air. Generally, if the biochar dust settles quickly out of the air, there is little explosion risk, but any biochar dust needs to be respected since it consists of a fuel that can react with oxygen in air and might subsequently deflagrate (burn via combustion in a progressive flame) or detonate (explode with an accompanying shock wave).

Other post-pyrolysis steps involve crushing and sieving the biochar into particle size ranges, blending additional ingredients such as rock dust and clays to provide additional benefits when added to the soil, or other steps that modify the biochar for the end use, including mixing with compost or other crop fertilizer sources. A popular and beneficial practice is to introduce biochar at the start of the composting process to improve nutrient capture during composting and stabilizing the biochar by allowing the compost microbes to consume excess soluble sugars and volatile organics present in the biochar.

7.4 Pyrolysis processing of biomass into biochar

While the pre-pyrolysis steps are generally directed at improving the biomass properties to allow better and faster conversion to biochar, and the post-processing is focused on handling and downstream requirements, it is the pyrolysis step that converts biomass into biochar. Furthermore, during that conversion, new carbon structures are created from the biomass that was formed when the plants grew by consuming carbon dioxide from the atmosphere. As will be discussed, pyrolysis is essentially the only process where many of the unique properties of biochar can be created and possibly destroyed. Once pyrolysis creates the graphitic portion of the biochar, virtually nothing can modify it since it is a three-dimensional (3D) solid that is nonreactive once cooled from pyrolysis temperatures. As such, especially at ambient temperatures, the graphitic backbone of biochar is essentially indestructable under the conditions that biochar subsequently encounters. While the graphitic backbone is stable against chemical changes, it can and does still participate in interactions occurring in the soil where the biochar is present.

In addition to converting the biomass to biochar, the pyrolysis processing is an opportunity to add modifiers to the biomass. The modifiers can be integrated into the biochar and subsequently improve the performance in the soil. The specific modifiers are dictated by the characteristics of the climate, soil, and crops involved. As such, the permutations will not be elaborated here, but the biochar producer should not overlook the opportunities to augment biochar properties during pyrolysis. In addition to chemically blending the additives into the biochar, the pyrolysis conditions offer the possibility of heating the additives and, also, exposing them to the pyroligneous acid vapors. A recent publication by Singh et al. (2023) provided an overview of modified biochar methods and possibilities.

7.4.1 Qualitative changes when biomass becomes biochar

It is insightful to remember that virtually all living biomass, including plants, exists within a very narrow temperature range throughout its life—between 0°C, where water freezes, and 50°C, where most of the enzymes have denatured such that no more metabolism is possible. As such, when biomass is subjected to pyrolysis temperatures, which basically start above 100°C, there are really no requirements that structures be preserved to allow for propagation of the species. During pyrolysis, biology steps aside and chemistry takes over. Since most biochar is made from woody and grassy biomass, or from wastes derived from these previously living sources, it is helpful to visualize the molecular level spatial interlocking of the three main plant building blocks of lignin, hemicellulose, and cellulose, as shown in Fig. 7.1. As can be seen, these materials are extensively interwoven and create a myriad of interfaces and composites at the molecular scale.

Fig. 7.2 summarizes the sequence of qualitative changes that occur in the three major constituents of biomass with increasing HTT. While the sequence is the same, each constituent transforms over different temperature ranges. The torrefaction range, characterized by increasing shades of brown, as seen in coffee roasting, features localized mixtures of volatiles and tars, accompanied by the onset of graphene structures. Above 300°C, the composite solid turns black, but the evolution of the chemical and mechanical structures in the biochar continues as the HTT increases. In addition, the biochar present continues to interact with the vapor atmosphere in the pyrolyzer.

As noted, above 300°C, the solids turn black and the properties are increasingly dominated by the newly created graphene structures. Fig. 7.3 shows the developing graphene sheets, which are connected by residual biomass and partially pyrolyzed tar-like materials that are not shown.

At higher temperatures, the remaining tars are converted to more vapors and additional graphene structures, creating a rigid 3D solid. This solid is composed of mostly graphene sheets with additional organic fragments on the edges and diminishing levels of tars connecting and adsorbed on the graphene surfaces.

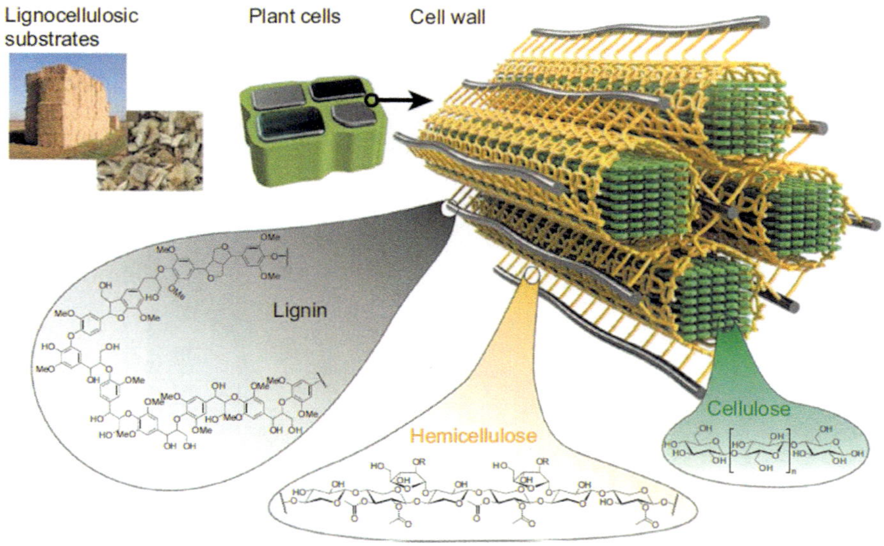

FIGURE 7.1 Artist depiction of plant structures at the molecular scale.

The three main constituents of lignocellulosic materials are interwoven to create the plant structures, with many interfaces of differing materials (Brethauer et al., 2020).

From Brethauer, S., Shahab, R. L., & Studer, M. H. (2020). Impacts of biofilms on the conversion of cellulose. Applied Microbiology and Biotechnology, *104(12), 5201–5212. https://doi.org/10.1007/s00253-020-10595-y.*

7.4.2 Quantitative properties—pyrolysis processing of biomass into biochar

While biochar and charcoal are both produced by pyrolysis processes, the main difference is that biochar is produced at much higher temperatures, or HTT, than charcoal, as shown in Fig. 7.4. As shown, charcoal has a much higher yield at the lower HTT, retaining more energy when used as a fuel, but biochar requires higher HTT to achieve adsorption properties. Much of the foundation for this section is presented by Mclaughlin and Shields (2010). Section 6.3.2 provides additional discussion of adsorption capacity and R134a, which is quantified on the right-side axis of Fig. 7.4. The reader is referred to the original work for the additional discussions of the underlying trends and implications.

The development of adsorption capacity above 300°C is attributed to the formation of localized microporous graphene regions within the char as the residual solid becomes progressively more carbon-rich. The decline after 600°C is attributed to consolidation of the developed graphene plates into denser, but less

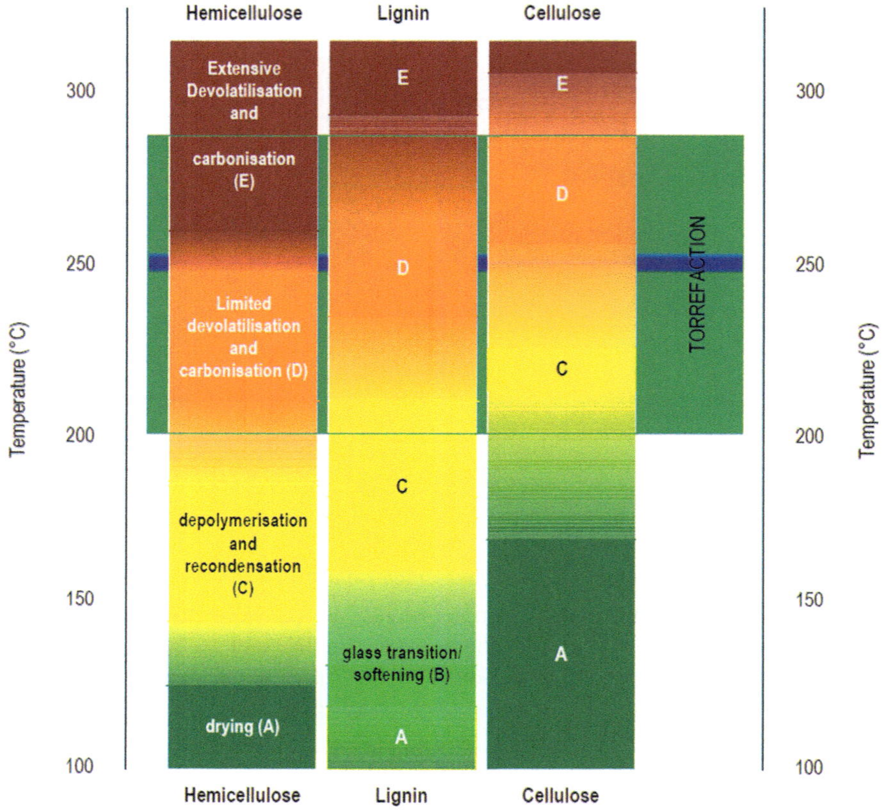

FIGURE 7.2 Transition temperatures for biomass constituents.

All components of lignocellulosic materials transform between drying and char formation, but individual materials transform through stages A to E at different temperature ranges (Bergman et al., 2005).

From Bergman, P., Boersma, A., Zwart, R., & Kiel, J. (2005). Torrefaction for biomass co-firing in existing coal-fired power stations. *Technical report. Energy Research Centre of the Netherlands.*

microporous, regions of amorphous graphite. As the biochar, initially formed at 300°C, is heated to progressively higher HTTs, the internal graphitic domains grow in size and the residual nongraphite organics continue to decompose into volatiles and additional graphitic backbone. This sequence is accompanied by the external shrinking of the biochar particle, in addition to changes in the relative amounts of C, H, and O in the composite biochar material (Antal & Grønli, 2003). Fig. 7.5 shows the progression for nut shells and woody biomass, two of the most prevalent biochar starting materials (in addition to grassy biomass such as corn stover, miscanthus, bagasse, and grain straws).

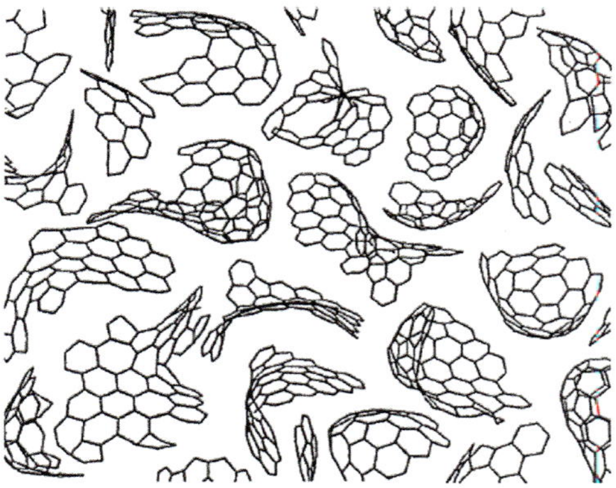

FIGURE 7.3 Curved individual graphene sheets, floating in space.

Developing individual graphene sheets, curved with 5, 6, and 7 member rings, and connected by tars (not shown) (Harris, 2013).

From Harris, P. J. F. (2013). Rosalind Franklin's work on coal, carbon, and graphite. Interdisciplinary Science Reviews, 26*(3), 204–210. https://doi.org/10.1179/030801801679467.*

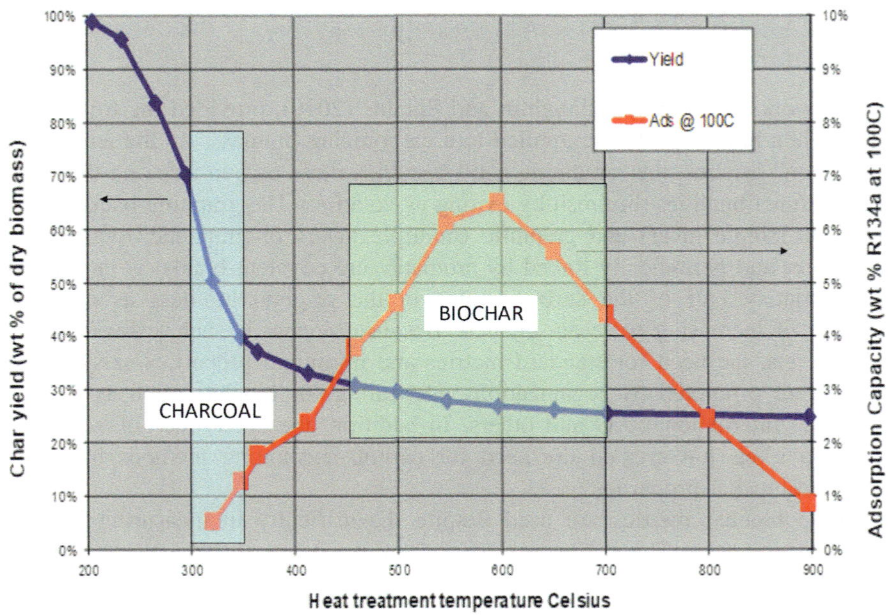

FIGURE 7.4 Trends in conversion and adsorption capacity versus temperatures.

HTT difference for charcoal versus biochar, yield and adsorption capacity (Mclaughlin & Shields, 2010).

FIGURE 7.5 Biomass changes as temperature of pyrolysis increases.

Shrinkage due to consolidation of the graphitic backbone with increasing HTT.

7.4.3 Chemical and structural changes—pyrolysis processing of biomass into biochar

Earlier works, including Mclaughlin and Shields (2010), provided the trends that occur when higher HTTs are applied to a carbonizing biomass. As the term "carbonization" implies, the remaining solid becomes more carbon-rich and the other organic functionalities diminish by exiting as volatiles. The graphitic backbone of graphene (single layer) and graphitic (multiple layers of graphene, overlapping each other and periodically linked by aromatic and covalent bonds) is the fate of approximately half of the carbon atoms in the original biomass as aromatic domains of increasing scale are created. Because biochar is such a diverse class of materials, the need for standard metrics and minimum properties exists. Such metrics are promoted by organizations advocating for biochar, such as Carbon Standards International AG and others. In addition, the emergence of voluntary carbon markets has created the need for carbon accounting methods based on measured biochar properties.

These biochar metrics are used despite the difficulty in measuring biochar chemical properties. The challenge derives from the range of properties and impurities that interfere with analytical methods, typically developed for other materials, such as coals. To date, only H/Corg (the molar ratio of hydrogen to organic carbon atoms) and the mass fraction of carbon atoms in a sample have taken on any widespread utilization. In addition to the current commercial valuation of biochar, there have been significant advances in measuring the intrinsic properties of biochars. The current understanding of biochar chemical and structural properties is detailed in the two publications by Wood et al. (2023). The first part focuses on individual chemical properties, and the challenges and limitations of measuring them, and the second part assembles the properties into a holistic molecular model called "The Molecular Dynamic Simulation Model." The research is a collaboration between the Department of Chemistry and the UK Biochar Research Centre, both at the University of Edinburgh.

"The Molecular Dynamic Simulation Model" functions by allowing the user to adjust biochar input parameters until the model predictions match experimental measurements over a reasonable range of conditions, typically HTT. Once agreement is reached, there is the dangerous tendency to conclude that the simulation has utilized the same mechanism as the experimental studies, which is overlaying conjecture onto an experimental outcome and not easy to prove or disprove. However, it is one of the few available steps in the right direction. The true measure of the validity of a modeling effort is whether it can predict properties that are not directly controlled and optimized in the development of the model, such as adsorption, physical properties such as bulk density, friability and mechanical strength, and unique application considerations like water holding capacity. Unfortunately, beyond the commercial value assigned to biochars by the carbon content and H/Corg metrics, there has been sporadic and erratic progress relating measured biochar characteristics and properties with soil applications of repeatable and quantifiable value.

7.5 Measuring biochar properties to quantify amount of biochar constituents

Biochar is basically a mixture of inorganic compounds, often measured together and termed "ash," and carbon-rich structures formed from the biomass and created by the pyrolysis conditions. The organic structures are usually classified as "stable" if dominated by graphitic domains. Unstable organic entities include a much wider class of substances that range from small volatile compounds that are water soluble to larger tar-like compounds that slowly break down in soils through consumption by microbes or gradual dissolution in groundwater over time. Over time, perhaps a decade or more, the stable organics remain as solids and are fully integrated into the soil, while the unstable organics have disappeared. In addition to ash and organic compounds, biochar contains two other substances that are equally important in the unique role that biochar plays in the soil: moisture, in the form of liquid water, and air, which fills the voids not filled with water. Since biochar is mostly volume that is not occupied by solid matter (ash and organics), the voidage, as it is termed, is a dynamic volume that varies from full of liquid water to full of air depending on the conditions of the soil. Many benefits to the crops and soil properties come from both the water storage and modulation by biochar and the ability of the porosity to prevent anaerobic soil conditions that inhibit beneficial microbial populations.

7.5.1 Measuring biochar water content by removing it correctly

Biochar is a complex material with a variety of internal structures and local chemical functionalities. As a result, biochars contain both *bulk water*, also called *free water*, and *adsorbed water*, which is contained in biochar micropores. In order to remove both kinds of water, it is necessary to dry the biochar at 150°C for a few hours. Biochar will lose the bulk water at 105°C, the typical laboratory drying temperature for glassware and nonadsorbing materials, but it may be several days to reach a stable endpoint. Enders et al. (2012) used 105°C and defined the dried weight at this temperature after 18 hours of drying. The 150°C drying temperature is adopted from the ASTM methods for activated carbons, which share many properties with biochar.

7.5.2 Measuring biochar ash by oxidizing at an intermediate temperature

ASTM proximate analysis methods were developed for fuels, especially coals, and use 750°C as the temperature for oxidizing away the organic compounds. The resulting measurement reflects the results of combusting the material at temperatures of 1000°C or higher. Since biochar is not a fuel, there is a need to modify the ASTM method to accurately measure the ash content in the biochar. Many biochars contain calcium and magnesium in the form of carbonates. The ashing temperature is lowered to 450°C–550°C to minimize the extent that the carbonates decompose and lose carbon dioxide, which would result in less weight in the subsequent ash determination.

Since the recommended ashing temperature of biochar is lower, the destruction of the organics by oxidation is slower. Thus the ashing is done with finely ground material and continued until the residue converts to a uniform off-white material. One proven method utilizes an inverted lid from a 30-mL porcelain crucible, filled with about 1 g of dry ground biochar and weighed to the nearest 0.1 mg. The samples are heated in a Muffle Furnace set to 550°C until uniformly converted to ash, visually confirmed by periodically opening the furnace door to view the progression of transformation. After ashing, the samples are allowed to cool and weighed to measure the ash content, reported as a weight percentage of the original dry biochar.

The ash in most biochars consists either of plant nutrients, such as phosphorus (P), potassium (K), calcium (Ca), and magnesium (Mg), or inert materials, typically sand or clay, that were collected with the incoming biomass. Therefore there may be merit in having the biochar tested by a standard fertilizer assay to account for the amount of plant fertilizer being provided by the biochar, especially for biochars with levels of ash above 10 wt.% or when the biochar dosage is significant, such as above 25 tons/ha. Due to the pyrolysis conversion step, the "total nitrogen" measured in biochar is usually not plant-available nor biodegradable, meaning the biochar will not be a source of nitrogen for the crops. During pyrolysis, any remaining nitrogen is incorporated into carbon-rich aromatic structures and becomes a permanent component of the stable graphene structures. If the biochar is tested for "plant-available nitrogen," that portion can be considered a source of plant nutrients. However, the plant nutrients in biochar are largely consumed in one or two growing seasons, whereas the stable carbon in the biochar is likely to persist for centuries.

7.5.3 Partitioning the organics into stable and unstable

With the advent of carbon credits, predicting the amount of stable carbon in a biochar has become an issue of significant economic interest. While the valuation of biochar is not quite as speculative as cryptocurrency, it is similar in that its value is determined by what others are willing to assign to it, rather than being an intrinsic property of the material that can be measured and quantified.

Mclaughlin and Shields (2010) contains a dated but coherent set of pyrolysis runs, known as a "homologous series," where HTT is varied, and the resulting chars are characterized by the modified proximate analyses approach. The relative relationships between the stable (termed "resident") versus unstable (termed "mobile") portions of biochar, with each portion divided into carbon (measured directly, MW = 12) and the oxygen and hydrogen assumed to be present as H's and OH's in equal portions (= H_2O, mw = 18), are shown in Fig. 7.6. This data set is representative of many biochars, but not quantitatively exact due to effects related to the distribution of local HTT's in commercial pyrolysis reactors. However, the sharp transition of biochar properties between 300°C and 400°C is apparent and widespread among all biochars, irrespective of starting biomass.

Incremental progress is being made in the analytical options for partitioning biochar into stable and unstable portions, but the current metric of H/Corg seems to be the most widely accepted metric for assigning carbon sequestration valuation (Fawzy et al., 2022). Unfortunately, the determination of Corg is basically calculated from total carbon less the fraction of inorganic carbon contained in carbonates. While total carbon is accurately measured by analytical instruments, the methods for quantifying inorganic carbon are significantly less precise and contain the bulk of the error in the determination of the organic carbon fraction. As such, biochars with elevated ash fractions represent a significant analytical challenge. It is anticipated that more accurate methods will be developed and adopted for measuring stable carbon contents in biochars that are being valued and transacted for carbon credits.

In addition to the potential carbon credit value, the stable carbon portion persists in the soil for a very long time and becomes a dynamic component of a

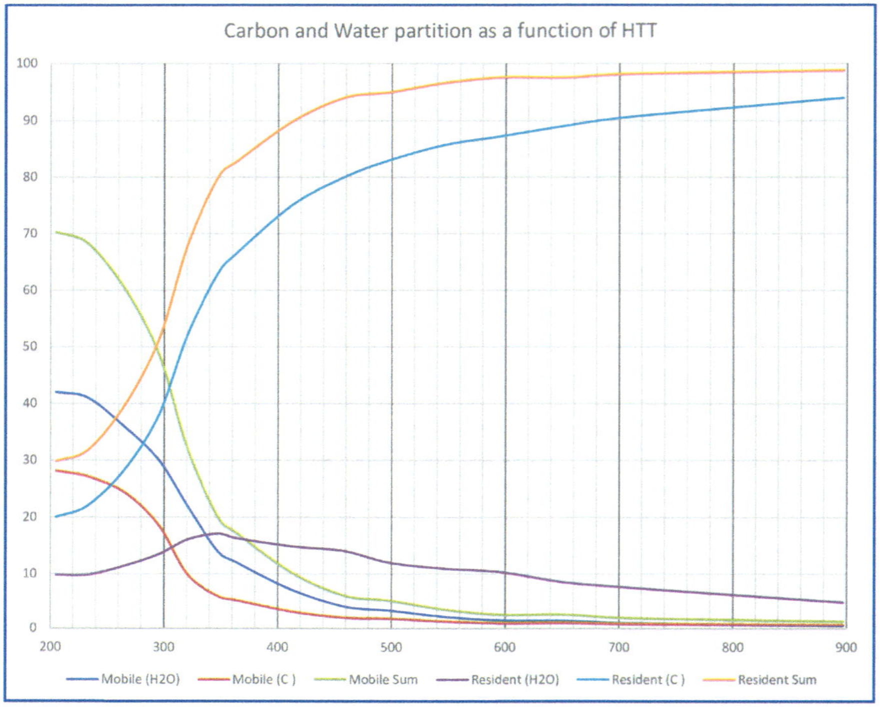

FIGURE 7.6 Trends in biochar constituents as temperature increases.

Homologous series of biochars (Mclaughlin & Shields, 2010).

living soil. In contrast, the unstable portion may produce several different impacts depending on the nature of the organics and the soil conditions. In most cases, the unstable organics are a source of food for soil microbes, which also require nitrogen for growth. In many situations, the soil microbes out-compete the plants for the available nitrogen in the soil until the excess unstable carbon is consumed. This effect causes the observed crop growth delay or inhibition, as compared with the "no biochar" control in many biochars research studies.

Some biochars may contain phytotoxins and polycyclic aromatic hydrocarbons, usually measured by extraction of the biochar with a strong solvent like toluene. For any properly made biochar, which will have significant adsorption capacity (discussed in depth later), such toxins are not bio-available or are desorbed into the groundwater at levels that do not cause unacceptable effects and are consumed by soil microbes. As such, toxicity is rarely a legitimate constraint in biochar utilization, although it may represent a barrier if regulations prevent specific materials or applications due to inferred or implied risks that are unlikley to exist in real world applications.

One benefit of unstable soluble organics that leach into the soil is stimulating development of microbial populations in soils that are sterile or have minimal organic matter. A living soil works with the crops to provide the plants with moisture and plant nutrients in exchange for the plant exudates that feed the microbes. When soils lack microbial populations, there is no role or "market" for the plant exudates and the plant suffers accordingly. Excess soluble carbon, provided as unstable organics in the biochar, allows the soil to build microbial populations and restores one natural mechanism by which the soil interacts with the crops. Sandy arid climates often have this paradigm, as encountered in Australia and many desert climates.

7.5.4 Partitioning the biochar into relative volumes

Biochar is a low bulk density material with an internal skeleton of rigid structures. The graphitic skeleton density of biochars has been measured at about 1.5 g/cm^3. The typical bulk density of dry low-ash biochar is 0.15–0.25 g/cm^3. For the less-dense end of the range, 1 L (1000 cm^3) of biochar would have 150 g of skeleton, which would only require 100 cm^3 of the volume, leaving 90% as voidage. For the high end of the density range, 250 g of skeleton at 1.5 g/cm^3 is 167 cm^3, leaving 83% voidage. Since most of the ash in higher ash biochars is either within the graphite skeleton or present as grains of sand with a density of 2.5 g/cm^3 or higher, virtually any biochar is at least 80% voidage. About 40% of the 80% voidage is between particles and the rest is inside the biochar.

Let's consider a liter of uniform granular 0.15 g/cm^3 biochar with 100 cm^3 of carbon skeleton and 40%, or 400 cm^3, of volume external to the biochar particles. That leaves 500 cm^3 of volume inside the biochar particles. If all the internal volume is filled with water, there would be 500 additional grams of water and 3.33 g of water for each gram of biochar. Since some water is also retained on the outer surfaces and between the particles of the biochar, it is not unusual for four to five times the weight of biochar to be retained as "held water" when biochar is in the soil and the soil is flooded by excess precipitation.

The voidage of biochar is occupied by either liquid water or air, with the sum of the two volumes being constant. Biochar does not significantly swell or shrink when wetted, although interparticle surface water may make the biochar "fluff up" when mixed with water and shrink when that water is removed by evaporation in dry conditions. The solid skeleton is a range of interconnected volumes, often termed pores, which do not have any specific shape or geometry and range from visible to the human eye to the scale of atoms. Virtually all the unique properties of biochar adsorption phenomena are associated with the smallest sites, called micropores. Biochar micropores are created during the pyrolysis step and are not significantly modified over time. It is best to measure the micropore properties of a biochar before it is integrated in the soil. This unique aspect of biochar is discussed in greater detail in Section 7.6.3.2.

7.6 Measuring biochar properties to predict biochar performance

The goal of the previous discussion was to partition the biochar into four separate constituents, in addition to air, each of which has a very different contribution to the value of the biochar. The water is present but of little value to the purchaser and will certainly not be the last water the biochar encounters. The ash may have some value depending on the plant fertilizer profile, but it is valued as a fertilizer, consumed in one or two growing seasons, and does not act as a permanent soil modifier. Unstable organics are present and will have a short-term impact, but their contribution is more accommodated than coveted in most applications. The stable portion of the biochar is the portion that qualifies for carbon credits and imparts the unique and long-lasting changes that modify and improve the soil properties and crop performance. This section will explore biochar properties that can be measured and used to predict biochar impact in soil applications. Needless to say, the biochar impact is highly dependent on the soil characteristics, the climate during the growing season, and the crop requirements for growth and quality. However, biochar is basically an inert material with predictable contributions to the soil, and those contributions will be discussed.

7.6.1 Revisiting pH and soil buffering concerns and constraints

Biochar pH is used to predict how the biochar will influence the soil pH and whether any change will impact crop performance. pH is how acidic or basic a solution is, typically measured by wetting a portion of biochar in deionized or neutral (pH = 6.5–7.5) water. Since the soil has also a pH before the biochar addition, there is a melding when the two are mixed, and the blend depends on properties and relative amounts of each. Buffering capacity refers to the ability of a solution to resist changes in pH when a strong acid or base is added. The pH in biochars is due to both inorganic compounds, which are part of the ash and usually strong bases, and organic functionalities found in both the stable and unstable carbon compounds, which are weak acids and bases. The same is true of soils, where soil organic matter and minerals can participate in the pH tug-of-war. The surest way to predict the pH of the biochar–soil mixture is to make a mixture and test it for pH. Thus, for a proposed biochar addition to a known soil, simply make up the anticipated blend, add an excess of deionized or reasonably pure water and measure the pH, with a simple pH paper strip being sufficiently accurate. The goal is to make sure the resulting pH is not outside the range recommended for the crop, with each crop's acceptable pH range being available in the literature or on the web.

The biggest driver in biochar pH is the highest temperature, HTT, the biochar reached during pyrolysis. Above 600°C, calcium carbonate in the biochar loses carbon dioxide and converts to calcium oxide, which converts to calcium hydroxide upon wetting, which raises the pH. Magnesium has a similar sequence of transformations at the same temperatures. It is not uncommon for biochar to have both carbonate and oxide forms due to the HTT range during pyrolysis. Since both forms are measured as part of the ash; however, it is hard to know how much of each is present. Still, the higher the ash level, the more inorganic material that can impact the pH and the greater the need to confirm that the biochar–soil blend will provide the appropriate range for the intended crop. Complicating the pH balancing act is the ability of calcium and magnesium hydroxides to react with atmospheric carbon dioxide to return to the corresponding carbonate, or even the bicarbonate form in aqueous solutions. Low-ash biochars with <10% ash on a dry weight basis are not a pH concern in most applications. Still, it is easy to confirm the pH of the biochar–soil blend and avoid subjecting the crop to conditions that slow or inhibit the many positive benefits of biochar addition.

7.6.2 Measuring biochar density and properties on a weight and volume basis

Every biochar arrives with an unknown amount of water and ash, but these are easy to measure by the methods of Sections 5.1 and 5.2. When biochar is received, especially in bulk shipments, the volume of the shipment should be measured or estimated, and a small known volume of representative material weighed to provide the "as received" volumetric density (lb/ft^3, lb/yd^3, or kg/m^3). A portion of the "retain sample" is then tested for wt.% water and wt.% ash. These weight percentages are multiplied times the volumetric density to calculate the weight of water and ash for any volume of biochar. Most biochar is handled and dispensed on a volume basis, so it is important to know the properties on a volume basis and not use the reported weight-based measurements that may come from an analytical lab.

7.6.3 Biochar properties that depend on the stable carbon characteristics

The unique aspect of biochar is its stable graphitic backbone, which is resistant to microbial breakdown and forms the micropores that provide adsorption sites. Notably, the graphitic backbone of biochar is significantly more complex than the similar structures found in most activated carbon products. This difference stems from additional chemical functionalities attached to the edges of the biochar graphene sheets. Unlike activated carbons, the edges of biochar can have a wide range of organic functional groups directly bonded to the graphitic backbone. An important functionality is the carboxylic acid group, which is largely responsible for cationic exchange capacity (CEC). In addition, biochar can adsorb humic and fulvic acids, providing additional CEC. These two remaining biochar attributes, along with adsorption capacity, are challenging to measure analytically. Conceptually, the former is the extent to which biochar behaves like ion-exchange resin and the latter is the extent that biochar performs like activated carbon. Many biochars exhibit significant and measurable amounts of CEC and adsorption capacity, and these properties provide significant long-term improvements in the soil properties.

7.6.3.1 Measuring cation exchange capacity

Biochar CEC can be measured by taking a sample of char and converting all the cations to one form, then displacing them with another cation, and finally quantifying the displaced cations to measure the CEC. CEC is not a very common analytical test, and exact procedures can vary from laboratory to another. As a result, values reported in the literature are highly variable and reproducibility is low, mainly due to differences in methods used (Munera-Echeverri et al., 2018). An example of the CEC method is the following procedure:

A sample of dried char is shaken, centrifuged, and drained three times with sodium acetate solution, followed by shaken, centrifuged, and drained three times with 2-propanol. The alcohol rinses remove excess cations present in solution but not bound to the char. The sodium-loaded char is then shaken, centrifuged, and drained with an ammonium acetate solution three times. The total solution from the three ammonia acetate rinses is measured for sodium level and the CEC calculated in milli-equivalents per 100 g of dry starting char. Predicting the value of CEC in a specific biochar is further complicated by the creation of *more CEC* over time, within the soil and depending on soil conditions, as presented by Cheng et al. (2006). Because of these post-pyrolysis oxidation effects, the measurement of CEC of a char determines the current level of the CEC property at the time of measurement but does not indicate what additional CEC may develop in the future.

7.6.3.2 Adsorption capacity

Adsorption capacity is another property that is poorly understood and underappreciated in biochar. Because adsorption capacity is a property of the graphene portion of the char itself, the property is formed at the time the char is created. The adsorption capacity of a char cannot be accurately measured after the biochar is introduced in the soil, since something will likely be occupying the adsorption sites. As such, measuring the adsorption capacity of a freshly made char yields the most accurate measurement of the maximum adsorption capacity available over the life of the biochar.

In terms of what biochar contributes to soil dynamics, adsorption capacity is understood to provide moisture retention under desiccating conditions and the ability to adsorb and desorb soluble organic compounds. As such, these characteristics may be pivotal in the enhancement of the microbial populations in the soil by stabilizing the minimum moisture level and assuring residual carbon sources in the soil, thereby elevating microbial survival rates during times of drought and shortages of other soluble carbon sources. Adsorption capacity is measured by "challenging" the char with a known substance, usually an organic vapor, and measuring the extent of uptake of the challenge gas under controlled conditions. The test is not a routine analytical method and the closest historic analytical method is the Brunauer, Emmett and Teller (BET) surface area assay, named after the inventors BET. BET-N2 calculates a metric of m2/gm of adsorbent, which does not predict adsorption performance in real world applications. The BET method is performed under conditions far removed from what occurs in the soil, with the BET method measuring the uptake of nitrogen vapor at liquid nitrogen temperatures (−196°C). As such, BET measurements do not accurately predict or differentiate the adsorption capacity of chars in typical biochar applications (Maziarka et al., 2021).

For the purposes of comparing chars, it is sufficient to subject all chars to the same adsorption conditions and measure the extent of adsorption. There are many historic standardized adsorption methods, such as butane activity, iodine number, and CTC Number, that provide a single numerical measure of the biochar adsorption, with higher numbers being better. Unfortunately, biochars possess a wide range of adsorption sites, extending from very high adsorption energy sites to sites with insufficient adsorption energy to have any utility. Since it is difficult to accurately characterize any distribution with a single metric, methods that provide measurements over a range of adsorption energies or multiple measurements by different assays provide more insight into a specific biochar's adsorption abilities.

One newer adsorption assay is known as "GACS" or gravimetric adsorption capacity scan. The GACS assay is performed on a modified thermogravimetric analyzer (TGA) and measures all the adsorption behavior of chars and activated carbons over a wide range of adsorption conditions. The GACS method measures a wide range of adsorption energies of practical utility by measuring the gravimetric uptake of R134a, a hydrofluorocarbon, from 300°C down to 100°C. It is possible to modify existing commercial TGAs to generate the same range of data or custom instruments can be created from any analytical balance with below the balance weighing capability. Additional information on the GACS method is available in Appendix B of *Schenkel and Shenxue Revisited* by Mclaughlin and Shields (2010). The method was also adopted by Licht et al. (2014).

7.6.4 Measuring biochar impact on soil water management

One of the greatest benefits of biochar when added to existing soils is to improve the ability of soil to capture, retain and deliver water to the crops. Laboratory studies have shown several water management improvements with biochar, including adsorption of water vapor under desiccating conditions and improved retention of bulk water when excess water is drained. Conceptually, whether the crops are grown with natural precipitation or irrigation, the goal is to have sufficient available water. This requires that the moisture be retained by the soil and made available to the crop as it grows. There are many methods for measuring the properties of soils, yet little consensus on how to predict the benefits to the crop. Unfortunately, while it is possible to measure water retention properties of the pure biochar, it is much harder to predict the performance of the soil when amended with biochar beyond the general direction of "improved." This is not surprising, since the biochar amendment is only a minor fraction of the composite soil, usually $<10\%$ by volume of the root zone. As such, the impact of any biochar addition depends predominantly on the characteristics of the original soil.

A recent meta-analysis by Wei et al. (2023) provides an overview of the many factors involved in measuring soil water dynamics and summarizing trends that appear in multiple studies. The review concludes with one set of guidance that seems well-supported and non-obvious: "Our meta-analysis showed that biochar applications increased capillary pores in coarse-textured soils and aeration pores in fine-textured soils, ultimately resulting in a significant increase in AWC (Available Water Capacity)." As the climate changes and natural precipitation becomes less predictable, and supplemental irrigation becomes more costly and less available, biochar represents one option for improving the moisture management properties of a wide range of soils in the future.

7.7 How biochar addresses deficiencies in soils

Section 7.6.4 mentioned how biochar has been shown to improve two different soil types by two different mechanisms. Since biochar co-evolved with soils, including soil microbiology and crops successfully adapted to localized climates, it is logical that various properties of biochars provide vital and different properties in many growing situations. Many crop deficiencies result from modern agricultural practices, where synthetic fertilizers have replaced natural systems involving plant exudates and soil microbiology, which results in loss of soil organic matter and reduced moisture retention. While the short-term increases in crop yields are undeniable with chemical fertilization, the long-term deterioration of soil properties ultimately results in the failure of the soil-crop performance in the absence of excess fossil carbon-based fertilizers and supplemental irrigation. As climate disruption due to elevated temperatures and water shortages increase, the viability of continued fossil carbon-based practices decreases and the merits of returning to natural soil dynamics increase. Biochar addition to existing soils provides a faster option for recovering natural soil properties such as cover crops, fallowing, or allowing natural biochar sources, such as wildfires, to restore soil carbon inventories. Principal soil deficiencies that biochar addresses are moisture management and improving soil aeration, which involve the rigid yet porous volume provided by the biochar stable carbon backbone. The stable volume alternates between moisture storage and aeration as water inventories fluctuate. With the stabilization of the soil moisture levels and suppression of anaerobic conditions, the stage is set for crop exudates to stimulate soil microbiology, which promotes crop health and productivity.

Another unique contribution of biochar is lowering the level of toxicity in the soil groundwater. Toxicity in this instance refers to a broad class of chemical compounds that inhibit both microbial and crop health. One obvious source of toxicity is residual pesticides from historic applications or contaminated irrigation water, typically from rivers receiving runoff from industrial processes or chemical intensive agricultural practices. Other, less obvious sources are allelochemicals, which are "a toxic chemical produced by a plant in order to defend itself against herbivores or competing plants" (https://www.merriam-webster.com/dictionary/allelochemical). The biochar serves to remove the toxins, when present, by adsorption from the soil groundwater, thereby lowering any effect and enhancing soil microbiology and crop viability. Subsequently, in periods when the level of toxins is lower, the biochar desorbs the toxins at reduced levels into the groundwater. This reversable adsorption–desorption cycle restores adsorption capacity in the biochar and provides sources of carbon for soil microbes while destroying the toxins. Biochar provides a similar buffering role by accumulating soluble humic and fulvic acids during periods of decaying plant residues or detritus. Subsequently, during periods of less soluble carbon in the soil, the humic and fulvic acids are desorbed and serve as maintenance food sources for the microbiology.

When toxicity is the leading shortcoming in an existing soil, relatively small amounts of biochar can effectively address this constraint. When using a biochar with reasonable adsorption capacity, with at least 25 wt.% of the adsorption capacity of activated carbon, 1 vol.% of biochar will usually remove the toxicity. Furthermore, targeting the toxicity by locating the biochar nearer the seedlings, by banding or adding biochar below the seedling when planting, will further improve the efficacy and reduce the biochar requirement. The net effect of the biochar in the soil is to improve the soil "health" (Schmidt et al., 2021), enabling the microbes to interact with the crops synergistically. Two significant advantages of elevated soil microbiology are the ability to adjust soil pH by excreting or consuming organic acids and the ability to reduce nutrient leaching by having the microbes consume soluble nutrients and incorporate them into non-soluble microbial populations.

7.7.1 How biochar reduces plant stress in growing systems

The concept of plant stress is central to understanding how crops grow and thrive. One popular definition, from Lichtenthaler (1996), is "Any unfavorable condition or substance that affects or blocks a plant's metabolism, growth or development, is to be regarded as stress." With this framework, it is easy to see that biochar can help with many of the typical sources of plant stress, as discussed in the previous section, including increased moisture retention, soil aeration, and toxicity reduction. One way of gauging crop productivity is to start with the commodity that the plant creates via photosynthesis, being simple sugars. The crop will make as much of these sugars as possible based on the amount of sunlight, nutrients, and water available, assuming the plants are healthy and growing in a suitable climate. The plants will metabolize sugars to obtain the necessary amount of water, which can range from very little if water is plentiful, to essentially all the plant's energy if water is so scarce that the plant reaches its wilting point. The plant also uses the sugars for root exudates, to feed soil microbes in exchange for plant nutrients and micronutrients. If the plant expends less sugars obtaining necessary nutrients, then more is available for plant growth. This is how biochar and a living microbial soil benefit the plant the most, by providing an environment where the plant can efficiently obtain necessary water and nutrients.

However, too much of a good thing has its downside with respect to water and plant nutrients. Excess water may cause anaerobic zones in the soil, which is a hostile environment for the roots and soil microbes. Biochar, as discussed, helps keep soil moisture at the levels that promote crop growth. Excess nutrients, especially soluble forms of nitrogen and phosphorus typically from chemical fertilizers, result in the plant not wasting sugars on root exudates, which may reduce the health of the soil microbiology. If excessive plant nutrients are provided, the soil microbiology level can deteriorate such that the crop loses its partnership with the soil microbiology, resulting in the issues associated with sterile soils such as poor moisture management and nutrient leaching of those same excess soluble fertilizers.

While excess water and nutrients cause problems, biochar addition to the soil can be beneficial by compensating for soil deficiencies. Biochar is a permanent soil amendment that integrates into the texture of the healthy living soil unless lost due to soil erosion. By lowering the requirements of water and plant nutrients, biochar addition benefits the growing system for decades. The reduction in annual nutrient and water inputs offset the cost of the biochar by allowing crops to grow more efficiently year after year. This dynamic presents itself in the literature in studies where tropical applications of biochar showed dramatic improvements and temperate climates showed little to none (Jeffery et al., 2017; Novak et al., 2019). This partitioning is attributed to the practice of providing supplemental fertilizers and irrigation in temperate climates versus greater reliance on available, potentially minimal, soil fertility, and natural precipitation in tropical agriculture. Significant progress has been made in understanding biochar properties and the interactions with soils and growing systems since the earlier works that divided the tropical and temperate worlds into two different biochar universes. What has emerged is a holistic model that accounts for soil microbiology, physical properties, and nutrient additions to understand how biochar fixes soil deficiencies and reduces plant stress. This model enhances nutrient and water use efficiency and contributes to more sustainable crop production systems.

7.8 Conclusion and future perspectives

Biochar is a highly variable material, with the characteristics of the end product depending on the type of biomass, the pyrolysis temperature and duration, and any post-pyrolysis modifications. Due to this variability, it does not lend itself to definitive and absolute guidance when integrated into a growing system of soil, crops, and climate. As with any living system, there are many factors to be balanced and optimized, including the cost of inputs and labor, the variability of the climate, and the markets for the crops. A few prudent concepts and guidelines that can be applied when using biochar agronomically are summarized below:

1. Biochar is a natural component of living soil, not an artificial modification.
2. Biochar works on a natural time scale and may take several growing seasons to fully integrate into the soil–crop partnership.
3. The only major requirement when applying biochar is that it stays where intended, so avoid applying it when it might either blow away or wash away before the growing season.
4. Multiple smaller additions of biochar seem to work better than one large addition because it avoids any shock due to the biochar suddenly changing the soil properties or taking up nutrients.
5. Avoid adding raw biochar directly to growing crops unless it is top-dressed to allow integration into the soil over time.
6. Biochar has quantifiable properties and qualities—measure them and learn what works and what does not.
7. While biochar is a climate-positive action, its major value is in the improved crop performance and reduced crop inputs.

Acknowledgment

The authors wish to thank Tera Lewandowski, PhD, for reviewing the manuscript.

Author contribution

Hugh McLaughlin: writing—original draft, writing—reviewing and editing; Samer Fawzy: writing—original draft, writing—reviewing and editing.

Conflict of interest

The authors declare no conflict of interest.

Statement on the use of AI tools for writing

The authors declare that AI tools were not utilized at any stage of preparing this manuscript.

References

Antal, M. J., & Grønli, M. (2003). The art, science, and technology of charcoal production. *Industrial and Engineering Chemistry Research*, *42*(8), 1619–1640. Available from https://doi.org/10.1021/ie0207919, http://pubs.acs.org/journal/iecred.

Bergman, P., Boersma, A., Zwart, R., & Kiel, J. (2005). *Torrefaction for biomass co-firing in existing coal-fired power stations*. Technical report. Energy Research Centre of the Netherlands.

Brethauer, S., Shahab, R. L., & Studer, M. H. (2020). Impacts of biofilms on the conversion of cellulose. *Applied Microbiology and Biotechnology*, *104*(12), 5201–5212. Available from https://doi.org/10.1007/s00253-020-10595-y, http://link.springer.de/link/service/journals/00253/index.htm.

Cheng, C. H., Lehmann, J., Thies, J. E., Burton, S. D., & Engelhard, M. H. (2006). Oxidation of black carbon by biotic and abiotic processes. *Organic Geochemistry*, *37* (11), 1477–1488. Available from https://doi.org/10.1016/j.orggeochem.2006.06.022.

Enders, A., Hanley, K., Whitman, T., Joseph, S., & Lehmann, J. (2012). Characterization of biochars to evaluate recalcitrance and agronomic performance. *Bioresource Technology*, *114*, 644–653. Available from https://doi.org/10.1016/j.biortech.2012.03.022.

Fawzy, S., Osman, A. I., Doran, J., & Rooney, D. W. (2020). Strategies for mitigation of climate change: A review. *Environmental Chemistry Letters*, *18*(6), 2069–2094. Available from https://doi.org/10.1007/s10311-020-01059-w, http://springerlink.metapress.com/app/home/journal.asp?wasp = d86tgdwvtg0yvw9gvkwp&referrer = parent&backto = browsepublicationsresults,140,541;.

Fawzy, S., Osman, A. I., Mehta, N., Moran, D., Al-Muhtaseb, A. H., & Rooney, D. W. (2022). Atmospheric carbon removal via industrial biochar systems: A techno-economic-environmental study. *Journal of Cleaner Production*, *371*. Available from https://doi.org/10.1016/j.jclepro.2022.133660, https://www.journals.elsevier.com/journal-of-cleaner-production.

Fawzy, S., Osman, A. I., Yang, H., Doran, J., & Rooney, D. W. (2021). Industrial biochar systems for atmospheric carbon removal: A review. *Environmental Chemistry Letters*, *19*(4), 3023–3055. Available from https://doi.org/10.1007/s10311-021-01210-1, http://springerlink.metapress.com/app/home/journal.asp?wasp = d86tgdwvtg0yvw9gvkwp& referrer = parent&backto = browsepublicationsresults,140,541;.

Harris, P. J. F. (2013). Rosalind Franklin's work on coal, carbon, and graphite. *Interdisciplinary Science Reviews*, *26*(3), 204–210. Available from https://doi.org/10.1179/030801801679467.

Jeffery, S., Abalos, D., Prodana, M., Bastos, A. C., Van Groenigen, J. W., Hungate, B. A., & Verheijen, F. (2017). Biochar boosts tropical but not temperate crop yields. *Environmental Research Letters*, *12*(5). Available from https://doi.org/10.1088/1748-9326/aa67bd, http://iopscience.iop.org/article/10.1088/1748-9326/aa67bd/pdf.

Licht, J., McLaughlin, H., Burns, C., & Shields, F. (2014). Can biochar come to the rescue of coastal barren species? A controlled study reports on the impact of biochar amendment on their survival. *BioResources*, *9*(4), 6214–6226. Available from https://doi.org/10.15376/biores.9.4.6214-6226, http://www.ncsu.edu/bioresources/BioRes_09/BioRes_09_4_6214_Licht_Biochar_Rescue_Soil_Inoculation_5496.pdf.

Lichtenthaler, H. K. (1996). Vegetation stress: An introduction to the stress concept in plants. *Journal of Plant Physiology*, *148*(1-2), 4–14. Available from https://doi.org/10.1016/s0176-1617(96)80287-2.

Maziarka, P., Wurzer, C., Arauzo, P. J., Dieguez-Alonso, A., Mašek, O., & Ronsse, F. (2021). Do you BET on routine? The reliability of N2 physisorption for the quantitative assessment of biochar's surface area. *Chemical Engineering Journal*, *418*. Available from https://doi.org/10.1016/j.cej.2021.129234, http://www.elsevier.com/inca/publications/store/6/0/1/2/7/3/index.htt.

Mclaughlin, H., & Shields, F. (2010). Schenkel and Shenxue revisited-implications on char production and biochar properties. *Initiative conference* (pp. 1–16).

Munera-Echeverri, J. L., Martinsen, V., Strand, L. T., Zivanovic, V., Cornelissen, G., & Mulder, J. (2018). Cation exchange capacity of biochar: An urgent method modification. *Science of the Total Environment*, *642*, 190–197. Available from https://doi.org/10.1016/j.scitotenv.2018.06.017, http://www.elsevier.com/locate/scitotenv.

Novak, J. M., Sigua, G. C., Ducey, T. F., Watts, D. W., & Stone, K. C. (2019). Designer biochars impact on corn grain yields, biomass production, and fertility properties of a highly-weathered ultisol. *Environments – MDPI*, *6*(6). Available from https://doi.org/10.3390/environments6060064, https://www.mdpi.com/2076-3298/6/6/64/pdf.

Osman, A. I., Fawzy, S., Farghali, M., El-Azazy, M., Elgarahy, A. M., Fahim, R. A., Maksoud, M. I. A. A., Ajlan, A. A., Yousry, M., Saleem, Y., & Rooney, D. W. (2022). Biochar for agronomy, animal farming, anaerobic digestion, composting, water treatment, soil remediation, construction, energy storage, and carbon sequestration: A review. *Environmental Chemistry Letters*, *20*(4), 2385–2485. Available from https://doi.org/10.1007/s10311-022-01424-x, http://springerlink.metapress.com/app/home/journal.asp?wasp = d86tgdwvtg0yvw9gvkwp&referrer = parent&backto = browsepublicationsresults,140,541;.

Schmidt, H. P., Kammann, C., Hagemann, N., Leifeld, J., Bucheli, T. D., Sánchez Monedero, M. A., & Cayuela, M. L. (2021). Biochar in agriculture – a systematic review of 26 global meta-analyses. *GCB Bioenergy*, *13*(11), 1708–1730. Available from https://doi.org/10.1111/gcbb.12889, http://onlinelibrary.wiley.com/journal/10.1111/(ISSN)1757-1707.

Singh, P., Rawat, S., Jain, N., Bhatnagar, A., Bhattacharya, P., & Maiti, A. (2023). A review on biochar composites for soil remediation applications: Comprehensive solution to contemporary challenges. *Journal of Environmental Chemical Engineering*, *11*(5), 110635. Available from https://doi.org/10.1016/j.jece.2023.110635.

Wei, B., Peng, Y., Lin, L., Zhang, D., Ma, L., Jiang, L., Li, Y., He, T., & Wang, Z. (2023). Drivers of biochar-mediated improvement of soil water retention capacity based on soil texture: A meta-analysis. *Geoderma*, *437*, 116591. Available from https://doi.org/10.1016/j.geoderma.2023.116591.

Wood, R., Mašek, O., & Erastova, V. (2023). Biochars at the molecular level. Part 2: Development of realistic molecular models of biochars. *arXiv*. Available from https://doi.org/10.48550/arXiv.2303.09907, https://arxiv.org.

PART 3

Biochar for enhancing soil fertility, functioning, & plant responses

CHAPTER

8 Recent trends in multifarious benefits of biochar application to soil

Sara de Jesus Duarte[1], Gertrudis Pentón Fernández[2,3] and Odelin Brea Maure[4]

[1]*Latin America Institute of Biochar, Ciudad de Mexico, Mexico*

[2]*Indio Hatuey Pastures and Forages Experimental Station, University of Matanzas, Matanzas, Cuba*

[3]*Latin American Institute of Biochar, Mexico City, Mexico*

[4]*Agricultural Faculties, University of Matanzas, Matanzas, Cuba*

8.1 Introduction

The sustainable management of agricultural production should be a priority in current times. It requires processes based on natural laws that, when used rationally, result in the production of healthy and abundant food at a low cost, without environmental impact, and with a significant contribution to human well-being. From ancient civilizations that used biochar to improve soil fertility to its modern resurgence, it has become a versatile amendment, encapsulating the transformative potential of carbon (C)-rich substances. It has transcended geographical borders and found a place in agricultural landscapes worldwide. Scientific and technological research on the characterization of biochar and the optimization of its agricultural use is numerous, varied, and contextualized. However, these efforts are still insufficient, and recent knowledge production is distinguished by its novelty and relevance.

The capacity of biochar to retain and release moisture and nutrients, the modifications to the physical and biological characteristics of the soil, and the use of by-products from technological processes, such as sugarcane bagasse, agave, tree pruning residues, enriched with biofertilizers and high-nutrient organic substances, continue to have a high scientific and technical impact. The role of final uses is emphasized, and the alterations in physicochemical properties when produced at different pyrolysis temperatures and particle sizes play a leading role. These characteristics alter interactions with the soil, affecting its chemical, physical, and biological properties, as well as its application in water treatment. In this chapter, we aim to achieve a better understanding of the relationships between biochar raw material, pyrolysis temperature, and their effects on the chemical, physic-hydraulic, and biological properties of the soil.

Biochar Ecotechnology for Sustainable Agriculture and Environment. DOI: https://doi.org/10.1016/B978-0-443-29855-4.00010-2

8.2 The properties of biochar vary with different pyrolysis temperatures, influencing its impact on soil chemical and physic-hydraulic properties

Certainly, the choice of feedstock and pyrolysis temperature stands as the primary determinant of biochar's ultimate characteristics. Biochar, with diverse chemical and physical properties, can distinctly impact soil chemistry, physics, and hydraulics properties. It becomes imperative to carefully select biochar tailored to specific end-use requirements. To achieve this, attention must first be directed toward the biochar feedstock, followed by the selection of an appropriate pyrolysis temperature. This is crucial as feedstocks upon undergoing the pyrolysis process, each nutrient exhibiting unique behavior, and as the particle size and porosity changes had direct effects on soil chemical physical and hydraulic properties.

The characteristics of biochar and its response to pyrolysis temperature are contingent upon the type of biomass employed as the raw material. Key indicators commonly used for its characterization and certification include elemental composition (total C, organic C, hydrogen (H), nitrogen (N), oxygen (O), sulfur (S), and ashes), as well as physical-structural parameters (electrical conductivity, salt content, pH, and water-soluble carbohydrates [WSCs]) (Table 8.1). The O/C and (N + O)/C ratios, indicating a reduction with rising temperature, signify an increase in aromaticity and a decline in the polarity of biochar, reflecting an enhanced sorption capacity of the biochar (Blanco-Canqui et al., 2020; Figure 10.1). The fixed or recalcitrant C varies depending on the type of raw material and the presence of labile C fractions, being a nonlinear function of the H/C ratio. A standard value

Table 8.1 Elemental composition and physical-structural parameters of biochar for agricultural use.

General comment	Authors
The properties of biochar, as well as its response to pyrolysis temperature, depend on the type of biomass used as raw material.	Bednik et al. (2022), Gaffar et al. (2021), Masís-Meléndez et al. (2020), Batista et al. (2018)
The relationship between fixed C and labile C varies depending on the pyrolysis temperature and the type of raw material, constituting a nonlinear function of the H/C ratio.	Kwaku et al. (2022), Alfattani et al. (2022), Chen et al. (2022), Gaffar et al. (2021), Shen et al. (2022)
The qualities of biochar determine its potential uses in agriculture, whether for bioremediation, the reduction of greenhouse gas emissions, or as a restorer of soil physical structure and/or carrier of nutrients, biofertilizers, and biopesticides.	Campos et al. (2020), EBC (2021), Gaffar et al. (2021), Hagemann et al. (2018), Kwaku et al. (2022), López-Cano et al. (2018), Pandit et al. (2020), Pérez-Cabrera et al. (2022), Qayyum et al. (2019), Schmidt et al. (2017), Shen et al. (2022)

Modified from Sara de Jesus Duarte, Gertrudis Fernández y Odelin Maure.

of 0.7 has been established for the H/C ratio for biochar used in agriculture. The European Biochar Certificate (EBC, 2021) also established that the best pyrolyzed organic biomass is one that achieves more than 50% C content.

An important aspect regarding the qualities of biochar and its potential uses in agriculture for bioremediation, reduction of greenhouse gas emissions, or as a restorer of soil physical structure and carrier of nutrients, biofertilizers, and biopesticides is the content of organic or labile C versus C capture (Table 8.1). During the pyrolysis process, organic matter begins to thermally decompose at temperatures above 120°C. Hemicelluloses degrade between 200°C and 260°C, cellulose between 240°C and 350°C, and lignin between 280°C and 500°C (Kwaku et al., 2022). The increase (Table 8.1) in pyrolysis temperature from 350°C to 500°C could result in up to a 70% increase in the fixed C content in biochar, with a significant reduction in labile or mineralizable C at temperatures between 500°C and 700°C (Bednik et al., 2022).

Low pyrolysis temperatures in biomass such as sugarcane bagasse can be explained by a low lignin content compared to denser woody species. Raw materials with high cellulose content generate more recalcitrant char, while labile C, which is more related to WSCs content, represents less C fixation. Therefore it is considered important to include this aspect in discussions about the biochar–soil interaction (Table 8.1).

8.3 Biochar pyrolysis temperature and its effects on soil chemical properties

Biochar produced at low pyrolysis temperatures (200°C–400°C) is characterized by an abundance of oxygen-containing functional groups, including –COOH, –OH, C = O, phenolic –OH, and –CHO groups (Fig. 8.1). These groups stimulate nutrient exchange and enhance soil fertility (Mandal, Pu et al., 2019). A higher C/O ratio in biochar, associated with lower pyrolysis temperatures, suggests the presence of more oxygenated functional groups, such as hydroxyl, carboxylate, and carbonyl. This contributes to high cation exchange capacity (CEC) values ranging between 50 and 200 mmol/kg (Graber et al., 2017), indicating a more negatively charged surface for biochar. It means that biochar produced at lower temperatures could contribute to increasing soil fertility more rapidly than that produced at higher pyrolysis temperatures. High-temperature biochar might exhibit high ionic binding strength through physical adsorption, potentially locking up nutrients in unavailable forms. In contrast, low-temperature biochar, with chemisorption and ion exchange capacity associated with surface functional groups, could provide a more efficient pathway for reserving available phosphorus (Ngatia et al., 2017; Fig. 8.1). Meaning that high pyrolysis temperature could contribute to increase some soil nutrients content in unavailable forms and low pyrolysis temperature increase some soil nutrients availability (Fig. 8.1).

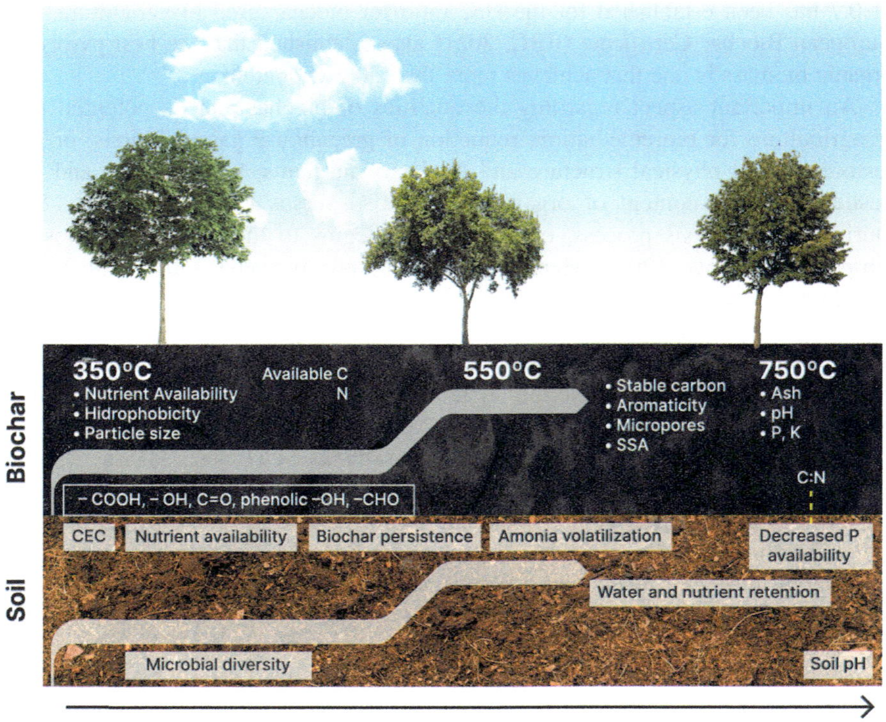

FIGURE 8.1 Biochar pyrolysis temperature and its effects on soil chemical, physical-hydric properties, and microbial association.

The pyrolysis temperature of biochar impacts soil porosity, particle size, bulk density, nutrient retention and availability, and directly influences microbial diversity. Biochar produced between 400°C and 700°C boosts microbial diversity, at lower temperatures enhance nutrient availability and soil fertility, while higher temperatures increase pH and porosity, improving soil water and nutrient retention.

Data from Mandal, Pu et al. (2019), Graber et al. (2017), Ngatia et al. (2017), Bilias et al. (2023), Leng et al. (2020), Xiao et al. (2018), Xu et al. (2016), Yang et al. (2018), Hossain et al. (2020), Pallarés et al. (2018), Keiluweit et al. (2010), Ren et al. (2023). Costumized credit line from Sara de Jesus Duarte and Rafael Gamboa.

The behavior of macronutrient content, particularly N in comparison to phosphorus (P) and potassium (K), exhibits distinct patterns with increasing pyrolysis temperature. As the temperature rises, the N content decreases, while the P and K content increases (Fig. 8.1). The decline in N content is ascribed to the conversion of amino acids into pyridine-N and pyrrolic-N, accompanied by the loss of ammoniacal nitrogen (NH_4^+-N) through volatilization as NH_3 during pyrolysis (Leng et al., 2020). Conversely, the elevation of P and K content in biochar with rising pyrolytic temperature can be attributed to the "concentration effect," a phenomenon resulting from decreased biochar yield as temperature increases (Xiao et al., 2018). The behavior of N, P, and K when pyrolysis temperature increases is shown as follows:

8.3.1 Nitrogen

It is crucial to consider that, despite the reduction in N as the pyrolysis temperature increases (Fig. 8.1), the loss of N through NH_3 emission is highly influenced by the biochar pH. Biochar with slightly acidic or near-neutral pH has been shown to decrease NH_3 volatilization from the soil (Mandal, Donner et al., 2019). Consequently, biochar produced at higher temperatures contributes to increase the pH and it may contribute to an increase in NH_3 volatilization, thereby reducing N concentration in the soil. Numerous studies have highlighted that the application of biochar can elevate soil total N (TN) content, and escalating application rates result in additional soil TN accumulation compared to control conditions (Wu et al., 2020; Zhao et al., 2023).

The nutrient content of biochar does not guarantee its immediate availability for plants. For instance, despite the N content of biochar ranging from <0.10% to 9%, its actual availability is influenced by the rate, type of biochar, soil type, and biochar C-to-N (C:N) ratio. In this chapter, our focus will be on feedstock and the biochar C:N ratio, and their influence on biochar N availability. Biochar derived from feedstocks high in carbohydrates, such as crop residues and manure, contains higher levels of labile C, while lignin-rich wood biochar tends to have relatively low levels (Bednik et al., 2022).

Although biochar C:N ratios can be relatively high, a significant portion of the C is considered resistant to microbial decomposition, limiting its ability to stimulate microbial N immobilization (Cross & Sohi, 2011). Alternatively, biochar can adsorb high C:N organic molecules from the soil solution, potentially increasing mineralization (Gundale & DeLuca, 2007). Consequently, the addition of biochar to soil may result in no change or a slight enhancement in mineral N concentrations under specific biochar C:N subgroups. Notably, a reduction in inorganic N following biochar additions is more pronounced for biochar produced under low temperatures or derived from low C:N feedstocks, such as manure or crop residue. The presence of labile C in biochar also plays a role in influencing N availability. The introduction of additional degradable C to the soil through biochar is likely to induce microbial N immobilization. This process results in a decrease in soil inorganic N as soil microorganisms utilize the additional label C. In such scenarios, it becomes necessary to supplement N in agricultural soil to meet the demands of plants. For example, combining biochar applications with organic fertilizer has demonstrated significant potential for improving inorganic N availability (Gao et al., 2019).

8.3.2 Phosphorus

Phosphorus (P) availability in soils is more pronounced when biochar is produced at lower pyrolysis temperatures (Fig. 8.1), given that at a higher pyrolysis temperatures more stable P species could be formed where the presence of poly-P, crandallite ($CaAl_3(OH)5(PO_4)_2$), and wavellite ($Al_3(OH)3(PO_4)_2$) were observed to be at greater concentrations in high-temperature biochar regardless of the feedstock (Xu et al., 2016). Moreover, an increase in temperature contributes to elevated stable C levels, resulting in high C:N ratio (Fig. 8.1), the application of a relatively high C:N ratio biochar would be predicted to enhance microbial N demand, N mobilization, and relative N limitation (Cleveland & Liptzin, 2007). In turn, conditions of N scarcity would be predicted to reduce the microbial demand for P, induce declines in microbial P, and contribute to net increases in P mineralization (Gao et al., 2019).

To enhance soil-available P, it is advisable to produce biochar with a low C:N ratio, ideally within the range of 15–45. To achieve this objective, selecting a biochar derived from manure, crop residues, or other materials with a naturally low C:N ratio and producing the biochar at a low pyrolysis temperature (below 450°C), or apply a biochar to slight acidic to neutral soils it is possible to make a biochar with potential to contribute to up to 45% available P (Gao et al., 2019). As a conclusion, soils amended with biochar would be more likely to exhibit an increase in soil P with decreased soil N.

8.3.3 Potassium

Phosphorus and K exhibit a similar behavior in response to pyrolysis temperature. Higher temperatures (>500°C) tend to increase the total K content, whereas lower temperatures enhance the water-soluble and exchangeable K fractions. Biochar demonstrates considerable potential for improving K availability through various mechanisms, ultimately leading to increased K uptake by plants. Indirect mechanisms involve high K retention capacity based on biochar properties such as CEC, porosity, and specific surface area. Simultaneously, direct K supply can be facilitated by producing biochar from K-rich sources (Bilias et al., 2023).

The levels of available K in the soil show significant increases even with low levels of biochar addition. This phenomenon is attributed to the generation of free nutrient cations during biochar pyrolysis, which contributes to enhancing the availability of cations such as K into the soil. As the biochar concentration increases, there is a notable rise in both soluble and available K content in the soil. The availability of K is directly linked to the soil CEC witch predominant in biochar produced at low pyrolysis temperature (Fig. 8.1). Soils with higher CEC exhibit more ability to adsorb ions such as Ca, K, NH_4^+, and Mg. This enhanced adsorption improves the utilization of nutrient ions in the soil, reduces nutrient loss, and increases nutrient supply for plants, thereby minimizing leaching.

8.3.4 pH

The pyrolysis temperature of biochar significantly influences soil pH, and these pH changes directly impact nutrient availability. Biochar inherently possesses an alkaline nature, with pH values ranging from 6.52 to 12.64, attributed to the presence of base cations within biochar (Fidel et al., 2017). Biochar produced at elevated pyrolysis temperatures is more likely to exhibit a higher pH (Fig. 8.1) due to the retention of alkali and alkaline metals in the feedstocks, which are not volatilized during pyrolysis (Hossain et al., 2020; Yang et al., 2018). Additionally, acidic functional groups in organic material are often removed through volatilization on high pyrolysis temperature (Zhao et al., 2023). Upon application to soil, biochar can either increase or decrease soil pH, depending on the initial soil properties (e.g., pH, texture) and the pH and alkalinity of the biochar (Hossain et al., 2020). Given that most biochars are alkaline, possessing an acid-neutralizing capacity of up to 33% of agricultural lime (Van Zwieten et al., 2010), the application of alkaline biochar tends to elevate the pH of acidic and neutral soils (Buss et al., 2016). However, it's worth noting if older biochar may have a lower pH (Li et al., 2019).

8.4 Biochar pyrolysis temperature and its effects on soil physical and hydraulic properties

Porosity, specific surface area, and hydrophobicity are the main factors that control its effects on soil physical and hydraulic properties. Under pyrolysis temperature ($>550°C$), biochar is characterized by having a high aromaticity, porosity, and specific surface area (Edeh et al., 2020; Fig. 8.1). When pyrolysis temperature is under 400°C, biochar tends to be more hydrophobic with low specific surface area and total pore volume (Fig. 8.1). Low temperature is not able to provide conditions for complete devolatilization and consequently, they may bring about the blockage of some pores and hinder the formation of new pores (Pallarés et al., 2018).

The drastic increase starts at 400°C–500°C; resulting in an increase in the specific surface area, total pore volume and reduction of bulk density in biochar associated with the decomposition of cellulose and hemicellulose, removal of pore-blocking substances, destruction of aliphatic alkyls and ester groups, exposure of the aromatic lignin core, thermal cracking, and the formation of vesicular structures or channel networks (Keiluweit et al., 2010). Regarding this, the results found do not allow for the establishment of a clear trend and could be heavily influenced by the raw material. As the biochar porosity increases due to the temperature or feedstock, the reduction on soil bulk density is higher from 1.44 ± 0.10 to 1.38 ± 0.06 g/cm after orchard pruning biochar application in various soil types (Leng et al., 2021).

As pyrolysis temperature increases, the hydrophobicity reduces resulting in an increase in water retention (Hussaina et al., 2020; Fig. 8.1). It happens because low temperature of biochar production is more hydrophobic initially due to accumulation of aliphatic compounds in pores and on the surface; such compounds are usually lost during pyrolysis at higher temperatures. Hydrophobicity can inhibit water uptake by biochar particles (Gray et al., 2014), but this effect dissipates over time, normally after 1 year of biochar application into the soil. It means that biochar produced at low pyrolysis temperatures normally did not contribute to retain water into the soil, but this effect is dependent on the feedstock. Feedstock with good capacity to retain water could retain it even when produced at low pyrolysis temperature. de Jesus Duarte et al. (2023) studying different feedstocks pyrolyzed at low pyrolysis temperature found that *Pennisetum purpureum* could retain a higher quantity of water even when produced at low pyrolysis temperature (380°C). Another key factor influencing biochar performance is particle size, specific surface area, and porosity indicating that both soil–biochar interparticle and biochar intraparticle pores are important factors. To achieve optimum soil water relations, the biochar pore diameter range 1–8 nm is ideal (Turunen et al., 2020) and to make it available to the plants the biochar particle size range ($\sim 0.5 - 1$ mm) is optimal for plant performance (Thomas, 2021).

8.5 The impact of biochar on soil microbial properties is influenced by changes in pyrolysis temperature

The application of biochar produced at temperatures between 400°C and 700°C, enhances the microbial diversity in the rhizosphere at the phylum, order, and genus levels. The contribution rates of fungi tend to be higher than those of bacteria in general (Ren et al., 2023). Among the variation factors related to biochar and its effects on soil microbiota, notable factors include pH, porosity, water storage capacity, redox potential, labile organic matter content, alkaline hydrolyzable N, available phosphorus and potassium, and exchangeable calcium and magnesium. There is consensus in the consulted literature on these aspects, as well as on the fact that pyrolysis temperature and the source material are among the determining factors (Table 8.2). Temperatures exceeding 700°C result in biochar with a low quantity of N, H, oxygen, and functional groups on the surface, which may reduce nutrient sorption for microbial growth (Table 8.2; Tomczyk et al., 2020). Under such conditions, Sun et al. (2020) reported more pronounced positive effects on the functioning of N-fixing bacteria.

The reviewed literature suggests that pyrolysis temperatures below 700°C stimulate the abundance of microorganisms and are associated with the decomposition of labile fractions of biochar (aliphatic compounds and carbohydrates such as cellulose). Phares et al. (2022) found that in biochar pyrolyzed at 450°C, the activity of *Bacillus* spp. and *Pseudomonas* spp. was related to a pH of 9.8, a total C content of 208 g/kg, 0.52% TN, and 0.17% total P. Husna et al. (2019) discovered that in biochar pyrolyzed at low temperatures, with moisture content of 26.86%, a pH of 7.74, and a pore size with an average of 6.59 μM, the survival capacity of the phosphate-solubilizing microorganism inoculated increased for up

Table 8.2 Effect of biochar on soil microbiological properties.

General comment	Authors
The pH, porosity, water storage capacity, redox potential, labile organic matter content, alkaline hydrolyzable nitrogen, available phosphorus and potassium, and exchangeable calcium and magnesium of biochar influence changes in rhizosphere microbiota.	Cowie (2021), Cuthbertson (2018), Figueiredo et al. (2019), Hansen et al. (2017), Husson (2013), Lacomino et al. (1996), Lehmann et al. (2021), Pentón et al. (2021), Sashidhar et al. (2020), Schmidt et al. (2021), Yang et al. (2019)
Biochar enhances the competitive capacity of beneficial microorganisms such as plant growth promoting rhizobacteria, *Bacillus subtilis, Bacillus megaterium*, *Trichoderma harzianum*, *Pseudomonas* spp., against pathogenic bacteria.	Akhter et al. (2016), Bolan et al. (2023), Egamberdieva et al. (2018), Jing et al. (2022), Pavlicevic et al. (2023), Poveda et al. (2021), Wu et al. (2020)

Modified from Gertrudis Fernández.

to 6 months. Figueiredo et al. (2019) observed that biochar obtained at a temperature of 300°C exhibited higher H/C ratios and lower C/N ratios than that obtained at a pyrolysis temperature of 500°C. This stimulated the presence of arbuscular mycorrhizal fungi (AMF). When biochar reaches a very high pH and an excessive amount of ashes, it can be toxic and/or provide no benefit to acid-loving bacteria (Xiang et al., 2022).

An interesting aspect corroborated by the consulted literature is that, although enriching biochar with acidic nutrients or beneficial native microorganisms such as IHPLUSBF maintains the redox value above 300 mV and pH close to neutrality (Milera et al., 2021), enriched biochar can alter the correlations between bacteria and fungi, driven by the increased pH that alleviates bacterial acid stress (Chen et al., 2021). In experiments on soil modified with biochar pyrolyzed at 700°C, this author observed negative effects on the relative abundance of fungi but positive effects on bacteria. For instance, the addition of biochar reduced root-associated fungi (16:1ω5c) and the assimilated C13 by these fungi (from 71.2 to 26.3 ng C/g soil). The unfavorable relation to fungi led to a decrease in aggregation and stability because the fungi were less entangled with the aggregates, exposing rhizodeposits to bacterial decomposition (actinomycetes).

On the other hand, Xu et al. (2021) indicated that the abundance of AMF, microbial biomass, and functional richness increased with the addition of biochar, regardless of the dosage and time elapsed since application. However, AMF abundance was highly correlated with pH, specific surface area, soil aggregate geometric mean diameter, and soil aggregate size. In turn, these parameters were negatively correlated with glomalin production, which showed a positive correlation with bulk density (Barna et al., 2020).

The interactions among microorganisms induced by root exudates constitute a crucial mechanism in the combined effect of biochar and soil microorganisms

(Gu et al., 2017; Jaiswal et al., 2020; Wu et al., 2016; Xu et al., 2021). Biochar significantly determines the metabolomic distribution in the rhizosphere, encompassing primary and secondary metabolites, mainly acids, alcohols, esters, amines, amino acids, sterols, and sugars (Table 8.2). Ren et al. (2023) detected significant differences between the absence and presence of biochar in 10 metabolic pathways, including thiamine metabolism, arginine and proline metabolism, glutathione metabolism, ATP-binding cassette transporters, butanoate metabolism, cyan amino acid metabolism, tyrosine metabolism, phenylalanine metabolism, phosphotransferase system, and lysine degradation. In this regard, it is known that secondary metabolites secreted by plants have a chemotaxis function; through which, bacteria are able to communicate among themselves and with other organisms through stimuli released into the environment, and bacterial cells near the induction translate and inform other bacterial cells in response to it. This conditions a greater microbiological diversity in the soil by "recruiting" beneficial bacteria for plants. The metabolomic analysis conducted by Wu et al. (2020) indicated that biochar significantly influenced metabolic processes in the rhizosphere, the modification of root exudates, soil properties, and nutrient availability.

8.5.1 Dynamics of beneficial fungi and their extracellular enzymatic activities

The simultaneous application of biochar and the inoculation of AMF conditions a significant increase in organic C, enzymatic activities of microbial biomass, organic phosphorus, and alkaline and acid respiration. Vahedi et al. (2021) found that AMF combined with biochar increased microbial respiration levels and substrate-induced respiration in the rhizosphere by 13.06% and 7.95%, respectively, compared to the nonrhizosphere zone. Li and Cai (2021) discovered that the addition of biochar increased the colonization rate of AMF and had significant effects on root morphology. Its beneficial effect on phosphatase activity and microbial activity was not direct but rather through AMF inoculation.

A metagenomic study conducted by Cortes et al. (2021) demonstrated that the addition of AMF with bamboo biochar (*Bambusa vulgaris*) modified the prokaryotic diversity of the soil. This is an interesting topic to delve into in new research, relating it to factors such as pyrolysis temperature and the origin of biochar. The study revealed that the three most abundant phyla in the treated soils were Proteobacteria, Acidobacteria, and Actinobacteria. The relative abundance of Proteobacteria and Acidobacteria increased with the treatments, while Actinobacteria decreased. Biochar increased the genera not classified under the Acidobacteria and Chromatia orders. On the other hand, AMF increased the abundance of an unclassified genus under the Xanthomonada order. The genus *Rhodoplanes* was the most dominant in soils treated with AMF + 15% bamboo biochar.

8.5.2 Dynamics of beneficial bacteria and their extracellular enzymatic activities

The response of bacteria to the presence of biochar depends largely on the soil condition (Guo et al., 2021; Xu et al., 2021), which, in turn, can be modified with organic amendments. The high porosity of biochar allows rhizobacteria to remain viable and metabolically active for extended periods (Ajeng et al., 2020; Rafique et al., 2017; Zheng et al., 2019), intensifying their natural activity (Chuaphasuk & Prapagdee, 2019; Egamberdieva et al., 2018). In such conditions, concentrations of available forms of P, Fe, and Zn increase (Araujo et al., 2020; Vahedi et al., 2022). For instance, Phares et al. (2022) observed that the improvement in dehydrogenase activity and the metabolic quotient rate is associated with the synergistic interaction between biochar and plant growth promoting bacteria, *Bacillus*, and *Pseudomonas* spp. They found a positive correlation between the metabolic quotient and dehydrogenase enzyme activity with CEC, pH, C, and an inverse correlation with soil bulk density.

Javeed et al. (2021) highlighted that the addition of *Bacillus* and *Pseudomonas* to biochar provided a superior positive effect by forming a synergistic interaction and efficiently promoting the utilization of NPK fertilizer through their mechanisms of solubilizing N, P, and K. A similar phenomenon occurred with *Azotobacter vinelandii*, which can convert atmospheric N into NH_3 utilized by plants (Plunkett et al., 2020). It is well established that biochar increases the competitive capacity of beneficial microorganisms such as Plant growth promoting rhizobacteria, *Bacillus subtilis*, *Bacillus megaterium*, *Trichoderma harzianum*, *Pseudomonas* spp., against pathogenic bacteria. The mechanisms of action are associated with the release hormones of plant such as salicylic acid and jasmonic acid, enzymes such as peroxidases, polyphenoloxidases, glucanases, chitinases, lipoxygenases, and antifungal compounds such as tetraamine saccharose polyphosphates, hexadecanoic acid methyl ester, and tetracosane with antifungal activity (Table 8.2).

8.6 Case study: physicochemical characteristics of biochar derived from *Saccharum officinarum* with different particle sizes

The size of particles and the choice of raw materials directly impact the physicochemical properties of biochar, influencing its effects not just in soil but also in water treatment. This case study aims to assess the quality of biochar sourced from *Saccharum officinarum* bagasse in comparison to biochar from woody species, considering various particle sizes for the filtration of water and nutrient solution for IHPLUS BF native efficient microorganisms. Biochar was generated using the kon-tiki technique through slow pyrolysis (Pentón et al., 2021).

Subsequently, these biochars underwent pretreatment with particle sizes of 1 and 2 mm, immersed either in drinking water (H_2O), and 50% solution of IHPLUSBF efficient microorganisms for 24 hours, followed by filtration for 1 hour. The experimental design employed a completely randomized approach with 5 repetitions and 20 treatments, encompassing five raw materials: *S. officinarum*, *Vitex parviflora*, *Acacia auriculiformis*, *Morus alba*, and *B. vulgaris*. The treatments also included two particle sizes (1 and 2 mm) and two filtration tests (IHPLUSBF and drinking water).

Assessment parameters included the storage index in biochar (g of moisture/g of oven-dried biochar), pH, electrical conductivity (mS/cm), and redox potential [Eh (pH7)1(mV)]. K, NH_3-N, and organic acids were quantified both before and after filtration, and the volume of these nutrients retained in the biochar was estimated. Instruments such as the Digital ExStik pH meter from EXTECH, GREISINGER Type K thermometer (NiCr-Ni), ORPMETER YK-23RP, LUTRON ELECTRONIC, and HACH Dr3900 spectrophotometer were utilized for these analyses.

Statistical analysis encompassed descriptive statistics and ANOVA, with mean comparisons conducted through Tukey's test at a 0.05 probability level. The statistical software Infostat 2008 (Di Rienzo et al., 2008) was applied for data analysis.

8.6.1 Physicochemical characteristics of biochars from *Saccharum officinarum* bagasse versus *Vitex parviflora*, *Acacia auriculiformis*, *Morus alba*, and *Bambusa vulgaris* with different particle sizes

8.6.1.1 Electrical conductivity, pH, and redox potential

No factorial interaction effect was observed. *Morus alba*-embedded biochar was characterized by higher electrical conductivity (2.28 mS/cm), followed by *S. officinarum* biochar with 0.85 mS/cm. The pH was higher in *M. alba* biochar (8.20), followed by *V. parviflora* biochar (7.50), and lower in *S. officinarum* biochar (6.63). The redox potential reached values between 499 and 546 mV in *S. officinarum, M. alba*, and *V. parviflora* biochars, which are considered optimal according to Husson (2013). Particle size conditioned significant variations in the conductivity of embedded biochars; being higher with 1 mm (0.94 mS/cm) compared to 2 mm (0.73 mS/cm).

The enrichment substances (water vs IHPLUSBF) marked significant variations in pH and redox potential; higher values were observed in water-treated biochar (8.28 and 581.16), while lower values were noted with IHPLUSBF (6.41 and 540.93 mV). The significant decrease in pH in biochar when IHPLUSBF was added was related to the higher content of soluble solids that could be retained in its structure. It is noteworthy that IHPLUSBF is a solution with high levels of metabolites produced by various microorganisms such as photosynthetic bacteria, lactic acid bacteria, actinomycetes, yeasts, and fungi, cohabiting in a pH range between 3.2 and 3.8, with a high load of acidic functional groups (IHPLUS®BF, 2020).

8.6.1.2 Water and nutrient storage capacity in biochar

The water and substance storage capacity showed no factorial interaction. The storage index quadrupled the weight of *S. officinarum* biochar on a dry basis. *M. alba, V. parviflora*, and *A. auriculiformis* biochars doubled their weight, with significant differences compared to *B. vulgaris* biochar. The 2 mm particle size exceeded the storage index of the 1 mm particle size by more than half a unit (g). Schmidt et al. (2021) suggested that the storage capacity in biochar is affected by both pore structure and surface chemistry. An increase in porosity is correlated with elevated catalytic activity, adsorption capacity, and storage (Kwaku et al., 2022) This phenomenon is further linked to the zeta potential and CEC. Studies have demonstrated that *S. officinarum* exhibits notable surface energy and high levels of polar and dispersive components (Batista et al., 2018; Fig. 8.2).

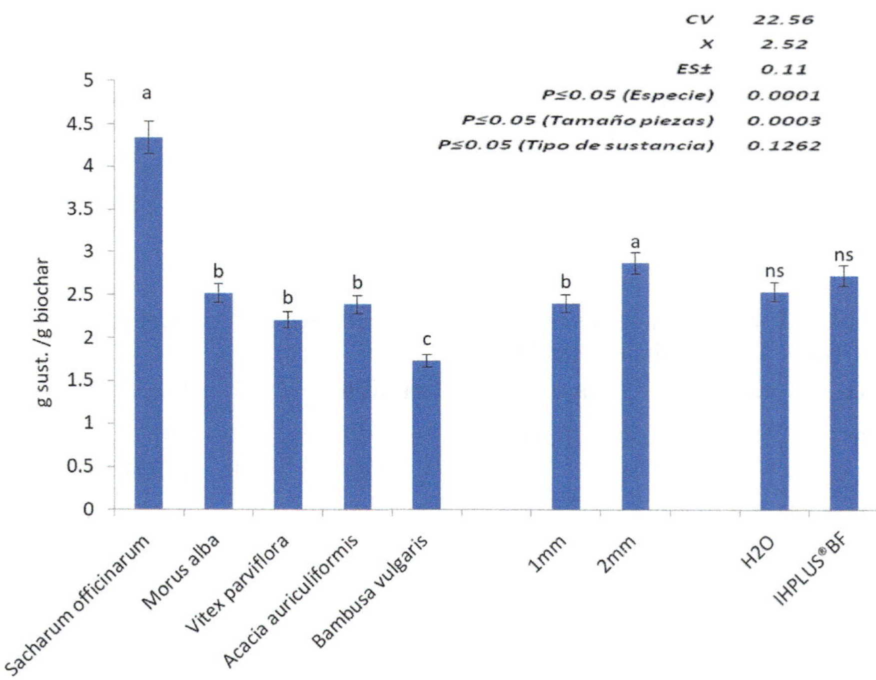

FIGURE 8.2 Storage capacity of biochar produced from different raw materials and particle sizes.

The biochar feedstock, particle size, and enrichment solution influence biochar's capacity to store water and retain or release substances and ions. The storage index was higher in *Saccharum officinarum* biochar, with a particle size of 2 mm and when using IHPLUS BF microorganisms.

Data from Gertrudis Fernández experiment located at Indio Hatuey Pastures and Forages Experimental Station, University of Matanzas, Cuba.

8.6.2 Characteristics of biochar from *Saccharum officinarum* bagasse regarding the retention versus release of NH_4-N, K, and organic acids

The release of K from the filtration of water and IHPLUS BF in the biochar from *S. officinarum* bagasse confirms the high K contents present in the ash generated during pyrolysis and the low retention of the element in its structures. K was notably and consistently leached, regardless of the initial K content in the filtered substances (Table 8.3). Ammoniacal-N was retained, with significant variations between water filtration and IHPLUSBF retained more than 75% of the initially detected ion in the solution. Regarding organic acids, IHPLUSBF made a notable contribution compared to water; and in both substances, high percentages were retained, ranging from 81% to 98% (Table 8.3). The increase in pH and conductivity of the filtered substances corresponded to the release of potassium ions from the biochar.

The descriptive statistical analysis of filtered substances (before and after) and the release or retention of substances and anions revealed that K was released, while NH4-N and total sugars were retained. The letters B and A denote the condition before and after filtration. Values in parentheses indicate the percentage represented by the indicator (K, N-NH_3, organic acids). X: mean, St: standard deviation; Se: standard error.

Table 8.3 Retention versus release of potassium (K), ammoniacal nitrogen (NH_4-N), and organic acids in biochar.

Indicator	Treatment	H_2O		IHPLUS BF				Se +
1 mm	2 mm	1 mm	2 mm			X	St	(%)
K	B	0.9	0.9	51.9	51.9	26.4	29.44	14.72 (55.76)
	A	265.0	365.3	431.1	443.4	376.2	81.68	40.84 (10.86)
	Release (mg/L)	264.1	364.4	379.2	391.5	349.8	58.2	29.10 (8.32)
N-NH_3	B	0.08	0.08	3.55	3.55	1.82	2.00	1.0 (55.1)
	A	0.04	0.03	0.86	0.83	0.46	0.45	0.22 (47.83)
	Retetion (mg/L)	0.04 (50)	0.05 (62.5)	2.69 (75.77)	2.72 (76.62)	1.36	1.56	0.78 (57.56)
Organic acids	B	1214	1214	15,254	15,254	8234	8106	4053 (49.22)
	A	46.9	22.2	2394	2894	1339.28	1520.37	760.18 (56.76)
	Retention (mg/L)	1167.1 (96.14)	1191.8 (98.17)	12,860 (84.31)	12,360 (81.03)	6894.73	6602.59	3301.3 (47.88)

8.7 Conclusions

Biochar produced at lower pyrolysis temperatures tends to enhance nutrient availability such as P and K and improve soil fertility more rapidly than biochar produced at higher pyrolysis temperatures. Nonetheless, the rise in pH at higher pyrolysis temperatures elevates soil pH and enhances nutrient availability for plants. Biochar produced at temperatures exceeding 500°C tends to exhibit increased porosity, a higher specific surface area, and reduced hydrophobicity, contributing to a reduction in soil bulk density and an enhancement in soil water and nutrient retention. Biochar produced at temperatures between 400°C and 700°C stimulates microbial diversity in the rhizosphere, the abundance of beneficial or antagonistic microorganisms, and their effects on the biological quality and health of the soil. The changes in the dynamics of fungi and bacteria are influenced by variations in the physicochemical characteristics of both the biochar and the soil in its presence. The incorporation of biochar into the soil determines the metabolomic distribution in the soil, and a crucial mechanism in the interaction between biochar and soil microorganisms is the modification of root exudates and those generated in the rhizosphere, inducing the action of microorganisms. The ability of biochar to store water and retain or release substances and ions was significantly influenced by the source material, charcoal particle size, and the properties of the enrichment solution. The storage index was notably higher in *S. officinarum* biochar, particularly with a particle size of 2 mm and when using microoganisms IHPLUS BF. The marked reduction in pH and redox potential observed in biochar, enriched with IHPLUSBF, was consistent with its strong capacity to retain NH_4-N, and organic acids.

Acknowledgments

We extend our gratitude to the Latin American Institute of Biochar (ILBI), represented by Patrick Atanasije Pineda, Luisa Marin de Block, and Rafael Gamboa. Additionally, special thanks to the members of the research team in biofertilizers and plant nutrition at EEPFIH/UM, including MSc. Esther Beatríz Norda Castro, Ing. Julio Brunet Zulueta, Yania Susnay Oropeza Scool, Ing. Jairo Urguellez, MSc. Aramis Soto Díaz, MSc. Leyma Rodríguez Navarro, and Leydi Laura Pentón Arias.

Credit author statement

Sara de Jesus Duarte: conceptualization, methodology, investigation, data curation, writing—original draft, writing—review and editing, visualization, supervision, project administration. Gertrudis Pentón Fernández: conceptualization, methodology, formal analysis, investigation, resources, data curation, writing—

original draft, writing—review and editing, visualization, funding acquisition. Odelin Brea Maure: data curation, investigation, writing—original draft, writing—review and editing.

Conflict of interest

We confirm that we have no conflicts of interest regarding this book chapter. We are not affiliated with any organization or entity that could influence our impartiality or judgment in this matter.

AI disclosure

This chapter does not contain any content created by AI or other automated technologies.

References

Ajeng, A. A., Abdullah, R., Ling, T. C., Ismail, S., Lau, B. F., Ong, H. C., Chew, K. W., Show, P. L., & Chang, J. S. (2020). Bioformulation of biochar as a potential inoculant carrier for sustainable agriculture. *Environmental Technology and Innovation*, *20*. Available from https://doi.org/10.1016/j.eti.2020.101168, http://www.journals.elsevier.com/environmental-technology-and-innovation/.

Akhter, A., Hage-Ahmed, K., Soja, G., & Steinkellner, S. (2016). Potential of Fusarium wilt-inducing chlamydospores, in vitro behaviour in root exudates and physiology of tomato in biochar and compost amended soil. *Plant and Soil*, *406*(1-2), 425–440. Available from https://doi.org/10.1007/s11104-016-2948-4.

Alfattani, R., Shah, M. A., Siddiqui, M. I. H., Ali, M. A., & Alnaser, I. A. (2022). Bio-char characterization produced from walnut shell biomass through slow pyrolysis: Sustainable for soil amendment and an alternate bio-fuel. *Energies*, *15*(1), 1. Available from https://doi.org/10.3390/en15010001.

Araujo, J., Díaz-Alcántara, C.-A., Urbano, B., & González-Andrés, F. (2020). Inoculation with native Bradyrhizobium strains formulated with biochar as carrier improves the performance of pigeonpea (*Cajanus cajan* L.). *European Journal of Agronomy*, *113*, 125985. Available from https://doi.org/10.1016/j.eja.2019.125985.

Barna, G., Makó, A., Takács, T., Skic, K., Füzy, A., & Horel, Á. (2020). Biochar alters soil physical characteristics, arbuscular mycorrhizal fungi colonization, and glomalin production. *Agronomy*, *10*(12), 1933. Available from https://doi.org/10.3390/agronomy10121933.

Batista, E. M. C. C., Shultz, J., Matos, T. T. S., Fornari, M. R., Ferreira, T. M., Szpoganicz, B., De Freitas, R. A., & Mangrich, A. S. (2018). Effect of surface and porosity of biochar on water holding capacity aiming indirectly at preservation of the

Amazon biome. *Scientific Reports*, *8*(1). Available from https://doi.org/10.1038/s41598-018-28794-z, http://www.nature.com/srep/index.html.

Bednik, M., Medyńska-Juraszek, A., & Ćwieląg-Piasecka, I. (2022). Effect of six different feedstocks on biochar's properties and expected stability. *Agronomy*, *12*(7), 1525. Available from https://doi.org/10.3390/agronomy12071525.

IHPLUS® BF. (2020). *Dictamen de registro No. RCF 039/20*. República de Cuba. *Ministerio de la Agricultura*.

Bilias, F., Kalderis, D., Richardson, C., Barbayiannis, N., & Gasparatos, D. (2023). Biochar application as a soil potassium management strategy: A review. *Science of the Total Environment*, *858*, 159782. Available from https://doi.org/10.1016/j.scitotenv.2022.159782.

Blanco-Canqui, H., Laird, D. A., Heaton, E. A., Rathke, S., & Acharya, B. S. (2020). Soil carbon increased by twice the amount of biochar carbon applied after 6 years: Field evidence of negative priming. *GCB Bioenergy*, *12*(4), 240–251. Available from https://doi.org/10.1111/gcbb.12665, http://onlinelibrary.wiley.com/journal/10.1111/(ISSN)1757-1707.

Bolan, S., Hou, D., Wang, L., Hale, L., Egamberdieva, D., Tammeorg, P., Li, R., Wang, B., Xu, J., Wang, T., Sun, H., Padhye, L. P., Wang, H., Siddique, K. H. M., Rinklebe, J., Kirkham, M. B., & Bolan, N. (2023). The potential of biochar as a microbial carrier for agricultural and environmental applications. *Science of the Total Environment*, *886*, 163968. Available from https://doi.org/10.1016/j.scitotenv.2023.163968.

Buss, W., Graham, M. C., Shepherd, J. G., & Mašek, O. (2016). Suitability of marginal biomass-derived biochars for soil amendment. *Science of the Total Environment*, *547*, 314–322. Available from https://doi.org/10.1016/j.scitotenv.2015.11.148, http://www.elsevier.com/locate/scitotenv.

Campos, P., Miller, A. Z., Knicker, H., Costa-Pereira, M. F., Merino, A., & De la Rosa, J. M. (2020). Chemical, physical and morphological properties of biochars produced from agricultural residues: Implications for their use as soil amendment. *Waste Management*, *105*, 256–267. Available from https://doi.org/10.1016/j.wasman.2020.02.013, http://www.elsevier.com/locate/wasman.

Chen, Z., Kumar, A., Fu, Y., Singh, B. P., Ge, T., Tu, H., Luo, Y., & Xu, J. (2021). Biochar decreased rhizodeposits stabilization via opposite effects on bacteria and fungi: diminished fungi-promoted aggregation and enhanced bacterial mineralization. *Biology and Fertility of Soils*, *57*(4), 533–546. Available from https://doi.org/10.1007/s00374-020-01539-9.

Chen, H., Gao, Y., Li, J., Fang, Z., Bolan, N., Bhatnagar, A., Gao, B., Hou, D., Wang, S., Song, H., Yang, X., Shaheen, S. M., Meng, J., Chen, W., Rinklebe, J., & Wang, H. (2022). Engineered biochar for environmental decontamination in aquatic and soil systems: a review. *Carbon Research*, *1*(1). Available from https://doi.org/10.1007/s44246-022-00005-5, http://www.springer.com/journal/44246.

Chuaphasuk, C., & Prapagdee, B. (2019). Effects of biochar-immobilized bacteria on phytoremediation of cadmium-polluted soil. *Environmental Science and Pollution Research*, *26*(23), 23679–23688. Available from https://doi.org/10.1007/s11356-019-05661-6.

Cleveland, C. C., & Liptzin, D. (2007). C:N:P stoichiometry in soil: Is there a "Redfield ratio" for the microbial biomass? *Biogeochemistry*, *85*(3), 235–252. Available from https://doi.org/10.1007/s10533-007-9132-0.

Cortes, A. D., Aggangan, N. S., & Opulencia, R. B. (2021). Taxonomic microbiome profiling and abundance patterns in the cacao (*Theobroma cacao* L.) rhizosphere treated with arbuscular mycorrhizal fungi and bamboo biochar. *Philippine Agricultural Scientist*, *104*(1), 19–33. Available from https://pas.cafs.uplb.edu.ph/download/

taxonomic-microbiome-profiling-and-abundance-patterns-in-the-cacao-theobroma-cacao-l-rhizosphere-treated-with-arbuscular-mycorrhizal-fungi-and-bamboo-biochar/.

Cowie, A. L. (2021). *Meeting the Net Zero challenge: A key role for the land sector.*

Cross, A., & Sohi, S. P. (2011). The priming potential of biochar products in relation to labile carbon contents and soil organic matter status. *Soil Biology and Biochemistry*, *43* (10), 2127–2134. Available from https://doi.org/10.1016/j.soilbio.2011.06.016.

Cuthbertson, D. M. (2018). *The production of pyrolytic biochar for addition in value-added composite material* (Doctoral dissertation).

Di Rienzo, Casanoves, F., Balzarini, M. G., Gonzalez, L., Tablada, M., & Robledo, C. W. (2008). *Infostat: Estadística descriptiva, inferencial y multivariada*. Argentina: Grupo Infostat, FCA, Universidad Nacional de Córdoba.

EBC. (2021). *Certification of the carbon sink potential of biochar*. Version 2.1E.

Edeh, I. G., Mašek, O., & Buss, W. (2020). A meta-analysis on biochar's effects on soil water properties – new insights and future research challenges. *Science of the Total Environment, 2020.*

Egamberdieva, D., Hua, M., Reckling, M., Wirth, S., & Bellingrath-Kimura, S. D. (2018). Potential effects of biochar-based microbial inoculants in agriculture. *Environmental Sustainability*, *1*(1), 19–24. Available from https://doi.org/10.1007/s42398-018-0010-6.

Fidel, R. B., Laird, D. A., Thompson, M. L., & Lawrinenko, M. (2017). Characterization and quantification of biochar alkalinity. *Chemosphere*, *167*, 367–373. Available from https://doi.org/10.1016/j.chemosphere.2016.09.151, http://www.elsevier.com/locate/chemosphere.

de Figueiredo, C. C., Monteiro Farias, W., Coser, T. R., Monteiro de Paula, A., Sartori da Silva, M. R., & Paz-Ferreiro, J. (2019). Sewage sludge biochar alters root colonization of mycorrhizal fungi in a soil cultivated with corn. *European Journal of Soil Biology*, *93,* 103092. Available from https://doi.org/10.1016/j.ejsobi.2019.103092.

Gaffar, S., Dattamudi, S., Baboukani, A. R., Chanda, S., Novak, J. M., Watts, D. W., Wang, C., & Jayachandran, K. (2021). Physiochemical characterization of biochars from six feedstocks and their effects on the sorption of atrazine in an organic soil. *Agronomy*, *11*(4). Available from https://doi.org/10.3390/agronomy11040716, https://www.mdpi.com/2073-4395/11/4/716/pdf.

Gao, S., DeLuca, T. H., & Cleveland, C. C. (2019). Biochar additions alter phosphorus and nitrogen availability in agricultural ecosystems: A meta-analysis. *Science of the Total Environment*, *654*, 463–472. Available from https://doi.org/10.1016/j.scitotenv.2018.11.124, http://www.elsevier.com/locate/scitotenv.

Graber, E. R., Singh, B., Hanley, K., & Lehmann, J. (2017). *Determination of cation exchange capacity in biochar. Biochar: A guide to analytical methods* (2017, pp. 74–84). CRC Press.

Gray, M., Johnson, M. G., Dragila, M. I., & Kleber, M. (2014). Water uptake in biochars: The roles of porosity and hydrophobicity. *Biomass and Bioenergy*, *61*, 196–205. Available from https://doi.org/10.1016/j.biombioe.2013.12.010.

Gu, Y., Hou, Y., Huang, D., Hao, Z., Wang, X., Wei, Z., Jousset, A., Tan, S., Xu, D., Shen, Q., Xu, Y., & Friman, V.-P. (2017). Application of biochar reduces Ralstonia solanacearum infection via effects on pathogen chemotaxis, swarming motility, and root exudate adsorption. *Plant and Soil*, *415*(1-2), 269–281. Available from https://doi.org/10.1007/s11104-016-3159-8.

Gundale, M. J., & DeLuca, T. H. (2007). Charcoal effects on soil solution chemistry and growth of Koeleria macrantha in the ponderosa pine/Douglas-fir ecosystem.

Biology and Fertility of Soils, *43*(3), 303–311. Available from https://doi.org/10.1007/s00374-006-0106-5.

Guo, S., Liu, X., Zhao, H., Wang, L., & Tang, J. (2021). High pyrolysis temperature biochar reduced the transport of petroleum degradation bacteria Corynebacterium variabile HRJ4 in porous media. *Journal of Environmental Sciences*, *100*, 228–239. Available from https://doi.org/10.1016/j.jes.2020.07.012.

Hagemann, N., Subdiaga, E., Orsetti, S., de la Rosa, J. M., Knicker, H., Schmidt, H. P., Kappler, A., & Behrens, S. (2018). Effect of biochar amendment on compost organic matter composition following aerobic compositing of manure. *Science of the Total Environment*, *613-614*, 20–29. Available from https://doi.org/10.1016/j.scitotenv.2017.08.161, http://www.elsevier.com/locate/scitotenv.

Hansen, V., Müller-Stöver, D., Imparato, V., Krogh, P. H., Jensen, L. S., Dolmer, A., & Hauggaard-Nielsen, H. (2017). The effects of straw or straw-derived gasification biochar applications on soil quality and crop productivity: A farm case study. *Journal of Environmental Management*, *186*, 88–95. Available from https://doi.org/10.1016/j.jenvman.2016.10.041, https://www.sciencedirect.com/journal/journal-of-environmental-management.

Hossain, M. Z., Bahar, M. M., Sarkar, B., Donne, S. W., Ok, Y. S., Palansooriya, K. N., Kirkham, M. B., Chowdhury, S., & Bolan, N. (2020). Biochar and its importance on nutrient dynamics in soil and plant. *Biochar*, *2*(4), 379–420. Available from https://doi.org/10.1007/s42773-020-00065-z. Available from:, springer.com/journal/42773.

Husna, N., Budianta, D., Munandar., & Napoleon, A. (2019). Evaluation of several biochar types as inoculant carrier for indigenous phosphate solubilizing microoorganism from acid sulphate soil. *Journal of Ecological Engineering*, *20*(6), 1–8. Available from https://doi.org/10.12911/22998993/109078, http://www.jeeng.net/pdf-109078-40028?filename = Evaluation%20of%20Several.pdf.

Hussaina, R. K., Ravia, K., & Ankit, G. (2020). Influence of biochar on the soil water retention characteristics (SWRC): Potential application in geotechnical engineering structures. *Soil Tillage Res*, *204*.

Husson, O. (2013). Redox potential (Eh) and pH as drivers of soil/plant/microorganism systems: A transdisciplinary overview pointing to integrative opportunities for agronomy. *Plant and Soil*, *362*(1-2), 389–417. Available from https://doi.org/10.1007/s11104-012-1429-7, http://www.wkap.nl/journalhome.htm/0032-079X.

Jaiswal, A. K., Alkan, N., Elad, Y., Sela, N., Philosoph, A. M., Graber, E. R., & Frenkel, O. (2020). Molecular insights into biochar-mediated plant growth promotion and systemic resistance in tomato against Fusarium crown and root rot disease. *Scientific Reports*, *10*(1). Available from https://doi.org/10.1038/s41598-020-70882-6. Available from:, http://www.nature.com/srep/index.html.

Javeed, H. M. R., Ali, M., Ahmed, I., Wang, X., Al-Ashkar, I., Qamar, R., Ibrahim, A., Habib-Ur-rahman, M., Ditta, A., & Sabagh, A. E. L. (2021). Biochar enriched with buffalo slurry improved soil nitrogen and carbon dynamics, nutrient uptake and growth attributes of wheat by reducing leaching losses of nutrients. *Land*, *10*(12). Available from https://doi.org/10.3390/land10121392, https://www.mdpi.com/2073-445X/10/12/1392/pdf.

de Jesus Duarte, S., Cerri, C. E. P., Rittl, T. F., Abbruzzin, T. F., & Pano, B. L. P. (2023). Biochar physical and hydrological characterization to improve soil attributes for plant production. *Journal of Soil Science and Plant Nutrition*, *23*(3), 3051–3057. Available from https://doi.org/10.1007/s42729-023-01273-9, https://www.springer.com/journal/42729.

Jing, F., Sun, Y., Liu, Y., Wan, Z., Chen, J., & Tsang, D. C. W. (2022). Interactions between biochar and clay minerals in changing biochar carbon stability. *Science of the Total Environment*, *809,* 151124. Available from https://doi.org/10.1016/j.scitotenv.2021.151124.

Keiluweit, M., Nico, P. S., Johnson, M., & Kleber, M. (2010). Dynamic molecular structure of plant biomass-derived black carbon (biochar). *Environmental Science and Technology*, *44*(4), 1247–1253. Available from https://doi.org/10.1021/es9031419.

Kwaku, A., Tobias, C., & Kwasi, F.-M. (2022). Physical and mechanical properties of *Holarrhena floribunda* and *Nesogordonia papaverifera* branch wood in Ghana. *African Journal of Plant Science*, *16*(3), 29–33. Available from https://doi.org/10.5897/ajps2022.2228.

Lacomino, G., Sarker, T. C., Ippolito, F., Bonanomi, G., Vinale, F., Staropoli, A., & Idbella, M. (1996). Biochar and compost application either alone or in combination affects vegetable yield in a volcanic Mediterranean soil. *Agronomy*, *12*(9).

Lehmann, J., Cowie, A., Masiello, C. A., Kammann, C., Woolf, D., Amonette, J. E., Cayuela, M. L., Camps-Arbestain, M., & Whitman, T. (2021). Biochar in climate change mitigation. *Nature Geoscience*, *14*(12), 883–892. Available from https://doi.org/10.1038/s41561-021-00852-8, http://www.nature.com/ngeo/index.html.

Leng, L., Xiong, Q., Yang, L., Li, H., Zhou, Y., Zhang, W., Jiang, S., Li, H., & Huang, H. (2021). An overview on engineering the surface area and porosity of biochar. *Science of the Total Environment*, *763,* 144204. Available from https://doi.org/10.1016/j.scitotenv.2020.144204.

Leng, L., Xu, S., Liu, R., Yu, T., Zhuo, X., Leng, S., Xiong, Q., & Huang, H. (2020). Nitrogen containing functional groups of biochar: An overview. *Bioresource Technology*, *298,* 122286. Available from https://doi.org/10.1016/j.biortech.2019.122286.

Li, H., Lu, X., Xu, Y., & Liu, H. (2019). How close is artificial biochar aging to natural biochar aging in fields? A meta-analysis. *Geoderma*, *352*, 96–103. Available from https://doi.org/10.1016/j.geoderma.2019.06.006.

Li, M., & Cai, L. (2021). Biochar and arbuscular mycorrhizal fungi play different roles in enabling maize to uptake phosphorus. *Sustainability*, *13*(6), 3244. Available from https://doi.org/10.3390/su13063244.

López-Cano, I., Cayuela, M., Sánchez-García, M., & Sánchez-Monedero, M. (2018). Suitability of different agricultural and urban organic wastes as feedstocks for the production of biochar—Part 2: Agronomical evaluation as soil amendment. *Sustainability*, *10*(6), 2077. Available from https://doi.org/10.3390/su10062077.

Mandal, S., Donner, E., Smith, E., Sarkar, B., & Lombi, E. (2019). Biochar with near-neutral pH reduces ammonia volatilization and improves plant growth in a soil-plant system: A closed chamber experiment. *Science of the Total Environment*, *697,* 134114. Available from https://doi.org/10.1016/j.scitotenv.2019.134114.

Mandal, S., Pu, S., He, L., Ma, H., & Hou, D. (2019). Biochar induced modification of graphene oxide & nZVI and its impact on immobilization of toxic copper in soil. *Environmental Pollution*, 113851. Available from https://doi.org/10.1016/j.envpol.2019.11385.

Masís-Meléndez, F., Segura-Chavarría, D., García-González, C. A., Quesada-Kimsey, J., & Villagra-Mendoza, K. (2020). Variability of physical and chemical properties of TLUD stove derived biochars. *Applied Sciences*, *10*(2), 507. Available from https://doi.org/10.3390/app10020507.

Milera, M., Pentón, G., Schmidt, H. P., Machado, H., Miranda, T., Martín, G. J., et al. (2021). Manejo agroecológico de los sistemas agropecuarios. *Usos del suelo con abonos y biochar*, *978*(2021).

Ngatia, L. W., Hsieh, Y. P., Nemours, D., Fu, R., & Taylor, R. W. (2017). Potential phosphorus eutrophication mitigation strategy: Biochar carbon composition, thermal stability and pH influence phosphorus sorption. *Chemosphere*, *180*, 201–211. Available from https://doi.org/10.1016/j.chemosphere.2017.04.012.

Pallarés, J., González-Cencerrado, A., & Arauzo, I. (2018). Production and characterization of activated carbon from barley straw by physical activation with carbon dioxide and steam. *Biomass and Bioenergy*, *115*, 64–73. Available from https://doi.org/10.1016/j.biombioe.2018.04.015.

Pandit, N. R., Schmidt, H. P., Mulder, J., Hale, S. E., Husson, O., & Cornelissen, G. (2020). Nutrient effect of various composting methods with and without biochar on soil fertility and maize growth. *Archives of Agronomy and Soil Science*, *66*(2), 250–265. Available from https://doi.org/10.1080/03650340.2019.1610168, http://www.tandf.co.uk/journals/titles/03650340.asp.

Pavlicevic, M., Elmer, W., Zuverza-Mena, N., Abdelraheem, W., Patel, R., Dimkpa, C., 'Keefe, T. O., Haynes, C. L., Pagano, L., Caldara, M., Marmiroli, M., Maestri, E., Marmiroli, N., & White, J. C. (2023). Nanoparticles and biochar with adsorbed plant growth-promoting rhizobacteria alleviate Fusarium wilt damage on tomato and watermelon. *Plant Physiology and Biochemistry*, *203*, 108052. Available from https://doi.org/10.1016/j.plaphy.2023.108052.

Pentón, G., Milera, M. D. L. C., & Schmidt, H. P. (2021). *Manual para la elaboración de biochar y microorganismos eficientes IHPLUS® BF. Estación Experimental Indio Hatuey* (26, p. 978).

Pérez-Cabrera, C. A., Juárez-López, P., Anzaldo-Hernández, J., Alia-Tejacal, I., Valdez-Aguilar, L. A., Alejo-Santiago, G., Castro-Brindis, R., López-Martínez, V., & Alvarado-Camarillo, D. (2022). Biochar from sugarcane apices as a soil conditioner for greenhouse cultivation of *Ocimum basilicum var. thyrsif lora. Terra Latinoamericana*, *40*. Available from https://doi.org/10.28940/terra.v40i0.1077, https://terralatinoamericana.org.mx/index.php/terra/article/view/1077.

Phares, C. A., Amoakwah, E., Danquah, A., Afrifa, A., Beyaw, L. R., & Frimpong, K. A. (2022). Biochar and NPK fertilizer co-applied with plant growth promoting bacteria (PGPB) enhanced maize grain yield and nutrient use efficiency of inorganic fertilizer. *Journal of Agriculture and Food Research*, *10*, 100434. Available from https://doi.org/10.1016/j.jafr.2022.100434.

Plunkett, M. H., Knutson, C. M., & Barney, B. M. (2020). Key factors affecting ammonium production by an Azotobacter vinelandii strain deregulated for biological nitrogen fixation. *Microbial Cell Factories*, *19*, 1–12.

Poveda, J., Martínez-Gómez, Á., Fenoll, C., & Escobar, C. (2021). The use of biochar for plant pathogen control. *Phytopathology®*, *111*(9), 1490–1499. Available from https://doi.org/10.1094/phyto-06-20-0248-rvw.

Qayyum, M. F., Abdullah, M. A., Rizwan, M., Haider, G., Ali, M. A., Zafar-ul-Hye, M., & Abid, M. (2019). Different nitrogen and biochar sources' application in an alkaline calcareous soil improved the maize yield and soil nitrogen retention. *Arabian Journal of Geosciences*, *12*(21). Available from https://doi.org/10.1007/s12517-019-4846-6, http://www.springer.com/geosciences/journal/12517?cm_mmc = AD-_-enews-_-PSE1892-_-0.

Rafique, M., Sultan, T., Ortas, I., & Chaudhary, H. J. (2017). Enhancement of maize plant growth with inoculation of phosphate-solubilizing bacteria and biochar amendment in soil. *Soil Science and Plant Nutrition*, *63*(5), 460–469. Available from https://doi.org/10.1080/00380768.2017.1373599, http://www.tandfonline.com/toc/tssp20/current.

Ren, H., Guo, H., Shafiqul Islam, M., Zaki, H. E. M., Wang, Z., Wang, H., Qi, X., Guo, J., Sun, L., Wang, Q., Li, B., Li, G., & Radwan, K. S. A. (2023). Improvement effect of biochar on soil microbial community structure and metabolites of decline disease bayberry. *Frontiers in Microbiology*, *14*. Available from https://doi.org/10.3389/fmicb.2023.1154886.

Sashidhar, P., Kochar, M., Singh, B., Gupta, M., Cahill, D., Adholeya, A., & Dubey, M. (2020). Biochar for delivery of agri-inputs: Current status and future perspectives. *Science of the Total Environment*, *703,* 134892. Available from https://doi.org/10.1016/j.scitotenv.2019.134892.

Schmidt, H. P., Kammann, C., Hagemann, N., Leifeld, J., Bucheli, T. D., Sánchez Monedero, M. A., & Cayuela, M. L. (2021). Biochar in agriculture – a systematic review of 26 global meta-analyses. *GCB Bioenergy*, *13*(11), 1708–1730. Available from https://doi.org/10.1111/gcbb.12889, http://onlinelibrary.wiley.com/journal/10.1111/(ISSN)1757-1707.

Schmidt, H. P., Pandit, B. H., Cornelissen, G., & Kammann, C. I. (2017). Biochar-based fertilization with liquid nutrient enrichment: 21 field trials covering 13 crop species in Nepal. *Land Degradation and Development*, *28*(8), 2324–2342. Available from https://doi.org/10.1002/ldr.2761, http://onlinelibrary.wiley.com/journal/10.1002/(ISSN)1099-145X.

Shen, M., Song, W., Shi, X., Wang, S., Wang, H., Liu, J., Jin, W., Fan, S., & Cao, Z. (2022). New insights into physicochemical properties of different particulate size-fractions and dissolved organic matter derived from biochars and their sorption capacity for phenanthrene. *Journal of Hazardous Materials*, *434,* 128867. Available from https://doi.org/10.1016/j.jhazmat.2022.128867.

Sun, Q., Liu, Y., Liu, H., & Dumroese, R. K. (2020). Interaction of biochar type and rhizobia inoculation increases the growth and biological nitrogen fixation of robinia pseudoacacia seedlings. *Forests*, *11*(6). Available from https://doi.org/10.3390/f11060711, https://res.mdpi.com/d_attachment/forests/forests-11-00711/article_deploy/forests-11-00711-v2.pdf.

Thomas, S. C. (2021). Post-processing of biochars to enhance plant growth responses: a review and meta-analysis. *Biochar*, *3*(4), 437–455. Available from https://doi.org/10.1007/s42773-021-00115-0, springer.com/journal/42773.

Tomczyk, A., Sokołowska, Z., & Boguta, P. (2020). Biochar physicochemical properties: pyrolysis temperature and feedstock kind effects. *Reviews in Environmental Science and Biotechnology*, *19*(1), 191–215. Available from http://www.kluweronline.com/issn/1569-1705, 10.1007/s11157-020-09523-3.

Turunen, M., Hyväluoma, J., Heikkinen, J., Keskinen, R., Kaseva, J., Hannula, M., & Rasa, K. (2020). Quantifying the pore structure of different biochars and their impacts on the water retention properties of *Sphagnum moss* growing media. *Biosystems Engineering*, *191*, 96–106. Available from https://doi.org/10.1016/j.biosystemseng.2020.01.006.

Vahedi, R., Rasouli-Sadaghiani, M. H., Barin, M., & Vetukuri, R. R. (2022). Effect of biochar and microbial inoculation on P, Fe, and Zn bioavailability in a calcareous soil. *Processes*, *10*(2), 343. Available from https://doi.org/10.3390/pr10020343.

Vahedi, R., Rasouli-Sadaghiani, M. H., Barin, M., & Vetukuri, R. R. (2021). Interactions between biochar and compost treatment and mycorrhizal fungi to improve the qualitative properties of a calcareous soil under rhizobox conditions. *Agriculture*, *11*(10), 993. Available from https://doi.org/10.3390/agriculture11100993.

Van Zwieten, L., Kimber, S., Morris, S., Chan, K. Y., Downie, A., Rust, J., Joseph, S., & Cowie, A. (2010). Effects of biochar from slow pyrolysis of papermill waste on agronomic performance and soil fertility. *Plant and Soil*, *327*(1-2), 235–246. Available from https://doi.org/10.1007/s11104-009-0050-x.

Wu, H., Zeng, G., Liang, J., Chen, J., Xu, J., Dai, J., Li, X., Chen, M., Xu, P., Zhou, Y., Li, F., Hu, L., & Wan, J. (2016). Responses of bacterial community and functional marker genes of nitrogen cycling to biochar, compost and combined amendments in soil. *Applied Microbiology and Biotechnology*, *100*(19), 8583–8591. Available from https://doi.org/10.1007/s00253-016-7614-5.

Wu, G. S., Peng, M. Y. P., Chen, Z., Du, Z., Anser, M. K., & Zhao, W. X. (2020). The effect of relational embeddedness, absorptive capacity, and learning orientation on SMEs' competitive advantage. *Frontiers in Psychology*, *11*. Available from https://doi.org/10.3389/fpsyg.2020.01505, http://www.frontiersin.org/Psychology.

Xiang, L., Harindintwali, J. D., Wang, F., Redmile-Gordon, M., Chang, S. X., Fu, Y., He, C., Muhoza, B., Brahushi, F., Bolan, N., Jiang, X., Ok, Y. S., Rinklebe, J., Schaeffer, A., Zhu, Y.-guan, Tiedje, J. M., & Xing, B. (2022). Integrating biochar, bacteria, and plants for sustainable remediation of soils contaminated with organic pollutants. *Environmental Science & Technology*, *56*(23), 16546–16566. Available from https://doi.org/10.1021/acs.est.2c02976.

Xiao, X., Chen, B., Chen, Z., Zhu, L., & Schnoor, J. L. (2018). Insight into multiple and multilevel structures of biochars and their potential environmental applications: A critical review. *Environmental Science & Technology*, *52*(9), 5027–5047. Available from https://doi.org/10.1021/acs.est.7b06487.

Xu, G., Zhang, Y., Sun, J., & Shao, H. (2016). Negative interactive effects between biochar and phosphorus fertilization on phosphorus availability and plant yield in saline sodic soil. *Science of the Total Environment*, *568*, 910–915. Available from https://doi.org/10.1016/j.scitotenv.2016.06.079.

Xu, W., Whitman, W. B., Gundale, M. J., Chien, C. -C., & Chiu, C. -Y. (2021). Functional response of the soil microbial community to biochar applications. *GCB Bioenergy*, *13*(1), 269–281. Available from https://doi.org/10.1111/gcbb.12773.

Yang, X., Igalavithana, A. D., Oh, S.-E., Nam, H., Zhang, M., Wang, C.-H., Kwon, E. E., Tsang, D. C. W., & Sik Ok, Y. (2018). Characterization of bioenergy biochar and its utilization for metal/metalloid immobilization in contaminated soil. *Science of the Total Environment*, 704–713. Available from https://doi.org/10.1016/j.scitotenv.2018.05.298.

Yang, X. Y., Chang, K. H., Kim, Y. J., Zhang, J., & Yoo, G. (2019). Effects of different biochar amendments on carbon loss and leachate characterization from an agricultural soil. *Chemosphere*, *226*, 625–635. Available from https://doi.org/10.1016/j.chemosphere.2019.03.085, http://www.elsevier.com/locate/chemosphere.

Zhao, Y., Li, X., Li, Y., Bao, H., Xing, J., Zhu, Y., Nan, J., & Xu, G. (2023). Biochar acts as an emerging soil amendment and its potential ecological risks: A review. *Energies*, *16*(1), 410. Available from https://doi.org/10.3390/en16010410.

Zheng, B. X., Ding, K., Yang, X. R., Wadaan, M. A. M., Hozzein, W. N., Peñuelas, J., & Zhu, Y. G. (2019). Straw biochar increases the abundance of inorganic phosphate solubilizing bacterial community for better rape (*Brassica napus*) growth and phosphate uptake. *Science of the Total Environment*, *647*, 1113–1120. Available from https://doi.org/10.1016/j.scitotenv.2018.07.454, http://www.elsevier.com/locate/scitotenv.

CHAPTER

Biochar for climate change mitigation and soil health management

9

Daniel Guarín[1], Alejandro Fernandez[2], Jhony Benavides[3] and Nicolás Rangel[4]

[1]*Department of Ecosystem Science and Management, Pennsylvania State University, University Park, PA, United States*

[2]*Department of Earth and Environmental Sciences, University of Minnesota, Minneapolis, MN, United States*

[3]*School of Natural Resources and Environmental Engineering—EIDENAR, College of Engineering, Universidad del Valle, Cali, Colombia*

[4]*Geothermal Energy and Geofluids, Institute of Geophysics, ETH Zürich, Zurich, Switzerland*

9.1 Introduction

Biochar is a highly porous and carbon-rich charcoal-like material derived from the combustion of biomass under limited oxygen. It can represent a pivotal carbon-negative strategy for climate change mitigation with multiple cobenefits promoting enhanced soil health. During photosynthesis, plants naturally capture atmospheric carbon dioxide and store it in their live tissues. However, at the end of their life or after a harvest cycle, plant material slowly decomposes, and the stored carbon is released back into the atmosphere as carbon dioxide (CO_2) along with methane, a greenhouse gas, and the second largest contributor to global warming after CO_2 (Saunois et al., 2020).

The production of biochar prevents the decomposition of organic material and therefore averts the carbon captured by plants from returning to the atmosphere. This carbon is locked away in a stable and solid form for up to hundreds of years, given its highly recalcitrant nature (Woolf et al., 2010). Recent research has shown that biochar permanence can extend to a geological timeframe of thousands to millions of years as a highly stable mineralized organic carbon polymer called inertinite if produced under the right circumstances (Sanei et al., 2024). In terms of its potential as a tool to mitigate the effects of climate change, biochar has the potential to remove up to 12% of anthropogenic annual greenhouse gas emissions (Woolf et al., 2010) and can remove up to 3 gigatons of CO_2 equivalents by 2050 (Project Drawdown, 2023). Given its high potential for scalability, low cost of production, and many cobenefits, biochar has been placed as one of the main nature-based methods for carbon dioxide removal for climate change mitigation (Li et al., 2021).

Beyond climate change mitigation, biochar's potential lies in its many cobenefits and positive effects on soil fertility. Biochar application can enhance plant growth,

Biochar Ecotechnology for Sustainable Agriculture and Environment. DOI: https://doi.org/10.1016/B978-0-443-29855-4.00011-4

FIGURE 9.1 UN SDGs promoted by biochar.

United Nations Sustainable Development Goals that are benefited by the adoption of biochar. https://www.un.org/sustainabledevelopment.

which means more CO_2 is captured by photosynthesis (Schulz et al., 2013). Its highly porous structure reduces the need for chemical fertilizers by retaining and preventing nutrient loss from the soil (Neogi et al., 2022) while lowering greenhouse gas emissions from fertilizer manufacturing. Biochar fosters a favorable habitat for soil microbes (Schulz et al., 2013), promoting a nutrient-rich soil environment and enhancing plant growth and crop yields. Additionally, biochar's water retention ability increases soil moisture and prevents water stress for plants in fast-draining soils (Woolf et al., 2010), while reducing bulk density and aiding aeration in compacted soils, promoting healthier root structures (Neogi et al., 2022). The use of biochar, especially in developing nations, can help accomplish multiple sustainable development goals, such as no poverty, zero hunger, good health and well-being, gender equality, decent work and economic growth, and life on land, among others (for a detailed description on how biochar helps accomplish every goal, visit https://www.dutchcarboneers.com/projects; Fig. 9.1).

These properties of biochar make it a vital tool in the efforts to mitigate climate change while benefiting agricultural productivity through healthier and more resilient soils. This chapter examines the underlying mechanisms of biochar's dual role as a climate change mitigation tool and soil health enhancer, contributing to further understanding of its effectiveness and expansion potential in the light of 2023 taking the dishonorable prize as the warmest year since global records began in 180 and 2024 building up to surpass 2023 (NOAA, 2024).

9.2 Biochar for climate change mitigation

The latest assessment report (AR6) of the Intergovernmental Panel on Climate Change (IPCC) estimated global warming due to human activities since the start of the Industrial Revolution at approximately 1.1°C (IPCC, 2022). Since global human emissions continue to rise despite the pressing need for abatement to reach

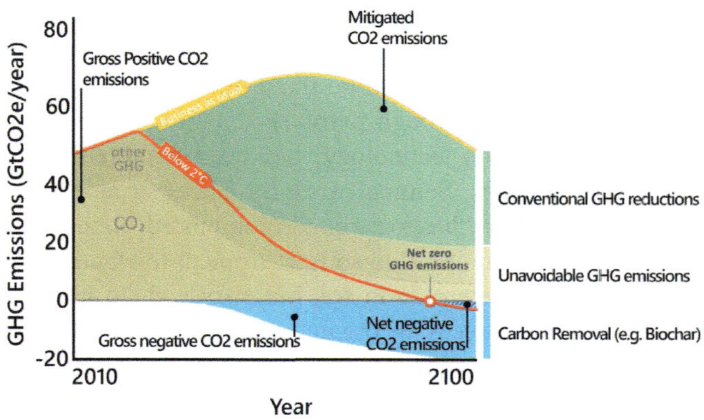

FIGURE 9.2 Net zero pathways.

Projected pathways to achieve net negative GHG emissions under a business-as-usual scenario and a set of policies and methodologies to limit global warming to 2°C by 2100. Note the paramount need for CDR technologies to compensate for unavoidable GHG emissions.

Modified from UNEP. (2017). https://www.unep.org/resources/emissions-gap-report-2017. Pathways to net zero in 2100.

1.5°C warming goals set during the Paris Agreement, the need for technologies and practices capable of decreasing the concentration of greenhouse gases (GHG) in the atmosphere is paramount (Fig. 9.2).

Although the idea of burning biomass for the improvement of soil characteristics and fuel production has been around for centuries—reaching as far back as communities in the pre-Columbian Amazon rainforest as well as ancient Egypt and China (Zhang et al., 2019)—recent scientific literature has been flooded with studies that prove the potential of biochar as a climate mitigation technique (Schmidt et al., 2021; Woolf et al., 2010). The production of biochar in local-to-regional scales is thought to serve as a Carbon sink (Glaser et al., 2001) by transforming decay-prone biomass into a more stable, less reactive form of carbon that can be effectively removed from the carbon cycle (Lehmann et al., 2021), thereby representing an important climate mitigation tool. Although this transformation of carbon fluxes is in itself a potential climate mitigation strategy, the factor that makes biochar stand out as a desirable and viable solution is its many cobenefits. This section evaluates the state of the science of biochar as a carbon sink, in what timescales it can be considered to store carbon, what mechanisms drive this storage, and why it's a strong candidate for aiding in the fight against the climate crisis.

9.2.1 Photosynthesis captures and organic matter decomposition releases CO_2

Anthropogenic climate change has dramatically increased the amount of CO_2 in the atmosphere available for photosynthetic organisms to utilize. However, the biological carbon cycle is a constant balance between carbon uptake by plants and its return to the atmosphere through respiration and decay. Biochar represents a way to bypass the recycling of the carbon fixed by photosynthesizers by making it effectively less bioavailable (through pyrolysis). Its potential as a climate solution, or a CO_2 removal (CDR) technology, has been explored and documented over the last couple of decades. Seminal work by Glaser et al. (2001), for example, suggested the presence of "black carbon" (combusted organic matter) is key in the productivity in otherwise infertile soils of tropical lowlands, which not only enhances water and nutrient retention in soil but also leads to longer permanence of carbon. By preventing the decay of organic matter, biochar effectively extends the time it takes for photosynthetically stored carbon to be recycled back into the atmosphere.

For the process of pyrolysis of biomass to be considered a form of carbon capture, two conditions must be met (Lehmann & Joseph, 2015): (1) the rate of plant matter production must be equal to that of pyrolysis because the carbon sequestration occurs during plant growth—through photosynthesis—and preventing annually-formed biomass from decomposing diverts that carbon from the biological cycle; and (ii) the stability of the biochar must be significantly greater than that of the biomass it is made from.

Since the rate of decay of biomass varies considerably according to climatic conditions, type of biomass, and numerous other factors, no avoided emissions can be considered to have happened through the application or production of biochar until after the time in which the unpyrolyzed matter would have decayed (Woolf et al., 2010). Thus pyrolysis of durable forms of biological matter, such as timber, might lead to an increase in GHG during biological cycle timescales that must be accounted for. Because of this, any estimate of negative emissions or carbon storage estimates must take into account the concept of carbon debt and carbon-payback time, which consider land-use changes, use of feedstocks with long life expectancy, and other positive carbon fluxes that may be incurred in when producing biochar (Woolf et al., 2010).

Though estimates vary, the approximate amount of available biomass is around 100 billion metric tons of carbon per year (Fahmy et al., 2020), half of which is on land. Woolf et al. (2010) estimated that global implementation of biochar to manage biomass could (as an upper limit) offset $\sim 12\%$ of current annual global emissions, only accounting for the carbon being sequestered. To this, one could add the amount of emissions offset if the biochar was to be sustainably converted to a clean energy source. In a systematic assessment of negative emissions technologies, Werner et al. (2018) showed the potential of pyrolysis-based technologies to aid in reaching climate targets, with the caveat that the amount of land that would need to be converted would possibly endanger biosphere integrity. It should be noted that the amount of carbon sequestered or prevented from being recycled back into the atmosphere depends on many factors, such as the type and fertility of the soil amended, the biomass feedstocks, and the specific characteristics of the recalcitrant nature of the product (Zhang et al., 2019).

9.2.2 How pyrolysis prevents the carbon from being decomposed and locks it away

To unlock the potential climate benefits of biochar production and application, the process of pyrolysis must be attentively defined and standardized. In a review by Fahmy et al. (2020), the process of pyrolysis is defined as "destructive distillation" wherein organic material undergoes thermochemical decomposition under high temperatures and in the absence of oxygen. Biochar is namely the principal solid product of pyrolysis and contains between 40% and 80% carbon (Blenis et al., 2023). There are several byproducts of this distillation of biomass such as bio-oils and tars, gases such as syngas and methane, and solid impurities in the form of aromatic compounds.

In terms of its long-term climate benefits, an important characteristic of the pyrolysis process is that it transforms the forms of carbon in the parent biomass into fused aromatic ring structures, which are more resistant to mineralization or adsorption (Lehmann & Joseph, 2015). Thus biochar is enriched in carbon and depleted in oxygen and hydrogen compared to the feedstock biomass, making it a chemically stable and manageable carbon sink. Due to the high chemical complexity and variety of biochar material—depending on the feedstock, temperature of pyrolysis, time of burn (also known as residence time), among other factors—the physicochemical characteristics of biochar are usually put into a continuum or categories, instead of being specifically defined (Keiluweit et al., 2010). Although most carbon in biochar is in aromatic compounds and biologically unavailable, there is a labile fraction (<10%) which depends on the conditions of pyrolysis, and which can be emitted back to the atmosphere as a metabolic byproduct as it is used by soil microorganisms (Hue, 2020).

The review paper by Fahmy et al. (2020) summarizes the classification of pyrolysis for biochar processes as follows: (1) slow pyrolysis is done in periods of up to several days, usually does not surpass 500°C and heating rates range from 0.1°C to 0.2°C per second. The extended period of heating allows for repolymerization reactions, which lead to a lower proportion of gas and volatile products, and the resulting material is mostly char and/or tar; (2) fast pyrolysis that lasts several seconds and yields a higher proportion of liquid, and gaseous products such as bio-oil and syngas, with heating rates of 10°C–200°C per second and maximum temperatures of thousands of degree Celsius; and (3) flash pyrolysis is similar to fast pyrolysis with shorter residence times. This process is defined by the fastest heating rates, reaching up to 2500°C/s, and taking fractions of a second. In short, lower temperatures of pyrolysis generally lead to a higher yield of biochar, although the nature of the material has been proposed to be more bioavailable, and therefore higher temperatures have been associated with more recalcitrant biochar (Zimmermann et al., 2012).

The choice of the type and specific parameters of the pyrolysis process is a compromise between the amount of biochar yield and the potential residence time in the soil when it is applied (Werner et al., 2018). A study by Safarian (2023) showcased the dependency of the study site to determine what temperature and pyrolysis factors lead to the most optimal emissions reductions using lifecycle assessments. In their study, they recommend different pyrolysis temperatures that lead to a compromise between abated GHG emissions versus optimal soil fertility and carbon sequestration potentials. This is because, as a general rule, the climate benefits of biochar application to soils are directly related to its recalcitrant characteristics which determine its residence time and the extent to its cobenefits.

The most important characteristic of pyrolyzed organic matter, which makes biochar an attractive carbon sequestration and storage medium, is its recalcitrance. Due to the physicochemical processes that result from the anoxic burning of biomass, including the development of aromatic structures (Lehmann & Joseph, 2015), biochar particles remain in the soil for a much longer time than any other form of organic carbon (Coppola et al., 2018). The aromatic carbon compounds resulting from pyrolysis become arranged into large sheets that give biochar particles a relatively large specific surface area (SSA) and porosity (Blenis et al., 2023; Hue, 2020), thereby allowing soil amendment through the inclusion of biochar particles.

The slow decomposition is characterized by endothermic drying and depolymerization, followed by an exothermic phase releasing volatile gases, resulting in a largely amorphous structure (Joseph et al., 2021). Persistence is notably greater in soils dominated by minerals forming stable aggregates, such as kaolinite and sesquioxides in Oxisols and Ultisols (Fungo et al., 2014; Joseph et al., 2021). While some evidence suggests a potential decrease in biochar persistence with increasing ambient temperature, the movement of biochar through the soil profile may enhance persistence in certain soil types (Wang et al., 2016).

9.2.3 Climate change mitigation potential of biochar

Evidence is robust that biochar can act as a carbon sink by displacing carbon from the biological cycle to longer timescales and effectively removing it as a flux. Several reviews and studies have tried to untangle the approximate quantity of equivalent CO_2 that could be annually offset through biochar application. In their review paper, Lehmann et al. (2021) show that the climate impact of biochar is twofold, balancing emissions offsets through the use of biochar as a pathway to different energy sources and carbon removal through the sequestration of atmospheric CO_2 in the biochar itself. The authors estimate an emissions reduction of 3.4–6.3 gigatons of CO_2 equivalent, half of which corresponds to direct carbon removal.

Multiple studies have estimated the CDR potential of biochar based on different conditions and assumptions. Wang et al. (2023) estimated that biochar application could mitigate as much as 2.56 gigatons of CO_2 equivalent emissions, which accounts for ~5% of total global emissions. Likewise, another generalized framework study estimated that ~6% of total GHG emissions were potentially offset (Lefebvre et al., 2023). Using a global multisector model, Bergero et al. (2022)

estimate a carbon sink of up to 2.8 gigatons of CO_2 equivalent per year. In a more integrated assessment of carbon dioxide removal technologies, Fuhrman et al. (2023) put biochar in the hundreds of megatons of CO_2 equivalent, representing a lower bound to the emissions offset that biochar could provide. They highlight the potential biochar has in helping some regions in particular—including those most vulnerable to climate change—build resilience while storing carbon. Further research on the long-term permanence and decomposition of biochar is still needed as the heterogeneous nature of the soil environment prevents a one-size-fits-all framework for its large-scale application.

Perhaps of more importance is the underlying need to pair biochar with other carbon removal strategies (such as direct air carbon, enhanced rock weathering, and biomass burial, among others) and a considerable reduction in emissions altogether. It is important to clarify that the amount of emissions that could be curtailed by biochar deployment, along with any other type of CDR technology, must not be considered to be alternative pathways to decarbonization and climate remediation. Although some level of CDR technology will be essential in reaching climate goals (IPCC, 2022), emissions reductions must still be the focus point as they are the most efficient and actionable paths to reducing GHG concentrations in the atmosphere (Dooley et al., 2022). It has been estimated that by 2050, when net-zero emissions must have been reached to keep the Paris Agreement goals, only about 6% of the emissions reductions will have been achieved by carbon removal (Rockström et al., 2017). Indeed, there exists a sizable gap between individual countries' plans for CDR deployment, and what is needed to keep warming below 1.5°C (Smith et al., 2023). Additionally, there is an ongoing debate in the literature about the moral hazard and so-called mitigation deterrence that focusing on CDR and emissions offsets might have in efficiently reaching net zero (Carton et al., 2023; Grant et al., 2021). Although researching and implementing technologies that may offset emissions from hard-to-abate sectors (e.g., cement, steel, air travel) is worthwhile, emissions reductions remain the most important priority in combating the climate crisis.

9.3 Biochar cobenefits: effects on soil health

9.3.1 Enhancing plant growth

Soils require a balance of aggregates, solids, and organic matter to create an optimal growing medium for plants. Biochar's varying degree of porosity and large surface area, especially when compared to sandy soils, remediates the physical structure of nutrient depleted soils (Allohverdi et al., 2021). The germination and early seedling development are enhanced in soils with biochar, promoting the release of salts, phytotoxins, and germination-inducing hormones, and enhancing water-holding capacity (WHC) and soil aeration (Joseph et al., 2021). Furthermore, the increase in ethylene production, a phytohormone that regulates growth and senescence, has been evidenced in soils with biochar amendments, ultimately leading to increased crop yields (Allohverdi et al., 2021; Fig. 9.3).

Biochar has been reported to increase crop yields significantly, especially in tropical, highly weathered, and degraded soils. For instance, in a study applying biochar at 68 t/ha to highly weathered soils in Zambia, rice and cowpea biomass increased by 20% and 50%, respectively (Glaser et al., 2002). At higher application rates (136.75 t/ha), cowpea biomass doubled. Another study analyzing the effects of Eucalyptus-derived biochar in highly weathered and degraded soils in Kenya saw corn yields doubling after application (Kimetu et al., 2008). In tropical climates like the Philippines and Thailand, rice-husk biochar improved yields by 16%–35% in dry, poor, nonacidic soils, attributed to enhanced water retention and increased availability of potassium and phosphorus (Haefele et al., 2011). Elias et al. (2020) conducted a study at the Crops for the Future Field Research

FIGURE 9.3 Artisanal biochar in Odisha 1.

Artisanal biochar production by a community in Odisha, India. In this region, soil is highly weathered and depleted of nutrients. Most farmers in this area depend on rice as their main cash crop, while a garden provides their food supply. Biochar in this project is projected to increase crop yield, nutritional value of food produced, and resistance to insects and drought. Image and testimony courtesy of Carboneers, a certified Carbon Sink Manager, working with ground partners and rural farmer communities in the global South. https://www.dutchcarboneers.com/#home. Artisanal biochar in Odisha, India.

Center in Malaysia to explore the potential of Leucaena biochar. The results demonstrate a positive crop yield response to Leucaena biochar, particularly when combined with fertilizer additions on very strongly to strongly acidic tropical soils. The study underscores the economic feasibility of biochar production from local biomass or waste residues using low-cost pyrolysis technology.

Biochar application at high rates highlights the potential to alleviate the adverse effects of salt stress on plant growth, as evidenced by studies on *Abutilon theophrasti* and *Prunella vulgaris*. Notably, the application of biochar at 50 t/ha has proven effective in mitigating salt-induced mortality, maintaining growth rates comparable to plants without salt addition (Thomas et al., 2013). Similarly, in Ghana, the application of Sorghum and rice-husk biochar (2 t/ha) led to a significant increase in maize yield along with improved soil pH, soil OC, and the availability of nitrogen, phosphorous, and potassium (Calys-Tagoe et al., 2019).

While tropical regions show an average 25% increase in crop yield, temperate regions exhibit lower effects, possibly due to the already high fertility, cation exchange capacity (CEC), water retention capacity, and nutrient availability in temperate soils (Jeffery et al., 2017). The liming effect (raising the pH in acidic soils), increased WHC, and enhanced nutrient retention and availability are identified as key factors contributing to the yield increase in tropical conditions (Vijay et al., 2021). The positive residual effects of biochar extend to reducing Na^+ uptake in subsequent crops, demonstrating its novel technique in addressing salinity-induced reductions in mineral uptake (Akhtar et al., 2015).

It is noteworthy that field studies have reported delayed positive effects on crop yield, with increases observed in the second or subsequent years rather than immediately after biochar application. Long-term trials in Colombia and Nepal demonstrated a gradual increase in maize and soybean yield in the second, third, and fourth years, emphasizing the long-term impact of biochar (Major et al., 2010; Pandit et al., 2018). The high surface area and CEC of biochar can lead to the need for charging or inoculation before benefits are seen. This process can be done before application, by mixing the biochar with compost or fertilizer mix, or by waiting for the biochar to charge naturally (Glaser et al., 2015). Biochar aging (i.e., natural inoculation) has shown enhanced nutrient retention and subsequent crop yield improvement (Jin et al., 2019; Xu et al., 2023).

As biochar undergoes aging processes in the soil, disturbances such as cultivation, wetting–drying cycles, and freeze–thaw cycles, along with ingestion by soil fauna, contribute to the evolution of biochar particles over time (Wang et al., 2020). Recent investigations have identified the formation of porous organo-mineral heterogeneous microagglomerates with mineral phases consisting of Fe, Al, Si oxides, phosphates (Ca/Fe/Al), carbonates (Ca/Mg), and chlorides (K, Na), bound together by organic compounds and bonded to the biochar surface (Kavitha et al., 2018; Kocsis et al., 2022). As these microagglomerates age, they increase their surface area, CEC, and anion exchange capacity (Dong et al., 2017). This evolution contributes to the prolonged increase in P availability where enhancement was observed to last up to 5 years (Glaser & Lehr, 2019).

9.3.2 Enhancing nutrient retention capacity

Biochar application serves as an effective strategy to mitigate nutrient leaching, particularly during the monsoon season in tropical regions, where leaching is prominent and accelerates soil acidification, leading to decreased crop yield and heightened fertilizer demands (Luo et al., 2020). Biochar application can significantly reduce the leaching of essential nutrients such as N, NO^{3-}, K, P, Mg, Na, and Ca from the soil (Major et al., 2012). Biochar has also demonstrated effectiveness in reducing N leaching, leading to increased N levels in the soil (Lu et al., 2020). A study applying wheat straw biochar applied at rates ranging from 0.5% to 1%, saw a decrease in $NH^{4\ +}$ and NO^{3-} leaching ranging from 11.6% to 24% and 13.2%–29.7%, respectively (Sun et al., 2017). The physicochemical attributes of biochar significantly influence its adsorption capabilities. Acidic functional groups present in biochar have presented an increased affinity for ammonium (NH_4^+) adsorption, an essential source of nitrogen to plants (Spokas et al., 2012).

Nutrient adsorption by biochar is highly dependent on the pyrolysis temperature and feedstock. Lower pyrolysis temperatures (e.g., 400°C–500°C) tend to facilitate nutrient adsorption For instance, NH_4^+ adsorption rates of pepperwood biochar produced at 450°C were higher than those produced at 300°C and 600°C while biochar produced at 450°C from sugar cane bagasse showed higher phosphate adsorption rates compared to biochar produced at 300°C and 600°C. Meanwhile, biochar produced at higher pyrolysis temperatures demonstrated an improved ability to retain nutrients, reducing leaching in soil (Ding et al., 2017).

The CEC serves as a measurement of the ability of soils to hold and provide essential nutrients to plants. The CEC of biochar serves as a crucial indicator of its capacity to absorb essential cations like NH^{+4} and Ca^{2+}, which are vital nutrients for plant growth (Hue, 2020). By increasing the number of cation exchange sites and creating functional groups on its surface, such as –COOH and –OH, biochar enhances the cation adsorption capacity of soils (Ramezanzadeh et al., 2023). Within the diverse fractions generated during pyrolysis, the biochar's nutrient retention capacity becomes evident (Kocsis et al., 2022). The persistent fraction, with its amorphous structure and graphite-like crystalline formations, serves as a reservoir for essential nutrients while exhibiting resistance to degradation, while the labile fraction, comprising mainly cellulose and hemicellulose, undergoes rapid mineralization upon soil application, swiftly releasing nutrients for crop uptake (Kocsis et al., 2022). This dynamic nutrient release mechanism contributes to the biochar's potential to enhance soil fertility and nutrient availability.

Biochar plays a crucial role in soil nutrient retention, and one key aspect is its pH, typically ranging from 7 to 10, although variations can occur based on pyrolysis temperature and feedstock type (Muhammad et al., 2020). Increased pH facilitates the retention and availability of essential nutrients such as P, K, and micronutrients (Vijay et al., 2021). The pH values of biochars derived from different feedstocks and pyrolysis temperatures highlight the diversity in nutrient retention characteristics. Woody biomass-derived biochars displayed average pH values two units lower than nonwoody biomass-derived biochars under similar carbonization conditions (Qiu et al., 2022). The alkaline nature of biochar is often associated with components such as −O− and −COO groups, carbonate content, and oxygen-containing functionalities such as quinine, diketone, chromene, and pyrone (Kumari et al., 2023). The carbonization process during biochar production contributes to the formation of oxygen-rich functionalities, and the increase in pH values may be linked to the concentration of nonpyrolyzed inorganic components and the decomposition of the organic matrix (Xiao et al., 2023). Moreover, biochars prepared at higher pyrolysis temperatures tend to have greater pH values, attributed to the release of alkali salts from organic biomass substances (Kamarudin et al., 2022).

9.3.3 Reducing fertilizer needs

Overcoming the challenge of boosting yields by lowering or eliminating the need for synthetic fertilizers is a promising effect of the agronomic use of biochar. Biochar application shows promise in enhancing fertilizer use efficiency, thereby reducing the overall need for fertilizers in agricultural practices. Chan et al. (2007) demonstrated that biochar application alone did not increase radish yields, but in the presence of N fertilizer, it significantly improved radish yields by 42% and 96% with biochar application at 10 and 50 t/ha, respectively (Chan et al., 2008). This improvement in yields was attributed to the enhanced availability of N facilitated by biochar application. Similarly, Zhang et al. (2010) reported a 12.1% increase in rice yield and a substantial improvement in agronomic N use efficiency (from 1.3 to 5.3 kg of increased grain production per kg of N fertilizer) with biochar application at 40 t/ha in N-fertilized soil.

The use of biochar can improve fertilizer use efficiency and soil fertility (Fig. 9.4). Biochar's large surface area and diverse functional groups contribute to enhanced nutrient adsorption, reducing nutrient leaching and promoting slow-release fertilizer capabilities (Ding et al., 2017). Biochar's effects on nutrient cycling and microbial activity contribute to increased P availability, particularly in acidic or neutral soils (Jin et al., 2019) by enhancing mycorrhizal colonization (Graber et al., 2015). A study reported that biochar applied with iron (Fe)-P and enriched with arbuscular mycorrhizal (AM) hyphae enhanced AM colonization by 6% and improved AM-related Fe-P uptake by 12% (Vanek & Lehmann, 2015). The soluble P located on biochar enhanced total plant and microbial P, and biochar application reversed reductions in specific root length induced by AM, enhancing AM's access to sparingly soluble P and root/microbial access to soluble P (Vanek & Lehmann, 2015).

Acharya et al. (2023) focused on the potential of biochar-based organic and inorganic fertilizers, the research aims to evaluate their impact on soil fertility and Okra yield in a greenhouse pot experiment. The biochar-blended

FIGURE 9.4 Artisanal biochar in Odisha 2.

Artisanal biochar manufacturing in Odisha, India. This project is part of a decentralized effort to expand biochar in remote communities with no access to financial or infrastructure resources for a high-tech pyrolysis plant. This approach allows farmer communities to reap the benefits of both the biochar application to their soils and the financial influx from carbon credit sales without an elevated upfront investment. Image and testimony courtesy of Carboneers, a certified Carbon Sink Manager, working with ground partners and rural farmer communities in the global South. https://www.dutchcarboneers.com/#home. Artisanal biochar in Odisha, India.

formulations involved mixing biochar with mineral NPK fertilizer, vermicompost, and goat manure, while biochar enrichment formulations included biochar enriched with cow urine and mineral NPK fertilizer in an aqueous solution. Results indicated that all biochar-amended treatments exhibited improved soil chemical properties, including higher pH, OC, total N, and available P and K compared with nonbiochar control plots. Notably, biochar blended with goat manure showed the most significant impact on soil fertility and fruit yield, increasing Okra yield by 89% over the control with no biochar and no fertilizers (CK) and by 88% over the plot with only recommended NPK fertilizer and no biochar (F). Similarly, cow urine–enriched biochar demonstrated positive effects, enhancing fruit yield by 29% and 28% compared to CK and F, respectively.

9.3.4 Lowering N_2O emissions

An intersection of biochar's potential as a soil enhancer and a tool to capture atmospheric carbon is in its ability to lower nitrous oxide (N_2O) and methane (CH_4) emissions, two potent GHGs. Recent meta-analyses, incorporating a larger number of field studies and realistic biochar application rates, found average reductions of 12% in soil N_2O emissions in field studies (Verhoeven et al., 2017) and 38% in laboratory and field studies (Borchard et al., 2019). The optimal effect is observed with biochar doses of 10–20 t/ha, and the effectiveness might diminish with time as biochar ages in soil (Cayuela et al., 2014) which suggests periodic reapplications might be needed.

The ability of biochar to restrict nitrate (NO_3) within the rhizosphere, reduce leaching, and improve N use efficiency contributes to lower N_2O emissions (Fungo et al., 2014). One key mechanism involves the inhibition of stages in both nitrification and denitrification processes (Weldon et al., 2019). This inhibition plays a crucial role in diminishing the overall source capacity of the soil for N_2O by impeding the activities of anaerobic microorganisms involved in denitrification, and biochar application. Moreover, biochar-induced improvement in soil aeration reduces denitrification by inhibiting anaerobic microorganisms, resulting in the positive impact of biochar on soil gas diffusivity and subsequent N_2O emission reduction (Joseph et al., 2021). Additionally, biochar application facilitates microbial immobilization of available N in the soil. decreasing the source capacity of the soil for N_2O by limiting the availability of N substrates, contributing to overall emission reduction (Song et al., 2019).

The increased pH of the soil following biochar application is associated with the conversion of N_2O to N_2, further contributing to the mitigation of N_2O emissions (Song et al., 2019). Gases such as NH_3, N_2O, and CH_4 produced during biotic and abiotic reactions in soils can diffuse into the nanopores of biochar where they undergo reactions with oxidants and reductants, especially in the presence of water, potentially reducing N loss and greenhouse gas emissions (Quin et al., 2015). Additionally, N_2O can be transformed into NH_3, pyridine, or pyrrole compounds on biochar surfaces, further decreasing N_2O emissions (Quin et al., 2015).

Biochar application has shown promise in significantly lowering N_2O emissions in rice paddies and other crop ecosystems. Rice paddies, due to alternative wetting and drying conditions, are major contributors to N_2O emissions (Verhoeven et al., 2006). Alternative studies have reported up to a 50% reduction in N_2O emissions in biochar-amended acidic soils in soybean plots in the eastern Colombian Plains (Major et al., 2012), and reductions of about 85% in rewetted soils with 10% biochar compared to soils without biochar (Yanai et al., 2007). The effectiveness of biochar in mitigating N_2O emissions varies with feedstock source and pyrolysis temperature (Fungo et al., 2014). For instance, maize stover biochar showed 17% less N_2O emissions than Eucalyptus wood biochar, and biochar pyrolyzed at 350°C exhibited 3% lower emissions compared to that at 550°C (Fungo et al., 2014).

9.3.5 Improving soil microbial habitat

Soil microbial communities play a crucial role in soil health, impacting soil respiration, organic carbon content, nutrient cycling, and overall crop productivity (Joseph et al., 2021). Biochar, often considered less available to soil microbes, has demonstrated positive impacts on soil microbial biomass, growth, and activity (Li et al., 2020). Sorption and subsequent inactivation of growth-inhibiting materials by biochar modulate the abundance of soil biota in biochar-amended soils (Li et al., 2020). The porous structure and high SSA of biochar contribute to nutrient and water retention, establishing habitats that promote microbial activity (Manirakiza et al., 2019). Additionally, biochar surfaces contain partially soluble N and carbon sources that are suitable for microbial activities (Ammal et al., 2020). Multiple studies highlight the positive impact of biochar on soil enzyme activities. For instance, rice-husk–derived biochar at 12 t/ha significantly improved alkaline phosphatase, invertase, catalase, and urease activities. The enhanced enzyme activities are attributed to increased organic substrates such as soil organic carbon (SOC), N pools, and microbial biomass carbon (Krahn et al., 2023).

Biochar's impact on soil microbial habitat is particularly pronounced through the formation of persistent and labile fractions during pyrolysis (Kocsis et al., 2022). The persistent fraction, characterized by an amorphous structure with graphite-like crystalline formations, provides a conducive environment for beneficial soil microorganisms. In contrast, the labile fraction, composed of cellulose and hemicellulose, supports rapid mineralization upon soil application, fostering a dynamic microbial habitat. Likewise, biochar's porous nature provides surface protection for microorganisms, safeguarding against physical damage from soil compaction or changes in soil fauna nutrition (Kocsis et al., 2022).

The enrichment of biochar has been found to particularly benefit mycorrhizal and rhizobial populations at the root level, fostering a symbiotic relationship with plants (Allohverdi et al., 2021). This interaction occurs at the interface between root hairs and microbes in the soil matrix, facilitating the sorption of various organic compounds bonded to biochar structures that can be utilized by plants. Likewise, biochar application in the rhizosphere increases microbe numbers, favoring the presence and biomass of common microsymbionts such as AM and ectomycorrhiza (Videgain-Marco et al., 2021). Studies have reported a 30%–40% higher soil OC content with varying biochar application rates (Qianqian et al., 2022). This increase is attributed to the recalcitrant nature of biochar's high carbon content, along with cumulative effects from microbial activity, rhizosphere decomposition, and root exudates (Qianqian et al., 2022). The enhanced microbial activities stimulated by biochar play a crucial role in nutrient cycling. Biochar promotes increased microbial biomass C and enzymatic activities, which influence the availability of essential nutrients (Joseph et al., 2021). The interaction of fungi and bacteria with biochar's nutrient-rich pores and the subsequent release of root exudates and microbial metabolites contribute to elevated (SOC) levels (Zhang & Shen, 2022).

9.3.6 Increasing soil moisture retention

The ability of biochar to enhance water retention is attributed to its impact on soil structure, increasing microporosity and pore connectivity, which in turn, results in better WHC and reduced water drainage (Joseph et al., 2021). Additionally, biochar plays a crucial role in moderating the reduction in infiltration rate during high-intensity rainstorms, thereby reducing surface sealing and lowering runoff and erosion rates (Abrol et al., 2016).

Ndede et al. (2022) assessed the impact of different types and application rates of biochar on soil properties related to water retention, particularly in sandy soils prone to water deficiency. The results indicate that irrespective of biochar type, increasing the amendment rate led to a reduction in bulk density. The WHC increased with the application rate, while readily available water exhibited a peak at a 5% biochar content for most biochar types. Woodchip and waterweed biochars showed the highest readily available water increments, reaching 165% and 191%, respectively, at a 10% (v/v) rate. The study concludes that both biochar type and application rate significantly influence hydrological properties and the readily available water capacity of sandy soils (Ndede et al., 2022).

A recent study analyzed the impact of biochar, organic manure, and their combination on hydro-physical properties of sandy soils in a sandy loam acidic mountain soil of India, this study focuses on the effects after 6 years of maize-black gram cropping (Das & Ghosh, 2022). Biochar application rate positively influenced soil water content and volumetric water content, with the highest WHC observed in the vermicompost and biochar combination. Porosity increased with the biochar application rate, while hydraulic conductivity decreased with the biochar application rate. Additionally, the combination of biochar and organic manure positively influenced soil porosity, aggregate stability, mean weight diameter, WHC, and soil moisture content, while decreasing soil bulk density and hydraulic conductivity (Das & Ghosh, 2022).

9.3.7 Reducing soil bulk density

The addition of biochar to agricultural soils influences soil properties, including bulk density. By reducing soil bulk density, biochar enhances soil structure and aeration, promoting better root growth, and nutrient uptake (Hussain et al., 2017). Soil compaction, characterized by increased bulk and particle density, poses significant challenges to root growth, particularly within the top 30 cm of soil. High bulk density (>1.7 t/m^3) results in poor aeration, physical resistance, and reduced root growth, impacting plant growth due to hindered soil pore penetration (Xiang et al., 2017). Compacted soils exhibit lower pore space, restricting oxygen diffusion and creating anaerobic conditions that substantially decrease plant growth (Wong et al., 2022). Notably, biochar incorporation has been shown to reduce soil compaction by over 10%, contributing to a physical loosening of the soil (Peake et al., 2014). Various investigations underscore the positive impact of biochar on soil bulk density, with reductions ranging from 3% to 31%, leading to improved soil porosity by 14%–64% (Blanco-Canqui, 2017). The enhancement of soil structure, as demonstrated by decreased bulk density and increased porosity, facilitates the transport of water, gases, and heat in soils (Liang et al., 2006).

9.3.8 Improving soil aeration and root growth

Intrapores, interpores, particle size distribution, and adsorption/absorption properties of biochar contribute to changes in porosity, water retention, hydraulic conductivity, and air permeability in biochar-amended soils (Vijay et al., 2021). Field studies consistently report improved soil porosity and WHC with biochar application (Randolph et al., 2017). A meta-analysis involving 74 publications revealed significant enhancements in soil porosity (8.4%), WHC (15.1%), and saturated hydraulic conductivity (25.2%) with biochar application (Omondi et al., 2016). In specific scenarios, such as clayey soils, biochar's introduction prompts the creation of larger pores, exceeding the main pores of the clay, improving aeration in compacted soils (Kocsis et al., 2022). Conversely, in sandy soils, biochar proves beneficial for augmenting WHC and nutrient adsorption rather than enhancing aeration.

9.3.9 Applications for remediation of soils contaminated with heavy metals

Biochar demonstrates promise in remediating soils contaminated with heavy metals (HMs). The large surface area, microporous structure, active functional groups, and high pH of biochars enable them to act as effective adsorbents (Paz-Ferreiro et al., 2014). Through cation exchange mechanisms and complex formation with surface functional groups, biochar adsorbs HM such as copper (Cu), lead (Pb), and cadmium (Cd) (Uchimiya et al., 2010). Traditional approaches for remediating soils contaminated with HM encompass physical, chemical, and physicochemical methods. These techniques, ranging from thermal treatment to soil replacement, leaching, and ion exchange, have demonstrated efficacy in mitigating contamination (Qayyum et al., 2020). Challenges such as high costs, inefficiency, and environmental unsustainability have hindered their widespread adoption (Khalid et al., 2017). Moreover, certain methods may compromise soil quality or lead to the loss of valuable metals, emphasizing the need for alternative strategies (Paz-Ferreiro et al., 2014).

Among the various pollutants, cadmium emerges as a critical threat due to its persistence, toxicity, and potential to contaminate food supplies (Guarín et al., 2023, 2024; Lewis et al., 2018; Ramtahal et al., 2019; Fig. 9.5). Cadmium, a highly toxic HM, can inflict severe harm even at minimal concentrations, raising concerns about its impact on human health and ecosystems (Lewis et al., 2018). Field studies have demonstrated the potential of biochar in reducing Cd in soils.

FIGURE 9.5 Cacao tree.

Cacao tree in northern Colombia. Research has shown that biochar is a cost-effective and easily available amendment strategy to mitigate soil contamination with heavy metals (especially cadmium).

Biochar addition to soils results in average decreases in plant tissue concentrations of Cd by 38% (Chen et al., 2018). Various mechanisms contribute to this remediation effect, including the immobilization of HM through surface O-functional groups on biochar, ion exchange, precipitation, cation and anion metal attraction, reduction, electron shuttling, and physisorption (Joseph et al., 2021).

One study, conducted in cacao farms in Ecuador with reported elevated bean Cd levels, focuses on the effectiveness of four biochars derived from coffee husk, quinoa straw, and oil palm residues (López et al., 2022). Results revealed that quinoa biochar at a 2% rate significantly reduced bioavailable Cd in both moderately acidic and slightly alkaline soils by approximately 71% while leaf-Cd was reduced by 48%. Biochar characteristics, including ash content, basic cation content, and surface functional groups, were identified as indicators for selecting biochars to mitigate Cd uptake by cacao. Furthermore, the application of quinoa-derived biochar provided essential nutrients such as P and K, potentially enhancing productivity and offsetting mitigation costs (López et al., 2022). Other studies on Cd accumulation in cacao plants demonstrated a 35% reduction in live leaf tissues in East Trinidad (Ramtahal et al., 2019), 90% in cacao pod waste in northern

Colombia (Pinzon-Nuñez et al., 2022), and 50% in live tissues when coapplied with dolomitic lime in southwest Colombia (López & Saldarriaga, 2023).

9.4 Biochar challenges and opportunities

Challenges lie ahead for the future of biochar production and expansion, particularly as the adoption of CDR methods grows. A significant obstacle is the acquisition of feedstocks, as competition for biomass is anticipated to increase (IPCC, 2022). While certain feedstocks, such as municipal biomass waste, logging residues, or crop leftovers, may pose fewer economic concerns, others, like switchgrass or willow, may involve added expenses related to crop cultivation and harvesting (IBI, 2022). However, some feedstocks may be obtained at a negative price if they constitute a waste stream that would otherwise necessitate costly treatment, such as municipal solid waste. Feedstock acquisition is a critical factor in biochar's cost–benefit analysis. To achieve profitability, it may be necessary to shift away from marketable feedstocks like corn stover and brewery grain and instead focus on waste stream feedstocks, such as forest thinning or yard waste (Zilberman et al., 2023).

Another challenge is the cost of transporting feedstock, which can be high when waste biomass is sourced from a location far away from where it will be pyrolyzed. In some cases, it may be beneficial to densify the biomass through chipping or pelletizing before transportation (IBI, 2022). Additionally, if the distance between the feedstock source and the pyrolysis facility is significant, the greenhouse gas emissions from transporting the biomass may outweigh the carbon captured (Schmidt et al., 2021). Furthermore, the physical and chemical properties of biochar in soils change over time after application. An improved understanding of how these factors impact end uses in different climates and soil types could aid in identifying situations where biochar application will benefit crops while maximizing carbon capture, leading to biochar expansion (Amonette et al., 2021). Lastly, there are additional opportunities associated with the potential impact of biochar on crop insurance and farmer loans. If biochar can be shown to consistently reduce production risks and adapt to climate variability, especially in the future agronomic uncertainty under climate change, farmers who use biochar could potentially receive discounted crop insurance rates, which could incentivize adoption (Amonette et al., 2021).

9.5 Conclusions and takeaways

CDR techniques, such as biochar, can effectively remove and long-term store carbon dioxide from the atmosphere. To attain global and national goals of net zero CO_2 and greenhouse gas emissions, the implementation and growth of biochar is crucial. It is an essential component in all models that restrict global warming to 2°C or below by 2100. Biochar is not a replacement for deep emissions reductions, but it is a crucial tool that should be utilized in conjunction with other mitigation methods. Biochar is a cost-effective and low-entry barrier method for capturing and storing atmospheric CO_2 in a stable solid state for centuries to millennia. In addition to its primary purpose of carbon sequestration, biochar offers a range of cobenefits that make it an appealing option for policymakers, farmers, and consumers. Biochar's physical and chemical properties promote healthy soil and enhanced plant growth. Proven cobenefits include higher seedling survival rates, increased plant growth, higher soil CEC, improved water and nutrient retention, and a thriving microbial and mycorrhizal community. Even in heavily polluted soils, biochar is a cost-effective and promising mitigation strategy. Overall, biochar is a scalable, equitable, and permanent nature-based carbon removal method that leverages naturally occurring photosynthesis to sequester carbon. It provides an opportunity for farmers of all sizes and incomes to enhance crop yields and productivity while contributing to global efforts to combat climate change.

Acknowledgments

The authors would like to thank the editors for inviting us to contribute to this important book. Their insightful edits and valuable suggestions significantly strengthened the manuscript. Special thanks to Carboneers for sharing stories and pictures that show the potential of biochar for carbon removal and community impact.

Author contributions

Daniel Guarín: Writing—review and editing, writing—original draft, visualization, validation, methodology, investigation, formal analysis, data curation. Alejandro Fernandez: writing—original draft, visualization, validation, methodology, investigation, formal analysis, data curation. Jhony Benavides: writing—original draft, visualization, validation, methodology, investigation, formal analysis, data curation. Nicolás Rangel: writing—original draft, visualization, validation, methodology, investigation, formal analysis, data curation.

Conflict of interest

The authors declare that they have no known competing financial interests or personal relationships that could have appeared to influence the work reported in this paper.

AI disclosure

During the preparation of this work the author(s) used Chat GPT in order to check for spelling and grammar. After using this tool/service, the author(s) reviewed and edited the content as needed and take(s) full responsibility for the content of the publication.

References

Abrol, V., Ben-Hur, M., Verheijen, F. G. A., Keizer, J. J., Martins, M. A. S., Tenaw, H., Tchehansky, L., & Graber, E. R. (2016). Biochar effects on soil water infiltration and erosion under seal formation conditions: Rainfall simulation experiment. *Journal of Soils and Sediments*, *16*(12), 2709–2719. Available from https://doi.org/10.1007/s11368-016-1448-8.

Acharya, N., Vista, S. P., Shrestha, S., Neupane, N., & Pandit, N. R. (2023). Potential of biochar-based organic fertilizers on increasing soil fertility, available nutrients, and okra productivity in slightly acidic sandy loam soil. *Nitrogen (Switzerland)*, *4*(1), 1–15. Available from https://doi.org/10.3390/nitrogen4010001, http://www.mdpi.com/journal/nitrogen.

Akhtar, S. S., Andersen, M. N., & Liu, F. (2015). Residual effects of biochar on improving growth, physiology and yield of wheat under salt stress. *Agricultural Water Management*, *158*, 61–68. Available from https://doi.org/10.1016/j.agwat.2015.04.010, http://www.journals.elsevier.com/agricultural-water-management/.

Allohverdi, T., Mohanty, A. K., Roy, P., & Misra, M. (2021). A review on current status of biochar uses in agriculture. *Molecules (Basel, Switzerland)*, *26*(18). Available from https://doi.org/10.3390/molecules26185584, https://www.mdpi.com/1420-3049/26/18/5584/pdf.

Ammal, A., Abunyewa, A. A., & Yeboah, E. (2020). Influence of integrated soil fertility management on the vegetative growth parameters of zea mays in the guinea savanna eco-zone of ghana. *Journal of Agricultural Sciences (Belgrade*, *65*(2), 187–197. Available from https://doi.org/10.2298/JAS2002187A, http://scindeks.ceon.rs/article.aspx?query = ISSID%26and%2615114&page = 6&sort = 8&stype = 0&backurl = %2fissue.aspx%3fissue%3d15114.

Amonette, J. E., Archuleta, J. G., Fuchs, M. R., Hills, K. M., Yorgey, G. G., Flora, G., Hunt, J., Han, H. S., Jobson, B. T., Miles, T. R., Page-Dumroese, D. S., Thompson, S., Trippe, K. M., Wilson, K., Baltar, R., Carloni, K., Christoforou, C., Collins, D. P., Dooley, J., ... Wheeler, E. (2021). *Biomass to biochar: Maximizing the carbon value*. Center for Sustaining Agriculture and Natural Resources, Washington State University.

Bergero, C., Wise, M., Lamers, P., Wang, Y., & Weber, M. (2022). Biochar as a carbon dioxide removal strategy in integrated long-run climate scenarios. *Research Square*. Available from https://doi.org/10.21203/rs.3.rs-1805544/v1, https://www.researchsquare.com/browse.

Blanco-Canqui, H. (2017). Biochar and soil physical properties. *Soil Science Society of America Journal*, *81*(4), 687–711. Available from https://doi.org/10.2136/sssaj2017.01.0017.

Blenis, N., Hue, N., Maaz, T. M. C., & Kantar, M. (2023). Biochar production, modification, and its uses in soil remediation: A review. *Sustainability (Switzerland)*, *15*(4). Available from https://doi.org/10.3390/su15043442, http://www.mdpi.com/journal/sustainability/.

Borchard, N., Schirrmann, M., Cayuela, M. L., Kammann, C., Wrage-Mönnig, N., Estavillo, J. M., Fuertes-Mendizábal, T., Sigua, G., Spokas, K., Ippolito, J. A., & Novak, J. (2019). Biochar, soil and land-use interactions that reduce nitrate leaching and N_2O emissions: A meta-analysis. *Science of the Total Environment*, *651*, 2354–2364. Available from https://doi.org/10.1016/j.scitotenv.2018.10.060, http://www.elsevier.com/locate/scitotenv.

Calys-Tagoe, E., Sadick, A., Yeboah, E., & Amoah, B. (2019). Biochar effect on maize yield in selected farmers fields in the northern and upper east regions of Ghana. *Journal of Experimental Agriculture International*, *30*(6), 1–9. Available from https://doi.org/10.9734/jeai/2019/44168.

Carton, W., Hougaard, I.-M., Markusson, N., & Lund, J. F. (2023). Is carbon removal delaying emission reductions?Is carbon removal delaying emission reductions? *WIREs Climate Change*, *14*(4), 1–18. Available from https://doi.org/10.1002/wcc.826, https://onlinelibrary.wiley.com/doi/abs/10.1002/wcc.826.

Cayuela, M. L., Van Zwieten, L., Singh, B. P., Jeffery, S., Roig, A., & Sánchez-Monedero, M. A. (2014). Biochar's role in mitigating soil nitrous oxide emissions: A review and meta-analysis. *Agriculture, Ecosystems & Environment*, *191*, 5–16. Available from https://doi.org/10.1016/j.agee.2013.10.009.

Chan, K. Y., Van Zwieten, L., Meszaros, I., Downie, A., & Joseph, S. (2007). Agronomic values of greenwaste biochar as a soil amendment. *Australian Journal of Soil Research*, *45*(8), 629–634. Available from https://doi.org/10.1071/SR07109.

Chan, K. Y., Van Zwieten, L., Meszaros, I., Downie, A., & Joseph, S. (2008). Using poultry litter biochars as soil amendmentsUsing poultry litter biochars as soil amendments. *Australian Journal of Soil Research*, *46*(5), 437–444. Available from https://doi.org/10.1071/SR08036.

Chen, D., Liu, X., Bian, R., Cheng, K., Zhang, X., Zheng, J., Joseph, S., Crowley, D., Pan, G., & Li, L. (2018). Effects of biochar on availability and plant uptake of heavy metals – a meta-analysis. *Journal of Environmental Management*, *222*, 76–85. Available from https://doi.org/10.1016/j.jenvman.2018.05.004.

Coppola, A. I., Wiedemeier, D. B., Galy, V., Haghipour, N., Hanke, U. M., Nascimento, G. S., Usman, M., Blattmann, T. M., Reisser, M., Freymond, C. V., Zhao, M., Voss, B., Wacker, L., Schefuß, E., Peucker-Ehrenbrink, B., Abiven, S., Schmidt, M. W. I., & Eglinton, T. I. (2018). Global-scale evidence for the refractory nature of riverine black carbon. *Nature Geoscience*, *11*(8), 584–588. Available from https://doi.org/10.1038/s41561-018-0159-8, http://www.nature.com/ngeo/index.html.

Das, S. K., & Ghosh, G. K. (2022). Soil hydro-physical properties affected by biomass-derived biochar and organic manure: A low-cost technology for managing acidic mountain sandy soils of north eastern region of India. *Biomass Conversion and Biorefinery*. Available from https://doi.org/10.1007/s13399-022-03107-7, http://www.springer.com/engineering/energy + technology/journal/13399.

Ding, Y., Liu, Y., Liu, S., Huang, X., Li, Z., Tan, X., Zeng, G., & Zhou, L. (2017). Potential benefits of biochar in agricultural soils: A review. *Pedosphere*, *27*(4), 645–661. Available from https://doi.org/10.1016/S1002-0160(17)60375-8, http://pedosphere.issas.ac.cn.

Dong, X., Li, G., Lin, Q., & Zhao, X. (2017). Quantity and quality changes of biochar aged for 5 years in soil under field conditions. *Catena*, *159*, 136–143. Available from https://doi.org/10.1016/j.catena.2017.08.008.

Dooley, K., Nicholls, Z., & Meinshausen, M. (2022). Carbon removals from nature restoration are no substitute for steep emission reductions. *One Earth*, *5*(7), 812–824. Available from https://doi.org/10.1016/j.oneear.2022.06.002, https://www.sciencedirect.com/science/article/pii/S2590332222003232.

Elias, D. M. O., Ooi, G. T., Ahmad Razi, M. F., Robinson, S., Whitaker, J., & Mcnamara, N. P. (2020). Effects of Leucaena biochar addition on crop productivity in degraded tropical soils. *Biomass and Bioenergy*, *142*. Available from https://doi.org/10.1016/j.biombioe.2020.105710, http://www.journals.elsevier.com/biomass-and-bioenergy/.

Fahmy, T. Y. A., Fahmy, Y., Mobarak, F., El-Sakhawy, M., & Abou-Zeid, R. E. (2020). Biomass pyrolysis: past, present, and future. *Environment, Development and Sustainability*, *22*(1), 17–32. Available from https://doi.org/10.1007/s10668-018-0200-5, http://www.wkap.nl/journalhome.htm/1387-585X.

Fuhrman, J., Bergero, C., Weber, M., Monteith, S., Wang, F. M., Clarens, A. F., Doney, S. C., Shobe, W., & Mcjeon, H. (2023). Diverse carbon dioxide removal approaches could reduce impacts on the energy–water–land system. *Nature Climate Change*, *13* (4), 341–350. Available from https://doi.org/10.1038/s41558-023-01604-9.

Fungo, B., Guerena, D., Thiongo, M., Lehmann, J., Neufeldt, H., & Kalbitz, K. (2014). N_2O and CH_4 emission from soil amended with steam-activated biochar. *Journal of Plant Nutrition and Soil Science*, *177*(1), 34–38. Available from https://doi.org/10.1002/jpln.201300495.

Glaser, B., Lehmann, J., & Zech, W. (2002). Ameliorating physical and chemical properties of highly weathered soils in the tropics with charcoal – a review. *Biology and Fertility of Soils*, *35*(4), 219–230. Available from https://doi.org/10.1007/s00374-002-0466-4.

Glaser, B., & Lehr, V. I. (2019). Biochar effects on phosphorus availability in agricultural soils: A meta-analysis. *Scientific Reports*, *9*(1). Available from https://doi.org/10.1038/s41598-019-45693-z, http://www.nature.com/srep/index.html.

Glaser, B., Wiedner, K., Seelig, S., Schmidt, H.-P., & Gerber, H. (2015). Biochar organic fertilizers from natural resources as substitute for mineral fertilizers. *Agronomy for Sustainable Development*, *35*(2), 667–678. Available from https://doi.org/10.1007/s13593-014-0251-4.

Glaser, B., Haumaier, L., Guggenberger, G., & Zech, W. (2001). The 'Terra Preta' phenomenon: a model for sustainable agriculture in the humid tropics. *Die Naturwissenschaften*, *88*(1), 37–41. Available from https://doi.org/10.1007/s001140000193.

Graber, E. R., Tsechansky, L., Mayzlish-Gati, E., Shema, R., & Koltai, H. (2015). A humic substances product extracted from biochar reduces Arabidopsis root hair density and length under P-sufficient and P-starvation conditions. *Plant and Soil*, *395*(1-2), 21–30. Available from https://doi.org/10.1007/s11104-015-2524-3.

Grant, N., Hawkes, A., Mittal, S., & Gambhir, A. (2021). Confronting mitigation deterrence in low-carbon scenarios. *Environmental Research Letters*, *16*(6), 64099. Available from https://doi.org/10.1088/1748-9326/ac0749, https://doi.org/10.1088/1748-9326/ac0749.

Guarín, D., Martín-López, J. M., Libohova, Z., Benavides-Bolaños, J., Maximova, S. N., Guiltinan, M. J., Spargo, J., Da Silva, M., Fernandez, A., & Drohan, P. (2024). Accumulation of cadmium in soils, litter and leaves in cacao farms in the North Sierra Nevada de Santa Marta, Colombia. *Geoderma Regional*, *36*, e00762. Available from https://doi.org/10.1016/j.geodrs.2024.e00762.

Guarín, D., Hamamura, N. R., Cortez, J. B., Benavides, J., Spargo, J., Guiltinan, M., Maximova, S., De La Cruz, O. A., & Drohan, P. J. (2023). Cadmium contamination in cacao farms of Piura, North Peru: A comprehensive assessment of geogenic and anthropogenic sources and implications for future production. *Environmental Challenges, 13*, 100765. Available from https://doi.org/10.1016/j.envc.2023.100765.

Haefele, S. M., Konboon, Y., Wongboon, W., Amarante, S., Maarifat, A. A., Pfeiffer, E. M., & Knoblauch, C. (2011). Effects and fate of biochar from rice residues in rice-based systems. *Field Crops Research, 121*(3), 430–440. Available from https://doi.org/10.1016/j.fcr.2011.01.014.

Hue, N. (2020). *Biochar for maintaining soil health* (pp. 21–46). Springer Science and Business Media LLC. Available from 10.1007/978-3-030-44364-1_2.

Hussain, M., Farooq, M., Nawaz, A., Al-Sadi, A. M., Solaiman, Z. M., Alghamdi, S. S., Ammara, U., Ok, Y. S., & Siddique, K. H. M. (2017). Biochar for crop production: Potential benefits and risks. *Journal of Soils and Sediments, 17*(3), 685–716. Available from https://doi.org/10.1007/s11368-016-1360-2.

IBI. (2022). *Biochar feedstocks. How to make biochar.*

IPCC. (2022). *Summary for policymakers climate change 2022: Impacts, adaptation, and vulnerability. Contribution of working group II to the sixth assessment report of the intergovernmental panel on climate change* (pp. 3–33). Cambridge University Press. Available from 10.1017/9781009325844.001.

Jeffery, S., Abalos, D., Prodana, M., Bastos, A. C., Van Groenigen, J. W., Hungate, B. A., & Verheijen, F. (2017). Biochar boosts tropical but not temperate crop yields. *Environmental Research Letters, 12*(5). Available from https://doi.org/10.1088/1748-9326/aa67bd, http://iopscience.iop.org/article/10.1088/1748-9326/aa67bd/pdf.

Jin, Z., Chen, C., Chen, X., Jiang, F., Hopkins, I., Zhang, X., Han, Z., Billy, G., & Benavides, J. (2019). Soil acidity, available phosphorus content, and optimal biochar and nitrogen fertilizer application rates: A five-year field trial in upland red soil, China. *Field Crops Research, 232*, 77–87. Available from https://doi.org/10.1016/j.fcr.2018.12.013.

Joseph, S., Cowie, A. L., Van Zwieten, L., Bolan, N., Budai, A., Buss, W., Cayuela, M. L., Graber, E. R., Ippolito, J. A., Kuzyakov, Y., Luo, Y., Ok, Y. S., Palansooriya, K. N., Shepherd, J., Stephens, S., Weng, Z., & Lehmann, J. (2021). How biochar works, and when it doesn't: A review of mechanisms controlling soil and plant responses to biochar. *GCB Bioenergy, 13*(11), 1731–1764. Available from https://doi.org/10.1111/gcbb.12885, http://onlinelibrary.wiley.com/journal/10.1111/(ISSN)1757-1707.

Kamarudin, N. S., Dahalan, F. A., Hasan, M., An, O. S., Parmin, N. A., Ibrahim, N., Hamdzah, M., Zain, N. A. M., Muda, K., & Wikurendra, E. A. (2022). Biochar: A review of its history, characteristics, factors that influence its yield, methods of production, application in wastewater treatment and recent development. *Biointerface Research in Applied Chemistry, 12*(6), 7914–7926. Available from https://doi.org/10.33263/BRIAC126.79147926, https://biointerfaceresearch.com/wp-content/uploads/2021/12/20695837126.79147926.pdf.

Kavitha, B., Reddy, P. V. L., Kim, B., Lee, S. S., Pandey, S. K., & Kim, K. H. (2018). Benefits and limitations of biochar amendment in agricultural soils: A review. *Journal of Environmental Management, 227*, 146–154. Available from https://doi.org/10.1016/j.jenvman.2018.08.082, https://www.sciencedirect.com/journal/journal-of-environmental-management.

Keiluweit, M., Nico, P. S., Johnson, M. G., & Kleber, M. (2010). Dynamic molecular structure of plant biomass-derived black carbon (biochar). *Environmental Science & Technology, 44*(4), 1247–1253. Available from https://doi.org/10.1021/es9031419.

Khalid, S., Shahid, M., Niazi, N. K., Murtaza, B., Bibi, I., & Dumat, C. (2017). A comparison of technologies for remediation of heavy metal contaminated soils. *Journal of Geochemical Exploration*, *182*, 247–268. Available from https://doi.org/10.1016/j.gexplo.2016.11.021.

Kimetu, J. M., Lehmann, J., Ngoze, S. O., Mugendi, D. N., Kinyangi, J. M., Riha, S., Verchot, L., Recha, J. W., & Pell, A. N. (2008). Reversibility of soil productivity decline with organic matter of differing quality along a degradation gradient. *Ecosystems*, *11*(5), 726–739. Available from https://doi.org/10.1007/s10021-008-9154-z.

Kocsis, T., Ringer, M., & Biró, B. (2022). Characteristics and applications of biochar in soil–plant systems: A short review of benefits and potential drawbacks. *Applied Sciences (Switzerland)*, *12*(8). Available from https://doi.org/10.3390/app12084051, https://www.mdpi.com/2076-3417/12/8/4051/pdf.

Krahn, K. M., Cornelissen, G., Castro, G., Arp, H. P. H., Asimakopoulos, A. G., Wolf, R., Holmstad, R., Zimmerman, A. R., & Sørmo, E. (2023). Sewage sludge biochars as effective PFAS-sorbents. *Journal of Hazardous Materials*, *445*. Available from https://doi.org/10.1016/j.jhazmat.2022.130449, http://www.elsevier.com/locate/jhazmat.

Kumari, K., Kumar, R., Bordoloi, N., Minkina, T., Keswani, C., & Bauddh, K. (2023). Unravelling the recent developments in the production technology and efficient applications of biochar for agro-ecosystems. *Agriculture (Switzerland)*, *13*(3). Available from https://doi.org/10.3390/agriculture13030512, http://www.mdpi.com/journal/agriculture.

Lefebvre, D., Fawzy, S., Aquije, C. A., Osman, A. I., Draper, K. T., & Trabold, T. A. (2023). Biomass residue to carbon dioxide removal: quantifying the global impact of biochar. *Biochar*, *5*(1). Available from https://doi.org/10.1007/s42773-023-00258-2.

Lehmann, J., Cowie, A., Masiello, C. A., Kammann, C., Woolf, D., Amonette, J. E., Cayuela, M. L., Camps-Arbestain, M., & Whitman, T. (2021). Biochar in climate change mitigation. *Nature Geoscience*, *14*(12), 883–892. Available from https://doi.org/10.1038/s41561-021-00852-8.

Lehmann, J., & Joseph, S. (2015). *Biochar for environmental management: An introduction* (pp. 33–46). Informa UK Limited. Available from 10.4324/9780203762264-8.

Lewis, C., Lennon, A. M., Eudoxie, G., & Umaharan, P. (2018). Genetic variation in bioaccumulation and partitioning of cadmium in *Theobroma cacao* L. *Science of the Total Environment*, *640-641*, 696–703. Available from https://doi.org/10.1016/j.scitotenv.2018.05.365.

Li, T., Kirk, D. W., & Jia, C. Q. (2021). Monolithic wood biochar as functional material for sustainability. *Canadian Journal of Chemical Engineering*, *99*(3), 640–656. Available from https://doi.org/10.1002/cjce.23944, http://onlinelibrary.wiley.com/journal/10.1002/(ISSN)1939-019X.

Li, X., Wang, T., Chang, S. X., Jiang, X., & Song, Y. (2020). Biochar increases soil microbial biomass but has variable effects on microbial diversity: A meta-analysis. *Science of the Total Environment*, *749*, 141593. Available from https://doi.org/10.1016/j.scitotenv.2020.141593.

Liang, B., Lehmann, J., Solomon, D., Kinyangi, J., Grossman, J., O'neill, B., Skjemstad, J. O., Thies, J., Luizão, F. J., Petersen, J., & Neves, E. G. (2006). Black carbon increases cation exchange capacity in soils. *Soil Science Society of America Journal*, *70*(5), 1719–1730. Available from https://doi.org/10.2136/sssaj2005.0383.

López, J. E., Arroyave, C., Aristizábal, A., Almeida, B., Builes, S., & Chavez, E. (2022). Reducing cadmium bioaccumulation in *Theobroma cacao* using biochar: Basis for scaling-up to field. *Heliyon*, *8*(6). Available from https://doi.org/10.1016/j.heliyon.2022.e09790, http://www.journals.elsevier.com/heliyon/.

López, J. E., & Saldarriaga, J. F. (2023). Surface co-application of dolomitic lime with either biochar or compost changes the fractionation of Cd in the soil and its uptake by cacao seedlings. *Journal of Soil Science and Plant Nutrition*, *23*(4), 4926–4936. Available from https://doi.org/10.1007/s42729-023-01469-z.

Lu, H., Yan, M., Wong, M. H., Mo, W. Y., Wang, Y., Chen, X. W., & Wang, J.-J. (2020). Effects of biochar on soil microbial community and functional genes of a landfill cover three years after ecological restoration. *Science of the Total Environment*, *717*, 137133. Available from https://doi.org/10.1016/j.scitotenv.2020.137133.

Luo, C., Yang, J., Chen, W., & Han, F. (2020). Effect of biochar on soil properties on the Loess Plateau: Results from field experiments. *Geoderma*, *369*. Available from https://doi.org/10.1016/j.geoderma.2020.114323.

Major, J., Rondon, M., Molina, D., Riha, S. J., & Lehmann, J. (2010). Maize yield and nutrition during 4 years after biochar application to a Colombian savanna oxisol. *Plant and Soil*, *333*(1), 117–128. Available from https://doi.org/10.1007/s11104-010-0327-0.

Major, J., Rondon, M., Molina, D., Riha, S. J., & Lehmann, J. (2012). Nutrient leaching in a Colombian savanna oxisol amended with biochar. *Journal of Environmental Quality*, *41*(4), 1076–1086. Available from https://doi.org/10.2134/jeq2011.0128.

Manirakiza, E., Ziadi, N., St. Luce, M., Hamel, C., Antoun, H., & Karam, A. (2019). Nitrogen mineralization and microbial biomass carbon and nitrogen in response to co-application of biochar and paper mill biosolids. *Applied Soil Ecology*, *142*, 90–98. Available from https://doi.org/10.1016/j.apsoil.2019.04.025.

Muhammad, N., Nafees, M., Khan, M. H., Ge, L., & Lisak, G. (2020). Effect of biochars on bioaccumulation and human health risks of potentially toxic elements in wheat (*Triticum aestivum* L.) cultivated on industrially contaminated soil. *Environmental Pollution*, *260*, 113887. Available from https://doi.org/10.1016/j.envpol.2019.113887.

Ndede, E. O., Kurebito, S., Idowu, O., Tokunari, T., & Jindo, K. (2022). The potential of biochar to enhance the water retention properties of sandy agricultural soils. *Agronomy*, *12*(2). Available from https://doi.org/10.3390/agronomy12020311, https://www.mdpi.com/2073-4395/12/2/311/pdf.

Neogi, S., Sharma, V., Khan, N., Chaurasia, D., Ahmad, A., Chauhan, S., Singh, A., You, S., Pandey, A., & Bhargava, P. C. (2022). Sustainable biochar: A facile strategy for soil and environmental restoration, energy generation, mitigation of global climate change and circular bioeconomy. *Chemosphere*, *293*, 133474. Available from https://doi.org/10.1016/j.chemosphere.2021.133474.

NOAA. (2024). *2023 was the world's warmest year on record.*

Omondi, M. O., Xia, X., Nahayo, A., Liu, X., Korai, P. K., & Pan, G. (2016). Quantification of biochar effects on soil hydrological properties using meta-analysis of literature data. *Geoderma*, *274*, 28–34. Available from https://doi.org/10.1016/j.geoderma.2016.03.029, http://www.elsevier.com/inca/publications/store/5/0/3/3/3/2.

Pandit, N. R., Mulder, J., Hale, S. E., Zimmerman, A. R., Pandit, B. H., & Cornelissen, G. (2018). Multi-year double cropping biochar field trials in Nepal: Finding the optimal biochar dose through agronomic trials and cost-benefit analysis. *Science of the Total Environment*, *637-638*, 1333–1341. Available from https://doi.org/10.1016/j.scitotenv.2018.05.107, http://www.elsevier.com/locate/scitotenv.

Paz-Ferreiro, J., Lu, H., Fu, S., Méndez, A., & Gascó, G. (2014). Use of phytoremediation and biochar to remediate heavy metal polluted soils: A review. *Solid Earth*, *5*(1), 65–75. Available from https://doi.org/10.5194/se-5-65-2014.

Peake, L. R., Reid, B. J., & Tang, X. (2014). Quantifying the influence of biochar on the physical and hydrological properties of dissimilar soils. *Geoderma*, *235-236*, 182–190. Available from https://doi.org/10.1016/j.geoderma.2014.07.002, http://www.elsevier.com/inca/publications/store/5/0/3/3/3/2.

Pinzon-Nuñez, D. A., Adarme-Durán, C. A., Vargas-Fiallo, L. Y., Rodriguez-Lopez, N., & Rios-Reyes, C. A. (2022). Biochar as a waste management strategy for cadmium contaminated cocoa pod husk residues. *International Journal of Recycling of Organic Waste in Agriculture*, *11*(1), 101–115. Available from https://doi.org/10.30486/ijrowa.2021.1920124.1192, http://ijrowa.khuisf.ac.ir/article_686239_48-c5e2fdd1ea04ec0e297d8c47a10009.pdf.

Project Drawdown. (2023). *Biochar production*. Drawdown Solutions Library. https://drawdown.org/solutions/biochar-production#skip-link.

Qayyum, S., Khan, I., Meng, K., Zhao, Y., & Peng, C. (2020). A review on remediation technologies for heavy metals contaminated soil. *Central Asian Journal of Environmental Science and Technology Innovation*, *1*, 21–29. Available from https://doi.org/10.22034/CAJESTI.2020.01.03.

Qianqian, M., Haider, F. U., Farooq, M., Adeel, M., Shakoor, N., Jun, W., Jiaying, X., Wang Wang, X., Panjun, L., & Cai, L. (2022). Selenium treated foliage and biochar treated soil for improved lettuce (*Lactuca sativa* L.) growth in Cd-polluted soil. *Journal of Cleaner Production*, *335*. Available from https://doi.org/10.1016/j.jclepro.2021.130267.

Qiu, M., Liu, L., Ling, Q., Cai, Y., Yu, S., Wang, S., Fu, D., Hu, B., & Wang, X. (2022). Biochar for the removal of contaminants from soil and water: A review. *Biochar*, *4*(1). Available from https://doi.org/10.1007/s42773-022-00146-1, springer.com/journal/42773.

Quin, P., Joseph, S., Husson, O., Donne, S., Mitchell, D., Munroe, P., Phelan, D., Cowie, A., & Van Zwieten, L. (2015). Lowering N_2O emissions from soils using eucalypt biochar: The importance of redox reactions. *Scientific Reports*, *5*(1). Available from https://doi.org/10.1038/srep16773.

Ramezanzadeh, H., Zarehaghi, D., Baybordi, A., Bouket, A. C., Oszako, T., Alenezi, F. N., & Belbahri, L. (2023). The impacts of biochar-assisted factors on the hydrophysical characteristics of amended soils: A review. *Sustainability (Switzerland)*, *15*(11). Available from https://doi.org/10.3390/su15118700, http://www.mdpi.com/journal/sustainability/.

Ramtahal, G., Umaharan, P., Hanuman, A., Davis, C., & Ali, L. (2019). The effectiveness of soil amendments, biochar and lime, in mitigating cadmium bioaccumulation in *Theobroma cacao* L. *Science of the Total Environment*, *693*, 133563. Available from https://doi.org/10.1016/j.scitotenv.2019.07.369.

Randolph, P., Bansode, R. R., Hassan, O. A., Rehrah, D., Ravella, R., Reddy, M. R., Watts, D. W., Novak, J. M., & Ahmedna, M. (2017). Effect of biochars produced from solid organic municipal waste on soil quality parameters. *Journal of Environmental Management*, *192*, 271–280. Available from https://doi.org/10.1016/j.jenvman.2017.01.061.

Rockström, J., Gaffney, O., Rogelj, J., Meinshausen, M., Nakicenovic, N., & Schellnhuber, H. J. (2017). A roadmap for rapid decarbonization. *Science (New York, N.Y.)*, *355* (6331), 1269–1271. Available from https://doi.org/10.1126/science.aah3443, https://www.science.org/doi/10.1126/science.aah3443.

Safarian, S. (2023). Climate impact comparison of biomass combustion and pyrolysis with different applications for biochar based on LCA. *Energies*, *16*(14), 5541. Available from https://doi.org/10.3390/en16145541.

Sanei, H., Rudra, A., Przyswitt, Z. M. M., Kousted, S., Sindlev, M. B., Zheng, X., Nielsen, S. B., & Petersen, H. I. (2024). Assessing biochar's permanence: An inertinite benchmark. *International Journal of Coal Geology*, *281*, 104409. Available from https://doi.org/10.1016/j.coal.2023.104409.

Saunois, M., Stavert, A. R., Poulter, B., Bousquet, P., Canadell, J. G., Jackson, R. B., Raymond, P. A., Dlugokencky, E. J., Houweling, S., Patra, P. K., Ciais, P., Arora, V. K., Bastviken, D., Bergamaschi, P., Blake, D. R., Brailsford, G., Bruhwiler, L., Carlson, K. M., Carrol, M., ... Zhuang, Q. (2020). The global methane budget 2000–2017. *Earth System Science Data*, *12*(3), 1561–1623. Available from https://doi.org/10.5194/essd-12-1561-2020.

Schmidt, H. P., Kammann, C., Hagemann, N., Leifeld, J., Bucheli, T. D., Sánchez Monedero, M. A., & Cayuela, M. L. (2021). Biochar in agriculture – a systematic review of 26 global meta-analyses. *GCB Bioenergy*, *13*(11), 1708–1730. Available from https://doi.org/10.1111/gcbb.12889, http://onlinelibrary.wiley.com/journal/10.1111/(ISSN)1757-1707.

Schulz, H., Dunst, G., & Glaser, B. (2013). Positive effects of composted biochar on plant growth and soil fertility. *Agronomy for Sustainable Development*, *33*(4), 817–827. Available from https://doi.org/10.1007/s13593-013-0150-0.

Smith, S. M., Geden, O., Nemet, G. F., Gidden, M. J., Lamb, W. F., Powis, C., Bellamy, R., Callaghan, M. W., Cowie, A., Cox, E., Fuss, S., Gasser, T., Grassi, G., Greene, J., Lück, S., Mohan, A., Müller-Hansen, F., Peters, G. P., Pratama, Y., Repke, T., Riahi, K., Schenuit, F., Steinhauser, J., Strefler, J., Valenzuela, J. M., & Minx, J. C. (2023). The State of Carbon Dioxide Removal 1st edition. *The State of Carbon Dioxide Removal*. Available from https://doi.org/10.17605/OSF.IO/W3B4Z.

Song, Y., Li, Y., Cai, Y., Fu, S., Luo, Y., Wang, H., Liang, C., Lin, Z., Hu, S., Li, Y., & Chang, S. X. (2019). Biochar decreases soil N_2O emissions in Moso bamboo plantations through decreasing labile N concentrations, N-cycling enzyme activities and nitrification/denitrification rates. *Geoderma*, *348*, 135–145. Available from https://doi.org/10.1016/j.geoderma.2019.04.025.

Spokas, K. A., Novak, J. M., & Venterea, R. T. (2012). Biochar's role as an alternative N-fertilizer: Ammonia capture. *Plant and Soil*, *350*(1-2), 35–42. Available from https://doi.org/10.1007/s11104-011-0930-8.

Sun, H., Lu, H., Chu, L., Shao, H., & Shi, W. (2017). Biochar applied with appropriate rates can reduce N leaching, keep N retention and not increase NH3 volatilization in a coastal saline soil. *Science of the Total Environment*, *575*, 820–825. Available from https://doi.org/10.1016/j.scitotenv.2016.09.137.

Thomas, S. C., Frye, S., Gale, N., Garmon, M., Launchbury, R., Machado, N., Melamed, S., Murray, J., Petroff, A., & Winsborough, C. (2013). Biochar mitigates negative effects of salt additions on two herbaceous plant species. *Journal of Environmental Management*, *129*, 62–68. Available from https://doi.org/10.1016/j.jenvman.2013.05.057, https://www.sciencedirect.com/journal/journal-of-environmental-management.

Uchimiya, M., Lima, I. M., Klasson, K. T., & Wartelle, L. H. (2010). Contaminant immobilization and nutrient release by biochar soil amendment: Roles of natural organic matter. *Chemosphere*, *80*(8), 935–940. Available from https://doi.org/10.1016/j.chemosphere.2010.05.020.

Vanek, S. J., & Lehmann, J. (2015). Phosphorus availability to beans via interactions between mycorrhizas and biochar. *Plant and Soil*, *395*(1-2), 105–123. Available from https://doi.org/10.1007/s11104-014-2246-y, http://www.wkap.nl/journalhome.htm/0032-079X.

Verhoeven, E., Pereira, E., Decock, C., Suddick, E., Angst, T., & Six, J. (2017). Toward a better assessment of biochar–nitrous oxide mitigation potential at the field scale. *Journal of Environmental Quality*, *46*(2), 237–246. Available from https://doi.org/10.2134/jeq2016.10.0396.

Verhoeven, J. T. A., Arheimer, B., Yin, C., & Hefting, M. M. (2006). Regional and global concerns over wetlands and water quality. *Trends in Ecology and Evolution*, *21*(2), 96–103. Available from https://doi.org/10.1016/j.tree.2005.11.015.

Videgain-Marco, M., Marco-Montori, P., Martí-Dalmau, C., Jaizme-Vega, M. del C., Manyà-Cervelló, J. J., & García-Ramos, F. J. (2021). The effects of biochar on indigenous arbuscular mycorrhizae fungi from agroenvironments. *Plants*, *10*(5), 950. Available from https://doi.org/10.3390/plants10050950.

Vijay, V., Shreedhar, S., Adlak, K., Payyanad, S., Sreedharan, V., Gopi, G., Sophia Van Der Voort, T., Malarvizhi, P., Yi, S., Gebert, J., & Aravind, P. V. (2021). Review of large-scale biochar field-trials for soil amendment and the observed influences on crop yield variations. *Frontiers in Energy Research*, *9*. Available from https://doi.org/10.3389/fenrg.2021.710766.

Wang, J., Xiong, Z., & Kuzyakov, Y. (2016). Biochar stability in soil: Meta-analysis of decomposition and priming effects. *GCB Bioenergy*, *8*(3), 512–523. Available from https://doi.org/10.1111/gcbb.12266, http://onlinelibrary.wiley.com/journal/10.1111/(ISSN)1757-1707.

Wang, L., O'connor, D., Rinklebe, J., Ok, Y. S., Tsang, D. C. W., Shen, Z., & Hou, D. (2020). Biochar aging: Mechanisms, physicochemical changes, assessment, and implications for field applications. *Environmental Science and Technology*, *54*(23), 14797–14814. Available from https://doi.org/10.1021/acs.est.0c04033, http://pubs.acs.org/journal/esthag.

Wang, L., Deng, J., Yang, X., Hou, R., & Hou, D. (2023). Role of biochar toward carbon neutrality. *Carbon Research*, *2*(1). Available from https://doi.org/10.1007/s44246-023-00035-7.

Weldon, S., Rasse, D. P., Budai, A., Tomic, O., & Dörsch, P. (2019). The effect of a biochar temperature series on denitrification: Which biochar properties matter? *Soil Biology and Biochemistry*, *135*, 173–183. Available from https://doi.org/10.1016/j.soilbio.2019.04.018.

Werner, C., Schmidt, H.-P., Gerten, D., Lucht, W., & Kammann, C. (2018). Biogeochemical potential of biomass pyrolysis systems for limiting global warming to 1.5°C. *Environmental Research Letters*, *13*(4), 044036. Available from https://doi.org/10.1088/1748-9326/aabb0e.

Wong, J. T. F., Chow, K. L., Chen, X. W., Ng, C. W. W., & Wong, M. H. (2022). Effects of biochar on soil water retention curves of compacted clay during wetting and drying. *Biochar*, *4*(1). Available from https://doi.org/10.1007/s42773-021-00125-y, springer.com/journal/42773.

Woolf, D., Amonette, J. E., Street-Perrott, F. A., Lehmann, J., & Joseph, S. (2010). Sustainable biochar to mitigate global climate change. *Nature Communications*, *1*(5). Available from https://doi.org/10.1038/ncomms1053, http://www.nature.com/ncomms/index.html.

Xiang, Y., Deng, Q., Duan, H., & Guo, Y. (2017). Effects of biochar application on root traits: A meta-analysis. *GCB Bioenergy*, *9*(10), 1563–1572. Available from https://doi.org/10.1111/gcbb.12449.

Xiao, B., Jia, J., Wang, W., Zhang, B., Ming, H., Ma, S., Kang, Y., & Zhao, M. (2023). A review on magnetic biochar for the removal of heavy metals from contaminated soils: Preparation, application, and microbial response. *Journal of Hazardous Materials Advances*, *10*, 100254. Available from https://doi.org/10.1016/j.hazadv.2023.100254.

Xu, P., Gao, Y., Cui, Z., Wu, B., Yan, B., Wang, Y., Zaitongguli, K., Wen, M., Wang, H., Jing, N., Wang, Y., Chao, C., & Xue, W. (2023). Research progress on effects of biochar on soil environment and crop nutrient absorption and utilization. *Sustainability (Switzerland)*, *15*(6). Available from https://doi.org/10.3390/su15064861, http://www.mdpi.com/journal/sustainability/.

Yanai, Y., Toyota, K., & Okazaki, M. (2007). Effects of charcoal addition on N_2O emissions from soil resulting from rewetting air-dried soil in short-term laboratory experiments: Original article. *Soil Science and Plant Nutrition*, *53*(2), 181–188. Available from https://doi.org/10.1111/j.1747-0765.2007.00123.x.

Zhang, A., Cui, L., Pan, G., Li, L., Hussain, Q., Zhang, X., Zheng, J., & Crowley, D. (2010). Effect of biochar amendment on yield and methane and nitrous oxide emissions from a rice paddy from Tai Lake plain, China. *Agriculture, Ecosystems & Environment*, *139*(4), 469–475. Available from https://doi.org/10.1016/j.agee.2010.09.003.

Zhang, C., Zeng, G., Huang, D., Lai, C., Chen, M., Cheng, M., Tang, W., Tang, L., Dong, H., Huang, B., Tan, X., & Wang, R. (2019). Biochar for environmental management: Mitigating greenhouse gas emissions, contaminant treatment, and potential negative impacts. *Chemical Engineering Journal*, *373*, 902–922. Available from https://doi.org/10.1016/j.cej.2019.05.139.

Zhang, J., & Shen, J. L. (2022). Effects of biochar on soil microbial diversity and community structure in clay soil. *Annals of Microbiology*, *72*(1). Available from https://doi.org/10.1186/s13213-022-01689-1, https://rd.springer.com/journal/13213.

Zilberman, D., Laird, D., Rainey, C., Song, J., & Kahn, G. (2023). Biochar supply-chain and challenges to commercialization. *GCB Bioenergy*, *15*(1), 7–23. Available from https://doi.org/10.1111/gcbb.12952.

Zimmermann, M., Bird, M. I., Wurster, C., Saiz, G., Goodrick, I., Barta, J., Capek, P., Santruckova, H., & Smernik, R. (2012). Rapid degradation of pyrogenic carbon. *Global Change Biology*, *18*(11), 3306–3316. Available from https://doi.org/10.1111/j.1365-2486.2012.02796.x.

CHAPTER

10 Biochar as an emerging green and sustainable solution for amendment of degraded soil

Kushalindu Biswas[1,2], Manish Kumar[1,2], Arati Ray[1,2], Sibadutta Mohanty[1,2], Akshyurna Pattnaik[1,2] and Chandrama Dipa[1,2]

[1]*Academy of Scientific and Innovative Research (AcSIR), Ghaziabad, Uttar Pradesh, India*

[2]*CSIR—Institute of Minerals and Materials Technology, Bhubaneswar, Odisha, India*

10.1 Introduction

Soil is considered one of the most inevitable resources supporting life on this planet. Through its physical, chemical, and biological entities, it supports the entire human population with food resources, while also providing anchorage to plants. It is considered as natural and nonrenewable resource as soil formation through geological processes is highly complex which takes large time scale for its formation. In one hand, soil formation takes millions of years, but its degradation occuring at an accelerated rate posing a serious threat to its functioning. Soil degradation can be defined as reduction or impaired capacity of soil to restore its original form due to natural and anthropogenic factors (Osman, 2014). Degraded soil immensely affects the soil ecosystem by incorporation of contamination in soil system, which interrupt the entire structure, including soil air, water, organic matter, and microbial world. Other factors, such as deforestation, pollutant load, microbial activity, climate change, and land use patterns also significantly alter the soil characteristics. A number of conventional methods and measures have been undertaken to improve the degraded soil properties such as application of chemical pesticides, fertilizer, and liming treatment (Drobnik et al., 2018). But regular application and higher doses of these additives abruptly pollute the soil system and resulted into metal toxicity, soil erosion, loss of soil fertility, and high alkalinity/acidity of soil.

To dealing with these constraints, recently biochar is getting tremendous attention toward amendment of degraded soil. Biochar is a carbonaceous material produced through pyrolysis process in an oxygen-limited environment using biomass waste as feedstocks (Ding et al., 2016). Wide variety of biomass feedstocks ranging from different animal waste, plant wastes, sludge from municipal wastewater treatment plants and organic wastes generated from various sources can be used

Biochar Ecotechnology for Sustainable Agriculture and Environment. DOI: https://doi.org/10.1016/B978-0-443-29855-4.00034-5

for preparation of biochar with minimal processing (Kamali et al., 2022). The scope of tuning the reaction parameters during pyrolysis such as temperature, heating rate, and residence time enables the generation of biochar with desirable properties. Due to the presence of higher carbon content, nutrients, cation exchange capacity, and porous structure of biochar along with its high surface area made it suitable as green and sustainable material for multitude applications related to treatment of soil. Biochar is regarded as green and sustainable solution toward degraded soil treatment, as it belongs to organic material derived from biomass wastes, it may lack any other inorganic contaminants. Simultaneously pyrolysis of biomass generates huge amount of heat, which can be used to reenergize pyrolysis process as a reverse mechanism. The produced biochar is utilized for soil application, where it acts as nutrient source for plant growth. Biochar also acts as substitute of chemical fertilizers and pesticides due to its organic nature. At the same time when biochar used as per optimum dose and type, it does not cause any detrimental toxicity impact like lime and pesticides.

Different lab scale and field trial studies proofed that biochar has the potential for soil amelioration and successfully alters the chemical, physical, and biological characteristics of soil (Masud et al., 2023). It serves as a long-term amendment strategy and alternative for improvement of degraded soil. The first known instance of biochar application was to enhance the soil profile by native groups of Amazonian rain forest (Glaser & Birk, 2012). Since then, biochar has been extensively used to correct the physicochemical and biological properties of degraded soil (Allohverdi et al., 2021). As far the soil physical properties are concerned, biochar shows improved porosity, soil structure, water-holding capacity, aggregation, and lower the bulk density. In terms of chemical properties, biochar has been used to enhance the soil pH in case of acidic soils. It is also known to improve the cation exchange capacity and electrical conductivity of degraded soil (Fig. 10.1). Biochar has also been explored to enhance the soil microbial biomass thereby facilitating these organisms to perform their metabolism action. Similarly, biochar application significantly improves water retentio capacity and facilitates availability of essential nutrients to soil. It also has the capacity to enhances the ability of soil to capture excessive atmospheric C and can bring down the greenhouse gas emission from degraded soil significantly (Elkhlifi et al., 2023). Biochar, either in its pristine form or in a modified form it has the ability to enhance soil productivity without the aid of synthetic fertilizers or other agrochemicals.

Compared to the pristine biochar, composite form of biochar has shown promising impact in improving overall soil health by improving multiple nutrient availability or enrichment, facilitate microbial activity, heavy metal removal, and promoting C sequestration in degraded soil. Biochar composites are usually developed using different types of biomasses or pretreatment of raw biomasses before pyrolysis or posttreatment of pristine biochar. Compared to the pristine biochar, the composite biochar has shown greater potential in terms of restoration of degraded soil (Wang et al., 2022). The properties of composite biochar, including

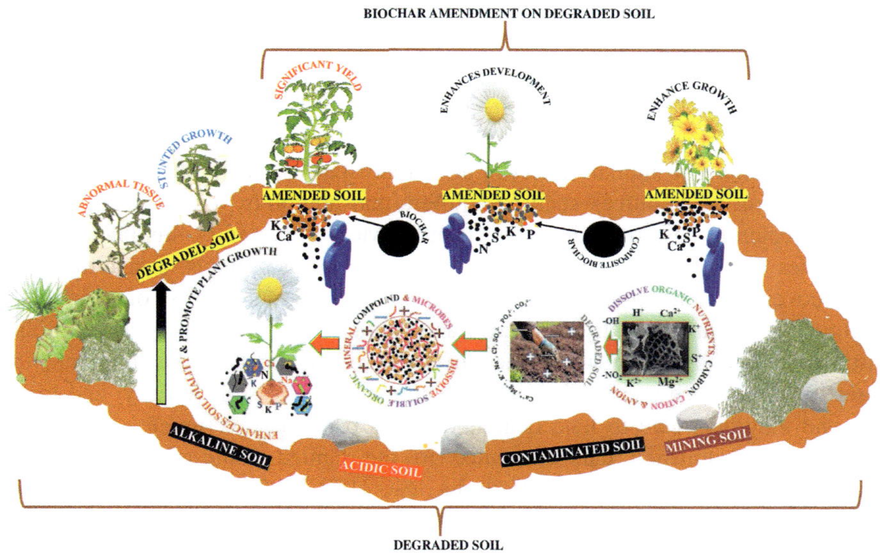

FIGURE 10.1

Schematic representation of the impact of biochar amendment on degraded soil conditions.

greater surface area, availability of multiple nutrients, improved water holding capacity, high surface charge, abundance of diverse functional groups, and greater pore size, give it an extra edge over pristine biochar, making it a fine solution for degraded soil health improvement and remediation of contaminants.

10.2 Major factors contributing toward soil degradation

10.2.1 Erosion

Soil erosion is a natural process but the intervention of human activities such as mining, uneven land use, deforestation, and quarrying severely increases the rate of erosion and which indirectly enhances the degradation of soil and it mostly affects all landforms (Balasubramanian, 2017). During erosion, mostly the top layer of the soil gets affected and partially it gets contaminated by the pollutants generated by human activities. Other secondary factors such as poor drainage patterns, transportation, and construction activities are also significantly contribute toward soil attrition processes. Erosion not only affects the upper capping layer of the soil but also contaminates the inner soil layer with time due to continuous erosion patterns over a long period. Erosion is one of the key challenges for agricultural practices, and proper management of this valuable resource is vital to sustaining soil health.

10.2.2 Desertification

The interruption of human intervening circumstances, including deforestation, overutilization of resources such as aquifers and aquitards, and climatic factors, which accelerate desertification significantly. Most importantly this phenomenon suppresses the ability of the soil to support life such as animals, agriculture, and green cover. Desertification may be a temporary or permanent loss of soil health. Eventually vegetation in drylands, arid and semiarid regions faces grasslands and shrublands disappearance with time due to desertification. In the long run, the desertification process can affect the moisture content of soil, fertility, and farming, and may also lead to land degradation. As per, the government's guidelines and recently published data presented to the United Nations Convention to Combat Desertification (UNCCD), India lost 31%, or 5.65 million hectares of grassland in a decade and the total area reduced to 12.3 million hectares from 18 million hectares in between 2005 and 2015 (Pandey, 2019).

10.2.3 Compaction

Compaction phenomenon of soil occurs due to overutilization of land where soil particles are got compressed due to internal injury to the soil system and lack of internal space and water (Hamza & Anderson, 2005). Ongoing extensive construction or agricultural activity, where heavy machinery tools, frequent tillage, and excessive use of pesticides immensely damage soil construction. Compaction mainly causes choking of the soil passages where the movement of water or infiltration rate is restricted to the soil system. Because of insufficient moister, nutrient, and pore space over time, compaction processes will accelerate, affecting the subsoil and potentially impacting porosity as well.

10.2.4 Deforestation

Loss of natural forest in the tropical region alters the soil physicochemical characteristics and adversely impacts soil microclimate, hydrological regime, and enhances soil erodibility (Sahani & Behera, 2001). Ongoing expansion of population along with industrialization together demands extensive use of land resources, which resulted in the clearing of forest and vegetation cover. With respect to time, modernization also affected huge agricultural land and arable land to nonviable land resources. Deforestation causes serious environmental damage such as loss of biodiversity and natural habitat loss. As a result of deforestation, the fertile layer or the top layer of the soil is directly get exposed to outside habitat, making it prone to pollutants, along with depletion of nutrient, moisture, and organic matter content.

10.2.5 Sealing

Soil sealing are the persistent covering over land and soil by impermeable artificial material like concrete and plastering objects (EU., 2012). Sealing is one of the critical land degradation condition where it disrupts the structure and permeability of land cover. It is a type of soil degradation where the surface is contaminated or capped due to construction, debris, pollutants, and impermeable materials, which combinedly restrict the percolation of water. In most of the urban and semiurban areas, land is covered with buildings, roads, and industries which leads up to 50%–60% of land sealing. This impermeable covering caused destruction of the ground cover and also simultaneously affected the subsoil. Growing urbanization and economic activities significantly affect the land covering by replacing the trees and shrubs with concrete, roads, buildings, and developed land.

10.2.6 Soil alkalinization

Soil alkalinization is a phenomenon where the deposition of weakly acidic bases such as bicarbonates, carbonates, and lime, leads to rise in soil pH, which indirectly affects soil health (Zewd & Siban, 2021). Extreme alkaline conditions can abruptly affect the nutrient mobility in soil and may likely to reduce the nutrient pathway to plant nutrient dynamics. At higher alkaline conditions soil nutrients bind to elemental forms, making them unavailable to soil and posing serious toxicity risks. Earlier studies reported that alkalinity significantly impacts soil physicochemical properties, for example, pH, cation exchange capacity, and nutrient availability.

10.2.7 Soil temperature

Soil temperature regulates the physiochemical and biological processes which can interrupt seed germination, seedling, emergence, growth, microbial function, and extractable nutrients (onwuka & Mang, 2018). To support the growth of any plant or microbes in soil environment, it needs optimal internal temperature (20°C–30°C or 68°F–86°F) and facilitate mutualism phenomenon. Soil temperature is a causative role behind the alteration of organic matter decomposition and mineralization of organic fractions in soil (Probert, 2022). Whereas with time developmental activities such as mining, deforestation, industrialization, and pollution, which actively disrupt the temperature profile and affect the optimal condition of soil. Various reports and articles indicate that soil temperature regulates biochemical processes, such as aeration, and affects soil–water interactions, as well as gaseous and fluid exchange metabolism.

10.2.8 Soil acidification

Soil acidification is an inherent process, which has been accelerated due to the rise in sulfur, nitrogen, and particulate matter pollutant deposition by the rising of human activities (Meng et al., 2019). In most of the studies, it was seen that soil acidification critically reduces soil pH. Factors such as intense agricultural practices, application of chemical fertilizer, pesticides, and herbicides, which contributed immensely toward acidification of soil. It is mainly caused by the accumulation of soluble inorganic and organic fractions of contaminants, which are released as free H^+ ions in the soil system. Acidification not only affects the soil, but it also alters the biochemical processes such as nutrient uptake, cation exchange capacity, conductivity, biological process, microbial interactions and enzymatic activity.

10.2.9 Salinization

Salinization impact negatively toward soil degradation to plant development. It directly affects the agricultural productivity and which ultimately affect the socio-economic scenario. Proper management of salinity at earlier stages may helps to restrict the degradation and may restore it. Salinization is a detrimental effect of degraded soil where the accumulation of salt occurs due to factors like excessive irrigation, synthetic pesticide/fertlizer application, soil drainage, and evaporation. The incoming irrigated water and pesticides contain fractions of soluble salt, while these water evaporates it leaves behind salt that causes salt stress across soil surface. It was expected that if the salinization problem could not be addressed this impact may even rise to 50% of the global irrigated area (Singh, 2021). Salinization process is also dependent on the parent soil and topography that determine the nature and characteristics of soil.

10.2.10 Nutrient depletion

Nutrient depletion is one of the important indicators of degraded soil which adversely influences soil quality and crop yield. Factors such as inappropriate use of fertilizer, mismanaged farming, overirrigation, continuous cropping, and excessive land use all significantly affect the nutrient concentration of soil and eventually lead to soil erosion as well. Continuously nutrient loosing from soil can cause toxicity stress to soil and may indirectly lead to agricultural constraints (Tan et al., 2005). According to a report from The Economics of Land Degradation (ELD Initiative 2015), approximately 20% of our land has undergone degradation, and on average, 52% of agricultural soil has been moderately affected by ongoing nutrient depletion and land degradation (Noel, 2015). With respect to several research studies, authors concluded about nutrient concentration was being declined by 42%, as such S content reduced by 33% and P reduced by 27% roughly within the past 45–60 years interval (Kopittke et al., 2017).

10.2.11 Loss of soil biodiversity

It is the deterioration of life forms in soil such as microbes, bacteria, fungi, and a range of species, which can significantly disrupt the functionality of soil. It provides an environment for the large number of diverse microbes assembled in complex and varied communities. Among biological entities, the presence of invasive species in soil systems can pose a serious threat to soil biodiversity. When invasive species were contaminated it modifies the local habitat by importing pathogens and competing with native species, thereby modifying the microclimate of the soil. Factors such as climate change, pollution, forest fire, intense solar insolation, and environmental degradation all these instances immensely alter the soil health degradation process.

10.2.12 Declining of soil organic matter

Organic matter is regarded as one of the important components of soil health, as it helps to maintain soil fertility. However, the reduction of organic matter suppresses the infiltration rate and increases runoff. The organic matter content of soil decreases due to the presence of decomposing organisms and contaminants that increase the rate of decaying and result into degradation of soil and emits harmful gases. Decreasing of organic matter content limits the ability of soil to supply nutrients to support plant growth. Factors such as biotic, climatic, vegetation, and temperature were critically responsible for the presence/generation of organic matter.

10.2.13 Soil borne pathogens

Affected soils contain a substantial number of phytopathogenic microbes that are capable of devastating soil crops. This includes pathogens such as fungi, bacteria, nematodes, and viruses, which can disrupt overall soil productivity and plant growth metabolism (Dixon & Tilston, 2010). Pathogens can also abruptly consume or drain micro and macronutrients from the soil. Pathogens are carriers that infect other components of soil and target physicochemical factors through airborne, erosion, runoff, and pollution. They most often affect the soil's biotic component by lowering organic matter, causing plant root wilting, stunted growth, and root decay, compared to the abiotic status.

10.2.14 Greenhouse gas emission

Greenhouse gases are types of atmospheric gases that trap the incoming solar insolation, warms the atmosphere and at same time altering the land surface temperature. These rising of soil internal temperature, which may directly affects the rates of biological, physical, and chemical characteristics of soil. Studies show that limited fluctuations in temperature range may influence physicochemical parameters (Howe & Smith, 2021). As soil is contributing one of the most important sources of carbon sink but simultaneously here vegetation cover plays a critical role in balancing the overall temperatures. Most importantly, these gases which genrated due to rise in atmospheric temperature can immensely disrupt precipitation patterns, weather conditions and altering the environmental phenomenon. All this constructive event can eventually lead to cause floods, landslide, and accelerate soil degradation.

10.2.15 Imbalance nutrient cycling

Nutrient cycling is a cyclic process where it flows across the physical environment, living organisms, and back to the natural environment through the carbon, nitrogen, and phosphorus cycles. It may impact on farming, food security and devoid of essential micronutrients such as iron, copper, and zinc that leads to decreasing yield parameter by causing plant disease (Lal, 2009). This cyclic process is mostly affected by human activities such as deforestation, mining, industrialization, and application of chemical pesticides and fertilizers. All these activities critically affect and trigger the continuing degradation process. Furthermore, natural phenomenon such as high rainfall, flood, and climatic conditions is partially exploiting the nutrient cycle up to some extent. Several studies and data represent that burning of fossil fuels, along with vehicular and industrial emissions, also significantly deteriorates the nutrient cycle. Most of these emission event alters the CO_2 misbalancing in the atmosphere, which indirectly affects nutrient uptake in soil and plant systems.

10.3 Potential mitigation strategies of degraded soil through biochar amendment

10.3.1 Impact on physical properties

10.3.1.1 Texture/particle size

Biochar amendment of extreme textural (sand, slit, and clay) soil conditions shows effective improvement in associated parameters such as water-holding capacity, nutrient retention, and structural development. They can refill the disrupted pore spaces of the soil particles, which can restrict or reduce water flow and additionally hold higher moister content. Earlier studies reported that biochar derived from wood waste can improve water infiltration and reduce aggregate formation (Reddy et al., 2015), and at same treatment it also decreases the plasticity of clayey soil (Zong et al., 2016). The mechanism of improving the particle size of soil partially depends on the initial surface area of biochar related to pyrolysis temperature, and feedstock type. Higher surface area and microparticles of biochar can alter the volume of soil pores and improve the particle size of degraded soil.

10.3.1.2 Bulk density and particle density

Biochar application to soil reduces both bulk density and particle density, thereby increasing porosity and water-holding capacity within soil matrix (Blanco-Canqui, 2017). Porosity directly influences root penetration depth, further access to available water. As density of the biochar particles is lower compared to that of soil, where biochar can easily incorporate into the soil system through effectively pushing soil particles apart and reduces overall soil bulk/particle density. It also interacts with soil particles to restore soil structure while simultaneously restrict particles compaction.

10.3.1.3 Macro and microporosity

Micro and macropores are the critical component of soil pore structural function. Macropores consist of larger pore sizes, while micropores are smaller. Biochar particles adhere to the unfilled pore spaces in the soil, by improving the amended soil porosity characteristics (Yu & Lu, 2020). Previous studies have reported that biochar production at higher pyrolysis temperatures and extended residence times leads to increased surface area and porosity (Tomczyk et al., 2020). The porous matrix of biochar provides microspace to its inner and outer matrix of biochar, which helps in absorbing moister, pollutants, and heavy metals in biochar amended soil pores. Biochar also contains substantial amounts of nutrients and mineral compounds that leach into biochar-amended soil, thereby improving soil pores.

10.3.1.4 Water-holding capacity

The ability of soil to support crop production and plant growth is highly dependent on the soil's water holding capacity. Biochar possess a high specific surface area, and its porous nature, which facilitates the retention of more water per unit volume of soil. The pores present in the biochar form an interconnected channel allowing water to pass through them. The inherent porous nature of biochar further enhances microbial activity providing these organisms with suitable microhabitats. These microorganisms further boost soil health and moisture retention. Thus the role of biochar in improving soil's water-holding capacity can prove essential in dry regions or regions with scarce water availability (Kabir et al., 2023).

10.3.2 Impact on chemical properties

10.3.2.1 pH

Biochar amended soil significantly boosts soil pH and reduces soil acidification. In biochar applied soil system, free bases of Ca, Mg, and K, along with dissolved organic compounds which are readily released into the soil, resulting in a net increase in soil pH and also enhancement in electrical conductivity (Chintala et al., 2014). At an optimum level of pH in biochar-amended soil, conditions can ameliorate and support to plant life; however, under extreme alkaline/acidic conditions pants may not survive. Biochar-induced soil pH found to be beneficial in terms of improving microbial function and growth in degraded soil to restore its fertility and productivity.

10.3.2.2 Cation exchange capacity

Biochar based soil amendment can improve the cation exchange capacity of soil particles, as biochar has negatively charged with higher surface area that help store and absorb nutrients in soil pores due to improved binding sites. After the introduction of biochar into the soil system, its functional groups, such as carboxylic and phenolic groups, promotes electrostatic attraction on biochar surface, which enhances oxidation and increases cation exchange. Previous experimental studies (Chathurika et al., 2016; Dey et al., 2023) reported about the enhancement of cation exchange capacity in biochar-treated soil. Similarly few studies found that biochar produces from cellulosic based biomass waste can enhance cation exchange capacity in soil.

10.3.2.3 Carbon content

However, various review and studies depict biochar is a carbonaceous material that can enhance soil carbon sequestration in amended soil. In case of degraded soil, the initial carbon content was found to be considerably minimal due to usual events of decomposition, agricultural practices, harvesting and erosion. However, further application of biochar could significantly boost the carbon fraction (Chagas et al., 2022). Biochar is a carbonaceous and porous material, with carbon content varying from 20% to 65%, depending on the nature of biomass and pyrolysis parameter. The presence of biochar restrict the organic carbon mineralization and partially accumulate nitrogen. Studies have found that biochar produced at lower pyrolysis temperatures contains more labile carbon and increases the organic carbon fraction in soil compared to biochar produce at higher temperatures (Tomczyk et al., 2020).

10.3.2.4 Nitrogen content

Biochar application shows positive interactions with nitrogen asimilation along with other nutrients in soil. Most importantly biochar produces through lower pyrolysis temperature/intermediate temperature can significantly contribute to nitrogen source in degraded soil. Number of demonstrative studies are undertaken into investigate the effects of biochar on soil available inorganic nitrogen such as ammonium and nitrate (Nguyen et al., 2017). With respect to time, it was observed initially higher biochar application rate may suppress the nitrogen metabolism and simultaneously leads to nitrogen toxicity by nitrate reductase. Whereas combined application of biochar along with nitrogen may shows synergetic effect on growth and sorption of excess nitrogen in degraded soil (Khan et al., 2022). Similarly biochar derive from wood waste can limit the reduction of nitrogen in degraded soil and absorb nitrate from soil (Nguyen et al., 2017).

10.3.2.5 Phosphorus content

Phosphorus is one of the important micronutrients, which essential for balancing the soil biochemical process, where ongoing process of erosion, runoff, leaching and plant uptake severely impact the available phosphorus in soil. Biochar application can significantly address the phosphorus availability in degraded soil due to the mechanism of adsorption, immobilization, and mineralization process. The available form of phosphorus content in biochar outer/inner matrix can be directly avail phosphorus in degraded soil. However, biochar produced from higher temperature pyrolysis ($>$700°C) can limit the release of available phosphorus as experimental studies have found that biochar produced at lower temperatures shows a higher contribution of available phosphorus (Zhang et al., 2016). Optimum phosphorus availability to contaminated/degraded soil through biochar treatment depends on the specific properties/types of biochar and nature of the degraded soil. Experimental trial observe biochar ash fraction can also enhance the phosphorus content in degrades soil and additionally, extreme alkaline conditions may affect phosphorus availability.

10.3.2.6 Absorption of heavy metals

Contamination of heavy metals such as Cr, Cd, As, and Zn from anthropogenic source, which are resistant to degrade and can persist in soil over a long time. Biochar found to be potential solution to immobilize those heavy metal in degraded soil and reducing the risk of pollution. The presence of immobilizing and reducing agent in biochar helps decrease metal contamination, mobility, and bioavailability due to their characteristics such as higher surface area, carbon, cation exchange capacity, and pH. The absorption mechanism of biochar depends on the biochar type and specific soil, studies find biochar produces at higher temperature are generally effective in adsorb contaminants in affected soil (Ahmad et al., 2014). The experimental evidences found that biochar with a higher content of oxygen containing functional groups can effectively stabilize heavy metals in acid-degraded soil with low organic carbon. Similarly, biochar with a more alkaline pH contributes to metal stabilization.

10.3.2.7 Absorption of pesticides

The excessive use of pesticides in soil reduces the potential of the soil to restore its productivity. Some of the applied conventional methods such as chemical remediation and destruction are less suitable option while compared to biochar amendment. Experimental trial found biochar treatment played significant role in remediation of pesticide contaminated soil and diffuse herbicide pollution (Ouyang et al., 2016). Biochar is also observed as strong adsorbent for hydrophobic organic compounds like pesticides, fertilizer and simultaneously it reduces leaching. Several comparative studies are being carried out to evaluate the efficiency of biochar in relation to the fate of pesticides. Ouyang et al. (2016) found rice straw and corn straw biochar are significantly remove herbicides in clay loam soil. The efficiency of biochar in adsorbing or decontaminating pesticides is often significantly greater when biochar is produced at higher temperatures will enhances its surface area and matrix.

10.3.3 Impact on biological properties

10.3.3.1 Impact on microbial community and abundance

Biochar treated soil acts as a substrate where it provides a space for growing of microbes, at same time it facilitates nutrients for their growth and abundance. Biochar mostly composed of carbon molecules and where soil microbes commonly utilizes those labile form of carbon and nutrient from biochar (Yin et al., 2021). Biochar application was found to be improved the diversity of the microbial community, partially reduces microbes uniformity and these diversity reflected as may change the soil macroenvironment. The experimental study shows biochar derived from green waste feedstock at 700°C pyrolysis results into positive abundance of bacteria (*Pseudomonas* and *Pedobactor*) in degraded soil (sandy loam soil). Similarly, biochar produces from different feedstock and production method can alter properties of biochar which can impact soil microbes diversity and richness (Kolton et al., 2017). However, biochar provides a potential supplement for improving microbiomes in affected soil and simultaneously enhancing soil quality.

10.3.3.2 Impact on pathogens

Biochar application suppresses the growth of pathogens in soil system by encountering the growth of beneficial bacteria and fungi to improve degraded soil properties. Experimental study found that biochar produced from eucalyptus and greenhouse-based waste, applied at rates of 0.5%, 1%, and 3% (v/v) showed a reduction in pathogen levels in tomato plants (test plant) compared to untreated soil (Lehmann et al., 2015). Degraded soil treated with rice straw biochar was found to be provide better resistance against tobacco bacterial wilt, while simultaneously enhancing bacterial richness and abundance compared to unamended soil (Zhang et al., 2017). In pathogen affected soil, microbes can become resistant to soil pathogens; however, biochar application can significantly improve the density of soil bacteria and suppress pathogenic activity.

10.3.3.3 Effect on flora and fauna

It is a critical component of soil structure, acting as the backbone for improving soil health/fertility but due to natural and anthropogenic circumstances eventually they disappears with time. Where in case of biochar application on these contaminated soil or degraded soil could provide a supplement to support and restore flora and fauna in the affected soil system. Biochar also acts as a source of nutrients/food for organisms/pants present in the soil or indirectly in the form of soil organic fraction and it provide microhabitat environment within matrix of soils (Lehmann et al., 2011). Due to its porous and carbonaceous nature, biochar provides a suitable site for micro species to thrive and support microflora/fauna and also defending against predatory species. Additionally biochar treated soil could provide essential available form of nutrients such as nitrogen, potassium, phosphorus, and carbon to improve flora and fauna of the degraded soil (Fig. 10.2).

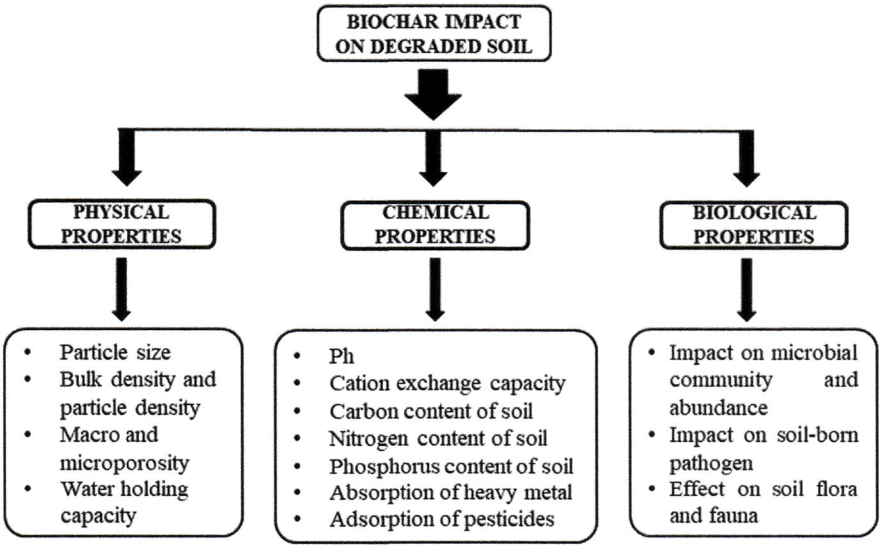

FIGURE 10.2 Impact of biochar amendment on degraded soil properties **(Wang et al., 2022).**

10.4 Recent trends of composite biochar production toward restoration of degraded soil

Biochar gains multifarious scope for application as high end products, soil application, and carbon mitigation by its utility for degraded soils (Table 10.1). It has some limitations for the specific constraints such as deficiency of required nutrients, multiple metals adsorption, carbon sequestration, soil plant productivity, and others supporting physicochemical parameters, which can be addressed in improved manner by composite biochar. Composite biochar produced from nutrient-rich multiple feedstocks and with scope to address multinutriens deficiency of degraded soils, an advantage over pristine biochar. Composite biochar enhances physical, chemical, and biodynamics of soil and remediates through immobilization of contaminants along with improvising soil carbon and restoring contaminated soil (Singh et al., 2023). Biochar composite can be modified by optimizing the process parameters such as pyrolysis conditions, change of feedstocks, and composite ratio. These biochar when applied, loaded with immobilization and degradation factor, assure its long term remediation effects on soil. Various studies are undertaken to explore the production of composite biochar from poultry litter, banana peduncle, phosphogypsum and its wastewater, effluent sludge etc. (Karim et al., 2019; Ray et al., 2023; Vimal et al., 2022), for preparation of phosphorous, sulfur, potassium multinutrient enriched biochar. In the

Table 10.1 Biochar application studies for management of degraded soil.

Feedstocks	Production technique and conditions	Soil application	Salient result	References
Rice straw	N_2 inert pyrolysis (300°C–700°C)	Soil amelioration	Biochar at 400°C used as better soil amendment and fertilizer	Wu et al. (2012)
Poultry litter	Continuous pyrolysis unit (450°C and 550°C)	Treating alfisol soil	Increases dry matter yield of radish and soil quality improvement	Chan et al. (2008)
Wheat straw	Continuous vertical kiln pyrolysis (450°C)	Decreasing bioavailability and sorption of Cd in affected soils	Cd removal up to 90% in soil	Cui et al. (2019)
Water hyacinth	Gasification	Biochar efficacy on soil amendment	Improve nutrient recovery and carbon mitigation	He et al. (2022)
Corn cob	Regulated pyrolysis (500°C–550°C)	Treatment of tropical sandy loam soils	Increases humic acid, fulvic acid, and TOC concentration of soil	Amoakwah et al. (2020)
Cow manure	Regulated pyrolysis (400°C)	Treatment of contaminated soil	Reduces the heavy metals bioavailability	Eissa (2019)
Sugarcane bagasse	Earthen kiln pyrolysis (600°C)	Carbon sequestration and growth of plant in degraded floodplain soil	Sustain soil organic matter, soil fertility, and enhance crop yield	Rahman et al. (2022)
Canola straw	Regulated pyrolysis (600°C)	Treating aluminum phytotoxicity in acid soil	Mitigation of Al (III) phytotoxicity up to 60%	Dong et al. (2022)
Switchgrass	Bench furnace pyrolysis 350°C and 800°C	Enhancement of microbial nitrogen cycling in acidic subsoil	10% treated soil enhances nitrogen fixation and moderate denitrification	Ducey et al. (2013)
Woodchip	Control pyrolysis (290°C and 700°C)	Enhancement of soil properties	5% 700 °C biochar reduces (47%–58%) CO_2 release and improves soil properties	Lai et al. (2013)

(Continued)

Table 10.1 Biochar application studies for management of degraded soil. *Continued*

Feedstocks	Production technique and conditions	Soil application	Salient result	References
Biosolid	Muffle furnace (450°C)	Biosolid biochar for amelioration of soil	Recovery of NH_3-N up to 5.3 mg/g in biochar amended soil	Carey et al. (2015)
Coconut fiber	Muffle furnace (500°C)	Immobilization Pb in soil	The Pb immobilization fraction increased by 29.5–33.5%	Li et al. (2019)
Paper sludge	Electric furnace (300°C, 400°C, and 500°C)	Valorization of heavy metal contaminated soil	Addition of biochar 400°C and 500°C reduced the amount of mobile Zn	Paz-Ferreiro et al. (2017)
Olive waste	Gasifier pyrolysis (400°C–450°C)	In situ remediation of metal contaminated soil	90 days of equilibration study found immobilization of Pb, Cd, and Zn in soil with ageing	Hmid et al. (2015)
Pine sawdust	Control pyrolysis (300°C and 550°C)	Treatment of As and Pb contaminated soil	Biochar 550°C reduced phytoavailable As and Pb	Beiyuan et al. (2017)

TOC, *total organic carbon.*

backdrop of biomass generation report by UNEP (2009), numerous disposal and environmental concern are envisaged. Therefore, their recycling and reuse to high value sustainable solution as composite biochar would provide substitute for fertilizer and conditioners for soil plant system. Several case studies reviewed the methods of optimization, preparation and applications of biochar composites for degraded soils such as heavy metals contaminated, nutrients deficit and saline-alkaline soil. So, preparation of innovative composites of biochar and potential application dynamics in degraded soil are highlighted (Table 10.2).

10.5 Challenges and future prospectives of biochar amendment for degraded soil

Biochar is a carbonaceous material once applied to soil system, cannot be isolated or separated again. Due to diverse characteristics, critical analysis is required

Table 10.2 Biochar composite application studies for degraded soil amendment.

Types of feedstocks	Production technique and conditions	Applications	Salient results	References
MgO-peanut	Muffle furnace pyrolysis in N_2 inert condition at 600°C for 4 h	Phosphate adsorption in saline-alkaline soils	MgO-biochar adsorbed phosphate 20% better than biochar	Wu et al. (2019)
Bottom ash-animal waste	Furnace-based pyrolysis at 500°C for 4 h	Cd immobilization in paddy soil	Cd concentration in rice decreases to 0.11 and 0.12 mg/kg in early and late production	Lei et al. (2020)
Iron-modified composite biochar	Regulated pyrolysis at 300°C for 12 h	Removal of Cd and As in paddy soil	Reduces the availability of As and Cd in soil	Pan et al. (2019)
Fe-palm-rice straw	Muffle pyrolysis at 600°C for 6 h	Decontamination of Cd and As in agricultural soil	Concentration of available Cd and As decreased by 74% and 14% and boost crop yield	Tang et al. (2020)
K-Fe-rice straw	Pyrolysis at 400°C and 600°C for 2 h	For sorption of NH_4^+, PO_4^{3-}, and NO_3	Sorption of NH_4^+, PO_4^{3-}, and NO_3 and availability of K, Ca, N, and P	Chandra et al. (2020)
Fe-Mn-corn straw	Slow pyrolysis at 600°C for 2 h	Increasing soil redox potential and reduced available As	Alter soil physicochemical character and reduced bioavailable As	Lin et al. (2019)
Banana peduncle-poultry litter-phosphogypsum composite	Tube furnace pyrolysis at 600°C for 1 h	Production of composite biochar fertilizer for soil application	Second-order kinetic model depicts slow release of P	Vimal et al. (2022)
Banana peduncle and phosphogypsum	Plasma pyrolysis	Generation of K–S rich nutrient for soil for degraded soil	The presence of plasmagen gases showed a significant effect on the biochar characteristics which may helps its various use for degraded soils	Karim, Kumar, Mohapatra, Singh, et al. (2019)
Phosphogypsum pond waste and banana peduncle	Regulated pyrolysis at 400°C, 600°C, and 800°C for 1 h	Recovery of essential K–P–S rich nutrient for mine and degraded soil	Potential application of biochar exhibited water-soluble nutrients and may be suitable for degraded soil	Ray et al. (2023)

before its application in various soil type. Recently, biochar founds as an emerging sustainable material for soil amelioration, and at the same time, it also acts as a dynamic material for adsorption and reduction of contaminants present in the degraded soil. Many reviews and articles have already discussed about the existing drawbacks for large scale application of biochar, since lab scale and mid scale experiments investigated the applicability of biochar treated soil for crop growth and decontamination. Contrasting literature reported about efficacy of mixed wood biochar on soil application: small scale laboratory based study showed reduction of CO_2 and N_2O by suppression of gasses from 38% to 56%, whereas only CO_2 was reduced significantly for long term field study (Fidel et al., 2019). Earlier study illustrated biochar as progressive material to decrease N_2O emission which is primarily dependent on properties of soil. Mostly, related trial studies are application oriented work carried as lab scale and bench scale; but real time field trial studies are still scarce in literature. Upcoming studies are recommended to inculcate the prior knowledge, identified gaps of lab to pilot scales and real time field applications, to achieve significant outcomes. Such practical applicable studies are essential for correlating the multiple dynamics of biochar production environment, feedstock types, properties of biochar and their impact on soil characteristics of degraded soil, which will help improving the degree of reliability of biochar amendments for problematic and degraded soil applications. Grip on thermochemical process, manifesting positive significance of composite biochar production for recycle and reuse aspects of biomass and other wastes toward management of degraded soil. Even with large benefits of biochar application, still some concerns unresolved due to different kind and complexities of degraded soil, about its response and nature of implementation. However, the utilization of biochar for improving degraded soil and decontamination of land resources, ultimately benefits the soil environment and ecosystem restoration.

10.6 Conclusions

Biochar acts as a sustainable substitute and additive to chemical fertilizers toward degraded soil amendment. It provides multiple benefits with respect to different constraints of degraded and adversely affected soil with deviations in their physical, chemical, and biological parameters. Many experimental studies revealed that application of biochar to degraded soil results in a synergistic effect on soil health improvement, leading to its amelioration, reduced pollutant loads and higher crop yield. Depending on soil conditions, nature of nutrient deficit, and other associated factors of degraded soil, selection of feedstocks and methods of biochar production are tailormade to mitigate reason of soil degradation. However, specific biochar may not show positive response on all type of degraded soil application because of its specific nature. Currently, researchers are exploring the desired properties of biochar through modifying production parameters (types of feedstocks used, temperature, inert condition, and residence time etc.) to treat specific type of soil degradation. Furthermore, long term demonstrative studies needs to be conducted for simultaneously underlining the positive and potential aspects with biochar soil interaction for degraded soil improvement.

Acknowledgments

Kushalindu Biswas (NTA Ref. No. 190510222943) cordially thanks to University Grants Commission (UGC), India for awarding Junior Research Fellowship (JRF) and Senior Research Fellowship (SRF) for pursuing PhD research work under AcSIR. Manish Kumar acknowledge funding support from DST-India and Paradeep Phosphates Limited, Paradeep, Odisha and CSIR-Institute of Minerals and Material Technology, Bhubaneswar [(DST/ TDT/ WM/2019/51 (G))]. The co-authors sincerely thanks to CSIR-Institute of Minerals and Materials Technology and Environmental and Sustainability Department for providing all the support and resources for scripting the book chapter. Also, the authors obliged for continuous support from editors and reviwers, which helped in improvement of content and quality of the chapter.

Author contributions

Conceptualization: Manish Kumar, Kushalindu Biswas; Initial scripting: Kushalindu Biswas, Sibadutta Mohanty, Arati Ray, Akshyurna Pattnaik, Chandrama Dipa and Manish Kumar; Correction: Manish Kumar, Kushalindu Biswas, Sibadutta Mohanty, Akshyurna Pattnaik, Arati Ray and Chandrama Dipa; Editing: Manish Kumar, Kushalindu Biswas

Conflict of interest

The authors report there are no conflict of interest.

Statement on the use of AI tools for writing

For the manuscript writeup and content, AI tools were not used.

References

Ahmad, M., Rajapaksha, A. U., Lim, J. E., Zhang, M., Bolan, N., Mohan, D., Vithanage, M., Lee, S. S., & Ok, Y. S. (2014). Biochar as a sorbent for contaminant management in soil and water: A review. *Chemosphere*, *99*, 19–33. Available from https://doi.org/10.1016/j.chemosphere.2013.10.071, http://www.elsevier.com/locate/chemosphere.

Allohverdi, T., Mohanty, A. K., Roy, P., & Misra, M. (2021). A review on current status of biochar uses in agriculture. *Molecules (Basel, Switzerland)*, *26*(18), 5584. Available from https://doi.org/10.3390/molecules26185584.

Amoakwah, E., Arthur, E., Frimpong, K. A., Parikh, S. J., & Islam, R. (2020). Soil organic carbon storage and quality are impacted by corn cob biochar application on a tropical sandy loam. *Journal of Soils and Sediments*, *20*(4), 1960–1969. Available from https://doi.org/10.1007/s11368-019-02547-5, http://www.springerlink.com/content/1439-0108.

Balasubramanian, A. (2017). *Soil erosion—causes and effects. Centre for Advanced Studies in Earth Science*. University of Mysore, Mysore.

Beiyuan, J., Awad, Y. M., Beckers, F., Tsang, D. C. W., Ok, Y. S., & Rinklebe, J. (2017). Mobility and phytoavailability of As and Pb in a contaminated soil using pine sawdust biochar under systematic change of redox conditions. *Chemosphere*, *178*, 110–118. Available from https://doi.org/10.1016/j.chemosphere.2017.03.022, http://www.elsevier.com/locate/chemosphere.

Blanco-Canqui, H. (2017). Biochar and soil physical properties. *Soil Science Society of America Journal*, *81*(4), 687–711. Available from https://doi.org/10.2136/sssaj2017.01.0017.

Carey, D. E., McNamara, P. J., & Zitomer, D. H. (2015). Biochar from pyrolysis of biosolids for nutrient adsorption and turfgrass cultivation. *Water Environment Research*, *87* (12), 2098–2106. Available from https://doi.org/10.2175/106143015x14362865227391.

Chagas, J. K. M., Figueiredo, C. Cd, & Ramos, M. L. G. (2022). Biochar increases soil carbon pools: Evidence from a global meta-analysis. *Journal of Environmental Management*, *305*. Available from https://doi.org/10.1016/j.jenvman.2021.114403, https://www.sciencedirect.com/science/journal/03014797.

Chan, K. Y., Van Zwieten, L., Meszaros, I., Downie, A., & Joseph, S. (2008). Using poultry litter biochars as soil amendments. *Soil Research*, *46*(5), 437. Available from https://doi.org/10.1071/sr08036.

Chandra, S., Medha, I., & Bhattacharya, J. (2020). Potassium-iron rice straw biochar composite for sorption of nitrate, phosphate, and ammonium ions in soil for timely and controlled release. *Science of the Total Environment*, *712*136337. Available from https://doi.org/10.1016/j.scitotenv.2019.136337.

Chathurika, J. A. S., Kumaragamage, D., Zvomuya, F., Akinremi, O. O., Flaten, D. N., Indraratne, S. P., & Dandeniya, W. S. (2016). Woodchip biochar with or without synthetic fertilizers affects soil properties and available phosphorus in two alkaline, chernozemic soils. *Canadian Journal of Soil Science*, *96*(4), 472–484. Available from https://doi.org/10.1139/cjss-2015-0094, http://pubs.aic.ca/toc/cjss/current.

Chintala, R., Mollinedo, J., Schumacher, T. E., Malo, D. D., & Julson, J. L. (2014). Effect of biochar on chemical properties of acidic soil. *Archives of Agronomy and Soil Science*, *60*(3), 393–404. Available from https://doi.org/10.1080/03650340.2013.789870, http://www.tandf.co.uk/journals/titles/03650340.asp.

Cui, L., Noerpel, M. R., Scheckel, K. G., & Ippolito, J. A. (2019). Wheat straw biochar reduces environmental cadmium bioavailability. *Environment International*, *126*, 69–75. Available from https://doi.org/10.1016/j.envint.2019.02.022, http://www.elsevier.com/locate/envint.

Dey, S., Purakayastha, T. J., Sarkar, B., Rinklebe, J., Kumar, S., Chakraborty, R., Datta, A., Lal, K., & Shivay, Y. S. (2023). Enhancing cation and anion exchange capacity of rice straw biochar by chemical modification for increased plant nutrient retention. *Science of the Total Environment, 886*. Available from https://doi.org/10.1016/j.scitotenv.2023.163681, http://www.elsevier.com/locate/scitotenv.

Ding, Y., Liu, Y., Liu, S., Li, Z., Tan, X., Huang, X., Zeng, G., Zhou, L., & Zheng, B. (2016). Biochar to improve soil fertility. A review. *Agronomy for Sustainable Development*, *36*(2). Available from https://doi.org/10.1007/s13593-016-0372-z, http://www.springerlink.com/content/1773-0155.

Dixon, G. R., & Tilston, E. L. (2010). *Soil-borne pathogens and their interactions with the soil environment. Soil microbiology and sustainable crop production* (pp. 197–271). Springer. Available from http://www.springerlink.com/openurl.asp?genre = book&isbn = 978-90-481-9478-0, https://doi.org/10.1007/978-90-481-9479-7_6.

Dong, Y., Yu, Y., Wang, R., Chang, E., Hong, Z., Hua, H., Liu, H., Jiang, J., & Xu, R. (2022). Insights on mechanisms of aluminum phytotoxicity mitigation by canola straw biochars from different regions. *Biochar*, *4*(1). Available from https://doi.org/10.1007/s42773-022-00179-6, springer.com/journal/42773.

Drobnik, T., Greiner, L., Keller, A., & Grêt-Regamey, A. (2018). Soil quality indicators – from soil functions to ecosystem services. *Ecological Indicators*, *94*, 151–169. Available from https://doi.org/10.1016/j.ecolind.2018.06.052, http://www.elsevier.com/locate/ecolind.

Ducey, T. F., Ippolito, J. A., Cantrell, K. B., Novak, J. M., & Lentz, R. D. (2013). Addition of activated switchgrass biochar to an aridic subsoil increases microbial nitrogen cycling gene abundances. *Applied Soil Ecology*, *65*, 65–72. Available from https://doi.org/10.1016/j.apsoil.2013.01.006.

Eissa, M. A. (2019). Effect of cow manure biochar on heavy metals uptake and translocation by zucchini (*Cucurbita pepo* L). *Arabian Journal of Geosciences*, *12*(2). Available from https://doi.org/10.1007/s12517-018-4191-1, http://www.springer.com/geosciences/journal/12517?cm_mmc = AD-_-enews-_-PSE1892-_-0.

Elkhlifi, Z., Iftikhar, J., Sarraf, M., Ali, B., Saleem, M. H., Ibranshahib, I., Bispo, M. D., Meili, L., Ercisli, S., Torun Kayabasi, E., Alemzadeh Ansari, N., Hegedűsová, A., & Chen, Z. (2023). Potential role of biochar on capturing soil nutrients, carbon sequestration and managing environmental challenges: A review. *Sustainability*, *15*(3), 2527. Available from https://doi.org/10.3390/su15032527.

EU. (2012). *Guidelines on best practice to limit, mitigate or compensate soil sealing*. Luxembourg: European Union, Luxembourg European Union. Available from http://ec.europa.eu/environment/soil/pdf/guidelines/pub/soil_en.pdf.

Fidel, R. B., Laird, D. A., & Parkin, T. B. (2019). Effect of biochar on soil greenhouse gas emissions at the laboratory and field scales. *Soil Systems*, *3*(1), 1–18. Available from https://doi.org/10.3390/soilsystems3010008, https://www.mdpi.com/2571-8789/3/1/8/pdf.

Glaser, B., & Birk, J. J. (2012). State of the scientific knowledge on properties and genesis of anthropogenic dark earths in Central Amazonia (terra preta de índio). *Geochimica et Cosmochimica Acta*, *82*, 39–51. Available from https://doi.org/10.1016/j.gca.2010.11.029.

Hamza, M. A., & Anderson, W. K. (2005). Soil compaction in cropping systems: A review of the nature, causes and possible solutions. *Soil and Tillage Research*, *82*(2), 121–145. Available from https://doi.org/10.1016/j.still.2004.08.009, http://www.elsevier.com/inca/publications/store/5/0/3/3/1/8.

He, X., Wang, Y., Tai, M. H., Lin, A., Owyong, S., Li, X., Leong, K., Yusof, M. L. M., Ghosh, S., & Wang, C. H. (2022). Integrated applications of water hyacinth biochar: A circular economy case study. *Journal of Cleaner Production*, *378*. Available from https://doi.org/10.1016/j.jclepro.2022.134621, https://www.journals.elsevier.com/journal-of-cleaner-production.

Hmid, A., Al Chami, Z., Sillen, W., De Vocht, A., & Vangronsveld, J. (2015). Olive mill waste biochar: A promising soil amendment for metal immobilization in contaminated soils. *Environmental Science and Pollution Research*, *22*(2), 1444–1456. Available from https://doi.org/10.1007/s11356-014-3467-6, https://link.springer.com/journal/11356.

Howe, J. A., & Smith, A. P. (2021). *The soil habitat principles and applications of soil microbiology* (3rd ed, pp. 23–55). Elsevier. Available from https://www.sciencedirect.com/book/9780128202029, 10.1016/B978-0-12-820202-9.00002-2.

Kabir, E., Kim, K. H., & Kwon, E. E. (2023). Biochar as a tool for the improvement of soil and environment. *Frontiers in Environmental Science*, *11*. Available from https://doi.org/10.3389/fenvs.2023.1324533, journal.frontiersin.org/journal/environmental-science.

Kamali, M., Sweygers, N., Al-Salem, S., Appels, L., Aminabhavi, T. M., & Dewil, R. (2022). Biochar for soil applications-sustainability aspects, challenges and future prospects. *Chemical Engineering Journal*, *428*131189. Available from https://doi.org/10.1016/j.cej.2021.131189.

Karim, A. A., Kumar, M., Mohapatra, S., & Singh, S. K. (2019). Nutrient rich biomass and effluent sludge wastes co-utilization for production of biochar fertilizer through different thermal treatments. *Journal of Cleaner Production*, *228*, 570–579. Available from https://doi.org/10.1016/j.jclepro.2019.04.330, https://www.journals.elsevier.com/journal-of-cleaner-production.

Karim, A. A., Kumar, M., Mohapatra, S., Singh, S. K., & Panda, C. R. (2019). Co-plasma processing of banana peduncle with phosphogypsum waste for production of lesser toxic potassium–sulfur rich biochar. *Journal of Material Cycles and Waste Management*, *21*(1), 107–115. Available from https://doi.org/10.1007/s10163-018-0769-7, http://link.springer.de/link/service/journals/10163/index.htm.

Khan, Z., Zhang, K., Khan, M. N., Bi, J., Zhu, K., Luo, L., & Hu, L. (2022). How biochar affects nitrogen assimilation and dynamics by interacting soil and plant enzymatic activities: Quantitative assessment of 2 years potted study in a rapeseed-soil system. *Frontiers in Plant Science*, *13*. Available from https://doi.org/10.3389/fpls.2022.853449, https://www.frontiersin.org/journals/plant-science.

Kolton, M., Graber, E. R., Tsehansky, L., Elad, Y., & Cytryn, E. (2017). Biochar-stimulated plant performance is strongly linked to microbial diversity and metabolic potential in the rhizosphere. *New Phytologist*, *213*(3), 1393–1404. Available from https://doi.org/10.1111/nph.14253, http://onlinelibrary.wiley.com/journal/10.1111/(ISSN)1469-8137.

Kopittke, P. M., Dalal, R. C., Finn, D., & Menzies, N. W. (2017). Global changes in soil stocks of carbon, nitrogen, phosphorus, and sulphur as influenced by long-term agricultural production. *Global Change Biology*, *23*(6), 2509–2519. Available from https://doi.org/10.1111/gcb.13513.

Lai, W. Y., Lai, C. M., Ke, G. R., Chung, R. S., Chen, C. T., Cheng, C. H., Pai, C. W., Chen, S. Y., & Chen, C. C. (2013). The effects of woodchip biochar application on crop yield, carbon sequestration and greenhouse gas emissions from soils planted with rice or leaf beet. *Journal of the Taiwan Institute of Chemical Engineers*, *44*(6), 1039–1044. Available from https://doi.org/10.1016/j.jtice.2013.06.028.

Lal, R. (2009). Soil degradation as a reason for inadequate human nutrition. *Food Security*, *1*(1), 45–57. Available from https://doi.org/10.1007/s12571-009-0009-z.

Lehmann, J., Kuzyakov, Y., Pan, G., & Ok, Y. S. (2015). Biochars and the plant-soil interface. *Plant and Soil*, *395*(1-2). Available from https://doi.org/10.1007/s11104-015-2658-3, http://www.wkap.nl/journalhome.htm/0032-079X.

Lehmann, J., Rillig, M. C., Thies, J., Masiello, C. A., Hockaday, W. C., & Crowley, D. (2011). Biochar effects on soil biota – a review. *Soil Biology and Biochemistry*, *43*(9), 1812–1836. Available from https://doi.org/10.1016/j.soilbio.2011.04.022.

Lei, S., Shi, Y., Xue, C., Wang, J., Che, L., & Qiu, Y. (2020). Hybrid ash/biochar biocomposites as soil amendments for the alleviation of cadmium accumulation by *Oryza sativa* L. in a contaminated paddy field. *Chemosphere*, *239*124805. Available from https://doi.org/10.1016/j.chemosphere.2019.124805.

Li, J., Wang, S. L., Zheng, L., Chen, D., Wu, Z., Xie, Y., Wu, W., Niazi, N. K., Ok, Y. S., Rinklebe, J., & Wang, H. (2019). Sorption of lead in soil amended with coconut fiber biochar: Geochemical and spectroscopic investigations. *Geoderma*, *350*, 52–60. Available from https://doi.org/10.1016/j.geoderma.2019.05.008, http://www.elsevier.com/inca/publications/store/5/0/3/3/3/2.

Lin, L., Li, Z., Liu, X., Qiu, W., & Song, Z. (2019). Effects of Fe-Mn modified biochar composite treatment on the properties of As-polluted paddy soil. *Environmental Pollution*, *244*, 600–607. Available from https://doi.org/10.1016/j.envpol.2018.10.011, http://www.elsevier.com/inca/publications/store/4/0/5/8/5/6.

Masud, M. A. A., Shin, W. S., Sarker, A., Septian, A., Das, K., Deepo, D. M., Iqbal, M. A., Islam, A. R. M. T., & Malafaia, G. (2023). A critical review of sustainable application of biochar for green remediation: Research uncertainty and future directions. *Science of the Total Environment*, *904*. Available from https://doi.org/10.1016/j.scitotenv.2023.166813, http://www.elsevier.com/locate/scitotenv.

Meng, C., Tian, D., Zeng, H., Li, Z., Yi, C., & Niu, S. (2019). Global soil acidification impacts on belowground processes. *Environmental Research Letters*, *14*(7)074003. Available from https://doi.org/10.1088/1748-9326/ab239c.

Nguyen, T. T. N., Xu, C. Y., Tahmasbian, I., Che, R., Xu, Z., Zhou, X., Wallace, H. M., & Bai, S. H. (2017). Effects of biochar on soil available inorganic nitrogen: A review and meta-analysis. *Geoderma*, *288*, 79–96. Available from https://doi.org/10.1016/j.geoderma.2016.11.004, http://www.elsevier.com/inca/publications/store/5/0/3/3/3/2.

Noel, S. (2015). *Reaping economic and environmental benefits from sustainable land management, Economics of land degradation initiative: Report for policy and decision makers* (2015). ELD.

onwuka, B., & Mang, B. (2018). Effects of soil temperature on some soil properties and plant growth. *Advances in Plants & Agriculture Research*, *8*(1). Available from https://doi.org/10.15406/apar.2018.08.00288.

Osman, K. T. (2014). Soil degradation, conservation and remediation. *Soil Degradation, Conservation and Remediation*, 1–237. Available from https://doi.org/10.1007/978-94-007-7590-9, http://doi.org/10.1007/978-94-007-7590-9.

Ouyang, W., Zhao, X., Tysklind, M., & Hao, F. (2016). Typical agricultural diffuse herbicide sorption with agricultural waste-derived biochars amended soil of high organic matter content. *Water Research*, *92*, 156–163. Available from https://doi.org/10.1016/j.watres.2016.01.055, http://www.elsevier.com/locate/watres.

Pan, D., Liu, C., Yu, H., & Li, F. (2019). A paddy field study of arsenic and cadmium pollution control by using iron-modified biochar and silica sol together. *Environmental*

Science and Pollution Research, *26*(24), 24979–24987. Available from https://doi.org/10.1007/s11356-019-05381-x, http://www.springerlink.com/content/0944-1344.

Pandey, K. (2019). India lost 31% of grasslands in a decade. *Down To Earth, 2019.*

Paz-Ferreiro, J., Plasencia, P., Gascó, G., & Méndez, A. (2017). Biochar from pyrolysis of deinking paper sludge and its use in the remediation of Zn-polluted soils. *Land Degradation and Development*, *28*(1), 355–360. Available from https://doi.org/10.1002/ldr.2597, http://onlinelibrary.wiley.com/journal/10.1002/(ISSN)1099-145X.

Probert, R. J. (2022). *The role of temperature in the regulation of seed dormancy and germination* (pp. 261–292). CABI Publishing. Available from 10.1079/9780851994321.0261.

Rahman, M. A., Jahiruddin, M., Kader, M. A., Islam, M. R., & Solaiman, Z. M. (2022). Sugarcane bagasse biochar increases soil carbon sequestration and yields of maize and groundnut in charland ecosystem. *Archives of Agronomy and Soil Science*, *68*(10), 1338–1351. Available from https://doi.org/10.1080/03650340.2021.1892651, http://www.tandf.co.uk/journals/titles/03650340.asp.

Ray, A., Kumar, M., Karim, A. A., Biswas, K., Mohanty, S., Shadangi, K. P., Kumar, S., & Sarkar, B. (2023). Potassium-phosphorus-sulfur augmented biochar production from potentially toxic elements abated gypsum pond wastewater of phosphate fertilizer industry. *Journal of Environmental Chemical Engineering*, *11*(5). Available from https://doi.org/10.1016/j.jece.2023.110404, http://www.journals.elsevier.com/journal-of-environmental-chemical-engineering/.

Reddy, K. R., Yaghoubi, P., & Yukselen-Aksoy, Y. (2015). Effects of biochar amendment on geotechnical properties of landfill cover soil. *Waste Management and Research*, *33*(6), 524–532. Available from https://doi.org/10.1177/0734242X15580192, wmr.sagepub.com/.

Sahani, U., & Behera, N. (2001). Impact of deforestation on soil physicochemical characteristics, microbial biomass and microbial activity of tropical soil. *Land Degradation & Development*, *12*(2), 93–105. Available from https://doi.org/10.1002/ldr.429.

Singh, A. (2021). Soil salinization management for sustainable development: A review. *Journal of Environmental Management*, *277*111383. Available from https://doi.org/10.1016/j.jenvman.2020.111383.

Singh, P., Rawat, S., Jain, N., Bhatnagar, A., Bhattacharya, P., & Maiti, A. (2023). A review on biochar composites for soil remediation applications: Comprehensive solution to contemporary challenges. *Journal of Environmental Chemical Engineering*, *11*(5)110635. Available from https://doi.org/10.1016/j.jece.2023.110635.

Tan, Z. X., Lal, R., & Wiebe, K. D. (2005). Global soil nutrient depletion and yield reduction. *Journal of Sustainable Agriculture*, *26*(1), 123–146. Available from https://doi.org/10.1300/j064v26n01_10.

Tang, X., Shen, H., Chen, M., Yang, X., Yang, D., Wang, F., Chen, Z., Liu, X., Wang, H., & Xu, J. (2020). Achieving the safe use of Cd- and As-contaminated agricultural land with an Fe-based biochar: A field study. *Science of the Total Environment*, *706*135898. Available from https://doi.org/10.1016/j.scitotenv.2019.135898.

Tomczyk, A., Sokołowska, Z., & Boguta, P. (2020). Biochar physicochemical properties: Pyrolysis temperature and feedstock kind effects. *Reviews in Environmental Science and Biotechnology*, *19*(1), 191–215. Available from https://doi.org/10.1007/s11157-020-09523-3, http://www.kluweronline.com/issn/1569-1705.

UNEP. (2009). *Converting waste agricultural biomass into a resource-compendium of technologies*. Author.

Vimal, V., Karim, A. A., Kumar, M., Ray, A., Biswas, K., Maurya, S., Subudhi, D., & Dhal, N. K. (2022). Nutrients enriched biochar production through co-pyrolysis of poultry litter with banana peduncle and phosphogypsum waste. *Chemosphere*, *300*. Available from https://doi.org/10.1016/j.chemosphere.2022.134512, http://www.elsevier.com/locate/chemosphere.

Wang, L., Ok, Y. S., Tsang, D. C. W., Alessi, D. S., Rinklebe, J., Mašek, O., Bolan, N. S., & Hou, D. (2022). Biochar composites: Emerging trends, field successes and sustainability implications. *Soil Use and Management*, *38*(1), 14–38. Available from https://doi.org/10.1111/sum.12731, http://onlinelibrary.wiley.com/journal/10.1111/(ISSN)1475-2743.

Wu, L., Wei, C., Zhang, S., Wang, Y., Kuzyakov, Y., & Ding, X. (2019). MgO-modified biochar increases phosphate retention and rice yields in saline-alkaline soil. *Journal of Cleaner Production*, *235*, 901–909. Available from https://doi.org/10.1016/j.jclepro.2019.07.043, https://www.journals.elsevier.com/journal-of-cleaner-production.

Wu, W., Yang, M., Feng, Q., McGrouther, K., Wang, H., Lu, H., & Chen, Y. (2012). Chemical characterization of rice straw-derived biochar for soil amendment. *Biomass and Bioenergy*, *47*, 268–276. Available from https://doi.org/10.1016/j.biombioe.2012.09.034, http://www.journals.elsevier.com/biomass-and-bioenergy/.

Yin, D., Li, H., Wang, H., Guo, X., Wang, Z., Lv, Y., Ding, G., Jin, L., & Lan, Y. (2021). Impact of different biochars on microbial community structure in the rhizospheric soil of rice grown in albic soil. *Molecules (Basel, Switzerland)*, *26*(16), 4783. Available from https://doi.org/10.3390/molecules26164783.

Yu, X., & Lu, S. (2020). Double effects of biochar in affecting the macropore system of paddy soils identified by high-resolution X-ray tomography. *Science of the Total Environment*, *720*137690. Available from https://doi.org/10.1016/j.scitotenv.2020.137690.

Zewd, I., & Siban, M. (2021). The effects of alkalinity on physical and chemical properties of soil. *Journal of Plant Biology and Agriculture Science.*, *3*, 1–5.

Zhang, C., Lin, Y., Tian, X., Xu, Q., Chen, Z., & Lin, W. (2017). Tobacco bacterial wilt suppression with biochar soil addition associates to improved soil physiochemical properties and increased rhizosphere bacteria abundance. *Applied Soil Ecology*, *112*, 90–96. Available from https://doi.org/10.1016/j.apsoil.2016.12.005, http://www.elsevier.com/inca/publications/store/5/2/4/5/1/8/index.htt.

Zhang, H., Chen, C., Gray, E. M., Boyd, S. E., Yang, H., & Zhang, D. (2016). Roles of biochar in improving phosphorus availability in soils: A phosphate adsorbent and a source of available phosphorus. *Geoderma*, *276*, 1–6. Available from https://doi.org/10.1016/j.geoderma.2016.04.020, http://www.elsevier.com/inca/publications/store/5/0/3/3/3/2.

Zong, Y., Xiao, Q., & Lu, S. (2016). Acidity, water retention, and mechanical physical quality of a strongly acidic Ultisol amended with biochars derived from different feedstocks. *Journal of Soils and Sediments*, *16*(1), 177–190. Available from https://doi.org/10.1007/s11368-015-1187-2, http://www.springerlink.com/content/1439-0108.

CHAPTER

11

Stability and reactivity of biochar in regulating nutrient dynamics and its supplies in agricultural ecosystems

Bayu Dume[1], Abebe Nigussie[1], Melkamu Mamuye[1] and Milkias Ahmed[2]

[1]*Department of Natural Resource managment, College of Agriculture and Vetrnary Medicine, Jimma University, Jimma, Ethiopia*

[2]*Department of Horticulture and Plant Science, College of Agriculture and Vetrnary Medicine, Jimma University, Jimma, Ethiopia*

11.1 Introduction

Nutrient management in agricultural systems is a critical and multifaceted practice that plays a pivotal role in ensuring sustainable and productive food production while safeguarding the environment (Selim, 2020). It involves the careful planning, application, and monitoring of essential nutrients, primarily nitrogen (N), phosphorus (P), and potassium (K), as well as other micronutrients, to optimize crop growth and minimize negative environmental impacts (Paramesh et al., 2023). The importance of nutrient management in agriculture cannot be overstated, as it directly influences crop yields, soil health, water quality, and the overall sustainability of farming practices (Pawlak & Kołodziejczak, 2020). Here are some key facets of its significance:

Adequate and balanced nutrient management is essential for achieving optimal crop yields. Different crops have varying nutrient requirements, and understanding these needs allows farmers to apply the right nutrients in the right amounts at the right time. This enhances crop growth, quality, and overall productivity. Nutrient management practices contribute to maintaining soil health and fertility. Over time, improper nutrient application can lead to soil depletion, erosion, and reduced agricultural productivity. By replenishing nutrients and maintaining soil pH levels, farmers can ensure the long-term sustainability of their land. Uncontrolled nutrient runoff from farms can have detrimental effects on the environment. Excessive nutrients, particularly nitrogen and phosphorus, can pollute water bodies, leading to issues like algal blooms, dead zones, and contamination of drinking water sources (Leslie et al., 2017). Proper nutrient management helps

Biochar Ecotechnology for Sustainable Agriculture and Environment. DOI: https://doi.org/10.1016/B978-0-443-29855-4.00013-8

minimize these environmental risks, protecting aquatic ecosystems and human health. Effective nutrient management can also lead to cost savings for farmers. By avoiding over-application of expensive fertilizers and reducing nutrient waste, farmers can optimize their resource use and improve their economic viability (Wurtsbaugh et al., 2019).

Many regions have regulations and guidelines in place to manage nutrient runoff from agricultural operations (Guo et al., 2021). Compliance with these regulations is crucial to avoid penalties and maintain a positive public image for the farming community. Nutrient management practices help farmers meet these requirements. Nutrient management can enhance the resilience of agricultural systems to the impacts of climate change. Balanced nutrient levels can improve crop tolerance to environmental stressors such as drought or extreme temperatures, helping farmers adapt to changing climate conditions (Malhi et al., 2021). Advances in technology, such as precision agriculture and soil testing, have revolutionized nutrient management. These tools enable farmers to make data-driven decisions, optimizing nutrient application and minimizing waste (Garg et al., 2023). Nutrient management is a cornerstone of sustainable agriculture. It promotes the efficient use of resources, reduces negative environmental externalities, and contributes to the long-term viability of farming practices (Tuğrul, 2019). In general, nutrient management in agricultural systems is an essential practice with far-reaching implications for food security, environmental conservation, and economic sustainability. As global food demand continues to rise, adopting effective nutrient management strategies is imperative to ensure that agriculture remains productive, resilient, and environmentally responsible in the face of evolving challenges.

Biochar is a solid product obtained from thermochemical conversion of biomass under limited oxygen supply (Lehmann & Joseph, 2009). It is a stable, carbon-rich material that can be used as a soil amendment to improve soil fertility and health (Novak et al., 2009). Biochar can also be used to improve water quality, reduce greenhouse gas emissions, and sequester carbon in the soil (Woolf et al., 2010). Biochar has a number of properties that make it a beneficial soil amendment. It is a good source of organic matter, which helps to improve soil structure and water retention. Biochar also contains nutrients that are essential for plant growth, such as nitrogen, phosphorus, and potassium. Additionally, biochar can help to suppress soil borne pests and diseases. Biochar can be applied to soil in a number of ways. It can be mixed into the soil at planting time, or it can be applied as a top dressing. Biochar can also be used to create biochar-based composts and mulches (Glaser et al., 2002). The amount of biochar that should be applied to soil will vary depending on the type of soil, the climate, and the desired benefits (Laird, 2008). In general, it is recommended to apply 1–2 tons of biochar per hectare of land (Pandit et al., 2018). Biochar is a safe and effective soil amendment that can improve soil fertility, water quality, and greenhouse gas emissions. It is a sustainable alternative to synthetic fertilizers and pesticides (Semida et al., 2019).

Biochar has been used as a soil amendment for centuries. It was first used by indigenous peoples in the Amazon rainforest, and it is still used today by farmers in many parts of the world (Lehmann, 2007). Biochar is a natural product that is made from plant material that has been heated to high temperatures in the absence of oxygen. This process, called pyrolysis, converts the plant material into a solid, charcoal-like material. Biochar has a number of properties that make it a beneficial soil amendment. It is a good source of organic matter, which helps to improve soil structure and water retention. Biochar also contains nutrients that are essential for plant growth, such as nitrogen, phosphorus, and potassium. Here are some of the potential benefits of using biochar as a soil amendment: Improves soil structure and water retention; increases nutrient availability for plants; suppresses soil borne pests and diseases; reduces greenhouse gas emissions; improves water quality; enhances crop yields; and provides a long-term soil amendment. Biochar is a versatile and effective soil amendment that can be used to improve soil health and productivity. It is a sustainable alternative to synthetic fertilizers and pesticides, and it has the potential to reduce greenhouse gas emissions (Lehmann et al., 2011). Therefore the objective of this chapter is to explore the stability and reactivity of biochar and its role in regulating nutrient dynamics in agricultural ecosystems.

11.2 Biochar production

There are several methods for producing biochar, each with its own approach to converting biomass into a carbon-rich and stable material. Some of the key methods include:

11.2.1 Pyrolysis

Pyrolysis involves heating biomass in a low or zero oxygen environments to break it down into biochar, bio-oil, and syngas (Lehmann & Joseph, 2009). The temperature range for pyrolysis typically varies from 300°C to 700°C. The resultant biochar is typically stable and retains much of the carbon originally present in the biomass (Novak et al., 2009).

11.2.2 Gasification

Gasification also involves heating biomass, but in a high-temperature environment with a controlled amount of oxygen (Woolf et al., 2010). This process usually occurs at temperatures between 700°C and 1500°C and produces synthesis gas (syngas) consisting of carbon monoxide and hydrogen, along with biochar as a by-product (Lehmann & Joseph, 2009).

11.2.3 Hydrothermal carbonization

Hydrothermal carbonization (HTC) uses high temperature and pressure in a liquid environment to convert biomass into biochar. HTC uses high temperature (180°C–250°C) and pressure (2–10 MPa) in a liquid environment to convert biomass into biochar (Glaser et al., 2002).

11.2.4 Slow pyrolysis

Slow pyrolysis is a method that focuses on the slow heating of biomass to produce biochar, bio-oil, and syngas. This process typically operates at temperatures ranging from 400°C to 600°C over a longer duration, resulting in biochar with specific properties based on the temperature and duration of pyrolysis (Lehmann, 2007).

11.2.5 Fast pyrolysis

Fast pyrolysis involves rapidly heating biomass at high temperatures in the absence of oxygen, resulting in the production of bio-oil, syngas, and biochar. Fast pyrolysis involves rapidly heating biomass at high temperatures, typically

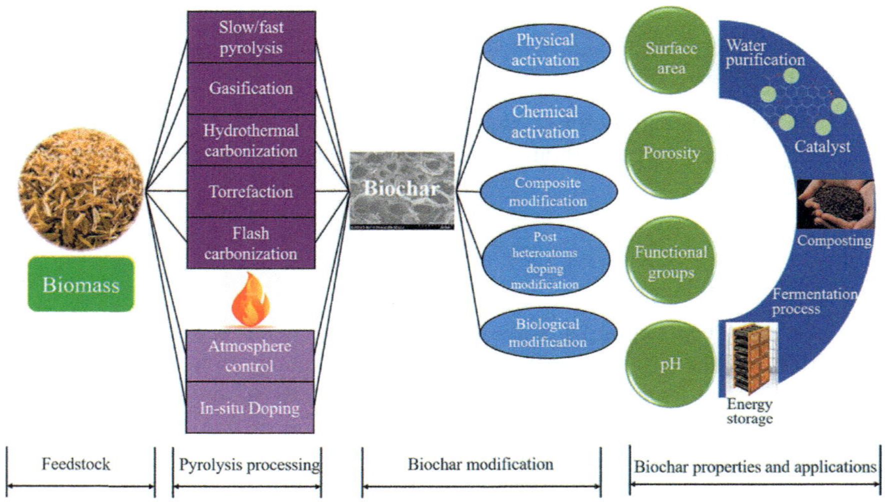

FIGURE 11.1 Production and utilization of biochar via selective pyrolysis of biomass (Li et al., 2020).

Illustrating the production and utilization of biochar through the selective pyrolysis of biomass, and detailing the properties of the resulting biochar. Biochar production and utilization.

From Li, Y., Xing, B., Ding, Y., Han, X., & Wang, S. (2020). A critical review of the production and advanced utilization of biochar via selective pyrolysis of lignocellulosic biomass. Bioresource Technology, 312, *123614. https://doi.org/10.1016/j.biortech.2020.123614.*

between 450°C and 600°C, in the absence of oxygen. This method is known for its efficiency in generating bio-oil suitable for fuel production, as well as syngas and biochar (Spokas, 2010). The production and utilization of biochar via selective pyrolysis of biomass are shown in Fig. 11.1. Each of these production methods offers specific advantages and yields biochar with distinct properties, making them suitable for various applications such as soil improvement, carbon sequestration, and renewable energy production.

11.3 Biochar physicochemical properties

11.3.1 Surface area

Biochar typically has a high surface area due to its porous nature, which allows for increased interactions with soil, water, and other environmental components (Jeffery et al., 2011). This high surface area can enhance its reactivity and its ability to adsorb nutrients and pollutants.

11.3.2 Porosity

Biochar possesses a porous structure, providing a network of nooks and crannies (Laird, 2008). This porosity influences the water-holding capacity of biochar and its ability to retain nutrients and microorganisms in the soil. It also affects the biochar's reactivity by offering sites for chemical reactions and microbial colonization.

11.3.3 Functional groups

Biochar contains various functional groups, such as hydroxyl, carboxyl, and phenolic groups, which contribute to its reactivity and stability (Semida et al., 2019). These functional groups can interact with nutrients and contaminants in the environment, influencing the biochar's ability to improve soil health and sequester carbon. The combination of these physicochemical properties plays a significant role in determining biochar's stability and reactivity, making it a valuable tool for applications in agriculture, environmental remediation, and carbon sequestration.

11.4 Biochar's impact on nutrient retention and availability

11.4.1 Nutrient adsorption

Biochar has a high surface area and pore structure, which allows it to adsorb and retain nutrients such as nitrogen, phosphorus, and potassium (Jeffery et al., 2011).

This adsorption capacity helps prevent nutrient leaching and makes them available to plants over an extended period.

11.4.2 Cation exchange capacity

Biochar's porous structure and negative surface charge contribute to its high cation exchange capacity (CEC; Laird, 2008). This property enables biochar to attract and retain positively charged nutrient ions, including potassium (K^+), calcium (Ca^{2+}), and magnesium (Mg^{2+}) (Semida et al., 2019). As a result, biochar-amended soil can hold onto these nutrients for plant uptake.

11.4.3 Nutrient stabilization

Biochar helps stabilize nutrients in soil, reducing their susceptibility to loss through leaching or volatilization (Lehmann & Joseph, 2009). This stabilization is particularly important for nutrients like phosphorus, which can easily bind with soil particles and become less available to plants (Novak et al., 2009). Biochar's carbonaceous structure can form strong bonds with phosphorus, making it less prone to runoff or fixation.

11.4.4 pH modification

Biochar has the potential to alter soil pH due to its alkaline nature (Spokas, 2010). This pH modification can influence nutrient availability, helping to adjust soil conditions to optimize nutrient uptake by plants (Woolf et al., 2010). While biochar exhibits positive effects on nutrient retention and availability, it's important to note that the exact impact may vary depending on factors such as feedstock type, pyrolysis conditions, soil characteristics, and management practices. It is advisable to consider specific studies and conduct site-specific assessments to understand the particular effects of biochar in different agricultural or environmental contexts. Fig. 11.1 illustrates the production and utilization of biochar through selective pyrolysis of biomass, as well as the properties of the resulting biochar.

Biochar efficiently retains nitrogen, phosphorus, and potassium nutrients in soil, as indicated by a review on different amounts of biochar (Lehmann et al., 2011). A meta-analysis study highlights short-term biochar impacts on crop performance and soil quality, emphasizing biochar's role in converting compounds into plant-available nutrients, including nitrogen, phosphorus, and potassium. Adsorption and desorption of nitrogen, phosphorus, and potassium in soils are influenced by biochar, with factors such as pH, organic matter, and cation exchange playing a role. Biochar interacts with nitrogen, phosphorus, and potassium through electrostatic interaction, complexation, and mineralization, enhancing fertilizer production (Shackley et al., 2012). Biochar's properties influence environmental pollutants, carbon and nitrogen cycling, and plant health, highlighting its crucial role in paddy soils. Biochar, enriched with organic carbon and containing nitrogen, phosphorus, and potassium, serves as a microbial carrier, emphasizing its potential in nutrient retention and soil improvement (Al-Wabel et al., 2013).

11.5 Explanation of the mechanisms involved in nutrient sorption and release by biochar

Nutrient sorption and release by biochar involve several key mechanisms, including physical and chemical processes that influence the interactions between biochar and nutrients (Fig. 11.2).

11.5.1 Physical adsorption

Biochar possesses a high surface area and a porous structure, allowing it to physically adsorb nutrients such as nitrogen, phosphorus, and potassium onto its

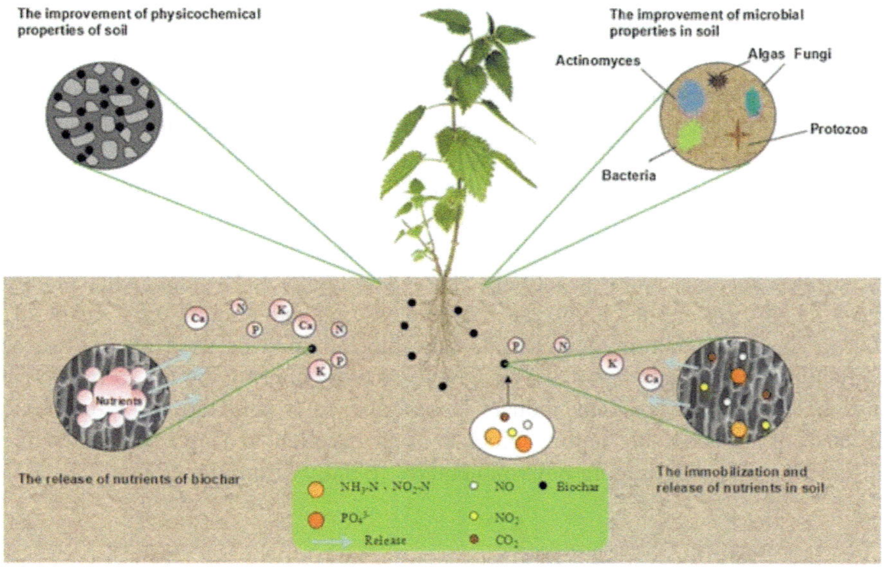

FIGURE 11.2 The possible mechanisms of biochar for improving soil fertility (Ding et al., 2016).

The potential mechanisms by which biochar enhances soil fertility encompass various physical, chemical, and biological processes. The potential mechanisms by which biochar enhances soil fertility.

From Ding, Y., Liu, Y., Liu, S., Li, Z., Tan, X., Huang, X., Zeng, G., Zhou, L., & Zheng, B. (2016). Biochar to improve soil fertility. A review. Agronomy for Sustainable Development, 36*(2). https://doi.org/10.1007/s13593–016-0372-z.*

surfaces. This mechanism involves the attraction and retention of nutrients through van der Waals forces, hydrogen bonding, and electrostatic interactions (Lehmann et al., 2011).

11.5.2 Chemical adsorption

Chemical reactions between biochar and nutrients contribute to nutrient retention. Functional groups on the biochar surface, such as hydroxyl, carboxyl, and phenolic groups, can form chemical bonds with nutrients, promoting their adsorption and retention (Shackley et al., 2012).

11.5.3 Ion exchange

Biochar's surface properties allow it to undergo ion exchange with nutrients in the soil solution. CEC and anion exchange capacity play crucial roles in the exchange of positively and negatively charged ions such as calcium, magnesium, nitrate, and phosphate (Al-Wabel et al., 2013).

11.5.4 pH-mediated reactions

The sorption and release of nutrients can be greatly influenced by the pH-dependent nature of biochar. Because of variations in surface charge and chemical speciation, biochar has the ability to either release or adsorb nutrients at different pH values. This fluctuation stems from the surface chemistry of the biochar changing with pH, which impacts how it interacts with soil nutrients. For instance, biochar surfaces may become positively charged at lower pH values, which would increase the adsorption of anions like phosphate. On the other hand, the surface becomes more negatively charged at higher pH values, which facilitates the adsorption of cations such as potassium and ammonium (Xu et al., 2023). These interactions demonstrate how biochar may be used to modify the availability of nutrients in soils. This can be done by altering the pH to maximize nutrient use efficiency and reduce losses due to volatilization or leaching. Because of its adaptable nature, biochar is an excellent resource for enhancing soil fertility and encouraging environmentally friendly farming methods.

11.5.5 Microbial interactions

Biochar can interact with soil microorganisms, influencing nutrient availability and uptake. Microbial activity can be influenced by biochar's physical and chemical properties, further impacting nutrient dynamics in the soil. Regarding nutrient release, biochar can act as a slow-release source of nutrients due to its ability to retain and gradually release nutrients over time, influenced by environmental conditions such as pH, moisture, and microbial activity. These mechanisms collectively contribute to the multifaceted role of biochar in nutrient sorption and release, offering potential benefits for soil fertility, nutrient management, and sustainable agriculture practices (Gorovtsov et al., 2020). Fig. 11.2 shows the possible mechanisms of biochar for improving soil fertility.

11.6 Comparison of different feedstock sources and their influence on nutrient retention capabilities

Different feedstock sources used to produce biochar can significantly influence its nutrient retention capabilities. Here are some general comparisons:

11.6.1 Wood-based biochar

Biochar derived from wood feedstock sources often exhibits good nutrient retention capabilities (Lehmann et al., 2011). It typically has a high surface area and porosity, offering favorable conditions for nutrient adsorption. Wood-based biochar may contain lower levels of nitrogen and phosphorus, but it can effectively adsorb and retain these nutrients from the surrounding environment.

11.6.2 Crop residue-based biochar

The ability of biochar made from crop leftovers, such as straw, husks, and stalks, to retain nutrients varies depending on the particular crop source. The original nutrient content of the feedstock can affect the amount of nutrients retained in crop residue-based biochar, which may have higher amounts of some nutrients than wood-based biochar. Due to the natural nutritious content of the feedstock materials employed, recent research have demonstrated that biochar generated from agricultural leftovers frequently has higher quantities of key elements like potassium, phosphate, and nitrogen (Brtnicky et al., 2021). For example, it has been discovered that biochar derived from rice husks has a higher silica concentration, which may help to improve the structure of the soil and the availability of nutrients. Similar to this, straw-derived biochar may include high concentrations of phosphate and potassium, which improves the soil's capacity to promote plant development. These variances highlight how crucial it is to choose the right feedstock materials depending on the particular nutritional requirements of the soil and the crops that are being grown (Dai et al., 2020).

11.6.3 Manure-based biochar

Manure-based biochar has the potential to contain significant levels of nutrients, such as potassium, phosphate, and nitrogen. As a result, biochar has better capacity to retain nutrients, particularly phosphate and nitrogen. The production method and the existence of additional pollutants, however, may have an impact on the efficiency of nutrient retention. Recent research has demonstrated that the richer nutritional profile of manure-based biochar can greatly increase soil fertility. For example, compared to plant-based biochars, biochar made from poultry litter or dairy manure frequently has higher quantities of important nutrients, which improves soil nutrient availability and retention (Elkhlifi et al., 2023). The final nutrient content and retention efficiency of manure-based biochar are significantly influenced by the production method, which includes the temperature and duration of the pyrolysis process. Although it might make some nutrients less available, high-temperature pyrolysis can strengthen the carbon's stability in biochar. Furthermore, the safety and effectiveness of using manure-based biochar as a soil supplement might be impacted by the presence of pollutants such heavy metals (Quilliam et al., 2012).

11.6.4 Nut shell–based biochar

Biochar produced from nut shells, such as coconut or almond shells, often displays good nutrient retention properties. Nut shell–based biochar can have a high surface area and porosity, providing favorable conditions for nutrient adsorption and retention. In general, the nutrient retention capabilities of biochar from different feedstock sources depend on factors such as the initial nutrient content of the feedstock, the production process used to create the biochar, as well as the physical and chemical properties of the resulting biochar. Additionally, factors such as pH, CEC, and surface functional groups contribute to the overall nutrient retention characteristics of biochar. Understanding these factors is essential for selecting the most effective biochar for specific nutrient management and soil improvement applications (Elkhlifi et al., 2023).

11.7 Influence of biochar on soil microbial communities

The influence of biochar on soil microbial communities is an active area of research within the field of soil science. Biochar, a form of charcoal produced from organic materials, has been shown to have various impacts on soil microorganisms. These impacts may include changes in microbial biomass, community composition, and activity. Research suggests that biochar application can influence soil microbial communities by altering the availability of nutrients, modifying soil pH, and providing a habitat for microbial colonization. Additionally, biochar can affect the interactions between soil microorganisms and plant roots, potentially influencing plant-microbe relationships. Further studies are needed to better understand the specific mechanisms underlying the influence of biochar on soil microbial communities and to assess the long-term impacts on soil health and ecosystem functioning (Lehmann et al., 2011).

11.7.1 Examination of research studies on the interactions between biochar and soil microbes

Numerous scientific studies have examined the interactions between biochar and soil microbes. Here are some common findings and insights from such research:

11.7.1.1 Microbial activity and metabolism

Studies have shown that biochar can influence soil microbial activity by altering the soil environment, such as pH and nutrient availability. The presence of biochar can impact the abundance, diversity, and function of soil microbial communities (Lehmann et al., 2011). Biochar can significantly influence microbial metabolism and nutrient cycling in the soil. By providing a porous structure and a large surface area, biochar serves as a habitat for soil microorganisms, enhancing their activity and diversity. This can lead to increased decomposition of organic matter, as microbes break down complex organic compounds into simpler substances. Additionally, biochar can adsorb nutrients, making them more accessible to microbes and plants. Studies have shown that biochar amendments can enhance microbial respiration rates and enzyme activities, thereby accelerating nutrient cycling processes. For instance, a study by Li et al. (2020) demonstrated that biochar application increased soil microbial biomass and enzyme activities related to carbon, nitrogen, and phosphorus cycling.

11.7.1.2 Soil carbon sequestration

Research has indicated that biochar amendments can impact soil carbon sequestration by altering microbial community activities involved in carbon turnover. Biochar can stabilize organic matter, reducing its decomposition rate and enhancing carbon storage in the soil. This is achieved through physical protection within biochar pores and chemical interactions that make organic carbon less accessible to decomposers. Moreover, biochar can influence the microbial processes of decomposition and stabilization of organic matter, promoting the formation of stable soil organic carbon compounds. According to Wang et al. (2021), biochar addition to soil significantly increased the microbial-derived carbon fraction, which is more resistant to decomposition, thereby contributing to long-term carbon sequestration.

11.7.1.3 Nutrient dynamics

Interactions between biochar and soil microbes can significantly affect nutrient cycling, particularly nitrogen dynamics. Biochar's porous structure can adsorb ammonium and nitrate, influencing their availability and transformation in the soil. This can alter microbial processes such as nitrification and denitrification, affecting nitrogen availability and losses. For example, biochar can provide niches for nitrifying bacteria, enhancing nitrification rates. Conversely, it can also reduce nitrogen losses through leaching and gaseous emissions. A study by Liu et al. (2021) found that biochar amendments improved nitrogen use efficiency by increasing soil microbial biomass and altering the community structure of nitrogen-cycling microbes, leading to reduced nitrogen leaching and nitrous oxide emissions.

11.7.1.4 Biodegradation of organic pollutants

Biochar has been investigated for its potential to enhance the biodegradation of organic pollutants in the soil. It can adsorb pollutants, reducing their bioavailability and toxicity to microorganisms, thus providing a more conducive environment for microbial degradation. Biochar's porous structure also supports the growth and activity of pollutant-degrading microbes. For instance, biochar can promote the activity of specific microbial consortia capable of breaking down complex organic pollutants such as polycyclic aromatic hydrocarbons (PAHs) and pesticides. Research by Chen et al. (2022) demonstrated that biochar application increased the degradation rates of PAHs in contaminated soils by enhancing the microbial population and enzymatic activities involved in pollutant breakdown.

11.7.1.5 Microbial community composition

Some studies have explored the impact of biochar on soil microbial community composition, revealing changes in the relative abundance of different microbial taxa. Biochar can selectively stimulate or inhibit the growth of certain microbial groups, leading to shifts in community structure. These changes can influence soil functions related to nutrient cycling, organic matter decomposition, and disease suppression. For example, biochar has been shown to increase the abundance of beneficial microbes such as plant growth-promoting rhizobacteria (PGPR) and mycorrhizal fungi, while suppressing pathogenic organisms. A study by Zhang et al. (2023) found that biochar amendments led to a significant increase in the relative abundance of PGPR and mycorrhizal fungi, which contributed to improved plant health and soil fertility. These findings collectively demonstrate the complex and multifaceted nature of the interactions between biochar and soil microbes. The specific outcomes of these interactions can vary depending on factors such as biochar properties (e.g., feedstock type, pyrolysis temperature), soil type, environmental conditions, and the composition of soil microbial communities. Understanding these interactions is crucial for optimizing biochar applications to enhance soil health and ecosystem services.

11.7.2 Discussion of how biochar affects microbial activity and diversity

Biochar can significantly influence microbial activity and diversity in soil due to its physical, chemical, and biological properties. Here's a discussion of how biochar affects microbial activity and diversity:

11.7.2.1 Physicochemical properties

Porosity and surface area: Biochar's porous structure and large surface area significantly contribute to its role in soil health. The high porosity of biochar provides numerous microhabitats for soil microorganisms, promoting microbial colonization and growth. These pores offer protection and a stable environment for microbes, enhancing their activity and diversity. Studies have shown that the enhanced surface area and porosity of biochar can improve microbial community structure by offering more sites for attachment and colonization. For example, Lehmann and Joseph (2015) reported that biochar's porous nature increases microbial biomass and activity, supporting more robust soil microbial ecosystems.

Water retention: Biochar's ability to retain water is another critical property that impacts soil microbial activity. By improving soil moisture content, biochar creates microhabitats that support diverse microbial communities. This increased water retention capability helps maintain adequate moisture levels during dry periods, thereby sustaining microbial processes and enhancing soil fertility. Research by Lyu et al. (2020) demonstrated that biochar amendments significantly improved soil water retention, which in turn supported higher microbial biomass and diversity, particularly under drought conditions.

pH moderation: Biochar can also moderate soil pH, a vital factor influencing microbial distribution and activity. Many soil microorganisms thrive within specific pH ranges, and biochar's pH-buffering capacity can help create optimal conditions for a diverse array of microbes. By neutralizing acidic soils or buffering alkaline soils, biochar enhances the microbial habitat, promoting greater microbial activity and diversity. For instance, a study by Liu et al. (2021) showed that biochar application adjusted soil pH to a more neutral range, which favored the growth of beneficial microbial taxa while inhibiting harmful ones.

Nutrient availability: Biochar's ability to adsorb and release nutrients such as nitrogen and phosphorus is crucial for influencing microbial nutrient availability. This nutrient cycling capability affects microbial populations and diversity by providing essential nutrients in a more accessible form. Biochar can adsorb excess nutrients, preventing their loss through leaching, and gradually release them, making them available to microbes and plants over time. According to a study by Zhang et al. (2021), biochar amendments increased the availability of phosphorus and nitrogen in the soil, which in turn boosted microbial activity and biomass, enhancing overall soil fertility and health. These physicochemical properties of biochar collectively enhance soil quality by creating a favorable environment for microbial growth and activity. The improvements in porosity, water retention, pH moderation, and nutrient availability contribute to more resilient and productive soil ecosystems.

11.7.2.2 Biological properties

Microbial habitats: Biochar can significantly influence the development of microbial habitats by providing numerous attachment sites and shelters for soil microorganisms. The porous structure of biochar provides an ideal environment for microbial colonization, enhancing the spatial distribution and diversity of microbial communities in the soil. This structure not only protects microbes from predation and environmental stresses but also creates microenvironments that support diverse microbial metabolic activities. Studies have shown that biochar amendments can alter microbial metabolic processes, affecting the decomposition of organic matter, nutrient cycling, and the balance of microbial populations. For instance, Wang et al. (2020) observed that biochar application increased the abundance and activity of soil microbes involved in the decomposition of organic matter, leading to enhanced nutrient cycling and improved soil fertility.

The impact of biochar on microbial activity and diversity is highly context-dependent, influenced by factors such as biochar type, application rate, soil type, and environmental conditions. Different types of biochar, produced from various feedstocks and under different pyrolysis conditions, have distinct physical and chemical properties that can differentially affect soil microbes. Additionally, the rate at which biochar is applied can also play a crucial role; too much or too little can lead to varying effects on microbial populations and activity. Soil type and environmental conditions further modulate biochar's impact. For example, biochar may enhance microbial activity and diversity in nutrient-poor soils by providing additional nutrients and improving soil structure, while its effects might be less pronounced in already fertile soils. Moreover, biochar's influence on microbial communities can depend on the existing soil microbial structure, nutrient availability, and specific needs of different microbial taxa. For instance, the presence of biochar can boost the activity of specific beneficial microbes, such as nitrogen-fixing bacteria, while potentially suppressing harmful pathogens.

Understanding the intricate interactions between biochar and soil microbes is essential for leveraging its potential benefits in promoting soil health and fertility. Biochar's ability to enhance microbial habitats and influence microbial processes can lead to improved soil structure, increased nutrient availability, and greater soil organic matter stabilization. For example, recent research by Chen et al. (2022) highlighted that biochar amendments not only increased microbial biomass but also shifted the microbial community composition toward more beneficial and efficient decomposers, resulting in enhanced soil fertility and crop productivity. However, to maximize these benefits, it is crucial to tailor biochar applications to specific soil and environmental conditions. This involves selecting the appropriate type and amount of biochar, considering the soil's existing microbial community, and understanding the nutrient dynamics within the soil. By doing so, biochar can be effectively used as a soil amendment to promote sustainable agriculture and soil health.

11.7.3 Implications for nutrient cycling and overall soil health

Biochar's influence on nutrient cycling and overall soil health is multifaceted, with both direct and indirect implications. Here are some key implications to consider:

11.7.3.1 Nutrient retention and availability

Biochar, distinguished by its porous structure and extensive surface area, significantly enhances nutrient adsorption and retention. This attribute mitigates nutrient leaching and ensures the prolonged availability of essential nutrients to plants, thereby augmenting soil fertility and promoting robust plant growth (Lehmann et al., 2011). Furthermore, biochar contributes substantially to carbon sequestration and stability. Its carbon-rich composition enables long-term carbon storage in soils, aiding in climate change mitigation. This stable form of soil organic carbon supports a balanced carbon cycle, enhancing soil health and long-term sustainability. Additionally, biochar provides a conducive environment for soil microorganisms, influencing their activity, diversity, and functionality. This interaction bolsters nutrient cycling processes, including organic matter decomposition, nitrogen fixation, and nutrient mineralization, which are crucial for maintaining overall soil health. Biochar also has the potential to moderate soil pH, reducing both acidity and alkalinity, thereby creating a more favorable environment for plant nutrient uptake and beneficial microbial activity. Moreover, its interaction with soil colloids enhances soil structure, promoting improved aeration, water retention, and root penetration. The structural improvements and nutrient retention capabilities of biochar also contribute to reducing nutrient runoff and soil erosion, conserving nutrients within the root zone and minimizing environmental pollution. Additionally, biochar-amended soils exhibit greater resilience to environmental stresses such as drought and salinity. This increased resilience is attributed to improved water-holding capacity, nutrient retention, and favorable microbial activity, collectively enhancing the soil's ability to withstand adverse conditions. In summary, the application of biochar has profound positive implications for nutrient cycling and overall soil health. It enhances soil fertility, improves microbial dynamics, sequesters carbon, and increases resilience to environmental perturbations. However, the specific impact of biochar on nutrient cycling and soil health may vary based on its properties, soil types, and regional conditions, necessitating further research and tailored applications.

11.8 Effects of biochar application on crop growth and yield

Biochar application can have various effects on crop growth and yield. Research indicates that biochar can improve soil structure, water retention, and nutrient availability, contributing to enhanced plant growth (Lehmann et al., 2011). Additionally, biochar may help mitigate nutrient leaching, thereby increasing nutrient retention and availability for plant uptake. By promoting beneficial microbial activity and enhancing soil fertility, biochar has the potential to positively impact crop productivity. However, the effects of biochar on crop growth and yield can be influenced by factors such as soil type, biochar properties, crop type, and environmental conditions. It is essential to consider these factors when evaluating the potential benefits of biochar application and to conduct site-specific investigations to optimize its use for achieving improved crop performance.

11.9 Challenges and limitations of biochar application

Biochar is a solid product obtained from thermochemical conversion of biomass under limited oxygen supply. It is a stable, carbon-rich material that can be used in a variety of applications, including soil amendment, water filtration, and renewable energy production. It can improve soil fertility and reduce erosion, filter water and remove pollutants, serve as a renewable energy source, and help mitigate climate change (Lehmann et al., 2011). However, there are also some challenges and limitations associated with biochar application. Here are some additional details on the challenges and limitations of biochar application: Biochar is produced by heating biomass in the absence of oxygen, a process known as pyrolysis. This process requires a significant amount of energy, which can come from a variety of sources, including fossil fuels, biomass, or solar energy. The use of fossil fuels to produce biochar can contribute to greenhouse gas emissions, while the use of biomass can compete with food production.

The cost of producing biochar varies depending on the type of biomass used, the pyrolysis process, and the location of the production facility. In general, biochar is more expensive than other soil amendments, such as compost. The use of biochar can alter soil chemistry and may not be suitable for all soils: biochar can alter soil pH, nutrient availability, and water retention. In some cases, biochar can improve soil health, but in other cases, it can have negative impacts. It is important to test the soil before applying biochar to ensure that it is compatible with the soil type. Biochar can release greenhouse gases when it is burned: biochar is a stable carbon-rich material, but it can release greenhouse gases when it is burned. This is because biochar is not a fossil fuel, but rather a product of biomass decomposition. When biochar is burned, the carbon that it contains is released into the atmosphere as carbon dioxide. The long-term effects of biochar application are not fully understood: biochar has been shown to have a number of potential benefits, but the long-term effects of biochar application are not fully understood. More research is needed to understand how biochar affects soil health, water quality, and greenhouse gas emissions over time.

11.9.1 Compilation of agricultural studies showcasing the impact of biochar on plant growth and yield

Numerous research studies have examined the impact of biochar on plant growth and yield across various crop types and growing conditions. Common findings from these studies highlight several key benefits of biochar application. Firstly, biochar amendments have been shown to improve soil fertility and nutrient availability, leading to enhanced plant growth and productivity. This is particularly evident in the increased availability of essential nutrients such as nitrogen, phosphorus, and potassium (Jeffery et al., 2011). Additionally, biochar enhances soil moisture retention, which can significantly improve plant growth and yield, especially in regions prone to drought or water stress (Lehmann et al., 2011). Furthermore, biochar's ability to moderate soil pH creates a more favorable environment for nutrient uptake and overall plant growth, particularly in acidic soils. Improved root development is another benefit observed in biochar-amended soils, with increased root proliferation and development facilitating better nutrient and water uptake, ultimately enhancing plant growth and yield. Some studies also report that biochar application may reduce pest pressure and disease incidence, indirectly benefiting plant growth and yield (Rajkovich et al., 2012). Moreover, biochar contributes to long-term carbon sequestration and stable soil organic matter accumulation, providing a beneficial environment for plant growth and yield. However, it is important to note that the specific effects of biochar on plant growth and yield can vary based on factors such as biochar type, application rate, soil type, crop species, and environmental conditions. While many studies have demonstrated the positive impacts of biochar on plant performance, further research is needed to elucidate the mechanisms underlying these effects and to optimize biochar application for specific agricultural scenarios. In conclusion, biochar has significant potential to enhance soil fertility, improve plant growth and yield, and support sustainable agricultural practices, though its application must be tailored to specific conditions to maximize its benefits.

11.9.2 Discussion of how nutrient availability facilitated by biochar influences crop productivity

The increased nutrient availability facilitated by biochar profoundly impacts crop productivity across various soil types and crop varieties (Basak et al., 2022). Biochar's porous structure and high surface area create a favorable environment for microbial activity, enhancing nutrient cycling and promoting the availability of essential elements for plant growth. Studies have shown that biochar-amended soils exhibit improved levels of nutrients such as nitrogen, phosphorus, and potassium, which are vital for crop development (Yuan et al., 2023). This enhanced nutrient availability contributes to healthier plant growth, improved stress tolerance, and ultimately higher crop yields. Additionally, biochar's ability to reduce nutrient leaching helps to maintain a more stable nutrient supply, ensuring sustained plant nutrition throughout the crop cycle (Bolan et al., 2023).

Case studies have demonstrated the positive effects of biochar on various crops and soil types. For instance, in acidic soils, biochar has been found to increase soil pH, reducing toxic aluminum concentrations and improving the availability of essential nutrients (Tsolis & Barouchas, 2023). In sandy soils, biochar has shown to enhance water and nutrient retention, mitigating the challenges of low water-holding capacity (Mukherjee et al., 2022). Furthermore, research on different crop types, including maize, rice, and vegetables, has revealed consistent improvements in growth and yield when biochar is incorporated into the soil (Mukherjee et al., 2022). These case studies collectively illustrate the positive impact of biochar on nutrient availability and crop productivity across diverse soil types and crop species, highlighting its potential as a sustainable strategy for enhancing agricultural output.

11.9.3 Discussion of potential drawbacks and challenges related to biochar use in agriculture

The use of biochar in agriculture presents several potential drawbacks and challenges that warrant consideration (Yadav & Ramakrishna, 2023). Firstly, the production of biochar often involves high temperatures, which can result in energy-intensive processes and greenhouse gas emissions if not managed efficiently. Additionally, the transportation and application of biochar may contribute to carbon emissions, particularly if large distances are involved. Another important consideration is the potential variability in biochar quality and characteristics, which can influence its effectiveness in different soil and crop systems (Mukherjee et al., 2022). In some cases, low-quality or improperly produced biochar may introduce contaminants or unwanted substances into the soil, posing risks to both plant and environmental health (Premalatha et al., 2023). Moreover, while biochar can enhance nutrient retention in soils, its effects on nutrient availability and plant growth may vary depending on factors such as soil type, climate, and management practices (Yadav & Ramakrishna, 2023). This variability may necessitate careful assessment and optimization to ensure that biochar application aligns with specific agricultural objectives.

Furthermore, the long-term stability and persistence of biochar in soil remain under investigation (Bogale et al., 2023). Understanding its interactions with native soil components, as well as its potential to impact soil microbial communities and ecosystem functions, is crucial for assessing its overall environmental implications. Lastly, the scalability and economics of biochar production and application in agriculture present challenges (Kochanek et al., 2022). The cost-effectiveness of large-scale biochar deployment, as well as the integration of biochar into existing agricultural systems, requires careful evaluation and investment. Addressing these drawbacks and challenges through research, regulation, and innovative practices will be essential in maximizing the benefits of biochar while minimizing potential risks in agricultural contexts.

11.9.4 Addressing issues like feedstock variability, production scalability, and economic considerations

Addressing issues such as feedstock variability, production scalability, and economic considerations is crucial for the widespread adoption of biochar in agriculture (Li et al., 2023).

11.9.4.1 Feedstock variability

Given the diverse range of feedstocks used in biochar production, it's essential to establish standards and best practices for feedstock selection and processing. Research into feedstock characteristics and their impact on biochar properties can help identify optimal sources for different soil and cropping systems (Bach et al., 2016). Moreover, promoting the use of sustainable and locally available feedstocks can minimize logistical challenges and reduce environmental footprints associated with transportation and sourcing (Li et al., 2023).

11.9.4.2 Production scalability

Developing scalable and efficient biochar production technologies is vital for widespread adoption. Innovation in pyrolysis and gasification processes can improve energy efficiency and reduce emissions. Additionally, exploring decentralized production models closer to biomass sources can minimize transport costs and facilitate smaller-scale agricultural applications (Rex et al., 2023).

11.9.4.3 Economic considerations

Assessing the economic viability of biochar production and application is critical. Encouraging investment in biochar research and development, as well as exploring potential carbon market opportunities, can help incentivize adoption. Furthermore, cost-benefit analyses and life cycle assessments can offer insights into the economic and environmental trade-offs associated with biochar use (Kumari et al., 2023).

11.9.4.4 Soil and ecosystem effects

Research on biochar-soil interactions and their long-term impacts on soil health and biodiversity is essential to understand potential consequences (Brtnicky et al., 2021). Monitoring studies can provide valuable insights into the persistence and effects of biochar on soil properties and microbial communities (Tsolis & Barouchas, 2023). Evaluating the net effects of biochar production and application on greenhouse gas emissions is also crucial (Yuan et al., 2023). Assessing carbon sequestration potential and emissions reductions can help determine the overall climate benefits of biochar use. Additionally, developing supportive policies and standards for biochar production and application can help mitigate potential environmental risks (Dwibedi et al., 2022). Establishing guidelines for feedstock sourcing, production practices, and application methods is necessary to ensure environmental integrity. Addressing these considerations through research, innovation, and policy support can advance the agricultural use of biochar in a sustainable and responsible manner.

11.10 Best practices and recommendations for biochar application

There are also some potential drawbacks and challenges associated with biochar use in agriculture (Hossain et al., 2020). These include: the production of biochar can be energy-intensive (Joseph et al., 2021), the high cost of producing biochar (Jatav et al., 2021), potential alterations to soil chemistry that may make biochar unsuitable for all soils (Wang et al., 2023), the release of greenhouse gases when biochar is burned (Mohammadi et al., 2020), and the incomplete understanding of the long-term effects of biochar application (Joseph et al., 2021).

Despite these challenges, biochar has the potential to be a valuable tool for sustainable agriculture. More research is needed to understand the full range of benefits and risks associated with biochar application (Jatav et al., 2021), but it has the potential to address a number of pressing environmental challenges. Here are some best practices and recommendations for biochar application: Test the soil before applying biochar: biochar can alter soil chemistry and may not be suitable for all soil types. It is important to test the soil before applying biochar to ensure that it is compatible with the soil type. Apply biochar at the correct rate: the rate of biochar application will vary depending on the soil type, the crop being grown, and the desired outcome. It is important to apply biochar at the correct rate to avoid negative impacts on the soil. Incorporate biochar into the soil: biochar should be incorporated into the soil to ensure that it is evenly distributed and to prevent it from being washed away by rain or irrigation. Use biochar in combination with other soil amendments: biochar can be used in combination with other soil amendments, such as compost, to improve soil health and crop yields. Monitor the results of biochar application: It is important to monitor the results of biochar application to ensure that it is having the desired impact on the soil and the crops being grown (Jatav et al., 2021). Overall, biochar has the potential to be a valuable tool for sustainable agriculture (Ding et al., 2016). However, more research is needed to understand the full range of benefits and risks associated with biochar application. It is important to test the soil, apply biochar at the correct rate, incorporate it into the soil, use it in combination with other soil amendments, and monitor the results of biochar application to ensure that it is having the desired impact on the soil and the crops being grown.

11.10.1 Compilation of guidelines for optimal biochar application in different agricultural contexts

Here are some specific guidelines for biochar application in different agricultural contexts: In conventional agriculture, biochar can be applied to the soil as a top dressing or incorporated into the soil at planting. The rate of application will vary depending on the soil type, the crop being grown, and the desired outcome. In general, biochar should be applied at a rate of 1–2 t/ha (Kloss et al., 2014). In organic agriculture, biochar can be used as a soil amendment or as a component of compost. Biochar can also be used to make biochar-based fertilizers. The rate of application will vary depending on the soil type, the crop being grown, and the desired outcome. In general, biochar should be applied at a rate of 0.5–1 t/ha. Furthermore, in regenerative agriculture, biochar can be used to improve soil health and to sequester carbon. Biochar can also be used to make biochar-based fertilizers. The rate of application will vary depending on the soil type, the crop being grown, and the desired outcome. In general, biochar should be applied at a rate of 1–2 t/ha (Montanarella & Lugato, 2013). It is important to note that these are just general guidelines and that the specific rate of biochar application will vary depending on the specific agricultural context. It is always best to consult with a soil scientist or other agricultural expert to determine the best rate of biochar application for our specific situation.

11.10.2 Integration of biochar with other soil management practices

Despite these challenges, biochar has the potential to be a valuable tool for sustainable agriculture (Kabir et al., 2023). More research is needed to understand the full range of benefits and risks associated with biochar application, but it has the potential to address a number of pressing environmental challenges. One way to integrate biochar with other soil management practices is to use it in combination with compost (Ayaz et al., 2021). Compost is a mixture of organic materials, such as manure, leaves, and food scraps that has been decomposed by microorganisms. Compost can improve soil fertility, water retention, and drainage. When biochar is added to compost, it can help to stabilize the compost and make it more resistant to degradation. This can extend the benefits of compost and make it a more effective soil amendment. Another way to integrate biochar with other soil management practices is to use it in combination with cover crops (Yadav & Ramakrishna, 2023). Cover crops are plants that are grown between cash crops. They can help to improve soil health, reduce erosion, and suppress weeds. When biochar is added to cover crops, it can help to improve the growth of the cover crops and make them more effective at improving soil health. Biochar can also be used in combination with other soil amendments, such as manure and rock phosphate. Manure is a nutrient-rich material that can improve soil fertility. Rock phosphate is a mineral that can provide phosphorus to plants. When biochar is added to manure or rock phosphate, it can help to improve the availability of nutrients to plants.

11.10.3 Sustainability aspects and future research directions

Biochar has the potential to be a sustainable soil amendment because it can improve soil health and crop yields, while also sequestering carbon from the atmosphere (Oni et al., 2019). However, there are some potential drawbacks and challenges associated with biochar use, such as the energy required to produce it and the potential for it to alter soil chemistry. More research is needed to understand the full range of benefits and risks associated with biochar use, but it has the potential to be a valuable tool for sustainable agriculture. Some of the sustainability aspects of biochar include: Biochar can improve soil health by increasing water retention, nutrient availability, and microbial activity (Yuan et al., 2023); biochar can sequester carbon from the atmosphere, which can help to mitigate climate change; biochar can be produced from a variety of biomass sources, including agricultural waste, forestry waste, and urban waste; biochar can be used in a variety of applications, including soil amendment, water filtration, and renewable energy production. Some of the future research directions for biochar include: understanding the long-term effects of biochar application on soil health and crop yields (Joseph et al., 2021); developing more efficient methods for biochar production (Mukherjee et al., 2022); exploring the use of biochar in other applications, such as water filtration and renewable energy production; and evaluating the economic and social impacts of biochar use.

11.11 Conclusion

The discussion on biochar's potential reveals its capacity to enhance soil fertility by improving water retention, increasing nutrient availability, and promoting microbial activity. Additionally, biochar's ability to sequester carbon from the atmosphere positions it as a valuable tool in mitigating climate change. The diverse sources from which biochar can be derived, including agricultural waste, forestry waste, and urban waste, underline its versatility and potential for sustainable resource utilization. Moreover, the multiple applications of biochar, such as soil amendment, water filtration, and renewable energy production, accentuate its adaptability across various environmental and agricultural contexts. The significance of biochar stability and reactivity in nutrient management underscores two critical aspects. Firstly, biochar stability determines its longevity in the soil and consequently, the duration for which its benefits are sustained. Moreover, biochar reactivity influences its interactions with soil and nutrients, ultimately determining its effectiveness in improving soil fertility and supporting plant growth. Understanding and optimizing these characteristics are crucial for maximizing the benefits of biochar application in agricultural systems. While biochar exhibits promise as a sustainable solution for agricultural ecosystems, it is important to note that further research is needed to comprehensively grasp its full range of benefits and associated risks. Nevertheless, the potential impact of biochar in enhancing soil health, reducing erosion, and sequestering carbon from the atmosphere highlights its significance in advancing sustainable agricultural practices. This underscores the necessity for continued exploration and understanding of biochar's potential as a transformative tool for promoting agricultural sustainability.

Acknowledgments

The authors express their gratitude to Jimma University College of Agriculture and Veterinary Medicine for providing essential technical support and reading materials. They also extend their thanks to the anonymous reviewers and Abhay Kumar for their constructive comments on an earlier version of this chapter.

Authors' contributions

Bayu Dume, Abebe Nigussie, Melkamu Mamuye, and Milkyas Ahmed conceived the idea. All authors contributed equally in literature review, data extraction, writing the manuscript draft, and review and editing the final draft. All authors approved the final version for publication.

Conflict of interest

The authors declare that they have no known competing financial interests or personal relationships that could have appeared to influence the work reported in this paper.

AI disclosure

During the preparation of this work the author(s) used Semantic Scholar, ResearchRabbit, GPT-4, Grammarly, and ProWritingAid in order to Literature search and review and for writing, English grammar correction, and improved readability. After using this tool/service, the author(s) reviewed and edited the content as needed and take(s) full responsibility for the content of the publication.

References

Al-Wabel, M. I., Al-Omran, A., El-Naggar, A. H., Nadeem, M., & Usman, A. R. A. (2013). Pyrolysis temperature induced changes in characteristics and chemical composition of biochar produced from conocarpus wastes. *Bioresource Technology*, *131*, 374–379. Available from https://doi.org/10.1016/j.biortech.2012.12.165, http://www.elsevier.com/locate/biortech.

Ayaz, M., Feizienė, D., Tilvikienė, V., Akhtar, K., Stulpinaitė, U., & Iqbal, R. (2021). Biochar role in the sustainability of agriculture and environment. *Sustainability*, *13*(3), 1330. Available from https://doi.org/10.3390/su13031330.

Bach, M., Wilske, B., & Breuer, L. (2016). Current economic obstacles to biochar use in agriculture and climate change mitigation. *Carbon Management*, *7*(3-4), 183–190.

Available from https://doi.org/10.1080/17583004.2016.1213608, http://www.tandfonline.com/loi/tcmt20.

Basak, B. B., Sarkar, B., Saha, A., Sarkar, A., Mandal, S., Biswas, J. K., Wang, H., & Bolan, N. S. (2022). Revamping highly weathered soils in the tropics with biochar application: What we know and what is needed. *Science of the Total Environment*, *822*. Available from https://doi.org/10.1016/j.scitotenv.2022.153461, http://www.elsevier.com/locate/scitotenv.

Bogale, A. A., Melash, A. A., & Percze, A. (2023). Symbiotic and asymmetric causality of the soil tillage system and biochar application on soil carbon sequestration and crop production. *Soil Systems*, *7*(2). Available from https://doi.org/10.3390/soilsystems7020048, http://www.mdpi.com/journal/soilsystems.

Bolan, S., Hou, D., Wang, L., Hale, L., Egamberdieva, D., Tammeorg, P., Li, R., Wang, B., Xu, J., Wang, T., Sun, H., Padhye, L. P., Wang, H., Siddique, K. H. M., Rinklebe, J., Kirkham, M. B., & Bolan, N. (2023). The potential of biochar as a microbial carrier for agricultural and environmental applications. *Science of the Total Environment*, *886*, 163968. Available from https://doi.org/10.1016/j.scitotenv.2023.163968.

Brtnicky, M., Datta, R., Holatko, J., Bielska, L., Gusiatin, Z. M., Kucerik, J., Hammerschmiedt, T., Danish, S., Radziemska, M., Mravcova, L., Fahad, S., Kintl, A., Sudoma, M., Ahmed, N., & Pecina, V. (2021). A critical review of the possible adverse effects of biochar in the soil environment. *Science of the Total Environment*, *796*. Available from https://doi.org/10.1016/j.scitotenv.2021.148756, http://www.elsevier.com/locate/scitotenv.

Chen, L., Zhang, S., Zhao, J., & Wang, X. (2022). Effects of biochar on soil properties and crop growth in sustainable agriculture: A review. *Agriculture, Ecosystems & Environment*, *325*, 107742. Available from https://doi.org/10.1016/j.agee.2021.107742.

Dai, Y., Zheng, H., Jiang, Z., & Xing, B. (2020). Comparison of different crop residue-based technologies for their energy production and air pollutant emission. *Science of the Total Environment*, *707*, 136122. Available from https://doi.org/10.1016/j.scitotenv.2019.136122.

Ding, Y., Liu, Y., Liu, S., Li, Z., Tan, X., Huang, X., Zeng, G., Zhou, L., & Zheng, B. (2016). Biochar to improve soil fertility. A review. *Agronomy for Sustainable Development*, *36*(2). Available from https://doi.org/10.1007/s13593-016-0372-z, http://www.springerlink.com/content/1773-0155.

Dwibedi, S. K., Pandey, V. C., Divyasree, D., & Bajpai, O. (2022). Biochar-based land development. *Land Degradation and Development*, *33*(8), 1139−1158. Available from https://doi.org/10.1002/ldr.4185, http://onlinelibrary.wiley.com/journal/10.1002/(ISSN)1099-145X.

Elkhlifi, Z., Iftikhar, J., Sarraf, M., Ali, B., Saleem, M. H., Ibranshahib, I., Bispo, M. D., Meili, L., Ercisli, S., Kayabasi, E. T., Ansari, N. A., Hegedűsová, A., & Chen, Z. (2023). Potential role of biochar on capturing soil nutrients, carbon sequestration and managing environmental challenges: A review. *Sustainability*, *15*(3), 2527. Available from https://doi.org/10.3390/su15032527.

Garg, S., Rumjit, N. P., & Roy, S. (2023). Smart agriculture and nanotechnology: Technology, challenges, and new perspective. *Advanced Agrochem*. Available from https://doi.org/10.1016/j.aac.2023.11.001, http://www.keaipublishing.com/en/journals/advanced-agrochem/.

Glaser, B., Lehmann, J., & Zech, W. (2002). Ameliorating physical and chemical properties of highly weathered soils in the tropics with charcoal — a review. *Biology and Fertility of Soils*, *35*(4), 219–230. Available from https://doi.org/10.1007/s00374-002-0466-4.

Gorovtsov, A. V., Minkina, T. M., Mandzhieva, S. S., Perelomov, L. V., Soja, G., Zamulina, I. V., Rajput, V. D., Sushkova, S. N., Mohan, D., & Yao, J. (2020). The mechanisms of biochar interactions with microorganisms in soil. *Environmental Geochemistry and Health*, *42*(8), 2495–2518. Available from https://doi.org/10.1007/s10653-019-00412-5, http://www.wkap.nl/journalhome.htm/0269-4042.

Guo, T., Campbell-Arvai, V., & Cardinale, B. J. (2021). Why does the public support or oppose agricultural nutrient runoff regulations? The effects of political orientation, environmental worldview, and policy specific beliefs. *Journal of Environmental Management*, *279*. Available from https://doi.org/10.1016/j.jenvman.2020.111708, https://www.sciencedirect.com/journal/journal-of-environmental-management.

Hossain, M. Z., Bahar, M. M., Sarkar, B., Donne, S. W., Ok, Y. S., Palansooriya, K. N., Kirkham, M. B., Chowdhury, S., & Bolan, N. (2020). Biochar and its importance on nutrient dynamics in soil and plant. *Biochar*, *2*(4), 379–420. Available from https://doi.org/10.1007/s42773-020-00065-z, http://springer.com/journal/42773.

Jatav, H. S., Rajput, V. D., Minkina, T., Singh, S. K., Chejara, S., Gorovtsov, A., Barakhov, A., Bauer, T., Sushkova, S., Mandzhieva, S., Burachevskaya, M., & Kalinitchenko, V. P. (2021). Sustainable approach and safe use of biochar and its possible consequences. *Sustainability*, *13*(18), 10362. Available from https://doi.org/10.3390/su131810362.

Jeffery, S., Verheijen, F. G. A., van der Velde, M., & Bastos, A. C. (2011). A quantitative review of the effects of biochar application to soils on crop productivity using meta-analysis. *Agriculture, Ecosystems and Environment*, *144*(1), 175–187. Available from https://doi.org/10.1016/j.agee.2011.08.015, http://www.elsevier.com/inca/publications/store/5/0/3/2/9/8.

Joseph, S., Cowie, A. L., Van Zwieten, L., Bolan, N., Budai, A., Buss, W., Cayuela, M. L., Graber, E. R., Ippolito, J. A., Kuzyakov, Y., Luo, Y., Ok, Y. S., Palansooriya, K. N., Shepherd, J., Stephens, S., Weng, Z., & Lehmann, J. (2021). How biochar works, and when it doesn't: A review of mechanisms controlling soil and plant responses to biochar. *GCB Bioenergy*, *13*(11), 1731–1764. Available from https://doi.org/10.1111/gcbb.12885, http://onlinelibrary.wiley.com/journal/10.1111/(ISSN)1757-1707.

Kabir, M. J., Hasan, M. M., Hossain, M. A., Islam, M. S., & Alam, M. Z. (2023). Biochar application improves soil health and enhances crop productivity: A meta-analysis. *Soil & Tillage Research*, *229*, 105668. Available from https://doi.org/10.1016/j.still.2022.105668.

Kloss, S., Zehetner, F., Wimmer, B., Buecker, J., Rempt, F., & Soja, G. (2014). Biochar application to temperate soils: Effects on soil fertility and crop growth under greenhouse conditions. *Journal of Plant Nutrition and Soil Science*, *177*(1), 3–15. Available from https://doi.org/10.1002/jpln.201200282.

Kochanek, J., Soo, R. M., Martinez, C., Dakuidreketi, A., & Mudge, A. M. (2022). Biochar for intensification of plant-related industries to meet productivity, sustainability and economic goals: A review. *Resources, Conservation and Recycling*, *179*. Available from https://doi.org/10.1016/j.resconrec.2021.106109, http://www.elsevier.com/locate/resconrec.

Kumari, K., Kumar, R., Bordoloi, N., Minkina, T., Keswani, C., & Bauddh, K. (2023). Unravelling the recent developments in the production technology and efficient applications of biochar for agro-ecosystems. *Agriculture*, *13*(3), 512. Available from https://doi.org/10.3390/agriculture13030512.

Laird, D. A. (2008). The charcoal vision: A win-win-win scenario for simultaneously producing bioenergy, permanently sequestering carbon, while improving soil and water quality. *Agronomy Journal*, *100*(1), 178–181. Available from https://doi.org/10.2134/agronj2007.0161.

Lehmann, J. (2007). Bio-energy in the black. *Frontiers in Ecology and the Environment*, *5*(7), 381–387. Available from https://doi.org/10.1890/1540-9295(2007)5[381:BITB]2.0.CO;2.

Lehmann, J., & Joseph, S. (Eds.), (2009). *Biochar for Environmental Management: Science and Technology* (1st ed.). Routledge. Available from https://doi.org/10.4324/9781849770552.

Lehmann, J., & Joseph, S. (Eds.), (2015). *Biochar for environmental management: Science, technology and implementation* (2nd ed.). Routledge. Available from https://doi.org/10.4324/9780203762264.

Lehmann, J., Rillig, M. C., Thies, J., Masiello, C. A., Hockaday, W. C., & Crowley, D. (2011). Biochar effects on soil biota – a review. *Soil Biology and Biochemistry*, *43*(9), 1812–1836. Available from https://doi.org/10.1016/j.soilbio.2011.04.022.

Leslie, J. E., Weersink, A., Yang, W., & Fox, G. (2017). Actual versus environmentally recommended fertilizer application rates: Implications for water quality and policy. *Agriculture, Ecosystems and Environment*, *240*, 109–120. Available from https://doi.org/10.1016/j.agee.2017.02.009, http://www.elsevier.com/inca/publications/store/5/0/3/2/9/8.

Li, S., Galoustian, T., & Trejo, H. (2023). Biochar pyrolyzed with concentrated solar radiation for enhanced nitrate adsorption. *Journal of Analytical and Applied Pyrolysis*, *174*, 106131. Available from https://doi.org/10.1016/j.jaap.2023.106131.

Li, Y., Xing, B., Ding, Y., Han, X., & Wang, S. (2020). A critical review of the production and advanced utilization of biochar via selective pyrolysis of lignocellulosic biomass. *Bioresource Technology*, *312*, 123614. Available from https://doi.org/10.1016/j.biortech.2020.123614.

Liu, C., Li, J., Xu, Y., Guo, S., Pan, G., & Wang, X. (2021). Biochar's effect on soil organic carbon sequestration in agricultural soils: A meta-analysis. *Science of The Total Environment*, *755*, 142531. Available from https://doi.org/10.1016/j.scitotenv.2020.142531.

Lyu, H., He, Y., Liu, H., Fan, J., & Hu, L. (2020). Biochar improves soil carbon sequestration and soil properties in drought-prone regions. *Agricultural Water Management*, *241*, 106379. Available from https://doi.org/10.1016/j.agwat.2020.106379.

Malhi, G. S., Kaur, M., & Kaushik, P. (2021). Impact of climate change on agriculture and its mitigation strategies: A review. *Sustainability*, *13*(3), 1318. Available from https://doi.org/10.3390/su13031318.

Mohammadi, A., Khoshnevisan, B., Venkatesh, G., & Eskandari, S. (2020). A critical review on advancement and challenges of biochar application in paddy fields: Environmental and life cycle cost analysis. *Processes*, *8*(10), 1275. Available from https://doi.org/10.3390/pr8101275.

Montanarella, L., & Lugato, E. (2013). The application of biochar in the eu: Challenges and opportunities. *Agronomy*, *3*(2), 462–473. Available from https://doi.org/10.3390/agronomy3020462, https://www.mdpi.com/2073-4395/3/2/462/pdf.

Mukherjee, S., Sarkar, B., Aralappanavar, V. K., Mukhopadhyay, R., Basak, B. B., Srivastava, P., Marchut-Mikołajczyk, O., Bhatnagar, A., Semple, K. T., & Bolan, N. (2022). Biochar-microorganism interactions for organic pollutant remediation: Challenges and perspectives. *Environmental Pollution, 308*. Available from https://doi.org/10.1016/j.envpol.2022.119609, https://www.journals.elsevier.com/environmental-pollution.

Novak, J. M., Lima, I., Xing, B., Gaskin, J. W., Steiner, C., Das, K. C., Ahmedna, M., Rehrah, D., Watts, D. W., Busscher, W. J., & Schomberg, H. (2009). Characterization of designer biochar produced at different temperatures and their effects on a loamy sand. *Annals of Environmental Science, 3*.

Oni, B. A., Oziegbe, O., & Olawole, O. O. (2019). Significance of biochar application to the environment and economy. *Annals of Agricultural Sciences, 64*(2), 222–236. Available from https://doi.org/10.1016/j.aoas.2019.12.006, http://www.elsevier.com/journals/annals-of-agricultural-sciences/0570-1783/.

Pandit, N. R., Mulder, J., Hale, S. E., Martinsen, V., Schmidt, H. P., & Cornelissen, G. (2018). Biochar improves maize growth by alleviation of nutrient stress in a moderately acidic low-input Nepalese soil. *Science of the Total Environment, 625*, 1380–1389. Available from https://doi.org/10.1016/j.scitotenv.2018.01.022, http://www.elsevier.com/locate/scitotenv.

Paramesh, V., Kumar, R., Meena, H. M., & Singh, S. (2023). Enhancing soil fertility and crop yield with biochar: A sustainable approach. *Journal of Cleaner Production, 384*, 135646. Available from https://doi.org/10.1016/j.jclepro.2022.135646.

Pawlak, K., & Kołodziejczak, M. (2020). The role of agriculture in ensuring food security in developing countries: Considerations in the context of the problem of sustainable food production. *Sustainability, 12*(13), 5488. Available from https://doi.org/10.3390/su12135488.

Premalatha, R. P., Poorna Bindu, J., Nivetha, E., Malarvizhi, P., Manorama, K., Parameswari, E., & Davamani, V. (2023). A review on biochar's effect on soil properties and crop growth. *Frontiers in Energy Research, 11*. Available from https://doi.org/10.3389/fenrg.2023.1092637, http://www.frontiersin.org/Energy_Research/.

Quilliam, R. S., Marsden, K. A., Gertler, C., Rousk, J., DeLuca, T. H., & Jones, D. L. (2012). Nutrient dynamics, microbial growth and weed emergence in biochar amended soil are influenced by time since application and reapplication rate. *Agriculture, Ecosystems and Environment, 158*, 192–199. Available from https://doi.org/10.1016/j.agee.2012.06.011.

Rajkovich, S., Enders, A., Hanley, K., Hyland, C., Zimmerman, A. R., & Lehmann, J. (2012). Corn growth and nitrogen nutrition after additions of biochars with varying properties to a temperate soil. *Biology and Fertility of Soils, 48*(3), 271–284. Available from https://doi.org/10.1007/s00374-011-0624-7.

Rex, P., Mohammed Ismail, K. R., Meenakshisundaram, N., Barmavatu, P., & Sai Bharadwaj, A. V. S. L. (2023). Agricultural biomass waste to biochar: A review on biochar applications using machine learning approach and circular economy. *ChemEngineering, 7*(3). Available from https://doi.org/10.3390/chemengineering7030050, http://www.mdpi.com/journal/ChemEngineering.

Selim, M. M. (2020). Introduction to the integrated nutrient management strategies and their contribution to yield and soil properties. *International Journal of Agronomy, 2020*. Available from https://doi.org/10.1155/2020/2821678, http://www.hindawi.com/journals/ija/.

Semida, W. M., Beheiry, H. R., Sétamou, M., Simpson, C. R., Abd El-Mageed, T. A., Rady, M. M., & Nelson, S. D. (2019). Biochar implications for sustainable agriculture and environment: A review. *South African Journal of Botany*, *127*, 333–347. Available from https://doi.org/10.1016/j.sajb.2019.11.015, http://www.elsevier.com.

Shackley, S., Carter, S., Knowles, T., Middelink, E., Haefele, S., Sohi, S., Cross, A., & Haszeldine, S. (2012). Sustainable gasification-biochar systems? A case-study of rice-husk gasification in Cambodia, Part I: Context, chemical properties, environmental and health and safety issues. *Energy Policy*, *42*, 49–58. Available from https://doi.org/10.1016/j.enpol.2011.11.026.

Spokas, K. A. (2010). Review of the stability of biochar in soils: Predictability of O:C molar ratios. *Carbon Management*, *1*(2), 289–303. Available from https://doi.org/10.4155/cmt.10.32.

Tsolis, V., & Barouchas, P. (2023). Biochar as soil amendment: The effect of biochar on soil properties using VIS-NIR diffuse reflectance spectroscopy, biochar aging and soil microbiology—a review. *Land*, *12*(8), 1580. Available from https://doi.org/10.3390/land12081580.

Tuğrul, K. M. (2019). Soil management in sustainable agriculture. In *Soil Management and Plant Nutrition for Sustainable Crop Production*, IntechOpen. Available from https://doi.org/10.5772/intechopen.88319.

Wang, L., Deng, J., Yang, X., Hou, R., & Hou, D. (2023). Role of biochar toward carbon neutrality. *Carbon Research*, *2*(1). Available from https://doi.org/10.1007/s44246-023-00035-7, http://www.springer.com/journal/44246.

Wang, J., Zhang, M., Xiong, Z., Liu, P., Pan, G., & Lev, S. (2021). Biochar addition to soil significantly increased the microbial-derived carbon fraction, which is more resistant to decomposition, thereby contributing to long-term carbon sequestration. *Journal of Soils and Sediments*, *21*(3), 1568–1581. Available from https://doi.org/10.1007/s11368-020-02872-5.

Wang, P., Zhang, W., Xue, Y., Liu, H., & Wu, J. (2020). The effect of biochar on soil properties and plant growth: A comprehensive review. *Environmental Science and Pollution Research*, *27*(29), 36230–36239. Available from https://doi.org/10.1007/s11356-020-09356-9.

Woolf, D., Amonette, J. E., Street-Perrott, F. A., Lehmann, J., & Joseph, S. (2010). Sustainable biochar to mitigate global climate change. *Nature Communications*, *1*(5). Available from https://doi.org/10.1038/ncomms1053, http://www.nature.com/ncomms/index.html.

Wurtsbaugh, W. A., Paerl, H. W., & Dodds, W. K. (2019). Nutrients, eutrophication and harmful algal blooms along the freshwater to marine continuum. *Wiley Interdisciplinary Reviews: Water*, *6*(5). Available from https://doi.org/10.1002/WAT2.1373, http://wires.wiley.com/WileyCDA/WiresJournal/wisId-WAT2.html.

Xu, P., Gao, Y., Cui, Z., Wu, B., Yan, B., Wang, Y., Zaitongguli, K., Wen, M., Wang, H., Jing, N., Wang, Y., Chao, C., & Xue, W. (2023). Research progress on effects of biochar on soil environment and crop nutrient absorption and utilization. *Sustainability*, *15*(6), 4861. Available from https://doi.org/10.3390/su15064861.

Yadav, R., & Ramakrishna, W. (2023). Biochar as an environment-friendly alternative for multiple applications. *Sustainability*, *15*(18), 13421. Available from https://doi.org/10.3390/su151813421.

Yuan, Y., Liu, Q., Zheng, H., Li, M., Liu, Y., Wang, X., Peng, Y., Luo, X., Li, F., Li, X., & Xing, B. (2023). Biochar as a sustainable tool for improving the health of salt-affected soils. *Soil & Environmental Health*, *1*(3), 100033. Available from https://doi.org/10.1016/j.seh.2023.100033.

Zhang, C., Li, S., Lin, Q., & Wei, J. (2021). Biochar effects on soil organic carbon storage and microbial activity in agricultural soils. *Soil & Tillage Research*, *208*, 104882. Available from https://doi.org/10.1016/j.still.2020.104882.

Zhang, Y., Liu, B., Chen, W., & Zhao, Y. (2023). Biochar's impact on carbon sequestration and soil health: A global synthesis. *Environmental Research*, *214*, 113929. Available from https://doi.org/10.1016/j.envres.2022.113929.

CHAPTER

Prospects of biochar for boosting crop productivity and soil sustainability

12

Joao Antonangelo[1], Mahesh Pallikonda[2] and Tarah Sullivan[1]

[1]*Department of Crop and Soil Sciences, Washington State University, Pullman, WA, United States*

[2]*Department of Engineering Technology, Austin Peay State University, Clarksville, TN, United States*

12.1 Introduction

12.1.1 Overview of the significance of biochar in modern agriculture and soil sustainability

Biochar, an organic substance, offers a versatile array of applications in agriculture, encompassing carbon storage, heightened plant yields, soil pH adjustment, and acting as a growth medium for microbes (Chausali et al., 2021; Wang et al., 2017). Its integration into agricultural soil not only curtails the need for lime, owing to its pH-balancing potential in acidic soils (Antonangelo & Zhang, 2019; Antonangelo et al., 2019) but also minimizes reliance on commercial fertilizers (Luque et al., 2023). Furthermore, biochar's introduction enhances macronutrients availability, notably phosphorus (P) and potassium (K), while simultaneously reducing the necessity for nitrogen (N) application, thus mitigating N leaching into groundwater (Li et al., 2023). The literature also underscores biochar's role in elevating soil pH, augmenting organic carbon (OC) storage, and enhancing cation exchange capacity (Chan et al., 2008).

12.1.2 Production of biochar: methods and considerations

Biochar is majorly produced through pyrolysis or gasification. Pyrolysis involves heating organic materials in the absence (or low presence) of oxygen, with feedstock selection impacting biochar properties. Temperature control during pyrolysis dictates stability, while additional activation processes enhance porosity and nutrient retention. Rigorous quality assessment ensures suitability for agricultural use, considering pH, surface area, and contaminants. Integration strategies focus on optimal application rates and compatibility with farming practices. Gasification, an alternative process, involves partial combustion and offers versatility in feedstock

Biochar Ecotechnology for Sustainable Agriculture and Environment. DOI: https://doi.org/10.1016/B978-0-443-29855-4.00014-X

utilization. Operating parameters in gasification, such as temperature and residence time, influence biochar properties, necessitating distinct considerations in quality assessment. Integrating insights from both processes enhances our understanding of diverse biochar production avenues, fostering a comprehensive approach to sustainable soil management in agriculture.

12.2 Different methods of biochar production

12.2.1 Factors influencing biochar properties: feedstock type, temperature, residence time, and heating rate

As the world faces increasing environmental challenges, the need for sustainable practices in agriculture, waste management, and carbon sequestration has become paramount. Biochar, a carbon-rich material produced through the pyrolysis of biomass, has emerged as a promising tool with the potential to address multiple sustainability goals (Anand et al., 2023). The properties of biochar are highly influenced by several key factors throughout the production process. This section explores the major factors that play a crucial role in shaping biochar characteristics, including feedstock type, temperature, residence time, heating rate, and pyrolysis method.

12.2.1.1 Feedstock type

The selection of feedstock in biochar production is a critical determinant of the final biochar properties. Different biomass sources contribute unique chemical compositions, physical structures, and functionalities to the biochar. The chemical composition of the feedstock significantly influences the elemental makeup of the resulting biochar. For instance, feedstocks with high lignin content, such as wood and certain crop residues, tend to yield biochars with greater aromaticity and recalcitrance (Lehmann & Joseph, 2015). On the other hand, feedstocks rich in cellulose and hemicellulose, like straw or grass, may produce biochar with higher oxygen content. The physical structure of the feedstock influences the surface area and porosity of the resulting biochar. Feedstocks with fibrous or porous structures, such as coconut shells or certain agricultural residues, can produce biochars with higher surface area and porosity. This is crucial for applications like water filtration and adsorption of contaminants. Different feedstocks contribute varying amounts of nutrients to the biochar. Manure-based feedstocks, for example, often result in biochars with higher nitrogen content, making them valuable for soil amendment (Spokas et al., 2012). The nutrient content of biochar derived from plant-based feedstocks depends on the original nutrient content of the biomass. When biochar is applied to soils as a soil amendment, the feedstock type can affect soil properties. For example, biochars derived from poultry litter may contribute to increased soil fertility due to their nutrient content (Holatko et al.,

2022). On the other hand, biochar from woody feedstocks may enhance soil structure and water retention.

12.2.1.2 Temperature

Pyrolysis temperature is a critical parameter in the production of biochar, influencing its physical, chemical, and functional properties (Sharma et al., 2024). The temperature at which biomass is subjected to pyrolysis determines the extent of decomposition and the nature of the resulting biochar. Pyrolysis temperature significantly affects the porosity and surface area of biochar. Higher temperatures generally result in biochars with increased porosity and larger surface areas. This is crucial for applications like water filtration, where higher surface areas enhance adsorption capacity. Lower pyrolysis temperatures may lead to biochars with higher bulk density, while higher temperatures often result in lighter, more porous biochars. Density impacts the physical handling of biochar and its ability to integrate effectively into soils. The pH of biochar is influenced by pyrolysis temperature, with higher temperatures typically yielding biochar with higher pH values. This can impact soil acidity and nutrient availability, especially in acidic soils (Spokas et al., 2012). Understanding the impact of pyrolysis temperature on biochar properties is crucial for tailoring biochar to specific soil amendment needs. For instance, biochars with higher nutrient content and lower recalcitrance may be desirable for promoting plant growth. The stability of biochar in soils, influenced by pyrolysis temperature, is central to its role in carbon sequestration. Biochars produced at optimal temperatures can potentially sequester carbon for extended periods, contributing to climate change mitigation (Lehmann & Joseph, 2015).

12.2.1.3 Residence time

Residence time, the duration of biomass exposure to pyrolysis conditions, is a crucial factor influencing the properties of biochar. The time biomass spends in the pyrolysis reactor affects the extent of decomposition, carbonization, and the resulting biochar characteristics (Sharma et al., 2024). Longer residence times generally lead to biochars with increased porosity and surface area. The extended exposure allows for more complete pyrolysis, resulting in a porous structure that enhances biochar's adsorption capacity. Longer residence times promote greater carbonization of the biomass, leading to biochars with higher carbon content and enhanced stability in soils. This increased carbonization is often associated with reduced volatile matter content. Shorter residence times typically result in biochars with higher volatile matter content. This may contribute to biochars with greater nutrient availability upon soil application. Biochars produced with longer residence times may have a higher potential to enhance soil fertility due to increased stability, nutrient retention, and improved physical structure.

12.2.1.4 Pyrolysis method

The choice of pyrolysis method is a critical factor in biochar production, influencing its yield, composition, and properties. Different thermochemical processes, including slow pyrolysis, fast pyrolysis, gasification, and torrefaction, have distinct effects on biomass, influencing its physical, chemical, and thermal properties (Anand et al., 2023). The effects of slow pyrolysis, fast pyrolysis, gasification, and torrefaction on biomass are diverse and application-dependent. Understanding these effects allows for the tailored production of biochars, bio-oils, syngas, or torrefied biomass to meet specific needs, contributing to the development of sustainable and efficient biomass utilization systems.

12.2.1.5 Heating rate

The heating rate during pyrolysis influences the kinetics of the thermochemical conversion process, impacting biochar properties. Faster heating rates often result in higher biochar yields, but the properties of the biochar may differ from those produced under slower heating conditions. Heating rate can influence the physical structure of biochar particles, including factors such as particle size, density, and surface morphology. The speed at which pyrolysis occurs can affect the chemical composition of biochar (Angın, 2013), influencing its nutrient content and stability.

12.2.2 Electrochemical properties of biochar

Biochar, a carbonaceous material derived from the pyrolysis or gasification of biomass, exhibits intriguing electrochemical properties that have garnered increasing attention for applications in energy storage, environmental remediation, and electrocatalysis. This section explores the electrochemical aspects of biochar, delving into its capacitive properties, conductivity, redox activity, and potential applications.

12.2.2.1 Capacitive properties

Capacitive property refers to the ability of a material or a device to store electrical energy in the form of an electric charge. This property is often associated with capacitors, which are electronic components designed to store and release electrical energy. One of the key factors contributing to the capacitive behavior of biochar is its high specific surface area. The porous structure, influenced by production parameters such as pyrolysis temperature and residence time, provides a large surface area for electrolyte interaction. This property is crucial for electrochemical capacitors, where efficient charge storage is essential. The porosity of biochar, determined by factors like residence time and heating rate during production, enhances its capacitive performance. The availability of pores allows for easy access of ions to the electrode surface, facilitating rapid charge/discharge cycles in supercapacitor applications.

12.2.2.2 Conductivity

Electrical conductivity (EC) is a fundamental property of materials that describes their ability to conduct an electric current. It is a measure of how easily electric charge (carried by electrons) can move through a material when a voltage is applied across it. The pyrolysis temperature significantly affects the EC of biochar. Higher temperatures generally result in biochars with enhanced conductivity. This is attributed to the graphitization and carbonization processes, which create a more conductive carbon matrix. The conductivity is crucial for applications where efficient electron transport is essential, such as in electrodes for energy storage devices. The type of feedstock used in biochar production also influences its conductivity. Different feedstocks contribute to biochars with distinct elemental compositions, which, in turn, affect their electrical properties. For example, biochars derived from certain wood types may exhibit higher conductivity compared to those from agricultural residues (Gabhi et al., 2024).

12.2.2.3 Redox activity

Redox activity refers to the ability of a substance to undergo redox (reduction–oxidation) reactions, which involve the transfer of electrons between chemical species. These reactions are fundamental in various chemical and biological processes, playing a crucial role in energy storage, corrosion, combustion, and biological redox signaling. Biochar's redox activity is linked to the presence of surface functional groups, such as hydroxyl, carbonyl, or quinone groups. These groups are influenced by factors like feedstock type and pyrolysis temperature. The redox activity enables biochar to undergo electron transfer reactions, making it suitable for applications involving redox processes. The redox activity of biochar has environmental implications, particularly in soil remediation. Biochar's ability to facilitate redox reactions can influence nutrient availability, metal mobility, and the degradation of pollutants in soils. This property enhances its role as a soil amendment and environmental remediation agent.

12.3 Explanation of biochar's electrochemical characteristics

12.3.1 Importance of surface area, porosity, and functional groups for electrochemical performance

The electrochemical performance of materials is significantly influenced by their surface area, porosity, and the presence of functional groups. These characteristics play a crucial role in determining how efficiently a material can participate in electrochemical reactions, making them key factors in various applications such as energy storage, catalysis, and sensing. The interplay of surface area, porosity, and functional groups is pivotal for optimizing electrochemical performance in various applications. Materials engineered to maximize these characteristics demonstrate improved efficiency in charge storage, catalysis, and sensing. Tailoring these features enables the design of advanced materials that meet the specific requirements of diverse electrochemical systems, contributing to the development of more efficient and sustainable technologies.

12.3.2 Surface area

The surface area of a material is directly proportional to its electrochemically active sites. A higher surface area provides more sites for electrochemical reactions to occur, facilitating greater interaction with ions or molecules in the surrounding environment. In energy storage devices like supercapacitors and batteries, a higher surface area allows for more charge storage sites. Electrodes with increased surface area enhance the capacity of the device to store electrical energy. Catalysts with higher surface areas provide more active sites for chemical reactions. This is particularly important in electrocatalysis, where the efficiency of catalysts in facilitating electrochemical reactions is directly related to their surface area. Finally, in electrochemical sensors, a larger surface area allows for a more significant interaction with analyte molecules, leading to improved sensitivity and detection limits.

12.3.3 Functional groups

Functional groups are specific arrangements of atoms within a molecule that confer distinct chemical properties. The presence of functional groups on a material's surface influences its reactivity and interaction with ions or molecules in electrochemical processes. Functional groups with redox-active sites can contribute to the electrochemical activity of a material. This is exploited in applications such as supercapacitors and batteries, where redox reactions contribute to energy storage. Certain functional groups act as catalytic sites, enhancing the electrocatalytic activity of materials. This is particularly important in fuel cells, where functional groups facilitate the electrochemical reactions involved in energy conversion. Functional groups with specific chemical affinities can be employed in electrochemical sensors to selectively interact with target analytes, improving the sensor's specificity and sensitivity.

12.3.4 Porosity

Porosity refers to the presence of pores or void spaces within a material. Porous materials have a larger effective surface area, and the interconnected pore structure enhances mass transport and ion diffusion. Porous materials facilitate the diffusion of ions, improving the kinetics of electrochemical reactions. This is particularly crucial in energy storage devices, where rapid ion transport is essential for high charge/discharge rates. The porous structure allows electrolyte access to a larger portion of the material's surface, ensuring more active sites are available for electrochemical reactions. This is vital for devices like fuel cells and batteries. Porous catalysts provide more sites for reactants to adsorb and participate in electrochemical reactions. This enhances the catalytic activity and efficiency of the material.

12.3.5 Potential applications of electrochemically active biochars, such as in microbial fuel cells or electrochemical remediation

Electrochemically active biochars can serve as anodes in microbial fuel cells (Mishra et al., 2024), providing a conductive surface for microbial colonization and facilitating the extracellular electron transfer process. Biochars, especially those with redox-active functional groups, contribute to the electrochemical reactions at the anode, leading to improved power generation in microbial fuel cells. Additionally, biochars can be employed as anodes or cathodes in electrochemical cells for pollutant degradation. Biochars can be used in electrochemical systems to induce either reduction or oxidation reactions, depending on the targeted contaminants. The use of biochars, especially those derived from sustainable sources, aligns with environmentally friendly remediation practices.

12.3.6 Enhancing nutrient use efficiency

Enhancing nutrient use efficiency in agriculture is crucial for sustainable and environmentally responsible farming practices. The integration of precision agriculture, controlled-release fertilizers, nutrient stabilizers, such as biochars (Zhang et al., 2024), cover crops, biological approaches, and effective nutrient monitoring contributes to optimizing nutrient utilization, promoting crop health, and minimizing the environmental impact of agricultural activities. Adopting these strategies collectively can lead to more resilient and sustainable agricultural systems.

12.4 Role of biochar in improving nutrient retention and availability in soil

12.4.1 Mechanisms behind biochar's ability to adsorb and release nutrients slowly over time

Industrialization, deforestation, and overgrazing have depleted soil nutrients, leading to reduced crop yields and food security (Ramadan et al., 2019). Excessive use of chemical fertilizers has caused nitrogen (N) leaching and phosphorus (P) runoff, posing environmental risks and diminishing soil nutrient availability. Conversely, biochar, with its high nutrient-carrying capacity due to its dense charge density, enhances soil fertility and boosting crop yields (Liu et al., 2016). Biochar serves as a soil enhancer by increasing P and N availability while reducing environmental losses (Dong et al., 2017; Pan et al., 2017; Rawat et al., 2019; Zeng et al., 2013). Recent studies suggest that combining biochar with advanced technologies, like biochar-based nanocomposites, could yield even better results (Bopp et al., 2016; Saxena et al., 2014). These materials, including nano-derived ones, help retain nutrients in the soil for longer periods by leveraging biochar's functional groups, pore structure, and high surface area (Gale et al., 2021). However, it remains uncertain whether biochar or nanobiochar effectively remediates excessive nutrient levels from past fertilizer overuse. Therefore a pragmatic approach might involve the simultaneous use of biochar and commercial fertilizers to enhance nitrogen, phosphorus, and potassium use efficiency.

12.4.2 Studies and examples demonstrating enhanced nutrient use efficiency with biochar amendment

Biochar, when enriched with external nutrients like N, P, and K, serves as an effective slow-release fertilizer for plants (Liu et al., 2023). This is particularly advantageous when biochar is derived from low-nutrient lignocellulosic biomass, as it can be conveniently impregnated with external nutrients and formed into pellets (Kim et al., 2014). The resulting product regulates nutrient release at a reduced rate, minimizing nutrient loss and enhancing nutrient utilization efficiency (Kim et al., 2014). Combining nano-structured biochar with commercial N fertilizer urea has shown significant improvements in crop yield, reducing the need for urea input (Joseph et al., 2013; Qian et al., 2014). Additionally, biochar containing phosphorus serves as a valuable source of nutrients for long-term soil fertility improvement as a slow-release fertilizer. Mixing charcoal with ashes and infusing nutrients like N, P, and K into wood residue yields slow-release K and N fertilizers (Liu et al., 2023). These biochar-based slow-release fertilizers exhibit promise in sustainable modern agriculture due to their nutrient retention capabilities, although comprehensive evaluation remains necessary (Liu et al., 2023).

Biochar-based products enhance soil nutrient availability due to their electrochemical properties and inherent nutrient content. They are rich in carbon and boast higher levels of key macronutrients such as N, P, K, Mg, Ca, and S, alongside essential micronutrients such as Cu, Mn, Fe, and Zn, as well as ash (Hossain et al., 2011), making them potent organic fertilizers. Studies consistently reveal rapid nutrient release upon biochar incorporation into the soil (Liu et al., 2023). Notably, animal-sourced feedstock biochar, e.g., broiler litter and sewage sludge, exhibits higher total P and N content compared to plant-based biomass biochar like wood and green waste (Bopp et al., 2016). In unpublished research by Antonangelo, poultry litter–derived biochar (PLB) significantly outperforms switchgrass-derived biochar (SGB) in increasing P and K availability, with a 20-fold difference (Fig. 12.1), attributed to higher nutrient levels in animal-derived biochar and its ash content compared to green waste-derived biochar (Antonangelo et al., 2019). However, caution is warranted as excessive PLB application rates (2%–4%) can lead to nutrient accumulation (Fig. 12.1), potentially harmful to the environment. Lower PLB rates (0.5% and 1%) suffice to achieve optimal perennial ryegrass (*Lolium perenne*) yields, with excessive rates negatively impacting crop productivity and growth (Antonangelo & Zhang, 2019, 2021).

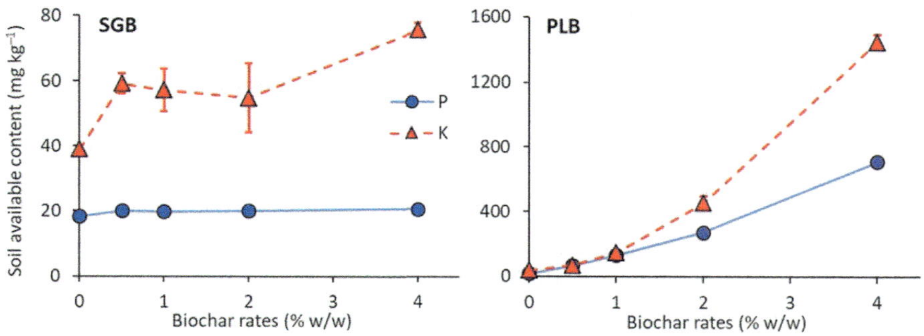

FIGURE 12.1 The increment of soil nutrients as a function of biochar application rates.

The method for soil available-P and -K is Mehlich-3 (Antonangelo et al., 2022). *PLB*, poultry litter–derived biochar; *SGB*, switchgrass-derived biochar. Bars are the standard deviation of replicates ($n = 3$). Note the difference between the graphs' scale on the y-axis.

Combining biochar with other fertilizers offers a more feasible and sustainable approach in agriculture to enhance soil health and boost crop productivity. This is primarily because it reduces the required application rates for both materials. Additionally, biochar composites containing minerals and organic substances provide further benefits to fertilizer utilization. Biochar-derived nanocarbon has garnered significant interest due to its capacity to retain nutrients, making it suitable for use as nanofertilizers. Studies have demonstrated substantial improvements in wheat crop development and productivity through the application of biochar-carbon nanoparticles (Rose et al., 2011; Zhang et al., 2020). The phosphate retention and rice yields were enhanced in saline-alkaline soil with the use of magnesium oxide-modified biochar (MgO-modified biochar) (Wu et al., 2019). Various nanocarbon and nanobiochar applications and their positive impact on plant growth and fertilizer efficiency have been shown elsewhere in the literature (Chausali et al., 2021).

12.5 Explanation of the pH-neutralizing capacity of biochar

12.5.1 Interaction between biochar and soil acidity, including buffering effects

The ideal soil pH range for agricultural lands typically falls between 5.5 and 7.0. This range is favored because it maximizes the availability of essential nutrients crucial for robust plant growth. Consequently, when soil pH levels drop below 5.5, creating acidic conditions, the application of biochar and modified biochar emerges as a valuable strategy. Biochar possesses significant potential to counteract acidity, thereby enhancing soil fertility and overall soil health. In essence, biochar functions as a natural liming material, effectively raising soil pH levels.

The final pH of biochar can vary depending on the feedstock and the pyrolysis temperature employed in its production (Antonangelo et al., 2019). Nevertheless, most biochar products tend to exhibit higher pH values, making them a valuable tool for ameliorating acidic soils (Dai et al., 2021). Notably, biochar's alkalinity increases as the pyrolysis temperature rises. Furthermore, biochar derived from manure sources typically boasts a higher pH than its counterparts derived from lignocellulose or green waste (Antonangelo et al., 2019). This discrepancy can be attributed to the enhanced electrochemical properties of manure-based biochar, which contribute to its greater buffering capacity. Fig. 12.2 provides a visual representation of this concept, illustrating the buffer capacity of biochars produced from

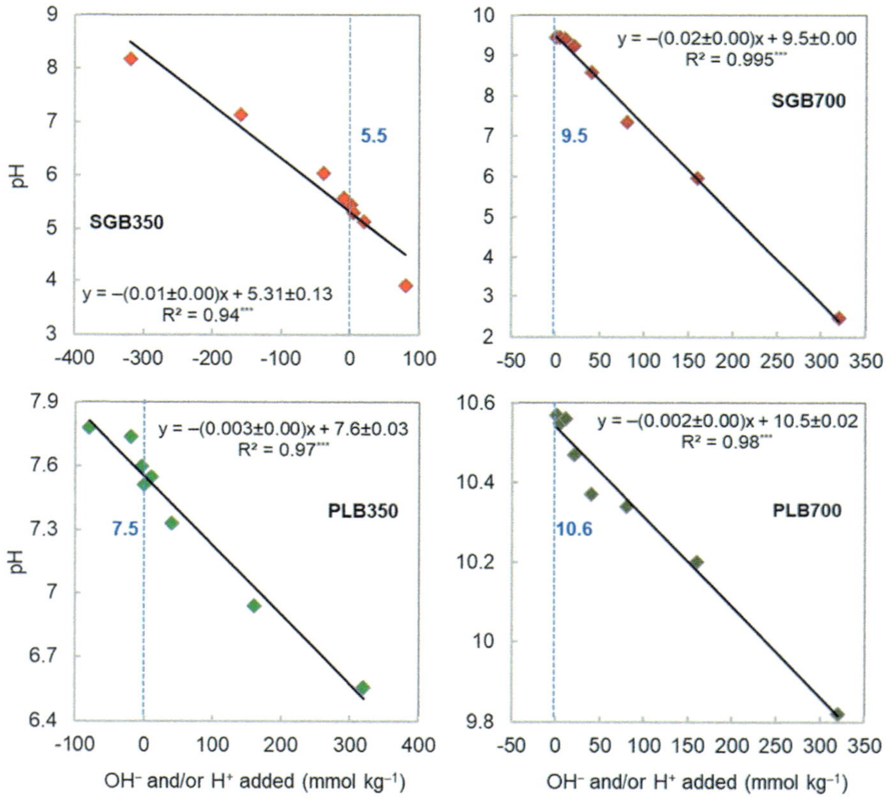

FIGURE 12.2 The linear response of biochar pH to the addition of H^+ and OH^- ions as HCl and NaOH.

Negative values indicate the addition of NaOH (or hydroxyl ions OH^-). The " ± " symbol indicates the standard deviation of linear regression coefficients. ***Models are significant at $P<.0001$. *PLB*, poultry litter–derived biochar; *SGB*, switchgrass-derived biochar. The blue dashed line indicates the initial biochar pH. Note the difference in the pH ranges after the addition of known concentrations of acid and alkali reagents.

various feedstocks and pyrolysis temperatures when subjected to the addition of alkali and acid materials, as adapted from the work of these authors.

12.5.2 Case studies: illustrating how biochar application can mitigate soil acidity issues

The rise in soil pH resulting from the addition of biochar has a direct positive impact on crop yields. The increase in maize yield, ranging from 18.92% to 27.67%, was primarily attributed to the elevated soil pH following biochar application (Liu et al., 2023). Furthermore, when bark biochar was applied to fertilized acidic soils in Indonesia, it led to increased peanut and corn production (Yamato et al., 2006). This phenomenon can be attributed to biochar's ability to neutralize soil pH, essentially creating a liming effect. Research indicates that the influence of biochar addition on soil pH elevation is more significant than other practices, such as traditional liming (Lu et al., 2015).

Attention must be given to the quantity of biochar applied in agricultural practices. When the ash content in the biochar material is significantly high, it leads to an increase in EC. Consequently, an excessive application of biochar can pose a significant threat to the growth and development of cash crops, as the soil's EC may exceed the threshold tolerated by certain plant species, even as the pH of the soil continues to rise. For instance, Antonangelo's unpublished data reveals that the optimal yields of perennial ryegrass were attained with the application of up to 1% (w/w) of PLB, as illustrated in Fig. 12.3. This level of application

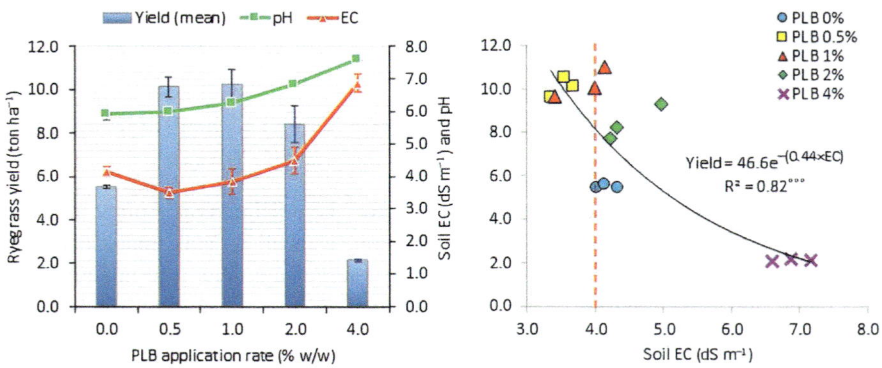

FIGURE 12.3 Ryegrass yield, soil pH, and electrical conductivity (EC) as a function of biochar application rates (left); and the negative exponential relationship between soil EC and ryegrass yield (right).

The red dashed line indicates the EC threshold for ryegrass cultivation. *PLB*, poultry litter–derived biochar. Bars are the standard deviation of replicates ($n = 3$). The average $\pm$ (standard deviation) of the whole dataset of measurements ($n = 15$) for pH and EC are, respectively, 6.5 $\pm$ (0.7) and 4.6 $\pm$ (1.3). ***$P < .001$.

effectively maintained the soil pH at an ideal range for nutrient availability without surpassing the EC threshold conducive to ryegrass production, as shown in Fig. 12.3. As PLB application rates increase beyond this point, a negative exponential relationship between EC and plant yield becomes apparent (Fig. 12.3).

A substantial presence of oxygen-containing functional groups, such as carboxylic acids (−COOH) and phenols (−OH), on the surface of biochar plays a pivotal role in mitigating soil acidity. This effect is further amplified when biochar is subjected to modifications involving nanoparticle loading (Hafeez et al., 2022). Additionally, the creation of nanobiochar also contributes to this enhancement.

In a separate investigation, the application of acid-modified straw biochar was found to enhance the pH buffering capacity and resistance to acidification in paddy soil (He et al., 2022). The researchers attributed this improvement to the heightened presence of surface functional groups, which, through the dissociation of these functional groups, resulted in the generation of protonation of organic anions. This process effectively slowed down the decline in soil pH during acidification. The study posited that these modified biochars exhibited a notable increase in carboxyl functionality, thereby enhancing soil resistance to acidification.

Another study demonstrated a rise in soil pH with the application of manganese oxide (MnO)-modified biochar, a phenomenon akin to the effects of alkaline-modified biochar (Yu et al., 2015). In summary, acid-modified biochar enhances pH buffering capacity and soil acidification resistance by promoting greater protonation of organic anions, while metal oxide-modified biochar displays a pronounced capacity for elevating soil pH (Hafeez et al., 2022).

12.6 Discussion on the potential of biochar to sequester carbon in soil

12.6.1 Importance of biochar as a tool for carbon sequestration and climate change mitigation

Soil organic matter comprises carbon-based materials, encompassing plant and animal remnants in various stages of decomposition, as well as soil microbe tissues and compounds synthesized by these microbes. It serves a crucial role in supporting both fauna and flora (Hafeez et al., 2022) and in driving soil nutrient cycling and carbon storage. Understandably, soil organic carbon (SOC) has emerged as a key quantifiable element in soil health assessments, forming the basis for soil quality categorization. The introduction of biochar not only affects soil nutrients, as previously detailed but also contributes to the stabilization of carbon sequestered within the soil carbon pool, thereby playing a vital role in carbon sequestration and the reduction of carbon dioxide (CO_2) emissions. One contributing factor is the micrometric porosity of biochar, which enhances soil moisture retention, subsequently reducing the degradation of SOC (Hussein et al., 2022) and consequently, mitigating CO_2 emissions.

12.6.2 Factors affecting biochar's stability and persistence in soil

Biochar is distinguished as a recalcitrant carbon source due to its lower availability of carbon, resulting in a notably higher carbon-to-nitrogen (C:N) ratio compared to its original feedstocks. This elevated C:N ratio, coupled with the absence of nitrogen supply, renders biochar resistant to decomposition by soil microbes. Consequently, it fosters the formation of a more stable OC pool, thus promoting improved soil carbon sequestration, a concept intricately tied to the C:N ratio. Consequently, biochar stands out as a carbon-negative energy resource that concurrently aids in carbon sequestration. To further augment this perspective, it is worth noting that carbon nanomaterials naturally occur in biochar. Given the inherent resilience of carbon against decay, it is believed that this sequestration can persist for an extensive timeframe, spanning centuries to millennia when applied to agricultural lands (Chausali et al., 2021).

In Antonangelo's unpublished research, the utilization of SGB, biochar sourced from green waste with a total carbon content of 31.4 ± 1.3% and minimal total nitrogen (0.7 ± 0.0%), demonstrated remarkable efficacy. Specifically, when applied at the highest rate of 4% (w/w) in a potted experiment, it was associated with elevated soil moisture levels. Consequently, these treated pots required less frequent watering to maintain the desired moisture content, resulting in a significant increase in SOC to 5.3 ± 0.2%. This represented a striking 152.4% surge compared to the control group devoid of biochar (SOC = 2.1 ± 0.1%). The

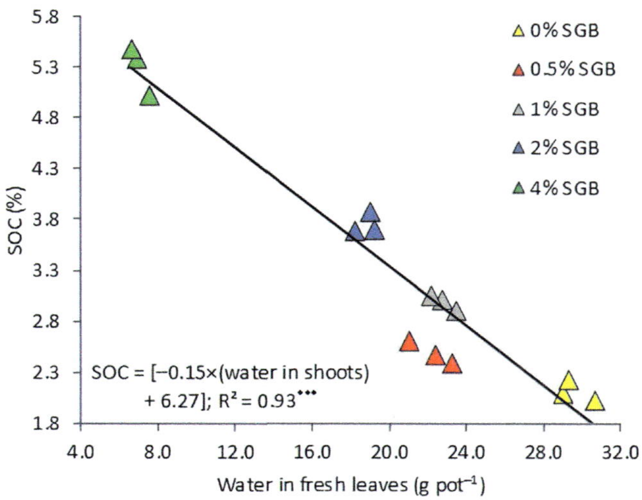

FIGURE 12.4 The inverse relationship between soil organic carbon (SOC) and water content in ryegrass fresh leaves.

SGB, switchgrass-derived biochar. Results comprise the whole dataset of measurements ($n = 15$). ***$P < .001$.

investigation delved deeper, establishing a direct correlation between water content in fresh ryegrass leaves and SOC (Fig. 12.4). Notably, due to the comparatively lower water content in the shoots of these biochar-treated pots, attributed to the higher soil moisture content, it was apparent that the plants exhibited reduced water use efficiency. Consequently, this combination of factors led to the highest SOC levels at the maximum biochar application rate and the lowest water levels in plant shoots, as illustrated in Fig. 12.4.

Nano-modified biochar has demonstrated its efficacy in bolstering SOC content (Wen et al., 2021). Moreover, metal-binding biochar, specifically cadmium (Cd-binding biochar), contributed to elevated soil C:N ratios and subsequently led to an increase in SOC (Li et al., 2020). Additionally, the utilization of iron-zinc oxide ($Fe_2H_2O_4Zn$) composite-modified corn straw, as opposed to pristine biochar, was found to result in higher organic matter content (Yang et al., 2021). Furthermore, various nanocarbons, including black, activated, or engineered nanomaterial carbon, have been recognized for their capacity to adsorb natural organic matter (Shen et al., 2020). In summation, mineral-modified biochar emerges as a pivotal player in the enhancement of SOC (Hafeez et al., 2022). Consequently, biochar nanocomposites and nanobiochar hold immense potential in sustaining soil carbon and soil health. Nonetheless, there remains a compelling need for further research to garner more comprehensive insights.

12.7 Exploration of how biochar affects soil microbial communities

12.7.1 Impacts of biochar on microbial diversity, activity, and nutrient cycling

Out of many others, one pivotal biochar's influence centers around its effects on soil microbial communities. Numerous studies indicate that biochar can wield a positive influence on microbial diversity by furnishing a stable habitat conducive to various microbial species (Li et al., 2018; Palansooriya et al., 2019). A study revealed that biochar amendments not only elevated microbial species richness but also augmented diversity within the soil, cultivating a more robust and adaptable microbial community (Lehmann & Joseph, 2015). Furthermore, the porous structure of biochars has been associated with heightened microbial activity, serving as an ideal substrate for microbial growth and colonization (Jeffery et al., 2011). This structural attribute of biochar is instrumental in stimulating microbial enzymatic activity, thereby amplifying nutrient turnover and availability.

Moreover, the impact of biochar on nutrient cycling unfolds in a multifaceted manner. As previously highlighted, biochar has demonstrated its prowess in enhancing nutrient retention within the soil, thereby curbing nutrient leaching. Additionally, biochar exerts a modulating effect on soil pH, thereby influencing the availability of nutrients to plants. From a microbial standpoint, findings underscored the ability of biochar to reshape nitrogen cycling, thereby influencing microbial nitrogen transformations in the soil (Spokas et al., 2012).

12.7.2 Balance between positive and potentially negative effects on soil microbial ecology

While biochar boasts various positive effects on soil microbial ecology, it is imperative to weigh these advantages against potential drawbacks. On the positive side, biochar proves instrumental in enhancing soil microbial ecology by establishing a stable and enriched habitat. Functioning as a carbon source, it fosters the proliferation of beneficial microorganisms and facilitates nutrient cycling, thereby engendering a positive feedback loop that contributes to overall soil health improvement. Nonetheless, it is crucial to acknowledge the potential negative consequences stemming from biochar mismanagement. The elevated pH levels associated with certain biochars can alter microbial community composition, impacting the abundance of specific microbial groups. Additionally, biochar's capacity to adsorb nutrients may affect their availability to microorganisms. Recognizing these potential challenges, it must be underscored the significance of considering biochar feedstock and production conditions as crucial factors in mitigating possible adverse effects (Lehmann & Joseph, 2015).

12.7.3 Field studies and practical applications

Field studies play a crucial role in evaluating the practical implications of biochar amendments on soil microbial communities. A comprehensive investigation through field trials revealed a significant enhancement in microbial biomass and activity in soils amended with biochar (Bolan et al., 2023). These findings underscore the potential of biochar to positively influence microbial communities in real-world agricultural settings.

It must be stressed that the effective application of biochar in agriculture hinges on a multitude of factors, encompassing soil type, climate, and crop variety. Previous research underscores the importance of considering these variables when implementing biochar amendments for sustainable agriculture (Agegnehu et al., 2016). The implementation of biochar should be tailored to specific agroecosystems to optimize the advantages of biochar while mitigating potential drawbacks.

In conclusion, the impact of biochar on soil microbial communities is intricate and contingent on the context. Unraveling the intricate dynamics governing the soil microecosystem poses a formidable challenge, primarily stemming from the absence of a unified agreement on how soil microbiota responds to biochar amendment (Xu et al., 2021). While biochar shows promise in enhancing microbial diversity, activity, and nutrient cycling, achieving its full potential for sustainable agriculture necessitates careful consideration of factors such as feedstock, production methods, and field conditions.

12.8 Overview of field trials and experiments showcasing the effects of biochar on crop productivity and soil sustainability

12.8.1 Consideration of different soil types, climates, and crop varieties

12.8.1.1 Soil types

Field trials have investigated the impact of biochar on different soil types, including sandy, loamy, and clayey soils. Results have shown that biochar can enhance water retention in sandy soils and improve drainage in clayey soils. Its porous structure helps in nutrient retention and exchange, promoting a healthier soil environment.

12.8.1.2 Climates

Experiments have been conducted in various climates, ranging from tropical to temperate regions. Biochar has been shown to have positive effects on soil microbial activity and nutrient availability across diverse climatic conditions. In tropical soils, biochar has been found to mitigate nutrient leaching and enhance nutrient retention.

12.8.1.3 Crop varieties

Studies have explored the effects of biochar on a wide range of crop varieties, including cereals, legumes, and vegetables. Biochar application has demonstrated increased crop yields, improved nutrient uptake, and enhanced resistance to certain diseases. The specific response varies among crops, with some showing more substantial benefits than others.

12.8.2 Practical recommendations for farmers and land managers regarding biochar application techniques

In agricultural practices, various application techniques are employed to effectively integrate biochar into the soil and harness its benefits. One common method involves the incorporation of biochar into the soil during land preparation. This ensures close contact with plant roots and facilitates the seamless integration of biochar particles into the soil matrix. Another approach is surface application, where biochar is spread over the soil surface. This technique enhances water retention and improves nutrient availability, although it may be less effective for long-term soil carbon sequestration.

An innovative method involves the combination of biochar with compost, known as biochar-amended compost. This synergistic blend not only enhances nutrient content but also promotes microbial activity, offering a balanced approach to overall soil improvement. Additionally, precision application has emerged as a crucial strategy, tailoring biochar application rates based on specific soil characteristics, crop types, and climate conditions. This personalized approach ensures optimal benefits from biochar while minimizing potential drawbacks. By adopting these diverse application techniques, farmers and land managers can harness the full potential of biochar to enhance soil quality and promote sustainable agriculture.

12.8.3 Challenges and future directions

Challenges in the widespread adoption of biochar revolve around economic viability and long-term stability. The production and application of biochar can be economically intensive, posing a hurdle to its utilization, particularly in regions with limited resources. Additionally, the long-term stability and persistence of biochar in soils remain subjects of investigation, necessitating a deeper understanding of degradation rates and potential environmental impacts.

Looking toward future directions, ongoing research is focused on integrating biochar into sustainable agriculture practices. This includes exploring synergies with techniques such as cover cropping and crop rotation. A holistic approach to assessments is also recommended, urging future studies to not only consider crop productivity but also evaluate the overall impact on soil health, water quality, and greenhouse gas emissions. Furthermore, customization of biochar production is seen as a promising avenue. Tailoring biochar properties to specific soil and crop requirements could significantly enhance its effectiveness. This involves optimizing pyrolysis conditions and carefully selecting appropriate feedstocks for a more targeted and efficient application in agriculture.

The field trials and experiments conducted so far indicate the potential of biochar to positively influence soil productivity and sustainability across diverse conditions. However, addressing challenges and refining application practices are essential for realizing the full benefits of biochar in agriculture. Ongoing research and collaboration between scientists, farmers, and policymakers will contribute to a more comprehensive understanding of biochar's role in sustainable agriculture.

12.9 Discussion of challenges in scaling up biochar production and application

The challenges associated with scaling up biochar production are deeply embedded in scientific intricacies. Diverse feedstocks wield a profound influence on biochar properties, and the availability of suitable biomass emerges as a critical limiting factor. The intricate dynamics of production costs, particularly in the context of slow pyrolysis, necessitate a meticulous optimization approach to alleviate economic burdens and curtail energy inputs (Mukherjee & Lal, 2013). Moreover, the ambitious scale-up demands not only the optimization of production methods but also the orchestration of an efficient and scalable technological infrastructure.

It is imperative to develop sophisticated systems capable of accommodating large-scale biochar production, coupled with strategic considerations for the transportation of biochar (Jeffery et al., 2011).

12.9.1 Potential limitations and risks associated with biochar use

Turning to the application of biochar, the challenges extend to achieving a uniform distribution in the field and determining optimal application rates. These tasks are not merely logistical hurdles but demand a nuanced understanding of the intricate interactions between biochar and soil properties (Spokas et al., 2012). The longevity of biochar in soils, a pivotal factor in its overall effectiveness, introduces another layer of complexity. There is a need for a comprehensive understanding of degradation processes and the long-term stability of biochar. As we delve into the potential limitations and risks associated with biochar use, the specter of contaminants looms large. Certain biochar production processes may introduce contaminants, necessitating a deep understanding and robust mitigation strategies (Beesley et al., 2011). Additionally, the impact of biochar on soil microbial communities adds a layer of complexity, where both positive and negative influences need to be carefully balanced for sustainable soil management (Kuzyakov et al., 2009).

12.9.2 Research gaps and future research directions in understanding and optimizing biochar's role in agriculture and soil sustainability

Looking forward, addressing research gaps and defining future research directions emerges as a critical imperative. The intricate web of biochar-soil-plant interactions beckons for a more nuanced understanding, particularly in terms of how biochar influences nutrient cycling and plant growth (van Zwieten et al., 2010). Simultaneously, the environmental fate of biochar takes center stage, demanding focused research on potential leaching and runoff effects, as well as a comprehensive assessment of the overall environmental impact (Bol et al., 2016). Furthermore, the integration of biochar with other soil management practices, such as cover cropping and conservation tillage, unveils synergistic opportunities for enhancing the sustainability of agricultural systems, a domain ripe for exploration in future research endeavors.

12.10 Conclusion

12.10.1 Summary of key findings and implications of using biochar for enhancing crop productivity and ensuring soil sustainability

This chapter highlights the remarkable potential of biochar and its nano-derived counterparts in enhancing soil quality, fostering robust plant growth, and boosting crop yields. This efficacy stems from the electrochemical attributes of biochar, which impart advantageous thermal, mechanical, and electrical enhancements to the soil. When blended with conventional fertilizers, biochar, and its nanocomposites not only serve as a rich source of soil nutrients but also amplify fertilizer efficiency. Through the augmentation of specific surface area, porosity, and surface adsorption characteristics by nanocomposites, biochar takes center stage in facilitating the gradual release of essential nutrients and OC, all while contributing to carbon sequestration and acting as a pH regulator for acidic soils.

Without a doubt, nanobiochar exhibits a diverse array of applications, including significant yield improvement, nutrient recycling, optimization of fertilizer application, and its role as a soil pH modifier. Encouragingly, the exploration of biochar, biochar-based nanocomposites, and nanobiochar continues to unfold. However, it's crucial to underscore that their impact on soil quality varies depending on soil type, crop selection, and nutrient balance. Thus future research endeavors must delve deeper into understanding the mechanisms governing the nano-additive influence of biochar on plant growth, ultimately contributing to the sustainability of agriculture in the long run.

12.10.2 Emphasis on the importance of sustainable agricultural practices and the potential of biochar as a valuable tool in achieving these goals

In conclusion, the imperative to adopt sustainable agricultural practices has never been more urgent, given the escalating challenges posed by climate change and environmental degradation. This chapter has underscored the pivotal role that biochar can play in fostering agricultural sustainability. As a carbon-negative soil amendment, biochar not only enhances soil fertility and water retention but also mitigates greenhouse gas emissions. Its potential to sequester carbon and promote resilient crop yields presents a compelling case for its incorporation into mainstream agricultural practices. The journey toward a more sustainable future demands a paradigm shift, and biochar stands as a beacon of innovation, offering a promising and practical solution to cultivate a harmonious coexistence between agriculture and the environment. As we navigate the complexities of modern agriculture, the integration of biochar represents a transformative step toward cultivating a resilient and regenerative agricultural system that nourishes both the land and future generations.

Author contribution

JAA: conceptualization, methodology, validation, formal analysis, investigation, resources, data curation, writing—original draft, writing—review and editing, visualization, and supervision. MKP: conceptualization, investigation, writing—original draft, and writing—review and editing. TS: investigation, writing—original draft, and writing—review and editing.

Conflict of interest

The authors declare and certify that there are not any actual or potential conflicts of interests.

AI disclosure

During the preparation of this work the author(s) used ChatGPT in order to assist with keeping a good reading flow for the readers. After using this tool/service, the author(s) reviewed and edited the content as needed and take(s) full responsibility for the content of the publication.

References

Agegnehu, G., Bass, A. M., Nelson, P. N., & Bird, M. I. (2016). Benefits of biochar, compost and biochar-compost for soil quality, maize yield and greenhouse gas emissions in a tropical agricultural soil. *Science of the Total Environment*, *543*, 295–306. Available from https://doi.org/10.1016/j.scitotenv.2015.11.054, http://www.elsevier.com/locate/scitotenv.

Anand, A., Gautam, S., & Ram, L. C. (2023). Feedstock and pyrolysis conditions affect suitability of biochar for various sustainable energy and environmental applications. *Journal of Analytical and Applied Pyrolysis*, *170*, 105881. Available from https://doi.org/10.1016/j.jaap.2023.105881.

Angın, D. (2013). Effect of pyrolysis temperature and heating rate on biochar obtained from pyrolysis of safflower seed press cake. *Bioresource Technology*, *128*, 593–597. Available from https://doi.org/10.1016/j.biortech.2012.10.150.

Antonangelo, J. A., & Zhang, H. (2019). Heavy metal phytoavailability in a contaminated soil of northeastern Oklahoma as affected by biochar amendment. *Environmental Science and Pollution Research*, *26*(32), 33582–33593. Available from https://doi.org/10.1007/s11356-019-06497-w, https://link.springer.com/journal/11356.

Antonangelo, J. A., Zhang, H., Sun, X., & Kumar, A. (2019). Physicochemical properties and morphology of biochars as affected by feedstock sources and pyrolysis temperatures. *Biochar*, *1*(3), 325–336. Available from https://doi.org/10.1007/s42773-019-00028-z, http://springer.com/journal/42773.

Antonangelo, J. A., Souza, J. L. B., Whitaker, A., Arnall, B., & Zhang, H. (2022). Evaluation of Mehlich-3 as a multi-element extractant of micronutrients and sulfur in a soil–ryegrass system amended with varying biochar rates from two feedstocks. *Land*, *11*(11), 1979. Available from https://doi.org/10.3390/land11111979.

Antonangelo, J., & Zhang, H. (2021). Influence of biochar derived nitrogen on cadmium removal by ryegrass in a contaminated soil. *Environments*, *8*(2), 11. Available from https://doi.org/10.3390/environments8020011.

Beesley, L., Moreno-Jiménez, E., Gomez-Eyles, J. L., Harris, E., Robinson, B., & Sizmur, T. (2011). A review of biochars' potential role in the remediation, revegetation and restoration of contaminated soils. *Environmental Pollution*, *159*(12), 3269–3282. Available from https://doi.org/10.1016/j.envpol.2011.07.023.

Bol, R., Julich, D., Brödlin, D., Siemens, J., Kaiser, K., Dippold, M. A., Spielvogel, S., Zilla, T., Mewes, D., von Blanckenburg, F., Puhlmann, H., Holzmann, S., Weiler, M., Amelung, W., Lang, F., Kuzyakov, Y., Feger, K. H., Gottselig, N., Klumpp, E., ... Hagedorn, F. (2016). Dissolved and colloidal phosphorus fluxes in forest ecosystems—an almost blind spot in ecosystem research. *Journal of Plant Nutrition and Soil Science*, *179*(4), 425–438. Available from https://doi.org/10.1002/jpln.201600079, http://onlinelibrary.wiley.com/journal/10.1002/(ISSN)1522-2624.

Bolan, S., Hou, D., Wang, L., Hale, L., Egamberdieva, D., Tammeorg, P., Li, R., Wang, B., Xu, J., Wang, T., Sun, H., Padhye, L. P., Wang, H., Siddique, K. H. M., Rinklebe, J., Kirkham, M. B., & Bolan, N. (2023). The potential of biochar as a microbial carrier for agricultural and environmental applications. *Science of the Total Environment*, *886*, 163968. Available from https://doi.org/10.1016/j.scitotenv.2023.163968.

Bopp, C., Christl, I., Schulin, R., & Evangelou, M. W. H. (2016). Biochar as possible long-term soil amendment for phytostabilisation of TE-contaminated soils. *Environmental Science and Pollution Research*, *23*(17), 17449–17458. Available from https://doi.org/10.1007/s11356-016-6935-3, http://www.springerlink.com/content/0944-1344.

Chan, K. Y., Van Zwieten, L., Meszaros, I., Downie, A., & Joseph, S. (2008). Using poultry litter biochars as soil amendments. *Soil Research*, *46*(5), 437. Available from https://doi.org/10.1071/sr08036.

Chausali, N., Saxena, J., & Prasad, R. (2021). Nanobiochar and biochar based nanocomposites: Advances and applications. *Journal of Agriculture and Food Research*, *5*, 100191. Available from https://doi.org/10.1016/j.jafr.2021.100191.

Dai, Z., Xiong, X., Zhu, H., Xu, H., Leng, P., Li, J., Tang, C., & Xu, J. (2021). Association of biochar properties with changes in soil bacterial, fungal and fauna communities and nutrient cycling processes. *Biochar*, *3*(3), 239–254. Available from https://doi.org/10.1007/s42773-021-00099-x, http://springer.com/journal/42773.

Dong, X., Li, G., Lin, Q., & Zhao, X. (2017). Quantity and quality changes of biochar aged for 5 years in soil under field conditions. *Catena*, *159*, 136–143. Available from https://doi.org/10.1016/j.catena.2017.08.008, http://www.elsevier.com/inca/publications/store/5/2/4/6/0/9.

Gabhi, R., Tan, K., Feng, T., Kirk, D. W., Giorcelli, M., Tagliaferro, A., & Jia, C. Q. (2024). Intrinsic electrical conductivity of monolithic biochar. *Biomass and Bioenergy*, *181*, 107051. Available from https://doi.org/10.1016/j.biombioe.2024.107051.

Gale, M., Nguyen, T., Moreno, M., & Gilliard-Abdulaziz, K. L. (2021). Physiochemical properties of biochar and activated carbon from biomass residue: Influence of process conditions to adsorbent properties. *ACS Omega*, *6*(15), 10224–10233. Available from https://doi.org/10.1021/acsomega.1c00530, http://pubs.acs.org/journal/acsodf.

Hafeez, A., Pan, T., Tian, J., & Cai, K. (2022). Modified biochars and their effects on soil quality: A review. *Environments*, *9*(5), 60. Available from https://doi.org/10.3390/environments9050060.

He, X., Hong, Z.-N., Shi, R.-Y., Cui, J.-Q., Lai, H.-W., Lu, H.-L., & Xu, R.-K. (2022). The effects of H_2O_2- and HNO_3/H_2SO_4-modified biochars on the resistance of acid paddy soil to acidification. *Environmental Pollution*, *293*, 118588. Available from https://doi.org/10.1016/j.envpol.2021.118588.

Holatko, J., Hammerschmiedt, T., Kucerik, J., Baltazar, T., Radziemska, M., Havlicek, Z., Kintl, A., Jaskulska, I., Malicek, O., & Brtnicky, M. (2022). Soil properties and maize yield improvement with biochar-enriched poultry litter-based fertilizer. *Materials*, *15*(24), 9003. Available from https://doi.org/10.3390/ma15249003.

Hossain, M. K., Strezov Vladimir, V., Chan, K. Y., Ziolkowski, A., & Nelson, P. F. (2011). Influence of pyrolysis temperature on production and nutrient properties of wastewater sludge biochar. *Journal of Environmental Management*, *92*(1), 223–228. Available from https://doi.org/10.1016/j.jenvman.2010.09.008, https://www.sciencedirect.com/journal/journal-of-environmental-management.

Hussein, B. A., Mahdi, A. B., Izzat, S. E., Dwijendra, N. K. A., Parra, R. M. R., Arenas, L. A. B., Mustafa, Y. F., Yasin, G., Hammid, A. T., & Kianfar, E. (2022). Production, structural properties nano biochar and effects nano biochar in soil: A review. *Egyptian Journal of Chemistry*, *65*(12), 607–618. Available from https://doi.org/10.21608/EJCHEM.2022.131162.5772, https://ejchem.journals.ekb.eg/article_238521_6c7c0522b7-f3773ecfc3007eef9d75e9.pdf.

Jeffery, S., Verheijen, F. G. A., van der Velde, M., & Bastos, A. C. (2011). A quantitative review of the effects of biochar application to soils on crop productivity using meta-analysis. *Agriculture, Ecosystems and Environment*, *144*(1), 175–187. Available from https://doi.org/10.1016/j.agee.2011.08.015, http://www.elsevier.com/inca/publications/store/5/0/3/2/9/8.

Joseph, S., Graber, E. R., Chia, C., Munroe, P., Donne, S., Thomas, T., Nielsen, S., Marjo, C., Rutlidge, H., Pan, G. X., Li, L., Taylor, P., Rawal, A., & Hook, J. (2013). Shifting paradigms: Development of high-efficiency biochar fertilizers based on nano-structures and soluble components. *Carbon Management*, *4*(3), 323–343. Available from https://doi.org/10.4155/cmt.13.23.

Kim, P., Hensley, D., & Labbé, N. (2014). Nutrient release from switchgrass-derived biochar pellets embedded with fertilizers. *Geoderma*, *232-234*, 341–351. Available from https://doi.org/10.1016/j.geoderma.2014.05.017, http://www.elsevier.com/inca/publications/store/5/0/3/3/3/2.

Kuzyakov, Y., Subbotina, I., Chen, H., Bogomolova, I., & Xu, X. (2009). Black carbon decomposition and incorporation into soil microbial biomass estimated by 14C labeling. *Soil Biology and Biochemistry*, *41*(2), 210–219. Available from https://doi.org/10.1016/j.soilbio.2008.10.016.

Lehmann, J., & Joseph, S. (2015). *Biochar for environmental management: An introduction* (pp. 33–46). Informa UK Limited. Available from https://doi.org/10.4324/9780203762264-8.

Li, D., Manu, M. K., Varjani, S., & Wong, J. W. C. (2023). Role of tobacco and bamboo biochar on food waste digestate co-composting: Nitrogen conservation, greenhouse gas emissions, and compost quality. *Waste Management*, *156*, 44–54. Available from https://doi.org/10.1016/j.wasman.2022.10.022, http://www.elsevier.com/locate/wasman.

Li, K., Yin, G., Xu, Q., Yan, J., Hseu, Z.-Y., Zhu, L., & Lin, Q. (2020). Influence of aged biochar modified by Cd2 + on soil properties and microbial community. *Sustainability*, *12*(12), 4868. Available from https://doi.org/10.3390/su12124868.

Li, Y., Li, Y., Chang, S. X., Yang, Y., Fu, S., Jiang, P., Luo, Y., Yang, M., Chen, Z., Hu, S., Zhao, M., Liang, X., Xu, Q., Zhou, G., & Zhou, J. (2018). Biochar reduces soil heterotrophic respiration in a subtropical plantation through increasing soil organic carbon recalcitrancy and decreasing carbon-degrading microbial activity. *Soil Biology and Biochemistry*, *122*, 173–185. Available from https://doi.org/10.1016/j.soilbio.2018.04.019.

Liu, S. B., Tan, X. F., Liu, Y. G., Gu, Y. L., Zeng, G. M., Hu, X. J., Wang, H., Zhou, L., Jiang, L. H., & Zhao, B. B. (2016). Production of biochars from Ca impregnated ramie biomass (*Boehmeria nivea* (L.) Gaud.) and their phosphate removal potential. *RSC*

Advances, *6*(7), 5871–5880. Available from https://doi.org/10.1039/c5ra22142k, http://pubs.rsc.org/en/journals/journalissues.

Liu, T., Awasthi, S. K., Zhou, Y., Varjani, S., Zhang, Z., Pandey, A., Ngo, H. H., & Awasthi, M. K. (2023). *Biochar for sustainable agriculture. Current developments in biotechnology and bioengineering: Biochar towards sustainable environment* (pp. 299–331). Elsevier. Available from https://www.sciencedirect.com/book/9780323918732, https://doi.org/10.1016/B978-0-323-91873-2.00009-1.

Lu, H., Li, Z., Fu, S., Méndez, A., Gascó, G., & Paz-Ferreiro, J. (2015). Effect of biochar in cadmium availability and soil biological activity in an anthrosol following acid rain deposition and aging. *Water, Air, and Soil Pollution*, *226*(5). Available from https://doi.org/10.1007/s11270-015-2401-y, http://www.kluweronline.com/issn/0049-6979/.

Luque, M. P. A., López, J. E., Henao, N., & Saldarriaga, J. F. (2023). Remediation of Pb-contaminated soil using biochar-based slow-release P Fertilizer and biomonitoring employing bioindicators. *Scientific Reports*, *13*. Available from https://www.researchsquare.com/browse, https://doi.org/10.21203/rs.3.rs-2289314/v1.

Mishra, R., Gollakota, A. R. K., & Shu, C. M. (2024). Cultivating eco-advantages: Unleashing the distinctive potential of biochar in microbial fuel cells. *Process Safety and Environmental Protection*, *185*, 614–631. Available from https://doi.org/10.1016/j.psep.2024.02.084, https://www.sciencedirect.com/science/journal/09575820.

Mukherjee, A., & Lal, R. (2013). Biochar impacts on soil physical properties and greenhouse gas emissions. *Agronomy*, *3*(2), 313–339. Available from https://doi.org/10.3390/agronomy3020313, https://www.mdpi.com/2073-4395/3/2/313/pdf.

Palansooriya, K. N., Wong, J. T. F., Hashimoto, Y., Huang, L., Rinklebe, J., Chang, S. X., Bolan, N., Wang, H., & Ok, Y. S. (2019). Response of microbial communities to biochar-amended soils: A critical review. *Biochar*, *1*(1), 3–22. Available from https://doi.org/10.1007/s42773-019-00009-2.

Pan, F., Chapman, S. J., Li, Y., & Yao, H. (2017). Straw amendment to paddy soil stimulates denitrification but biochar amendment promotes anaerobic ammonia oxidation. *Journal of Soils and Sediments*, *17*(10), 2428–2437. Available from https://doi.org/10.1007/s11368-017-1694-4, http://www.springerlink.com/content/1439-0108.

Qian, L., Chen, L., Joseph, S., Pan, G., Li, L., Zheng, J., Zhang, X., Zheng, J., Yu, X., & Wang, J. (2014). Biochar compound fertilizer as an option to reach high productivity but low carbon intensity in rice agriculture of China. *Carbon Management*, *5*(2), 145–154. Available from https://doi.org/10.1080/17583004.2014.912866, http://www.future-science.com/loi/cmt.

Ramadan, M. M., Asran-Amal., & Abd-Elsalam, K. A. (2019). *Micro/nano biochar for sustainable plant health: Present status and future prospects. Carbon nanomaterials for agri-food and environmental applications* (pp. 323–357). Elsevier. Available from http://www.sciencedirect.com/science/book/9780128197868, https://doi.org/10.1016/B978-0-12-819786-8.00016-5.

Rawat, J., Saxena, J., & Sanwal, P. (2019). *Biochar: A sustainable approach for improving plant growth and soil properties*. IntechOpen. Available from https://doi.org/10.5772/intechopen.82151.

Rose, P., Bhat, M., Vidhani, K., Ajmeri, N., Gole, A., & Ghaisas, S. (2011). Intelligent informatics platform for nano-agriculture. *Proceedings of the IEEE Conference on Nanotechnology*, 916–919. Available from https://doi.org/10.1109/NANO.2011.6144536.

Saxena, M., Maity, S., & Sarkar, S. (2014). Carbon nanoparticles in 'biochar' boost wheat (*Triticum aestivum*) plant growth. *RSC Advances*, *4*(75), 39948. Available from https://doi.org/10.1039/c4ra06535b.

Sharma, T., Hakeem, I. G., Gupta, A. B., Joshi, J., Shah, K., Vuppaladadiyam, A. K., & Sharma, A. (2024). Parametric influence of process conditions on thermochemical techniques for biochar production: A state-of-the-art review. *Journal of the Energy Institute*, *113*, 101559. Available from https://doi.org/10.1016/j.joei.2024.101559.

Shen, Y., Tang, H., Wu, W., Shang, H., Zhang, D., Zhan, X., & Xing, B. (2020). Role of nano-biochar in attenuating the allelopathic effect from: Imperata cylindrica on rice seedlings. *Environmental Science: Nano*, *7*(1), 116–126. Available from https://doi.org/10.1039/c9en00828d, http://pubs.rsc.org/en/journals/journal/en.

Spokas, K. A., Cantrell, K. B., Novak, J. M., Archer, D. W., Ippolito, J. A., Collins, H. P., Boateng, A. A., Lima, I. M., Lamb, M. C., McAloon, A. J., Lentz, R. D., & Nichols, K. A. (2012). Biochar: A synthesis of its agronomic impact beyond carbon sequestration. *Journal of Environmental Quality*, *41*(4), 973–989. Available from https://doi.org/10.2134/jeq2011.0069, https://www.agronomy.org/publications/jeq/pdfs/41/4/973.

Wang, S., Dai, G., Yang, H., & Luo, Z. (2017). Lignocellulosic biomass pyrolysis mechanism: A state-of-the-art review. *Progress in Energy and Combustion Science*, *62*, 33–86. Available from https://doi.org/10.1016/j.pecs.2017.05.004.

Wen, E., Yang, X., Chen, H., Shaheen, S. M., Sarkar, B., Xu, S., Song, H., Liang, Y., Rinklebe, J., Hou, D., Li, Y., Wu, F., Pohořelý, M., Wong, J. W. C., & Wang, H. (2021). Iron-modified biochar and water management regime-induced changes in plant growth, enzyme activities, and phytoavailability of arsenic, cadmium and lead in a paddy soil. *Journal of Hazardous Materials*, *407*, 124344. Available from https://doi.org/10.1016/j.jhazmat.2020.124344.

Wu, L., Wei, C., Zhang, S., Wang, Y., Kuzyakov, Y., & Ding, X. (2019). MgO-modified biochar increases phosphate retention and rice yields in saline-alkaline soil. *Journal of Cleaner Production*, *235*, 901–909. Available from https://doi.org/10.1016/j.jclepro.2019.07.043, https://www.journals.elsevier.com/journal-of-cleaner-production.

Xu, W., Whitman, W. B., Gundale, M. J., Chien, C. C., & Chiu, C. Y. (2021). Functional response of the soil microbial community to biochar applications. *GCB Bioenergy*, *13* (1), 269–281. Available from https://doi.org/10.1111/gcbb.12773, http://onlinelibrary.wiley.com/journal/10.1111/(ISSN)1757-1707.

Yamato, M., Okimori, Y., Wibowo, I. F., Anshori, S., & Ogawa, M. (2006). Effects of the application of charred bark of Acacia mangium on the yield of maize, cowpea and peanut, and soil chemical properties in South Sumatra, Indonesia. *Soil Science and Plant Nutrition*, *52*(4), 489–495. Available from https://doi.org/10.1111/j.1747-0765.2006.00065.x.

Yang, T., Xu, Y., Huang, Q., Sun, Y., Liang, X., Wang, L., Qin, X., & Zhao, L. (2021). An efficient biochar synthesized by iron-zinc modified corn straw for simultaneously immobilization Cd in acidic and alkaline soils. *Environmental Pollution*, *291*, 118129. Available from https://doi.org/10.1016/j.envpol.2021.118129.

Yu, Z., Zhou, L., Huang, Y., Song, Z., & Qiu, W. (2015). Effects of a manganese oxide-modified biochar composite on adsorption of arsenic in red soil. *Journal of Environmental Management*, *163*, 155–162. Available from https://doi.org/10.1016/j.jenvman.2015.08.020, https://www.sciencedirect.com/journal/journal-of-environmental-management.

Zeng, Z., Zhang, S. D., Li, T. Q., Zhao, F. L., He, Z. L., Zhao, H. P., Yang, X. E., Wang, H. L., Zhao, J., & Rafiq, M. T. (2013). Sorption of ammonium and phosphate from aqueous solution by biochar derived from phytoremediation plants. *Journal of Zhejiang University: Science B*, *14*(12), 1152–1161. Available from https://doi.org/10.1631/jzus.B1300102.

Zhang, K., Khan, Z., Khan, M. N., Luo, T., Luo, L., Bi, J., & Hu, L. (2024). The application of biochar improves the nutrient supply efficiency of organic fertilizer, sustains soil quality and promotes sustainable crop production. *Food and Energy Security*, *13*(1). Available from https://doi.org/10.1002/fes3.520, http://onlinelibrary.wiley.com/journal/10.1002/(ISSN)2048-3694.

Zhang, P., Huang, P., Xu, X., Sun, H., Jiang, B., & Liao, Y. (2020). Spectroscopic and molecular characterization of biochar-derived dissolved organic matter and the associations with soil microbial responses. *Science of the Total Environment*, *708*, 134619. Available from https://doi.org/10.1016/j.scitotenv.2019.134619.

van Zwieten, L., Kimber, S., Morris, S., Chan, K. Y., Downie, A., Rust, J., Joseph, S., & Cowie, A. (2010). Effects of biochar from slow pyrolysis of papermill waste on agronomic performance and soil fertility. *Plant and Soil*, *327*(1), 235–246. Available from https://doi.org/10.1007/s11104-009-0050-x.

CHAPTER

13 Biochar's potential for enhancing soil functions, nutrient balance, and crop productivity in Mediterranean and tropical regions

Taoufik El Rasafi[1], Ahmed El Moukhtari[2], Wassila Bouta[3,4], Aziz Ziouti[1], Abdelmajid Haddioui[3] and Abdallah Oukarroum[5]

[1]Laboratory of Health and Environment, Faculty of Sciences Ain Chock, Hassan II University, Casablanca, Morocco

[2]Laboratory of Ecology and Environment, Faculty of Sciences Ben Msik, Hassan II University, Casablanca, Morocco

[3]Laboratory of Agro-Industrial and Medical Biotechnologies, Faculty of Science and Techniques, University of Sultan Moulay Slimane, Beni Mellal, Morocco

[4]Office of Professional Training and Employment Promotion, Agricultural Center, City of Trades and Skills (CMC-BMK), Beni Mellal, Morocco

[5]AgroBioSciences Department (AgBS), University Mohammed VI Polytechnic (UM6P), Ben Guerir, Morocco

13.1 Introduction

Abiotic stresses such as drought, salinity, extreme temperature (high/low), metal contamination, and nutrient efficiency pose serious threats to plant productivity and food security (Debnath et al., 2019; El Rasafi et al., 2022; Maiti & Satya, 2014; Nguyen et al., 2016), especially in Mediterranean and tropical lands. They can gradually lead to a high decline in water and soil quality and severely affect all plant attributes in every development stage, reducing the quality and quantity of food production.

Organic amendments like biochar have gained much interest in the improvement of soil properties such as physicochemical characteristics of soil, enhancing soil fertility, amelioration of plant growth, and mitigation of stresses (Hamid et al., 2019; Kabir et al., 2023; Rizwan, Ali, Abbas, et al., 2016; Wang et al., 2017). Fig. 13.1 shows the major roles of biochar in Mediterranean and tropical soil. Previous investigations showcased that biochar was frequently used as a soil

Biochar Ecotechnology for Sustainable Agriculture and Environment. DOI: https://doi.org/10.1016/B978-0-443-29855-4.00015-1

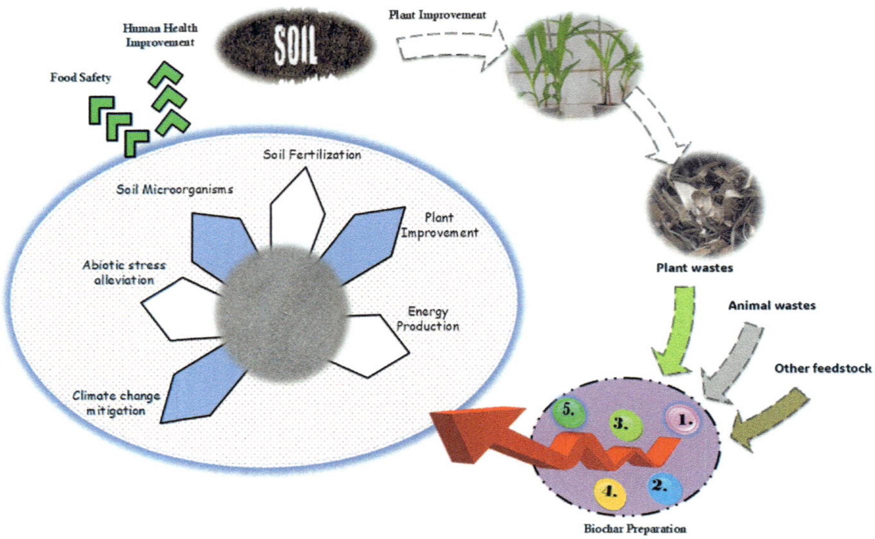

FIGURE 13.1

Role of biochar amendment in improving plant growth and development. Numbers from 1 to 5 indicate the biochar production steps. The figure shows the major roles of biochar in Mediterranean and tropical region.

amendment to improve soil productivity (Kim et al., 2015; Park et al., 2011). This practice has recently changed and many recent studies have investigated the role of biochar in plant tolerance to problems related to water (drought or salinity), saline soils, and varying temperatures (cold/chilling) (Faloye et al., 2019; Farhangi-Abriz & Torabian, 2018; Ghassemi-Golezani & Farhangi-Abriz, 2019; Jia et al., 2019; Lu et al., 2017; Prasad & Kumar, 2023; Singh et al., 2020; Xu et al., 2016). However, to our knowledge, no comprehensive summary of the potential of biochar on enhancing soil functions, nutrient balance, and crop productivity in both Mediterranean and tropical regions has been investigated so far. The present work summarizes (1) feedstock, production, and characteristics of biochar, (2) biochar's roles in enhancing soil functions, nutrient balance, and crop productivity in Mediterranean and tropical regions, and (3) legislation and use policies of biochar in both Mediterranean and tropical regions.

13.2 Climate and situation of the Mediterranean and tropical regions

The Mediterranean region consists of the region located around the Mediterranean Sea and belongs to Africa, Europe, and Asia (Fig. 13.2). This region extends on 2,073,806 km^2 (Guarino et al., 2020) characterized by diversified geographical and topographical features including mountains, peninsulas, and forests. The climate in

FIGURE 13.2

Map of the countries belonging to the Mediterranean region. Countries of the Mediterranean region. The Mediterranean region consists of the region located around the Mediterranean sea and belongs to Africa, Europe, and Asia. Map lines delineate study areas and do not necessarily depict accepted national boundaries.

the Mediterranean region is a dry and hot summer and a cool and rainy winter, which lies between 30 degrees north to 45 degrees south of the equator (Guarino et al., 2020; Ukhurebor et al., 2021). Interestingly, this climate is sometimes found in other regions around the world, including California, the South African Cape Province, south and southwestern Australia, and several parts of central Chile (Fady & Médail, 2004; Galindo et al., 2018). The Mediterranean region is characterized by a large plant biodiversity and soil type variety (e.g., Calcisol, Cambiosols; Acrisol, Oxisols), which is largely affected by the scarcity of water, dry seasons, and nutrient availability.

On the other hand, the tropical region is the area located between the Tropic of Cancer, situated at 23.5°N latitude, and the Tropic of Capricorn at 23.5°S (Henry, 2005). The tropical climate is characterized by an average annual temperature of 20°C and frequent rainfall throughout the year (Henry, 2005; Ukhurebor et al., 2021).

13.3 Biochar: properties and composition

13.3.1 Definition and proprieties

Biochar is the result of the pyrolysis of biomass in the absence or limited supply of oxygen (Bopp et al., 2016; Rizwan et al., 2017; Xu et al., 2016) using various biological materials (Beesley et al., 2011; Bopp et al., 2016; Tang et al., 2016) such as crop residues, animal manures, crop straws, wood chips, sludge, and other biomass residues from agricultural production (Qambrani et al., 2017; Rizwan, Ali, Qayyum, et al., 2016; Xu et al., 2016; Yang et al., 2016; Zhang et al., 2013). Biochar production involves several steps starting with biological material selection to separation (Tan et al., 2017).

Fig. 13.3 shows the main production steps of biochar. Several previous studies have reported that biochar can be produced at varying temperatures and durations, depending on the type of biomass used (Table 13.1). In general, biochar may be produced using two important procedures: slow and fast pyrolysis (Rawat et al., 2019; Yu et al., 2017) depending on the heating time and the temperature applied (Oliveira et al., 2017; Qambrani et al., 2017; Saletnik et al., 2019; Weber & Quicker, 2018; Yuan et al., 2019). It has been reported that the temperature, pyrolysis time, oxygen levels, and the nature of the biological materials used are the major factors affecting the quality of the produced biochar (Hussain et al., 2017; Liu et al., 2018; Qambrani et al., 2017; Tan et al., 2017). Biochar is rich in stable carbon (for long periods reaching thousands of years) and both macro and microelements, as well as a high number of micropores (Agegnehu et al., 2017; Dissanayake et al., 2020; Qambrani et al., 2017; Tan et al., 2017). In fact, biochar porosity and surface area both greatly influence the absorption capacity of biochar

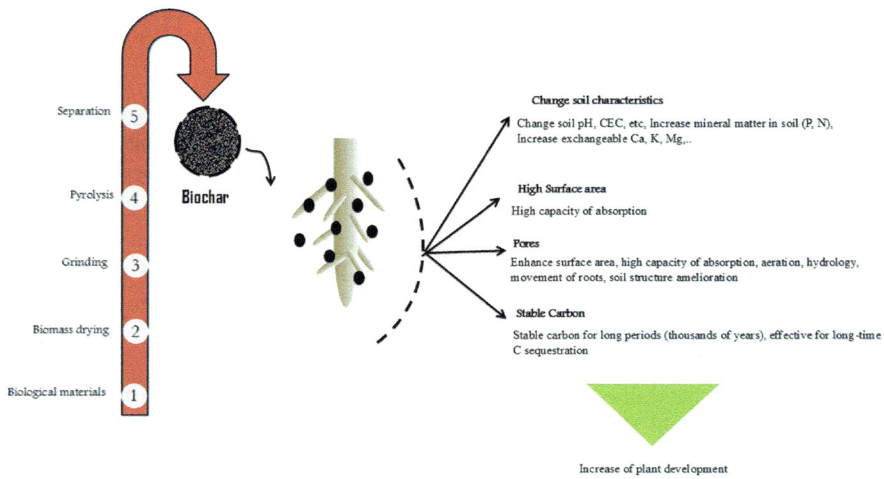

FIGURE 13.3

Biochar production steps and its advantages. Step 1: selection of the feedstock. Step 2: remove the material moisture to enhance combustion and reduce energy consumption. Step 3: reduce the size of the particles for facilitating their use. Step 4: use of energy to transform the biological material to ash. Step 5: classification of the resulted ash depending on the particle size (fine particles have higher surface area, then high reactivity). The figure shows the steps of the production of biochar and its advantages.

Table 13.1 Temperature, time, and feedstock used in the production of biochar.

Production temperature (°C)	Feedstock	Time (min)	References
350–550	Ramie stick and rice straw	60	Shen et al. (2016)
350, 450, 550	Wheat straw	120	Bian et al. (2019)
400	Rice straw and swine manure	120	Meng et al. (2018)
450	Cotton stalks	NS	Younis et al. (2016)
450	Cotton stick and wheat straw	NS	Zia ur Rehman et al. (2021)
450	Rice-straw	120	Abbas et al. (2017)
450	*Eucalyptus saligna* wood	NS	Rehman et al. (2016)
450–600	Rabbit manure	60	Gascó et al. (2019)
500	Hardwood biomass (Quercus, Carpinus, Fagus)	180	Lebrun et al. (2019)
500	Rice hull	NS	Kim et al. (2015)
500	Chicken and duck manure, barley straw and tomato green waste	120	Yasmin Khan et al. (2017)
500	Hard-wood	NS	Khan et al. (2018)
500	Green garden waste	120	Hussain et al. (2018)
500	Hard-wood of platelets and chips of oak, beech and charm	NS	Norini et al. (2019)
500, 550, 600	Bamboo hardwoods	120, 300	Wu et al. (2019)
550	Coniferous tree wood chips	NS	Bopp et al. (2016)
550	Rice-straw	60	Dai et al. (2018)
550	Eucalyptus biochar	NS	Dang et al. (2018)
550	Southern Yellow hardwood chips	NS	Xin et al. (2019)
560	Maple residues (*Acer pseudoplatanus* L.)	360	Farhangi-Abriz and Torabian (2017)
500–600	Cacao shell and rice husk	210	Smebye et al. (2016)
500–600	Straw	NS	Cen et al. (2021)
600	Cattle manure	240	Gavili et al. (2019)
600–700	Shrubby wood	NS	Pandit et al. (2018)
700	Sugar cane straw	60	Puga et al. (2015)
700	Alkaline pine biochar	60	Rezaei Rashti et al. (2019)
750	Bamboo biochar	180	Xu et al. (2016)
900	Walnut shell	NS	Wang et al. (2016)
1200	Maize silage	40	Paetsch et al. (2018)

NS, *not specified.*

(Tang et al., 2013; Tareq et al., 2019) due to the presence of various sites such as carbonyl, carboxylic, phenolic, and hydroxyl groups (Yuan et al., 2019). Surface area may increase with the increase of residence time (Tan et al., 2017).

The pH is considered as one of the important characteristics of biochar (Table 13.2). This parameter plays a crucial role in potential trace elements (PTE)

Table 13.2 Physicochemical characteristics of biochar derived from different feedstock.

Feedstock	pH	EC (dS/m)	CEC (cmolc/kg)	C (%)	N	Ash	SSA (m^2/g)	Pore volume (cm^3/g)	Cd (mg/kg)	Pb	Ca	Fe	Mn	Mg	P	K	References
Rice hull	10.2	0.82	50.4	20.5	0.26	–		–	0.02	0.48	0.72	–	–	0.65	–	1.12	Kim et al. (2015)
Ramie stick	10.02		90.5		1.34	–	5.2	–	0.84	11.41	–	–	–	–	1770	20,700	Shen et al. (2016)
Rice straw	8.99		90.3		1.55	–	3.4	–	17.66	15.6	–	–	–	–	1680	79,100	Shen et al. (2016)
Barley straw	9.51	–	–	53	1.2	–	12.02	–	–	–	–	–	–	–	–	–	Yasmin Khan et al. (2017)
Chicken manure	7.81	–	–	11	0.9	–	57.56	–	–	–	–	–	–	–	–	–	Yasmin Khan et al. (2017)
Duck manure	8.70	–	–	35	1.9	–	19.95	–	–	–	–	–	–	–	–	–	Yasmin Khan et al. (2017)
Tomato green waste	8.92	–	–	42.5	2.01	–	65.85	–	–	–	–	–	–	–	–	–	Yasmin Khan et al. (2017)
Swine manure	8.74	–	–	19.3	1.57	–	48.07	–	–	–	–	–	–	–	–	–	Yasmin Khan et al. (2017)
Bamboo	8.86	0.537	14.25	73	0.713	–	–	–	–	0.06	–	–	–	–	–	–	Ali et al. (2017)
Rice straw	9.5	3.36	64.76	29.3	0.183	–	14.5	15.3	<0.005		1.34	–	–	0.62	1430	18,900	Zhang et al. (2017)
Dairy manure	10.7	–	–	51.1	2.9	37.2	4.26	–	–	–	3.336	0.466	0.082	1.882	3.068	1.021	Xu et al. (2018)
Sawdust	8.2	–	–	65.6	0.6	7.05	40.1	–	–	–	2.090	0.108	0.009	0.348	0.061	1.189	Xu et al. (2018)
Coffee husk	10.31		22.54	67.11	2.05	–	244	–	–	–	0.14	–	–	0.12	470.65	22.17	Lima et al. (2018)

Coffee grounds	9.65		5.56	68.81	4.30	–	23.5	–	–	–	1.56	–	–	0.72	311.36	2.68	Lima et al. (2018)
Coconut shell	10.55	0.025	42.5	736		–	–	–	nd	16.10	–	–	–	–	–	–	Liu et al. (2018)
Honey beeswax waste	9.42	3.96	8.79	22	3.47	69	–	–	–	–	–	–	–	–	28.31	384	Moradi et al. (2019)
Peanut shell	9.65	–	–	0.95	1.56	10.6	10.1	–	–	–	12.4	2.4	0.2	–	1.5	16.1	Xu et al. (2019)
Plantain peel	10.3	>17.02		18.1	0.6	77.5			nd	0.043		669.12	–	–	–	–	Nzediegwu et al. (2019)
Wood	10.43	0.39	11.8	6.01	0.12	11.97	362.4	–	nd	nd	–	–	–	–	230	–	Wang et al. (2019)
Bamboo	9.27	0.1	16.2	88	0.51	12.4	918.7	–	nd	nd	–	–	–	–	240	–	Wang et al. (2019)
Rice straw	9.71	0.16	46.7	51.9	1.72	41.6	38.1	–	nd	nd	–	–	–	–	270	–	Wang et al. (2019)
Chinese walnut shell	9.12	0.08	13.5	6.94	0.16	2.7	271.5	–	nd	nd	–	–	–	–	180	–	Wang et al. (2019)
Rabbit manure	10.88	0.46	132	24.99	0.38	–	35.96	2.1239	0.34	11	–	–	–	–	752	11.6	Gascó et al. (2019)
Rice husk	9.10	0.16×10^{-3}	–	–	–	49.52	60.18	0.05	0.48	–	–	–	–	–	–	–	Liu et al. (2020)
Codonopsis pilosula (Dangshen) and *Angelica sinensis* (Danggui)	9.75	–	–	79.09	1.51	57.65	85.3	8.03×10^{-2}	–	–	–	–	–	–	–	–	Wei et al. (2020)

nd, *not determined;* SSA, *specific surface area.*

adsorption, leading to a decrease in their availability in soil solutions (Dai et al., 2018). Various reports have indicated the increase of soil pH due to the application of biochar as soil amendment (El-Naggar et al., 2020; Yin et al., 2016; Zhang et al., 2017). The observed increase in soil pH is mostly due to the alkaline pH of biochar. Generally, all derived biochar showed high values of pH (alkaline) resulting from the high amount of ash containing different forms of mineral elements (oxides and carbonates) and decomposition of functional groups (−COOH, −OH, and −CHO) as a result of increasing pyrolysis temperature (Tareq et al., 2019; Weber & Quicker, 2018). Gascó et al. (2019) found high values of pH of rabbit manure derived biochar at high pyrolysis temperature ($pH_{600°C} = 10.88$) comparing with lower temperature ($pH_{450°C} = 10.55$). Similar results have been reported by Ibn Ferjani et al. (2019) showing that the increase of temperature has increased the biochar pH. Table 13.2 presents the physicochemical characteristics of biochar derived from diverse feedstock, indicating that the pH ranges between 7.81 and 10.88. Feedstock such as rabbit manure and dairy manure presents higher values of pH while other feedstock from vegetal material such as coconut shell, wood, coffee husk, plantain peel, rice hull, and ramie stick showed alkaline pH. As with other characteristics, surface area and biochar porosity are highly important in the sorption of organic contaminants that are related to the feedstock and pyrolysis temperature used (Liu et al., 2018). Wang et al. (2019) demonstrated that bamboo biochar had a high specific surface area (SSA) compared with wood biochar, Chinese walnut shell biochar, and rice-straw biochar, with values of (m^2/g) 918.7, 362.4, 271.5, and 38.1, respectively. Therefore biochar amendment of PTE-contaminated soil may help reduce their bioavailability due to the binding of these contaminants to the numerous functional groups on the biochar surface. This binding can be enhanced with increases in the biochar's surface area and pore volume (Han et al., 2020; Nigam et al., 2019; Yilu et al., 2018).

13.3.2 Composition of biochar

Derived biochar composition is highly varied depending on the feedstock used in biochar production and pyrolysis conditions (Table 13.2). Pandey et al. (2020) have reported that different characteristics have been affected by feedstock type such as ash content and carbon percentage (Liu et al., 2018; Pandey et al., 2020). Al-Wabel et al. (2015) demonstrated that conocarpus wastes exposed to 800°C of pyrolysis showed higher carbon content (85%) compared with 200°C (64%). Biochar mineral content is affected also by feedstock and pyrolysis temperature (Vijayaraghavan, 2021). Different studies have reported that derived biochar from different feedstock contain high amounts of several mineral elements such as K, Ca, Mg, Na, Fe, N, and P (Lima et al., 2018; Nigam et al., 2019; Nzediegwu et al., 2019; Omoni et al., 2020; Xu et al., 2018; Yilu et al., 2018). The presence of such elements has a beneficial role in improving soil characteristics and plant growth and development (El-Naggar et al., 2019; Han et al., 2020; Lima et al., 2018; Nigam et al., 2019).

13.4 Effect of biochar on the Mediterranean soil

Soil quality is a major factor influencing plant growth and development (Zafar et al., 2019). Soils with low organic matter, lack of nutrients, and extreme pH may lead to reduced plant growth, loss of vegetation, and degradation of soil (Abbasi & Anwar, 2015; Agegnehu et al., 2017; Montiel-Rozas et al., 2016; Virto et al., 2015), especially in Mediterranean soils. Among the soil quality improvement techniques, the application of organic amendments, such as biochar, is very efficient.

13.4.1 Soil proprieties and soil functions

Soil properties refer to the physical, chemical, and biological characteristics of soil, such as texture, structure, pH, organic matter content, nutrient availability, and water-holding capacity (WHC; Centenaro et al., 2023). However, soil functions refer to the roles that soil plays in supporting life and ecosystem processes. These functions include providing a medium for plant growth, regulating water flow and storage, filtering and purifying water, cycling nutrients, storing carbon, supporting biodiversity, and serving as a habitat for various organisms (Schulte et al., 2018). Soil proprieties are highly linked to soil functions and affecting them is automatically leading to soil functions perturbation.

Amendment with biochar has been reported to affect positively the soil properties such as physicochemical characteristics (pH, cation exchange capacity (CEC)) in the presence of various feedstock such as hardwood, bagasse, rice husk, and maize comb waste (Khan et al., 2018); physical proprieties (e.g., soil porosity, bulk density) in the presence of straw miscanthus and winter wheat (300°C per 15 minutes) (Głąb et al., 2016), and soil biochemical properties (e.g., soil respiration, microbial biomass, soil enzyme activities) in the presence of eucalyptus (at 600°C) and poultry litter (at 400°C) (Lu et al., 2015). In Mediterranean soil, amendments with several biochar types derived from poplar and pinewood splinters or thermally dried municipal sewage sludge have influenced both the chemical and physical properties of the soil (Ojeda et al., 2015). Biochar can help maintain a more stable and optimal pH range for plant growth in Mediterranean soil that is prone to pH fluctuations. It can have a liming effect on acidic soils, helping to raise the pH toward neutrality. It has been shown that the application of holm oak biochar increased the pH of an acidic Acrisol with the increase of biochar doses; however, the biochar showed an opposite effect on the pH of the alkaline Calcisol without differences between treatments (Teutscherova et al., 2018). Similarly, coniferous tree woods biochar amended to Moroccan soil has increased soil pH under barley (El Rasafi & Haddioui, 2020) and maize (EL Rasafi, Oukarroum, & Haddioui, 2023) growth by almost 1 pH unit. This pH regulation is particularly important in Mediterranean regions where soils can be highly alkaline, as it allows for better nutrient availability and uptake by plants. Biochar also affects other soil characteristics such as electrical conductivity (EC). In a field experiment in Tarragona, Spain, amendment with maize corn cob rachis derived biochar increased soil pH, soil C/N ratio, and EC after 2 years of application (Andrés et al., 2019). In general, and due to biochar's properties, biochar amendments may improve several other properties of the Mediterranean soil, such as improving its aggregation, creating a more stable structure that prevents soil compaction, and enhancing its aeration due to its porous structure, which leads to better air circulation and water infiltration.

13.4.2 Soil water status

The amendments of soil with biochar have been reported to have beneficial effects on plant and soil water status. WHC is considered a potential indicator of soil quality and productivity. WHC of soil has been increased after biochar amendments, which may lead to an increase in water availability for crops. Plants fight water loss by closing their stomata, which affects CO_2 use by leaves leading to a negative effect on photosynthesis. This inhibition of photosynthetic enzymes ultimately leads to a noticeable reduction in crop productivity (Fischer et al., 2019; Zhang et al., 2020). In Mediterranean soil, biochar positively affects soil water status by increasing WHC and water content due to its capture in biochar surface pores, biochar external surface, and biochar-soil particles (Fig. 13.4). This may lead to help plants to increase plant productivity and resist drought stress in Mediterranean regions where water availability is limited and droughts are

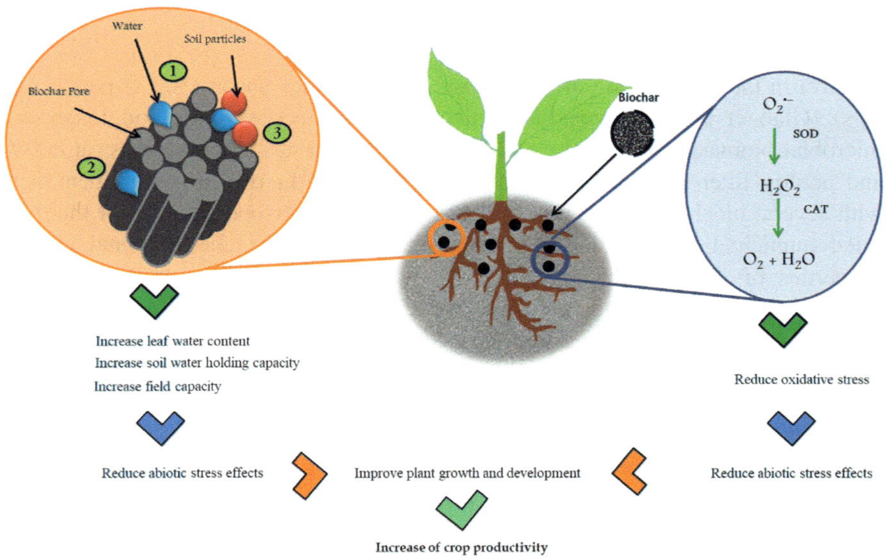

FIGURE 13.4

Effect of biochar on reduction of abiotic stress and soil water status. *CAT*, catalase; *SOD*, superoxide dismutase. The figure shows the effect of biochar on reduction of abiotic stress and water status in soil.

common. For example, it has been shown that biochar amended to sandy soil in Tortona, Italy, improved tomato seedlings' productivity under a deficit of water and increased their resistance to shortage of water (Mulcahy et al., 2013). Similarly, biochar derived from wheat bran (800°C/3 hours) amended to a silty clay loam soil in Parma, Northern Italy, at a rate of 14 t/ha was found to increase soil water retention properties by increasing various parameters such as field capacity and available water capacity as compared to the control (Andrenelli et al., 2016). In field study conducted by Paneque et al. (2016) where sunflowers grown in Calcic Cambisol in Spain under no irrigation regime, biochar obtained from pine wood, paper sludge, and grapevine wood increased WHC up to 7%. Maize-cop residues biochar increased the WHC up to 9% at 50 kPa of water tension (Faloye et al., 2019). The observed increase of soil WHC under abiotic stresses such as drought is likely due to the changes of soil structural characteristics by biochar such as soil porosity and agglomerate level (Tan et al., 2017) as well as higher CEC and porous structure of biochar (Ali et al., 2017).

13.4.3 Soil biodiversity

Besides its crucial role in the improvement of the soil water status of Mediterranean soil, biochar has important benefits for the development of the soil habitat community. Biochar as an organic amendment was reported to stimulate soil microbial activity (Semida et al., 2019). For example, Lehmann et al. (2011) mentioned the positive effect of biochar on bacteria and both ectomycorrhizal and arbuscular mycorrhizal fungi. This may affect soil microbiological properties. Luis Moreno et al. (2022) tested the effect of biochar (holm oak chips/600°C) on Mediterranean soil enzyme activities under various climate manipulations such as ambient conditions, rain exclusion, and rain exclusion + warming. Results indicated that rain exclusion + warming increased the activity of soil dehydrogenase and urease and decreased the activity of β-glucosidase and protease, while no effect was observed on alkaline phosphor-mono-esterase. Moreover, an increase in microbial basal respiration was reported after the application of pine chip (*Pinus pinaster* + *Pinus radiata*) biochar to soil collected from Cerdanyola del Vallès, Catalonia, NE Spain, where barely plants were grown (Ribas et al., 2019). Similar results were found by Teutscherova et al. (2018), where holm oak residues pyrolyzed at 600°C increased the soil biota activities especially basal respiration, dehydrogenase, and urease activity in an acid sandy Acrisol in the first 6 weeks of testing. The same findings were found by Abou Jaoude et al. (2020) in a Lebanon subalkaline soil amended with 3% of biochar (organic municipal solid waste/400°C/2 hours). The authors found that the application of biochar increased the metabolic potential and the use of carboxylic acids and polymers by the soil microbial communities, as well as urease, β-glucosidase, and soil dehydrogenase activities.

In contrast, other works have reported that wood gasification biochar at a rate of 60 Mg/ha amended to subalkaline soil under a Mediterranean climate has no effect on soil enzymatic activities and wheat yield parameters (Baldoni et al., 2023). Moreover, soil microbial biomass was reported to be reduced after biochar application, while no effect was observed on soil microarthropods in a Mediterranean vineyard (Andrés et al., 2019). It is clearly observed that the effect of biochar on soil biota is highly linked to the soil type (Teutscherova et al., 2018). In general, biochar has shown great potential to improve the soil microorganisms' diversity and abundance (Gul et al., 2015). Wheat straw (520°C) at a rate of 1.5% was found to increase the bacterial community richness (Zhang et al., 2019). This may be due to the crucial role of biochar pores that offer favorable conditions for their growth, propagation, and protection against microorganism competition (Rawat et al., 2019; Tan et al., 2017). Moreover, biochar could affect soil microorganisms directly by toxicity and emission of volatile organic compounds or indirectly by modifying soil characteristics, regulating the nutrient availability and changing enzymatic activities (Gorovtsov et al., 2020; Semida et al., 2019).

13.4.4 Biochar and metal contamination and drought stress

Biochar has proven to have a good effect on mitigating various abiotic stresses (e.g., drought, salinity, heavy metals pollution, low or high temperatures, and floods) in different climates conditions such as arid, semiarid, equatorial, and temperate. Similarly, biochar has been reported to combat stresses in Mediterranean soils. Amendments of rabbit manure biochar in the mining area of Riotinto, Spain have reduced the concentration of various elements such as Cu, Co, Cr, Se, and Pb in the soil where *Brassica napus* were grown (Gascó et al., 2019). In addition, a field test was conducted in Arganda del Rey, Spain by Moreno-Jiménez et al. (2016), where barley seeds were grown in contaminated soil with multimetals and amended with holm oak chips (600°C/slow pyrolysis). Results indicated that Pb and Cd were stabilized after biochar treatment and had minimal effect on Zn, Cu, and Ni grain fortification. In France, a pot experiment was conducted using mine soil formed by Ag–Pb ore mining, amended with biochar derived from platelets and chips of oak, beech, and charm where ryegrass (*Lolium perenne*) seeds and willow (*Salix viminalis* L.) where grown for 5 and 7 weeks, respectively (Norini et al., 2019). The authors found that soil nutrient status and the development of both plant roots were improved, and Pb and Zn soluble fractions were reduced while increased that of Ba. Moreover, biochar amendment has decreased both the mobility and bioavailability of Pb and increased the Sb and As mobility after 15 days of beans (*Phaseolus vulgaris* L.) in a gold mine located in Massif central of France (Lomaglio et al., 2017). In Lebanon, biochar was found to increase the now-extractable fractions of Ni, Cr, As, and Sb, while decrease their water soluble and exchangeable fractions (Abou Jaoude et al., 2020).

Biochar is reported to be a good tool to mitigate the negative effects of metals in other Mediterranean regions. Application of coniferous tree woods biochar improved soil proprieties and plant growth traits of barley (El Rasafi & Haddioui, 2020) and maize (EL Rasafi, Oukarroum, & Haddioui, 2023) grown in multicontaminated soil originated from a former iron mine site in the massif central of Morocco. Moreover, in a pot experiment conducted by Eissa (2019), the effect of maize stalk biochar (400°C/3 hours) on accumulation of *Atriplex nummularia* Lindl is tested in contaminated soil with Cd, Pb, Cu, and Zn originating from Illwan, Assiut, Egypt. The results showed that application of biochar helped the old man saltbush to stabilize metals in roots by reducing their shoot accumulation by 44%, 38%, and 30% of Pb, Cu, and Zn, respectively, in comparison to the control.

Besides its role in metals remediation, it can be a useful tool to face various stresses in Mediterranean soils such as drought stress. Various biochar proprieties such as the high number of pores, SSA, and surface functions, biochar can retain water helping to the improvement of WHC in the soil. For example, Paneque et al. (2016) found that biochar at a rate of 15 t/ha applied to a Calcic Cambisol in Spain increased the WHC of the soil after lack of rain. Moreover, a 2-year field experiment was conducted by combining sprinkler irrigation, conventional tillage, and first-year holm oak prunings biochar (550°C/48 hours) application has led to reduced weed pressure, improvement of soil proprieties, and increase of water productivity (Rato-Nunes et al., 2023). Biochar proved to be effective in storing water and making water available for plants in the dry Mediterranean environment, which maintains their productivity. In the Mediterranean region, the application of biochar reduces the effect of environmental problems, promotes long-term sustainability and resilience in agricultural systems, forestry practices, and other land uses, and leads to the conservation of soil resources.

13.5 Beneficial effect of biochar on soil nutrient balance

13.5.1 Carbon sequestration

The soil nutrient balance refers to the differences between both nutrient inputs and outputs. This means the difference between nutrients entering the soil via fertilization or livestock manure and the nutrients leaving the soil for plants and pasture production. Various stresses are highly expressed under the Mediterranean climate, such as drought, soil nutrient depletion (Chtouki et al., 2024), and drastic soil organic carbon (SOC) losses may lead to land degradation (Zabaniotou & Stamou, 2020). This severely affects the different biogeochemical processes in the soil affecting in the end plant growth, productivity, and plant diversity. To face these problems, biochar considered as beneficial strategy for C storage, regulation of nutrient imbalance and counteract soil degradation (El Rasafi et al., 2023), especially in Mediterranean dryland environments (Giannetta et al., 2023; Nogués et al., 2023; Plaza et al., 2018). Biochar is a promising strategy for carbon sequestration, which refers to the capture and storage of carbon in the form of

CO_2 or other forms of carbon for longtime period (El Rasafi et al., 2023; Kabir et al., 2023). The increase of C amount in soil after biochar application is mainly due to the biochar itself since it is rich carbon material and the resistance of C to chemical and biological degradation, which explain its ability to remain in soil for a longtime period (El Rasafi et al., 2023).

In a 2 years field experiment conducted in Pistoia, Italy, biochar at a rate of 30 and 60 t/ha was applied in the first year only while durum wheat (*Triticum durum* L.) was sustained for two consecutive seasons (Vaccari et al., 2011). Results indicated that biochar affects positively the biomass production and yield of durum wheat during both seasons. Moreover, biochar was very effective in atmospheric CO_2 sequestration showing that 30–60 t/ha of biochar was equivalent to 92–184 t of atmospheric CO_2 that was taken from the atmosphere by plants and transferred into the soil. In addition, and not far to Pistoia, biochar derived from orchard pruning feedstock was applied in single and double applications in a field vineyard located in Montepulciano Central Italy for 10 years. Results showed that SOC was increased after double application to 23.1 and 17.3 g/kg after single application as compared to the control (12.7 g/kg) (Idbella et al., 2024). In Southwest Spain, five types of biochar were produced from pine wood, paper sludge, sewage sludge, old vineyard, and a mix of wood chips. These biochars were amended to a field with Calcic Cambisol over a 24-month period (de la Rosa et al., 2018). The authors found that biochar increased the slow C pool and increased the physical fragmentation values of all types of biochar over time at field conditions.

In another experiment, cumulative application of biochar combined with other amendments such as municipal solid waste compost and sewage sludge for 9 years affected positively the carbon storage, by increasing its content especially in the particulate organic matter fraction by three times and with 1.5 in the mineral-associated organic C content (Giannetta et al., 2024). The stability of biochar in soil (from 100 to 4000 years) is mainly due to its proprieties with an estimation of 80% of stability in the first 100 years (El Rasafi et al., 2023; Nogués et al., 2023). Carbon sequestration by biochar is mainly due to its recalcitrant nature, reduction of mineralization of native SOC, and the increase of the atmospheric C capture by plant (Yin et al., 2022). Nogués et al. (2023) reported that in Mediterranean soils, the great proportion of recalcitrant carbon is behind the great ability of carbon storage after biochar application.

13.5.2 Nutrients

To face nutrient imbalance, the application of biochar is considered a great solution due to its numerous advantages. As presented in Table 13.2, biochar is rich in nutrients such as N, P, K, Ca, Fe, and Mn depending on the feedstock used. High amounts of P and N are found in biochar derived from animal manures (Zabaniotou & Stamou, 2020). Adding biochar to Mediterranean soil could increase the nutrient inputs and improve nutrient retention and availability in soil. It has been shown that *Quercus ilex* wood biochar (400°C/8 hours) with high K, Cu, Ca, Mg, Na, and total carbon values enhanced soil nutrient availability (Forján et al., 2017). Biochar is found to enhance soil nutrient retention (Zabaniotou & Stamou, 2020), which reduces synthetic fertilization use. Moreover, various biochar characteristics such as cation exchange capacity, large SSA, and the high number of pores are behind the adsorption and nutrient retention leading to minimization of nutrient dynamics and leaching out of the soil. The leaching prevention may correct soil nutrient imbalances, improve nutrient-use efficiency, and minimize nutrient losses, leading to more sustainable and efficient nutrient management in agricultural systems. In terms of soil pH type, the increase in pH of acidic soil due to biochar application is reported to increase the availability of soil nutrients, while increasing the soil alkalinity may affect negatively nutrient availability. The increase of soil pH above 8 may lead to the precipitation of nutrients decreasing their availability for plants (Brtnicky et al., 2021).

13.6 Beneficial effect of biochar on plant growth and development

Biochar has been known for its beneficial effects on plants under different growth conditions, including the Mediterranean conditions. In fact, the positive effects of biochar can be seen at different growth stages, from germination to yield (Semida et al., 2019). These effects are mediated by several physiological, biochemical, and molecular changes.

13.6.1 Effect on plant growth

It has been reported that the application of biochar to degraded soils is one of the most sustainable strategies that can improve plant growth, especially under stressful conditions (Hasnain et al., 2023). In fact, when applied exogenously, biochar showed several benefits, such as improving seed germination, plant growth, and resilience. In Pakistan, 2% biochar was prepared (*via* pyrolysis) from dried leaves material, and sawdust improved significantly the germination percentage, seedling vigor index, root length, shoot length, and leaf area of wheat plants (Kanwal et al., 2018). Similarly, a pot experiment was conducted recently by Ullah et al. (2020) underwire house conditions in Pakistan to study the effect of 4% algal biochar on *Zea mays* growth. Results indicated that biochar was efficient in improving plant height, shoot fresh weight, shoot dry weight, root fresh weight, root dry weight, root length, number of leaves, and leaf area by 89%, 168%, 157%, 95%, 175%, 91%, 109%, and 151%, respectively. In another intriguing investigation conducted by Yue et al. (2019), rice-hull biochar application improved dry weight, plant height, and root parameters, including surface area, root volume, average diameter, root tips, and root crossing of rice plants. In the same line, the effect of peanut shell biochar on peanut (cultivar Middleton) growth in two different soil types (red ferrosol and redox-hydrosol) was studied (Xu et al., 2015). Results revealed a significant improvement in peanut biomass and pod yield up to two- and threefold, respectively, in red ferrosol and redox-hydrosol. More interestingly, in the model plant Arabidopsis, biochar application increased leaf area (130%), rosette diameter (61%), and root length (100%) when compared with the nontreated plants while increased significantly lettuce (*Lactuca sativa* L.) leaf cell expansion (Viger et al., 2015). Moreover, at the field scale, biochar applied at a rate of 14 t/ha in Parma (North, Italy) showed spectacular positive effects on root dry weight, root length, and plant total dry weight of tomato plants (Vaccari et al., 2015).

Despite the beneficial effects of biochar application on plant growth, several papers reported that its positive effect is dose-dependent; high doses could cause a toxic effect in plants. For example, Gonzaga et al. (2018) investigated the effects of biochar derived from coconut husk, orange bagasse, and pine wood chips at different rates (0, 5, 10, 20, and 60 t/ha) on *Z. mays* plants. The authors found that the effect was dependent upon the biochar type and application rate. In fact, soil treated with coconut husk biochar at an equivalent rate of 30 t/ha resulted in a 90% increase in *Z. mays* biomass and plant N and P concentrations of 0.88% and 0.12%, respectively, while the application of orange bagasse biochar (applied at a similar rate) or pine wood chip biochar to soil did not affect plant biomass or nutrition of *Z. mays* plants (Gonzaga et al., 2018). Similarly, in *Viola cornuta*, while low doses (10% and 25%) of biochar were effective in improving plant biomass, growth, root length, and flowering, 50% biochar was found to have negative effects on plant growth and flowering (Regmi et al., 2023). The above investigations evidenced that biochar application at suitable doses may improve plant growth, but other studies using different doses of different biochar types are still needed to better understand the relationship between biochar and plant growth.

13.6.2 Biochemical and physiological characteristics of plants

Biochar has repeatedly been found to have positive effects on the physiological and biochemical characteristics of plants. Wheat biochar increases relative water content (RWC), chlorophyll (Chl), carotenoid contents, and photosynthetic rate (Anwar et al., 2023). Investigations by Ullah et al. (2020) indicated that the application of algal biochar at 4% to *Z. mays* plants improved net photosynthesis and transpiration rate. Furthermore, biochar from rice hull was found to be efficient in improving total Chl and lowering lipid peroxidation and electrolyte leakage in *Oryza sativa* (Yue et al., 2019). In another investigation by Hafez et al. (2019), biochar increased Chl content, stomatal conductance, and RWC by up to 35%, 19%, and 10%, respectively, while reducing proline content by 39% relative to biochar-untreated rice plants. The same results were reported by Kanwal et al. (2018) in wheat plants, where proline content was decreased by 51% in response to a 2% biochar application.

Moreover, the effects of biochar on phytohormone biosynthesis in plants have been reviewed in several studies. Farhangi-Abriz & Ghassemi-Golezani (2022) showcased that biochar application to the soil at 25 g/kg increased indol acetic acid (IAA) and silicic acid (SA) in mint plants (*Mentha crispa* L.), with no significant effect on abscisic acid (ABA) and jasmonic acid. Similarly, in the study of Bibi et al. (2023), foliar spray of biochar (30 mg/L) allows a significant accumulation of IAA, gibberellic acid (GA), and SA in tomato plants. The ABA and ACC contents were not affected in plants treated by either 50 or 100 g biochar/kg soil whereas the IAA content and the IAA/ABA and IAA/ACC ratios were significantly improved (Nikpour-Rashidabad et al., 2019). Besides physiological and biochemical changes, biochar has also been found to modulate gene expression in several plant species. In Arabidopsis and lettuce plants, positive growth effects were accompanied by the downregulation of several plant defense genes, including the jasmonic acid biosynthetic pathway, defensins, and most categories of secondary metabolites (Viger et al., 2015).

13.6.3 Crop productivity

Previous investigations evidenced the beneficial effects of biochar on yield and productivity of several plant species. For example, Uzoma et al. (2011) evidenced a significant positive correlation between the rate (0, 10, 15, and 20 t/ha) of cow manure biochar and *Z. mays* yield. In fact, application of biochar at 15 and 20 t/ha mixing rates significantly increased plant grain yield by 150% and 98% relative to the control. In another interesting study, Hafez et al. (2019) tested the effect of biochar from slow pyrolysis of rice husk and corn stalk (1:1) at 350°C under an oxygen depleted condition on yield components of rice plants. Authors found that the applications of biochar improved 1000-grain weight, number of grains per panicle, and number of panicles per m^2 by 44%, 9%, and 5%, respectively, as compared with the untreated control plants. Authors also indicated that rice yield in terms of grain yield, straw yield, and harvest index were significantly improved when biochar was applied (Hafez et al., 2019). In a similar study, biochar application (at 14 t/ha) affects positively harvest parameters, including fresh fruit total production, fresh fruit marketable production, fresh fruit no marketable production, fresh fruit weight, fruit number per plant, fresh fruit optical residue, fresh fruit pH, and fresh fruit hunter scale of *Lycopersicon esculentum* plants grown under field condition in Parma (North, Italy) (Vaccari et al., 2015). Likewise, by evaluating the effect of different biochar rates (0, 2.5, 8, and 16 t/ha) on soybean yield, Arabi et al. (2018) showed that biochar improved number of pods per plant, yield, harvest index, oil content, and protein content when compared with plants without biochar.

13.7 Effect of biochar on the tropical regions

Similarly to the Mediterranean region, tropical soils suffer from various stresses such as urbanization, pollution, salinization, degradation, and soil fertility losses (Adekiya et al., 2023). These stresses severely affect plant growth, development, and yield, drastically declining food production and quality.

13.7.1 Beneficial effect of biochar on soil functions

Similarly to the Mediterranean soil, tropical soils' physicochemical properties are varied. Large tropical areas are acidic or moderately acidic, including regions in African such as Nigeria (Adekiya et al., 2024), Ethiopia (Agegnehu & Amede, 2017), and Zambia (Obia et al., 2016), as well as in Asia, including Nepal (Pandit et al., 2018), Thailand (Prakongkep et al., 2020), and China (Chen et al., 2018; Yan et al., 2021). Moreover, tropical soils are characterized by a low CEC (Glaser et al., 2022; Obia et al., 2016). Low pH in tropical lands may affect negatively crop growth and development, yield, and productivity. Biochar was reported to be more effective in enhancing tropical soil pH and promoting soil characteristics and plant growth. It has been shown that 5.0% of bamboo biochar (750°C–800°C) and rice biochar (550°C) increased soil pH from 4.3 to 5.2 and 5.3, respectively (Yan et al., 2021). In Brazil, increasing coffee husk biochar doses from 4 to 16 Mg/ha significantly raised soil pH from 5.5 to 7 (Lima et al., 2018). In addition, the authors observed increased soil water retention with higher biochar doses and a 50% improvement in water use efficiency with 16 Mg/ha of biochar. Biochar provides plants with carbon and nutrients, promoting their growth through its properties such as pH, CEC, WHC, and macro- and micronutrient content. Additionally, the surface area and pore structure of biochar influence soil compaction, water content, soil porosity, and permeability (Agegnehu et al., 2017; Qambrani et al., 2017).

13.7.2 Beneficial effect of biochar on nutrient balance

The tropical lands' high rainfall, high temperatures, SOC dynamics, intense weathering, deforestation, and uncontrolled agricultural practices lead to soil degradation. This is mostly due to nutrient depletion, leaching, and soil organic matter (Atkinson et al., 2010). Glaser et al. (2022) reported that tropical soil suffered from a lack of nutrient content and rapid mineralization of soil organic matter. In tropical soils, the topmost humic deposits contain high concentrations of nutrients that are quickly recycled (Arunrat et al., 2020). However, mobile nutrients such as K and NO^{3-} can be easily leaked due to high rainfall, affecting their availability (Glaser et al., 2022). Yan et al. (2021) reported that low nutrient availability is highly observed in most tea gardens in both tropical and subtropical areas. To face these problems, the application of biochar was reported to be very efficient in changing the nutrient status in tropical soils. An increase in nutrient availability was observed in a Colombian savanna Oxisol after the application of biochar for 4 years (Major et al., 2010). Similarly, Yamato et al. (2006) found that the application of bark charcoal of *Acacia mangium* increased both total N and P_2O_5 levels, indicating that biochar increased nutrient retention. Moreover, Lima et al. (2018) reported an increase in nutrient and nutrient use efficiency with increasing coffee husk biochar doses. Biochar is rich in nutrients such as P, N, and K (Chen et al., 2019). Its incorporation into tropical soil plays a crucial role in mitigating nutrient deficiency by increasing their supply, availability, and use efficiency, while also reducing nutrient leaching.

13.7.3 Beneficial effect of biochar on plant growth and development

In the tropics, where the fertility of the major agricultural soil is depleted, biochar application has been proposed as a sustainable approach to improve crop growth and productivity (Moiwo et al., 2019). In Riyadh city (Saudi Arabia), Ahmad et al. (2018) showcased that applying biochar to soil, especially at the highest rate of 3%, effectively enhanced parameters of *Z. mays* (such as shoot and root lengths and dry matter) and P uptake. Moreover, when applied at 50 g/kg to Cambodian soil, biochar from rice-husk char improved above ground biomass, below ground biomass, number of leaves, and stem length of both lettuce (*L. sativa*) and cabbage (*Brassica chinensis*) plants (Carter et al., 2013). More interestingly, nodule number of fenugreek (*Trigonella foenum-graecum*) plants was increased by 36% after the application of 1.0% of woody materials biochar (400°C–500°C) (Jabborova, Annapurna, Paul, et al., 2021). Biochar was also found to improve yield of several plant species. For example, to improve maize yield grown in the Colombian savanna, Major et al. (2010) conducted a 4-year field study using different biochar rates (0, 8, and 20 t/ha). Here the authors found that yield did not change during the first year, but increased significantly the second, third, and fourth years of the experiment, particularly in plants that received 20 t/ha of biochar, with increases of 28%, 30%, and 140%, respectively. Similarly, in the study of Agegnehu et al. (2016), biochar significantly impacted the yield of *Z. mays* grown in a tropical Ferralsol at Tolga, Australia. This was evidenced by increases of 29%, 18%, and 10% in grain yield, total biomass yield, and harvest index, respectively.

In the same line, a field trial was performed to evaluate the effect of two biochar amendment rates (2 and 40 t/ha) on the production of the super japonica rice variety grown under a semihumid temperate continental monsoon climate in Liaoning, China (Cui et al., 2017). Results showed that biochar amendment improved only seed setting rate with no significant effects on grain number per panicle, 1000-grain weight, number of effective panicles per hill and yield. Furthermore, a comparative study using bamboo biochar, rice straw biochar, and rice straw was established to study the effect of biochar on grain yield and yield components of rice in a cold waterlogged paddy (Liu et al., 2016). Rice-straw biochar resulted in the greatest improvement in grain yield, straw yield, thousand-grain weight, and productive tillers per plant, with increases of 11%, 13%, 5%, and 8%, respectively, compare to the control. Interestingly, in the intercropping system *Z. mays* and peanut (*Arachis hypogaea* L.), applying biochar at 10 and 20 g/kg increased N transfer from peanut-rhizobia symbiosis to maize plants by 33% and 36%, respectively, which allow an increase in maize yield. Biochar produced from *Acacia* spp., at a rate of 50 Mg/ha was found to improve yield of *Z. mays* by 1.2 Mg/ha and soybean by 0.4 Mg/ha when grown in rotation in Kenyan soil (Kätterer et al., 2019). However, in other crops such as corn, a negative effect on yield was observed with biochar application (Rajkovich et al., 2012). Indeed, an 80% decline in crop productivity was observed under food waste biochar amendment at a relatively high rate of 90 t/ha. Similarly, in the study of Liang et al. (2014), a 5-year field experiment was established in a calcareous soil in Shangzhuang, China to assess the effect of biochar application on winter wheat or summer *Z. mays* yield production grown in calcareous soils. Results revealed that biochar has no significant positive effect on annual yield of both species, while it significantly improved the cumulative yield over the first four growing seasons.

Indeed, in the study of Jabborova, Annapurna, Paul, et al. (2021), photosynthetic pigments, net photosynthesis, stomatal conductance, and transpiration rate were all improved in spinach plants upon biochar supply. In the symbiosis fenugreek-rhizobia, Jabborova, Annapurna, Choudhary, et al. (2021) showcased that biochar application resulted in a significant increase in Chl a by 18%, Chl b by 29%, total Chl by 18%, carotenoids by 48%, and RWC by up to 5% relative to the untreated control plants.

Susanti and Dewi (2023) investigated the effect of rice-husk biochar at different doses (0%, 3%, 5%, or 10%) on the endogenous hormones of Chinese cabbage grown in Yogyakarta soil in Indonesia. Results revealed an increase in the relative content of endogenous IAA, GA, cytokinin, jasmonic acid, and SA in plants that received 10% biochar treatment, whereas the ABA was drastically reduced. Moreover, in Chinese cherry (*Prunus pseudocerasus*), tobacco straw biochar addition mediated higher levels of IAA, GA, and zeatin (Yang et al., 2022). In contrast, lower levels of ABA were observed under biochar treatment. Interestingly, using RNA seq analysis, Yan et al. (2019) showcased that biochar application to tobacco plants grown in Ganzhou (Xinfeng) resulted in 20,249 differentially expressed genes, with 15,960 upregulated and 4289 downregulated. Interestingly, the Kyoto Encyclopedia of Genes and Genomes enrichment analysis revealed that 10 representative pathways were among the most significantly enriched, including circadian rhythm, proteasome, valine, leucine and isoleucine degradation, glyoxylate and dicarboxylate metabolism, pyrimidine metabolism, sesquiterpenoid and triterpenoid biosynthesis, carbon fixation in photosynthetic organisms, arginine and proline metabolism, biosynthesis of amino acids, and ribosome biogenesis in eukaryotes.

13.8 Legislation and policy

Biochar application has gained much interest due to its beneficial role in ameliorating soil quality, improving plant growth and development, mitigating various stress effects, and minimizing greenhouse gas emissions, which reduces climate change effects. However, its application is still not controlled, and in most countries, there is a lack of policies and legislation frameworks. Around the world, the application of biochar is not equally distributed depending on the continent. In Europe, biochar is widely used and accepted by farmers and policymakers who are highly convinced of biochar roles, which justify its implications in various sectors such as climate change policy, waste management, environment protection, agricultural policy, research policy, industry, and energy (Hammond et al., 2016; Montanarella & Lugato, 2013). This led to the creation of some laws and recommendations about biochar production and uses. Nogués et al. (2023) reported that the addition of biochar by the Commission Delegated Regulation (EU 2021/2088, 2021) into the current EU Fertilizer Regulation (EU/2019/1009) sets limit values for contaminants and heavy metals in EU fertilizing products, recategorizing the well-known fertilizer types into new product function categories. At national levels, only Italy from the EU Mediterranean countries regulated the application of biochar as a fertilizer (Garcia et al., 2022) similar to other European non-Mediterranean states including Germany, Denmark, Netherlands, Austria, Switzerland, and Norway (Meyer et al., 2017; Nogués et al., 2023).

To the best of our knowledge, no laws or directives have been set for biochar production and use in African countries within the Mediterranean or tropical areas. In these regions, especially in sub-Saharan African countries, most farmers and policymakers are unaware of biochar and lack a consistent strategy or regulatory foundation for biochar technologies (Gwenzi et al., 2015; Zanli et al., 2022). Omulo (2020) reported that policy and legal frameworks, institutional, socioeconomic, fiscal, ecological, health, and technical issues are the main obstacles in front of the biochar technology adoption in East Africa and sub-Saharan Africa.

13.9 Conclusion

In both Mediterranean and tropical regions, biochar application plays a key role in minimizing soil degradation, optimizing resource use, and maintaining ecosystem functions due to several beneficial effects such as water retention, nutrient cycling, and improvement of soil health. Moreover, biochar is very effective in the mitigation of various abiotic stresses, including heavy metal stress, salinity, and drought, as well as in the reduction of greenhouse gas emissions. However, biochar technology applications are still not well developed in a large number of countries in both Mediterranean and tropical regions, facing various challenges such as acceptability, legislation, ecological, socioeconomic, and technical issues.

More efforts are needed to increase farmers' and policymakers' knowledge of biochar's benefits, encourage scientists in universities and research institutes to be involved in biochar technology research, and facilitate the involvement of industrial investors in biochar production and commercialization.

Acknowledgment

The authors would like to thank the editors Prasad Majeti and Abhay Kumar, for the opportunity to contribute to elaborating of the book.

Authors' contribution

Taoufik El Rasafi and Ahmed El Moukhtari: conceptualization, writing—original draft preparation. Taoufik El Rasafi: tables and figures illustration, visualization. Wassila BoutaAziz Ziouti, Abdelmajid Haddioui, and Abdallah Oukarroum: writing—review and editing. All authors read and approved the manuscript.

References

Abbas, T., Rizwan, M., Ali, S., Zia-ur-Rehman, M., Qayyum, M. F., Abbas, F., Hannan, F., Rinklebe, J., & Sik Ok, Y. (2017). Effect of biochar on cadmium bioavailability and uptake in wheat (*Triticum aestivum* L.) grown in a soil with aged contamination. *Ecotoxicology and Environmental Safety*, *140*, 37–47. Available from https://doi.org/10.1016/j.ecoenv.2017.02.028.

Abbasi, M. K., & Anwar, A. A. (2015). Ameliorating effects of biochar derived from poultry manure and white clover residues on soil nutrient status and plant growth promotion – greenhouse experiments. *PLoS One*, *10*(6), 1–18. Available from https://doi.org/10.1371/journal.pone.0131592.

Abou Jaoude, L., Castaldi, P., Nassif, N., Pinna, M. V., & Garau, G. (2020). Biochar and compost as gentle remediation options for the recovery of trace elements-contaminated soils. *Science of the Total Environment*, *711*, 134511. Available from https://doi.org/10.1016/j.scitotenv.2019.134511.

Adekiya, A. O., Alori, E. T., Ogunbode, T. O., Sangoyomi, T., & Oriade, O. A. (2023). Enhancing organic carbon content in tropical soils: Strategies for sustainable agriculture and climate change mitigation. *Open Agriculture Journal*, *17*. Available from https://doi.org/10.2174/01187433152824762311240742O6, https://openagriculturejournal.com/VOLUME/17/ELOCATOR/e18743315282476/PDF/.

Adekiya, A. O., Ayorinde, B. B., & Ogunbode, T. (2024). Combined lime and biochar application enhances cowpea growth and yield in tropical Alfisol. *Scientific Reports*, *14*(1). Available from https://doi.org/10.1038/s41598-024-52102-7, https://www.nature.com/srep/.

Agegnehu, G., & Amede, T. (2017). Integrated soil fertility and plant nutrient management in tropical agro-ecosystems: A review. *Pedosphere*, *27*(4), 662−680. Available from https://doi.org/10.1016/s1002-0160(17)60382-5.

Agegnehu, G., Bass, A. M., Nelson, P. N., & Bird, M. I. (2016). Benefits of biochar, compost and biochar-compost for soil quality, maize yield and greenhouse gas emissions in a tropical agricultural soil. *Science of the Total Environment*, *543*, 295−306. Available from https://doi.org/10.1016/j.scitotenv.2015.11.054, http://www.elsevier.com/locate/scitotenv.

Agegnehu, G., Srivastava, A. K., & Bird, M. I. (2017). The role of biochar and biochar-compost in improving soil quality and crop performance: A review. *Applied Soil Ecology*, *119*, 156−170. Available from https://doi.org/10.1016/j.apsoil.2017.06.008, http://www.elsevier.com/inca/publications/store/5/2/4/5/1/8/index.htt.

Ahmad, M., Usman, A. R. A., Al-Faraj, A. S., Ahmad, M., Sallam, A., & Al-Wabel, M. I. (2018). Phosphorus-loaded biochar changes soil heavy metals availability and uptake potential of maize (*Zea mays* L.) plants. *Chemosphere*, *194*, 327−339. Available from https://doi.org/10.1016/j.chemosphere.2017.11.156, http://www.elsevier.com/locate/chemosphere.

Ali, A., Guo, D., Zhang, Y., Sun, X., Jiang, S., Guo, Z., Huang, H., Liang, W., Li, R., & Zhang, Z. (2017). Using bamboo biochar with compost for the stabilization and phytotoxicity reduction of heavy metals in mine-contaminated soils of China. *Scientific Reports*, *7*, 1−12. Available from https://doi.org/10.1038/s41598-017-03045-9.

Ali, S., Rizwan, M., Qayyum, M. F., Ok, Y. S., Ibrahim, M., Riaz, M., Arif, M. S., Hafeez, F., Al-Wabel, M. I., & Shahzad, A. N. (2017). Biochar soil amendment on alleviation of drought and salt stress in plants: A critical review. *Environmental Science and Pollution Research*, *24*(14), 12700−12712. Available from https://doi.org/10.1007/s11356-017-8904-x, http://www.springerlink.com/content/0944-1344.

Al-Wabel, M. I., Usman, A. R. A., El-Naggar, A. H., Aly, A. A., Ibrahim, H. M., Elmaghraby, S., & Al-Omran, A. (2015). Conocarpus biochar as a soil amendment for reducing heavy metal availability and uptake by maize plants. *Saudi Journal of Biological Sciences*, *22*(4), 503−511. Available from https://doi.org/10.1016/j.sjbs.2014.12.003, http://www.ksu.edu.sa/sites/Colleges/CollegeofScinces/BotanyDepartment/SJBS/Default.aspx.

Andrenelli, M. C., Maienza, A., Genesio, L., Miglietta, F., Pellegrini, S., Vaccari, F. P., & Vignozzi, N. (2016). Field application of pelletized biochar: Short term effect on the hydrological properties of a silty clay loam soil. *Agricultural Water Management*, *163*, 190−196. Available from https://doi.org/10.1016/j.agwat.2015.09.017.

Andrés, P., Rosell-Melé, A., Colomer-Ventura, F., Denef, K., Cotrufo, M. F., Riba, M., & Alcañiz, J. M. (2019). Belowground biota responses to maize biochar addition to the soil of a Mediterranean vineyard. *Science of the Total Environment*, *660*, 1522−1532. Available from https://doi.org/10.1016/j.scitotenv.2019.01.101, http://www.elsevier.com/locate/scitotenv.

Anwar, T., Munwwar, F., Qureshi, H., Siddiqi, E. H., Hanif, A., Anwaar, S., Gul, S., Waheed, A., Alwahibi, M. S., & Kamal, A. (2023). Synergistic effect of biochar-based compounds from vegetable wastes and gibberellic acid on wheat growth under salinity stress. *Scientific Reports*, *13*(1). Available from https://doi.org/10.1038/s41598-023-46487-0, https://www.nature.com/srep/.

Arabi, Z., Eghtedaey, H., Gharehchmaghloo, B., & Faraji, A. (2018). Effects of biochar and bio-fertilizer on yield and qualitative properties of soybean and some chemical properties of soil. *Arabian Journal of Geosciences*, *11*(21). Available from https://doi.org/10.1007/s12517-018-4041-1.

Arunrat, N., Kongsurakan, P., Sereenonchai, S., & Hatano, R. (2020). Soil organic carbon in sandy paddy fields of Northeast Thailand: A review. *Agronomy*, *10*(8), 1061. Available from https://doi.org/10.3390/agronomy10081061.

Atkinson, C. J., Fitzgerald, J. D., & Hipps, N. A. (2010). Potential mechanisms for achieving agricultural benefits from biochar application to temperate soils: A review. *Plant and Soil*, *337*(1), 1–18. Available from https://doi.org/10.1007/s11104-010-0464-5.

Baldoni, N., Francioni, M., Trozzo, L., Toderi, M., Fornasier, F., D'Ottavio, P., Corti, G., & Cocco, S. (2023). Effect of wood gasification biochar on soil physicochemical properties and enzyme activities, and on crop yield in a wheat-production system with subalkaline soil. *Biomass and Bioenergy*, *176*, 106914. Available from https://doi.org/10.1016/j.biombioe.2023.106914.

Beesley, L., Moreno-Jiménez, E., Gomez-Eyles, J. L., Harris, E., Robinson, B., & Sizmur, T. (2011). A review of biochars' potential role in the remediation, revegetation and restoration of contaminated soils. *Environmental Pollution*, *159*(12), 3269–3282. Available from https://doi.org/10.1016/j.envpol.2011.07.023.

Bian, R., Joseph, S., Shi, W., Li, L., Taherymoosavi, S., & Pan, G. (2019). Biochar DOM for plant promotion but not residual biochar for metal immobilization depended on pyrolysis temperature. *Science of the Total Environment*, *662*, 571–580. Available from https://doi.org/10.1016/j.scitotenv.2019.01.224.

Bibi, S., Ullah, R., Burni, T., Ullah, Z., & Kazi, M. (2023). Impact of resorcinol and biochar application on the growth attributes, metabolite contents, and antioxidant systems of tomato (*Lycopersicon esculentum* Mill.). *ACS Omega*, *8*(48), 45750–45762. Available from https://doi.org/10.1021/acsomega.3c06233.

Bopp, C., Christl, I., Schulin, R., & Evangelou, M. W. H. (2016). Biochar as possible long-term soil amendment for phytostabilisation of TE-contaminated soils. *Environmental Science and Pollution Research*, *23*(17), 17449–17458. Available from https://doi.org/10.1007/s11356-016-6935-3, http://www.springerlink.com/content/0944-1344.

Brtnicky, M., Datta, R., Holatko, J., Bielska, L., Gusiatin, Z. M., Kucerik, J., Hammerschmiedt, T., Danish, S., Radziemska, M., Mravcova, L., Fahad, S., Kintl, A., Sudoma, M., Ahmed, N., & Pecina, V. (2021). A critical review of the possible adverse effects of biochar in the soil environment. *Science of the Total Environment*, *796*, 148756. Available from https://doi.org/10.1016/j.scitotenv.2021.148756.

Carter, S., Shackley, S., Sohi, S., Suy, T. B., & Haefele, S. (2013). The impact of biochar application on soil properties and plant growth of pot grown lettuce (*Lactuca sativa*) and cabbage (*brassica chinensis*). *Agronomy*, *3*(2), 404–418. Available from https://doi.org/10.3390/agronomy3020404, https://www.mdpi.com/2073-4395/3/2/404/pdf.

Cen, R., Feng, W., Yang, F., Wu, W., Liao, H., & Qu, Z. (2021). Effect mechanism of biochar application on soil structure and organic matter in semi-arid areas. *Journal of Environmental Management*, *286*, 112198. Available from https://doi.org/10.1016/j.jenvman.2021.112198.

Centenaro, G., de Miguel, S., Amouzgar, L., Piñuela, Y., Son, D., Bonet, J. A., de Aragón, J. M., Dashevskaya, S., Castaño, C., & Alday, J. G. (2023). Silvicultural management and altitude prevail on soil properties and fungal community in shaping understorey plant communities in a Mediterranean pine forest. *Science of the Total Environment*, *858*. Available from https://doi.org/10.1016/j.scitotenv.2022.159860, http://www.elsevier.com/locate/scitotenv.

Chen, W., Wei, R., Yang, L., Yang, Y., Li, G., & Ni, J. (2019). Characteristics of wood-derived biochars produced at different temperatures before and after deashing: Their different potential advantages in environmental applications. *Science of the Total Environment*, *651*, 2762–2771. Available from https://doi.org/10.1016/j.scitotenv.2018.10.141.

Chen, X. W., Wong, J. T. F., Chen, Z. T., Tang, T. W. L., Guo, H. W., Leung, A. O. W., Ng, C. W. W., & Wong, M. H. (2018). Effects of biochar on the ecological performance of a subtropical landfill. *Science of the Total Environment*, *644*, 963–975. Available from https://doi.org/10.1016/j.scitotenv.2018.06.379, http://www.elsevier.com/locate/scitotenv.

Chtouki, M., Naciri, R., & Oukarroum, A. (2024). A review on phosphorus drip fertigation in the Mediterranean region: Fundamentals, current situation, challenges, and perspectives. *Heliyon*, *10*(3), e25543. Available from https://doi.org/10.1016/j.heliyon.2024.e25543.

Cui, Yf, Meng, J., Wang, Qx, Zhang, Wm, Cheng, Xy, & Chen, Wf (2017). Effects of straw and biochar addition on soil nitrogen, carbon, and super rice yield in cold waterlogged paddy soils of North China. *Journal of Integrative Agriculture*, *16*(5), 1064–1074. Available from https://doi.org/10.1016/S2095-3119(16)61578-2, http://www.elsevier.com/journals/journal-of-integrative-agriculture/2095-3119.

Dai, S., Li, H., Yang, Z., Dai, M., Dong, X., Ge, X., Sun, M., & Shi, L. (2018). Effects of biochar amendments on speciation and bioavailability of heavy metals in coal-mine-contaminated soil. *Human and Ecological Risk Assessment*, *24*(7), 1887–1900. Available from https://doi.org/10.1080/10807039.2018.1429250, http://www.tandf.co.uk/journals/titles/10807039.asp.

Dang, T., Marschner, P., Fitzpatrick, R., & Mosley, L. M. (2018). Assessment of the binding of protons, Al and Fe to biochar at different pH values and soluble metal concentrations. *Water (Switzerland)*, *10*(1). Available from https://doi.org/10.3390/w10010055, http://www.mdpi.com/2073-4441/10/1/55/pdf.

Debnath, M., Ashwath, N., & Midmore, D. J. (2019). Physiological and morphological responses to abiotic stresses in two cultivars of Stevia rebaudiana (Bert.) Bertoni. *South African Journal of Botany*, *123*, 124–132. Available from https://doi.org/10.1016/j.sajb.2019.01.025.

Dissanayake, P. D., You, S., Igalavithana, A. D., Xia, Y., Bhatnagar, A., Gupta, S., Kua, H. W., Kim, S., Kwon, J. H., Tsang, D. C. W., & Ok, Y. S. (2020). Biochar-based adsorbents for carbon dioxide capture: A critical review. *Renewable and Sustainable Energy Reviews*, *119*. Available from https://doi.org/10.1016/j.rser.2019.109582, https://www.journals.elsevier.com/renewable-and-sustainable-energy-reviews.

Eissa, M. A. (2019). Effect of compost and biochar on heavy metals phytostabilization by the halophytic plant old man saltbush [*Atriplex nummularia* Lindl]. *Soil and Sediment Contamination*, *28*(2), 135–147. Available from https://doi.org/10.1080/15320383.2018.1551325, http://www.tandfonline.com/toc/bssc20/current.

El-Naggar, A., El-Naggar, A. H., Shaheen, S. M., Sarkar, B., Chang, S. X., Tsang, D. C. W., Rinklebe, J., & Ok, Y. S. (2019). Biochar composition-dependent impacts on soil nutrient release, carbon mineralization, and potential environmental risk: A review. *Journal of Environmental Management*, *241*, 458–467. Available from https://doi.org/10.1016/j.jenvman.2019.02.044, https://www.sciencedirect.com/science/journal/03014797.

El-Naggar, A., Lee, M. H., Hur, J., Lee, Y. H., Igalavithana, A. D., Shaheen, S. M., Ryu, C., Rinklebe, J., Tsang, D. C. W., & Ok, Y. S. (2020). Biochar-induced metal immobilization and soil biogeochemical process: An integrated mechanistic approach. *Science of the Total Environment*, *698*. Available from https://doi.org/10.1016/j.scitotenv.2019.134112, http://www.elsevier.com/locate/scitotenv.

EU 2021/2088. (2021). Commission Delegated Regulation (EU) 2021/2088 of 7 July 2021 Amending Annexes II, III, and IV to Regulation (EU) 2019/1009 of the European Parliament and of the Council for the Purpose of Adding Pyrolysis and Gasification Materials as a Component Material Category in EU Fertilising Products. *Official Journal of the European Union*. Available from http://data.europa.eu/eli/reg_del/2021/2088/oj.

Fady, B., & Médail, F. (2004). Temperate and Mediterranean forests. *Mediterranean Forest Ecosystems*, 1403–1413.

Faloye, O. T., Alatise, M. O., Ajayi, A. E., & Ewulo, B. S. (2019). Effects of biochar and inorganic fertiliser applications on growth, yield and water use efficiency of maize under deficit irrigation. *Agricultural Water Management*, *217*, 165–178. Available from https://doi.org/10.1016/j.agwat.2019.02.044.

Farhangi-Abriz, S., & Ghassemi-Golezani, K. (2022). The modified biochars influence nutrient and osmotic statuses and hormonal signaling of mint plants under fluoride and cadmium toxicities. *Frontiers in Plant Science*, *13*. Available from https://doi.org/10.3389/fpls.2022.1064409.

Farhangi-Abriz, S., & Torabian, S. (2017). Antioxidant enzyme and osmotic adjustment changes in bean seedlings as affected by biochar under salt stress. *Ecotoxicology and Environmental Safety*, *137*, 64–70. Available from https://doi.org/10.1016/j.ecoenv.2016.11.029.

Farhangi-Abriz, S., & Torabian, S. (2018). Effect of biochar on growth and ion contents of bean plant under saline condition. *Environmental Science and Pollution Research*, *25* (12), 11556–11564. Available from https://doi.org/10.1007/s11356-018-1446-z.

Fischer, B. M. C., Manzoni, S., Morillas, L., Garcia, M., Johnson, M. S., & Lyon, S. W. (2019). Improving agricultural water use efficiency with biochar – a synthesis of biochar effects on water storage and fluxes across scales. *Science of the Total Environment*, *657*, 853–862. Available from https://doi.org/10.1016/j.scitotenv.2018.11.312, http://www.elsevier.com/locate/scitotenv.

Forján, R., Rodríguez-Vila, A., Cerqueira, B., & Covelo, E. F. (2017). Comparison of the effects of compost versus compost and biochar on the recovery of a mine soil by improving the nutrient content. *Journal of Geochemical Exploration*, *183*, 46–57. Available from https://doi.org/10.1016/j.gexplo.2017.09.013, http://www.sciencedirect.com/science/journal/03756742.

Galindo, A., Collado-González, J., Griñán, I., Corell, M., Centeno, A., Martín-Palomo, M. J., Girón, I. F., Rodríguez, P., Cruz, Z. N., Memmi, H., Carbonell-Barrachina, A. A., Hernández, F., Torrecillas, A., Moriana, A., & López-Pérez, D. (2018). Deficit irrigation and emerging fruit crops as a strategy to save water in Mediterranean semiarid agrosystems. *Agricultural Water Management*, *202*, 311–324. Available from https://doi.org/10.1016/j.agwat.2017.08.015, http://www.journals.elsevier.com/agricultural-water-management/.

Garcia, B., Alves, O., Rijo, B., Lourinho, G., & Nobre, C. (2022). Biochar: Production, applications, and market prospects in Portugal. *Environments - MDPI*, *9*(8). Available from https://doi.org/10.3390/environments9080095, http://www.mdpi.com/journal/environments.

Gascó, G., Álvarez, M. L., Paz-Ferreiro, J., & Méndez, A. (2019). Combining phytoextraction by *Brassica napus* and biochar amendment for the remediation of a mining soil in Riotinto (Spain). *Chemosphere*, *231*, 562–570. Available from https://doi.org/10.1016/j.chemosphere.2019.05.168.

Gavili, E., Moosavi, A. A., & Kamgar Haghighi, A. A. (2019). Does biochar mitigate the adverse effects of drought on the agronomic traits and yield components of soybean? *Industrial Crops and Products*, *128*, 445–454. Available from https://doi.org/10.1016/j.indcrop.2018.11.047, http://www.elsevier.com/inca/publications/store/5/2/2/8/2/5.

Ghassemi-Golezani, K., & Farhangi-Abriz, S. (2019). Biochar alleviates fluoride toxicity and oxidative stress in safflower (*Carthamus tinctorius* L.) seedlings. *Chemosphere*, *223*, 406–415. Available from https://doi.org/10.1016/j.chemosphere.2019.02.087.

Giannetta, B., Plaza, C., Cassetta, M., Mariotto, G., Benavente-Ferraces, I., García-Gil, J. C., Panettieri, M., & Zaccone, C. (2023). The effects of biochar on soil organic matter pools are not influenced by climate change. *Journal of Environmental Management*, *341*. Available from https://doi.org/10.1016/j.jenvman.2023.118092, https://www.sciencedirect.com/science/journal/03014797.

Giannetta, B., Plaza, C., Galluzzi, G., Benavente-Ferraces, I., García-Gil, J. C., Panettieri, M., Gascó, G., & Zaccone, C. (2024). Distribution of soil organic carbon between particulate and mineral-associated fractions as affected by biochar and its co-application with other amendments. *Agriculture, Ecosystems and Environment*, *360*. Available from https://doi.org/10.1016/j.agee.2023.108777, http://www.elsevier.com/inca/publications/store/5/0/3/2/9/8.

Glaser, B., Lehmann, J., & Zech, W. (2022). Ameliorating physical and chemical properties of highly weathered soils in the tropics with charcoal – a review. *Biology and Fertility of Soils*, *35*(4), 219–230. Available from https://doi.org/10.1007/s00374-002-0466-4.

Gonzaga, M. I. S., Mackowiak, C., de Almeida, A. Q., de Carvalho Junior, J. I. T., & Andrade, K. R. (2018). Positive and negative effects of biochar from coconut husks, orange bagasse and pine wood chips on maize (*Zea mays* L.) growth and nutrition. *Catena*, *162*, 414–420. Available from https://doi.org/10.1016/j.catena.2017.10.018, http://www.elsevier.com/inca/publications/store/5/2/4/6/0/9.

Gorovtsov, A. V., Minkina, T. M., Mandzhieva, S. S., Perelomov, L. V., Soja, G., Zamulina, I. V., Rajput, V. D., Sushkova, S. N., Mohan, D., & Yao, J. (2020). The mechanisms of biochar interactions with microorganisms in soil. *Environmental Geochemistry and Health*, *42*(8), 2495–2518. Available from https://doi.org/10.1007/s10653-019-00412-5, http://www.wkap.nl/journalhome.htm/0269-4042.

Guarino, R., Vrahnakis, M., Rojo, M. P. R., Giuga, L., & Pasta, S. (2020). Grasslands and shrublands of the Mediterranean region. *Encyclopedia of the World's Biomes*, *3-5*, 638–655. Available from https://doi.org/10.1016/B978-0-12-409548-9.12119-0, https://www.sciencedirect.com/referencework/9780128160978/encyclopedia-of-the-worlds-biomes.

Gul, S., Whalen, J. K., Thomas, B. W., Sachdeva, V., & Deng, H. (2015). Physico-chemical properties and microbial responses in biochar-amended soils: Mechanisms and future directions. *Agriculture, Ecosystems and Environment*, *206*, 46–59. Available from https://doi.org/10.1016/j.agee.2015.03.015, http://www.elsevier.com/inca/publications/store/5/0/3/2/9/8.

Gwenzi, W., Chaukura, N., Mukome, F. N. D., Machado, S., & Nyamasoka, B. (2015). Biochar production and applications in sub-Saharan Africa: Opportunities, constraints, risks and uncertainties. *Journal of Environmental Management*, *150*, 250–261. Available from https://doi.org/10.1016/j.jenvman.2014.11.027, https://www.sciencedirect.com/journal/journal-of-environmental-management.

Głąb, T., Palmowska, J., Zaleski, T., & Gondek, K. (2016). Effect of biochar application on soil hydrological properties and physical quality of sandy soil. *Geoderma*, *281*, 11–20. Available from https://doi.org/10.1016/j.geoderma.2016.06.028.

Hafez, E. M., Alsohim, A. S., Farig, M., Omara, A. E. D., Rashwan, E., & Kamara, M. M. (2019). Synergistic effect of biochar and plant growth promoting rhizobacteria on alleviation of water deficit in rice plants under salt-affected soil. *Agronomy*, *9*(12). Available from https://doi.org/10.3390/agronomy9120847, https://www.mdpi.com/2073-4395/9/12/847/pdf.

Hamid, Y., Tang, L., Yaseen, M., Hussain, B., Zehra, A., Aziz, M. Z., He, Zl, & Yang, X. (2019). Comparative efficacy of organic and inorganic amendments for cadmium and lead immobilization in contaminated soil under rice–wheat cropping system. *Chemosphere*, *214*, 259–268. Available from https://doi.org/10.1016/j.chemosphere.2018.09.113, http://www.elsevier.com/locate/chemosphere.

Hammond, J., Schmidt, H.-P., Scholl, van, L., Nelissen, V., Ibarrola, R., O'Toole, A., & Laer, van, T. (2016). *The legality of biochar use – regulatory requirements and risk assessment. Biochar in European soils and agriculture*. Routledge – Taylor & Francis Group. Available from https://doi.org/10.4324/97813158844622016.

Han, L., Sun, K., Yang, Y., Xia, X., Li, F., Yang, Z., & Xing, B. (2020). Biochar's stability and effect on the content, composition and turnover of soil organic carbon. *Geoderma*, *364*, 114184. Available from https://doi.org/10.1016/j.geoderma.2020.114184.

Hasnain, M., Munir, N., Abideen, Z., Zulfiqar, F., Koyro, H. W., El-Naggar, A., Caçador, I., Duarte, B., Rinklebe, J., & Yong, J. W. H. (2023). Biochar-plant interaction and detoxification strategies under abiotic stresses for achieving agricultural resilience: A critical review. *Ecotoxicology and Environmental Safety*, *249*. Available from https://doi.org/10.1016/j.ecoenv.2022.114408, http://www.elsevier.com/inca/publications/store/6/2/2/8/1/9/index.htt.

Henry, J. (2005). *Tropical and equatorial climates. Encyclopedia of earth sciences series* (pp. 742–750). Springer. Available from https://doi.org/10.1007/1-4020-3266-8_212, http://www.springer.com/series/5898.

Hussain, F., Hussain, I., Khan, A. H. A., Muhammad, Y. S., Iqbal, M., Soja, G., Reichenauer, T. G., Zeshan., & Yousaf, S. (2018). Combined application of biochar, compost, and bacterial consortia with Italian ryegrass enhanced phytoremediation of petroleum hydrocarbon contaminated soil. *Environmental and Experimental Botany*, *153*, 80–88. Available from https://doi.org/10.1016/j.envexpbot.2018.05.012, http://www.elsevier.com/inca/publications/store/2/6/7.

Hussain, M., Farooq, M., Nawaz, A., Al-Sadi, A. M., Solaiman, Z. M., Alghamdi, S. S., Ammara, U., Ok, Y. S., & Siddique, K. H. M. (2017). Biochar for crop production: potential benefits and risks. *Journal of Soils and Sediments*, *17*(3), 685–716. Available from https://doi.org/10.1007/s11368-016-1360-2, http://www.springerlink.com/content/1439-0108.

Ibn Ferjani, A., Jeguirim, M., Jellali, S., Limousy, L., Courson, C., Akrout, H., Thevenin, N., Ruidavets, L., Muller, A., & Bennici, S. (2019). The use of exhausted grape marc to produce biofuels and biofertilizers: Effect of pyrolysis temperatures on biochars properties. *Renewable and Sustainable Energy Reviews*, *107*, 425–433. Available from https://doi.org/10.1016/j.rser.2019.03.034.

Idbella, M., Baronti, S., Giagnoni, L., Renella, G., Becagli, M., Cardelli, R., Maienza, A., Vaccari, F. P., & Bonanomi, G. (2024). Long-term effects of biochar on soil chemistry, biochemistry, and microbiota: Results from a 10-year field vineyard experiment. *Applied Soil Ecology*, *195*, 105217. Available from https://doi.org/10.1016/j.apsoil.2023.105217.

Jabborova, D., Annapurna, K., Choudhary, R., Bhowmik, S. N., Desouky, S. E., Selim, S., Azab, I. H. E., Hamada, M. M. A., Nahhas, N. E., & Elkelish, A. (2021). Interactive impact of biochar and arbuscular mycorrhizal on root morphology, physiological properties of fenugreek (*Trigonella foenum-graecum* L.) and soil enzymatic activities. *Agronomy*, *11*(11). Available from https://doi.org/10.3390/AGRONOMY11112341, https://www.mdpi.com/2073-4395/11/11/2341.

Jabborova, D., Annapurna, K., Paul, S., Kumar, S., Saad, H. A., Desouky, S., Ibrahim, M. F. M., & Elkelish, A. (2021). Beneficial features of biochar and arbuscular mycorrhiza for improving spinach plant growth, root morphological traits, physiological properties, and soil enzymatic activities. *Journal of Fungi*, *7*(7). Available from https://doi.org/10.3390/jof7070571, https://www.mdpi.com/2309-608X/7/7/571/pdf.

Jia, W., Wang, C., Ma, C., Wang, J., Sun, H., & Xing, B. (2019). Mineral elements uptake and physiological response of *Amaranthus mangostanus* (L.) as affected by biochar. *Ecotoxicology and Environmental Safety*, *175*, 58–65. Available from https://doi.org/10.1016/j.ecoenv.2019.03.039.

Kabir, E., Kim, K.-H., & Kwon, E. E. (2023). Biochar as a tool for the improvement of soil and environment. *Frontiers in Environmental Science*, *11*. Available from https://doi.org/10.3389/fenvs.2023.1324533.

Kanwal, S., Ilyas, N., Shabir, S., Saeed, M., Gul, R., Zahoor, M., Batool, N., & Mazhar, R. (2018). Application of biochar in mitigation of negative effects of salinity stress in wheat (*Triticum aestivum* L.). *Journal of Plant Nutrition*, *41*(4), 526–538. Available from https://doi.org/10.1080/01904167.2017.1392568.

Khan, M. A., Ding, X., Khan, S., Brusseau, M. L., Khan, A., & Nawab, J. (2018). The influence of various organic amendments on the bioavailability and plant uptake of cadmium present in mine-degraded soil. *Science of the Total Environment*, *636*, 810–817. Available from https://doi.org/10.1016/j.scitotenv.2018.04.299, http://www.elsevier.com/locate/scitotenv.

Kim, H. S., Kim, K. R., Kim, H. J., Yoon, J. H., Yang, J. E., Ok, Y. S., Owens, G., & Kim, K. H. (2015). Effect of biochar on heavy metal immobilization and uptake by lettuce (*Lactuca sativa* L.) in agricultural soil. *Environmental Earth Sciences*, *74*(2), 1249–1259. Available from https://doi.org/10.1007/s12665-015-4116-1, http://www.springerlink.com/content/121380/.

Kätterer, T., Roobroeck, D., Andrén, O., Kimutai, G., Karltun, E., Kirchmann, H., Nyberg, G., Vanlauwe, B., & Röing de Nowina, K. (2019). Biochar addition persistently increased soil fertility and yields in maize-soybean rotations over 10 years in subhumid regions of Kenya. *Field Crops Research*, *235*, 18–26. Available from https://doi.org/10.1016/j.fcr.2019.02.015.

Lebrun, M., Miard, F., Nandillon, R., Scippa, G. S., Bourgerie, S., & Morabito, D. (2019). Biochar effect associated with compost and iron to promote Pb and As soil stabilization and *Salix viminalis* L. growth. *Chemosphere*, *222*, 810–822. Available from https://doi.org/10.1016/j.chemosphere.2019.01.188, http://www.elsevier.com/locate/chemosphere.

Lehmann, J., Rillig, M. C., Thies, J., Masiello, C. A., Hockaday, W. C., & Crowley, D. (2011). Biochar effects on soil biota – a review. *Soil Biology and Biochemistry*, *43*(9), 1812–1836. Available from https://doi.org/10.1016/j.soilbio.2011.04.022.

Liang, F., Li, Gt, Lin, Qm, & Zhao, Xr (2014). Crop yield and soil properties in the first 3 years after biochar application to a calcareous soil. *Journal of Integrative Agriculture*, *13*(3), 525–532. Available from https://doi.org/10.1016/S2095-3119(13)60708-X.

Lima, J. Rd. S., de Moraes Silva, W., de Medeiros, E. V., Duda, G. P., Corrêa, M. M., Martins Filho, A. P., Clermont-Dauphin, C., Antonino, A. C. D., & Hammecker, C. (2018). Effect of biochar on physicochemical properties of a sandy soil and maize growth in a greenhouse experiment. *Geoderma*, *319*, 14–23. Available from https://doi.org/10.1016/j.geoderma.2017.12.033, http://www.elsevier.com/inca/publications/store/5/0/3/3/3/2.

Liu, N., Liao, P., Zhang, J., Zhou, Y., Luo, L., Huang, H., & Zhang, L. (2020). Characteristics of denitrification genes and relevant enzyme activities in heavy-metal polluted soils remediated by biochar and compost. *Science of the Total Environment*, *739*, 139987. Available from https://doi.org/10.1016/j.scitotenv.2020.139987.

Liu, Y., Lonappan, L., Brar, S. K., & Yang, S. (2018). Impact of biochar amendment in agricultural soils on the sorption, desorption, and degradation of pesticides: A review. *Science of the Total Environment*, *645*, 60–70. Available from https://doi.org/10.1016/j.scitotenv.2018.07.099, http://www.elsevier.com/locate/scitotenv.

Liu, Y., Lu, H., Yang, S., & Wang, Y. (2016). Impacts of biochar addition on rice yield and soil properties in a cold waterlogged paddy for two crop seasons. *Field Crops Research*, *191*, 161–167. Available from https://doi.org/10.1016/j.fcr.2016.03.003.

Liu, H., Xu, F., Xie, Y., Wang, C., Zhang, A., Li, L., & Xu, H. (2018). Effect of modified coconut shell biochar on availability of heavy metals and biochemical characteristics of soil in multiple heavy metals contaminated soil. *Science of the Total Environment*, *645*, 702–709. Available from https://doi.org/10.1016/j.scitotenv.2018.07.115.

Lomaglio, T., Hattab-Hambli, N., Bret, A., Miard, F., Trupiano, D., Scippa, G. S., Motelica-Heino, M., Bourgerie, S., & Morabito, D. (2017). Effect of biochar amendments on the mobility and (bio) availability of As, Sb and Pb in a contaminated mine technosol. *Journal of Geochemical Exploration*, *182*, 138–148. Available from https://doi.org/10.1016/j.gexplo.2016.08.007, http://www.sciencedirect.com/science/journal/03756742.

Lu, H., Li, Z., Fu, S., Méndez, A., Gascó, G., & Paz-Ferreiro, J. (2015). Combining phytoextraction and biochar addition improves soil biochemical properties in a soil contaminated with Cd. *Chemosphere*, *119*, 209–216. Available from https://doi.org/10.1016/j.chemosphere.2014.06.024.

Lu, K., Yang, X., Gielen, G., Bolan, N., Ok, Y. S., Niazi, N. K., Xu, S., Yuan, G., Chen, X., Zhang, X., Liu, D., Song, Z., Liu, X., & Wang, H. (2017). Effect of bamboo and rice straw biochars on the mobility and redistribution of heavy metals (Cd, Cu, Pb and Zn) in contaminated soil. *Journal of Environmental Management*, *186*, 285–292. Available from https://doi.org/10.1016/j.jenvman.2016.05.068, https://www.sciencedirect.com/journal/journal-of-environmental-management.

Luis Moreno, J., Bastida, F., Díaz-López, M., Li, Y., Zhou, Y., López-Mondéjar, R., Benavente-Ferraces, I., Rojas, R., Rey, A., Carlos García-Gil, J., & Plaza, C. (2022). Response of soil chemical properties, enzyme activities and microbial communities to biochar application and climate change in a Mediterranean agroecosystem. *Geoderma*, *407*, 115536. Available from https://doi.org/10.1016/j.geoderma.2021.115536.

Maiti, R. K., & Satya, P. (2014). Research advances in major cereal crops for adaptation to abiotic stresses. *GM Crops & Food*, *5*(4), 259–279. Available from https://doi.org/10.4161/21645698.2014.947861.

Major, J., Rondon, M., Molina, D., Riha, S. J., & Lehmann, J. (2010). Maize yield and nutrition during 4 years after biochar application to a Colombian savanna oxisol. *Plant and Soil*, *333*(1), 117–128. Available from https://doi.org/10.1007/s11104-010-0327-0.

Meng, J., Tao, M., Wang, L., Liu, X., & Xu, J. (2018). Changes in heavy metal bioavailability and speciation from a Pb-Zn mining soil amended with biochars from co-pyrolysis of rice straw and swine manure. *Science of the Total Environment*, *633*, 300–307. Available from https://doi.org/10.1016/j.scitotenv.2018.03.199.

Meyer, S., Genesio, L., Vogel, I., Schmidt, H. P., Soja, G., Someus, E., Shackley, S., Verheijen, F. G. A., & Glaser, B. (2017). Biochar standardization and legislation harmonization. *Journal of Environmental Engineering and Landscape Management*, *25* (2), 175–191. Available from https://doi.org/10.3846/16486897.2016.1254640, http://www.tandfonline.com/loi/teel20.

Moiwo, J. P., Wahab, A., Kangoma, E., Blango, M. M., Ngegba, M. P., & Suluku, R. (2019). Effect of biochar application depth on crop productivity under tropical rainfed conditions. *Applied Sciences (Switzerland)*, *9*(13). Available from https://doi.org/10.3390/app9132602, https://res.mdpi.com/applsci/applsci-09-02580/article_deploy/applsci-09-02602.pdf?filename = &attachment = 1.

Montanarella, L., & Lugato, E. (2013). The application of biochar in the EU: Challenges and opportunities. *Agronomy*, *3*(2), 462–473. Available from https://doi.org/10.3390/agronomy3020462.

Montiel-Rozas, M. M., Madejón, E., & Madejón, P. (2016). Effect of heavy metals and organic matter on root exudates (low molecular weight organic acids) of herbaceous species: An assessment in sand and soil conditions under different levels of contamination. *Environmental Pollution*, *216*, 273–281. Available from https://doi.org/10.1016/j.envpol.2016.05.080.

Moradi, R., Pourghasemian, N., & Naghizadeh, M. (2019). Effect of beeswax waste biochar on growth, physiology and cadmium uptake in saffron. *Journal of Cleaner Production*, *229*, 1251–1261. Available from https://doi.org/10.1016/j.jclepro.2019.05.047.

Moreno-Jiménez, E., Fernández, J. M., Puschenreiter, M., Williams, P. N., & Plaza, C. (2016). Availability and transfer to grain of As, Cd, Cu, Ni, Pb and Zn in a barley agri-system: Impact of biochar, organic and mineral fertilizers. *Agriculture, Ecosystems and Environment*, *219*, 171–178. Available from https://doi.org/10.1016/j.agee.2015.12.001, http://www.elsevier.com/inca/publications/store/5/0/3/2/9/8.

Mulcahy, D. N., Mulcahy, D. L., & Dietz, D. (2013). Biochar soil amendment increases tomato seedling resistance to drought in sandy soils. *Journal of Arid Environments*, *88*, 222–225. Available from https://doi.org/10.1016/j.jaridenv.2012.07.012.

Nguyen, D., Rieu, I., Mariani, C., & van Dam, N. M. (2016). How plants handle multiple stresses: hormonal interactions underlying responses to abiotic stress and insect herbivory. *Plant Molecular Biology*, *91*(6), 727–740. Available from https://doi.org/10.1007/s11103-016-0481-8, http://www.wkap.nl/journalhome.htm/0167-4412.

Nigam, N., Khare, P., Yadav, V., Mishra, D., Jain, S., Karak, T., Panja, S., & Tandon, S. (2019). Biochar-mediated sequestration of Pb and Cd leads to enhanced productivity in Mentha arvensis. *Ecotoxicology and Environmental Safety*, *172*, 411–422. Available from https://doi.org/10.1016/j.ecoenv.2019.02.006.

Nikpour-Rashidabad, N., Tavasolee, A., Torabian, S., & Farhangi-Abriz, S. (2019). The effect of biochar on the physiological, morphological and anatomical characteristics of mung bean roots after exposure to salt stress. *Arhiv za bioloske nauke*, *71*(2), 321–327. Available from https://doi.org/10.2298/abs181005014n.

Nogués, I., Mazzurco Miritana, V., Passatore, L., Zacchini, M., Peruzzi, E., Carloni, S., Pietrini, F., Marabottini, R., Chiti, T., Massaccesi, L., & Marinari, S. (2023). Biochar soil amendment as carbon farming practice in a Mediterranean environment. *Geoderma Regional*, *33*, e00634. Available from https://doi.org/10.1016/j.geodrs.2023.e00634.

Norini, M. P., Thouin, H., Miard, F., Battaglia-Brunet, F., Gautret, P., Guégan, R., Le Forestier, L., Morabito, D., Bourgerie, S., & Motelica-Heino, M. (2019). Mobility of Pb, Zn, Ba, As and Cd toward soil pore water and plants (willow and ryegrass) from a mine soil amended with biochar. *Journal of Environmental Management*, *232*, 117–130. Available from https://doi.org/10.1016/j.jenvman.2018.11.021, https://www.sciencedirect.com/journal/journal-of-environmental-management.

Nzediegwu, C., Prasher, S., Elsayed, E., Dhiman, J., Mawof, A., & Patel, R. (2019). Effect of biochar on heavy metal accumulation in potatoes from wastewater irrigation. *Journal of Environmental Management*, *232*, 153–164. Available from https://doi.org/10.1016/j.jenvman.2018.11.013.

Obia, A., Mulder, J., Martinsen, V., Cornelissen, G., & Børresen, T. (2016). In situ effects of biochar on aggregation, water retention and porosity in light-textured tropical soils. *Soil and Tillage Research*, *155*, 35–44. Available from https://doi.org/10.1016/j.still.2015.08.002.

Ojeda, G., Mattana, S., Àvila, A., Alcañiz, J. M., Volkmann, M., & Bachmann, J. (2015). Are soil-water functions affected by biochar application? *Geoderma*, *249-250*, 1–11. Available from https://doi.org/10.1016/j.geoderma.2015.02.014, http://www.elsevier.com/inca/publications/store/5/0/3/3/3/2.

Oliveira, F. R., Patel, A. K., Jaisi, D. P., Adhikari, S., Lu, H., & Khanal, S. K. (2017). Environmental application of biochar: Current status and perspectives. *Bioresource Technology*, *246*, 110–122. Available from https://doi.org/10.1016/j.biortech.2017.08.122, http://www.elsevier.com/locate/biortech.

Omoni, V. T., Baidoo, P. K., Fagbohungbe, M. O., & Semple, K. T. (2020). The impact of enhanced and non-enhanced biochars on the catabolism of 14C-phenanthrene in soil. *Environmental Technology and Innovation*, *20*. Available from https://doi.org/10.1016/j.eti.2020.101146, http://www.journals.elsevier.com/environmental-technology-and-innovation/.

Omulo, G. (2020). *Biochar potential in improving agricultural production in East Africa*. IntechOpen. Available from http://doi.org/10.5772/intechopen.92195.

Paetsch, L., Mueller, C. W., Kögel-Knabner, I., Von Lützow, M., Girardin, C., & Rumpel, C. (2018). Effect of in-situ aged and fresh biochar on soil hydraulic conditions and microbial C use under drought conditions. *Scientific Reports*, *8*(1). Available from https://doi.org/10.1038/s41598-018-25039-x, http://www.nature.com/srep/index.html.

Pandey, D., Daverey, A., & Arunachalam, K. (2020). Biochar: Production, properties and emerging role as a support for enzyme immobilization. *Journal of Cleaner Production*, *255*, 120267. Available from https://doi.org/10.1016/j.jclepro.2020.120267.

Pandit, N. R., Mulder, J., Hale, S. E., Martinsen, V., Schmidt, H. P., & Cornelissen, G. (2018). Biochar improves maize growth by alleviation of nutrient stress in a moderately acidic low-input Nepalese soil. *Science of the Total Environment*, *625*, 1380–1389. Available from https://doi.org/10.1016/j.scitotenv.2018.01.022, http://www.elsevier.com/locate/scitotenv.

Paneque, M., De la Rosa, J. M., Franco-Navarro, J. D., Colmenero-Flores, J. M., & Knicker, H. (2016). Effect of biochar amendment on morphology, productivity and water relations of sunflower plants under non-irrigation conditions. *Catena*, *147*, 280–287. Available from https://doi.org/10.1016/j.catena.2016.07.037, http://www.elsevier.com/inca/publications/store/5/2/4/6/0/9.

Park, J. H., Lamb, D., Paneerselvam, P., Choppala, G., Bolan, N., & Chung, J. W. (2011). Role of organic amendments on enhanced bioremediation of heavy metal(loid) contaminated soils. *Journal of Hazardous Materials*, *185*(2-3), 549–574. Available from https://doi.org/10.1016/j.jhazmat.2010.09.082.

Plaza, C., Gascó, G., Méndez, A. M., Zaccone, C., & Maestre, F. T. (2018). *Soil organic matter in dryland ecosystems. The future of soil carbon: Its conservation and formation* (pp. 39–69). Elsevier Inc. Available from https://www.sciencedirect.com/book/9780128116876/the-future-of-soil-carbon, https://doi.org/10.1016/B978-0-12-811687-6.00002-X.

Prakongkep, N., Gilkes, R. J., Wisawapipat, W., Leksungnoen, P., Kerdchana, C., Inboonchuay, T., Delbos, E., Strachan, L.-J., Ariyasakul, P., Ketdan, C., & Hammecker, C. (2020). Effects of biochar on properties of tropical sandy soils under organic agriculture. *Journal of Agricultural Science*, *13*(1), 1. Available from https://doi.org/10.5539/jas.v13n1p1.

Prasad, M. N. V., & Kumar, C. (2023). *Agroecological approaches for sustainable soil management*. Wiley. Available from https://doi.org/10.1002/9781119911999.

Puga, A. P., Abreu, C. A., Melo, L. C. A., & Beesley, L. (2015). Biochar application to a contaminated soil reduces the availability and plant uptake of zinc, lead and cadmium. *Journal of Environmental Management*, *159*, 86–93. Available from https://doi.org/10.1016/j.jenvman.2015.05.036.

Qambrani, N. A., Rahman, M. M., Won, S., Shim, S., & Ra, C. (2017). Biochar properties and eco-friendly applications for climate change mitigation, waste management, and wastewater treatment: A review. *Renewable and Sustainable Energy Reviews*, *79*, 255–273. Available from https://doi.org/10.1016/j.rser.2017.05.057, https://www.journals.elsevier.com/renewable-and-sustainable-energy-reviews.

Rajkovich, S., Enders, A., Hanley, K., Hyland, C., Zimmerman, A. R., & Lehmann, J. (2012). Corn growth and nitrogen nutrition after additions of biochars with varying properties to a temperate soil. *Biology and Fertility of Soils*, *48*(3), 271–284. Available from https://doi.org/10.1007/s00374-011-0624-7.

El Rasafi, T., & Haddioui, A. (2020). Growth, survival and biomass production of barley in a polluted mine soil amended with biochar and animal manure. *Bulletin of Environmental Contamination and Toxicology*, *105*(1), 155–165. Available from https://doi.org/10.1007/s00128-020-02914-w.

El Rasafi, T., Haouas, A., Tallou, A., Chakouri, M., Aallam, Y., El Moukhtari, A., Hamamouch, N., Hamdali, H., Oukarroum, A., Farissi, M., & Haddioui, A. (2023). Recent progress on emerging technologies for trace elements-contaminated soil remediation. *Chemosphere*, *341*. Available from https://doi.org/10.1016/j.chemosphere.2023.140121.

EL Rasafi, T., Oukarroum, A., & Haddioui, A. (2023). Response of maize to coniferous tree woods biochar and sheep manure application to contaminated mine soil. *Biomass Conversion and Biorefinery*, *13*(5), 3917–3927. Available from https://doi.org/10.1007/s13399-021-01390-4.

El Rasafi, T., Oukarroum, A., Haddioui, A., Song, H., Kwon, E. E., Bolan, N., Tack, F. M. G., Sebastian, A., Prasad, M. N. V., & Rinklebe, J. (2022). Cadmium stress in plants: A critical review of the effects, mechanisms, and tolerance strategies. *Critical Reviews in Environmental Science and Technology*, *52*(5), 675–726. Available from https://doi.org/10.1080/10643389.2020.1835435, http://www.tandf.co.uk/journals/titles/10643389.asp.

Rato-Nunes, J. M., Martín-Franco, C., Peña, D., Terrón-Sánchez, J., Vicente, L. A., Fernández-Rodríguez, D., Albarrán, Á., & López-Piñeiro, A. (2023). Combined use of biochar and sprinkler irrigation may enhance rice productivity in water-stressed regions. *Annals of Agricultural Sciences*, *68*(1), 48–59. Available from https://doi.org/10.1016/j.aoas.2023.05.002, http://www.elsevier.com/journals/annals-of-agricultural-sciences/0570-1783/.

Rawat, J., Saxena, J., & Sanwal, P. (2019). *Biochar: A sustainable approach for improving plant growth and soil properties*. IntechOpen. Available from https://doi.org/10.5772/intechopen.82151.

Regmi, A., Poudyal, S., Singh, S., Coldren, C., Moustaid-Moussa, N., & Simpson, C. (2023). Biochar influences phytochemical concentrations of *Viola cornuta* flowers. *Sustainability*, *15*(5), 3882. Available from https://doi.org/10.3390/su15053882.

Rehman, M. Z. U., Rizwan, M., Ali, S., Fatima, N., Yousaf, B., Naeem, A., Sabir, M., Ahmad, H. R., & Ok, Y. S. (2016). Contrasting effects of biochar, compost and farm manure on alleviation of nickel toxicity in maize (*Zea mays* L.) in relation to plant growth, photosynthesis and metal uptake. *Ecotoxicology and Environmental Safety*, *133*, 218–225. Available from https://doi.org/10.1016/j.ecoenv.2016.07.023, http://www.elsevier.com/inca/publications/store/6/2/2/8/1/9/index.htt.

Rezaei Rashti, M., Esfandbod, M., Phillips, I. R., & Chen, C. (2019). Biochar amendment and water stress alter rhizosphere carbon and nitrogen budgets in bauxite-processing residue sand under rehabilitation. *Journal of Environmental Management*, *230*, 446–455. Available from https://doi.org/10.1016/j.jenvman.2018.09.093, https://www.sciencedirect.com/journal/journal-of-environmental-management.

Ribas, A., Mattana, S., Llurba, R., Debouk, H., Sebastià, M. T., & Domene, X. (2019). Biochar application and summer temperatures reduce N_2O and enhance CH_4 emissions in a Mediterranean agroecosystem: Role of biologically-induced anoxic microsites. *Science of the Total Environment*, *685*, 1075–1086. Available from https://doi.org/10.1016/j.scitotenv.2019.06.277.

Rizwan, M., Ali, S., Abbas, T., Zia-ur-Rehman, M., Hannan, F., Keller, C., Al-Wabel, M. I., & Ok, Y. S. (2016). Cadmium minimization in wheat: A critical review. *Ecotoxicology and Environmental Safety*, *130*, 43–53. Available from https://doi.org/10.1016/j.ecoenv.2016.04.001, http://www.elsevier.com/inca/publications/store/6/2/2/8/1/9/index.htt.

Rizwan, M., Ali, S., Adrees, M., Ibrahim, M., Tsang, D. C. W., Zia-ur-Rehman, M., Zahir, Z. A., Rinklebe, J., Tack, F. M. G., & Ok, Y. S. (2017). A critical review on effects, tolerance mechanisms and management of cadmium in vegetables. *Chemosphere*, *182*, 90–105. Available from https://doi.org/10.1016/j.chemosphere.2017.05.013, http://www.elsevier.com/locate/chemosphere.

Rizwan, M., Ali, S., Qayyum, M. F., Ibrahim, M., Zia-ur-Rehman, M., Abbas, T., & Ok, Y. S. (2016). Mechanisms of biochar-mediated alleviation of toxicity of trace elements in plants: A critical review. *Environmental Science and Pollution Research*, *23*(3), 2230–2248. Available from https://doi.org/10.1007/s11356-015-5697-7, http://www.springerlink.com/content/0944-1344.

de la Rosa, J. M., Rosado, M., Paneque, M., Miller, A. Z., & Knicker, H. (2018). Effects of aging under field conditions on biochar structure and composition: Implications for biochar stability in soils. *Science of the Total Environment*, *613-614*, 969–976. Available from https://doi.org/10.1016/j.scitotenv.2017.09.124, http://www.elsevier.com/locate/scitotenv.

Saletnik, B., Zagula, G., Bajcar, M., Tarapatskyy, M., Bobula, G., & Puchalski, C. (2019). Biochar as a multifunctional component of the environment – a review. *Applied Sciences (Switzerland)*, *9*(6). Available from https://doi.org/10.3390/app9061139, https://res.mdpi.com/applsci/applsci-09-01139/article_deploy/applsci-09-01139.pdf.

Schulte, R., O'Sullivan, L., & Creamer, R. (2018). *Soil functions—an introduction* (pp. 201–208). Springer Science and Business Media LLC. Available from https://doi.org/10.1007/978-3-319-71189-8_13.

Semida, W. M., Beheiry, H. R., Sétamou, M., Simpson, C. R., Abd El-Mageed, T. A., Rady, M. M., & Nelson, S. D. (2019). Biochar implications for sustainable agriculture and environment: A review. *South African Journal of Botany*, *127*, 333–347. Available from https://doi.org/10.1016/j.sajb.2019.11.015, http://www.elsevier.com.

Shen, X., Huang, D. Y., Ren, X. F., Zhu, H. H., Wang, S., Xu, C., He, Y. B., Luo, Z. C., & Zhu, Q. H. (2016). Phytoavailability of Cd and Pb in crop straw biochar-amended soil is related to the heavy metal content of both biochar and soil. *Journal of Environmental Management*, *168*, 245–251. Available from https://doi.org/10.1016/j.jenvman.2015.12.019, https://www.sciencedirect.com/journal/journal-of-environmental-management.

Singh, C., Tiwari, S., & Singh, J. S. (2020). *Biochar: A sustainable tool in soil pollutant bioremediation* (pp. 475–494). Springer Science and Business Media LLC. Available from http://doi.org/10.1007/978-981-13-3426-9_19.

Smebye, A., Alling, V., Vogt, R. D., Gadmar, T. C., Mulder, J., Cornelissen, G., & Hale, S. E. (2016). Biochar amendment to soil changes dissolved organic matter content and composition. *Chemosphere*, *142*, 100–105. Available from https://doi.org/10.1016/j.chemosphere.2015.04.087, http://www.elsevier.com/locate/chemosphere.

Susanti, N. L., & Dewi, K. (2023). The effect of rice husk biochar on the growth and phytohormone profile of Chinese cabbage under drought conditions. *Transactions of the Chinese Society of Agricultural. Machinery.*, *54*, 15–29.

Tan, Z., Lin, C. S. K., Ji, X., & Rainey, T. J. (2017). Returning biochar to fields: A review. *Applied Soil Ecology*, *116*, 1–11. Available from https://doi.org/10.1016/j.apsoil.2017.03.017, http://www.elsevier.com/inca/publications/store/5/2/4/5/1/8/index.htt.

Tang, J., Li, X., Luo, Y., Li, G., & Khan, S. (2016). Spectroscopic characterization of dissolved organic matter derived from different biochars and their polycylic aromatic hydrocarbons (PAHs) binding affinity. *Chemosphere*, *152*, 399–406. Available from https://doi.org/10.1016/j.chemosphere.2016.03.016.

Tang, J., Zhu, W., Kookana, R., & Katayama, A. (2013). Characteristics of biochar and its application in remediation of contaminated soil. *Journal of Bioscience and Bioengineering*, *116*(6), 653–659. Available from https://doi.org/10.1016/j.jbiosc.2013.05.035.

Tareq, R., Akter, N., & Azam, M. S. (2019). *Biochars and biochar composites* (pp. 169–209). Elsevier BV. Available from https://doi.org/10.1016/b978-0-12-811729-3.00010-8.

Teutscherova, N., Lojka, B., Houška, J., Masaguer, A., Benito, M., & Vazquez, E. (2018). Application of holm oak biochar alters dynamics of enzymatic and microbial activity in two contrasting Mediterranean soils. *European Journal of Soil Biology*, *88*, 15–26. Available from https://doi.org/10.1016/j.ejsobi.2018.06.002, https://www.journals.elsevier.com/european-journal-of-soil-biology.

Ukhurebor, K. E., Adetunji, C. O., Olugbemi, O. T., Nwankwo, W., Olayinka, A. S., Umezuruike, C., & Hefftg, D. I. (2021). *AI, edge and IoT-based smart agriculture climatic/environmental/telecommunication, 2021. Precision agriculture: Weather forecasting for future farming* (pp. 101–121). Elsevier. Available from https://doi.org/10.1016/B978-0-12-823694-9.00008-6.

Ullah, N., Ditta, A., Khalid, A., Mehmood, S., Rizwan, M. S., Ashraf, M., Mubeen, F., Imtiaz, M., & Iqbal, M. M. (2020). Integrated effect of algal biochar and plant growth promoting Rhizobacteria on physiology and growth of maize under deficit irrigations. *Journal of Soil Science and Plant Nutrition*, *20*(2), 346–356. Available from https://doi.org/10.1007/s42729-019-00112-0, https://www.springer.com/life + sciences/plant + sciences/journal/42729.

Uzoma, K. C., Inoue, M., Andry, H., Fujimaki, H., Zahoor, A., & Nishihara, E. (2011). Effect of cow manure biochar on maize productivity under sandy soil condition. *Soil Use and Management*, *27*(2), 205–212. Available from https://doi.org/10.1111/j.1475-2743.2011.00340.x.

Vaccari, F. P., Baronti, S., Lugato, E., Genesio, L., Castaldi, S., Fornasier, F., & Miglietta, F. (2011). Biochar as a strategy to sequester carbon and increase yield in durum wheat. *European Journal of Agronomy*, *34*(4), 231–238. Available from https://doi.org/10.1016/j.eja.2011.01.006.

Vaccari, F. P., Maienza, A., Miglietta, F., Baronti, S., Di Lonardo, S., Giagnoni, L., Lagomarsino, A., Pozzi, A., Pusceddu, E., Ranieri, R., Valboa, G., & Genesio, L. (2015). Biochar stimulates plant growth but not fruit yield of processing tomato in a fertile soil. *Agriculture, Ecosystems & Environment*, *207*, 163–170. Available from https://doi.org/10.1016/j.agee.2015.04.015.

Viger, M., Hancock, R. D., Miglietta, F., & Taylor, G. (2015). More plant growth but less plant defence? First global gene expression data for plants grown in soil amended with biochar. *GCB Bioenergy*, *7*(4), 658–672. Available from https://doi.org/10.1111/gcbb.12182, http://onlinelibrary.wiley.com/journal/10.1111/(ISSN)1757-1707.

Vijayaraghavan, K. (2021). The importance of mineral ingredients in biochar production, properties and applications. *Critical Reviews in Environmental Science and Technology*, *51*(2), 113–139. Available from https://doi.org/10.1080/10643389.2020.1716654.

Virto, I., Imaz, M. J., Fernández-Ugalde, O., Gartzia-Bengoetxea, N., Enrique, A., & Bescansa, P. (2015). Soil degradation and soil quality in Western Europe: Current situation and future perspectives. *Sustainability (Switzerland)*, *7*(1), 313–365. Available from https://doi.org/10.3390/su7010313.

Wang, D., Griffin, D. E., Parikh, S. J., & Scow, K. M. (2016). Impact of biochar amendment on soil water soluble carbon in the context of extreme hydrological events. *Chemosphere*, *160*, 287–292. Available from https://doi.org/10.1016/j.chemosphere.2016.06.100, http://www.elsevier.com/locate/chemosphere.

Wang, L., Ji, B., Hu, Y., Liu, R., & Sun, W. (2017). A review on in situ phytoremediation of mine tailings. *Chemosphere*, *184*, 594–600. Available from https://doi.org/10.1016/j.chemosphere.2017.06.025.

Wang, Y., Zhong, B., Shafi, M., Ma, J., Guo, J., Wu, J., Ye, Z., Liu, D., & Jin, H. (2019). Effects of biochar on growth, and heavy metals accumulation of moso bamboo

(*Phyllostachy pubescens*), soil physical properties, and heavy metals solubility in soil. *Chemosphere*, *219*, 510–516. Available from https://doi.org/10.1016/j.chemosphere.2018.11.159.

Weber, K., & Quicker, P. (2018). Properties of biochar. *Fuel*, *217*, 240–261. Available from https://doi.org/10.1016/j.fuel.2017.12.054.

Wei, L., Huang, Y., Huang, L., Li, Y., Huang, Q., Xu, G., Müller, K., Wang, H., Ok, Y. S., & Liu, Z. (2020). The ratio of H/C is a useful parameter to predict adsorption of the herbicide metolachlor to biochars. *Environmental Research*, *184*, 109324. Available from https://doi.org/10.1016/j.envres.2020.109324.

Wu, C., Shi, L., Xue, S., Li, W., Jiang, X., Rajendran, M., & Qian, Z. (2019). Effect of sulfur-iron modified biochar on the available cadmium and bacterial community structure in contaminated soils. *Science of the Total Environment*, *647*, 1158–1168. Available from https://doi.org/10.1016/j.scitotenv.2018.08.087.

Xin, D., Xian, M., & Chiu, P. C. (2019). New methods for assessing electron storage capacity and redox reversibility of biochar. *Chemosphere*, *215*, 827–834. Available from https://doi.org/10.1016/j.chemosphere.2018.10.080, http://www.elsevier.com/locate/chemosphere.

Xu, C. Y., Hosseini-Bai, S., Hao, Y., Rachaputi, R. C. N., Wang, H., Xu, Z., & Wallace, H. (2015). Effect of biochar amendment on yield and photosynthesis of peanut on two types of soils. *Environmental Science and Pollution Research*, *22*(8), 6112–6125. Available from https://doi.org/10.1007/s11356-014-3820-9, http://www.springerlink.com/content/0944-1344.

Xu, P., Sun, C. X., Ye, X. Z., Xiao, W. D., Zhang, Q., & Wang, Q. (2016). The effect of biochar and crop straws on heavy metal bioavailability and plant accumulation in a Cd and Pb polluted soil. *Ecotoxicology and Environmental Safety*, *132*, 94–100. Available from https://doi.org/10.1016/j.ecoenv.2016.05.031, http://www.elsevier.com/inca/publications/store/6/2/2/8/1/9/index.htt.

Xu, X., Huang, H., Zhang, Y., Xu, Z., & Cao, X. (2019). Biochar as both electron donor and electron shuttle for the reduction transformation of Cr(VI) during its sorption. *Environmental Pollution*, *244*, 423–430. Available from https://doi.org/10.1016/j.envpol.2018.10.068, http://www.elsevier.com/inca/publications/store/4/0/5/8/5/6.

Xu, Z., Xu, X., Tsang, D. C. W., & Cao, X. (2018). Contrasting impacts of pre- and post-application aging of biochar on the immobilization of Cd in contaminated soils. *Environmental Pollution*, *242*, 1362–1370. Available from https://doi.org/10.1016/j.envpol.2018.08.012, http://www.elsevier.com/inca/publications/store/4/0/5/8/5/6.

Yamato, M., Okimori, Y., Wibowo, I. F., Anshori, S., & Ogawa, M. (2006). Effects of the application of charred bark of *Acacia mangium* on the yield of maize, cowpea and peanut, and soil chemical properties in South Sumatra, Indonesia. *Soil Science and Plant Nutrition*, *52*(4), 489–495. Available from https://doi.org/10.1111/j.1747-0765.2006.00065.x.

Yan, P., Shen, C., Zou, Z., Fu, J., Li, X., Zhang, L., Zhang, L., Han, W., & Fan, L. (2021). Biochar stimulates tea growth by improving nutrients in acidic soil. *Scientia Horticulturae*, *283*, 110078. Available from https://doi.org/10.1016/j.scienta.2021.110078.

Yan, S., Niu, Z., Yan, H., Zhang, A., Liu, G., & Paz-Ferreiro, J. (2019). Transcriptome sequencing reveals the effect of biochar improvement on the development of tobacco plants before and after topping. *PLoS One*, *14*(10), e0224556. Available from https://doi.org/10.1371/journal.pone.0224556.

Yang, K., Jiang, Y., Wang, J., Cai, X., Wen, Z., Qiu, Z., & Qiao, G. (2022). Tobacco straw biochar improved the growth of Chinese cherry (*Prunus pseudocerasus*) via altering plant physiology and shifting the rhizosphere bacterial community. *Scientia Horticulturae*, *303*, 111244. Available from https://doi.org/10.1016/j.scienta.2022.111244.

Yang, X., Liu, J., McGrouther, K., Huang, H., Lu, K., Guo, X., He, L., Lin, X., Che, L., Ye, Z., & Wang, H. (2016). Effect of biochar on the extractability of heavy metals (Cd, Cu, Pb, and Zn) and enzyme activity in soil. *Environmental Science and Pollution Research*, *23*(2), 974–984. Available from https://doi.org/10.1007/s11356-015-4233-0, http://www.springerlink.com/content/0944-1344.

Yasmin Khan, K., Ali, B., Cui, X., Feng, Y., Yang, X., & Joseph Stoffella, P. (2017). Impact of different feedstocks derived biochar amendment with cadmium low uptake affinity cultivar of pak choi (*Brassica rapa ssb. chinensis* L.) on phytoavoidation of Cd to reduce potential dietary toxicity. *Ecotoxicology and Environmental Safety*, *141*, 129–138. Available from https://doi.org/10.1016/j.ecoenv.2017.03.020.

Yilu, X., Seshadri, B., Sarkar, B., Wang, H., Rumpel, C., Sparks, D., Farrell, M., Hall, T., Yang, X., & Bolan, N. (2018). Biochar modulates heavy metal toxicity and improves microbial carbon use efficiency in soil. *Science of the Total Environment*, *621*, 148–159. Available from https://doi.org/10.1016/j.scitotenv.2017.11.214.

Yin, D., Wang, X., Chen, C., Peng, B., Tan, C., & Li, H. (2016). Varying effect of biochar on Cd, Pb and As mobility in a multi-metal contaminated paddy soil. *Chemosphere*, *152*, 196–206. Available from https://doi.org/10.1016/j.chemosphere.2016.01.044.

Yin, J., Zhao, L., Xu, X., Li, D., Qiu, H., & Cao, X. (2022). Evaluation of long-term carbon sequestration of biochar in soil with biogeochemical field model. *Science of the Total Environment*, *822*, 153576. Available from https://doi.org/10.1016/j.scitotenv.2022.153576.

Younis, U., Malik, S. A., Rizwan, M., Qayyum, M. F., Ok, Y. S., Shah, M. H. R., Rehman, R. A., & Ahmad, N. (2016). Biochar enhances the cadmium tolerance in spinach (*Spinacia oleracea*) through modification of Cd uptake and physiological and biochemical attributes. *Environmental Science and Pollution Research*, *23*(21), 21385–21394. Available from https://doi.org/10.1007/s11356-016-7344-3, http://www.springerlink.com/content/0944-1344.

Yu, K. L., Lau, B. F., Show, P. L., Ong, H. C., Ling, T. C., Chen, W. H., Ng, E. P., & Chang, J. S. (2017). Recent developments on algal biochar production and characterization. *Bioresource Technology*, *246*, 2–11. Available from https://doi.org/10.1016/j.biortech.2017.08.009, http://www.elsevier.com/locate/biortech.

Yuan, P., Wang, J., Pan, Y., Shen, B., & Wu, C. (2019). Review of biochar for the management of contaminated soil: Preparation, application and prospect. *Science of the Total Environment*, *659*, 473–490. Available from https://doi.org/10.1016/j.scitotenv.2018.12.400.

Yue, L., Lian, F., Han, Y., Bao, Q., Wang, Z., & Xing, B. (2019). The effect of biochar nanoparticles on rice plant growth and the uptake of heavy metals: Implications for agronomic benefits and potential risk. *Science of the Total Environment*, *656*, 9–18. Available from https://doi.org/10.1016/j.scitotenv.2018.11.364.

Zabaniotou, A., & Stamou, K. (2020). Balancing waste and nutrient flows between urban agglomerations and rural ecosystems: Biochar for improving crop growth and urban air quality in the Mediterranean region. *Atmosphere*, *11*(5), 539. Available from https://doi.org/10.3390/atmos11050539.

Zafar, N., Zafar Iqbal, M., Shafiq, M., Kabir, M., & Farooqi, Z.-ur-R. (2019). Effects of exhaust pollutants and garden soil on seed germination and seedling growth of maize (*Zea mays* L.) and sunflower (*Helianthus annuus* L.). *Asian Journal of Research in Crop Science*, 1–7. Available from https://doi.org/10.9734/ajrcs/2019/v4i230065.

Zanli, B. L. G. L., Gbossou, K. C., Tang, W., Kamoto, M., & Chen, J. (2022). A review of biochar potential in Cote d'Ivoire in light of the challenges facing Sub-Saharan Africa. *Biomass and Bioenergy*, *165*. Available from https://doi.org/10.1016/j.biombioe.2022.106581, http://www.journals.elsevier.com/biomass-and-bioenergy/.

Zia ur Rehman, M., Waqar, M., Bashir, S., Rizwan, M., Ali, S., El Baroudy, A. A. E. F., Khalid, H., Ayub, M. A., Usman, M., & Jahan, S. (2021). Effect of biochar and compost on cadmium bioavailability and its uptake by wheat–rice cropping system irrigated with untreated sewage water: A field study. *Arabian Journal of Geosciences*, *14*(2). Available from https://doi.org/10.1007/s12517-020-06383-7, http://www.springer.com/geosciences/journal/12517?cm_mmc = AD-_-enews-_-PSE1892-_-0.

Zhang, M., Wang, J., Bai, S. H., Zhang, Y., Teng, Y., & Xu, Z. (2019). Assisted phytoremediation of a co-contaminated soil with biochar amendment: Contaminant removals and bacterial community properties. *Geoderma*, *348*, 115–123. Available from https://doi.org/10.1016/j.geoderma.2019.04.031.

Zhang, R. H., Li, Z. G., Liu, X. D., Wang, Bc, Zhou, G. L., Huang, X. X., Lin, C. F., Wang, Ah, & Brooks, M. (2017). Immobilization and bioavailability of heavy metals in greenhouse soils amended with rice straw-derived biochar. *Ecological Engineering*, *98*, 183–188. Available from https://doi.org/10.1016/j.ecoleng.2016.10.057, http://www.elsevier.com/inca/publications/store/5/2/2/7/5/1.

Zhang, X., Wang, H., He, L., Lu, K., Sarmah, A., Li, J., Bolan, N. S., Pei, J., & Huang, H. (2013). Using biochar for remediation of soils contaminated with heavy metals and organic pollutants. *Environmental Science and Pollution Research*, *20*(12), 8472–8483. Available from https://doi.org/10.1007/s11356-013-1659-0, https://link.springer.com/journal/11356.

Zhang, Y., Ding, J., Wang, H., Su, L., & Zhao, C. (2020). Biochar addition alleviate the negative effects of drought and salinity stress on soybean productivity and water use efficiency. *BMC Plant Biology*, *20*(1). Available from https://doi.org/10.1186/s12870-020-02493-2.

PART

4

Biochar in managing biogeochemical cycles & greenhouse gases

CHAPTER

Biochar application for sustainable soil carbon sequestration and greenhouse gas mitigation

14

Md. Rayhan Shaheb[1], Ayesha Sarker[2] and Shafiqur Rahman[3]

[1]*Department of Agricultural and Life Sciences, Central State University, Wilberforce, OH, United States*

[2]*Agricultural and Environmental Research Station, West Virginia State University, Institute, WV, United States*

[3]*Agricultural Research and Development Program, Central State University, Wilberforce, OH, United States*

14.1 Introduction

Biochar has been recently coined as a term that evolved in conjunction with renewable fuel, soil amelioration, carbon sequestration, greenhouse gas (GHG) emissions, and climate change. Biochar is produced from the thermochemical conversion of biomass, a process called pyrolysis (Lehmann & Joseph, 2012; Talberg, 2009), which takes place with limited oxygen delivery at temperatures <700°C (Clough & Condron, 2010; Kulyk, 2012; Wang et al., 2022). According to guidelines of the International Biochar Initiative (2015), "biochar is a solid material obtained from the thermochemical conversion of biomass in an oxygen-limited environment." Biochar includes char and charcoal, excluding fossil fuel products, produced by partially combusting carbonaceous organic materials like trees and plants (Warnock et al., 2007). Biochar is enriched with highly porous, stable carbon components, functions as a soil supplement, and can increase soil nutrient availability, improve soil fertility and productivity and enhance the soil and environmental quality (Clough & Condron, 2010; Wang et al., 2022). There are many sources of feedstock that are considered for biochar extraction. The more common sources are woody and dry matter, including agricultural wastes such as straw, husks, and nut shells (Yadav et al., 2023). Biochar production by utilizing agricultural and agro-industry residues can be an effective economic waste management (Parmar et al., 2014). It plays a vital role in amending soil, improving soil physicochemical properties, and promoting soil ecological functions that are pivotal in sustainable agriculture. The widespread application of agrochemicals such as fertilizers, pesticides, and herbicides in production agriculture often results in soil quality degradation and environmental contamination (Shaheb et al., 2024) This is, in fact, a great

Biochar Ecotechnology for Sustainable Agriculture and Environment. DOI: https://doi.org/10.1016/B978-0-443-29855-4.00035-7

concern for sustainable agriculture and the environment because it deteriorates water quality and the environment and impacts human health.

Agricultural waste and crop residues improve soil structure and health, but unmanaged residues are likely to contribute to GHG emissions, contributing to climate change. According to the US Environmental Protection Agency (EPA, 2024), annual GHG emissions from agriculture (e.g., livestock, agricultural soils, and rice production) in 2022 accounted for a little over 10% of total US emissions (not counting emissions from onsite energy use). Biochar can provide multiple environmental benefits and has shown considerable success in many recent studies addressing various agricultural and environmental problems. Key benefits include soil and water remediation, increased soil aggregation and structure, water retention, and pH. Further, it promotes microbial activities and can help improve soil C storage, as well as reduce soil emissions of GHG, such as nitrous oxide (N_2O) and methane (CH_4) (Lu et al., 2020; Montanarella & Lugato, 2013; Tan et al., 2015; Wang & Wang, 2018, 2019). Biochar in the soil can also adsorb contaminants such as pesticide residues, metalloids, microplastics, etc., and improve the overall soil and environmental quality, including climate change mitigation by reducing the net GHGs emissions to the environment (Lehmann & Joseph, 2012). The increase use of biochar in recent years is noticed as an application of biochar to improve soil health and increase sustainable agriculture production (Yadav et al., 2023). Fig. 14.1 shows an oxygen-enriched biochar chemical structure produced by applying air steam in mild temperature conditions.

However, the impact of biochar on agroecosystems depends on biochar sources and types, soil types, and other inherent characteristics, including agricultural land use, cropping systems, and soil and the environment. The differences in biochar chemical properties and overall effects on soil and environment also vary due to the differences in feedstock/sources, production conditions, pyrolysis temperature, and speed, including biochar products' posthandling and storage conditions (Brown, 2009; Joseph et al., 2009; Kulyk, 2012). A detailed understanding of this form of bioenergy functioning in agroecosystems will enable a full assessment of the economic and environmental benefits. Further, a thorough study of different sources of biochar's impact on soil biota, interaction with an inherent soil carbon stock and other carbon sources, carbon sequestration, GHG emissions, and soil–plant–environment quality and human health are essential. The present chapter highlights these issues, including the ecological functions of soil–plant–environmental systems and the impact on plant growth, development, agricultural productivity, and sustainability. A schematic flow diagram of biochar production, properties and applications, and effects on soil, crop, and environment is shown in Fig. 14.2.

14.2 Background on carbon sequestration and greenhouse gas emissions

Increasing GHG levels in Earth's atmosphere cause global warming, which is considered one of the most significant environmental problems in the modern

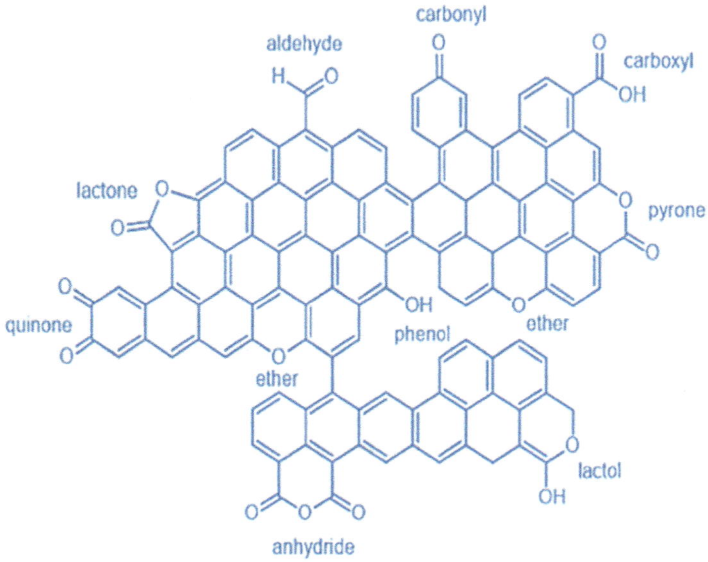

FIGURE 14.1 Oxygen-enriched biochar produced by in the presence of an air steam at mild temperature conditions.

Chemical structure of oxygen-enriched biochar (Conte et al., 2021).

From Conte, P., Bertani, R., Sgarbossa, P., Bambina, P., Schmidt, H. P., Raga, R., Lo Papa, G., Chillura Martino, D. F., & Lo Meo, P. (2021). Recent developments in understanding biochar's physical–chemistry. Agronomy, 11*(4). https://doi.org/10.3390/agronomy11040615.*

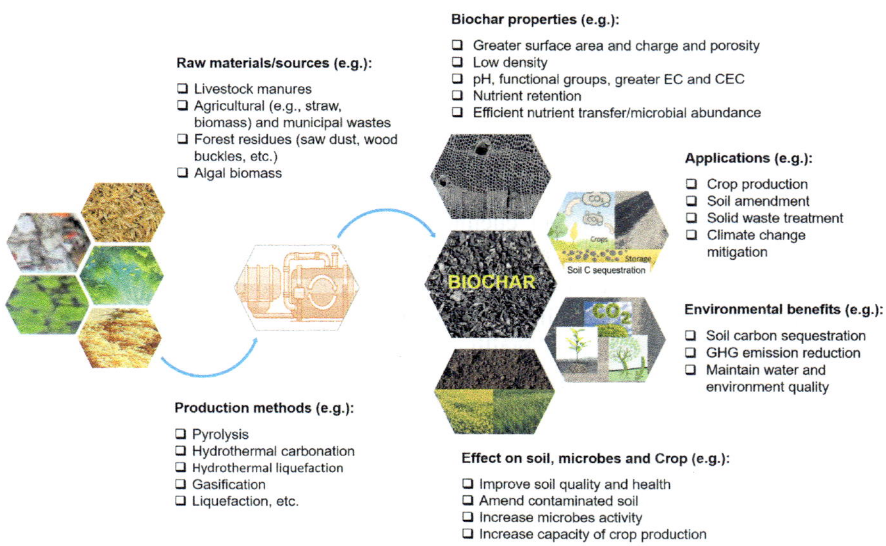

FIGURE 14.2 Biochar production, properties and applications and their effects on soil, crop, and environment.

Flow diagram of feedstocks, biochar production, and applications.

From Carbon Gold.COM (2024), Chalmers (2021), Lungkadee et al. (2023), and http://www.Vectormince.com.

history of the Earth. Global warming has been observed since the preindustrial period, between 1850 and 1900. It is unequivocal that human activities, such as burning fossil fuels and changing land use, increase GHG that traps long-term heating of the Earth's surfaces. According to Intergovernmental Panel on Climate Change (IPCC) Sixth Assessment Report, the earth's global average temperature is estimated to increase by about 1°C and is currently growing by more than 0.2°C per decade (Allen et al., 2012). Global mean CO_2, CH_4, and N_2O concentrations in 2019 increased about 47%, 156%, and 23%, respectively, above the levels in 1750. In reducing the emission of GHG, several approaches, such as adopting energy-efficient technology, renewable fuel or alternative fuel, combined heat and power, use of renewable energy, and zero waste are promising approaches (C2ES, 2024). It is also, therefore, urgent to adopt strategies to sequestrate carbon in soils from the atmosphere carbon (C) pool to offset this rapidly increasing concentration of CO_2. By offsetting anthropogenic emissions, soil C sequestration benefits ecosystem services such as food security, water quality, and biodiversity. Managing interventions toward organic and inorganic transformations through photosynthetic biomass humification and secondary carbonate formation enhances soil C pool. The soil C sink is a C deficit due to the transfer of soil C to the atmosphere. This can be filled and increase positive C and nutrient budgets by adapting management strategies such as reforestation and perennial vegetation cover, as well as changes to favorable soil temperate and moisture regimes following conservation agriculture. Therefore, understanding soil carbon sequestration and the greenhouse effect to address global warming and the processes, properties, and practices affecting the soil C pool and its dynamics is essential.

14.3 Brief overview of biochar and its potential role in greenhouse gas mitigation

Biochar possesses physical and chemical characteristics and can improve econatural quality. It can act as a catalyst in the degradation of contaminants; for example, it gathers transition metals (Lyu et al., 2020). Biochar contains a higher percentage of oxygen-containing carbon groups and greater pore structure and surface area. These properties facilitate biochar to adsorb carbon components, leading to increased potential in carbon sequestration and GHG emission reduction while improving soil physical structure, soil fertility, and crop productivity (Lu et al., 2020; Montanarella & Lugato, 2013; Tan et al., 2015; Wang & Wang, 2018, 2019).

Biochar management systems can lower GHG emissions and thus help in mitigating climate change. Appropriate organic waste management can indirectly reduce methane and other GHG emissions from landfills and industrial energy usage (Lehmann & Joseph, 2012). For example, biochar produced from 2.2 GtC of feedstock can contribute to removing 0.49 GtC per year from the atmosphere to soil (Montanarella & Lugato, 2013). Animal manure-based biochar can also reduce the risk of phosphates and nitrates leaching into the soil (Lehmann & Joseph, 2012; Parmar et al., 2014).

14.4 Biochar production methods

Biochar production depends on factors like production methods (e.g., pyrolysis, thermal carbonization, gasification, etc.), conditions (e.g., low temperature vs high temperature, slow vs fast, heating rate, etc.), and biomass sources. Three factors directly influence biochar production (Kulyk, 2012). These are manufacturing processes, which include methods and temperature; biomass types or sources such as straw, food wastes, and animal by-products; and solid wastes and production technologies such as pyrolysis, thermal carbonization, gasification, etc. There are many sources of feedstock to use to produce biochar. Woody feedstock (e.g., oak sawdust) is primarily used as a sustainable medium for biochar production (Ulusal et al., 2021). Biochar (e.g., charcoal from pyrolysis) contains a well-maintained porous structure with enough functional groups, different inorganic nutrients, and well-stable carbon components (Lehmann & Joseph, 2012; Talberg, 2009; Wang et al., 2022). Based on the source, biochar can be lower carbon ash-containing feedstock 3%–5%(e.g., bamboo, nut shells); ash-containing feedstock 3%–13% (e.g., agricultural residues, tree bark, green wastes); and ash-containing feedstock >13% (e.g., waste paper, manures, industrial effluent, municipal solid wastes) (Joseph & Taylor, 2014). Feedstock, like woody and dry matter, straw, husks, and nut shells, are ideal for biochar extraction (Ok et al., 2016; Parmar et al., 2014). Biochar products can be solid, liquid, and gas forms. Solid and liquid products of biochar can be maximized by using slow and fast pyrolysis (Noor et al., 2012; Ok et al., 2016). Several production methods for the extraction of biochar are briefly described below.

14.4.1 Pyrolysis

Pyrolysis is the process of thermal decomposition with a temperature ranging from 400°C to 600°C, and the heating rate ranges from 5°C to 25°C per minute applied to obtain high C-containing elements like biochar (Noor et al., 2012; Ok et al., 2016; Parmar et al., 2014; Yu et al., 2017). Pyrolysis produces syngas and bio-oil, fuels that can be used for heating or energy production, and a solid residual known as biochar. Pyrolysis depends on operating parameters, namely temperature, heating rate, and residence, and they directly affect the yield and composition of the final product. The increase in temperature and heating rate decreased the pyrolyzed product and vice versa (Yu et al., 2017). Several reactors, such as wagon reactors, agitated drums, kilns for sand rotating, and paddle pyrolysis kilns, are involved in this process. Pyrolysis can be performed incontinuous and batch modes. Improved final yield can be obtained through the continuous mode compared to the batch mode (Gwenzi et al., 2015). Batch mode includes earthen and mound kilns or brick, concrete, and metal kilns that can operate at temperatures ranging from 300°C to 500°C and are less expensive, more straightforward to design, and operate (Gwenzi et al., 2015; Panwar et al., 2019). Pyrolysis processes are of different types, such as dry, slow, fast, etc., depending on the heating rate, temperature, residence time, and pressure (Zhang et al., 2019). Tripathi et al. (2016) reported six subclasses of pyrolysis: slow pyrolysis, fast pyrolysis, flash pyrolysis, vacuum pyrolysis, intermediate pyrolysis, and hydropyrolysis. Some of the subclasses of the pyrolysis are briefly described below. A typical schematic flow diagram of the pyrolysis process is shown in Fig. 14.3.

14.4.1.1 Dry pyrolysis

Dry pyrolysis involves the thermal decomposition of the dry chemical-free substrate into biochar under an anaerobic condition at a high temperature. High temperature, heating, heat transfer rate, the interaction between vapor and solid, pressure, etc., are the key process-related parameters in this method (Libra et al., 2011). Both feedstock properties such as particle size and shape, lignin, cellulose, hemicellulose, etc., and temperature influence the properties of the final product or biochar, including content, structure, and morphology of biochar (Brown, 2009; Joseph et al., 2009). Biochar produced through dry pyrolysis can sequester carbon in the soil for long periods, thus mitigating climate change. However, if

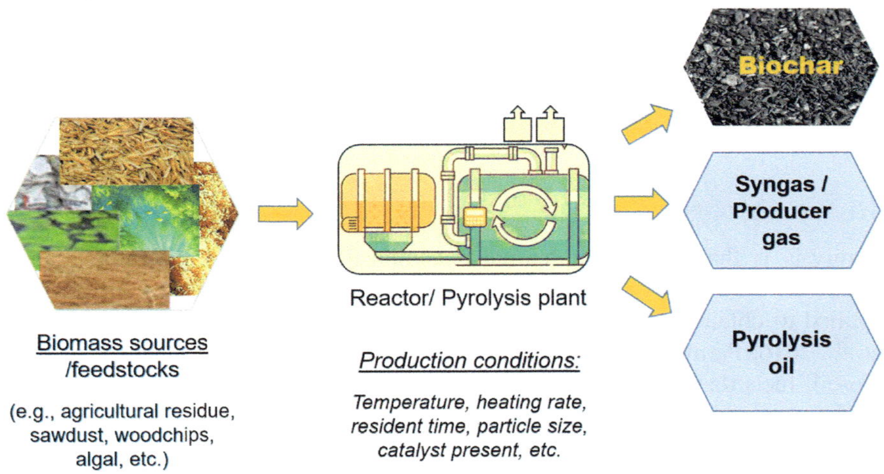

FIGURE 14.3 A typical schematic flow diagram of pyrolysis for biochar production.

Biochar production.

not properly managed, pyrolysis can produce emissions of volatile organic compounds and particulate matter.

14.4.1.2 Fast pyrolysis

Fast pyrolysis is conducted at temperatures 800°C–1300°C with a heating rate of 200°C per minute for no more than 10 seconds to improve bio-oil production by up to 75% (Elkhalifa et al., 2019; Zhang et al., 2019). Fast pyrolysis is generally used to maximize the liquid product. Biomass degradation occurs in anaerobic conditions at 10°C–100°C within 0.5–2 seconds (Sakhiya et al., 2020). Key reactors involved in this process are vacuum, bubbling fluidized beds, circulating beds, ablative reactors, entrained flow, and rotating cones (Elkhalifa et al., 2019; Zhang et al., 2019). Bio-oil can be used in boilers, furnaces, and engines for heat and power generation, while syngas can be used to generate heat or electricity, improving overall energy efficiency. However, challenges related to bio-oil and syngas quality, economic viability, and technical complexity need to be addressed to fully realize the potential of fast pyrolysis. Fig. 14.4 shows a typical fast pyrolysis process for biomass production.

14.4.1.3 Slow pyrolysis

Slow pyrolysis is a promising carbon-negative procedure to reduce the composition of atmosphere CO_2 and recalcitrant carbon into various products (Noor et al., 2012). It is performed at temperatures ranging from 350°C – 400°C to 600°C – 700°C with a heating rate of 5°C per minute (Lehmann & Joseph, 2012; Noor et al., 2012; Ok et al., 2016). This method is particularly focused on maximizing biochar production.

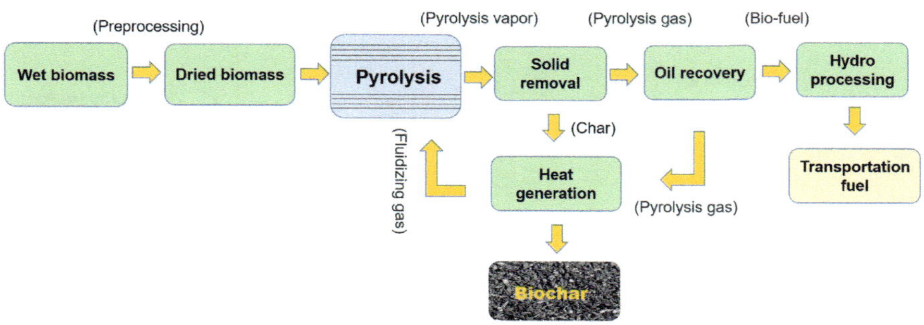

FIGURE 14.4 A flow diagram of fast pyrolysis of biochar production. Fast pyrolysis maximizes bio-oil production with biochar but has lower-yielding coproducts, such as pyrolysis gas.

A flow diagram of fast pyrolysis process of biochar production (Brown et al., 2011).

Adapted with permission from Brown, T. R., Wright, M. M., & Brown, R. C. (2011). Estimating profitability of two biochar production scenarios: Slow pyrolysis vs fast pyrolysis. Biofuels, Bioproducts and Biorefining, 5 (1), 54–68. https://doi.org/10.1002/bbb.254.

Reactors involved in slow pyrolysis are retorts, converters, and kilns (Sakhiya et al., 2020). Heat applied in slow pyrolysis decomposes the cellulose and hemicelluloses of the feedstuffs. Predrying treatment is conducted to remove the moisture of grounded materials and improve heating efficiency. During slow pyrolysis, prepyrolysis occurs before solid decomposition to obtain a product (Yu et al., 2017). It is estimated that 1 kg of the log in slow pyrolysis provides 30%–35% biochar, 40%–45% liquid product, and 20%–25% gaseous product. Slow pyrolysis techniques are suitable for generating biochar from agricultural wastes like corn stover, rice waste, and wood sawdust. Slow pyrolysis is used to maximize the solid product of biochar (Noor et al., 2012; Yu et al., 2017). Biochar produced through slow pyrolysis produces large pores and an increase in surface area that are beneficial for soil health, improving fertility, water retention, and microbial activity. A flow diagram of slow pyrolysis of biochar production is shown in Fig. 14.5.

14.4.2 Microwave-assisted pyrolysis

It is one of the most efficient thermochemical processes of biochar extraction (Abas & Ani, 2014; Yu et al., 2017). It provides microwave radiation for the heating process and converts electromagnetic energy to thermal energy. Both power output and radiation frequency are important, with a common frequency of 2.45 GHz, and temperature varies between 400°C and 600°C (Omoriyekomwan et al., 2017). This process involves pretreatment of the feedstuff before heating, which is not prevalent in conventional pyrolysis. It is an energy-saving, pollution-free, and highly efficient method (Abas & Ani, 2014), which is also known as an electric volumetric heating method. It improves the product's quality by preventing the formation of secondary reactors, which is an advantage over conventional

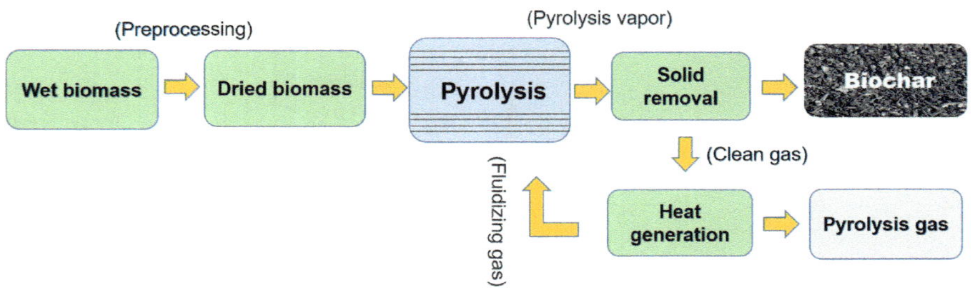

FIGURE 14.5 A flow diagram of slow pyrolysis of biochar production. Slow pyrolysis generally produces high char yields but has a relatively low-value energy product (pyrolysis gas).

A flow diagram of slow pyrolysis process of biochar production.

Adapted with permission from Brown, T. R., Wright, M. M., & Brown, R. C. (2011). Estimating profitability of two biochar production scenarios: Slow pyrolysis vs fast pyrolysis. Biofuels, Bioproducts and Biorefining, 5 (1), 54–68. https://doi.org/10.1002/bbb.254.

biochar production. Kostas et al. (2020) reported that microwaves influence the texture and properties of biochar, and approximately 78% carbon was obtained in the biochar from olive pomace using microwave pyrolysis.

14.4.3 Hydrothermal carbonization

Hydrothermal carbonization (HTC) is a thermal decomposition of substrate held in the subcritical aqueous substrate to obtain a high C-containing product (Parmar et al., 2014; Zhang et al., 2019). The decomposition in HTC, also called wet pyrolysis, is usually held at temperatures ranging from 180°C–250°C, with 2–10 MPa of pressure and water. The aqueous substrate is the main component in HTC in carbonizing and decomposing lignin, cellulose, and hemicelluloses (Parmar et al., 2014; Zhang et al., 2019). Feedstocks or waste residues with high moisture, such as algae aquatic plants, are suitable for processing in HTC (Cheng & Li, 2018) that produce biochar, liquid, and gaseous products. Glucose, levulinic acid, organic acid, hydro methyl furfural, and various organic acids are the chemical products found in HTC. HTC method is cost-effective and can be applicable for small-scale and farm-based biochar production.

14.4.4 Gasification

Gasification is a process of conversion of carbon sources into a gaseous product. In this process, a carbon source at a temperature <70°C in the presence of a control oxidizing agent (air, oxygen, and steam) is converted into a gaseous product or syngas (Elkhalifa et al., 2019; Zhang et al., 2019). Plant biomass is usually exposed to two sections in this process: gasification occurs, and cleaning and cooling of the produced syngas (Wiedner et al., 2013). Gasification applies to obtaining more syngas ($CO + H_2$) (Ok et al., 2016). The final yield is approximately 10% of the biomass. Thus, gasification is a less effective method compared to pyrolysis. Reactant/biomass ratio, temperature, residence duration, particle size, and pressure are the primary agents involved in this process. Auger reactor is utilized to continuously generate biochar production (Sakhiya et al., 2020).

14.4.5 Hydrothermal liquefaction

Hydrothermal liquefaction (HTL) is a thermochemical conversion process of biomass that undergoes depolymerization reactions that produce solids (biochar), liquid water-immiscible (bio-oil), liquid water-miscible (aqueous phase fraction), and gaseous materials (Toor et al., 2011). Biomass in HTL is treated at 5–20 MPa pressure and at temperatures between 200°C and 374°C (Ponnusamy et al., 2020). Depolymerization reactions in HTC involve fragmentation, hydrolysis, dehydration, deoxygenation, aromatization, and biomass depolymerization. Hydrothermal processes can be divided into HTC, HTL, and hydrothermal gasification. As per (Jain et al., 2016), their primary basis is mainly operating temperature. HTL is energy efficient and can recover more than 70% of the feedstock carbon as bio-oil or bio-char. HTC is, in fact, considered promising for biomass conversion into fuel.

14.4.6 Torrefaction

Torrefaction is the thermochemical conversion of feedstuffs into solid components at temperatures between 200°C and 300°C and atmospheric pressure in anaerobic conditions (Yu et al., 2017; Zhang et al., 2019). This method is often referred to as mild pyrolysis and mainly consists of dry and wet torrefaction methods. The dry torrefaction thermochemical conversion process occurs at temperatures between 200°C and 300°C in the presence of an inert gas for a 30–60 minutes residence time (Acharya & Dutta, 2013) with a heating rate of about 50°C/minute. Rago et al. (2018) examined the isothermal properties of torrefaction of food waste at similar temperatures (between 200°C and 300°C) for 60 and 180 minutes. They highlighted that the constant increase in temperature of 15°C per minute increases the amount of C, energy density, and caloric content of food waste. They mentioned that the reacting agent flushed with nitrogen gas in a small stream for 10 minutes helps remove any remaining O_2. To increase the product quality, pretreatment (dry or wet) is given to eliminate volatile substances such as moisture, CO_2, and O_2 through decomposition (Zhang et al., 2019). Wet torrefaction is the thermochemical conversion process of biomass in subcritical water at temperatures of 180°C–265°C for a resident time from 5 minutes to several hours (Acharya et al., 2015; Yan et al., 2009). Torrefaction is capable of processing lignocellulosic biomass such as hemicellulose, cellulose, lignin, water extractives, and ash and nonlignocellulosic biomass such as animal manure, animal fat, sewage sludge, etc. Various manufacturing companies have adopted several reactor types for dry torrefaction technologies. These are rotating drums, screw reactors, multiple hearth furnaces, turned and microwave reactors, compact moving beds, fixed beds, and belt dryers (Acharya et al., 2015). A typical flow diagram of torrefaction process technology is shown in Fig. 14.6.

Different biochar production technologies and their production conditions and product(s) yields are referenced in Section 14.4. Production methods have been described in Table 14.1.

14.5 Biochar's role in carbon sequestration and carbon stabilization in soil

Biochar application can improve soil quality and minimize GHG emissions, thus mitigating global climate change (Yadav et al., 2023). It contains a greater surface charge and higher surface area, a more significant percentage of oxygen-

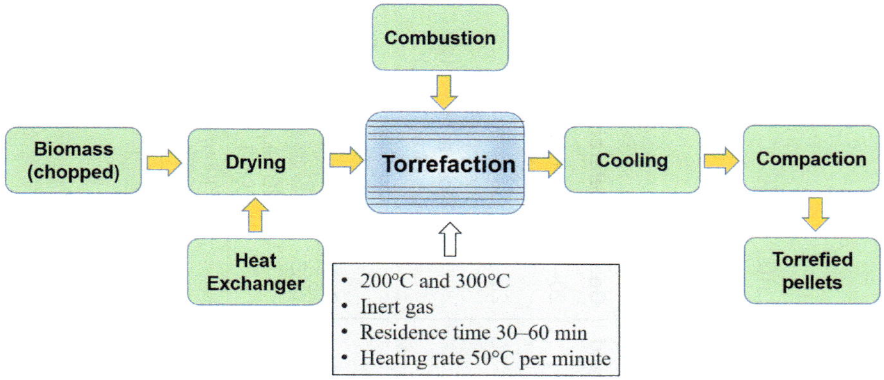

FIGURE 14.6 A schematic diagram of the dry correction biochar production method.

A schematic diagram illustrating the dry correction biochar production process.

Adapted from Acharya, B. (2017). Hybrid thermochemical and biochemical conversion of biomass for value added products *(PhD thesis). The University of Guelph.*

containing carbon groups, and inert carbon components. The pore structure of biochar is greater that leads to increased potential in carbon sequestration, reduced GHG emission, improved structure and soil fertility, and enhanced crop production (Lu et al., 2020; Montanarella & Lugato, 2013; Tan et al., 2015; Wang & Wang, 2018, 2019). Diversion of waste from the current waste disposal practices (Montanarella & Lugato, 2013) to biochar production is likely to resolve both disposal issues and climate change mitigation since the application of biochar can effectively store carbon in the soil in the long-term that would otherwise be burnt or composted away to create minerals during waste management techniques and strategies of biomass. Biochar contains relatively stable carbon components and is utilized as a soil amendment to store carbon, enhancing carbon sequestration in the soil (Bolan et al., 2022).

Biochar is considered a stable soil nutrient compared to other organic nutrients and has a profound role in adsorption and mineralization (Clough & Condron, 2010; Kulyk, 2012). Specific biochar-based composites may also improve biochar's stability and biomass carbon retention in soil compared to virgin biochar (Wang et al., 2022). Biochar production from agriculture and food processing industries leftover biomass can aid long-term carbon sequestration and improve soil and environmental quality (Parmar et al., 2014). Biochar, as a porous with high carbon content material, when integrated into fields, can sequester carbon, improve soil fertility and crop yield, decrease GHG (e.g., N_2O), and improve nitrogen retention in the soil while reducing the risks of nitrate leaching (Ahmed et al., 2019; Lehmann & Joseph, 2012; Sanford & Larson, 2020; Xu et al., 2016). A recent life cycle assessment for the CO_2 sequestration potential of sugarcane bagasse was conducted by Usevičiūtė and Baltrėnaitė-Gedienė (2021). The key considerations were the biochar life cycle, including processes such as

Table 14.1 Different biochar production technologies and their production conditions and product(s) yields.

Production methods	Feedstock pretreatment	Production conditions			Typical product yield (%)			
Operating temperature (°C)	**Residence time**	**Heating rate**	**Residence time**	**Heating rate**	**Solid**	**Liquid**	**Gases**	**References**
Dry-torrefaction		200–300	30 min–4 h	10°C–15°C/min	60–80	–	20–40	Yu et al. (2017), Zhang et al. (2019)
Slow pyrolysis	Dry biomass	300–650	5 min–12 h	10°C–30°C/min	25–35	20–30	25–35	Lehmann and Joseph (2012), Noor et al. (2012), Ok et al. (2016)
Fast pyrolysis	–	800°C–1300°C		200°C/min for no more than 10 s		up to 75%		Elkhalifa et al. (2019)
HTC	Wet biomass	200–374	5 min–12 h	5°C–10°C/min	45–70	5–25	2–5	Parmar et al. (2014)
Gasification		600–900	10–20 s	50°C–100°C/s	<10	<5	>85	Elkhalifa et al. (2019), Zhang et al. (2019)
Liquefaction	Wet biomass	260–400	1–120 min	10°C–100°C/min	–	–	–	Ponnusamy et al. (2020)

Notes: *Select table and click icon in the top toolbar to start the permission process or indicate no permission is required.*

transportation, biochar production methods such as pyrolysis, and farmland application. The study highlighted that approximately 50%–70% of the bagasse carbon can be stabilized as biochar, accounting for a total of 1200–1800 tCO_2, sequestered yearly from the available bagasse of 12,000 t/year.

14.6 Biochar's role in greenhouse gas mitigation

Biochar can be utilized to promote sustainable agricultural production, environmental mitigation, and GHG emission reduction (Wang et al., 2017). Biochars assist in improving soil quality and reducing pollutants and play a role in decreasing GHG emissions, thus mitigating global climate change (Bolan et al., 2022; Kwon et al., 2020; Lehmann et al., 2021). As an emerging technology, biochar production can enhance food security and promise to sequester carbon to slow climate change (Kuppusamy et al., 2016). Biochar alone can reduce GHG emissions; however, biochar composites or mixed with compost in the soil help mitigate climate change (Kulyk, 2012) by increasing the stable carbon in soil (Parmar et al., 2014) and increasing soil organic matter and decreasing GHG emission (Kulyk, 2012). For example, a recent study showed that biochar application decreases CH_4 emission from rice fields by increasing methanotroph (CH_4-consuming bacteria) and decreasing methanogens (CH_4-producing bacteria) communities in soil (Singh, Tiwari, et al., 2017). Another application of biochar is to support the ZVI (zerovalent iron) particles, which are environmentally and economically beneficial (Zhou et al., 2014). However, some studies reported no effect of biochar (Cheng & Li, 2018), on either decreased (Nelissen et al., 2014) or increased emissions (Clough et al., 2010); this implies that the impact of biochar may vary, depending on factors such as feedstock sources, pyrolysis temperatures, biochar properties, soil type, soil water content, land use, vegetation, etc.

The use of biochar has been seen in the remediation of contaminants. Biochar can potentially adsorb contaminants from soils and remove organic pollutants, heavy metals, and other contaminated sources (Tan et al., 2015; Yang et al., 2021) based on the application processes and rates (Kumar et al., 2016). Cost-effectiveness and the availability of abundant feedstock materials make biochar an effective alternative remediation agent for various environmental contaminants, including heavy metallic ions, organic pollutants, and even nutrients (Zhou et al., 2014).

14.7 Impact of biochar on soil microbe, gas dynamics, and soil fertility enhancement

As a stable soil nutrient, biochar can contribute to adsorption and mineralization (Clough et al., 2010; Kulyk, 2012; Lehmann et al., 2021). It removes various heavy metals from soils through adsorption (Mian & Liu, 2018; Wang & Wang, 2018). Adsorption eliminates toxic elements and organic contaminants in soil and plants (Wang & Wang, 2019). Soil contamination through industrial and domestic activity and the use of agricultural chemical compounds or substances directly or indirectly pollute soil, water, and the environment and affect microorganisms in soil (Akhil et al., 2021; Yadav et al., 2023). Biochar application in soil has been proven effective in cleaning up contaminated soils (Hale et al., 2015). Several researchers further documented that biochar can decompose or remove organic pollutants, including polychlorinated biphenyls, diethyl phthalate, chlorobenzene, *p*-nitrophenol, halo hydrocarbon, polycyclic aromatic hydrocarbons (PAHs), and 2-chlorobiphenyl (Deng et al., 2021; Lu et al., 2022; Lyu et al., 2020; Oni et al., 2019; Yaashikaa et al., 2019). With higher affordability and being an ecologically sustainable solution made from waste and greater surface area with a highly porous nature and water-holding ability, biochar reduces the pollutants and repairs the soils (Akhil et al., 2021; Yadav & Khare, 2020). Biochar from *Carya* spp. can significantly adsorb and reduce the risks of leaching of sodium bispyribac and clomazone in the soil (Wang & Wang, 2019). Using biochar has a significant positive impact as an immobilization strategy in remediating soils from heavy metals and metalloids (Shakya & Agarwal, 2020; Sharma & Irmak, 2020; Yaashikaa et al., 2019).

Microbial activity in soils directly affects soil processes such as organic matter decomposition, nutrient cycling, nutrient status and availability, and impact on crop production (Matykiewicz, 2020). Biochar application in soils can increase the activity of microorganisms (Karimi et al., 2020). Biochar can change soil's physical and chemical properties, act as a carrier or medium, and enhance the establishment of microorganisms and microbial growth in the soil (Hale et al., 2015; Parmar et al., 2014). Kannan et al. (2020) reported that biochar influences both functions and habitat of mycorrhizal fungi and other soil microbes, improving soil quality and health. Biochar's recalcitrant nature makes organic matter unavailable for microorganisms and other decomposers as substrate and has the potential for long-term carbon sequestration (Parmar et al., 2014). Biochar positively influences microbial growth due to the presence of a greater concentration of functional groups, especially those that are oxygenated at 400°C–450°C, which can easily dissolve in organic substances, and water facilitates greater germination and growth of microorganisms (Talberg, 2009). Biochar has a certain water-holding capability, surface area, and carbon:nitrogen concentration, which influences the survival of inoculant microorganisms in soil (Hale et al., 2015).

14.8 Biochar and soil health and the effect on crop production system

Biochar application in soils can improve soil structure, soil health, and crop productivity. Several studies reported that the application of biochar improves soil physiological, chemical, and biological properties in agricultural soils (Liang et al., 2017; Oliveira et al., 2017; Tisserant & Cherubini, 2019). Biochar increases the absorption capacity of nutrients in soils, and due to its greater charge density, it can transport large quantities of nutrients, altering the soil properties and thus enhancing crop performance (Chausali et al., 2021). Biochar has been proven to increase plant-soil nutrient availability while enhancing environmental quality (Clough & Condron, 2010). Other researchers indicated the benefits of biochar application in soils to increase water holding capacity, water filtration, available soil moisture, and plant nutrient retention (Schmidt et al., 2014; Yadav & Khare, 2020).

The use of biochar in agricultural soils can also minimize soil compaction. Soil compaction is identified as a limiting factor by producers in production agriculture (Shaheb et al., 2021). The problem has been reported in many areas of Europe and, more recently, in some parts of the United States. High gross weight vehicles/machinery traffic causes soil compaction, thus damaging soil structure and the environment critical for sustainable crop production (Keller et al., 2019; Shaheb et al., 2023, 2024). Although potential strategies, as reported by (Shaheb et al., 2021), several studies highlighted that biochar application ameliorated soil compaction and increased water infiltration and soil structure and water holding capacity (Acharya et al., 2024; Liu et al., 2017; Novak et al., 2016).

Mylavarapu et al. (2013) reported that biochar can maintain water and nutrients in the upper soil layer for an extended time, reducing nutrient loss risks from the crop root zone and enhancing crop performance. Biochar application in soils has been reported to have a 10% crop yield increment (Parmar et al., 2014; Singh, Rajawat, et al., 2017), possibly due to the changes in soil physicochemical characteristics, including soil structure, water retention, and nutrient availability. Several studies showed beneficial effects in reducing pests and diseases in crop fields. For example, Parmar et al. (2014) reported its role as promising to reduce fungal pathogens and pests growth, while others reported controlling weeds in fava beans and enhancing crop productivity (Saudy et al., 2021, 2022). Several studies conducted on a variety of crops such as *Phaseolus vulgaris, Cucumis sativus, Fragaria* × *ananassa, Solanum lycopersicum, Zea mays, Citrullus lanatus*, and *Piper nigrum*, and wheat and found that biochar application in soils improved soil physical conditions, nutrient availability and has greater impact on crop growth and yield (Das et al., 2023; Singh et al., 2019). Biochar can also reduce nitrogen leaching (Yadav & Khare, 2020), stimulate the soil's inorganic nitrogen, and increase the amount of accessible Fe, Cu, and Zn in calcareous soil (Karimi et al., 2020) with high adsorption capacity was reported by (Akhil et al., 2021) that permits the adsorption of potassium, nitrogen, organic matter, and phosphorous in the soil. Biochar properties and their effects on soil, plant, and environmental systems are described in Table 14.2.

14.9 Challenges and considerations

Biochar application in agricultural soils improves soil environmental conditions and has been proven a recyclable source of energy that also contributes to

Table 14.2 Biochar properties and effects on soil, plant, and environmental systems and climate change: a brief overview.

Biochar-source/feedstocks and association	Biochar properties	Key parameters studied/discussed	Effect on agriculture and environment	Effect on climate change	References
Varies. Crop residues, biodegradable waste, etc.	High carbon content; porous structure	Soil organic matter (OM); water holding capacity (WHC)	Soil OM (+); WHC (+)	CC (+)	Montanarella and Lugato (2013)
Rice/corn straw, wood, manure, etc.	Porous structure	Porosity (P), macroporosity (MP), hydrophilicity (HDP), water adsorption rate (AR)	P (+); MP (+); HPP (+); water AR (+)	N_2O emission (–)	Chausali et al. (2021)
Forestry, ag wastes, crop residues, switchgrass, etc., are discussed	Porous, specific surface area	Cation exchange capacity (CEC)	Soil properties (+); CEC (+); soil functions (+); soil compaction (±); limited research findings	CC (+); GHG (–)	Verheijen et al. (2010)
Ag wastes, wood chips, crop residues, manure, etc., are discussed.	Soil OM, pH, BD, CEC, P	Soil properties; crop yield (CY); nutrient availability	CY (+); chemical properties (+); physical properties (+) microbial properties (+); Nutrient availability (+)	CC (+) GHG (–)	Kuppusamy et al. (2016), Verheijen et al. (2010)
Different sources of biochar discussed	Color, P, nutrient cycling, biochar properties	Plant productivity (PP), CY, soil properties, GHG (CH_4, N_2O) emissions, carbon sequestration (CC), nutrient leaching (NL)	PP (+); CY (+)	CEC (+); CH_4 emissions (–); N_2O emissions (–); CC (+); NL (–)	Clough et al. (2010), Kannan et al. (2020)

Wood, manure, crop residues, etc.	Porus structure, higher carbon content, etc.	BD, aggregate stability (AS), P, WHC, saturated hydraulic conductivity (HC)	BD (−7.6%), AS (+8.2%), soil P (+8.4%); WHC (+15%), saturated HC (+25%)	GHG (−); CC (+)	Omondi et al. (2016)
Wood, manure, green waste, algae, and various crop residues and byproducts like straw, cobs, husks, bagasse, etc., are discussed	Greater surface area, porosity, etc.	Soil texture (ST), soil compaction (SC), water infiltration (WI), BD, saturated HC, SC, SA	SC ameliorated (+); WI (+), saturated HC (+), BD (−), SC (−), SA (+), WHC (+)	GHG (−); CC (−)	Acharya et al. (2024), Novak et al. (2016)
Biochar and mycorrhizal associations	–	Ecosystem services, soil properties, CC, PP	Mycorrhizal abundance and/or functioning. (+); PP (+)	Ecosystem restoration (+), soil CC (+)	Warnock et al. (2007)
Herbaceous, rice biochar, citrus juice sediments, etc.	Porous, H carbon content	Nutrient element; mycorrhizal properties	Plant phosphorus nutrition (+); mycorrhizal response (+610%)	–	Ishii and Kadoya (1994)
Agricultural residues, waste, animal manure, wood, etc.	Porous with high carbon content, low BD	Soil health and fertility, CY (+)	Soil amendment (+), soil health and fertility (+), CY (+)	GHG (e.g., N_2O) emissions; N retention (+); nitrate leaching (−)	Lehmann and Joseph (2015), Sanford and Larson (2020), Sanford et al. (2022)

enhancing carbon sequestration and reducing GHG emissions. However, biochar may contain some phytotoxic substances that affect seed germination and plant establishment (Kuppusamy et al., 2016). Although biochar has been promising in influencing soil physiochemical characteristics, the applied biochar can cause agrochemical binding and deactivation, such as herbicide, and has been found to slow the net CH oxidation rate in the soil (Spokas & Reicosky, 2009). The excess biochar applied to the soil can also impact the germination and biological processes in the soil due to electrical conductivity and pH (Kuppusamy et al., 2016; Pereira Lopes & Astruc, 2021). Poor mechanical strength and the dusty nature of biochar are other issues, especially when used in soils where erosion routinely happens (Mohammadi, 2021). Biochar may have a potential chance of releasing hazardous substances, such as heavy metals and PAHs (Kuppusamy et al., 2016). Variations in production processes and feedstocks sometimes result in biochar being less efficient in storing nutrients and is often vulnerable to microbial degradation (Kuppusamy et al., 2016; Mylavarapu et al., 2013).

Similarly, the application of biochar is very challenging. Typically, biochar is applied on the surface, followed by tilling. However, achieving uniform distribution can be difficult. For example, tilling can disturb soil structure and microbial communities and release N_2O. Also, due to the hydrophobic nature of biochar, it may not mix well with soil, creating pockets that resist wetting and impacting soil structure and porosity. Exploring the concept of waste valorization for the production of biochar and mechanical extrusion process, whether or not to improve biochar properties, and the molecular behavior in biochar needs more focus.

14.10 Case studies and research lessons

Case Study 1: A recent study conducted in Wisconsin, United States, in biochar-amended sandy loam soil columns at 5% (w/w) to evaluate the impact of repeated N applications on N soil retention, leaching, and gaseous emissions (Sanford & Larson, 2020). Corncob biochar is produced at a production temperature of 400°C (BC1) and 700°C (BC2) and oxidized at 700°C (BC3) for a holding time of 1 hour under oxygen-limiting conditions. Soil columns were received weekly at varying rates using organic N, ammonium (NH_4^+-N), or nitrate (NO_3^--N). The N retained in biochar-amended sandy loam soils, N leached, and the N_2O and ammonia (NH_3) emissions were quantified for approximately 3 months. N leaching in BC1 was reduced by 19% and 15% for ORG-N and $NH4_+$-N, respectively, compared to the control. BC2 biochar for NO_3^--N applications reduced leaching by 25% compared to the controls. Biochar amended columns enhanced soil retention of ORG-N and NO_3. Further, oxygenated functional groups and cationic minerals on the biochar surface, possibly through cationic bridging or forming an organomineral layer on the surface, likely improve nitrogen retention. A couple of other experimentations in Wisconsin, along with these findings, demonstrate the potential application of biochar in agricultural systems to improve soil health and quality, carbon sequestration, and reduce the risk of N leaching to groundwater sources (Lehmann & Joseph, 2015; Sanford & Larson, 2020; Sanford et al., 2022).

14.11 Future outlook

Biochar use is increasing rapidly for the sustainable production of agricultural goods and to improve food security (Chen et al., 2019; Parmar et al., 2014; Verheijen et al., 2010). The manufacturing procedure of biomass gives heterogeneous physical and chemical properties, usually pyrolyzed at different conditions. The utilization of biochar in improving the composition and texture of the soil, soil productivity, and sorption of nutrients directly enhances crop production. Biochar also helps to remove toxified soils, enhances photosynthesis in plants, improves carbon sequestration, minimizes GHG emissions, and reduces the island effect. Furthermore, biochar is cost-effective compared with chemical fertilizers and can be easily transported. Thus, greater emphasis must be placed on collaboration among farmers, soil scientists, researchers, and concerned authorities to utilize biochar in concerned fields promptly. Further research is required on how and what mechanisms biochar benefits soil fertility, crop physiology and productivity, and environmental sustainability.

Authors' contribution

Conceptualization—Md Rayhan Shaheb (M.R.S.); review of literature collections—M.R.S. and Ayesha Sarker (A.S.); analysis and interpretation—M.R.S. and Shafiqur Rahman (S.R.); writing original manuscript—M.R.S.; review and editing—all authors. All authors have read and agreed to the published version of the chapter.

Conflict of interest

The authors declare that there is no conflict of interest.

References

Abas, F. Z., & Ani, F. N. (2014). Comparing characteristics of oil palm biochar using conventional and microwave heating. *Jurnal Teknologi*, *68*(3). Available from https://doi.org/10.11113/jt.v68.2926.

Acharya, B., & Dutta, A. (2013). Characterization of torrefied willow for combustion application. *Journal of Biobased Materials and Bioenergy*, *7*(6), 667–674. Available from https://doi.org/10.1166/jbmb.2013.1372.

Acharya, B., Dutta, A., & Minaret, J. (2015). Review on comparative study of dry and wet torrefaction. *Sustainable Energy Technologies and Assessments*, *12*, 26–37. Available from https://doi.org/10.1016/j.seta.2015.08.003.

Acharya, B. S., Dodla, S., Wang, J. J., Pavuluri, K., Darapuneni, M., Dattamudi, S., Maharjan, B., & Kharel, G. (2024). Biochar impacts on soil water dynamics: Knowns, unknowns, and research directions. *Biochar*, *6*(1). Available from https://doi.org/10.1007/s42773-024-00323-4.

Ahmed, R., Li, Y., Mao, L., Xu, C., Lin, W., Ahmed, S., & Ahmed, W. (2019). Biochar effects on mineral nitrogen leaching, moisture content, and evapotranspiration after 15N urea fertilization for vegetable crop. *Agronomy*, *9*(6). Available from https://doi.org/10.3390/agronomy9060331, https://www.mdpi.com/2073-4395/9/6/331/pdf.

Akhil, D., Lakshmi, D., Kartik, A., Vo, D. V. N., Arun, J., & Gopinath, K. P. (2021). Production, characterization, activation and environmental applications of engineered biochar: A review. *Environmental Chemistry Letters*, *19*(3), 2261–2297. Available from https://doi.org/10.1007/s10311-020-01167-7.

Allen, S. K., Barros, V., Burton, I., Campbell-Lendrum, D., Cardona, O.-D., Cutter, S.L., Pauline Dube, O., Ebi, K.L., Field, C.B., Handmer, J.W., Lal, P.N., Lavell, A., Mach, K.J., Mastrandrea, M.D., McBean, G.A., Mechler, R., Mitchell, T., Nicholls, N., O'Brien, K.L., ... Wilbanks, T.J. (2012). Summary for policymakers. In Field CB, Barros V, Stocker TF, Dahe Q. (Eds.). Managing the risks of extreme events and disasters to advance climate change adaptation: Special report of the intergovernmental panel on climate change. Cambridge University Press. 3–22. Available from https://doi.org/10.1017/CBO9781139177245.003.

Bolan, N., Hoang, S. A., Beiyuan, J., Gupta, S., Hou, D., Karakoti, A., Joseph, S., Jung, S., Kim, K. H., Kirkham, M. B., Kua, H. W., Kumar, M., Kwon, E. E., Ok, Y. S., Perera, V., Rinklebe, J., Shaheen, S. M., Sarkar, B., Sarmah, A. K., ... Van Zwieten, L. (2022). Multifunctional applications of biochar beyond carbon storage. *International Materials Reviews*, *67*(2), 150–200. Available from https://doi.org/10.1080/09506608.2021.1922047.

Brown, R. (2009). Biochar production technology. In J. Lehmann, & S. Joseph (Eds.), *Biochar for environmental management* (1st ed., pp. 159–178). Routledge.

Brown, T. R., Wright, M. M., & Brown, R. C. (2011). Estimating profitability of two biochar production scenarios: Slow pyrolysis vs fast pyrolysis. *Biofuels, Bioproducts and Biorefining*, *5*(1), 54–68. Available from https://doi.org/10.1002/bbb.254.

C2ES. (2024). *Controlling industrial greenhouse gas emissions*. Center for Climate and Energy Solutions. https://www.c2es.org/content/regulating-industrial-sector-carbon-emissions/#: ~ :text = There%20are%20many%20ways%20to,use%20and%20recycling%20of%20materials

Chausali, N., Saxena, J., & Prasad, R. (2021). Nanobiochar and biochar based nanocomposites: Advances and applications. *Journal of Agriculture and Food Research*, *5*. Available from https://doi.org/10.1016/j.jafr.2021.100191.

Chen, W., Meng, J., Han, X., Lan, Y., & Zhang, W. (2019). Past, present, and future of biochar. *Biochar*, *1*(1), 75–87. Available from https://doi.org/10.1007/s42773-019-00008-3.

Cheng, F., & Li, X. (2018). Preparation and application of biochar-based catalysts for biofuel production. *Catalysts*, *8*(9). Available from https://doi.org/10.3390/catal8090346, http://www.mdpi.com/2073-4344/8/9/346/pdf.

Clough, T. J., Bertram, J. E., Ray, J. L., Condron, L. M., O'Callaghan, M., Sherlock, R. R., & Wells, N. S. (2010). Unweathered wood biochar impact on nitrous oxide emissions from a bovine-urine-amended pasture soil. *Soil Science Society of America Journal*, *74* (3), 852–860. Available from https://doi.org/10.2136/sssaj2009.0185.

Clough, T. J., & Condron, L. M. (2010). Biochar and the nitrogen cycle: Introduction. *Journal of Environmental Quality*, *39*(4), 1218–1223. Available from https://doi.org/10.2134/jeq2010.0204.

Conte, P., Bertani, R., Sgarbossa, P., Bambina, P., Schmidt, H. P., Raga, R., Lo Papa, G., Chillura Martino, D. F., & Lo Meo, P. (2021). Recent developments in understanding biochar's physical–chemistry. *Agronomy*, *11*(4). Available from https://doi.org/10.3390/agronomy11040615, https://www.mdpi.com/2073-4395/11/4/614/pdf.

Das, S. K., Ghosh, G. K., & Avasthe, R. (2023). Application of biochar in agriculture and environment, and its safety issues. *Biomass Conversion and Biorefinery*, *13*(2), 1359–1369. Available from https://doi.org/10.1007/s13399-020-01013-4.

Deng, S., Chen, J., & Chang, J. (2021). Application of biochar as an innovative substrate in constructed wetlands/biofilters for wastewater treatment: Performance and ecological benefits. *Journal of Cleaner Production*, *293*. Available from https://doi.org/10.1016/j.jclepro.2021.126156.

Elkhalifa, S., Al-Ansari, T., Mackey, H. R., & McKay, G. (2019). Food waste to biochars through pyrolysis: A review. *Resources, Conservation and Recycling*, *144*, 310–320. Available from https://doi.org/10.1016/j.resconrec.2019.01.024.

EPA. (2024). *Inventory of U.S. greenhouse gas emissions and sinks: 1999-2022*. U.S. Environmental Protection Agency, EPA 430-R-24-004. https://www.epa.gov/ghgemissions/inventory-us-greenhouse-gas-emissions-and-sinks-1990-2022.

Gwenzi, W., Chaukura, N., Mukome, F. N. D., Machado, S., & Nyamasoka, B. (2015). Biochar production and applications in sub-Saharan Africa: Opportunities, constraints, risks and uncertainties. *Journal of Environmental Management*, *150*, 250–261. Available from https://doi.org/10.1016/j.jenvman.2014.11.027.

Hale, L., Luth, M., & Crowley, D. (2015). Biochar characteristics relate to its utility as an alternative soil inoculum carrier to peat and vermiculite. *Soil Biology and Biochemistry*, *81*, 228–235. Available from https://doi.org/10.1016/j.soilbio.2014.11.023.

Ishii, T., & Kadoya, K. (1994). Effects of charcoal as a soil conditioner on citrus growth and vesicular-arbuscular mycorrhizal development. *Journal of the Japanese Society for Horticultural Science*, *63*(3), 529–535. Available from https://doi.org/10.2503/jjshs.63.529.

Jain, A., Balasubramanian, R., & Srinivasan, M. P. (2016). Hydrothermal conversion of biomass waste to activated carbon with high porosity: A review. *Chemical Engineering Journal*, *283*, 789–805. Available from https://doi.org/10.1016/j.cej.2015.08.014.

Joseph, S., Peacocke, C., Lehmann, J., & Munroe, P. (2009). Developing a biochar classification and test methods. *Biochar for Environmental Management: Science and Technology*, 107–126. Available from https://doi.org/10.4324/9781849770552.

Joseph, S., & Taylor, P. (2014). *The production and application of biochar in soils. Advances in biorefineries: Biomass and waste supply chain exploitation* (pp. 525–555). Elsevier Ltd.

Available from http://www.sciencedirect.com/science/book/9780857095213, https://doi.org/10.1533/9780857097385.2.525.

Kannan, P., Krishnaveni, D., & Ponmani, S. (2020). *Biochars and its implications on soil health and crop productivity in semi-arid environment. Biochar applications in agriculture and environment management* (pp. 99–122). Springer International Publishing. Available from http://www.springer.com/in/book/9783030409968, https://doi.org/10.1007/978-3-030-40997-5_5.

Karimi, A., Moezzi, A., Chorom, M., & Enayatizamir, N. (2020). Application of biochar changed the status of nutrients and biological activity in a calcareous soil. *Journal of Soil Science and Plant Nutrition*, *20*(2), 450–459. Available from https://doi.org/10.1007/s42729-019-00129-5.

Keller, T., Sandin, M., Colombi, T., Horn, R., & Or, D. (2019). Historical increase in agricultural machinery weights enhanced soil stress levels and adversely affected soil functioning. *Soil and Tillage Research*, *194*. Available from https://doi.org/10.1016/j.still.2019.104293.

Kostas, E. T., Durán-Jiménez, G., Shepherd, B. J., Meredith, W., Stevens, L. A., Williams, O. S. A., Lye, G. J., & Robinson, J. P. (2020). Microwave pyrolysis of olive pomace for bio-oil and bio-char production. *Chemical Engineering Journal*, *387*.

Kulyk, N. (2012). *Cost-benefit analysis of the biochar application in the U.S. cereal crop cultivation*. Center of Public Policy, University of Massachusetts. https://hdl.handle.net/20.500.14394/8676.

Kumar, A., Schreiter, I. J., Wefer-Roehl, A., Tsechansky, L., Schüth, C., & Graber, E. R. (2016). Production and utilization of biochar from organic wastes for pollutant control on contaminated sites. In M. N. V. Prasad, & K. Shih (Eds.), *Environmental materials and waste: Resource recovery and pollution prevention* (pp. 91–116). Elsevier Inc. Available from 10.1016/B978-0-12-803837-6.00005-6.

Kuppusamy, S., Thavamani, P., Megharaj, M., Venkateswarlu, K., & Naidu, R. (2016). Agronomic and remedial benefits and risks of applying biochar to soil: Current knowledge and future research directions. *Environment International*, *87*, 1–12. Available from https://doi.org/10.1016/j.envint.2015.10.018.

Kwon, G., Bhatnagar, A., Wang, H., Kwon, E. E., & Song, H. (2020). A review of recent advancements in utilization of biomass and industrial wastes into engineered biochar. *Journal of Hazardous Materials*, *400*. Available from https://doi.org/10.1016/j.jhazmat.2020.123242.

Lehmann, J., Cowie, A., Masiello, C. A., Kammann, C., Woolf, D., Amonette, J. E., Cayuela, M. L., Camps-Arbestain, M., & Whitman, T. (2021). Biochar in climate change mitigation. *Nature Geoscience*, *14*(12), 883–892. Available from https://doi.org/10.1038/s41561-021-00852-8.

Lehmann, J., & Joseph, S. (2015). Biochar for environmental management: An introduction. In J. Lehmann, & S. Joseph (Eds.), *Biochar for environmental management* (2nd ed, pp. 33–46). Routledge. Available from https://doi.org/10.4324/9780203762264-8.

Lehmann, J. S., & Joseph, S. (2012). *Biochar for environmental management*. Earthscan.

Liang, J., Yang, Z., Tang, L., Zeng, G., Yu, M., Li, X., Wu, H., Qian, Y., Li, X., & Luo, Y. (2017). Changes in heavy metal mobility and availability from contaminated wetland soil remediated with combined biochar-compost. *Chemosphere*, *181*, 281–288. Available from https://doi.org/10.1016/j.chemosphere.2017.04.081.

Libra, J. A., Ro, K. S., Kammann, C., Funke, A., Berge, N. D., Neubauer, Y., Titirici, M. M., Führer, C., Bens, O., Kern, J., & Emmerich, K. H. (2011). Hydrothermal carbonization of biomass residuals: A comparative review of the chemistry, processes and applications of wet and dry pyrolysis. *Biofuels*, *2*(1), 71–106. Available from https://doi.org/10.4155/bfs.10.81.

Liu, Q., Liu, B., Zhang, Y., Lin, Z., Zhu, T., Sun, R., Wang, X., Ma, J., Bei, Q., Liu, G., Lin, X., & Xie, Z. (2017). Can biochar alleviate soil compaction stress on wheat growth and mitigate soil N_2O emissions? *Soil Biology and Biochemistry*, *104*, 8–17. Available from https://doi.org/10.1016/j.soilbio.2016.10.006.

Lu, L., Yu, W., Wang, Y., Zhang, K., Zhu, X., Zhang, Y., Wu, Y., Ullah, H., Xiao, X., & Chen, B. (2020). Application of biochar-based materials in environmental remediation: From multi-level structures to specific devices. *Biochar*, *2*(1), 1–31. Available from https://doi.org/10.1007/s42773-020-00041-7.

Lu, Y., Cai, Y., Zhang, S., Zhuang, L., Hu, B., Wang, S., Chen, J., & Wang, X. (2022). Application of biochar-based photocatalysts for adsorption-(photo)degradation/reduction of environmental contaminants: Mechanism, challenges and perspective. *Biochar*, *4*(1). Available from https://doi.org/10.1007/s42773-022-00173-y.

Lyu, H., Zhang, Q., & Shen, B. (2020). Application of biochar and its composites in catalysis. *Chemosphere*, *240*. Available from https://doi.org/10.1016/j.chemosphere.2019.124842.

Matykiewicz, D. (2020). Biochar as an effective filler of carbon fiber reinforced bio-epoxy composites. *Processes*, *8*(6), 1–13. Available from https://doi.org/10.3390/pr8060724.

Mian, M. M., & Liu, G. (2018). Recent progress in biochar-supported photocatalysts: Synthesis, role of biochar, and applications. *RSC Advances*, *8*(26), 14237–14248. Available from https://doi.org/10.1039/c8ra02258e.

Mohammadi, A. (2021). Overview of the benefits and challenges associated with pelletizing biochar. *Processes*, *9*(9). Available from https://doi.org/10.3390/pr9091591, https://www.mdpi.com/2227-9717/9/9/1591/pdf.

Montanarella, L., & Lugato, E. (2013). The application of biochar in the eu: Challenges and opportunities. *Agronomy*, *3*(2), 462–473. Available from https://doi.org/10.3390/agronomy3020462, https://www.mdpi.com/2073-4395/3/2/462/pdf.

Mylavarapu, R., Nair, V., & Morgan, K. (2013). An introduction to biochars and their uses in agriculture. *EDIS*, *2013*(8). Available from https://doi.org/10.32473/edis-ss585-2013.

Nelissen, V., Saha, B. K., Ruysschaert, G., & Boeckx, P. (2014). Effect of different biochar and fertilizer types on N_2O and NO emissions. *Soil Biology and Biochemistry*, *70*, 244–255. Available from https://doi.org/10.1016/j.soilbio.2013.12.026.

Noor, N. M., Shariff, A., & Abdullah, N. (2012). Slow pyrolysis of cassava wastes for biochar production and characterization. *Iranica Journal of Energy & Environment*, *3*, 60–65. Available from https://doi.org/10.5829/idosi.ijee.2012.03.05.10.

Novak, J., Sigua, G., Watts, D., Cantrell, K., Shumaker, P., Szogi, A., Johnson, M. G., & Spokas, K. (2016). Biochars impact on water infiltration and water quality through a compacted subsoil layer. *Chemosphere*, *142*, 160–167. Available from https://doi.org/10.1016/j.chemosphere.2015.06.038.

Ok, Y. S., Uchimiya, S. M., Chang, S. X., & Bolan, N. (2016). *Biochar: Production, characterization, and applications*. CRC Press.

Oliveira, F. R., Patel, A. K., Jaisi, D. P., Adhikari, S., Lu, H., & Khanal, S. K. (2017). Environmental application of biochar: Current status and perspectives. *Bioresource*

Technology, *246*, 110–122. Available from https://doi.org/10.1016/j.biortech.2017.08.122.

Omondi, M. O., Xia, X., Nahayo, A., Liu, X., Korai, P. K., & Pan, G. (2016). Quantification of biochar effects on soil hydrological properties using meta-analysis of literature data. *Geoderma*, *274*, 28–34. Available from https://doi.org/10.1016/j.geoderma.2016.03.029.

Omoriyekomwan, J. E., Tahmasebi, A., Zhang, J., & Yu, J. (2017). Formation of hollow carbon nanofibers on bio-char during microwave pyrolysis of palm kernel shell. *Energy Conversion and Management*, *148*, 583–592. Available from https://doi.org/10.1016/j.enconman.2017.06.022.

Oni, B. A., Oziegbe, O., & Olawole, O. O. (2019). Significance of biochar application to the environment and economy. *Annals of Agricultural Sciences*, *64*(2), 222–236. Available from https://doi.org/10.1016/j.aoas.2019.12.006.

Panwar, N. L., Pawar, A., & Salvi, B. L. (2019). Comprehensive review on production and utilization of biochar. *SN Applied Sciences*, *1*(2). Available from https://doi.org/10.1007/s42452-019-0172-6.

Parmar, A., Nema, P. K., & Agarwal, T. (2014). Biochar production from agro-food industry residues: A sustainable approach for soil and environmental management. *Current Science*, *107*(10), 1673–1682. Available from http://www.currentscience.ac.in/Volumes/107/10/1673.pdf.

Pereira Lopes, R., & Astruc, D. (2021). Biochar as a support for nanocatalysts and other reagents: Recent advances and applications. *Coordination Chemistry Reviews*, *426*. Available from https://doi.org/10.1016/j.ccr.2020.213585.

Ponnusamy, V. K., Nagappan, S., Bhosale, R. R., Lay, C. H., Duc Nguyen, D., Pugazhendhi, A., Chang, S. W., & Kumar, G. (2020). Review on sustainable production of biochar through hydrothermal liquefaction: Physico-chemical properties and applications. *Bioresource Technology*, *310*. Available from https://doi.org/10.1016/j.biortech.2020.123414.

Rago, Y. P., Surroop, D., & Mohee, R. (2018). Assessing the potential of biofuel (biochar) production from food wastes through thermal treatment. *Bioresource Technology*, *248*, 258–264. Available from https://doi.org/10.1016/j.biortech.2017.06.108.

Sakhiya, A. K., Anand, A., & Kaushal, P. (2020). Production, activation, and applications of biochar in recent times. *Biochar*, *2*(3), 253–285. Available from https://doi.org/10.1007/s42773-020-00047-1.

Sanford, J., Aguirre-Villega, H., Larson, R., Sharara, M., Liu, Z., & Schott, L. (2022). *Biochar production through slow pyrolysis of animal manure*. University of Wisconsin-Madison. Available from https://tammi.tamu.edu/files/2022/03/mp-series-biochar.pdf.

Sanford, J. R., & Larson, R. A. (2020). Assessing nitrogen cycling in corncob biochar amended soil columns for application in agricultural treatment systems. *Agronomy*, *10*(7). Available from https://doi.org/10.3390/agronomy10070979, https://www.mdpi.com/2073-4395/10/7/979/pdf.

Saudy, H. S., El-Metwally, I. M., Telb, S. T. S., & Abd-Alwahed, S. H. A. A. (2022). Mycorrhiza, charcoal, and rocket salad powder as eco-friendly methods for controlling broomrape weed in inter-planted faba bean with flax. *Journal of Soil Science and Plant Nutrition*, *22*(4), 5195–5206. Available from https://doi.org/10.1007/s42729-022-00995-6.

Saudy, H. S., Hamed, M. F., El–Metwally, I. M., Ramadan, K. A., & Aisa, K. H. (2021). Assessing the effect of biochar or compost application as a spot placement on broomrape control in two cultivars of faba bean. *Journal of Soil Science and Plant Nutrition*, *21*(3), 1856–1866. Available from https://doi.org/10.1007/s42729-021-00485-1.

Schmidt, H. P., Kammann, C., Niggli, C., Evangelou, M. W. H., Mackie, K. A., & Abiven, S. (2014). Biochar and biochar-compost as soil amendments to a vineyard soil: Influences on plant growth, nutrient uptake, plant health and grape quality. *Agriculture, Ecosystems and Environment*, *191*, 117–123. Available from https://doi.org/10.1016/j.agee.2014.04.001.

Shaheb, M. R., Islam, T., Sarker, A., & Rahman, M. M. (2024). Biofertilizers: Catalysts for enhancing soil and plant health in pursuit of sustainable agriculture. In S. Dheeman, T. Islam, & D. Egamberdeiva (Eds.), *Soil bacteria – biofertilization and soil health* (pp. 3–42). Singapore: Springer Nature. Available from https://doi.org/10.1007/978-981-97-3473-3_1.

Shaheb, M. R., Misiewicz, P. A., Godwin, R. J., Dickin, E., White, D. R., & Grift, T. E. (2024). The effect of tire inflation pressure and tillage systems on soil properties, growth and yield of maize and soybean in a silty clay loam soil. *Soil Use and Management*, *40*(2)e13063. Available from https://doi.org/10.1111/sum.13063.

Shaheb, M. R., Misiewicz, P. A., Godwin, R. J., Dickin, E., White, D. R., Lowenberg-DeBoer, J., Shearer, S. A., & Grift, T. E. (2023). An economic appraisal of the effect of tire inflation pressure for alternative tillage systems on a silty clay loam soil. *Agronomy Journal*, *115*(6), 3144–3161. Available from https://doi.org/10.1002/agj2.21440.

Shaheb, M. R., Venkatesh, R., & Shearer, S. A. (2021). A review on the effect of soil compaction and its management for sustainable crop production. *Journal of Biosystems Engineering*, *46*(4), 417–439. Available from https://doi.org/10.1007/s42853-021-00117-7.

Shakya, A., & Agarwal, T. (2020). Potential of biochar for the remediation of heavy metal contaminated soil. In J. S. Singh, & C. Singh (Eds.), *Biochar applications in agriculture and environment management* (pp. 77–98). Springer International Publishing. Available from https://doi.org/10.1007/978-3-030-40997-5_4.

Sharma, V., & Irmak, S. (2020). Economic comparisons of variable rate irrigation and fertigation with fixed (uniform) rate irrigation and fertigation and pre-plant fertilizer management for maize in three soils. *Agricultural Water Management*, *240*106307. Available from https://doi.org/10.1016/j.agwat.2020.106307.

Singh, C., Tiwari, S., Boudh, S., & Singh, J. S. (2017). Biochar application in management of paddy crop production and methane mitigation. In J. Singh, & G. Seneviratne (Eds.), *Agro-environmental sustainability. 2* (pp. 123–145). Cham: Springer. Available from https://doi.org/10.1007/978-3-319-49727-3_7.

Singh, D., Rajawat, M. V. S., Kaushik, R., Prasanna, R., & Saxena, A. K. (2017). Beneficial role of endophytes in biofortification of Zn in wheat genotypes varying in nutrient use efficiency grown in soils sufficient and deficient in Zn. *Plant and Soil*, *416* (1-2), 107–116. Available from https://doi.org/10.1007/s11104-017-3189-x.

Singh, R., Singh, P., Singh, H., & Raghubanshi, A. S. (2019). Impact of sole and combined application of biochar, organic and chemical fertilizers on wheat crop yield and water productivity in a dry tropical agro-ecosystem. *Biochar*, *1*(2), 229–235. Available from https://doi.org/10.1007/s42773-019-00013-6.

Spokas, K. A., & Reicosky. (2009). Impacts of sixteen different biochars on soil greenhouse gas production. *Annals of Environmental Science*, *3*, 179–193.

Talberg, A. (2009). *The basics of biochar*. Parliamentary Library.

Tan, X., Liu, Y., Zeng, G., Wang, X., Hu, X., Gu, Y., & Yang, Z. (2015). Application of biochar for the removal of pollutants from aqueous solutions. *Chemosphere*, *125*, 70−85. Available from https://doi.org/10.1016/j.chemosphere.2014.12.058.

Tisserant, A., & Cherubini, F. (2019). Potentials, limitations, co-benefits, and trade-offs of biochar applications to soils for climate change mitigation. *Land*, *8*(12). Available from https://doi.org/10.3390/LAND8120179, https://www.mdpi.com/2073-445X/8/12/179.

Toor, S. S., Rosendahl, L., & Rudolf, A. (2011). Hydrothermal liquefaction of biomass: A review of subcritical water technologies. *Energy*, *36*(5), 2328−2342. Available from https://doi.org/10.1016/j.energy.2011.03.013.

Tripathi, M., Sahu, J. N., & Ganesan, P. (2016). Effect of process parameters on production of biochar from biomass waste through pyrolysis: A review. *Renewable and Sustainable Energy Reviews*, *55*, 467−481. Available from https://doi.org/10.1016/j.rser.2015.10.122.

Ulusal, A., Apaydın Varol, E., Bruckman, V. J., & Uzun, B. B. (2021). Opportunity for sustainable biomass valorization to produce biochar for improving soil characteristics. *Biomass Conversion and Biorefinery*, *11*(3), 1041−1051. Available from https://doi.org/10.1007/s13399-020-00923-7.

Usevičiūtė, L., & Baltrėnaitė-Gedienė, E. (2021). Dependence of pyrolysis temperature and lignocellulosic physical-chemical properties of biochar on its wettability. *Biomass Conversion and Biorefinery*, *11*(6), 2775−2793. Available from https://doi.org/10.1007/s13399-020-00711-3.

Verheijen, F., Jeffery, S., Bastos, A. C., Velde, M., der, van, & Diafas, I. (2010). EUR 24099 EN *Biochar application to soils : A critical scientific review of effects on soil properties, processes and functions*. Luxembourg: European Commission JRC55799.

Wang, C., & Wang, H. (2018). Pb(II) sorption from aqueous solution by novel biochar loaded with nano-particles. *Chemosphere*, *192*, 1−4. Available from https://doi.org/10.1016/j.chemosphere.2017.10.125.

Wang, J., & Wang, S. (2019). Preparation, modification and environmental application of biochar: A review. *Journal of Cleaner Production*, *227*, 1002−1022. Available from https://doi.org/10.1016/j.jclepro.2019.04.282.

Wang, B., Gao, B., & Fang, J. (2017). Recent advances in engineered biochar productions and applications. *Critical Reviews in Environmental Science and Technology*, *42*(22), 2158−2207. Available from https://doi.org/10.1080/10643389.2017.1418580.

Wang, L., Ok, Y. S., Tsang, D. C. W., Alessi, D. S., Rinklebe, J., Mašek, O., Bolan, N. S., & Hou, D. (2022). Biochar composites: Emerging trends, field successes and sustainability implications. *Soil Use and Management*, *38*(1), 14−38. Available from https://doi.org/10.1111/sum.12731.

Warnock, D. D., Lehmann, J., Kuyper, T. W., & Rillig, M. C. (2007). Mycorrhizal responses to biochar in soil − concepts and mechanisms. *Plant and Soil*, *300*(1-2), 9−20. Available from https://doi.org/10.1007/s11104-007-9391-5.

Wiedner, K., Rumpel, C., Steiner, C., Pozzi, A., Maas, R., & Glaser, B. (2013). Chemical evaluation of chars produced by thermochemical conversion (gasification, pyrolysis and hydrothermal carbonization) of agro-industrial biomass on a commercial scale. *Biomass and Bioenergy*, *59*, 264−278. Available from https://doi.org/10.1016/j.biombioe.2013.08.026.

Xu, N., Tan, G., Wang, H., & Gai, X. (2016). Effect of biochar additions to soil on nitrogen leaching, microbial biomass and bacterial community structure. *European Journal of Soil Biology*, *74*, 1−8. Available from https://doi.org/10.1016/j.ejsobi.2016.02.004.

Yaashikaa, P. R., Senthil Kumar, P., Varjani, S. J., & Saravanan, A. (2019). Advances in production and application of biochar from lignocellulosic feedstocks for remediation of environmental pollutants. *Bioresource Technology*, *292*. Available from https://doi.org/10.1016/j.biortech.2019.122030.

Yadav, S. P. S., Bhandari, S., Bhatta, D., Poudel, A., Bhattarai, S., Yadav, P., Ghimire, N., Paudel, P., Paudel, P., Shrestha, J., & Oli, B. (2023). Biochar application: A sustainable approach to improve soil health. *Journal of Agriculture and Food Research*, *11*. Available from https://doi.org/10.1016/j.jafr.2023.100498.

Yadav, V., & Khare, P. (2020). Impact of pyrolysis techniques on biochar characteristics: Application to soil. In J. Singh, & C. Singh (Eds.), *Biochar applications in agriculture and environment management* (pp. 33–52). Springer International Publishing. Available from 10.1007/978-3-030-40997-5_2.

Yan, W., Acharjee, T. C., Coronella, C. J., & Vásquez, V. R. (2009). Thermal pretreatment of lignocellulosic biomass. *Environmental Progress and Sustainable Energy*, *28*(3), 435–440. Available from https://doi.org/10.1002/ep.10385.

Yang, Y., Ye, S., Zhang, C., Zeng, G., Tan, X., Song, B., Zhang, P., Yang, H., Li, M., & Chen, Q. (2021). Application of biochar for the remediation of polluted sediments. *Journal of Hazardous Materials*, *404*. Available from https://doi.org/10.1016/j.jhazmat.2020.124052.

Yu, K. L., Lau, B. F., Show, P. L., Ong, H. C., Ling, T. C., Chen, W. H., Ng, E. P., & Chang, J. S. (2017). Recent developments on algal biochar production and characterization. *Bioresource Technology*, *246*, 2–11. Available from https://doi.org/10.1016/j.biortech.2017.08.009.

Zhang, Z., Zhu, Z., Shen, B., & Liu, L. (2019). Insights into biochar and hydrochar production and applications: A review. *Energy*, *171*, 581–598. Available from https://doi.org/10.1016/j.energy.2019.01.035.

Zhou, Y., Gao, B., Zimmerman, A. R., Chen, H., Zhang, M., & Cao, X. (2014). Biochar-supported zerovalent iron for removal of various contaminants from aqueous solutions. *Bioresource Technology*, *152*, 538–542. Available from https://doi.org/10.1016/j.biortech.2013.11.021.

CHAPTER

Mechanistic insights into biochar's influence on nitrogen cycle homeostasis, microbial dynamics, and crop productivity in fertilized agroecosystems

15

Paurabi Das[1] and Prosanta Saha[2]

[1]*Crop Production and Protection Division, CSIR—Central Institute of Medicinal and Aromatic Plants, Lucknow, Uttar Pradesh, India*

[2]*Department of Botany, Durgapur Government College, Durgapur, West Bengal, India*

15.1 Introduction

Nitrogen (N) fertilizers have been a staple in agricultural practices for a long time since they are one of the most important nutrients for plant growth and food production (Zhang et al., 2021). On the other hand, excess usage of N-based fertilizers imposes negative environmental impacts by releasing greenhouse gases into the atmosphere, and contaminating underground water, and surface water bodies (Abdalla et al., 2019). Global consumption of N-based fertilizers is predicted to reach 119.4 million tons, growing at a pace of 1.4% per year (Yang et al., 2017). Anthropogenic N fluxes are destined to rise significantly due to anticipated trends in population expansion, food demand, agricultural methods, and energy use. Nonetheless, low N usage efficiency has become a global concern that requires attention. As a result, some nations and regional organizations, including the European Union (EU), have enacted laws (the Groundwater Directive, 2006/118/EC, and the Nitrates Directive, 91/676/EEC) that track and manage the release of NO_3 into reservoirs (Gibrilla et al., 2020). The permissible drinking water standard for short-term exposure is 50 mg NO_3/L^{-1} and 3 mg NO_2/L, while for long-term exposure it is 0.2 mg N/L (a provisional number) (World Health Organization, 2002). Around 40% of EU groundwater monitoring stations and around 50% of EU surface water monitoring stations surpassed average values of 25 and 10 mg NO_3/L, respectively, based on a prior investigation (Serra et al., 2024). However, based on a recent EU assessment, 13.2% of groundwater resources were above the 50 mg/L level, with some regions, like Catalonia

Biochar Ecotechnology for Sustainable Agriculture and Environment. DOI: https://doi.org/10.1016/B978-0-443-29855-4.00016-3

(northeastern Spain), having more than 77% of their water bodies above the limit (Carrey et al., 2021). Comparably, the United States Environmental Protection Agency (US EPA, 2011) states that two-thirds of the country's estuaries are contaminated due to increased N inputs. Besides water body contamination, N_2O, which has 298 times the potential to cause global warming than CO_2, is one of the most significant greenhouse gases (Ramzan et al., 2020). Agricultural soils are the main source of atmospheric N_2O, contributing 4.1 Tg N/year to the global N_2O budget of approximately 14 Tg N/year (Dawar et al., 2021). In addition, NH_3 volatilization has several negative environmental effects and reduces the amount of N available for plants. According to a recent study, employing urease inhibitors to slow down urea hydrolysis can significantly reduce NH_3 volatilization by 24%, 16%, and 17% in the United States, Europe, and Eastern China (ECH) (averaged by 18%), respectively (Liu et al., 2020). However, various studies strive to establish sustainable and eco-friendly management strategies to reduce N losses, improve nitrogen utilization efficiency (NUE), and increase crop yield.

The application of biochar as a soil supplement has grown tremendously as it improves soil N retention and decreases N loss. Researchers have highlighted the potential of biochar to positively regulate soil N activities by studying its residence period in the soil, its impact on the N cycle, soil microbial dynamics, and crop productivity (Hu et al., 2023). Biochar is a solid carbon-rich organic material that is produced as an external input material by heating biomass in anoxic or low-oxygen environments. Previous research has shown that adding biochar can prevent N leaching which can explained due to the ability of biochar to raise the soil cation exchange capacity (CEC) and anion exchange capacity (AEC). While biochar is mainly inert, its large surface area (SA), porosity, and capacity to absorb soluble nutrients offer a favorable environment for soil microbes, enhance the physicochemical properties of soil, and facilitate the slow release of nutrients. Furthermore, the addition of biochar to the soil could alter the diversity of the soil microbial population (Ahmad et al., 2021).

According to some meta-analyses, applying biochar can significantly increase plant productivity and food production while reducing N losses including those brought about by N leaching and N_2O emission (Borchard et al., 2019; Gao et al., 2019; Liu et al., 2018; Yang et al., 2019). Soil microorganisms possess an essential function in directing the transformation and cycle of N in the soil. Through the expression of their functional genes and the activities of their extracellular enzymes, soil microbes predominantly regulate the transformation and cycle of soil N (Zhang et al., 2021). Through mineralization and hydrolysis, N-cycling enzymes in soil microorganisms control the availability of inorganic N. Ammonia-oxidizing genes play a crucial role in nitrification, facilitating the oxidation of ammonia. The nitrous oxide reductase gene (*nosZ*) is responsible for the complete denitrification of N_2O to N_2, while the nitrite reductase genes (*nirG*, *nirS*, and *nirK*) mediate the rate-limiting step of NO_3^- or NO_2^- to NO and N_2O. Lastly, N_2-fixing microbes can increase plant N absorption, enhancing root development and secretion, thus benefiting from N_2 utilization (Sun et al., 2024; Zhang et al., 2021). Researchers continued to be interested in the complex interactions between soil constitutions and biochar that alter the dynamics of accessible soil N. So, an in-depth comprehension of the soil N cycle, microbial dynamics, and crop productivity under the influence of biochar will be discussed in this chapter.

15.1.1 The influence of biochar on soil nitrogen dynamics

Biochar is a popular adsorbent in agri-environmental applications, with its effectiveness influenced by the type of feedstock and pyrolysis temperature. The typical feedstocks for biochar production namely agricultural residues, woody residues, municipal sludge and manures, animal wastes, and industrial wastes. The composition of these feedstocks affects their carbon and nutritional levels. Crop and wood-based biochar have higher carbon and lower ash content than those from manures and sludge (Das et al., 2021). Manure- and sludge-based biochar are rich in plant nutrients like N, phosphorus (P), and potassium (K). Biochar derived from agricultural crop remains also contains more nutrients than wood-derived biochar (Xi et al., 2020). Low-temperature biochar is more effective in maintaining nitrogenous components than higher-temperature biochar. High pyrolysis temperatures cause significant losses in N concentration due to the volatilization of N-containing molecules such as amino-N, amine-N, protein-N, amide-N, and inorganic N. The ideal temperature for biochar production is typically 300°C–700°C, with 450°C–550°C being used if the range is narrowed (Ahmad et al., 2021). The temperature during pyrolysis significantly impacts the SA and functional groups of biochar. As temperature increases, the aromatic CO– and phenolic –OH groups that shield the aromatic core are destroyed, while low temperatures remove –OH, aliphatic C–O, and ester C = O groups from the outer SA. However, higher temperatures enhance the SA of biochar (Das et al., 2021). The process also affects the chemical composition of biochar, such as the concentration of trace metals, pH value, C:N ratio, O:C ratio, H:C ratio, CEC, and electrical conductivity (EC). The most significant shift in C components occurs when the temperature of pyrolysis rises, causing an increase in aromaticity and a decrease in aliphatic and labile C fractions(Das et al., 2021). Therefore, a deliberate choice of feedstock and pyrolysis conditions is crucial to obtain the best output from biochar, considering the nature of the intended adsorbate and the unique characteristics of the biochar (Fig. 15.1).

15.1.2 Utilizing biochar to lessen N_2O and NH_3 agricultural emissions

Biochar is an alkaline substance that can increase the soil pH, affecting the carbon pool and subsequent N_2O emission. This alteration depends on the feedstock and application amount. As soil pH increases, N_2O emissions may decrease due to a shift in the N_2O:N_2 ratio (Lan et al., 2018). This increase in soil pH stimulates

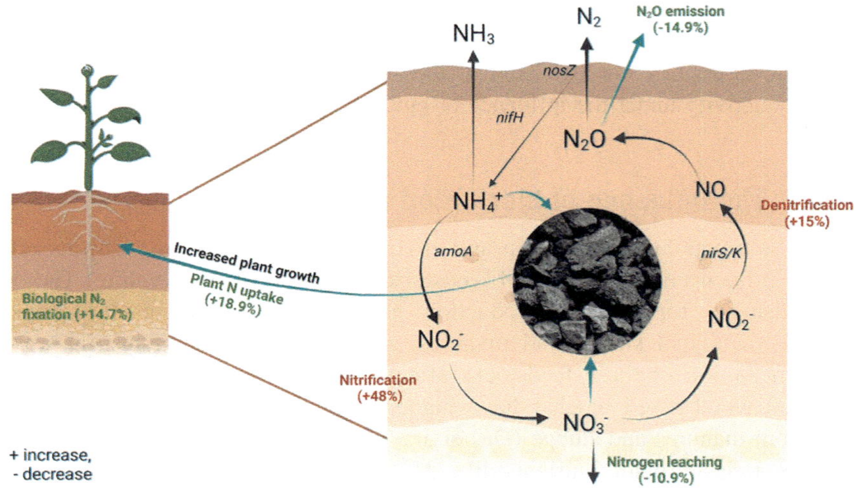

FIGURE 15.1

Conceptual impact of biochar on the soil nitrogen cycle. A schematic outline representing how biochar modulates N transformation in soil, plant N uptake, and corresponding functional microbial genes.

From Zhang, L., Jing, Y., Chen, C., Xiang, Y., Rezaei Rashti, M., Li, Y., Deng, Q., & Zhang, R. (2021). Effects of biochar application on soil nitrogen transformation, microbial functional genes, enzyme activity, and plant nitrogen uptake: A meta-analysis of field studies. GCB Bioenergy, 13*(12), 1859–1873.*

denitrifying enzymatic activities and promotes the formation of N_2 from N_2O. The molar H/Corg ratio of biochar is a crucial indicator of its ability to reduce N_2O emissions. A meta-analysis showed that biochar with a molar H/Corg ratio below 0.3 reduced N_2O emissions by 73% $\pm$ 7%, while those with a ratio above 0.5 were less effective (Cayuela et al., 2015). Biochar application to soil results in simultaneous changes in soil microclimate, such as soil aeration, porosity, moisture content, microbial growth, and surface chemistry of soil due to interaction with organic and inorganic compounds and microbes. Increased porosity and aeration may reduce denitrification by increasing oxygen availability, thereby enhancing soil health (Saffari et al., 2020). Biochar reduces N_2O emissions by weakening autotrophic nitrification and denitrification processes, which is possible by reducing ammonia-oxidizing bacteria (AOB) *amoA* and denitrifying bacterial ratio (*nirK* and *nirS*)/*nosZ* genes (Fan et al., 2020). Biochar application triggers the transformation of NO_3 into N_2 by increasing the abundance of *nosZ* and *nirK* genes, which stimulates N_2O reductases to catalyze the exergonic two-electron reduction. Biochar functions as an electron shuttle, mediating redox reactions of microbial strains (Aamer et al., 2020). Biochar can indirectly decrease N_2O emissions by inhibiting denitrification processes, which can be achieved by improving soil aeration. It can also directly reduce N_2O emissions by suppressing denitrification and inhibiting urease and protease activities (Wang et al., 2020). Biochar reduces NH_3 volatilization through adsorption, immobilization, and nitrification. The acidic groups present on the surface of biochar can protonate NH_3 gas to NH_4 ions. Biochar has been found to reduce NH_3 emissions by 32.92% by stimulating the abundance of the AOB *amoA* gene involved in nitrification (Yang et al., 2020). In another study, it was observed that reduction in NH_3 emission is feedstock-dependent, as manure and straw biochar stimulate net volatilization more than wood-based biochar due to their higher alkalinity, less functional groups, and less porosity structure (Li et al., 2019). However, biochar application at an appropriate rate (0.5%) can avoid dramatic NH_3 volatilization in rice paddy soils (Feng et al., 2017). Aged acidic biochar with higher oxygen-containing functionalities and less soil pH-raising effect could minimize NH_3 volatilization from terrestrial ecosystems (Esfandbod et al., 2017). Oxidized biochar from maple wood chips at 500°C retained higher amounts of NH_3 compared to unoxidized forms (Hestrin et al., 2020). A previous study where biochar mixed with cow dung was composted revealed that a higher application rate (10%) helps in increased oxygen content and faster rate of fermentation. This resulted in a higher concentration of total N, total P, and total K compared to control. The potential cause of the elevated total N levels in biochar (10%) treatment could be attributed to the adsorption of biochar on nitrogenous compounds such as ammonia (Yu et al., 2020). Similarly, as the application rates of bamboo biochar increased, NH_3 and N_2O emissions decreased. The reduction of NH_3 emission was due to strong adsorption affinity to the biochar surface while N_2O emission declined due to oxygenation on the surface of biochar (Awasthi et al., 2020).

15.1.3 Mitigation of N_2 runoff

The effectiveness of biochar production is determined by two key factors: the type of feedstock used and the temperature of pyrolysis during the thermochemical conversion process. Apart from these two factors, the particle size of biochar, application method, rate of application, soil type, and vegetation type greatly control N retention in soil (Ahmad et al., 2021). Water N runoff in agricultural fields is influenced by the water-soil-N nexus. Biochar can significantly modify soil water retention characteristics and reduce potential runoffs by limiting soil compaction (Rojimul Hussain et al., 2020). A study on lateritic red soil exposed to high rainfall intensity found that biochar incorporation into the soil surface layer maximized water-holding capacity and reduced total N, nitrate, and N losses by 31.1%, 68.3%, and 23.9%, respectively (Mai & Huang, 2021). Biochar effectively reduces surface runoff (N) and subsurface flow losses (NO_3), accounting for 63% of total runoff in sloping farmlands (Huang et al., 2020). The incorporation of biochar improved soil capacity to resist crusting and infiltration, resulting in a positive effect on water retention. Furthermore, 0.25 mm is the ideal size for maximum NH_4 adsorption, as smaller biochar particles have a higher NH_4 retention capacity than larger ones (Kizito et al., 2015). Due to the increased CEC of soil,

NH_4-N is held by biochar more firmly than NO_3-N. The extraordinary capacity of biochar to adsorb NH_4^+ and NO_3^- due to the presence of several functional groups (hydroxyl and carboxylic groups, lactones, ketones, and H-bonds) (Liao et al., 2021).

15.2 The role of biochar in stimulating biological nitrogen fixation

The term "biological N fixation" (BNF) refers to the process by which various bacteria use inert nitrogen (N_2), which is abundant in the atmosphere, to convert it into NH_3. BNF is limited to leguminous crops, where a molecular "crosstalk" between *Rhizobium* spp. and plant roots leads to the release of flavonoids from the roots, triggering the transcription of symbiotic N-fixing nod genes (Ajeng et al., 2020). Biochar amendment has been found to boost the population of *Bradyrhizobium* and *Frankiaceae*, suggesting that it can improve the microclimate of microbial growth. Additionally, free-living diazotrophs (such as *Azotobacter* sp., *Azospirillum*, and *Pseudomonas*) carry out nonsymbiotic BNF, which supplies external N inputs to terrestrial ecosystems (Gupta et al., 2019). The study compared the short-term effects of rice straw biochar application on diazotroph activity in alkaline soil using the *nifH* gene. Results showed that biochar application (20 t/ha) modified diazotrophic community structure, increased Sino-rhizobium abundance, and maximized N_2 fixation rate by 40.8%. This was also observed in pepper, with a 61% increase in BNF in biochar-amended treatment compared to unamended straw (Horel et al., 2019). Plants can absorb inorganic N in the form of nitrate or ammonium in soil. Nitrate is water-soluble, prone to evaporation, and does not form a link with the negatively charged sorption sites in the soil, while NH_4 being a cation can be sorbed onto negatively charged sites. When biochar is introduced to soil, the majority of recalcitrant fractions take a long time to decompose, while labile fractions mineralize quickly. An application of biochar accelerates the rate of nitrification, with a major contribution from the native organic N pools, leading to an increase in net N mineralization (Ahmad et al., 2021; Ramzan et al., 2020).

15.2.1 The influence of biochar on nitrogen use efficiency

NUE is the ratio of plant uptake of N to the total amount of N fertilizer applied. It depends on plant uptake, translocation, mobilization, and storage of N as well as the ability of soil for N supply (Ahmad et al., 2021). Due to its unique functionality, biochar can regulate the dynamics and transformations of available N in the soil and maximize plant uptake. It promotes the following outcomes: (1) maximizing available N contents by accelerating the rate of symbiotic and nonsymbiotic BNF and enhancing the rate of N mineralization; and (2) lowering the net amount of available N losses induced by volatilization/emission and runoff/leaching. Notably, biochar has a strong stimulating impact on BNF, both symbiotic and nonsymbiotic and the inert nature facilitates N mineralization (Li et al., 2019). Additionally, biochar improves soil AEC for NO_3 binding due to its high amount of pyridinium and oxonium groups, as well as the presence of protonated aromatic rings (Omara et al., 2020). Previous studies revealed that compared to 400°C–500°C and >500°C, the low pyrolysis temperature (300°C–400°C) proved to be highly effective at maximizing NUE over the control (without the use of biochar). Pyrolysis at high temperatures degrades several N molecules, including amino-N, amine-N, protein-N, amide-N, and inorganic N. Furthermore, biochar generated at high pyrolysis temperatures enhanced the C content of the soil, which accelerated the rate of N mobilization in the soil (Igalavithana et al., 2020).

15.3 Microorganism dynamics in biochar amendment soil

The high porosity of biochar allows microorganisms to thrive in it, providing a diverse range of niches for their growth. The pyrolysis process converts nutrients for microbial metabolism, while its pore structure allows for more chemical functional groups to adhere, increasing its ability to adsorb soil nutrients (Yan et al., 2021). Biochar also enhances the soil environment by increasing nutrient content, cation exchange capacity, and water-holding capacity, which indirectly affect the composition and variety of the soil microbial community (Yao et al., 2017). This makes biochar a valuable resource for microbial metabolism.

The N cycle in an agroecosystem is influenced by microbe-mediated pathways, including nitrification, which involves hydrolysis, mineralization, fixation, and denitrification (Yao et al., 2022). Among them, nitrification performed by nitrifiers in an aerobic condition is a crucial process because it is concerned with hydrolysis, mineralization, fixation, and denitrification. The first step of autotrophic nitrification is ammonia oxidation, catalyzed by AOB and ammonia-oxidizing archaea (AOA). AOB dominates ammonia oxidation in neutral and alkaline environments, while AOA is responsible in low-pH environments. Research shows that these pathways are sensitive to environmental variables (Hu et al., 2022; Li et al., 2019). Yao et al. (2022) found that biochar altered NH_4^+-N and NO_3^--N concentrations, consequently modulating nitrification potential and AOB population activity. AOA and AOB were less receptive to ammonia than comammox. AOA outnumbers AOB in low-nutrient habitats whereas AOB thrives in eutrophic environments. Li et al. (2019) stated the reaction of comammox clades A and B in response to N fertilizer application, revealing different ecotopic preferences in soil due to differences in NH_3 absorption systems. Soil pH, rate of biochar application, feedstock, and biochar residency duration in soil significantly impact soil microbial population dynamics.

15.3.1 Effect of biochar on nitrifiers

AOA are autotrophic ammonia-oxidizing microorganisms while AOB comprises heterotrophic ammonia-oxidizing microorganisms. After the addition of maize straw biochar, soil organic C, and N increased thereby favoring heterotrophic AOB activity (5.46×10^5 copies/g) compared to control soil. As the abundance of AOB increases, oxygen consumption also increases which leads to local anaerobiosis consequently limiting the activity of autotrophic AOA. This might be due to the high C/N ratio which accelerates N assimilation, thus providing an ample amount of substrates to nitrifiers (Abujabhah et al., 2018). The relative abundance of AOB (carrying the AOB amoA gene) was greatly increased in tea garden soil (TG) and Jerusalem artichoke cropping (JA) soil after the addition of biochar. Since biochar is alkaline in nature it provides a favorable environment for the growth of nitrifiers consequently facilitating N_2O production. Strikingly, the nitrification rate was increased with decreased N_2O emissions after biochar addition speculated that biochar-mediated soil nitrification was unlikely to contribute to N_2O emissions (Ji et al., 2020). At the maturity stage of wheat and maize, the expression of the AOA *amoA* gene was 4.1–19.9 times higher than that of the AOB genes across all biochar treatments. However, as the concentration of biochar increased, the abundance of AOA gradually decreased. A contrasting pattern of AOB abundance compared to AOA was observed. This microbial dynamics between AOA and AOB is correlated to soil pH which is modulated by biochar amendment (Xie et al., 2021). The application of cotton straw-derived biochar amendment at 2% and 5% affected the relative abundances of AOA and AOB (Liu, Lin, et al., 2021). AOA and AOB community were decreased by 27%–81% and 12%–62% across all treatments. In every treatment, the genus *Nitrososphaera* accounted for more than 93% of the AOA, and for AOB, the genus *Nitrosospira* accounted for 68%–80.3% (Liu, Lin, et al., 2021). Biochar coupled with N fertilizer significantly reduced the rate of nitrification and abundance of the *amoA*-AOB and *amoA*-AOA genes. The inhibitory effect of biochar became stronger with increased biochar rate. The autotrophic nitrification process was dominated by the *amoA*-AOB gene, and the abundance of the *amoA*-AOB gene was positively related to the nitrification rate. In contrast, a significant negative correlation was observed between the abundance of *amoA*-AOB and *amoA*-AOA genes and the concentration of nitrate-N. This adaptation was due to the shifting of soil microhabitat induced by biochar amendments (Yao et al., 2022).

The abundances of the AOA- and AOB-*amoA* genes varied between the two types of wheat straw biochar and soil. In acidic soil, a fresh biochar amendment somewhat reduced the abundance of the AOA-*amoA* gene; however, in alkaline soil, there was no change (in comparison to the control). Nonetheless, aged biochar increased the abundance of the AOA-*amoA* gene in both soils. AOB-*amoA* gene abundance in acidic soil was initially reduced by the addition of new biochar, but with time, a rising trend was seen. There was no apparent difference in the quantity of AOB-*amoA* genes in alkaline soil. On the other hand, aged biochar increased the abundance of the AOB-*amoA* gene in both soil types. However, the impact was more pronounced in acidic soil (Zhang, Duan, et al., 2019). Another study reported that the application of biochar inhibited the AOB community while dramatically increasing the AOA community by 143%–156% in comparison to the control. Nevertheless, in each treatment, the community of Comammox made a considerably higher contribution than AOA and AOB (Sun et al., 2024).

The impact of biochar on various soil types varies greatly depending upon the physicochemical properties of the soil which shape the microbial community. The brown sandy loam and red loam soils showed noticeable variations in bacterial diversity following the application of biochar, whereas the black clay loam soil was more adaptable. *Nitrospira* and nitrite-oxidizing bacteria (NOB) were more prevalent than typical AOB in all soils. The diversity of nitrifying bacteria in black clay loam soil increased with increased biochar application rates, whereas *Nitrospira* increased in abundance in brown sandy loam soil. The addition of biochar has a soil-dependent impact on the abundance of specific N_2 fixer groups. In the three soils, *Bradyrhizobium* spp., *Azosprillum* spp., *Rhizobium* spp., *Frankia* spp., and *Herbaspirillum* spp. were the most prevalent observable N-fixing bacterial groupings. *Rhizobium* and *Azospirillum* abundance showed strong positive relationships with higher biochar application rates in black clay loam soil. Additionally, higher biochar amendment in black clay loam soil dramatically raised the relative abundance of methanotrophs (Abujabhah et al., 2018). Apart from these nitrifiers and denitrifiers, *Proteobacteria* was the predominant phylum of bacteria present in all soil types, accounting for between 33% and 36% of the total. Actinobacteria (15%–22%), *Acidobacteria* (11%–14%), and *Verrucomicrobia* (4%–13%) were the other three predominant bacterial phyla that were found (Abujabhah et al., 2018). The addition of oilseed rape straws-derived biochar with larger SAs (2773 m^2/g) enhanced the relative abundances of bacteria that cycle N, including *Proteobacteria*, *Nitrospirae*, *Gemmatimonadetes*, *Rokubacteria*, and *Bacteroidetes* (Liao et al., 2021). According to Yuan et al. (2019) cellulose can be broken down by *Fibrobacteres*, nitrification activity is high in *Nitrospirae*, assimilative and dissimilative N processes are largely carried out by *Gemmatimonadetes*, and the production of secondary metabolites is aided by *Rokubacteria*.

The application of biochar with/or NPK enhanced the abundance of AOA compared to control which explains that the concentration of NH_4 acts as a limiting factor for the growth of AOA. AOB abundance where also increased in biochar with/or NPK treatments. However, the community structure of AOB and AOA greatly varied after the addition of biochar. Additionally, the Comammox *Nitrospira* clade A dominated usual ammonia oxidizers (AOA and AOB). Comammox *Nitrospira* clade A generally grows well under low concentrations of NH3. As a result, adding biochar and NPK together creates eutrophic conditions that greatly increase the abundance of Comammox *Nitrospira* clade A. Additionally, the addition of biochar markedly enhanced the abundance of *Nitrobacter*-like NOB rather than *Nitrospira*-like NOB, suggesting that biochar had a favorable influence on it (Sun et al., 2021). According to earlier research, high N concentrations could promote *Nitrobacter*-like NOB, which had a lower affinity for N substrate than *Nitrospira*-like NOB (Han et al., 2018).

Zhang, Sun, et al. (2019) reported that it typically outnumbered AOB in both acidic and alkaline soils, regardless of the crop kinds, fertilizer, and irrigation techniques used. In the wheat season, biochar derived from straw promoted the growth of AOA and AOB; however, in the rice season, there was no discernible variation in the fertilization treatments. The inconsistent results suggested that factors such as soil pH (acid or alkaline), crop type (vegetable or grain fields), oxygen content (flooded or dry soil), etc., may have an impact on how much biochar is used to reduce the number of ammonia oxidizers. AOB abundance was significantly impacted by the rise in exchangeable NH_4^+-N concentration following the addition of biochar. Exchangeable NH_4^+-N was a direct substrate resource for ammonia oxidizers (Zhang, Sun, et al., 2019). During the rice season, a negative correlation was found between AOA abundance and carbon/nitrogen (C/N). AOA can tolerate harsh conditions like low oxygen levels and switch between heterotrophic growth and chemoautotrophy in response to varying carbon supply conditions. In another study by Wu et al. (2020), AOB contributed ammonia oxidation up to 80% in alkaline soil. When acidic soil was treated with BE (biochar-enriched soil for 2–6 years) and FB (fresh biochar), AOB-driven N_2O production increased by 258.5% and 505.1%, respectively. However, in alkaline soil, the effects of BE and FB treatments resulted in a 38.3% and 13.5% decrease, respectively, in AOB-driven N_2O generation. In both soil types and all treatments—with a range of 78.2%–88.5%—the AOB was responsible for almost all of the NO production. AOB and AOA abundances in acidic soil were significantly lower than in alkaline soil, suggesting that higher soil pH levels could potentially result in more ammonia oxidation. Aged biochar, however, lost its ability to oxidize ammonia in comparison to fresh biochar, hence it did not affect the abundance of AOA and AOB (Wu et al., 2020).

15.3.2 Effect of biochar on denitrifiers

The process of denitrification releases NO and N_2O after converting NO_3^- to N_2. This conversion is the rate-limiting stage in the denitrification process which is catalyzed by nitrite reductase and regulated by the *nir* gene. Among the denitrifying genes, cytochrome cd1-containing reductase (*nirS*) and copper-containing reductase (*nirK*) are frequently regarded as representative molecular markers of the denitrification mechanism. The initial step of denitrification is the formation of NO_2 after the reduction of NO_3, which was catalyzed by the *narG* gene. The conversion of NO_2 to NO requires both the *nirK* and *nirS* genes and the *nosZ* gene is involved in the N_2O to N_2 conversion. There is a strong positive link between N_2O emissions and the abundance of *nirK* gene. The *nirK*-type denitrifiers are less impacted by soil type, while the community composition of *nirS*-type is more susceptible to it. Huang, Wang, et al. (2019) reported that the *nirS*, *nirK*, and *nosZ* gene copy numbers were reduced after the addition of biochar coupled with mineral fertilizer. In this study, *nirS* and *nosZ* gene copy numbers were lower under maize straw with mineral fertilizers (SF) and biochar with mineral fertilizer (BF) treatments compared to control. The BF treatment resulted in a decrease in the diversity of the *nirS* community while increasing that of the nirK especially *Rhizobium* and *nosZ* communities when compared to SF thus facilitating denitrification (Huang, Wang, et al., 2019). The relative abundance of denitrifiers (carrying the *nirK*, *nirS*, and *nosZ* genes) was greatly increased in TG and JA soil after the addition of biochar (Ji et al., 2020). Biochar limits denitrification by releasing polycyclic aromatic hydrocarbon compounds and by increasing the abundance of *nosZ* gene-carrying microorganisms (Xie et al., 2021). Additionally, the microsites within the soil particles formed after biochar addition resulted in elevated N_2O reductase activity. In contrast, biochar amendment enhanced N_2O emission in greenhouse vegetable soil though the abundance of *nosZ* gene-carrying microorganisms increased after biochar amendment. This can be due to distinct microbial communities in soil that shape the corresponding soil properties (Ji et al., 2020). In another study wheat straw-derived biochar, enhanced the *nirK*, *nirS*, and *nosZ* gene copy numbers by 111.4%, 41.9%, and 113.7%, respectively (Xie et al., 2021). The denitrifiers composition varied according to the biochar application rate. The genus *Bradyrhizobium* contained the highest proportion of the *nirK* denitrifier in each treatment, ranging from 31.9% to 46.6%. In this study, adding 2% biochar did not significantly alter the *nosZ* copies; however, adding 5% biochar resulted in a 12% increase in *nosZ* copies when compared to control, suggesting that applying additional biochar may have the ability to control the proliferation of *nosZ*-type bacteria. Under denitrification conditions, there was a considerable decrease in the relative abundance of *Burkholderia* and an increase in the relative abundance of *Cupriavidus* and *Pseudomonas*. When compared to the control condition, the heterotrophic nitrification treatments resulted in a considerable decline in the relative abundance of *Cupriavidus* (Liu, Lin, et al., 2021).

The modified biochar lowers NH_3 emissions by encouraging the conversion of NH_4^+-N to NO_3-N. The gene abundance of denitrifiers was increased by the addition of modified biochar. The modified biochar had a significant impact on *nirK* and *nosZ*, but not much on *nirS* (Zhou et al., 2021). Since biochar acts as an electron shuttle or donor and can create anoxic microsites, it enables variations in the abundance of the *nosZ* gene (Yu et al., 2020). Moreover, NH_3 emissions showed a positive correlation with *narG* and *nirS*, and modified biochar reduced their abundance (Zhou et al., 2021). Similarly, bamboo biochar amendment reduced *nirS* and *nirK* compared to control. However, *nosZ* abundance decreased after the addition of biochar. This might be due to the alkalinity of biochar which inhibited denitrification of oxidation reactions consequently affecting microbial activity indirectly (He et al., 2021). Green waste-derived biochar amendment significantly increased *nirK*, typical *nosZ*, and atypical *nosZ* gene copy numbers. As determined by the Shannon-Weaver diversity index, microbial communities comprising *nirK* were considerably less diverse than typical and atypical *nosZ* groups after biochar application. Unlike, the *nir* group, atypical *nosZ* communities lack *nir* genes. The genus *Mesorhizobium* and the order *Rhizobiales* dominated the *nirK*-expressing community. Alpha and betaproteobacteria as well as species from the genera *Oligotropha*, *Bradyrhizobium*, *Hyphomicrobium*, *Ensifer*, and *Azoarcus* dominated the usual *nosZ*-expressing soil community. *Chlorobi*, *Firmicutes*, *Chloroflexi*, and *Bacteroidetes* were among the many microbial species that made up the atypical *nosZ*-expressing community (Harter et al., 2017). Larger SAs in biochar led to a significant increase in *nosZ* copy number and ratio. The copy numbers of *nirK*, *nirS*, *nosZ*, and *nifH* accelerated by the addition of biochar with a SA of 2773 m^2/g compared to 1193 and 2023 m^2/g. Increased SA of biochar has larger pores, allowing for more microorganisms to thrive. The porosity of biochar may act as a habitat for nitrifiers and denitrifiers, encouraging the selection of N-related functional genes (Liao et al., 2021).

15.3.3 Effect of biochar on other microorganisms

The porous structure of biochar allows for the growth of microorganisms like filamentous microbes and fungi (Yin et al., 2021). The alkalinity of biochar can encourage the growth of both gram-positive and gram-negative bacteria (Kocsis et al., 2022). Factors like temperature, type, and carbon content during pyrolysis significantly influence microbial habitats, soil dehydrogenase activity, N immobilization, and mineralization (Murtaza et al., 2023). Biochar contains micro and mesopores that influence soil microbial composition and abundance. The high SA and porous nature of biochar attract soil microbes, which adhere to the micropores through hydrophobic attraction, electrostatic forces, or precipitate formation to prevent leaching and proliferate (Abhishek et al., 2022). Furthermore, the microbial composition and abundance in soils are changed by soluble chemicals found in the micropores and mesopores of biochar, including water, alcohol, sugars, acids, and ketone molecules (Adnan et al., 2020). It has been found to increase the microbial activity and population of filamentous fungi, *Bacillus* species, and *Pseudomonas* species in soils used for pepper cultivation. When applied at different concentrations, maize stalk biochar increased fungal abundance by 6.6%–31.22% compared to untreated soil (Yao et al., 2017). When corn residue biochar 1% and 2% (w/w) was added, soil microbial biomass increased by 20%–124% (Karimi et al., 2020). Biochar also increases the retention of N and P in the soil due to a shift in microbial community composition and increased activity. By controlling processes like nodule formation, immobilizing N, improving P supply, and changing soil pH, biochar is essential for BNF in legume crops (Semida et al., 2019). Competition among soil microbial communities is accelerated by changes in resources, physicochemical factors, water availability, or habitat access, resulting in altered community structure and composition. The functional groups present on the surface of biochar can bind enzymes and substrates, obstructing substrate diffusion (Osman et al., 2022). Biochar with higher porosity and SA may decrease extracellular enzymatic activity. Soil enzymes are crucial for soil quality, as they are directly linked to soil microbial activity and nutrient cycling. A recent global meta-analysis found that extracellular enzyme activities, such as phosphatase (25%), urease (23%), and dehydrogenase (20%), increased, but decreased β-glucosidase, acid phosphatase, and phenoloxidase (13%, 7%, and 6%, respectively) (Pokharel et al., 2020). Under environmental stress, *Ascomycota* dominated the fungal community, playing a crucial role in compost degradation. The addition of compost mixed with biochar increased the abundance of *Ascomycota*, while the relative abundance of *Basidiomycota* decreased due to their contrasting responses to stress. *Basidiomycota* can survive in anaerobic conditions, while soil amendments increase the abundance of *Mortierellomycota*. Soil amendments provided nutrient-rich habitats, benefiting *Mortierellomycota* as saprophytic fungi. Overall, compost along with biochar can enhance the response of the fungal community to environmental stress (Yan et al., 2021).

15.4 Impact of biochar on crop productivity

Rice is a major crop globally that generates a variety of solid wastes including bran, broken rice, husk, ash, and straw. Rice straw, which makes up 50% of its dry weight, is removed during harvest. The pyrolysis of rice husk and straw for biochar production varies from 15 minutes to 3 hours at 350–1100°C and from 30 minutes to 8 hours at 400–800°C, respectively (Bhatnagar et al., 2022). Wheat, another staple food crop, is widely cultivated and serves as the main dietary staple for 40% of the global population. Wheat straw pyrolysis varies from 30 minutes at 350°C to 8 hours at 750°C (Fermanelli et al., 2020). Maize is a significant source of carbohydrates in developing nations is also used for biochar production. Maize straw pyrolysis varies from 1 hour at 350°C to 30 minutes at 600°C (Ndiate et al., 2021). Other plant wastes are also used for biochar production, which accelerates plant growth (Table 15.1). Since these agricultural wastes are widely generated this section will focus on the effects of rice, wheat, maize, and aged biochar on crop productivity.

15.4.1 Effect of rice-based biochar on crop productivity

The application of biochar has an extensive impact on the soil's physicochemical properties and microbial diversity due to its large surface area (specific surface area), porosity, functional groups, and mineral content. After the addition of biochar, soil structural and functional properties are altered, reduction in nutrient leaching, enhancement of soil nutrient cycles, generation of labile carbon due to enhanced microbial activity, and increment of nutrient mineralization which consequently accelerate plant growth (Singh et al., 2022). Previous research indicates that the types of biochar, soil, and temperatures used to prepare the biochar all these factors greatly affect crop productivity. According to Luigi et al. (2022), the application of biochar improved soil pH and soil WHC, which appears to

Table 15.1 Effect of different feedstock biochar on plant growth.

Feedstock	Application rate (%, w/w)	Fabrication method (pyrolysis temperature, °C)	Residence time (min)	Crop	Mechanism of crop productivity	References
Rice straw	2.5	600	–	Ryegrass	Increased available P and K	Zhang et al. (2017)
	0, 2.25 and 11.3 Mg/ha	500	480	Rice and wheat	Slow release of macro- and micronutrients	Bi et al. (2019)
	2, 5, 10	450–550	–	*Phragmites communis*	Increased N and P uptake. Improved soil water content and photosynthetic rate	Huang, Li, et al. (2019)
	0, 1, 2	400	120	Maize	Increased P availability and soil pH	Baquy et al. (2020)
	3	450	120	Soybean	Increased nutrient uptake and chlorophyll, soluble proteins, and sugar contents	Mehmood et al. (2020)
	2.8 t/ha	500	120	Rice	Increased total N content and soil available K and Mg	Nan et al. (2020)
	2	400 and 800	–	Lettuce	increased soil available N and K, reduced Fe and Al	Xi et al. (2020)
	7.5 t/ha	450	360	Rice	Increase N uptake and N use efficiency	Zheng et al. (2020)
	15 t/ha	–	–	Faba bean	Reduced oxidative damage to leaf tissues and maintenance of the integrity of cell membranes	El Nahhas et al. (2021)
	1.0	450	120	Rice	Increased N use efficiency	Liu, Lin, et al. (2021)
	4.25 g/kg	500	300	Zucchini	Enhanced soil characteristics and increased nutrient availability	Farid et al. (2022)

Rice husk	2.5	Gasification (900–1100°C)		Rice	Improved water status	Fahad et al. (2016)
	20 t/ha	600	180	Rice	Increased N uptake and N use efficiency	Huang et al. (2018)
	21 g/kg	350–400	15	Rice	Increased soil total C, C/N ratio, exchangeable K, chlorophyll content, and nutrient uptake	Win et al. (2019)
	1–5 in water	Pyrolysis (400°C, 30 min) and liquid extraction with methanol	–	Rice	Increase gene and protein expression of ABP1	Yan et al. (2021)
	2.5, 5, and 7.5 t/ha	Modified biochar kiln (350°C)	–	Tomato	Improved soil physicochemical properties	Adebajo et al. (2022)
	0, 5, 20, and 80 g/kg	450	180	Chinese crab apple	Increased activity of soil enzymes and decreased abundance of *Fusarium solani*	Wang et al. (2019)
	3	450	–	Tomato	Increased expression of defense-related genes (PR-1b, JERF3)	Arshad et al. (2021)
Rice husks and maize stalk (1:1)	10 t/ha	350	180	Rice	Improved activity of CAT, APX, and POX	Hafez, Gowayed, et al. (2021)
	10 t/ha	350	180	Wheat	Increased chlorophyll, proline, and carotenoid content; increased relative water content and N, P, and K uptake	Hafez, Omara, et al. (2021)
	10	550	–	Barley	Increased P uptake	Shepherd et al. (2017)
	5 t ha^{-1}	600	180	Lentil	Increased the organic C content and improved other soil physicochemical properties	Safaei Khorram et al. (2018)

(*Continued*)

Table 15.1 Effect of different feedstock biochar on plant growth. *Continued*

Feedstock	Application rate (%, w/w)	Fabrication method (pyrolysis temperature, °C)	Residence time (min)	Crop	Mechanism of crop productivity	References
	8	350–550	–	Tomato	Na ions adsorption, and release of K, Ca, and Mg	She et al. (2018)
	1 t ha^{-1}	500–600		Wheat and rice	Improved soil aggregation and fungal community structure	Bai et al. (2019)
	2	500	240	Wheat	Reduced herbicide formesan uptake and increased microbial diversity	Meng et al (2019)
	1	350	30	Maize	Increased P, K, N, and microbial biomass in soil	Abbas et al. (2020)
	0.25	400	30	Rice	Increased root membrane potential resulting in an increased nutrient uptake	Chew et al. (2022)
	20–40 t ha^{-1}	450	–	Recan	Increased N and enzyme activities in soil	Hou et al. (2020)
	2	500	–	Tomato	Increased N and P uptake	Tian et al. (2021)
	2	400	–	Rice	Increased N, P, K, and Fe uptake	Chew et al. (2022)
	20 t ha^{-1}	450	–	Rice	Increased N and P uptake	Liu, Lin, et al. (2021)
	2	550	–	Tobacco	Increased P and K uptake	Liu et al. (2022)
A mixture of softwood and wheat straw	2	550	–	Tobacco	Improved soil hydrophysical properties	Liu, Wei, et al., (2021)

Maize cob	1	350	–	Quinoa	Improved the plant antioxidant defense machinery and enhanced nutrient uptake	Ramzani et al. (2017)
Maize cob and straw	2.5	400	–	Ryegrass	Increased P and K uptake	Zhang et al. (2017)
Maize straw	1	500	120	Rice	Increased soil P and Fe content by increasing Fe-reducing bacteria and phosphate-solubilizing bacteria	Xu et al. (2019)
	15.75–31.5 t/ha	500	–	Soybean and maize	Improved soil aggregation	Ji et al. (2020)
	9 kg/m^2	500	120	Soybean	Increased SOC content	Li et al. (2020)
	20–50 t/ha	500–600	–	Maize	Increased soil moisture and N uptake	Feng et al. (2021)
	5	350–500	60	Maize	Improved antioxidant defense machinery in plants and enhanced nutrient uptake	Ndiate et al. (2021)
Maize stalk	13.3 g/kg	400	90	Pepper	Increased soil organic matter, and N, P, and K content	Wang et al. (2020)
Barley straw	20 t/ha	400	–	Rice	Improved soil physical properties and increased soil chemical contents	Kang et al. (2019)
Sugarcane bagasse	2.5, 5.0, 7.5, and 10.0	450	30	Sugar beet	Increased in yield by increasing soil total C (by 27.8%)	Alves et al. (2021)
	5 and 10 t/ha	600	–	Maize and ground nut	Increased pH value, C, P, K, and S	Rahman et al. (2022)
	10 t/ha	–	–	Zucchini	Reduced carbon loss with increased significantly soil organic carbon content	Farid et al. (2022)

(Continued)

Table 15.1 Effect of different feedstock biochar on plant growth. *Continued*

Feedstock	Application rate (%, w/w)	Fabrication method (pyrolysis temperature, °C)	Residence time (min)	Crop	Mechanism of crop productivity	References
	10, 20, and 40 t/ha	Torrefaction (low-temperature pyrolysis; approximately 30 min residence time)	–	Cotton	Addition of nutrients like N, P, and K released by slow mineralization of biochar	Pinnamaneni, Lima, et al. (2023)
	0, 5, 10, 15, and 25 t/ha	–	–	*Zea mays*	Improved water availability in the root zone	Pinnamaneni, Lima, et al. (2023)
Popular wood and sugarcane bagasse	3, 7	550°C	480	Lettuce	Reduced Cr and Pb contamination with increased CEC and soil pH	Khan et al. (2020)
Pine-woodchip biochar	5 and 10 t/ha	450°C	–	Maize	Soil organic carbon and soil total nitrogen concentrations were significantly higher	Akbar et al. (2024)

Al, *aluminum;* APX, *ascorbate peroxidase;* Ca, *calcium;* CAT, *catalase;* CEC, *cation exchange capacity;* Cr, *chromium;* Fe, *iron;* K, *potassium;* Mg, *magnesium;* N, *nitrogen;* P, *phosphorus;* Pb, *lead;* POX, *peroxidase;* S, *sulfur;* SOC, *soil organic carbon.*

encourage the development of tomato seedlings, particularly at concentrations of 1%–20% (w/w with peat) by providing shelter to beneficial soil bacteria present in the rhizosphere. Biochar derived from rice husk when applied at a 2% (w/v) rate, the amount of N and K available in the soil led to the growth of taller lettuce plants with longer roots, stronger leaves and stems, and more leaf area. As biochar improves the inherent properties of soil it results in enhanced plant growth (Xi et al., 2020). Similarly, rice straw biochar improved overall soil N content, which led to enhanced *Phragmites communis* growth. On the other hand, rice biochar can also promote soil C and N cycling by altering the microbial community. For instance, improved soil pH, cation exchange capacity, and total C and N content, along with alterations in the microbiota, led to increased rape shoot biomass (from 2.31 to 4.23 g) following the application of rice biochar (Martínez-Gómez et al., 2022). In rice seedlings, liquor-based biochar (1%–5% (w/w) in water) increased plant height and root growth. The suggested mechanism of action was based on the accumulation of the protein product obtained from overexpression of the *ABP1* gene. With the help of molecular modeling, it was observed that a molecule on the surface of biochar interacted with the *abp1* protein stimulating plant growth (Gelová et al., 2021). Plant root, shoot, and leaf morphology were positively impacted by biochar as the available N was increased by 17%, and total C was also increased by 13%. Compared to untreated plants, biochar-treated plants had more leaves with increased leaf area by 26%–36%. Additionally, biochar amendment doubled the biomasses of the roots, stems, and leaves as well as the length and SA of the roots when compared to untreated plants (Simiele et al., 2022) (Fig. 15.2).

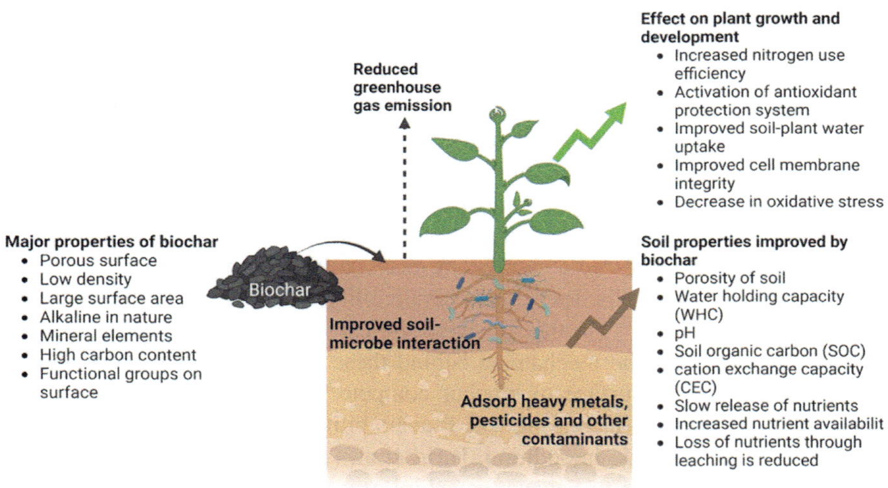

FIGURE 15.2

Biochar characteristics enhancing soil properties and boosting plant productivity. Diagram representing the effects of biochar on soil and plant productivity. The figure represents the characteristics of biochar which enhances soil properties and promotes plant productivity.

The addition of biochar boosted the yield, BNF, and nodulation of several legume species, such as red clover, alfalfa, and soybean (Lai et al., 2022). Potential mechanisms influencing the effects of biochar addition on BNF in legumes include (1) higher soil pH, which raises the availability of Mo, a nutrient that is essential to the BNF mechanism; (2) higher concentration of available N immobilization, which is associated with an improvement in BNF; (3) biochar has a strong influence on nodule production by utilizing the flavonoid and Nod features; and (4) adding biochar to P-deficient soils accelerates the concentration of available P (Chen et al., 2022). According to Simiele et al. (2022), the biochar-treated plants exhibited an increased quantity of flowers and fruits with increased levels of total soluble solids, titratable acidity, and *trans-* and *cis*-lycopene when compared to the untreated plants. Due to biochar amendment, high total P content in soil and high total N content in the roots may be responsible for the promotion of enhanced fruit quantity and quality. Moreover, additional reports indicate that when biochar is applied as a soil amendment, the phosphorous and N contents of growing substrates and plant tissues may be connected to the improvement of vegetative and reproductive processes that result in fruit production, lycopene, titratable acidity, and total soluble solids (Guo et al., 2021). In another study, continuous application of 20 t/ha rice straw–derived biochar increased plant growth and soil N uptake by 14%–26%, internal N use efficiency by 7%–11%, and grain yield by 6% in rice fields (Martínez-Gómez et al., 2022). In a 4-year field trial, Nan et al. (2020) observed a noticeable improvement in the relationships between soil bacteria following treatment with rice biochar. The rhizosphere bacterial community displayed increased complexity, probably due to an increase in the soil total N and total C contents along with soil available K and magnesium (Mg) contents which increased rice yield up to 14.5%. Yin et al. (2021) observed an increase in rice yield (38%–41%) following the application of N-enriched rice straw and waste wood biochar, which they attributed to higher soil C and N contents as well as increased availability of iron (Fe). Baquy et al. (2020) tested rice straw biochar combined with a P fertilizer (100 mg P/kg) in two different soils to investigate the mechanisms underlying the increase in soil P availability due to biochar. The results showed a significant increase in dry weight, and P uptake, in 2% biochar-amended soils. Increased P availability, a higher pH in the soil, and a lower exchangeable Al^{3+} in the soil were the causes of this dry weight increase. It was suggested that the mechanism behind these effects was that the functional group present on the surface of biochar compete with soil PO_4^{3-} to form complexes with Fe and Al, increasing soil P availability and a decrease of exchangeable Al^{3+} in the process. In addition, Farid et al. (2022) discussed the application of cocomposted biochar to enhance zucchini growth, which is composed of 70% rice straw, 15% farmyard manure, 10% rice straw biochar, and 5% mineral fertilizer. The cocomposted biochar increased plant height, dry weight, and chlorophyll content which may have been brought about by altered soil properties and enhanced nutrient availability. Additionally, rice biochar and fertilizers could be applied in combination to hasten ecosystem restoration.

15.4.2 Effect of biochar from wheat and maize on crop productivity

Wheat straw biochar when applied to rice field (20 t/ha), the yield increased by 17% as a result of increased N and P supply with an improvement in N use efficiency (Liu, Lin, et al., 2021; Liu et al., 2022). Liu et al. (2022) found that applying 2% (w/w) wheat straw biochar to tobacco plants, improved C and N content in rhizosphere level as well as P and K availability. Noticeably when a rice field is treated with wheat straw biochar represents a characteristic phenomenon that increases rice biomass along with N and P uptake. Due to greater negative electrical potential at the root surface compared to internal layers, these potential differences across the membranes act as catalysts for the uptake of minerals and other nutrients (Chew et al., 2022). The application of biochar with ammonium nitrate showed improvements in height, chlorophyll content, photosynthetic rate, and N, P, Fe, and K accumulation in *Carya illinoinensis* (pecan) (Hou et al., 2020). When tomato plants were treated with wheat biochar displayed improved yield, quality, and vegetative growth parameters when irrigated with saltwater because the adsorption of Na^+ ions released K^+, Ca^{+2}, and Mg^{+2} (She et al., 2018). The enhancement of soil hydro-physical characteristics such as bulk density, and water-holding capacity observed in tobacco plants is another way that wheat straw biochar can make plants more drought tolerant (Liu, Lin, et al., 2021). Hafeez et al. (2022) found that the addition of 20 t/ha of corn-derived biochar that had been pyrolyzed at 400°C did not significantly boost the growth of *Glycine max*.

15.5 Conclusion

There are many different types of biochar, and they can affect soil properties, crop growth, and yield in different ways. The characteristics of biochar and its impact on agricultural ecosystems are mostly determined by the feedstock and the parameters of the pyrolysis process. When applied at the right rate, biochar formulations can dramatically raise yields in situations with limited water, nutrients, and soil-specific restrictions. Pyrolyzed biochar at low temperatures has the potential to enhance crop yield and nutrient availability in both acidic and alkaline soil types. Conversely, biochar produced at high temperatures has the potential to raise soil carbon. Biochar speeds up microbial activity, which can improve nutrient mineralization and encourage plant uptake of minerals. The choice of biochar, how often it should be applied, and whether it works with cropping systems should all be taken into account before administering biochar. Large-scale carbon biochar sequestration lowers greenhouse gas emissions. Extensive insight into the diverse impacts of biochar on soil and plant systems might help in the fabrication of biochar for particular uses by selecting appropriate feedstock and modifying pre- or postproduction treatments of biochar that control its pH, nutrient availability, and adsorption capacity. Lastly, there should be developed guidelines for the selection and manufacture of biochar that take into account the specific limitations of the soil and environment.

Acknowledgment

Authors are grateful to the editors of the book for giving this opportunity.

Authors contribution

PD—conceptualization, investigation, resources, visualization, writing of the original draft, and review and editing; PS—review and editing of the original draft.

Conflict of interest

The authors declare no competing financial interest.

References

Aamer, M., Shaaban, M., Hassan, M. U., Guoqin, H., Ying, L., Hai Ying, T., Rasul, F., Qiaoying, M., Zhuanling, L., Rasheed, A., & Peng, Z. (2020). Biochar mitigates the N_2O emissions from acidic soil by increasing the nosZ and nirK gene abundance and soil pH. *Journal of Environmental Management*, *255*. Available from https://doi.org/10.1016/j.jenvman.2019.109891.

Abbas, A., Naveed, M., Azeem, M., Yaseen, M., Ullah, R., Alamri, S., Farooq, Q. U. A., & Siddiqui, M. H. (2020). Efficiency of wheat straw biochar in combination with compost and biogas slurry for enhancing nutritional status and productivity of soil and plant. *Plants*, *9*(11), 1–19. Available from https://doi.org/10.3390/plants9111516, https://www.mdpi.com/2223-7747/9/11/1516/pdf.

Abdalla, M., Hastings, A., Cheng, K., Yue, Q., Chadwick, D., Espenberg, M., Truu, J., Rees, R. M., & Smith, P. (2019). A critical review of the impacts of cover crops on nitrogen leaching, net greenhouse gas balance and crop productivity. *Global Change Biology*, *25*(8), 2530–2543. Available from https://doi.org/10.1111/gcb.14644.

Abhishek, K., Shrivastava, A., Vimal, V., Gupta, A. K., Bhujbal, S. K., Biswas, J. K., Singh, L., Ghosh, P., Pandey, A., Sharma, P., & Kumar, M. (2022). Biochar application for greenhouse gas mitigation, contaminants immobilization and soil fertility enhancement: A state-of-the-art review. *Science of the Total Environment*, *853*. Available from https://doi.org/10.1016/j.scitotenv.2022.158562.

Abujabhah, I. S., Doyle, R. B., Bound, S. A., & Bowman, J. P. (2018). Assessment of bacterial community composition, methanotrophic and nitrogen-cycling bacteria in three soils with different biochar application rates. *Journal of Soils and Sediments*, *18*(1), 148–158. Available from https://doi.org/10.1007/s11368-017-1733-1.

Adebajo, S. O., Oluwatobi, F., Akintokun, P. O., Ojo, A. E., Akintokun, A. K., & Gbodope, I. S. (2022). Impacts of rice-husk biochar on soil microbial biomass and agronomic performances of tomato (*Solanum lycopersicum* L.). *Scientific Reports, 12* (1). Available from https://doi.org/10.1038/s41598-022-05757-z.

Adnan, M., Fahad, S., Zamin, M., Shah, S., Mian, I. A., Danish, S., Zafar-Ul-hye, M., Battaglia, M. L., Naz, R. M. M., Saeed, B., Saud, S., Ahmad, I., Yue, Z., Brtnicky, M., Holatko, J., & Datta, R. (2020). Coupling phosphate-solubilizing bacteria with phosphorus supplements improve maize phosphorus acquisition and growth under lime induced salinity stress. *Plants*, *9*(7), 1−18. Available from https://doi.org/10.3390/plants9070900, https://www.mdpi.com/2223-7747/9/7/900/pdf.

Ahmad, Z., Mosa, A., Zhan, L., & Gao, B. (2021). Biochar modulates mineral nitrogen dynamics in soil and terrestrial ecosystems: A critical review. *Chemosphere*, *278*, 130378. Available from https://doi.org/10.1016/j.chemosphere.2021.130378.

Ajeng, A. A., Abdullah, R., Ling, T. C., Ismail, S., Lau, B. F., Ong, H. C., Chew, K. W., Show, P. L., & Chang, J. S. (2020). Bioformulation of biochar as a potential inoculant carrier for sustainable agriculture. *Environmental Technology and Innovation*, *20*. Available from https://doi.org/10.1016/j.eti.2020.101168.

Akbar, W. A., Rahim, H. U., Irfan, M., Alatalo, J. M., Mudasssir, M., & Bundschuh, J. (2024). Effects of incorporating pine-woodchip biochar with reduced NP fertilizer on calcareous soil characteristics, organic carbon, NPK availability, and maize productivity. *Journal of Soils and Sediments*, *24*(1), 177−188. Available from https://doi.org/10.1007/s11368-023-03617-5.

Alves, B. S. Q., Zelaya, K. P. S., Colen, F., Frazão, L. A., Napoli, A., Parikh, S. J., & Fernandes, L. A. (2021). Effect of sewage sludge and sugarcane bagasse biochar on soil properties and sugar beet production. *Pedosphere*, *31*(4), 572−582. Available from https://doi.org/10.1016/S1002-0160(21)60003-6.

Arshad, U., Azeem, F., Mustafa, G., Bakhsh, A., Toktay, H., McGiffen, M., Nawaz, M. A., Naveed, M., & Ali, M. A. (2021). Combined application of biochar and biocontrol agents enhances plant growth and activates resistance against Meloidogyne incognita in tomato. *Gesunde Pflanzen*, *73*(4), 591−601. Available from https://doi.org/10.1007/s10343-021-00580-4.

Awasthi, M. K., Duan, Y., Awasthi, S. K., Liu, T., & Zhang, Z. (2020). Influence of bamboo biochar on mitigating greenhouse gas emissions and nitrogen loss during poultry manure composting. *Bioresource Technology*, *303*. Available from https://doi.org/10.1016/j.biortech.2020.122952.

Bai, N., Zhang, H., Li, S., Zheng, X., Zhang, J., Zhang, H., Zhou, S., Sun, H., & Lv, W. (2019). Long-term effects of straw and straw-derived biochar on soil aggregation and fungal community in a rice−wheat rotation system. *PeerJ*, *6*(1), e6171. Available from https://doi.org/10.7717/peerj.6171.

Baquy, M. A. A., Jiang, J., & Xu, R. (2020). Biochars derived from crop straws increased the availability of applied phosphorus fertilizer for maize in Ultisol and Oxisol. *Environmental Science and Pollution Research*, *27*(5), 5511−5522. Available from https://doi.org/10.1007/s11356-019-06695-6.

Bhatnagar, A., Singhal, A., Tolvanen, H., Valtonen, K., Joronen, T., & Konttinen, J. (2022). Effect of pretreatment and biomass blending on bio-oil and biochar quality from two-step slow pyrolysis of rice straw. *Waste Management*, *138*, 298−307. Available from https://doi.org/10.1016/j.wasman.2021.12.013.

Bi, Y., Cai, S., Wang, Y., Xia, Y., Zhao, X., Wang, S., & Xing, G. (2019). Assessing the viability of soil successive straw biochar amendment based on a five-year column trial with six different soils: Views from crop production, carbon sequestration and net ecosystem economic benefits. *Journal of Environmental Management*, *245*, 173–186. Available from https://doi.org/10.1016/j.jenvman.2019.03.051.

Borchard, N., Schirrmann, M., Cayuela, M. L., Kammann, C., Wrage-Mönnig, N., Estavillo, J. M., Fuertes-Mendizábal, T., Sigua, G., Spokas, K., Ippolito, J. A., & Novak, J. (2019). Biochar, soil and land-use interactions that reduce nitrate leaching and N_2O emissions: A meta-analysis. *Science of the Total Environment*, *651*, 2354–2364. Available from https://doi.org/10.1016/j.scitotenv.2018.10.060.

Carrey, R., Ballesté, E., Blanch, A. R., Lucena, F., Pons, P., López, J. M., Rull, M., Solà, J., Micola, N., Fraile, J., Garrido, T., Munné, A., Soler, A., & Otero, N. (2021). Combining multi-isotopic and molecular source tracking methods to identify nitrate pollution sources in surface and groundwater. *Water Research*, *188*. Available from https://doi.org/10.1016/j.watres.2020.116537.

Cayuela, M. L., Jeffery, S., & van Zwieten, L. (2015). The molar H:Corg ratio of biochar is a key factor in mitigating N_2O emissions from soil. *Agriculture, Ecosystems & Environment*, *202*, 135–138. Available from https://doi.org/10.1016/j.agee.2014.12.015.

Chen, K., Li, N., Zhang, S., Liu, N., Yang, J., Zhan, X., & Han, X. (2022). Biochar-induced changes in the soil diazotroph community abundance and structure in a peanut field trial. *Biochar*, *4*(1). Available from https://doi.org/10.1007/s42773-022-00133-6.

Chew, J. K., Joseph, S., Chen, G., Zhang, Y., Zhu, L., Liu, M., Taherymoosavi, S., Munroe, P., Mitchell, D. R. G., Pan, G., Li, L., Bian, R., & Fan, X. (2022). Biochar-based fertiliser enhances nutrient uptake and transport in rice seedlings. *Science of the Total Environment*, *826*. Available from https://doi.org/10.1016/j.scitotenv.2022.154174.

Das, S. K., Ghosh, G. K., Avasthe, R. K., & Sinha, K. (2021). Compositional heterogeneity of different biochar: Effect of pyrolysis temperature and feedstocks. *Journal of Environmental Management*, *278*. Available from https://doi.org/10.1016/j.jenvman.2020.111501.

Dawar, K., Fahad, S., Jahangir, M. M. R., Munir, I., Alam, S. S., Khan, S. A., Mian, I. A., Datta, R., Saud, S., Banout, J., Adnan, M., Ahmad, M. N., Khan, A., Dewil, R., Habib-ur-Rahman, M., Ansari, M. J., & Danish, S. (2021). Biochar and urease inhibitor mitigate NH_3 and N_2O emissions and improve wheat yield in a urea fertilized alkaline soil. *Scientific Reports*, *11*(1). Available from https://doi.org/10.1038/s41598-021-96771-0.

El Nahhas, N., AlKahtani, M. D. F., Abdelaal, K. A. A., Al Husnain, L., AlGwaiz, H. I. M., Hafez, Y. M., Attia, K. A., El-Esawi, M. A., Ibrahim, M. F. M., & Elkelish, A. (2021). Biochar and jasmonic acid application attenuates antioxidative systems and improves growth, physiology, nutrient uptake and productivity of faba bean (Vicia faba L.) irrigated with saline water. *Plant Physiology and Biochemistry*, *166*, 807–817. Available from https://doi.org/10.1016/j.plaphy.2021.06.033.

Esfandbod, M., Phillips, I. R., Miller, B., Rezaei Rashti, M., Lan, Z. M., Srivastava, P., Singh, B., & Chen, C. R. (2017). Aged acidic biochar increases nitrogen retention and decreases ammonia volatilization in alkaline bauxite residue sand. *Ecological Engineering*, *98*, 157–165. Available from https://doi.org/10.1016/j.ecoleng.2016.10.077.

Fahad, S., Hussain, S., Saud, S., Hassan, S., Tanveer, M., Ihsan, M. Z., Shah, A. N., Ullah, A., Nasrullah., Khan, F., Ullah, S., Alharby, H., Nasim, W., Wu, C., & Huang, J. (2016). A combined application of biochar and phosphorus alleviates heat-induced adversities on physiological, agronomical and quality attributes of rice. *Plant*

Physiology and Biochemistry, *103*, 191–198. Available from https://doi.org/10.1016/j.plaphy.2016.03.001.

Fan, C., Duan, P., Zhang, X., Shen, H., Chen, M., & Xiong, Z. (2020). Mechanisms underlying the mitigation of both N_2O and NO emissions with field-aged biochar in an Anthrosol. *Geoderma*, *364*, 114178. Available from https://doi.org/10.1016/j.geoderma.2020.114178.

Farid, I. M., Siam, H. S., Abbas, M. H. H., Mohamed, I., Mahmoud, S. A., Tolba, M., Abbas, H. H., Yang, X., Antoniadis, V., Rinklebe, J., & Shaheen, S. M. (2022). Co-composted biochar derived from rice straw and sugarcane bagasse improved soil properties, carbon balance, and zucchini growth in a sandy soil: A trial for enhancing the health of low fertile arid soils. *Chemosphere*, *292*. Available from https://doi.org/10.1016/j.chemosphere.2021.133389.

Feng, W., Yang, F., Cen, R., Liu, J., Qu, Z., Miao, Q., & Chen, H. (2021). Effects of straw biochar application on soil temperature, available nitrogen and growth of corn. *Journal of Environmental Management*, *277*, 111331. Available from https://doi.org/10.1016/j.jenvman.2020.111331.

Feng, Y., Sun, H., Xue, L., Liu, Y., Gao, Q., Lu, K., & Yang, L. (2017). Biochar applied at an appropriate rate can avoid increasing NH3 volatilization dramatically in rice paddy soil. *Chemosphere*, *168*, 1277–1284. Available from https://doi.org/10.1016/j.chemosphere.2016.11.151.

Fermanelli, C. S., Córdoba, A., Pierella, L. B., & Saux, C. (2020). Pyrolysis and copyrolysis of three lignocellulosic biomass residues from the agro-food industry: A comparative study. *Waste Management*, *102*, 362–370. Available from https://doi.org/10.1016/j.wasman.2019.10.057.

Gao, S., DeLuca, T. H., & Cleveland, C. C. (2019). Biochar additions alter phosphorus and nitrogen availability in agricultural ecosystems: A meta-analysis. *Science of the Total Environment*, *654*, 463–472. Available from https://doi.org/10.1016/j.scitotenv.2018.11.124.

Gelová, Z., Gallei, M., Pernisová, M., Brunoud, G., Zhang, X., Glanc, M., Li, L., Michalko, J., Pavlovičová, Z., Verstraeten, I., Han, H., Hajný, J., Hauschild, R., Čovanová, M., Zwiewka, M., Hoermayer, L., Fendrych, M., Xu, T., Vernoux, T., & Friml, J. (2021). Developmental roles of auxin binding protein 1 in Arabidopsis thaliana. *Plant Science*, *303*, 110750. Available from https://doi.org/10.1016/j.plantsci.2020.110750.

Gibrilla, A., Fianko, J. R., Ganyaglo, S., Adomako, D., Anornu, G., & Zakaria, N. (2020). Nitrate contamination and source apportionment in surface and groundwater in Ghana using dual isotopes (15N and 18O-NO3) and a Bayesian isotope mixing model. *Journal of Contaminant Hydrology*, *233*. Available from https://doi.org/10.1016/j.jconhyd.2020.103658.

Guo, L., Yu, H., Kharbach, M., & Wang, J. (2021). The response of nutrient uptake, photosynthesis and yield of tomato to biochar addition under reduced nitrogen application. *Agronomy*, *11*(8), 1598. Available from https://doi.org/10.3390/agronomy11081598.

Gupta, V. V. S. R., Zhang, B., Penton, C. R., Yu, J., & Tiedje, J. M. (2019). Diazotroph diversity and nitrogen fixation in summer active perennial grasses in a Mediterranean region agricultural soil. *Frontiers in Molecular Biosciences*, *6*. Available from https://doi.org/10.3389/fmolb.2019.00115.

Hafeez, A., Pan, T., Tian, J., & Cai, K. (2022). Modified biochars and their effects on soil quality: A review. *Environments*, *9*, 60. Available from https://doi.org/10.3390/environments9050060, https://www.mdpi.com/2076-3298/9/5/60/pdf?version = 1652610423.

Hafez, E. M., Gowayed, S. M., Nehela, Y., Sakran, R. M., Rady, A. M. S., Awadalla, A., Omara, A. E. D., & Alowaiesh, B. F. (2021). Incorporated biochar-based soil amendment and exogenous glycine betaine foliar application ameliorate rice (*Oryza sativa* L.) tolerance and resilience to osmotic stress. *Plants*, *10*(9). Available from https://doi.org/10.3390/plants10091930, https://www.mdpi.com/2223-7747/10/9/1930/pdf.

Hafez, E. M., Omara, A. E. D., Alhumaydhi, F. A., & El-Esawi, M. A. (2021). Minimizing hazard impacts of soil salinity and water stress on wheat plants by soil application of vermicompost and biochar. *Physiologia Plantarum*, *172*(2), 587–602. Available from https://doi.org/10.1111/ppl.13261.

Han, S., Zeng, L., Luo, X., Xiong, X., Wen, S., Wang, B., Chen, W., & Huang, Q. (2018). Shifts in Nitrobacter- and Nitrospira-like nitrite-oxidizing bacterial communities under long-term fertilization practices. *Soil Biology and Biochemistry*, *124*, 118–125. Available from https://doi.org/10.1016/j.soilbio.2018.05.033.

Harter, J., El-Hadidi, M., Huson, D. H., Kappler, A., & Behrens, S. (2017). Soil biochar amendment affects the diversity of nosZ transcripts: Implications for N_2O formation. *Scientific Reports*, *7*(1), 3338.

He, X., Yin, H., Fang, C., Xiong, J., Han, L., Yang, Z., & Huang, G. (2021). Metagenomic and q-PCR analysis reveals the effect of powder bamboo biochar on nitrous oxide and ammonia emissions during aerobic composting. *Bioresource Technology*, *323*, 124567.

Hestrin, R., Enders, A., & Lehmann, J. (2020). Ammonia volatilization from composting with oxidized biochar. *Journal of Environmental Quality*, *49*(6), 1690–1702. Available from https://doi.org/10.1002/jeq2.20154.

Horel, Á., Gelybó, G., Potyó, I., Pokovai, K., & Bakacsi, Z. (2019). Soil nutrient dynamics and nitrogen fixation rate changes over plant growth in temperate soil. *Agronomy*, *9*(4), 179. Available from https://doi.org/10.3390/agronomy9040179.

Hou, Z., Tang, Y., Li, C., Lim, K. J., & Wang, Z. (2020). The additive effect of biochar amendment and simulated nitrogen deposition stimulates the plant height, photosynthesis and accumulation of NPK in pecan (*Carya illinoinensis*) seedlings. *AoB Plants*, *12*(4). Available from https://doi.org/10.1093/AOBPLA/PLAA035.

Hu, L., Dong, Z., Wang, Z., Xiao, L., & Zhu, B. (2022). The contributions of ammonia oxidizing bacteria and archaea to nitrification-dependent N_2O emission in alkaline and neutral purple soils. *Scientific Reports*, *12*(1). Available from https://doi.org/10.1038/s41598-022-23084-1.

Hu, T., Wei, J., Du, L., Chen, J., & Zhang, J. (2023). The effect of biochar on nitrogen availability and bacterial community in farmland. *Annals of Microbiology*, *73*(1). Available from https://doi.org/10.1186/s13213-022-01708-1.

Huang, M., Fan, L., Chen, J., Jiang, L., & Zou, Y. (2018). Continuous applications of biochar to rice: Effects on nitrogen uptake and utilization. *Scientific Reports*, *8*(1). Available from https://doi.org/10.1038/s41598-018-29877-7.

Huang, R., Gao, X., Wang, F., Xu, G., Long, Y., Wang, C., Wang, Z., & Gao, M. (2020). Effects of biochar incorporation and fertilizations on nitrogen and phosphorus losses through surface and subsurface flows in a sloping farmland of Entisol. *Agriculture, Ecosystems & Environment*, *300*, 106988. Available from https://doi.org/10.1016/j.agee.2020.106988.

Huang, R., Wang, Y., Liu, J., Li, J., Xu, G., Luo, M., Xu, C., Ci, E., & Gao, M. (2019). Variation in N_2O emission and N_2O related microbial functional genes in straw- and biochar-amended and non-amended soils. *Applied Soil Ecology*, *137*, 57–68. Available from https://doi.org/10.1016/j.apsoil.2019.01.010.

Huang, X. F., Li, S. Q., Li, S. Y., Ye, G. Y., Lu, L. J., Zhang, L., Yang, L. Y., Qian, X., & Liu, J. (2019). The effects of biochar and dredged sediments on soil structure and fertility promote the growth, photosynthetic and rhizosphere microbial diversity of *Phragmites communis* (Cav.) Trin. ex Steud. *Science of the Total Environment, 697*. Available from 10.1016/j.scitotenv.2019.134073.

Hussain, R., Ravi, K., & Garg, A. (2020). Influence of biochar on the soil water retention characteristics (SWRC): Potential application in geotechnical engineering structures. *Soil and Tillage Research, 204*, 104713. Available from https://doi.org/10.1016/j.still.2020.104713.

Igalavithana, A. D., Choi, S. W., Shang, J., Hanif, A., Dissanayake, P. D., Tsang, D. C. W., Kwon, J. H., Lee, K. B., & Ok, Y. S. (2020). Carbon dioxide capture in biochar produced from pine sawdust and paper mill sludge: Effect of porous structure and surface chemistry. *Science of the Total Environment, 739*. Available from https://doi.org/10.1016/j.scitotenv.2020.139845.

Ji, C., Li, S., Geng, Y., Miao, Y., Ding, Y., Liu, S., & Zou, J. (2020). Differential responses of soil N_2O to biochar depend on the predominant microbial pathway. *Applied Soil Ecology, 145*, 103348. Available from https://doi.org/10.1016/j.apsoil.2019.08.010.

Kang, S. W., Park, J. H., Kim, S. H., Seo, D. C., Ok, Y. S., & Cho, J. S. (2019). Establishment of optimal barley straw biochar application conditions for rice cultivation in a paddy field. *Environmental Geochemistry and Health, 41*(4), 1793–1803. Available from https://doi.org/10.1007/s10653-017-0009-4.

Karimi, A., Moezzi, A., Chorom, M., & Enayatizamir, N. (2020). Application of biochar changed the status of nutrients and biological activity in a calcareous soil. *Journal of Soil Science and Plant Nutrition, 20*(2), 450–459. Available from https://doi.org/10.1007/s42729-019-00129-5.

Khan, A. Z., Khan, S., Ayaz, T., Brusseau, M. L., Khan, M. A., Nawab, J., & Muhammad, S. (2020). Popular wood and sugarcane bagasse biochars reduced uptake of chromium and lead by lettuce from mine-contaminated soil. *Environmental Pollution, 263*. Available from https://doi.org/10.1016/j.envpol.2020.114446.

Kizito, S., Wu, S., Kipkemoi Kirui, W., Lei, M., Lu, Q., Bah, H., & Dong, R. (2015). Evaluation of slow pyrolyzed wood and rice husks biochar for adsorption of ammonium nitrogen from piggery manure anaerobic digestate slurry. *Science of The Total Environment, 505*, 102–112. Available from https://doi.org/10.1016/j.scitotenv.2014.09.096.

Kocsis, T., Ringer, M., & Biró, B. (2022). Characteristics and applications of biochar in soil–plant systems: A short review of benefits and potential drawbacks. *Applied Sciences, 12*(8). Available from https://doi.org/10.3390/app12084051, https://www.mdpi.com/2076-3417/12/8/4051/pdf.

Lai, H., Gao, F., Su, H., Zheng, P., Li, Y., & Yao, H. (2022). Nitrogen distribution and soil microbial community characteristics in a legume–cereal intercropping system: A review. *Agronomy, 12*(8), 1900. Available from https://doi.org/10.3390/agronomy12081900.

Lan, Z., Chen, C., Rezaei Rashti, M., Yang, H., & Zhang, D. (2018). High pyrolysis temperature biochars reduce nitrogen availability and nitrous oxide emissions from an acid soil. *GCB Bioenergy, 10*(12), 930–945. Available from https://doi.org/10.1111/gcbb.12529.

Li, C., Hu, H. W., Chen, Q. L., Chen, D., & He, J. Z. (2019). Comammox Nitrospira play an active role in nitrification of agricultural soils amended with nitrogen fertilizers. *Soil Biology and Biochemistry, 138*. Available from https://doi.org/10.1016/j.soilbio.2019.107609.

Li, Q., Wang, M., Fu, Q., Li, T., Liu, D., Hou, R., Li, H., Cui, S., & Ji, Y. (2020). Short-term influence of biochar on soil temperature, liquid moisture content and soybean growth in a seasonal frozen soil area. *Journal of Environmental Management*, *266*, 110609. Available from https://doi.org/10.1016/j.jenvman.2020.110609.

Liao, J., Hu, A., Zhao, Z., Liu, X., Jiang, C., & Zhang, Z. (2021). Biochar with large specific surface area recruits N_2O-reducing microbes and mitigate N_2O emission. *Soil Biology and Biochemistry*, *156*, 108212. Available from https://doi.org/10.1016/j.soilbio.2021.108212.

Liu, M., Lin, Z., Ke, X., Fan, X., Joseph, S., Taherymoosavi, S., Liu, X., Bian, R., Solaiman, Z. M., Li, L., & Pan, G. (2021). Rice seedling growth promotion by biochar varies with genotypes and application dosages. *Frontiers in Plant Science*, *12*. Available from https://doi.org/10.3389/fpls.2021.580462.

Liu, Q., Zhang, Y., Liu, B., Amonette, J. E., Lin, Z., Liu, G., Ambus, P., & Xie, Z. (2018). How does biochar influence soil N cycle? A meta-analysis. *Plant and Soil*, *426*(1-2), 211–225. Available from https://doi.org/10.1007/s11104-018-3619-4.

Liu, X., Ma, Y., Manevski, K., Andersen, M. N., Li, Y., Wei, Z., & Liu, F. (2022). Biochar and alternate wetting-drying cycles improving rhizosphere soil nutrients availability and tobacco growth by altering root growth strategy in Ferralsol and Anthrosol. *Science of the Total Environment*, *806*. Available from https://doi.org/10.1016/j.scitotenv.2021.150513.

Liu, X., Wei, Z., Ma, Y., Liu, J., & Liu, F. (2021). Effects of biochar amendment and reduced irrigation on growth, physiology, water-use efficiency and nutrients uptake of tobacco (*Nicotiana tabacum* L.) on two different soil types. *Science of the Total Environment*, *770*, 144769. Available from https://doi.org/10.1016/j.scitotenv.2020.144769.

Liu, L, Zhang, X, Xu, W, Liu, X, Li, Y, Wei, J, Gao, M, Bi, J, Lu, X, Wang, Z, & Wu, X (2020). Challenges for Global Sustainable Nitrogen Management in Agricultural Systems. *Journal of Agricultural and Food Chemistry*, *68*(11), 3354–3361. Available from https://doi.org/10.1021/acs.jafc.0c00273.

Luigi, M., Manglli, A., Dragone, I., Antonelli, M. G., Contarini, M., Speranza, S., Bertin, S., Tiberini, A., Gentili, A., Varvaro, L., Tomassoli, L., & Faggioli, F. (2022). Effects of biochar on the growth and development of tomato seedlings and on the response of tomato plants to the infection of systemic viral agents. *Frontiers in Microbiology*, *13*. Available from https://doi.org/10.3389/fmicb.2022.862075.

Mai, Y., & Huang, G. (2021). Hydrology and rainfall runoff pollutant removal performance of biochar-amended bioretention facilities based on field-scale experiments in lateritic red soil regions. *Science of the Total Environment*, *761*, 143252. Available from https://doi.org/10.1016/j.scitotenv.2020.143252.

Martínez-Gómez, Á., Poveda, J., & Escobar, C. (2022). Overview of the use of biochar from main cereals to stimulate plant growth. *Frontiers in Plant Science*, *13*. Available from https://doi.org/10.3389/fpls.2022.912264.

Mehmood, S., Ahmed, W., Ikram, M., Imtiaz, M., Mahmood, S., Tu, S., & Chen, D. (2020). Chitosan modified biochar increases soybean (*Glycine max* L.) resistance to salt-stress by augmenting root morphology, antioxidant defense mechanisms and the expression of stress-responsive genes. *Plants*, *9*(9), 1173. Available from https://doi.org/10.3390/plants9091173.

Meng, L., Sun, T., Li, M., Saleem, M., Zhang, Q., & Wang, C. (2019). Soil-applied biochar increases microbial diversity and wheat plant performance under herbicide Fomesafen stress. *Ecotoxicology and Environmental Safety*, *171*, 75–83. Available from https://doi.org/10.1016/j.ecoenv.2018.12.065.

Murtaza, G., Ahmed, Z., Eldin, S. M., Ali, B., Bawazeer, S., Usman, M., Iqbal, R., Neupane, D., Ullah, A., Khan, A., Hassan, M. U., Ali, I., & Tariq, A. (2023). Biochar-soil-plant interactions: A cross talk for sustainable agriculture under changing climate. *Frontiers in Environmental Science*, *11*. Available from https://doi.org/10.3389/fenvs.2023.1059449.

Nan, Q., Wang, C., Wang, H., Yi, Q., Liang, B., Xu, J., & Wu, W. (2020). Biochar drives microbially-mediated rice production by increasing soil carbon. *Journal of Hazardous Materials*, *387*, 121680. Available from https://doi.org/10.1016/j.jhazmat.2019.121680.

Ndiate, N. I., Saeed, Q., Haider, F. U., Liqun, C., Nkoh, J. N., & Mustafa, A. (2021). Co-application of biochar and Arbuscular mycorrhizal fungi improves salinity tolerance, growth and lipid metabolism of maize (*Zea mays* L.) in an alkaline soil. *Plants*, *10*(11). Available from https://doi.org/10.3390/plants10112490, https://www.mdpi.com/2223-7747/10/11/2490/pdf.

Omara, P., Aula, L., Oyebiyi, F. B., Eickhof, E. M., Carpenter, J., & Raun, W. R. (2020). Biochar application in combination with inorganic nitrogen improves maize grain yield, nitrogen uptake, and use efficiency in temperate soils. *Agronomy*, *10*(9). Available from https://doi.org/10.3390/agronomy10091241, https://www.mdpi.com/2073-4395/10/9/1241.

Osman, A. I., Fawzy, S., Farghali, M., El-Azazy, M., Elgarahy, A. M., Fahim, R. A., Maksoud, M. I. A. A., Ajlan, A. A., Yousry, M., Saleem, Y., & Rooney, D. W. (2022). Biochar for agronomy, animal farming, anaerobic digestion, composting, water treatment, soil remediation, construction, energy storage, and carbon sequestration: A review. *Environmental Chemistry Letters*, *20*(4), 2385–2485. Available from https://doi.org/10.1007/s10311-022-01424-x.

Pinnamaneni, S. R., Lima, I., Boone, S. A., Anapalli, S. S., & Reddy, K. N. (2023). Effect of continuous sugarcane bagasse-derived biochar application on rainfed cotton (*Gossypium hirsutum* L.) growth, yield and lint quality in the humid Mississippi Delta. *Scientific Reports*, *13*(1). Available from https://doi.org/10.1038/s41598-023-37820-8.

Pinnamaneni, S. R., Lima, I. M., Boone, S. A., Beacorn, J. A., & Bellaloui, N. (2023). Effects of pelleted sugarcane biochar applications on growth, yield and economics of rainfed corn (*Zea mays* L.) in the Mississippi Delta. *Industrial Crops and Products*, *204*. Available from https://doi.org/10.1016/j.indcrop.2023.117318.

Pokharel, P., Ma, Z., & Chang, S. X. (2020). Biochar increases soil microbial biomass with changes in extra- and intracellular enzyme activities: A global meta-analysis. *Biochar*, *2*(1), 65–79. Available from https://doi.org/10.1007/s42773-020-00039-1.

Rahman, M. A., Jahiruddin, M., Kader, M. A., Islam, M. R., & Solaiman, Z. M. (2022). Sugarcane bagasse biochar increases soil carbon sequestration and yields of maize and groundnut in charland ecosystem. *Archives of Agronomy and Soil Science*, *68*(10), 1338–1351. Available from https://doi.org/10.1080/03650340.2021.1892651.

Ramzan, S., Rasool, T., Bhat, R. A., Ahmad, P., Ashraf, I., Rashid, N., ul Shafiq, M., & Mir, I. A. (2020). Agricultural soils a trigger to nitrous oxide: A persuasive greenhouse gas and its management. *Environmental Monitoring and Assessment*, *192*(7). Available from https://doi.org/10.1007/s10661-020-08410-2.

Ramzani, P. M. A., Shan, L., Anjum, S., Khan, W. U. D., Ronggui, H., Iqbal, M., Virk, Z. A., & Kausar, S. (2017). Improved quinoa growth, physiological response, and seed nutritional quality in three soils having different stresses by the application of acidified biochar and compost. *Plant Physiology and Biochemistry*, *116*, 127–138. Available from 10.1016/j.plaphy.2017.05.003.

Safaei Khorram, M., Fatemi, A., Khan, M. A., Kiefer, R., & Jafarnia, S. (2018). Potential risk of weed outbreak by increasing biochar's application rates in slow-growth legume, lentil (*Lens culinaris* Medik. *Journal of the Science of Food and Agriculture*, *98*(6), 2080–2088. Available from https://doi.org/10.1002/jsfa.8689.

Saffari, N., Hajabbasi, M. A., Shirani, H., Mosaddeghi, M. R., & Mamedov, A. I. (2020). Biochar type and pyrolysis temperature effects on soil quality indicators and structural stability. *Journal of Environmental Management*, *261*, 110190. Available from https://doi.org/10.1016/j.jenvman.2020.110190.

Semida, W. M., Beheiry, H. R., Sétamou, M., Simpson, C. R., Abd El-Mageed, T. A., Rady, M. M., & Nelson, S. D. (2019). Biochar implications for sustainable agriculture and environment: A review. *South African Journal of Botany*, *127*, 333–347. Available from https://doi.org/10.1016/j.sajb.2019.11.015.

Serra, J., Marques-dos-Santos, C., Marinheiro, J., Cruz, S., Cameira, M. R., de Vries, W., Dalgaard, T., Hutchings, N. J., Graversgaard, M., Giannini-Kurina, F., Lassaletta, L., Sanz-Cobeña, A., Quemada, M., Aguilera, E., Medinets, S., Einarsson, R., & Garnier, J. (2024). Assessing nitrate groundwater hotspots in Europe reveals an inadequate designation of nitrate vulnerable zones. *Chemosphere*, *355*, 141830. Available from https://doi.org/10.1016/j.chemosphere.2024.141830.

She, D., Sun, X., Gamareldawla, A. H. D., Nazar, E. A., Hu, W., Edith, K., & Yu, S. (2018). Benefits of soil biochar amendments to tomato growth under saline water irrigation. *Scientific Reports*, *8*(1). Available from https://doi.org/10.1038/s41598-018-33040-7.

Shepherd, J. G., Joseph, S., Sohi, S. P., & Heal, K. V. (2017). Biochar and enhanced phosphate capture: Mapping mechanisms to functional properties. *Chemosphere*, *179*, 57–74. Available from https://doi.org/10.1016/j.chemosphere.2017.02.123.

Simiele, M., Argentino, O., Baronti, S., Scippa, G. S., Chiatante, D., Terzaghi, M., & Montagnoli, A. (2022). Biochar enhances plant growth, fruit yield, and antioxidant content of cherry tomato (*Solanum lycopersicum* L.) in a soilless substrate. *Agriculture*, *12*(8). Available from https://doi.org/10.3390/agriculture12081135.

Singh, H., Northup, B. K., Rice, C. W., & Prasad, P. V. V. (2022). Biochar applications influence soil physical and chemical properties, microbial diversity, and crop productivity: A meta-analysis. *Biochar*, *4*(1). Available from https://doi.org/10.1007/s42773-022-00138-1.

Sun, J., Rengel, Z., Zhou, Y., Li, H., & Zhang, A. (2024). Ammonia-oxidizing archaea bacteria (AOB) and comammox drive the nitrification in alkaline soil under long-term biochar and N fertilizer applications. *Applied Soil Ecology*, *193*, 105124. Available from https://doi.org/10.1016/j.apsoil.2023.105124.

Sun, P., Zhao, Z., Fan, P., Chen, W., Ruan, Y., & Wang, Q. (2021). Ammonia- and nitrite-oxidizing bacteria are dominant in nitrification of maize rhizosphere soil following combined application of biochar and chemical fertilizer. *Frontiers in Microbiology*, *12*. Available from https://doi.org/10.3389/fmicb.2021.715070.

Tian, J. H., Rao, S., Gao, Y., Lu, Y., & Cai, K. Z. (2021). Wheat straw biochar amendment suppresses tomato bacterial wilt caused by Ralstonia solanacearum: Potential effects of rhizosphere organic acids and amino acids. *Journal of Integrative Agriculture*, *20*(9), 2450–2462. Available from 10.1016/S2095-3119(20)63455-4.

US EPA. (2011). *Reactive nitrogen in the United States: An analysis of inputs, flows, consequences, and management options*. US EPA.

Wang, G., Ma, Y., Chenia, H. Y., Govinden, R., Luo, J., & Ren, G. (2020). Biochar-mediated control of phytophthora blight of pepper is closely related to the improvement of the rhizosphere fungal community. *Frontiers in Microbiology*, *11*. Available from https://doi.org/10.3389/fmicb.2020.01427.

Wang, Y., Ma, Z., Wang, X., Sun, Q., Dong, H., Wang, G., Chen, X., Yin, C., Han, Z., & Mao, Z. (2019). Effects of biochar on the growth of apple seedlings, soil enzyme activities and fungal communities in replant disease soil. *Scientia Horticulturae*, *256*, 108641. Available from https://doi.org/10.1016/j.scienta.2019.108641.

Win, K. T., Okazaki, K., Ookawa, T., Yokoyama, T., Ohwaki, Y., & Paz-Ferreiro, J. (2019). Influence of rice-husk biochar and Bacillus pumilus strain TUAT-1 on yield, biomass production, and nutrient uptake in two forage rice genotypes. *PLoS One*, *14* (7), e0220236. Available from https://doi.org/10.1371/journal.pone.0220236.

World Health Organization. (2002). *Guidelines for drinking-water quality*. World Health Organization.

Wu, Z., Zhang, Q., Zhang, X., Duan, P., Yan, X., & Xiong, Z. (2020). Biochar-enriched soil mitigated N_2O and NO emissions similarly as fresh biochar for wheat production. *Science of the Total Environment*, *701*, 134943. Available from https://doi.org/10.1016/j.scitotenv.2019.134943.

Xi, J., Li, H., Xi, J., Tan, S., Zheng, J., & Tan, Z. (2020). Effect of returning biochar from different pyrolysis temperatures and atmospheres on the growth of leaf-used lettuce. *Environmental Science and Pollution Research*, *27*(28), 35802−35813. Available from https://doi.org/10.1007/s11356-020-09840-8.

Xie, Y., Dong, C., Chen, Z., Liu, Y., Zhang, Y., Gou, P., Zhao, X., Ma, D., Kang, G., Wang, C., Zhu, Y., & Guo, T. (2021). Successive biochar amendment affected crop yield by regulating soil nitrogen functional microbes in wheat-maize rotation farmland. *Environmental Research*, *194*, 110671. Available from https://doi.org/10.1016/j.envres.2020.110671.

Xu, M., Gao, P., Yang, Z., Su, L., Wu, J., Yang, G., Zhang, X., Ma, J., Peng, H., & Xiao, Y. (2019). Biochar impacts on phosphorus cycling in rice ecosystem. *Chemosphere*, *225*, 311−319. Available from https://doi.org/10.1016/j.chemosphere.2019.03.069.

Yan, T., Xue, J., Zhou, Z., & Wu, Y. (2021). Biochar-based fertilizer amendments improve the soil microbial community structure in a karst mountainous area. *Science of the Total Environment*, *794*, 148757. Available from https://doi.org/10.1016/j.scitotenv.2021.148757.

Yang, E., Meng, J., Hu, H., Cheng, D., Zhu, C., & Chen, W. (2019). Effects of organic molecules from biochar-extracted liquor on the growth of rice seedlings. *Ecotoxicology and Environmental Safety*, *170*, 338−345. Available from https://doi.org/10.1016/j.ecoenv.2018.11.108.

Yang, X., Lu, Y., Ding, Y., Yin, X., Raza, S., & Tong, Y. 'an (2017). Optimising nitrogen fertilisation: A key to improving nitrogen-use efficiency and minimising nitrate leaching losses in an intensive wheat/maize rotation (2008−2014). *Field Crops Research*, *206*, 1−10. Available from https://doi.org/10.1016/j.fcr.2017.02.016.

Yang, Y., Kumar Awasthi, M., Wu, L., Yan, Y., & Lv, J. (2020). Microbial driving mechanism of biochar and bean dregs on NH_3 and N_2O emissions during composting. *Bioresource Technology*, *315*, 123829. Available from https://doi.org/10.1016/j.biortech.2020.123829.

Yao, Q., Liu, J., Yu, Z., Li, Y., Jin, J., Liu, X., & Wang, G. (2017). Three years of biochar amendment alters soil physiochemical properties and fungal community composition in a black soil of northeast China. *Soil Biology and Biochemistry*, *110*, 56–67. Available from https://doi.org/10.1016/j.soilbio.2017.03.005.

Yao, R., Li, H., Yang, J., Zhu, W., Yin, C., Wang, X., Xie, W., & Zhang, X. (2022). Combined application of biochar and N fertilizer shifted nitrification rate and amoA gene abundance of ammonia-oxidizing microorganisms in salt-affected anthropogenic-alluvial soil. *Applied Soil Ecology*, *171*, 104348. Available from https://doi.org/10.1016/j.apsoil.2021.104348.

Yin, D., Li, H., Wang, H., Guo, X., Wang, Z., Lv, Y., Ding, G., Jin, L., & Lan, Y. (2021). Impact of different biochars on microbial community structure in the rhizospheric soil of rice grown in albic soil. *Molecules (Basel, Switzerland)*, *26*(16), 4783. Available from https://doi.org/10.3390/molecules26164783.

Yu, H., Xie, B., Khan, R., Yan, H., & Shen, G. (2020). The changes in functional marker genes associated with nitrogen biological transformation during organic-inorganic co-composting. *Bioresource Technology*, *295*, 122197. Available from https://doi.org/10.1016/j.biortech.2019.122197.

Yuan, H., Zhang, Z., Li, M., Clough, T., Wrage-Mönnig, N., Qin, S., Ge, T., Liao, H., & Zhou, S. (2019). Biochar's role as an electron shuttle for mediating soil N_2O emissions. *Soil Biology and Biochemistry*, *133*, 94–96. Available from https://doi.org/10.1016/j.soilbio.2019.03.002.

Zhang, G., Guo, X., Zhu, Y., Han, Z., He, Q., & Zhang, F. (2017). Effect of biochar on the presence of nutrients and ryegrass growth in the soil from an abandoned indigenous coking site: The potential role of biochar in the revegetation of contaminated site. *Science of the Total Environment*, *601-602*, 469–477. Available from https://doi.org/10.1016/j.scitotenv.2017.05.218.

Zhang, H., Sun, H., Zhou, S., Bai, N., Zheng, X., Li, S., Zhang, J., & Lv, W. (2019). Effect of straw and straw biochar on the community structure and diversity of ammonia-oxidizing bacteria and archaea in rice-wheat rotation ecosystems. *Scientific Reports*, *9*(1). Available from https://doi.org/10.1038/s41598-019-45877-7.

Zhang, L., Jing, Y., Chen, C., Xiang, Y., Rezaei Rashti, M., Li, Y., Deng, Q., & Zhang, R. (2021). Effects of biochar application on soil nitrogen transformation, microbial functional genes, enzyme activity, and plant nitrogen uptake: A meta-analysis of field studies. *GCB Bioenergy*, *13*(12), 1859–1873. Available from https://doi.org/10.1111/gcbb.12898.

Zhang, X., Duan, P., Wu, Z., & Xiong, Z. (2019). Aged biochar stimulated ammonia-oxidizing archaea and bacteria-derived N_2O and NO production in an acidic vegetable soil. *Science of the Total Environment*, *687*, 433–440. Available from https://doi.org/10.1016/j.scitotenv.2019.06.128.

Zheng, Y., Han, X., Li, Y., Liu, S., Ji, J., & Tong, Y. (2020). Effects of mixed controlled release nitrogen fertilizer with rice straw biochar on rice yield and nitrogen balance in northeast China. *Scientific Reports*, *10*(1). Available from https://doi.org/10.1038/s41598-020-66300-6.

Zhou, S., Wen, X., Cao, Z., Cheng, R., Qian, Y., Mi, J., Wang, Y., Liao, X., Ma, B., & Zou, Y. (2021). Modified cornstalk biochar can reduce ammonia emissions from compost by increasing the number of ammonia-oxidizing bacteria and decreasing urease activity. *Bioresource Technology*, *319*, 124120.

CHAPTER

16 Investigation into pyrolysis impact on biochar traits, soil microbial community interaction, and nutrient dynamics: emission and leaching implications

Nur Aina Najwa Mohd Nor Azman[1], Mohd Asmadi[1] and Muzakkir Mohammad Zainol[2]

[1]*Chemical Reaction Engineering Group (CREG), Faculty of Chemical and Energy Engineering, Universiti Teknologi Malaysia, Johor Bahru, Malaysia*

[2]*School of Chemical Engineering, College of Engineering, Universiti Teknologi MARA, Shah Alam, Selangor, Malaysia*

16.1 Introduction

The massive production of biomass material arising from poor biomass waste management has created numerous disposal problems and governance issues. A majority of biomass wastes is left in the field to decompose naturally, discarded in landfills, or incinerated for energy supply. The lack of feasible and efficient approaches to recycling waste resources can lead to severe environmental pollution with greenhouse gas emissions and air quality deterioration. Since biodegradable biomass serves as a significant source of renewable organic matter, much interest and research have been encouraged to utilize this material using sustainable technologies. Thus, pyrolysis has emerged as one of the most efficient thermochemical technologies to recycle and recover biomass waste into value-added products. The pyrolysis processing method involves the thermal degradation of the biomass material in an oxygen-deficient condition and ambient pressure at a wide range of temperatures 400°C–1200°C producing biofuel products namely syngas, bio-oil, and biochar (Foong et al., 2020).

Biochar is a black, solid, and carbon-rich material produced from the pyrolysis process and has become a novel approach due to its wide range of agricultural benefits. Recent studies have elucidated that the high stability and biodegradability of biochar can be preserved for many years (> 1000 years) in the soil showing the high resistance of material to mineralization under abiotic and biotic stresses

Biochar Ecotechnology for Sustainable Agriculture and Environment. DOI: https://doi.org/10.1016/B978-0-443-29855-4.00018-7

(Tsolis & Barouchas, 2023). The characterized biochar products have a high porosity, pH, cation exchange capacity (CEC), and carbon content as illustrated in Fig. 16.1 to improve soil physicochemical properties. In general, biochar has a pH ranging from neutral to alkaline that is effectively used to ameliorate soil acidification by decreasing the hydrogen ions concentration (Geng et al., 2022). As the soil pH increases, the number of negative charges on the colloids increases thereby increasing CEC. Soils with a high CEC value are good indicators for soil rich in organic matter which also have a much lower percentage of cation in the soil water to prevent nutrient loss by leaching. The highly porous structure of biochar also contains an amount of extractable humic-like and fluvic-like substances with longitudinal pores of sizes ranging from micro- to macropores that can provide habitats for symbiotic microorganisms (Jatav et al., 2021). The addition of biochar interacts with soil particles to influence soil microbial communities by changing microhabitats and nutrient cycling. Hence, understanding the role of biochar traits on the responses of microbial communities is crucial to improve soil nutrients and fertility.

Soil microbial communities are highly complex ecological networks that serve in regulating nutrient cycling, carbon mineralization, and stabilization. A healthy soil microbial community is characterized by a large diversity of microbes which includes a different kinds range of unicellular organisms, bacteria, archaea, protists, and fungi (Chauhan et al., 2023). Multiple microbial interactions with plant roots can be enhanced with biochar addition owing to its soil improvement capabilities to increase soil pH, CEC, pore space, carbon, and water content. Moreover, the incorporation of biochar substrate imparts synergistic interactions

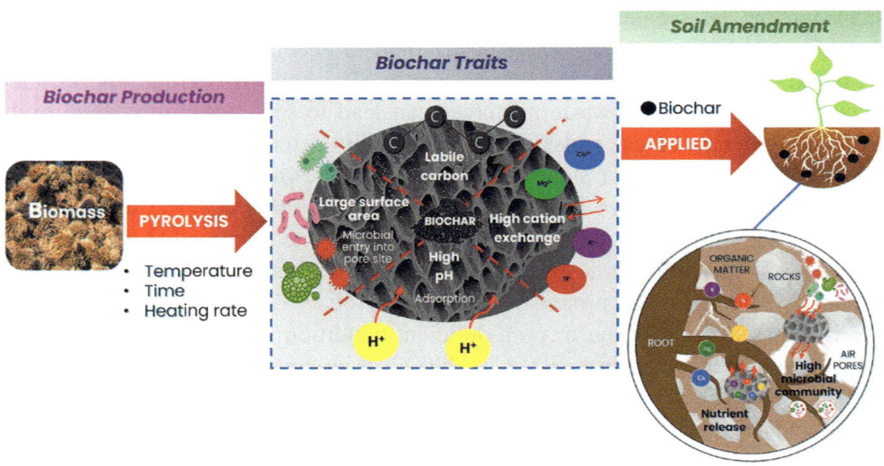

FIGURE 16.1 Direct effect of biochar amendment on microbial community and nutrient dynamics.

Biochar traits for a soil amendment application.

with microorganisms providing nutrients, nurturing colonization, and habitats while reducing contaminant toxicity (Li et al., 2023). However, the performance of biochar to enhance microbial community interaction varies depending on the produced biochar traits which require further study. Therefore this review emphasized the influence of feedstock and pyrolysis conditions on biochar production. The effect of biochar traits namely surface area, porosity, pH, CEC, and labile carbon on microbial community and nutrient dynamics was focused. The application of biochar as a soil amendment in Malaysia also was briefly discussed. Accordingly, a deep understanding of biomass selection and controlling pyrolysis conditions may help researchers produce high-quality biochar with specific traits to increase microbial activity and nutrient content for plant growth.

16.2 Effects of biomass feedstock and pyrolysis parameter on biochar traits

The selection of biomass feedstock and varied pyrolysis conditions influence the physical and chemical traits of the biochar formed. The characterization of biochar produced must satisfy the European Biochar Certificate (EBC) or International Biochar Initiative (IBI) guidelines to guarantee the sustainability and quality of biochar (EBC, 2012; IBI, 2013). The specific application of biochar may also depend on its physical and chemical characteristics. For such, a high degree of aromaticity and stability of biochar is favorable as a precursor for activated carbon and adsorbent application (Haeldermans et al., 2019). Meanwhile, biochar with a higher surface area ($>100\ m^2/g$) can be applied as a soil additive (Schimmelpfennig & Glaser, 2012). Hence, further studies on the formation of biochar under pyrolysis heating were focused. The pyrolysis operating conditions including pyrolysis temperature, time, and heating rate from different biomass feedstock were discussed to determine the optimum biochar production as a soil amendment.

16.2.1 Feedstock

Different biomass sources contain different polymeric components, primarily cellulose, hemicellulose, and lignin, which result in different biochar traits. Therefore the selection of biomass feedstock for biochar production was focused on classified biomass type: (1) woody biomass (tree parts, wood, and sawdust); (2) manure/waste biomass (sewage sludge); (3) agricultural and herbaceous wastes (straws, husks, shells, seeds); and (4) algae biomass (microalgae, macroalgae). Fig. 16.2 illustrates the morphology of Japanese cedar, poultry manure, sugarcane bagasse, and microalgae biochar pyrolyzed at 450°C (Kameyama et al., 2019; Michalak et al., 2019). The variability in the morphology of biochar structures was observed when different initial feedstock was used. The tubular

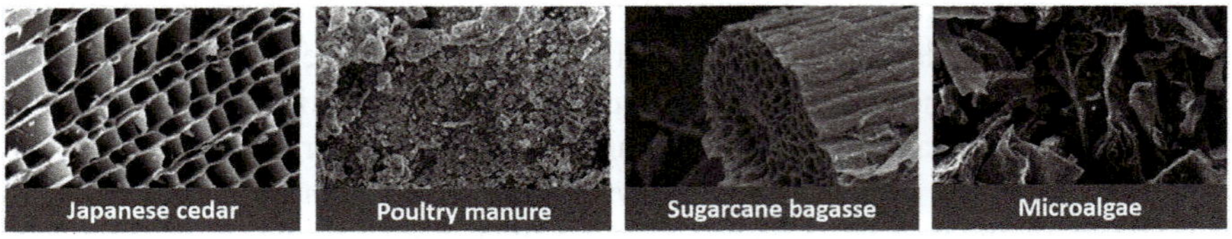

FIGURE 16.2 Scanning electron microscopy images of biochar produced from various biomass feedstock (Kameyama et al., 2019; Michalak et al., 2019).

Morphology of Japanese cedar, poultry manure, sugarcane bagasse, and microalgae biochar pyrolyzed at 450°C.

From Kameyama, K., Miyamoto, T., & Iwata, Y. (2019). The preliminary study of water-retention related properties of biochar produced from various feedstock at different pyrolysis temperatures. Materials, 12*(11). https://doi.org/10.3390/ma12111732. Michalak, I., Baśladyńska, S., Mokrzycki, J., & Rutkowski, P. (2019). Biochar from a freshwater macroalga as a potential biosorbent for wastewater treatment.* Water, 11*(7). https://doi.org/10.3390/w11071390.*

structure of lignocellulosic biomass such as Japanese cedar and sugarcane bagasse precisely favors microporosity development as compared to poultry manure and microalgae biochar. The structure of microalgae biochar shows an irregular shape structure covered with high roughness. Meanwhile, the poultry manure biochar appeared as lumps encrusted with fine and coarse powder. The high inorganic minerals (ash) presence in manure influenced the blockage of micropores resulting in the lower porosity and surface area of biochar. A meta-data analysis conducted by Ippolito et al. (2020) verified the high ash content of 44.6% in manure biochar produced a lower surface area of 52.2 m^2/g as compared to wood (184 m^2/g), crop waste (98.2 m^2/g), and grass (63.2 m^2/g) biochar. It was speculated the high fusion of molten ash can fill up pores of biochar, thereby reducing accessible surface area (Tomczyk et al., 2020). Meanwhile, the wood–biochar was observed to produce the lowest ash content of 10.2% than crop (21.1%) and grass (18%) biochar. This may be due to the high lignin content in wood material (20%–30%) which influences the preservation of the pore structure by the stability of lignin (Leng et al., 2021). Overall, the degradation mechanism of biomass polymers into biochar has important guiding significance for pore development.

The pyrolysis of the biomass material with a high ash content produces a biochar with a higher CEC and pH value. Ash acts as a heat-resistant component to protect organic compounds against degradation and hinder the formation of aromatic structures as charring intensity advances (Domingues et al., 2017). A study by Domingues et al. (2020) found a wide range of ash contents among the biochar samples, ranging from 1.1% to 56% in order of chicken manure (CM) > coffee husk (CH) > sugarcane bagasse (SB) > eucalyptus sawdust (ES) at 750°C. Consequently, the ash-rich feedstock of CM resulted in a higher CEC value of 21.3 cmol/kg than the biochar produced from CH (18.9 cmol/kg), SB (1.3 cmol/kg), and ES (1.4 cmol/kg). Available data suggest the concentrated potassium (K) element in the ash fraction of the CM materials stimulates carbonization reactions that may lead to higher thermal stability of carbon atoms during the pyrolysis process. Tag et al. (2016) also measured the high presence of ash content up to 40.0% in algal feedstock influenced the production of biochar with a higher CEC value of 57.5 cmol/kg. The concentrated alkali and alkali metal in algal material promoted the anion potential of the surface functional groups that occurred during the pyrolysis process, including hydroxyl (−OH), carboxyl (−COOH), ketone (−C(¼O) R), ester (−C(¼O) OR), aldehydes (−C(¼O) H), and amino ($-NH_2$) increasing the CEC value (Shaaban et al., 2013). Meanwhile, the orange pomace has the lowest CEC value of 29.9 cmol/kg and ash content of 14.2% due to the reduction of the oxygenated functional groups content on the biochar surface. Additionally, the ash contents also significantly influenced the pH properties of biochar. Li et al. (2023) found wood biochar had four to five times less ash content resulting in a lower pH of 8.83 in comparison to the corn stover (10.13) and switchgrass (9.00) biochar. A similar finding was reported by Wijitkosum (2022) with the lower pH of wood residue biochar derived from krachid (7.38) and rain tree (7.32) over corncob biochar (9.54). The increase in the pH value could be attributed to the concentration of non-pyrolyzed inorganic elements and the decomposition of the organic matrix (Garcia-Perez, 2008). The higher alkali salt content in corncob materials may also lead to the formation of anionic functional groups, while the acidic function groups are removed. Hence, corncob biochar is highly alkaline. The biochar CEC and pH are dependent on the feedstock selected. The complexity and diversity of components in biomass can result in biochar with varied traits.

16.2.2 Pyrolysis temperature

The pyrolysis temperature has a prominent effect on biochar traits despite the variety of initial biomass feedstock used. During the pyrolysis process, the release of volatile from the biochar's pore was accelerated with rising pyrolysis temperature, influencing the final biochar structure. Thus most studies mainly focused on biochar traits specifically based on high and low pyrolysis temperature principles at 500°C. Fig. 16.3 summarizes the scanning electron microscopy (SEM) micrographs of biochar derived from yak manure, rice straw, and palm fiber at pyrolysis temperatures of 300°C, 500°C, and 700°C (Phuong et al., 2016; Selvarajoo & Oochit, 2020; Zhang et al., 2018). The variation of shapes and sizes of pores structures was observed when the different pyrolysis temperature was applied to the biochar sample. Nonetheless, all biochar samples exhibit a similar trend with flat surfaces and no visible pore was produced at 300°C. The rising pyrolysis temperature from 300°C to 500°C and 700°C developed the porosity of biochar, increasing its surface area. Zhang et al. (2018) reported the surface area of yak manure biochar increased from 3.5 m^2/g (300°C) to 17.3 m^2/g (500°C) and 82.9 m^2/g (700°C). Similarly, Phuong et al. (2016) analyzed that straw biochar pyrolyzed at 700°C has a surface area 7.3 times larger than that biochar obtained at 300°C. The low pyrolysis temperature of 300°C may not be sufficient to induce

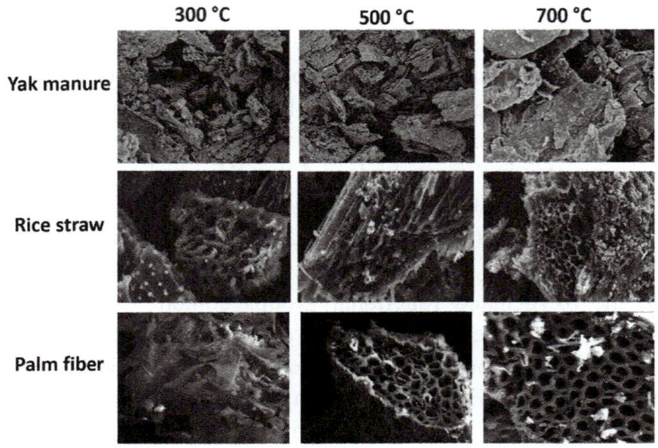

FIGURE 16.3

Scanning electron microscopy images of biochar produced from yak manure, rice straw, and palm fiber at pyrolysis temperature of 300°C, 500°C, and 700°C (Phuong et al., 2016; Selvarajoo & Oochit, 2020; Zhang et al., 2018).

a vigorous pyrolysis reaction that would trap more volatile constituents. allowing them to condense on the biochar pore's structure and hinder new pores from forming. Meanwhile, the rapid heating at a temperature >500°C releases volatile matter and pore-blocking substance on the surface of biochar increasing the externally accessible surface area (Tomczyk et al., 2020). However, operating the pyrolysis process beyond 700°C potentially shrinks and closes the pores of biochar (Phuong et al., 2016; Selvarajoo & Oochit, 2020). The increase of pyrolysis temperature to a sufficiently high value can lead to the dominant phenomenon "thermal deactivation", which influences the deformation or melting of biochar structures (Leng et al., 2021). To conclude, 500°C–700°C is an adequate pyrolysis temperature range for most biomass materials.

The pH and CEC of biochar varied considerably regulated by the pyrolysis temperature. (Domingues et al., 2017) assessed the CEC of CH biochar decreased as pyrolysis temperature increased. According to the Fourier-transform infrared spectra shown in Fig. 16.4A, the majority of the organic group assignments and bands responsible for generating negative charges were reduced at 750°C, indicating the removal of oxygen-containing functional groups from the biochar. Thus, a sharp decrease in CEC was observed from 450°C (72.0 cmol/kg) to 750°C (19 cmol/kg). However, the rising pyrolysis temperature will lead to higher ash content of up to 19% and total carbonate content increasing the pH of produced biochar. As the enrichment of basic cations in the ashes may be associated with alkaline species, such as carbonates, oxides, and hydroxides. The X-ray diffraction analysis in Fig. 16.4B shows the presence of kalicinite ($KHCO_3$) at 750°C was identified at 3.67, 2.97, 2.86, 2.63, and 2.37 Å. These peaks are sharp and

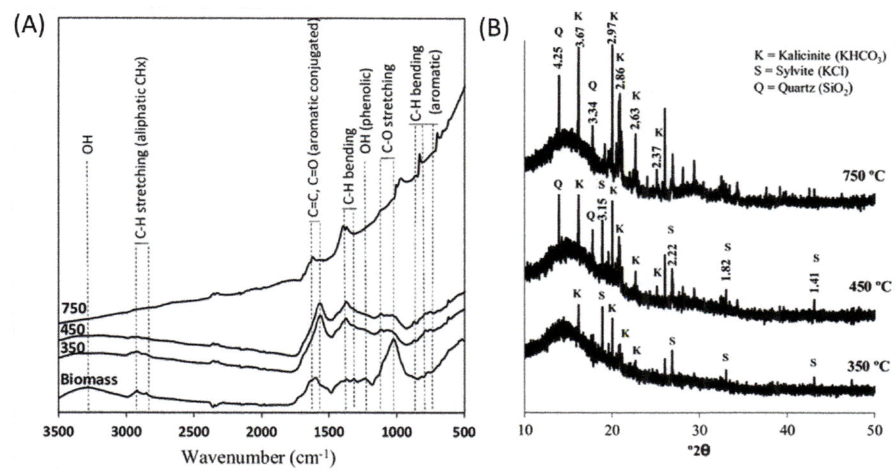

FIGURE 16.4

(A) Fourier-transform infrared spectra and (B) X-ray diffraction spectra of coffee husk biochar pyrolyzed at different temperatures of 350°C, 450°C, and 750°C (Domingues et al., 2017).

From Domingues, R. R., Trugilho, P. F., Silva, C. A., De Melo, I. C. N. A., Melo, L. C. A., Magriotis, Z. M., & Sánchez-Monedero, M. A. (2017). Properties of biochar derived from wood and high-nutrient biomasses with the aim of agronomic and environmental benefits. PLoS ONE, *12(5). https://doi.org/10.1371/journal.pone.0176884.*

intense indicating the presence of more alkaline crystalline minerals resulting in the high pH of biochar was produced. Similarly, Sun et al. (2017) reported the pH of plantain wood biochar had increased from 7.57 to 9.64 as the temperature rose from 300°C to 400°C. Askeland et al. (2019) and Song et al. (2023) also reported a higher pyrolysis temperature tends to favor the alkalinity of biochar due to an increase in inorganic carbonates. Evidently, the alkali metals became separated from the organic matrix at higher pyrolysis temperatures, resulting in biochar enriched with alkali metal salts.

16.2.3 Pyrolysis time

The pyrolysis process of biomass conversion occurs over shorter or longer durations depending on the pyrolysis mode used. The minimal effect of pyrolysis time on biochar pH and CEC was reported by Sun et al. (2017). Thus most studies have evaluated the contribution of varied pyrolysis time on biochar traits particularly on surface area and porosity of biochar. Wang et al. (2019) performed the co-pyrolysis of cotton stalks and sewage sludge in an electrical heating furnace at 600°C, with different pyrolysis times ranging from 30 to 150 minutes. It was observed the specific surface area of the biochar was increased as pyrolysis time increased from 30 minutes (9.17 m^2/g) to 90 minutes (31.33 m^2/g) with enhanced development pore volume from 0.040 cm^3/g to 0.065 cm^3/g. However, retaining the biochar for 150 minutes has reduced the surface area and pore volume biochar to 21.48 m^2/g and 0.056 cm^3/g. A similar result was obtained by Zhao et al. (2018), who found the increasing surface area of rapeseed biochar from 46.7 m^2/g to 98.4 m^2/g as pyrolysis time was extended from 10 minutes to 60 minutes. After 100 minutes pyrolysis process, the surface area of biochar decreased to 91.4 m^2/g. Leng et al. (2021) speculated the trend of an increase first then a decrease occurred if the optimal pyrolysis time has been selected with maximum surface area and pore volume. With sufficient reaction time, the release of volatile matter may be accelerated, boosting the development of rudimentary pore structures. However, excessive prolonged exposure to high temperatures may destroy the pore structure. Hence, the pyrolysis time is usually selected in the range of 5–60 minutes depending on the pyrolysis mode.

16.2.4 Heating rate

The heating rate of the pyrolysis process greatly affected the traits of biochar produced particularly on morphology and porosity properties. Shagali et al. (2021) found the limited distinct pore openings and tiny pores on the surface of walnut shell biochar produced at 10°C/min and 20°C/min as presented in Fig. 16.5A.

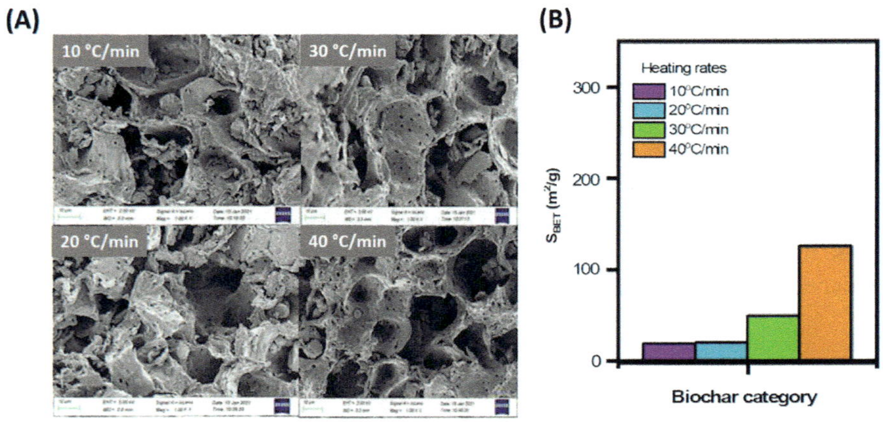

FIGURE 16.5

Structural analysis of biochar. (A) Scanning electron micrographs and (B) Brunauer–Emmett–Teller-specific surface area analysis of walnut shell biochar at varied heating rates (Shagali et al., 2021).

From Shagali, A. A., Hu, S., Wang, Y., Li, H., Wang, Y., Su, S., & Xiang, J. (2021). Comparative study on one-step pyrolysis activation of walnut shells to biochar at different heating rates. Energy Reports, *7, 388–396. https://doi.org/10.1016/j.egyr.2021.10.021.*

As the heating rate increased from 20°C/min to 30°C/min and 40°C/min, the porosity development in biochar sample was observed with a honeycomb-like appearance. This implies that higher heating rates affect the destruction of walnut shell cell structures by releasing more volatile matter and creating more pores. The Brunauer–Emmett–Teller results in Fig. 16.5B verified the fast-heating rates produced higher surface area and pore characteristics. The higher specific surface areas of biochar pyrolyzed at 40°C/min (126.50 m^2/g) than at 10°C/min (19.3 m^2/g). Zhao et al. (2018) also reported the surface area of biochar increased from 97.4 m^2/g to 115.6 m^2/g as the heating rate rose from 1°C/min to 20°C/min. Meanwhile, Zeng et al. (2015) analyzed the effect of heating rate was significantly correlated to pyrolysis temperature. The distinct biochar traits were more remarkable at a higher temperature with varied heating rates. The beech wood was slowly pyrolyzed in the Pyrex reactor at a temperature of 1200°C and 2000°C with heating rates from 5°C/s to 450°C/s. At 1200°C, the surface area and pore volume of biochar increased about 1.27 and 1.15 times when the heating rate increased from 50°C/s to 150°C/s and decreased about 1.1 and 1.2 times when the heating rate increased to 450°C/s. For comparison, the pyrolyzed biochar at 2000°C shows an increased surface area and pore volume by 3 and 2.5 times from 50°C/s to 150°C/s. Then, both surface area and pore volume decreased about 2 and 1.6 times when the heating rate reached 450°C/s. Thermal deactivation over 900°C may have caused the char to melt and pores to fuse, which led to a reduced surface area and higher pyrolysis temperature. A short diffusion time of volatiles at higher temperatures also causes the volatiles to accumulate between and within particles. Consequently, this accumulation of volatiles may lead to carbon deposits blocking pore entrances.

16.3 Biochar traits on soil microbiology community interaction and nutrient dynamics

Soil microorganism is an important functional group of soil organisms involved in many biogeochemical processes. It promotes the soil microcirculation ecosystem by increasing soil nutrient cycling, fertility maintenance, and crop production (Chauhan et al., 2023; Zhang & Shen, 2022). Soil microbial community also can be a key indicator to evaluate soil health and productivity. Several studies have emphasized that soil microbial communities are highly responsive to biochar amendment (Dai et al., 2017; Jaafar et al., 2015; Liu et al., 2022). The highly developed pore structure and specific surface area provide habitat and colonization of soil microorganisms increasing their activities and abundance. Besides, the application of biochar may directly or indirectly influence soil microbial status by changing the soil's physicochemical properties. However, the response of soil microbial communities to biochar is highly variable and influenced by the type and traits of biochar. Thus the correlation between the soil microbial community, nutrient dynamics, and biochar traits was further explored. Microbial activities were evaluated based on biochar's porosity, pH, CEC, and labile carbon.

16.3.1 Biochar surface area and porosity

Biochar exhibits porous aromatic structures with high surface areas to adsorb water, labile carbon, and mineral nutrients from adjacent soil. The large macropore structure of biochar can protect less competitive microorganisms from predators namely protists (Dai et al., 2021). Biochar particles also serve as a habitat for microbes to colonize, grow, and reproduce. The further growth and reproduction of microbes can promote the nutrient content. Jaafar et al. (2015) analyzed the rich microbial colonization on the porous structure of Simcoa biochar via fluorescence and SEM technique after 56 days of incubation in soil (Fig. 16.6A and B). It was observed these Hyphal networks were attached to the wall of larger biochar pores about 100 μm diameter. Meanwhile, smaller pores of biochar (20 μm diameter) in Fig. 16.6C and D were clogged by soil particles limiting the observation of fungal hyphae. Once the biochar particles are deposited in the soil, soil clumping and attachment may occur to the biochar particle by blocking smaller pores and altering the surface area and pore availability for microbial colonization. According to Quilliam et al. (2013), biochar pore diameter should ideally be two

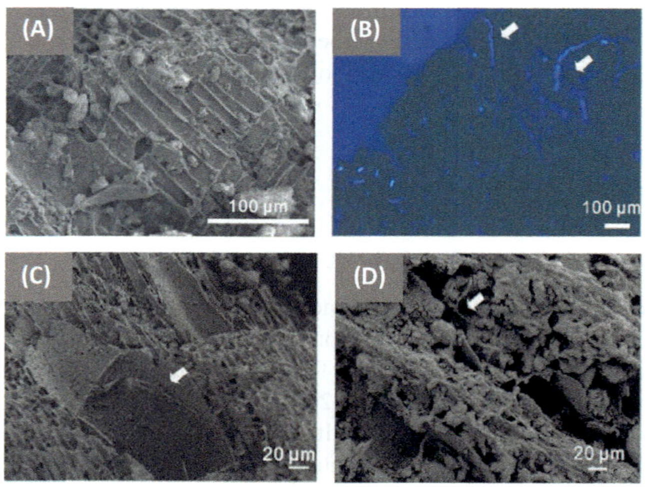

FIGURE 16.6

Scanning electron micrographs (A, C, and D) and fluorescent micrographs (B) of incubated (56 days) and colonized Simcoa biochar particles with soil particles and fungal networks (arrow) on biochar external surfaces (Jaafar et al., 2015).

From Jaafar, N. M., Clode, P. L., & Abbott, L. K. (2015). Soil microbial responses to biochars varying in particle size, surface and pore properties. Pedosphere, *25(5), 770–780. https://doi.org/10.1016/S1002-0160(15)30058-8.*

to five times more than the cell size of microorganisms for maximum microbial adsorption, and pore sizes of <1 μm (≥20% of total biochar pore volume) are considered uninhabitable for both bacteria and fungi. The high surface area of biochar is also crucial to provide more active sites for the growth and reproduction of microorganisms as well as nutrient retention. (Zhao et al., 2023) reported the amendment of biochar prepared at 650°C with the highest surface area of 46.12 m^2/g has increased soil nitrate and available potassium by 97.78% and 68.21%. It was deduced the available nutrients have increased attributed to the larger specific surface area and a greater number of micropores in biochar prepared at the higher temperature, leading to a stronger adsorption effect. The increase in microbial abundance particularly MND1 and mle1−7 (Nitrosomonadaceae), Arthrobacter and Bacillus also involved in the cycling and transformation of nitrogen and phosphorus in the soil. The higher porosity of biochar has a positive effect on soil microbial activity. Therefore the range of porosity of biochar must be explored to promote the survival and metabolism of microorganisms.

16.3.2 pH

Biochar is alkaline in nature (pH >7) with the level of alkalinity dependent on the feedstock source and pyrolysis conditions. Biochar is effectively applied as a liming material to reduce soil acidity creating a favorable environment where plants can thrive. The soil acidity neutralization mechanism involves the ability of biochar particles to absorb the H^+ ions. The changes in soil pH consequently influence the microbial biomass, activity, and community structure. Dai et al. (2017) found the manure biochar produced at 300°C with a pH value of 7.2 significantly increased the dissolved organic carbon content in comparison to the biochar amended soil pyrolyzed at 700°C (pH of 9.6) and unamended soil as summarized in Fig. 16.7C. Besides, the SEM spectra in Fig. 16.7B also revealed the community structure of bacteria dominantly by 71.8% and 55.7% of Actinobacteria extracted from Psammaquent (Ps) and Argiustoll (Ar) soil after 240-day incubation with biochar addition at 300°C. Meanwhile, biochar amendment at 700°C yields a small number of Actinobacteria of 40% and 33.2% in both Ps and Ar soil. The abundant growth of Actinobacteria at 300°C could be attributed to the easily mineralizable C than in high pyrolysis temperature of 700°C, which had more fused aromatic C as verified by the nuclear magnetic resonance (NMR) spectra in Fig. 16.7A. The results also revealed appropriate pH environment for microbial growth appeared in soils with near-neutral pHs of biochar. Contrarily, Wang et al. (2022) found that the high pH value of peanut shell biochar amendment at 700°C can increase the genus of bacteria that are relevant to nitrogen cycling, such as *Nitrospira*, *Nitrosotaleaceae*, and *Candidatus nitrosotalea* compared to other biochar condition of 300°C and 500°C. This indicates that different microbial communities thrive in different conditions, such as acidophilic or basophilic. For instance, acidophilic microbes tend to diminish with

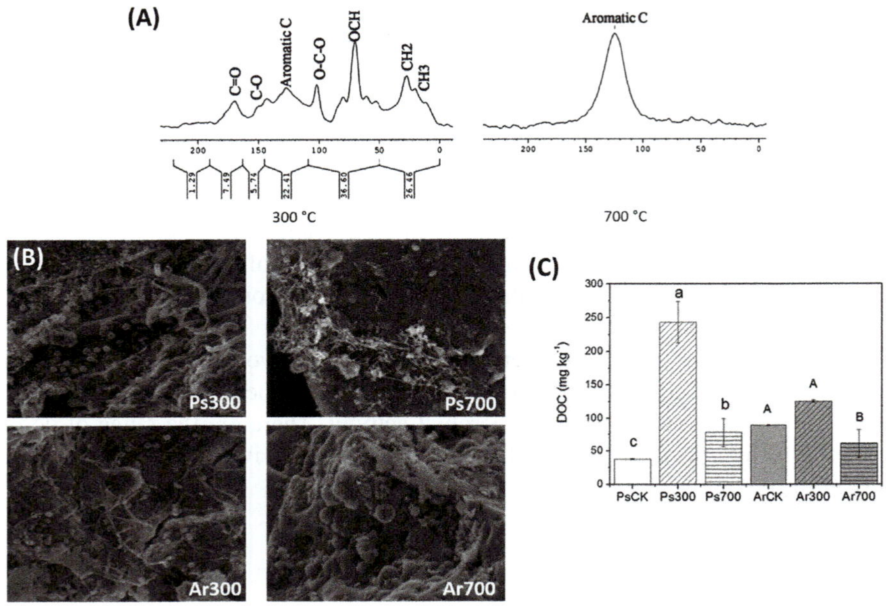

FIGURE 16.7

Microbial biomass and activity parameters of biochar produced at 300°C and 700°C. (A) Nuclear magnetic resonance spectra, (B) scanning electron micrograph, and (C) dissolved organic carbon concentrations. SEM images were taken at the end of the incubation experiment to show the microbial colonization in biochar particles that were extracted from soils (Dai et al., 2017).

From Dai, Z., Barberán, A., Li, Y., Brookes, P. C., & Xu, J. (2017). Bacterial community composition associated with pyrogenic organic matter (biochar) varies with pyrolysis temperature and colonization environment. mSphere, *2(2). https://doi.org/10.1128/mSphere.00085-17.*

increased soil alkalinity (Sheng & Zhu, 2018). Hence, the appropriate pH of biochar amendment must be identified to regulate the soils' pH accordance with the availability of soil microbes.

16.3.3 Cation exchange capacity

The CEC of biochar measures the ability to hold soil exchangeable cations such as Mg^{2+}, Ca^{2+}, K^{+}, and Na^{+}. The biochar incorporation may provide stronger adsorption capacity increasing the soil's adsorption capacity for mineral elements and the value of CEC. As CEC increases, more nutrients are bound to biochar and soil particles, which can then be transferred to the root surface for plant uptake. Zhang et al. (2023) found that the amendment of soybean straw biochar to the degraded black soil has positively increased soil CEC by 26.18%–43.56% much higher than the unamended soil. The increased soil CEC promoted the stability of soil aggregates and regulated the abundance of soil nutrient conversion-related microorganisms' activity. The functional growth of microbes in carbon fixation (Firmicutes and Bacteroidetes) and nitrification (Proteobacteria) was enriched after biochar addition, promoting C and N accumulation. The previous studies also focus on the influence of the pyrolysis process on the variability of biochar CEC produced that may impact the soil microbial community and nutrient dynamics. Tang et al. (2020) analyzed the biochar amendment at 400°C increased soil CEC from 15.44 cmol/kg to 22.99 cmol/kg, enhanced microbial activity, and produceded a higher available K content of 3327.20 mg/kg in comparison to the application of biochar at 600°C. Similarly, Domingues et al. (2020) evaluated the high CEC value of CH biochar obtained at 350°C (69.7 cmol/kg) than the biochar produced at 750°C (18.9 cmol/kg). After 9 months of incubation, a 20% rate of biochar at 350°C has significantly doubled the CEC value from 19.1 cmol/kg to 40.4 cmol/kg in Red Latosal (clayey texture) soil and by 10-fold from 2.3 to 23.1 Red-Yellow Latosol (medium texture) over the unamended and amended biochar pyrolyzed at 750°C. Approximately 75% of the original cellulose mass remains in matrices heated at temperatures up to 350°C, stimulating a greater microbial activity, potentially leading to the oxidation and formation of new functional groups in the structure of the biochar incubated in soil. The results show that soil CEC may relies on biochar CEC to increase microbial activity and retain more nutrients. The high soil CEC also can prevent the rapid drainage of soil water that can lead to the loss of nutrients beyond the root zone. Thus the high biochar CEC was preferred to sustain the soil nutrients.

16.3.4 Labile carbon

Biochar is highly stable, comprising more than 65% carbon (C) (Wijitkosum & Jiwnok, 2019). However, the varying forms of biochar structure yield a varied biochar component such as labile C (easy mineralizable C) or aromatic C/backbone (persistence C) that may result in different responses of the microbial community. A meta-analysis conducted by Dai et al. (2021) overviews the significant effects of labile C compared and aromatic C that can serve as soil C substrate. Since the labile C can directly provide available energy sources such as carbohydrates for microbial metabolism, growth, and reproduction. Liu et al. (2022) extracted the major fraction of labile and aromatic C in manure biochar using anhydrous acetone to analyze the effect of biochar structure on the microbial community. From Fig. 16.8A and B, the NMR of labile C of biochar was composed of aliphatic C such as alkyl and carbonyl, while aromatic C was mainly composed of aromatic rings. The 16S rRNA genes analysis shows that the microbial abundance of specialists responsive to labile C was 3.3 times larger than those responsive to the aromatic backbone, indicating the preferences of labile C specialists for energy sources. The Myxococcaceae and Flavisolibacter species are highly responsive to aromatic C as depicted in Fig. 16.8D. Meanwhile, the highest abundance of Ktedonobacteraceae and Bradyrhizobiaceae in Fig. 16.8C was identified in labile C. These species are effective organic matter decomposers

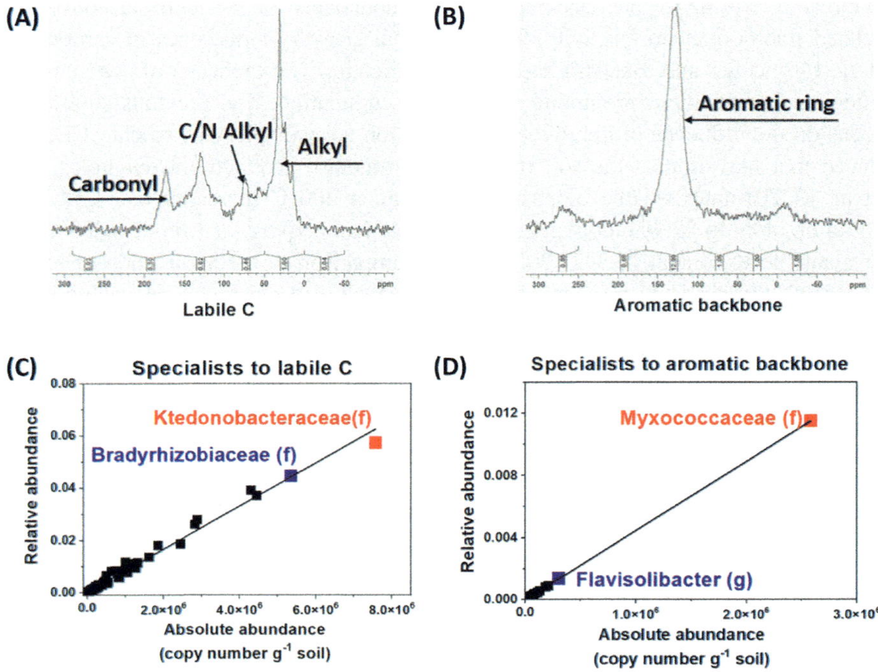

FIGURE 16.8

Carbon structure by nuclear magnetic resonance spectra (A and B) and (C and D) bacterial generalist in biochar materials and the specialist in the groups of labile C and aromatic backbone (Liu et al., 2022).

From Liu, H., Wang, X., Song, X., Leng, P., Li, J., Mazza Rodrigues, J. L., Hong, Z., Kuzyakov, Y., Xu, J., & Dai, Z. (2022). Generalists and specialists decomposing labile and aromatic biochar compounds and sequestering carbon in soil. Geoderma, 428. *https://doi.org/10.1016/j.geoderma.2022.116176.*

and have a good N_2 fixing capacity (Itakura et al., 2009). As a result, labile C has increased the respiration rate and verified enhanced microbial activity in the soil. Dai et al. (2019) also found that labile C in manure-based biochar produced higher microbial N cycling genes in comparison to the aromatic C and unamended soil. The rapid increases in microbial activity could favor denitrification by consuming O_2 and forming localized sites of anaerobiosis. This implies that the effects of labile C are crucial to increasing microbial activity.

16.4 Case study of biochar application in Malaysia

The commercialization of biochar in Malaysia is relatively new and still at an early stage. Previous studies have focused on biochar production from varied biomass materials using conventional (CV) heating method conditions (Bakhtiar et al., 2019; Noor et al., 2019; Shariff et al., 2016). The biochar from CV has been applied on tropical soil to maintain good soil pH, improve plant growth, and increase crop yield. However, the CV took a longer pyrolysis time to complete the pyrolysis process compared to the fast pyrolysis method. The fast pyrolysis method such as microwave-assisted pyrolysis (MWP) was performed with a higher heating rate and shorter pyrolysis time producing nearly the equivalent yield of biochar produced from the CV process. The microwave heating mechanism consists of an efficient volumetric heating system that can generate rapid heating within heated material from the core to the outer surface of the biomass sample (Miura et al., 2004). The enhanced biochar properties include a higher surface area and well-developed pores structure also can be produced from the MWP process (Halim et al., 2022). In Malaysia, the lack of literature regarding biomass decomposition under microwave heating shows the clear research gap present to assess the potential of microwaved biochar as a soil amender.

Recently, we evaluated the production of biochar from empty fruit bunch (EFB) material using different pyrolysis technologies of MWP and CV process at varied pyrolysis times. The MWP of EFB was performed at the desired power of 450 W with 1.25 g coconut shell–activated carbon addition and various irradiation times (5–20 minutes) under 1 L/min N_2 flow. Meanwhile, the biochar was prepared from CV pyrolysis under the following operating conditions: heating rate of 10°C/min; temperature of 550°C; N_2 gas of 1 L/min; and pyrolysis time at 5–120 minutes. The range of operating pyrolysis conditions was determined based on literature studies of biochar production using the MWP and CV heating methods (Azman et al., 2022; Haeldermans et al., 2019). Then, the selected biochar was characterized with N_2 adsorption–desorption isotherm at 77 K and field emission scanning electron microscopy analysis. It was observed the cross-sectional images of raw EFB as depicted in Fig. 16.9 exhibited a smooth surface with numerous microfibrils and sharp edges at the end. After 10 minutes MWP, the underlying layers of EFB become more exposed resulting honeycomb-like shape of biochar structure. The applied MWP heating also developed the new micropores and mesopores increasing the surface area of the raw EFB from 0.704 m^2/g to 337.40 m^2/g at 10 minutes MWP. Nonetheless, the surface area of biochar

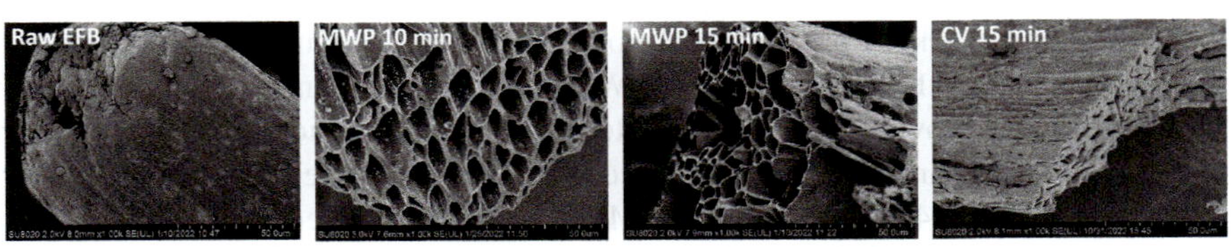

FIGURE 16.9 Field emission scanning electron microscopy images of raw empty fruit bunch and biochar pyrolyzed at different pyrolysis times under microwave and conventional heating.

was reduced to 98.43 m^2/g as pyrolysis time was extended to 15 minutes. The micrograph of the microwaved biochar at 15 minutes shows a much more complex and rougher surface appearance due to the fragility and fine particle size of the biochar. The results show that biochar pore structure can be destroyed by prolonged exposure to microwave heating at 450 W. Therefore it is reasonably expected the lower surface area and pore volume of biochar were obtained at 15 minutes.

Contrarily, the biochar produced via CV at 5–20 minutes has a lower total surface area of 0.66–1.40 m^2/g. Even though the pyrolysis time was extended to 120 minutes, the lower surface area of 0.99 m^2/g also was obtained, which is much lower as compared to the microwaved biochar. It was deduced that the heating mechanism of MWP and CV may influence the biochar traits formed. From Fig. 16.9, it was observed the CV heating resulted in an underdeveloped biochar pore structure, verifying the lower surface area of 1.40 m^2/g produced at 15 minutes. The CV biochar structure shows a slow development and may take much longer pyrolysis time to produce a fully developed biochar structure. The same results also were obtained by Fiore et al. (2018) and Sahoo et al. (2021) with underdeveloped biochar pore structures generated from the CV method. During CV heating, the materials are heated thoroughly as the heat is transmitted via conduction and/or convection from the surface into the inside of heated matter (Miura et al., 2004). In MWP, microwave heating could be generated directly within the EFB sample without the need for a transmission distance indicating the rapid heating of the pyrolysis process. The development of biochar porosity was enhanced under microwave heating due to direct heating penetration from the core of the EFB sample toward the outside surface of biochar resulting in a higher open structure and surface area. As previously discussed, the highly porous biochar structure has a high capacity to store nutrients by increasing the CEC of the soil and stimulating microbes' activities (Radin et al., 2018). The plant nutrient leaching losses from the soil can also be reduced. Hence, the microwaved biochar produced at 10 minutes with higher porosity was selected for further testing as a soil amendment.

For the soil amendment experiment, we applied okra seedlings due to the high germination rate of up to 80% (Nana et al., 2019). The biochar produced at the optimum MWP condition of 10 minutes was added at varied rates of 0.0 g, 2.5 g, 5 g, and 7.5 g to the 500 g sandy soil. The average plant height and leaf area were recorded weekly for 8 weeks. For comparison, the measurement of okra plant height and leaf area was taken at W-4, which corresponds to relatively early vegetative plant growth, and compared to the end of plantation period W-8. From Table 16.1, the unamended soil recorded the lowest average plant height of okra at 11.27 cm after a 4-week plantation. Meanwhile, the applied biochar at different rates of 2.5 g, 5 g, and 7.5 g enhances the plant growth by 12.40 cm, 13.67 cm, and 14.83 cm in the same week of okra plantation. The control soil grew taller by an average of 12.50 cm at 8 weeks of the plantation. Meanwhile, 7.5 g biochar addition shows the highest average plant growth at 16 cm as compared to control

Table 16.1 Effect of biochar application on average okra plant height and leaf area at W-4 and W-8.

Week	Biochar treatments (g)	Average okra plant height (cm)	Average leaf area (cm^2)
4	0	11.27	4.26
	2.5	12.40	5.63
	5	13.67	6.77
	7	14.83	7.48
8	0	12.50	5.71
	2.5	13.87	6.38
	5	15.17	7.15
	7	16.00	8.59

soils and different biochar rates. The average leaf width was observed to increase by 6.38–8.59 cm^2 with the biochar amendment compared with the control (5.71 cm^2) at the end of the planting period. The highest average leaf area was measured at 7.5 g showing that the higher biochar rate applied may promote the photosynthetic process improving plant productivity. As the plant grows, biochar treatments reduce the nutrients leaching and enhance soil nutrient availability through increasing microbial activity. The incorporation of biochar shows less soil compaction providing better aeration for root growth (Haque et al., 2021). The positive growth responses with biochar application were attributed to the improvement of the soil's physical and chemical properties.

At a 4-week okra plantation, we collected the soil sample and implemented a methodology reminiscent of our previous work to determine the soil's physical and chemical properties (Azman et al., 2023). As summarized in Table 16.2, it was observed that the highest biochar addition of 7.5 g promotes okra plant growth by improving the soil's physical and chemical properties compared to unamended soil and another biochar rate. The amendment of 7.5 g biochar causes the poorly aggregated soil to become less compacted decreasing the soil bulk density from 1.34 g/cm^3 to 1.30 g/cm^3 through mixing or dilution effect (Blanco-Canqui, 2017). The highly porous structure of biochar also modifies the pore system by contributing to extra pore formation by increasing the soil porosity from 49.43% to 51% (Dokoohaki et al., 2017). The greater number of pores also indicates more water can be retained. Thus the soil moisture content was increased from 22.82% to 27.04% with increasing soil porosity at a 7.5 g biochar rate. The improved soil physical properties provide better soil aeration affecting the availability of soil nutrients. Table 16.2 shows the lower bioavailability N presence in the control soil at 0.04%. Besides, the control soil has the lowest soil nutrient content at 0.06 mg/L P, 0.12 mg/L K, 0.28 mg/L Ca, and 0.26 mg/L Mg. The addition of 7.5 g of biochar gave higher soil nutrients by 0.62% N, 0.08 mg/L P, and 0.54 mg/L K. Meanwhile, the maximum of 0.40 mg/L Ca and 0.33 mg/L Mg

Table 16.2 Effect of biochar amendment on soil physical and chemical properties of okra plantation at W-4.

Parameter	Amount of biochar (g)			
	0	2.5	5	7.5
Physical				
Bulk density (g/cm^3)	1.34	1.31	1.31	1.30
Porosity (%)	49.43	50.47	50.56	51.00
Moisture content (%)	22.82	23.01	20.98	27.04
Chemical				
Nitrogen (N) (%)	0.04	0.25	0.29	0.62
Phosphorus (P) (mg/L)	0.06	0.08	0.12	0.08
Potassium (K) (mg/L)	0.12	0.27	0.50	0.54
Calcium (Ca) (mg/L)	0.28	0.37	0.40	0.33
Magnesium (mg/L)	0.26	0.31	0.33	0.28

was quantified from the 5 g biochar rate. The higher surface area and porosity of biochar may enhance the soil CEC to exchange mineral and metal elements with its environment by stimulating microbial activity. Therefore the high direct supply of nutrients was quantified from the biochar amendment. Our findings highlight the potential of microwaved biochar as a soil amendment to enhance plant growth by improving soil's physical and chemical properties. The high porosity and surface area of the microwaved biochar have proven to be a sustainable and effective method to remediate poor soil structure. The recommended biochar rate of 7.5 g was also suggested to ameliorate soil fertility with less soil compaction and high nutrient supplies from the biochar. However, the optimal biochar application may vary depending on the type of soil, crop variety, and environment. Thus further research and field trials are required to refine application methods and dosages for different contexts.

16.5 Future perspective and conclusion

This review has provided valuable insights into the impacts of feedstock and pyrolysis conditions on biochar traits, soil microbial community interaction, and nutrient dynamics. The selection of source feedstock and pyrolysis temperature are key parameters influencing the traits of the biochar. Studies show the decomposition of lignin-rich biomass at higher pyrolysis temperatures may produce a biochar with a larger specific surface area and higher C content. Meanwhile, biochar is derived from manures and sewage sludge that contain major inorganic elements resulting in higher pH and CEC value. The pyrolysis temperature parameter may also regulate biochar pH from acidic to alkaline at higher temperatures. The enrichment of basic cations in the ashes namely carbonates, oxides, and hydroxides cause the biochar pH to increase with rising temperature. The optimization of pyrolysis conditions using varied biomass materials was required to produce biochar with a strongly developed specific surface area, high pH, CEC, and labile C. The assessment of these traits is crucial to determine the effectiveness of biochar products as soil amendments. Since different biochar traits can create variability in biochar–microbe interaction. For such, the large quantities of pores of biochar can serve as a suitable habitat for many soil microbes and protection from predation. The labile C in biochar was prominent to directly increase the microbial communities by supplying energy sources such as carbohydrates. The incorporation of alkaline biochar with high CEC value was able to alter the soil pH and CEC. As soil CEC and pH increase, the growth and activities of microbes are enhanced enabling more nutrients can be transferred to the root surface for plant uptake. This shows biochar traits play a key role in maintaining soil fertility by increasing soil microbial interaction and nutrient dynamics.

A natural microbial community consists of several complex ecosystems with a variety of interactions. Several studies reported that different microbial communities thrive in different conditions, such as aerobic or anaerobic, acidophilic or basophilic, and large or small pores. Hence, suitable environmental conditions must be determined to promote the survival and metabolism of microorganisms. Future research should be taken into account (1) the extensive study on the influence of biochar traits on specific microbial functions such as nitrogen fixation and phosphorus solubilization. An in-depth understanding of the nutrient availability mechanism under different soil conditions helps to alter the biochar traits in targeted conditions. (2) The extensive study of the potential roles of co-composting biochar by blending biochar material with another soil additive such as compost or organic matter. The incorporation of compost substrate potentially modifies the biochar properties biotically and may influence microbial community and nutrient cycling. (3) The extensive study on the mechanism of biochar aging and stability over time. Assess the short- and long-term effects of the biochar–soil microbes' interaction that affect nutrient release. A more comprehensive understanding of biochar–microbe interaction will contribute to the development and best practices for optimizing biochar application in the agriculture field.

Acknowledgments

The authors would like to express their sincere gratitude to the Ministry of Higher Education (MOHE), Malaysia and Universiti Teknologi Malaysia (UTM) for the financial support via Fundamental Research Grant Scheme (FRGS) (FRGS/1/2020/TK0/UTM/02/1) and UTM fundamental Research Grant (Q.J130000.3846.22H57) and supporting one of the authors, Nur Aina Najwa Mohd Nor Azman.

CRediT author statement

Nur Aina Najwa Mohd Nor Azman: investigation, methodology, formal analysis, validation, and writing—original draft. Muzakkir Mohammad Zainol: supervision, resources, and funding acquisition. Mohd Asmadi: writing—review and editing, visualization, validation, supervision, resources, methodology, and investigation.

Declaration of competing interest

The authors declare that they have no known competing financial interest or personal relationships that could have appeared to influence the work reported in this paper.

References

Askeland, M., Clarke, B., & Paz-Ferreiro, J. (2019). Comparative characterization of biochars produced at three selected pyrolysis temperatures from common woody and herbaceous waste streams. *PeerJ*, *7*, e6784. Available from https://doi.org/10.7717/peerj.6784, https://pubmed.ncbi.nlm.nih.gov/31024777/.

Azman, N. A. N. M. N., Asmadi, M., Nawawi, M. A. S., Amin, N. A. S., Zakaria, Z. Y., Zainol, M. M., Lubes, Z. I. Z., & Phaiboonsilpa, N. (2022). Optimization of biochar production from slow pyrolysis of oil palm waste. *Chemical Engineering Transactions*, *97*, 163–168. Available from https://www.aidic.it/cet/22/97/028.pdf, https://doi.org/10.3303/CET2297028.

Azman, N. A. N. M. N., Khalid, P. I. M., Amin, N. A. S., Zakaria, Z. Y., Zainol, M. M., Ilham, Z., Phaiboonsilpa, N., & Asmadi, M. (2023). Effects of biochar, compost, and composted biochar soil amendments on okra plant growth. *Materials Today: Proceedings*, *2023*, 6–9.

Bakhtiar, M. H. A. B. M., Sari, N. B. A., Yaacob, A. B., Yunus, M. F. B. M., & Ismail, K. B. (2019). Characterization of oil palm empty fruit bunch (EFB) biochar activated with potassium hydroxide under different pyrolysis temperature. *Journal of Engineering Science and Technology*, *14*(5), 2792–2807. Available from http://jestec.taylors.edu.my/Vol%2014%20issue%205%20October%202019/14_5_25.pdf.

Blanco-Canqui, H. (2017). Biochar and soil physical properties. *Soil Science Society of America Journal*, *81*(4), 687–711. Available from https://doi.org/10.2136/sssaj2017.01.0017.

Chauhan, P., Sharma, N., Tapwal, A., Kumar, A., Verma, G. S., Meena, M., Seth, C. S., & Swapnil, P. (2023). Soil microbiome: Diversity, benefits and interactions with plants. *Sustainability (Switzerland)*, *15*(19). Available from https://doi.org/10.3390/su151914643, http://www.mdpi.com/journal/sustainability/.

Dai, Z., Barberán, A., Li, Y., Brookes, P. C., & Xu, J. (2017). Bacterial community composition associated with pyrogenic organic matter (biochar) varies with pyrolysis temperature and colonization environment. *mSphere*, *2*(2). Available from https://doi.org/10.1128/mSphere.00085-17, http://msphere.asm.org/content/msph/2/2/e00085-17.full.pdf.

Dai, Z., Li, Y., Zhang, X., Wu, J., Luo, Y., Kuzyakov, Y., Brookes, P. C., & Xu, J. (2019). Easily mineralizable carbon in manure-based biochar added to a soil influences N_2O

emissions and microbial-N cycling genes. *Land Degradation and Development*, *30*(4), 406−416. Available from https://doi.org/10.1002/ldr.3230, http://onlinelibrary.wiley.com/journal/10.1002/(ISSN)1099-145X.

Dai, Z., Xiong, X., Zhu, H., Xu, H., Leng, P., Li, J., Tang, C., & Xu, J. (2021). Association of biochar properties with changes in soil bacterial, fungal and fauna communities and nutrient cycling processes. *Biochar*, *3*(3), 239−254. Available from https://doi.org/10.1007/s42773-021-00099-x.

Dokoohaki, H., Miguez, F. E., Laird, D., Horton, R., & Basso, A. S. (2017). Assessing the biochar effects on selected physical properties of a sandy soil: An analytical approach. *Communications in Soil Science and Plant Analysis*, *48*(12), 1387−1398. Available from https://doi.org/10.1080/00103624.2017.1358742, http://www.tandfonline.com/loi/lcss20.

Domingues, R. R., Trugilho, P. F., Silva, C. A., De Melo, I. C. N. A., Melo, L. C. A., Magriotis, Z. M., & Sánchez-Monedero, M. A. (2017). Properties of biochar derived from wood and high-nutrient biomasses with the aim of agronomic and environmental benefits. *PLoS One*, *12*(5). Available from https://doi.org/10.1371/journal.pone.0176884, http://journals.plos.org/plosone/article/file?id = 10.1371/journal.pone.0176884&type = printable.

Domingues, R. R., Sánchez-Monedero, M. A., Spokas, K. A., Melo, L. C. A., Trugilho, P. F., Valenciano, M. N., & Silva, C. A. (2020). Enhancing cation exchange capacity ofweathered soils using biochar: Feedstock, pyrolysis conditions and addition rate. *Agronomy*, *10*(6). Available from https://doi.org/10.3390/agronomy10060824, https://www.mdpi.com/2073-4395/10/6/824/pdf.

EBC. (2012). *European biochar certificate-guidelines for a sustainable production of biochar* (pp. 1−18). European Biochar Certificate (EBC).

Fiore, S., Berruti, F., & Briens, C. (2018). Investigation of innovative and conventional pyrolysis of ligneous and herbaceous biomasses for biochar production. *Biomass and Bioenergy*, *119*, 381−391. Available from https://doi.org/10.1016/j.biombioe.2018.10.010.

Foong, S. Y., Liew, R. K., Yang, Y., Cheng, Y. W., Yek, P. N. Y., Wan Mahari, W. A., Lee, X. Y., Han, C. S., Vo, D. V. N., Van Le, Q., Aghbashlo, M., Tabatabaei, M., Sonne, C., Peng, W., & Lam, S. S. (2020). Valorization of biomass waste to engineered activated biochar by microwave pyrolysis: Progress, challenges, and future directions. *Chemical Engineering Journal*, *389*. Available from http://www.elsevier.com/inca/publications/store/6/0/1/2/7/3/index.htt, http://doi.org/10.1016/j.cej.2020.124401.

Garcia-Perez, M. (2008). *The formation of polyaromatic hydrocarbons and dioxins during pyrolysis: A review of the literature with descriptions of biomass composition, fast pyrolysis technologies and thermochemical reactions*. Washington State University.

Geng, N., Kang, X., Yan, X., Yin, N., Wang, H., Pan, H., Yang, Q., Lou, Y., & Zhuge, Y. (2022). Biochar mitigation of soil acidification and carbon sequestration is influenced by materials and temperature. *Ecotoxicology and Environmental Safety*, *232*, 113241. Available from https://doi.org/10.1016/j.ecoenv.2022.113241.

Haeldermans, T., Claesen, J., Maggen, J., Carleer, R., Yperman, J., Adriaensens, P., Samyn, P., Vandamme, D., Cuypers, A., Vanreppelen, K., & Schreurs, S. (2019). Microwave assisted and conventional pyrolysis of MDF − Characterization of the produced biochars. *Journal of Analytical and Applied Pyrolysis*, *138*, 218−230. Available from https://doi.org/10.1016/j.jaap.2018.12.027.

Halim, S. A., Mohd, N. A., & Razali, N. A. (2022). A comparative assessment of biofuel products from rice husk and oil palm empty fruit bunch obtained from conventional and microwave pyrolysis. *Journal of the Taiwan Institute of Chemical Engineers, 134*. Available from http://www.elsevier.com/wps/find/journaldescription.cws_home/715607/description#description, http://doi.org/10.1016/j.jtice.2022.104305.

Haque, A. N. A., Uddin, M. K., Sulaiman, M. F., Amin, A. M., Hossain, M., Aziz, A. A., & Mosharrof, M. (2021). Impact of organic amendment with alternate wetting and drying irrigation on rice yield, water use efficiency and physicochemical properties of soil. *Agronomy*, *11*(8). Available from https://doi.org/10.3390/agronomy11081529, https://www.mdpi.com/2073-4395/11/8/1529/pdf.

IBI. (2013). *Standardized product definition and product testing guidelines for biochar that is used in soil* (pp. 1–48). International Biochar Initiative (IBI).

Ippolito, J. A., Cui, L., Kammann, C., Wrage-Mönnig, N., Estavillo, J. M., Fuertes-Mendizabal, T., Cayuela, M. L., Sigua, G., Novak, J., Spokas, K., & Borchard, N. (2020). Feedstock choice, pyrolysis temperature and type influence biochar characteristics: A comprehensive meta-data analysis review. *Biochar*, *2*(4), 421–438. Available from https://doi.org/10.1007/s42773-020-00067-x, springer.com/journal/42773.

Itakura, M., Saeki, K., Omori, H., Yokoyama, T., Kaneko, T., Tabata, S., Ohwada, T., Tajima, S., Uchiumi, T., Honnma, K., Fujita, K., Iwata, H., Saeki, Y., Hara, Y., Ikeda, S., Eda, S., Mitsui, H., & Minamisawa, K. (2009). Genomic comparison of *Bradyrhizobium japonicum* strains with different symbiotic nitrogen-fixing capabilities and other Bradyrhizobiaceae members. *The ISME Journal*, *3*(3), 326–339. Available from https://doi.org/10.1038/ismej.2008.88.

Jaafar, N. M., Clode, P. L., & Abbott, L. K. (2015). Soil microbial responses to biochars varying in particle size, surface and pore properties. *Pedosphere*, *25*(5), 770–780. Available from https://doi.org/10.1016/S1002-0160(15)30058-8, http://pedosphere.issas.ac.cn.

Jatav, H. S., Rajput, V. D., Minkina, T., Singh, S. K., Chejara, S., Gorovtsov, A., Barakhov, A., Bauer, T., Sushkova, S., Mandzieva, S., Burachevskaya, M., & Kalinitchenko, V. P. (2021). Sustainable approach and safe use of biochar and its possible consequences. *Sustainability (Switzerland)*, *13*(18). Available from https://doi.org/10.3390/su131810362, https://www.mdpi.com/2071-1050/13/18/10362/pdf.

Kameyama, K., Miyamoto, T., & Iwata, Y. (2019). The preliminary study of water-retention related properties of biochar produced from various feedstock at different pyrolysis temperatures. *Materials*, *12*(11), 1732. Available from https://doi.org/10.3390/ma12111732.

Leng, L., Xiong, Q., Yang, L., Li, H., Zhou, Y., Zhang, W., Jiang, S., Li, H., & Huang, H. (2021). An overview on engineering the surface area and porosity of biochar. *Science of the Total Environment*, *763*, 144204. Available from https://doi.org/10.1016/j.scitotenv.2020.144204.

Li, L., Long, A., Fossum, B., & Kaiser, M. (2023). Effects of pyrolysis temperature and feedstock type on biochar characteristics pertinent to soil carbon and soil health: A meta-analysis. *Soil Use and Management*, *39*(1), 43–52. Available from https://doi.org/10.1111/sum.12848, http://onlinelibrary.wiley.com/journal/10.1111/(ISSN)1475-2743.

Liu, H., Wang, X., Song, X., Leng, P., Li, J., Mazza Rodrigues, J. L., Hong, Z., Kuzyakov, Y., Xu, J., & Dai, Z. (2022). Generalists and specialists decomposing labile and aromatic biochar compounds and sequestering carbon in soil. *Geoderma*, *428*, 116176. Available from https://doi.org/10.1016/j.geoderma.2022.116176.

Michalak, I., Baśladyńska, S., Mokrzycki, J., & Rutkowski, P. (2019). Biochar from a freshwater macroalga as a potential biosorbent for wastewater treatment. *Water*, *11*(7), 1390. Available from https://doi.org/10.3390/w11071390.

Miura, M., Kaga, H., Sakurai, A., Kakuchi, T., & Takahashi, K. (2004). Rapid pyrolysis of wood block by microwave heating. *Journal of Analytical and Applied Pyrolysis*, *71*(1), 187−199. Available from https://doi.org/10.1016/s0165-2370(03)00087-1.

Nana, R., Maïga, Y., Ouédraogo, R. F., Kaboré, W. G. B., Badiel, B., & Tamini, Z. (2019). Effect of water quality on the germination of okra (*Abelmoschus esculentus*) seeds. *International Journal of Agronomy*, *2019*. Available from http://www.hindawi.com/journals/ija/, https://doi.org/10.1155/2019/4938349.

Noor, N. M., Shariff, A., Abdullah, N., & Aziz, N. S. M. (2019). Temperature effect on biochar properties from slow pyrolysis of coconut flesh waste. *Malaysian Journal of Fundamental and Applied Sciences*, *15*(2), 153−158. Available from https://doi.org/10.11113/mjfas.v15n2.1015, https://mjfas.utm.my/index.php/mjfas/article/view/1015.

Phuong, D. T. M., Miyanishi, T., Okayama, T., & Kose, R. (2016). Pore characteristics & adsorption capacities of biochars derived from rice residues as affected by variety and pyrolysis temperature. *American Journal of Innovative Research & Applied Sciences*, *2*, 179−189.

Quilliam, R. S., Rangecroft, S., Emmett, B. A., Deluca, T. H., & Jones, D. L. (2013). Is biochar a source or sink for polycyclic aromatic hydrocarbon (PAH) compounds in agricultural soils? *GCB Bioenergy*, *5*(2), 96−103. Available from https://doi.org/10.1111/gcbb.12007, http://onlinelibrary.wiley.com/journal/10.1111/(ISSN)1757-1707.

Radin, R., Abu Bakar, R., Ishak, C. F., Ahmad, S. H., & Tsong, L. C. (2018). Biochar-compost mixture as amendment for improvement of polybag-growing media and oil palm seedlings at main nursery stage. *International Journal of Recycling of Organic Waste in Agriculture*, *7*(1), 11−23. Available from http://www.springer.com/environment/pollution + and + remediation/journal/40093, https://doi.org/10.1007/s40093-017-0185-3.

Sahoo, S. S., Vijay, V. K., Chandra, R., & Kumar, H. (2021). Production and characterization of biochar produced from slow pyrolysis of pigeon pea stalk and bamboo. *Cleaner Engineering and Technology*, *3*. Available from https://doi.org/10.1016/j.clet.2021.100101, https://www.journals.elsevier.com/cleaner-engineering-and-technology.

Schimmelpfennig, S., & Glaser, B. (2012). One step forward toward characterization: Some important material properties to distinguish biochars. *Journal of Environmental Quality*, *41*(4), 1001−1013. Available from https://doi.org/10.2134/jeq2011.0146.

Selvarajoo, A., & Oochit, D. (2020). Effect of pyrolysis temperature on product yields of palm fibre and its biochar characteristics. *Materials Science for Energy Technologies*, *3*, 575−583. Available from https://doi.org/10.1016/j.mset.2020.06.003.

Shaaban, A., Se, S. M., Mitan, N. M. M., & Dimin, M. F. (2013). Characterization of biochar derived from rubber wood sawdust through slow pyrolysis on surface porosities and functional groups. *Procedia Engineering*, *68*, 365−371. Available from https://doi.org/10.1016/j.proeng.2013.12.193, http://www.sciencedirect.com/science/journal/18777058.

Shagali, A. A., Hu, S., Wang, Y., Li, H., Wang, Y., Su, S., & Xiang, J. (2021). Comparative study on one-step pyrolysis activation of walnut shells to biochar at different heating rates. *Energy Reports*, *7*, 388−396. Available from https://doi.org/10.1016/j.egyr.2021.10.021.

Shariff, A., Aziz, M., Saleh, N. S., Ruzali, N., & I, N. S. (2016). The effect of feedstock type and slow pyrolysis temperature on biochar yield from coconut wastes. *International Journal of Chemical and Molecular Engineering*, *10*, 1410−1414.

Sheng, Y., & Zhu, L. (2018). Biochar alters microbial community and carbon sequestration potential across different soil pH. *Science of the Total Environment*, *622-623*,

1391–1399. Available from http://www.elsevier.com/locate/scitotenv, https://doi.org/10.1016/j.scitotenv.2017.11.337.

Song, S., Cong, P., Wang, C., Li, P., Liu, S., He, Z., Zhou, C., Liu, Y., & Yang, Z. (2023). Properties of biochar obtained from tropical crop wastes under different pyrolysis temperatures and its application on acidic soil. *Agronomy*, *13*(3), 921. Available from https://doi.org/10.3390/agronomy13030921.

Sun, J., He, F., Pan, Y., & Zhang, Z. (2017). Effects of pyrolysis temperature and residence time on physicochemical properties of different biochar types. *Acta Agriculturae Scandinavica, Section B—Soil & Plant Science*, *67*(1), 12–22. Available from https://doi.org/10.1080/09064710.2016.1214745.

Tag, A. T., Duman, G., Ucar, S., & Yanik, J. (2016). Effects of feedstock type and pyrolysis temperature on potential applications of biochar. *Journal of Analytical and Applied Pyrolysis*, *120*, 200–206. Available from https://doi.org/10.1016/j.jaap.2016.05.006.

Tang, J., Zhang, S., Zhang, X., Chen, J., He, X., & Zhang, Q. (2020). Effects of pyrolysis temperature on soil-plant-microbe responses to *Solidago canadensis* L.-derived biochar in coastal saline-alkali soil. *Science of the Total Environment*, *731*, 138938. Available from https://doi.org/10.1016/j.scitotenv.2020.138938.

Tomczyk, A., Sokołowska, Z., & Boguta, P. (2020). Biochar physicochemical properties: Pyrolysis temperature and feedstock kind effects. *Reviews in Environmental Science and Bio/Technology*, *19*(1), 191–215. Available from https://doi.org/10.1007/s11157-020-09523-3.

Tsolis, V., & Barouchas, P. (2023). Biochar as soil amendment: The effect of biochar on soil properties using VIS-NIR diffuse reflectance spectroscopy, biochar aging and soil microbiology—a review. *Land*, *12*(8), 1580. Available from https://doi.org/10.3390/land12081580.

Wang, Y., Shen, X., Bian, R., Liu, X., Zheng, J., Cheng, K., Xuhui, Z., Li, L., & Pan, G. (2022). Effect of pyrolysis temperature of biochar on Cd, Pb and As bioavailability and bacterial community composition in contaminated paddy soil. *Ecotoxicology and Environmental Safety*, *247*, 114237. Available from https://doi.org/10.1016/j.ecoenv.2022.114237.

Wang, Z., Liu, K., Xie, L., Zhu, H., Ji, S., & Shu, X. (2019). Effects of residence time on characteristics of biochars prepared via co-pyrolysis of sewage sludge and cotton stalks. *Journal of Analytical and Applied Pyrolysis*, *142*, 104659. Available from https://doi.org/10.1016/j.jaap.2019.104659.

Wijitkosum, S. (2022). Biochar derived from agricultural wastes and wood residues for sustainable agricultural and environmental applications. *International Soil and Water Conservation Research*, *10*(2), 335–341. Available from https://doi.org/10.1016/j.iswcr.2021.09.006.

Wijitkosum, S., & Jiwnok, P. (2019). Elemental composition of biochar obtained from agricultural waste for soil amendment and carbon sequestration. *Applied Sciences*, *9*(19), 3980. Available from https://doi.org/10.3390/app9193980.

Zeng, K., Minh, D. P., Gauthier, D., Weiss-Hortala, E., Nzihou, A., & Flamant, G. (2015). The effect of temperature and heating rate on char properties obtained from solar pyrolysis of beech wood. *Bioresource Technology*, *182*, 114–119. Available from http://www.elsevier.com/locate/biortech, https://doi.org/10.1016/j.biortech.2015.01.112.

Zhang, C., Zhao, X., Liang, A., Li, Y., Song, Q., Li, X., Li, D., & Hou, N. (2023). Insight into the soil aggregate-mediated restoration mechanism of degraded black soil via biochar addition: Emphasizing the driving role of core microbial communities and nutrient cycling. *Environmental Research*, *228*, 115895. Available from https://doi.org/10.1016/j.envres.2023.115895.

Zhang, J., & Shen, J. L. (2022). Effects of biochar on soil microbial diversity and community structure in clay soil. *Annals of Microbiology*, *72*(1). Available from https://doi.org/10.1186/s13213-022-01689-1, https://rd.springer.com/journal/13213.

Zhang, J., Huang, B., Chen, L., Li, Y., Li, W., & Luo, Z. (2018). Characteristics of biochar produced from yak manure at different pyrolysis temperatures and its effects on the yield and growth of highland barley. *Chemical Speciation & Bioavailability*, *30*(1), 57–67. Available from https://doi.org/10.1080/09542299.2018.1487774.

Zhao, B., O'Connor, D., Zhang, J., Peng, T., Shen, Z., Tsang, D. C. W., & Hou, D. (2018). Effect of pyrolysis temperature, heating rate, and residence time on rapeseed stem derived biochar. *Journal of Cleaner Production*, *174*, 977–987. Available from https://doi.org/10.1016/j.jclepro.2017.11.013.

Zhao, Y., Lu, Y., Zhuang, H., & Shan, S. (2023). In-situ retention of nitrogen, phosphorus in agricultural drainage and soil nutrients by biochar at different temperatures and the effects on soil microbial response. *Science of the Total Environment*, *904*, 166292. Available from https://doi.org/10.1016/j.scitotenv.2023.166292.

CHAPTER

Biochar augmentation on soil biotic processes and microbial structure and functions in fertilized tropical cropland

17

Chitranjan Kumar[1], Vashista Kotra[1], Ajay Tomar[1], Sangeeta Pandey[1] and Majeti Narasimha Vara Prasad[2]

[1]*Amity Institute of Organic Agriculture, Amity University, Noida, Uttar Pradesh, India*

[2]*Retired Professor, School of Life Sciences, University of Hyderabad, Hyderabad, Telangana, India*

17.1 Introduction

Biochar, a carbonaceous material derived from the pyrolysis of organic residues, has gained considerable attention as a soil amendment to enhance soil health and productivity in tropical croplands. The fertile landscapes of tropical croplands are host to intricate relationships between soil microorganisms, plants, and nutrient cycling, which sustain both ecosystems and humanity. However, these systems are facing disturbances due to intensifying agricultural practices, leading to concerns about soil degradation, reduced fertility, and environmental stress. In response, biochar, a carbon-rich material produced through pyrolysis, has emerged as a potential solution, offering a chance to restore balance by interacting with a fundamental component of these ecosystems: microbial communities (Yaashikaa et al., 2020). The exploration of biochar's interactions with soil microorganisms is akin to deciphering a cryptic code woven by nature herself. Biochar's significance lies in its versatility as a soil amendment, transcending conventional fertilization practices to reinvigorate tropical croplands (Kumar et al., 2022). This chapter embarks on a journey to unveil the intricate connections between biochar and soil microbial communities, revealing a network of relationships that shape nutrient cycling, structural integrity, and resilience. Within the soil matrix, biochar's introduction marks the beginning of a transformative process. This material, with its porous structure and high surface area, offers a sanctuary for microorganisms to flourish. Its sorption capacity also alters the availability of nutrients and organic compounds, influencing microbial functions and community structures (Kocsis et al., 2022). Tropical croplands provide a fertile ground for this exploration. With their warm climate, abundant rainfall, and diverse vegetation, they host a rich tapestry

Biochar Ecotechnology for Sustainable Agriculture and Environment. DOI: https://doi.org/10.1016/B978-0-443-29855-4.00019-9

of microbial life. The interplay between these systems sets the stage for biochar's potential to reintroduce balance to the equilibrium.

This study aims to unravel the complex interplay between biochar and soil microbial communities in fertilized tropical croplands. Through a comprehensive examination of literature and empirical evidence, the goal is to shed light on the mechanisms by which biochar influences microbial diversity, abundance, metabolic pathways, and structural organization. The ultimate objective is to uncover biochar's role as a catalyst for biotic processes that resonate with the intricate symphony of nature. To achieve this goal, this chapter follows a structured progression. It begins by establishing biochar's role as a catalyst for change within the soil matrix. From there, it delves into the context of tropical croplands and the dynamic environment that serves as the backdrop for these interactions. This chapter then transitions to explore how biochar transforms within the soil, setting the stage for its impact on microbial communities. Through a multidimensional lens, the study investigates how biochar influences microbial diversity, abundance, and metabolic pathways, drawing parallels to the nuanced processes of the natural world (Kolton et al., 2017). The significance of this study extends to the realm of sustainable agriculture. By deciphering the intricate relationship between biochar and soil microbes, we gain insights into enhancing soil health, nutrient cycling, and the resilience of ecosystems. This knowledge contributes to the growing field of ecological engineering, where human activities are harmonized with the intricacies of the natural world. In embarking on this exploration, we find ourselves tracing the paths of microorganisms, navigating a landscape characterized by complexity and interdependence.

17.2 Biochar's impact on microbial abundance: unveiling the microbial symphony

Soil ecosystems are intricately shaped by microbial abundance, which plays a crucial role in soil health and productivity. Biochar, a carbon-rich material from pyrolysis, is part of this complex web of interactions, creating a harmonious symphony of changes that affects microbial populations and their roles in nutrient cycling and ecosystem functions (Nascimento et al., 2023). In tropical croplands, microbial communities form the bedrock of soil vitality, and their abundance dictates the pace of organic matter decomposition, nutrient release, and soil structure resilience. Biochar's porous structure invites microbes to thrive, creating microenvironments that encourage microbial colonization and nutrient cycling (Kolton et al., 2017). Its porous structure also retains moisture, which enhances water-holding capacity and nutrient retention, leading to microbial proliferation and improved nutrient cycling.

Biochar's influence extends beyond immediate effects, leaving an enduring legacy. It sustains microhabitats over time, maintaining heightened microbial populations and leading to lasting shifts in the soil ecosystem's dynamics. Biochar's porous structure fosters diverse microenvironments, enabling the coexistence of varied microbial communities, mirroring nature's complex relationships. Biochar's power to enhance microbial abundance mirrors nature's complex balances, as diverse habitats create niches for distinct microbial communities. This alignment with nature's strategies presents a tangible solution for modern agricultural challenges. However, integrating biochar requires finesse, considering biochar type, application rates, and specific crop and ecosystem traits (Küçükağa et al., 2023). Biochar's influence extends beyond mere numbers, shaping community composition. Different biochar types influence specific microbial groups, resulting in shifts in community structure (J. Wang et al., 2023; Yin et al., 2021), which illuminates the broader ecological effects of biochar amendments. Biochar, a product of pyrolysis, has been explored for its potential to enhance soil health and productivity by altering the balance between nutrient cycling and microbial abundance. Its presence triggers a harmonious transformation that resonates throughout microbial communities, altering their numbers and roles in nutrient cycling and ecosystem functions. As biochar enters this intricate landscape, its presence triggers a harmonious transformation that resonates throughout microbial communities, altering their numbers and roles in nutrient cycling and ecosystem functions.

Tropical croplands are characterized by their rich soil ecosystems, where microbial abundance acts as a linchpin. These microorganisms are the conductors of essential biogeochemical processes, dictating organic matter decomposition, nutrient cycling, and the stability of soil structure (Gorovtsov et al., 2020). Biochar's impact on microbial populations is intricately linked to its physical properties. Its porous structure not only nurtures microorganisms but also enhances the soil's water-holding capacity, creating an environment conducive to microbial proliferation. This increased moisture availability, combined with improved nutrient retention, creates an environment conducive to microbial proliferation. As a result, microbial abundance swells in soils enriched with biochar. This surge in microbial populations resonates deeply within nutrient cycling processes (Giagnoni & Renella, 2022). Elevated microbial abundance accelerates the decomposition of organic matter, facilitating the release of essential nutrients for plant uptake. This intricate choreography of nutrient cycling mirrors the interconnectedness observed in natural ecosystems, where each component contributes to the harmonious functioning of the whole.

The implications of biochar's impact on microbial abundance reverberate throughout agricultural practices. By understanding biochar's role in nurturing microbial growth, farmers gain a tool to enhance soil fertility, amplify nutrient availability, and ultimately optimize crop yields. This alignment with sustainable strategies observed in nature offers a tangible solution to contemporary challenges in agriculture. However, the integration of biochar into soil management strategies requires finesse (Hardy et al., 2019). Balancing the benefits of heightened microbial populations with factors like biochar type, application rates, and specific crop and ecosystem attributes is essential for effective implementation (M. Wang et al., 2023). Further exploration unveils adjustments in functional groups within microbial communities. Specific microbial taxa may possess unique functional traits that thrive in biochar-fostered conditions. Exploring the implications of these shifts unravels the intricacies of biochar's impact on ecosystem functions.

Table 17.1 Exploring biochar–microbial aspects and their explanation for agricultural and ecological significance.

Aspect	Explanation	References
Biochar in soil ecosystems	Keystone element harmonizing biogeochemical processes and biodiversity	Smith et al. (2015) and Gorovtsov et al. (2020)
Biochar's introduction	Integration into soil matrix triggers transformative changes through the development of biochar-based soil amendments	Rombel et al. (2022)
Biochar under microbial abundance	Microbiome foundational for soil health and productivity	Suman et al. (2022)
Biochar–microbial interaction	Porous structure offers niches, fostering microorganism growth	Kocsis et al. (2022)
Shift in microbial equilibrium	Biochar initiates localized surges in microbial populations, and its application in circular bioeconomy	Yaashikaa et al. (2020)
Enhanced water-holding capacity	Biochar's porous structure enhances soil moisture retention in sandy agricultural soils	Ndede et al. (2022)
Nutrient cycling and soil C:N:P stoichiometry	Elevated microbial populations accelerate organic matter decomposition and C:N:P stoichiometry resists global change	Luo et al. (2020)
Elemental biochar composition	Physical structure of biochar sustains conducive conditions over time for carbon sequestration	Wijitkosum and Jiwnok (2019)
Habitat of diverse microbial environments	Biochar fosters diverse microhabitats, supporting varied communities facilitating contaminant biotransformation	Mukherjee et al. (2022) and Xiang et al. (2022)
Parallels with natural ecosystems	Biochar-generated microhabitats mimic natural ecological niches vital for sustainable agriculture	Chukwuka et al. (2020)
Implications for agriculture through biochar–microbe synergies	Biochar nurtures microbial growth, enhancing soil fertility in semiarid climate	Ahmad et al. (2020)
Balanced implementation	Considerations include biochar type, application rates, and ecosystem traits	Oni et al. (2019)
Shaping microbial community with functional adjustments and genetic richness	Biochar influences microbial community composition with positive effects on climate change mitigation	Xu et al. (2023)
Link to carbon sequestration	Microbial abundance interacts with carbon dynamics and their association with biochar increase C-use efficiency	Giagnoni and Renella (2022)

(*Continued*)

Table 17.1 Exploring biochar–microbial aspects and their explanation for agricultural and ecological significance. *Continued*

Aspect	Explanation	References
Plant–microbe interactions in nutrient cycling	Interplay between biochar, microbial abundance, and nutrient cycling	Yin et al. (2021)
Biochar for disease suppression	Augmented microbial abundance with biochar extends to disease suppression through inducting systemic resistance, sorption including induction detoxifying phytotoxic compounds, and enhancing rhizosphere competence of the microbial community	Iacomino et al. (2022)
Biochar–soil–plant interactions	Microbial populations indirectly impact soil's climate resilience through minimizing leaching losses and GHG emissions	Murtaza et al. (2023)
Multitrophic interactions	PGPR abundance, biochar, and other organic inputs affect soil sustainability and ecosystem functioning.	Kumar et al. (2022)
Complexity of microbial networks	Microbial interactions within ecosystems embody complexity which demand standardization of growth conditions for model communities	Bengtsson-Palme (2020)
Biochar for agricultural sustainability and environmental management	Biochar has potential for carbon sequestration, soil decontamination, and sustainable soil management	Prasad and Kumar (2023)

C, *carbon;* GHG, *greenhouse gas;* N, *nitrogen;* P, *phosphorus;* PGPR, *plant growth promoting rhizobacteria.*

The link between biochar and microbial abundance also extends to carbon sequestration. Microbial activity influences carbon turnover rates and the stabilization of organic matter. Biochar's provision of a substrate for microbial growth could potentially alter the balance between carbon storage and release, creating intricate connections. The interplay between biochar, microbial abundance, and nutrient cycling forms feedback loops. As microbial populations surge due to biochar, nutrient cycling rates may intensify, influencing microbial growth rates and fostering a dynamic interdependence. In conclusion, biochar's impact on microbial abundance is a complex and multifaceted phenomenon that can have significant implications for agriculture and ecosystems. Some key aspects of biochar–microbial interactions and their explanations are presented in Table 17.1.

17.3 Microbial diversity enhancement: painting a richer canvas

The exploration of soil ecosystems focuses on the intricacies of microbial diversity, which is a hallmark of thriving soil ecosystems. This intricate network of microorganisms, each with its distinct role and function, weaves the fabric of nutrient cycling, organic matter decomposition, and disease regulation. Biochar's presence introduces a subtle transformation in the composition of microbial communities within the soil matrix, creating specialized microenvironments that accommodate a diverse range of microbial taxa (Deshoux et al., 2023). Niche differentiation facilitated by biochar's three-dimensional matrix gives rise to microbial hotspots, which provide conditions conducive to the proliferation of specific microbial populations. Biochar's porous architecture fosters a playground of diversity, inviting numerous species to coexist harmoniously. It not only enriches microbial communities but also nurtures the growth of rare and unique microbial species, providing opportunities for less conspicuous microorganisms to flourish, contributing essential functions that underpin ecosystem stability. Microbial diversity extends beyond taxonomic variety to encompass functional genes that dictate microorganisms' capabilities. Biochar can influence the expression of these functional genes, potentially amplifying genes related to nutrient cycling, organic matter breakdown, and other crucial processes. Biochar's impact on microbial diversity unfolds over time, revealing a narrative of temporal dynamics. Initial shifts in microbial communities are followed by adaptations as microorganisms acclimatize to their new habitats. The long-term story of enhanced microbial diversity through biochar offers insights into soil ecosystems' stability and adaptability (Deshoux et al., 2023). Enhanced microbial diversity aligns with nature's strategy of fostering resilience in ecosystems. Diverse microbial communities possess greater resilience against environmental stressors such as temperature fluctuations, moisture changes, and nutrient variations. Biochar's role in diversifying microbial populations contributes to soil ecosystems' capacity to navigate challenges.

Biochar's potential to enrich microbial communities echoes the equilibrium observed in natural ecosystems. In untouched habitats, a multitude of species coexist, each contributing to the equilibrium of the ecosystem. By introducing diversity through biochar, we emulate nature's delicate balance in managed soils. The broadening of the microbial palette via biochar holds profound implications for agricultural sustainability. Diverse microbial communities enhance nutrient availability, bolster plant health, and suppress pathogens, aligning with regenerative agriculture principles (Deshoux et al., 2023). The impact of biochar on microbial diversity varies across different soil types, and understanding context-specific interactions is crucial for tailoring biochar applications to diverse agricultural settings.

Incorporating microbial diversity considerations into soil management strategies is paramount, as the benefits of increased microbial diversity must be balanced with other factors like nutrient management, water retention, and crop performance (Hardy et al., 2019). Microbial diversity plays a pivotal role in restoration ecology, supporting the recovery of degraded soils and ecosystems. By nurturing diverse microorganisms, biochar facilitates the establishment of functional ecosystems in areas facing ecological degradation. In conclusion, biochar emerges as a perpetual presence within soil ecosystems, enriching the landscape with a vibrant tapestry of life. The key functional aspects and impacts of biochar on microbial diversity are presented in Table 17.2.

17.4 Biochar–microbial interaction pathways: unveiling nature's blueprint

The exploration of soil ecosystems focuses on the intricate pathways of biochar's interaction with microbes. Biochar's porous structure provides a habitat for microbes, creating niches that shelter them from stressors and promote growth. This diversity mirrors nature's propensity to create diverse habitats. The carbon-rich composition of biochar fuels microorganisms' metabolic activities, reflecting the interconnectedness in ecosystems. Biochar's physical structure acts as a scaffold for microbial growth, allowing microorganisms to build complex communities within its matrix. The interaction between biochar and microbes can stimulate microbial activity, accelerating the breakdown of organic matter through microbial priming. This interaction aligns with species interactions in ecosystems. Biochar, as it ages, forms soil aggregates, protecting carbon, stabilizing rhizodeposits, and microbial products, and increasing pH, porosity, and water availability, promoting root development and microbiological function (Joseph et al., 2021).

Biochar enhances nutrient cycling within the soil ecosystem, intensifying nutrient turnover and promoting plant nutrient availability. The indirect effects of biochar on soil properties, such as pH and moisture retention, create a ripple effect that shapes microbial communities. Microorganisms engage in a biogeochemical partnership with biochar, reengineering its structure over time. Biochar triggers a delicate balance within microbial communities, with some flourishing in biochar-created niches while others compete for resources (Deshoux et al., 2023). The microbial community that colonizes biochar can influence its properties over time, creating a feedback loop that intertwines the destinies of biochar and microorganisms. As biochar prompts shifts in microbial roles and functions, genetic expression within microorganisms adapts to the presence of biochar, stimulating genes associated with nutrient acquisition and organic matter degradation. These interactions extend to higher trophic levels, revealing the complexity of the soil food web and the rhizosphere, where microbial interactions sustain plant growth. Biochar–microbial interactions along with their ecological implications and agricultural applications have been presented in Table 17.3.

Table 17.2 Functional aspects, significance, and impact of biochar on microbial communities and diversity.

Functional aspects	Description	Significance	Biochar's influence on microbial communities	Diversity's impacts	References
Microbial diversity enhancement	Within soil ecosystems, the focus shifts to microbial diversity. Biochar emerges as a transformative element, expanding the palette of microorganisms within tropical soils and imbuing the landscape with a more vibrant tapestry of life.	Microbial diversity is akin to a painter's palette, vital for thriving soil ecosystems. This intricate network of microorganisms underpins nutrient cycling, organic matter decomposition, and disease regulation, forming the bedrock of ecosystem resilience.	Biochar introduces subtle transformations in microbial communities. Its porous structure creates microenvironments, accommodating diverse microbial taxa. This spatial arrangement influences the composition and relative proportions of microorganisms.	Enhanced microbial diversity often leads to functional redundancy, bolstering ecosystem robustness and ensuring essential functions remain even if some microorganisms falter. The intricate interplay of cooperation and competition shapes the complex web of life beneath our feet.	Kolton et al. (2017)
Microbial diversity's significance	Microbial diversity is akin to a painter's palette, vital for thriving soil ecosystems. This intricate network of microorganisms underpins nutrient cycling, organic matter decomposition, and disease regulation, forming the bedrock of ecosystem resilience.	Microbial diversity serves as a valuable indicator of soil health and ecosystem functioning. Shifts in microbial communities triggered by biochar amendments provide insight into the ongoing ecological changes underway.	Biochar introduces subtle transformations in microbial communities. Its porous structure creates microenvironments, accommodating diverse microbial taxa. This spatial arrangement influences the composition and relative proportions of microorganisms.	Biochar's role in diversifying microbial communities aligns with nature's strategy of fostering resilience in ecosystems. Diverse microbial communities are better equipped to withstand environmental stressors and enhance the soil's ability to weather challenges.	Maron et al. (2018)

Biochar's influence on microbial communities	Biochar introduces subtle transformations in microbial communities. Its porous structure creates microenvironments, accommodating diverse microbial taxa. This spatial arrangement influences the composition and relative proportions of microorganisms.	Enhanced microbial diversity often leads to functional redundancy, bolstering ecosystem robustness and ensuring essential functions remain even if some microorganisms falter. The intricate interplay of cooperation and competition shapes the complex web of life beneath our feet.	Biochar introduces subtle transformations in microbial communities. Its porous structure creates microenvironments, accommodating diverse microbial taxa. This spatial arrangement influences the composition and relative proportions of microorganisms.	The interplay between biochar and soil organic matter further shapes microbial diversity. Biochar's role in enhancing the efficiency of organic matter decomposition contributes to nutrient cycling and soil fertility.	J. Zhang et al. (2022)
Niche differentiation and microbial hotspots	Biochar's three-dimensional matrix fosters niche differentiation and microbial hotspots. These microhabitats offer conditions suitable for specific microbial populations to thrive, creating a harmonious coexistence of diverse species.	Biochar's role in enhancing microbial diversity aligns with nature's strategy of fostering resilience in ecosystems. Diverse microbial communities are better equipped to withstand environmental stressors and enhance the soil's ability to weather challenges.	Biochar introduces subtle transformations in microbial communities. Its porous structure creates microenvironments, accommodating diverse microbial taxa. This spatial arrangement influences the composition and relative proportions of microorganisms.	The enhancement of microbial diversity through biochar resonates beyond the microbial realm. Microorganisms contribute to biodiversity conservation and overall ecosystem health, reflecting the interconnected relationships shaping nature's balance.	Yang et al. (2022)

(*Continued*)

Table 17.2 Functional aspects, significance, and impact of biochar on microbial communities and diversity. *Continued*

Functional aspects	Description	Significance	Biochar's influence on microbial communities	Diversity's impacts	References
Promoting rare and unique microbes	Biochar's presence nurtures the growth of rare and unique microbial species. This cooperation and competition lead to the flourishing of lesser-known microorganisms, enriching the microbial tapestry with valuable yet often overlooked contributors.	Enhanced microbial diversity often leads to functional redundancy, bolstering ecosystem robustness and ensuring essential functions remain even if some microorganisms falter. The intricate interplay of cooperation and competition shapes the complex web of life beneath our feet.	Biochar introduces subtle transformations in microbial communities. Its porous structure creates microenvironments, accommodating diverse microbial taxa. This spatial arrangement influences the composition and relative proportions of microorganisms.	Microbial diversity is a valuable indicator of soil health and ecosystem functioning. Monitoring these shifts offers a nuanced glimpse into the responses of soil ecosystems to biochar interventions.	H. Wang et al. (2022)
Functional redundancy and ecosystem robustness	Enhanced microbial diversity often leads to functional redundancy, bolstering ecosystem robustness and ensuring essential functions remain even if some microorganisms falter. The intricate interplay of cooperation and competition shapes the complex web of life beneath our feet.	The enhancement of microbial diversity through biochar resonates beyond the microbial realm. Microorganisms contribute to biodiversity conservation and overall ecosystem health, reflecting the interconnected relationships shaping nature's balance.	Biochar introduces subtle transformations in microbial communities. Its porous structure creates microenvironments, accommodating diverse microbial taxa. This spatial arrangement influences the composition and relative proportions of microorganisms.	The integration of microbial diversity into soil management strategies is vital. Balancing the benefits of elevated microbial populations with considerations such as biochar type, application rates, and specific crop and ecosystem characteristics is essential for effective implementation.	

Beyond taxonomic diversity: functional genes	Microbial diversity extends beyond taxonomic variety to encompass functional genes that dictate the capabilities of microorganisms. Biochar can influence the expression of these functional genes, potentially enhancing the presence of genes associated with nutrient cycling, organic matter degradation, and other critical processes.	Microbial diversity is a valuable indicator of soil health and ecosystem functioning. Monitoring these shifts offers a nuanced glimpse into the responses of soil ecosystems to biochar interventions.	Biochar introduces subtle transformations in microbial communities. Its porous structure creates microenvironments, accommodating diverse microbial taxa. This spatial arrangement influences the composition and relative proportions of microorganisms.	The integration of microbial diversity into soil management strategies is vital. Balancing the benefits of elevated microbial populations with considerations such as biochar type, application rates, and specific crop and ecosystem characteristics is essential for effective implementation.	Almendras et al. (2022)
Introducing microbial cooperation and competition	The intricate interplay of cooperation and competition is central to microbial diversity. Biochar introduces new inputs, altering the dynamics of interactions among microorganisms. Some species may form symbiotic partnerships with biochar, while others vie for dominance within the newly created microenvironments.	The broadening of the microbial palette through biochar holds implications for agricultural sustainability. Diverse microbial communities enhance nutrient availability, promote plant health, and suppress pathogens, aligning with principles of regenerative agriculture.	Biochar introduces subtle transformations in microbial communities. Its porous structure creates microenvironments, accommodating diverse microbial taxa. This spatial arrangement influences the composition and relative proportions of microorganisms.	The integration of microbial diversity into soil management strategies is vital. Balancing the benefits of elevated microbial populations with considerations such as biochar type, application rates, and specific crop and ecosystem characteristics is essential for effective implementation.	Lechón-Alonso et al. (2021)

(*Continued*)

Table 17.2 Functional aspects, significance, and impact of biochar on microbial communities and diversity. *Continued*

Functional aspects	Description	Significance	Biochar's influence on microbial communities	Diversity's impacts	References
Temporal dynamics: unveiling the story of microbial adaptability	The influence of biochar on microbial diversity unfolds over time, revealing a narrative of temporal dynamics. Initially, shifts in microbial communities may occur, followed by adjustments as microorganisms adapt to the new microhabitats. The long-term story of microbial diversity enhancement through biochar offers insights into the stability and adaptability of soil ecosystems.	The interplay between biochar and soil organic matter further shapes microbial diversity. Biochar's role in enhancing the efficiency of organic matter decomposition contributes to nutrient cycling and soil fertility.	Biochar introduces subtle transformations in microbial communities. Its porous structure creates microenvironments, accommodating diverse microbial taxa. This spatial arrangement influences the composition and relative proportions of microorganisms.	The integration of microbial diversity into soil management strategies is vital. Balancing the benefits of elevated microbial populations with considerations such as biochar type, application rates, and specific crop and ecosystem characteristics is essential for effective implementation.	Deshoux et al. (2023)
Fostering resilience in the agroecosystem and forestry on the face of stressors	Biochar's role in diversifying microbial communities aligns with nature's strategy of fostering resilience in ecosystems. Diverse microbial communities are better equipped to withstand environmental stressors and enhance the soil's ability to weather challenges.	The interplay between biochar and soil organic matter further shapes microbial diversity. Biochar's role in enhancing the efficiency of organic matter decomposition contributes to nutrient cycling and soil fertility.	Enhanced microbial diversity often leads to functional redundancy, bolstering ecosystem robustness and ensuring essential functions remain even if some microorganisms falter. The intricate interplay of cooperation and competition shapes the complex web of life beneath our feet.	The enhancement of microbial diversity through biochar resonates beyond the microbial realm. Microorganisms contribute to biodiversity conservation and overall ecosystem health, reflecting the interconnected relationships shaping nature's balance.	Sessions et al. (2019) and Bertola et al. (2021)

Table 17.3 Exploring biochar–microbial interactions: dynamics, ecological implications, and agricultural applications.

Interaction aspect	Description	Biochar–microbial dynamics	Ecological implications	Agricultural applications	References
Microbial niche creation	Biochar's porous structure offers microenvironments for colonization, mirroring natural habitats	Niche differentiation within biochar matrix	Supports rare and specialized microorganisms	Enhances nutrient availability and disease suppression	Wong and Ogbonnaya (2021) and Barone et al. (2023)
Habitat heterogeneity	Biochar introduces diverse niches within its matrix, supporting distinct microbial communities.	Diverse habitats for various microorganisms.	Promotes microbial diversity and functional redundancy.	Increases soil resilience and nutrient cycling.	Yan et al. (2021) and Bello et al. (2023)
Carbon as a microbial feast	Carbon-rich biochar serves as nourishment, fueling the metabolic activities of various microorganisms.	Microbes utilize biochar's carbon for growth.	Enhances microbial activity and nutrient cycling.	Improves soil fertility and organic matter decomposition.	Farrell et al. (2013) and Mansoor et al. (2021)
Chemical signaling	Biochar's chemical cues influence microbial behaviors, guiding them toward specific functions.	Chemical signals trigger shifts in microbial behavior.	Shapes nutrient cycling and disease suppression.	Promotes balanced microbial interactions and ecosystem stability.	Lu et al. (2020) and Afridi et al. (2022)
Structural support	Biochar's physical structure acts as a scaffold for microbial growth, fostering complex communities.	Microbes colonize biochar's intricate matrix.	Enhances microbial diversity and network complexity.	Improves soil structure and nutrient availability.	Yin et al. (2021)
Microbial priming	Biochar stimulates microbial activity, accelerating organic matter breakdown through microbial priming.	Microbial response to biochar's carbon source.	Accelerates organic matter decomposition and nutrient release.	Boosts nutrient availability and plant growth.	Zhu et al. (2019) and Rasul et al. (2022)

(Continued)

Table 17.3 Exploring biochar–microbial interactions: dynamics, ecological implications, and agricultural applications. *Continued*

Interaction aspect	Description	Biochar–microbial dynamics	Ecological implications	Agricultural applications	References
Microbial-guided nutrient cycling	Microbes perform nutrient cycling, amplified by biochar, intensifying nutrient turnover.	Microbial activity enhances nutrient cycling rates.	Improves nutrient availability for plants.	Optimizes nutrient management and reduces leaching.	Zhang M et al. (2022) and Shen et al. (2023)
Microbial adaptation	Microorganisms adapt to biochar's presence, leading to shifts in community composition and function.	Microbes adjust to new biochar-influenced environment.	Shapes microbial diversity and functional roles.	Enhances ecosystem resilience and stability.	Zhang et al. (2022) and Giagnoni and Renella (2022)
Symbiotic relationships	Biochar influences symbiotic interactions between microorganisms and plants, supporting plant health.	Enhances microbial interactions with plant roots.	Improves plant nutrient uptake and growth.	Facilitates sustainable and regenerative agriculture.	Harman et al. (2021) and Hou et al. (2022)
Soil health impacts	Biochar triggers cascading impacts on nutrient cycling, disease suppression, and organic matter decay.	Amplifies nutrient cycling and suppresses soil-borne pathogens.	Enhances overall soil health and functionality.	Improves crop productivity and reduces environmental impact.	de Medeiros et al. (2021) and Nepal et al. (2023)
Rhizosphere transformation	Biochar alters the rhizosphere, enhancing nutrient cycling and root health, benefiting plant growth.	Modifies rhizosphere microbial communities and functions.	Improves nutrient availability for plants and enhances root health.	Promotes optimal plant–microbe interactions and crop yield.	Mukherjee et al. (2022) and Evangelista et al. (2023)
Novel ecological connections	Biochar opens doors to novel ecological interactions, reshaping nutrient dynamics and carbon cycling.	Creates new microbial connections and feedbacks.	Influences nutrient cycling and ecosystem services.	Enhances ecosystem sustainability and agroecosystem functionality.	Zhang et al. (2021) and Prasad (2023)

Integration with management for safe and sustainable use	Balancing microbial benefits with other management considerations ensures sustainable biochar use.	Consideration of biochar in overall soil management.	Optimizes biochar application for desired outcomes.	Facilitates ecologically responsible and efficient land use.	Jatav et al. (2021) and Thakur et al. (2023)
Ecosystem-wide effects	Biochar's impact on microorganisms extends to broader ecosystem effects, influencing plant health and more.	Microbial interactions reverberate through entire ecosystems.	Shapes ecosystem functions and services.	Supports long-term sustainability and biodiversity.	Ren et al. (2020) and Yang et al. (2022)
Microbial feedback loops	Biochar–microbial interactions create feedback loops that shape both microbial growth and biochar.	Microbial colonization alters biochar's properties.	Influences biochar's longevity and impact on soil health.	Supports tailored soil management strategies.	Pang et al. (2021)

17.5 Structural adaptations of microbes: unveiling dynamic microbial ecosystems

The study of soil ecosystems focuses on the structural adaptations that emerge within microbial communities in response to biochar. Biochar acts as a catalyst, shaping the architecture of these microorganisms and creating a dynamic and interconnected ecosystem. The structural configurations of microbes serve as the architects that underlie soil health, encompassing networks, biofilms, and aggregates that carry out critical functions such as nutrient cycling, organic matter degradation, and disease suppression (Ajijah et al. 2023). Biochar's porous and intricate matrix introduces a new dimension to microbial architecture, offering a perfect platform for the assembly of microbial structures (Yin et al. 2021). Microorganisms cluster around biochar particles, giving rise to biofilms and aggregates that amplify their interactions and functional activities (Ajijah et al. 2023). These formations are akin to miniature cities, where various species come together to create cooperative habitats reminiscent of the interconnected relationships found in natural ecosystems.

The presence of biochar triggers a symphony of structural rearrangements within microbial communities, as microorganisms adapt their structures to thrive within the biochar matrix, much like nature's resilience. This adaptation results in a spatial heterogeneity that mirrors the diverse habitats and microclimates that characterize the natural world. In this adaptation process, the concept of microbial niche specialization comes to the forefront. The introduction of biochar prompts microorganisms to tailor their structures to specific microenvironments within the biochar matrix, echoes the vast diversity of habitats found in the natural world. The resulting microbial structures weave an intricate network that navigates the soil matrix, creating a dynamic web that mirrors the cooperative interactions fundamental to ecosystem resilience. Fungal hyphae explore the porous biochar matrix in search of nutrients and organic matter, while bacteria find a haven within the biochar matrix, occupying its pores and crevices. The structural adaptations brought about by biochar also contribute to the enhancement of soil structure, fostering biodiversity within the soil matrix and contributing to ecosystem resilience. As we peel back the layers of microbial structures influenced by biochar, we glimpse the intricate world of these architects that shape the very essence of soil ecosystems. The influence of biochar on microbial structure and ecosystem dynamics is presented in Table 17.4.

17.6 Biochar's role in nutrient cycling: unraveling the virtuous exchange

The study focuses on the role of biochar in soil ecosystems, specifically in nutrient cycling. Nutrient cycling is the core of soil ecosystems, a complex process

Table 17.4 Influence of biochar on microbial structure, adaptations, and ecosystem impacts.

Aspect of microbial structures	Biochar's influence	Microbial adaptations	Ecosystem impacts	Examples	References
Microbial architecture	Scaffold for assembly of structures	Formation of biofilms, aggregates	Enhanced nutrient cycling, disease suppression	Aggregation formations enhancing soil stability	Liang et al. (2023) and Ajijah et al. (2023)
Spatial variation	Creates spatial heterogeneity	Niche specialization	Diverse habitats, microclimates	Microbial adaptation to biochar matrix	Nakhli et al. (2021)
Functional zones	Leads to functional hotspots	Enhanced organic matter decomposition, nutrient cycling	Localized activity centers	Biofilm-driven nutrient cycling	Lyu et al. (2024)
Cooperative networks	Facilitates microbial communication	Sharing resources, information	Cooperative interactions	Biochar-enhanced microbial networks	Yan et al. (2023)
Mycorrhizal connectivity	Improves plant–microbe interactions	Enhanced nutrient uptake, plant health	Symbiotic relationships	Mycorrhizal networks and biochar	Meki et al. (2022)
Biodiversity enhancement	Promotes diverse microbial communities	Enhanced ecosystem resilience	Stability, adaptability	Diversity of species in microbial structures	H. Wang et al. (2022)

that replenishes essential elements for plant growth and ecosystem functions. Biochar, with its unique porous structure, acts as a nutrient reservoir, allowing it to retain and gradually release nutrients over time. This high cation exchange capacity mirrors the storage and release dynamics observed in natural soil horizons. The relationship between biochar and nutrients is a chemical ballet of adsorption, where biochar interacts with nutrients such as nitrogen, phosphorus, and potassium, temporarily adhering to its surfaces. This process mimics nature's ability to sequester elements in minerals and organic matter, ensuring the vitality of the ecosystem.

Biochar's influence extends beyond adsorption to controlled release of nutrients, which meet the specific needs of plants over time. This controlled release mechanism mirrors the efficient nutrient delivery systems in natural ecosystems. Microorganisms play a pivotal role in nutrient liberation from biochar, producing enzymes that break down organic matter attached to biochar, releasing bound nutrients in forms that plants can readily access. This microbial mediation echoes the cooperative dynamics that drive nutrient cycling in ecosystems, where species collaborate to recycle nutrients for the greater good. Biochar's presence also enhances nutrient retention within the soil matrix, reducing the risk of nutrient leaching and loss from the ecosystem. This aligns with nature's mechanisms for conserving resources within ecosystems, ensuring that precious nutrients remain available for plant uptake and other ecosystem processes. The synergy between biochar and nutrient cycling leads to improved nutrient use efficiency in plants, mitigating the risk of nutrient deficiencies and waste.

Biochar's introduction can alter soil pH, influencing the availability of certain nutrients. This modulation reflects the intricate pH-dependent nutrient dynamics observed in natural soil systems. It can promote the conversion of ammonium to nitrate, an essential step in nitrogen mineralization, and immobilize phosphorus, reducing its leaching potential. Additionally, biochar can adsorb potassium, potentially reducing its immediate availability, but microbial processes can release it back into the soil solution, reflecting nature's regulation of nutrient availability. Biochar's interaction with organic matter stabilizes nutrient cycling by protecting organic matter from rapid decomposition, allowing nutrients to be released gradually over time. This aligns with the principles of sustainable agricultural practices, where efficient nutrient use is key to ecosystem health. Biochar's role in nutrient cycling also aligns with organic farming principles, enhancing nutrient retention, controlled release, and efficiency (Gu et al., 2023). This harmony underscores biochar's potential as a tool for bridging agriculture with natural nutrient cycling processes that sustain soil fertility. Biochar revitalizes soil fertility, akin to the natural nutrient replenishment processes observed in undisturbed ecosystems. It enhances nutrient availability, promotes efficiency, reduces nutrient losses, and embodies nature's strategies for maintaining soil ecosystem balance. The exploration of biochar's impact on nutrient cycling offers valuable insight into soil health and ecosystem functioning (Maron et al., 2018). Changes in nutrient availability, cycling rates, and plant nutrient content serve as indicators of the influence of biochar amendments (Rombel et al., 2022). Biochar's role in nutrient cycling embodies the harmonious exchange of elements that sustains life beneath our feet. As we unravel the intricate mechanisms by which biochar enriches soil fertility, we will gain a deeper understanding of the metabolic harmonies, structural melodies, and functional rhythms that emerge from the interplay of biochar and nutrient dynamics.

17.7 Metabolic pathway modulation: unveiling biochar's influence

Biochar plays a crucial role in soil ecosystems by acting as a metabolic catalyst, fine-tuning the biochemical symphony beneath the soil's surface and shaping the complex metabolic pathways. Microbial metabolism is the pulse of ecosystems, driving the intricate machinery of soil life. Biochar's porous structure creates microenvironments that prompt shifts in microbial metabolism, fostering adaptations to optimize energy usage and resource utilization (Kolton et al., 2017). Biochar's influence extends to carbon cycling, a fundamental process dictating soil organic matter dynamics. It can accelerate the degradation of complex organic compounds (Xiang et al. 2022), enhancing the turnover of carbon. By interacting with microbial metabolism, biochar boosts the efficiency of energy utilization, as microorganisms adjust their metabolic pathways to suit the carbon-rich environment created by biochar. Biochar also guides nutrient transformations within microbial metabolism, guiding them to effectively utilize the nutrients available in biochar. This precision in nutrient utilization aligns with nature's emphasis on resource-focused metabolic strategies (Kolton et al., 2017; H. Wang et al., 2022). Biochar's influence extends to nitrogen cycles, as microorganisms within biochar enhance nitrogen mineralization, converting organic nitrogen into forms accessible to plants. Phosphorus metabolism also undergoes shifts due to biochar's influence, as microorganisms within biochar can impact the solubility of phosphorus compounds, influencing their availability to plants. Biochar's interaction with microbial metabolism fosters symbiotic relationships among microorganisms, enhancing nutrient cycling, carbon utilization, and organic matter degradation, reflecting the collaborative behaviors observed in natural ecosystems.

Biochar plays a crucial role in microbial metabolism, influencing enzyme activity, redox reactions, and the diversity of metabolic pathways within microbial communities. It can modulate enzyme activity, affecting rates of organic matter breakdown and nutrient transformation, and mirroring nature's ability to fine-tune metabolic processes (H. Wang et al., 2022). Microorganisms within biochar adjust their redox balance in response to biochar's electron dynamics, mirroring redox reactions in natural ecosystems. The interplay between biochar and microbial metabolism extends to the plant–microbe interface, where microorganisms within biochar influence nutrient cycling and availability for plants, fostering a dynamic exchange of elements. Biochar's effects ripple through soil enzyme dynamics, catalyzing reactions that influence nutrient availability and organic matter decomposition. Biochar's role in metabolic pathway modulation leads to shifts in microbial community compositions, with microorganisms that align their metabolic adaptations with biochar's influence potentially thriving. This flux mirrors nature's ability to shape communities based on resource availability.

Biochar's interaction sparks a molecular dialog, with chemical cues from biochar influencing microbial gene expression and enzyme production, guiding metabolic pathways. This dialog reflects the intricate communication observed in microbial communities within ecosystems. Biochar's modulation of microbial metabolism enhances carbon sequestration, vital for mitigating climate change, as microbes within biochar contribute to stable carbon storage and promote the formation of recalcitrant organic compounds. Biochar's role extends to the foundations of soil health, carbon cycling, which governs the fate of organic matter and profoundly impacts soil dynamics. As biochar entwines with microbial metabolism, it bestows enhanced efficiency upon the energy balance, enabling microorganisms to adjust their metabolic pathways to the carbon-rich niches fostered by biochar (H. Wang et al., 2022). This optimization reflects nature's underlying principle of resource allocation and reflects the harmony between microbes and their environment. Biochar also takes on the role of a guiding hand in nutrient transformations within microbial metabolism. As microorganisms adjust their metabolic pathways, they deftly harness the nutrients present in biochar, showcasing a precision that resonates with nature's meticulous focus on resource management. A striking interlude in the symphony is the acceleration of organic matter decomposition, where microbes residing within the porous matrix of biochar engage in breaking down complex organic molecules into simpler forms. This release of nutrients and energy echoes the rhythm of nature, where organic matter breakdown sustains the cycle of life.

Biochar's influence extends to nitrogen metabolism, a cornerstone of nutrient cycling, as microorganisms sheltered within the biochar matrix intensify the mineralization of nitrogen, converting organic nitrogen compounds into forms that nourish plants (Clough et al., 2013). This enhancement mirrors nature's innate ability to regulate nitrogen flows within ecosystems. Biochar also affects phosphorus metabolism, influencing the solubility of phosphorus compounds and directing the availability of this vital nutrient to plants. Enzymes, akin to conductors wielding their batons, are profoundly influenced by biochar's presence. A pivotal ensemble piece arises as biochar–microbe–plant connections form, facilitating nutrient exchange at the interface. This nexus reflects the intricate nutrient relationships that nourish plant growth in natural ecosystems. The effects of biochar resonate through soil enzyme dynamics, with enzymes secreted by microorganisms within biochar catalyzing reactions that influence nutrient availability and organic matter breakdown (Deshoux et al., 2023). Microbial communities are not static; they are in flux, adapting their compositions in response to biochar's modulation of metabolic pathways.

17.8 Biochar–mediated carbon sequestration: uncovering nature's locks

Biochar plays a crucial role in carbon sequestration, safeguarding atmospheric carbon, and contributing to the balance of carbon cycles. Its exceptional carbon storage capacity stems from organic matter that undergoes pyrolysis, transforming biomass into a stable form of carbon. This storage capacity mirrors nature's ability to accumulate carbon in long-lived organic forms (Li & Tasnady, 2023). Biochar's carbon sequestration mechanisms involve the stabilization of organic carbon, with its unique structure ensuring that stored carbon remains in the soil for extended periods. The presence of biochar can inhibit the decomposition of organic matter by reducing microbial access to substrates, mirroring the microbial regulation of organic matter decomposition in natural ecosystems. Biochar also serves as a habitat for carbon-storing microbes, with some microorganisms contributing to the formation of recalcitrant organic compounds, effectively storing carbon within the biochar matrix. The carbon priming effect, a dual-edged sword, reflects the complex interplay between carbon inputs and outputs.

Biochar's impact on soil aggregation contributes to carbon sequestration by trapping carbon within stable structures. Over time, the storage capacity of biochar may become saturated, leading to diminishing returns in terms of sequestration. This process reflects the balance and limits inherent in natural carbon sinks. Biochar's ability to sequester carbon for extended periods aligns with nature's long-term carbon persistence strategies, shielding stored carbon from rapid decomposition and locking it away from the atmosphere. This aligns with carbon farming, a regenerative agricultural approach aimed at increasing carbon storage in soils. Biochar amendments enhance soil organic carbon content, contributing to the accumulation of stable carbon stocks (Rombel et al., 2022). It plays a crucial role in carbon sequestration by enhancing soil aggregation, promoting microbial interactions, and stabilizing organic matter. This indirectly contributes to the formation of soil carbon sinks, mirroring the mechanisms that build resilient carbon stocks in natural ecosystems. Biochar's impact on microbial communities influences the fate of sequestered carbon, with microorganisms within biochar contributing to carbon storage and interactions with other microbial communities affecting decomposition rates. The exploration of biochar-mediated carbon sequestration holds implications for long-term carbon accounting, as the sequestration potential of biochar-amended soils contributes to the legacy of carbon storage, influencing carbon balance calculations over extended periods. This legacy reflects nature's imprint on carbon fluxes that shape global climate dynamics.

Biochar's role in carbon sequestration aligns with reforestation efforts, as it enhances soil carbon stocks through biochar amendments, complementing the carbon conservation potential of forests. Biochar's carbon sequestration facilitated by biochar holds promise for climate change mitigation, as it locks away atmospheric carbon in stable forms, contributing to reducing greenhouse gas concentrations. Biochar's carbon footprint is intertwined with its production process, with a comprehensive assessment including carbon emissions associated with biochar production. The potential for carbon sequestration through biochar has implications in carbon markets, where stored carbon represents a valuable asset that can be traded and monetized. Future research on biochar-mediated carbon sequestration will reveal the complexity of carbon cycling within soil ecosystems, reflecting nature's ever-evolving mechanisms for regulating carbon flows. Biochar's enduring legacy is carbon guardianship, safeguarding the Earth's carbon balance and shaping the trajectory of global environmental changes (Nepal et al., 2023). Biochar, a natural carbon storage system, has a long-standing legacy in mitigating climate change impacts. It captures carbon dioxide from the atmosphere and locks it away in stable forms, ensuring that it remains protected from decomposition and sequestered in the soil. This long-term storage mechanism mirrors nature's strategies for maintaining carbon reservoirs over geological time scales.

Biochar counters the anthropogenic release of carbon emissions by converting it into stable carbon forms, addressing the ongoing challenge of rising greenhouse gas concentrations. This counterbalance reflects nature's capacity for regulating carbon fluxes within ecosystems. Biochar's carbon guardianship is amplified through the enhancement of soil carbon sinks, which promote soil aggregation, microbial interactions, and organic matter stabilization. This indirectly contributes to the expansion of carbon storage in soils, reflecting nature's mechanisms for building resilient carbon stocks. Biochar's carbon guardianship is a gift to future generations, offering a pathway to mitigate climate change impacts. The legacy of sequestered carbon ensures that biochar-amended soils continue to provide benefits long after implementation. Biochar's carbon guardianship reinforces ecological resilience by contributing to soil fertility, water retention, and nutrient availability. It also contributes to the conservation of biodiversity by enhancing soil health and ecosystem functioning, creating habitats that support diverse plant and microbial communities (Maron et al., 2018). Biochar's carbon guardianship bridges traditional wisdom and modern science, as its practice dates back centuries and aligns with nature's intergenerational knowledge transmission. Carbon sequestration initiatives acknowledge biochar's potential to contribute to climate change mitigation, reflecting nature's principles of adopting measures that ensure long-term environmental sustainability. Biochar's carbon guardianship is a paradigm for sustainable land management, offering a bridge between carbon sources and sinks (Hardy et al., 2019). Biochar's role as a carbon guardian also contributes to climate resilience and adaptation by sequestering carbon and enhancing soil functions. This resilience reflects nature's capacity for adapting to changing environmental conditions.

Biochar's legacy fosters environmental stewardship, as communities embrace biochar for carbon sequestration and soil enhancement, creating a culture of responsible caretaking of ecosystems (Li & Tasnady, 2023). As climate challenges intensify, biochar stands as a beacon of hope in a changing world. It reflects humanity's capacity to work alongside nature to address pressing environmental issues. Biochar's carbon guardianship underscores the importance of collaborating with natural processes. By aligning human activities with the Earth's natural carbon cycling mechanisms, biochar offers a harmonious approach to carbon management. This collaboration reflects nature's principles of interdependence and synergy. The legacy of biochar extends beyond the present moment, leaving a positive imprint for future generations. As biochar-amended soils continue to sequester carbon and enhance ecosystem services, the impact of carbon guardianship endures. This lasting legacy reflects nature's gift of resilience and sustenance to those who follow. Biochar's legacy carries a powerful message to humanity, reminding us of our role as custodians of the Earth and echoes nature's call for conscious actions that uphold the health and longevity of our planet. By sequestering carbon, biochar exemplifies the equilibrium that nature seeks to maintain between carbon sources and sinks, reflecting nature's emphasis on harmony and equilibrium within the biosphere.

17.9 Elicitation of beneficial microbial consortia: unveiling symbiosis

Biochar plays a crucial role in soil ecosystems by fostering beneficial microbial consortia, which are collaborative assemblies of microorganisms that work together to achieve mutual benefits. These consortia can drive various soil processes, such as nutrient cycling, disease suppression, and plant growth promotion (Dorjee et al., 2024). Biochar's role as a microbial hub allows it to create microenvironments that attract and support diverse microorganisms, mirroring nature's creation of habitats that foster symbiotic interactions. The plant–microbe–biochar nexus is essential for nutrient exchange, as microorganisms within biochar can interact with plant roots, enhancing nutrient uptake and promoting plant growth. Mycorrhizal fungi also benefit from biochar, forming symbiotic partnerships with plants, enhancing nutrient and water uptake. Biochar–microbial synergy can help reduce disease incidence by antagonizing pathogenic organisms, creating a microbial balance that reduces disease incidence. This synergy reflects nature's strategies for maintaining equilibrium and health within ecosystems. Biochar's influence on microbial consortia can promote the proliferation of nitrogen-fixing bacteria, which convert atmospheric nitrogen into plant-available forms, enriching soil fertility. This encourages nature's utilization of microbial partners to enhance nutrient availability.

Biochar plays a crucial role in promoting beneficial fungal networks and cellulase activity (Yin et al., 2023), which can enhance nutrient uptake and disease resistance in soil ecosystems. It also facilitates the formation of biofilms, complex communities of microorganisms encased in a matrix, fostering cooperative interactions and resource sharing. Biochar's impact on beneficial microbial consortia amplifies functional diversity within soil ecosystems, allowing diverse microorganisms to perform various ecological roles. These partnerships contribute to the provision of ecosystem services, such as nutrient availability, disease suppression, and plant growth. Biochar's presence creates communication channels within microbial consortia, fostering cooperative behaviors and mirroring nature's intricate information exchange networks. The interactions facilitated by biochar can lead to beneficial feedback loops, as microorganisms within biochar enhance nutrient availability, plant health, and ecosystem functioning. This amplification reflects nature's ability to reinforce positive feedback mechanisms within ecosystems. Overall, biochar's role in microbial consortia and biofilm formation contributes to the resilience and sustainability of soil ecosystems.

Biochar plays a crucial role in soil resilience by fostering beneficial microbial communities that adapt to changing conditions, enhancing ecosystem stability and resistance to disturbances. This strengthens soil resilience, mirroring nature's creation of resilient communities that navigate environmental challenges. Biochar's influence on microbial consortia aligns with restoration ecology, a field focused on healing degraded ecosystems. The introduction of beneficial microbes through biochar can accelerate the recovery of ecosystem functions and biodiversity, reflecting nature's regenerative capacities within ecosystems. Biochar's interaction with microbial consortia extends to the soil food web, influencing trophic interactions and shaping the flow of energy and nutrients through the food web. This interaction resonates with nature's trophic dynamics that sustain ecosystem functioning (Maron et al., 2018). Biochar's influence on plant–microbe communication extends to the release of signaling compounds that affect plant responses to environmental cues.

Biochar's role in eliciting beneficial microbial consortia reflects the evolution of symbiotic relationships over time. Microbes within biochar adapt their strategies to collaborate and thrive within the biochar matrix, mirroring nature's evolutionary processes that give rise to cooperative associations. Biochar's impact on microbial consortia contributes to the harmonization of biogeochemical cycles, influencing nutrient cycling, carbon dynamics, and element interactions (Gorovtsov et al., 2020). In conclusion, biochar's role in eliciting beneficial microbial consortia reflects nature's capacity for balanced and integrated ecosystems.

17.10 Plant–microbe interplay in biochar-amended soils: a harmonious choreography

Biochar plays a crucial role in fostering harmonious growth, nutrient exchange, and ecosystem vitality through its interactions with plants and microbes. Plants and microbes are interdependent, with microbes contributing essential nutrients, protecting against pathogens, and enhancing stress tolerance, while plants provide sugars and organic compounds that fuel microbial activity. Biochar emerges as a mediator, bridging the gap between plants and microbes through its porous structure and chemical properties. Within biochar-amended soils, root zones become microbial enclaves, enhancing microbial diversity and abundance around plant roots. This enrichment reflects nature's strategy of cultivating microenvironments to optimize plant–microbe interactions (Xiang et al., 2022). Biochar's influence extends to nutrient exchange and mutualism, with beneficial microbes contributing to nutrient solubilization and mineralization, enhancing nutrient availability to plants. Biochar also amplifies mycorrhizal collaborations between plants and fungi, fostering mycorrhizal hyphal growth and efficient nutrient and water uptake by plants. This amplification reflects nature's mycorrhizal partnerships that optimize resource acquisition. In conclusion, biochar plays a vital role in the plant–microbe interplay, enhancing nutrient uptake by plants and promoting nutrient cycling pathways that sustain plant growth.

Biochar plays a crucial role in plant–microbe interactions, promoting disease resistance through the presence of beneficial microbes that act as allies, competing with pathogens and producing antimicrobial compounds. This defense strategy mirrors nature's use of microbial allies to protect plant health. Biochar's presence can also influence the production and availability of plant growth hormones, enhancing plant vigor and resilience. Its water-holding capacity and improved soil structure help mitigate drought and temperature extremes, supporting plant survival. Root exudates play a significant role in the plant–microbe interactions, shaping microbial communities that engage in nutrient cycling and disease suppression (Weller et al., 2002). Biochar's interaction with plant roots reinforces soil structure, creating stable aggregates that enhance soil porosity, aeration, and water infiltration. This contributes to nature's creation of soil architectures that support plant growth and microbial activities.

The interaction between plants, microbes, and biochar forms a feedback loop, influencing microbial communities within biochar-amended soils. This feedback reflects nature's use of cyclical relationships to drive ecosystem processes. The interaction enriches biodiversity and provides ecosystem services, promoting nutrient cycling, disease suppression, and organic matter decomposition. Biochar's role in carbon storage impacts plant growth dynamics, enhancing soil fertility and water-holding capacity. It also shapes root zone microbial ecology, creating unique microenvironments that cater to plant needs. This aligns with nature's design of microhabitats that accommodate specialized microbial communities. Biochar's role in agroecosystems is significant, as it enhances nutrient cycling, disease suppression, and plant growth, aligning with sustainable agricultural practices. The plant–microbe interactions showcase evolutionary strategies, as plants and microbes evolve mechanisms to communicate, exchange nutrients, and regulate each other's growth. This dialog reflects nature's commitment to shaping relationships for mutual benefit. Biochar plays a crucial role in enhancing plant–microbe interactions, leading to bountiful harvests and ecosystem resilience. It enhances nutrient availability and disease resistance, contributing to agricultural productivity and promoting dynamic equilibrium. In a changing climate, biochar functions harmonious plant–microbe interactions, improving nutrient cycling, water retention, and disease suppression. This harmony reflects nature's ability to adapt and balance in the face of environmental uncertainties. Biochar's role as a mediator and facilitator underscores its significance in fostering harmonious growth and nutrient exchange. In future chapters, we will explore metabolic dialogs, structural connections, and functional harmonies that arise from the intertwined relationships of plants, microbes, and biochar, guiding living organisms within the soil.

17.11 Biochar's impact on soil pH: unveiling nature's pH harmonizers

Biochar plays a crucial role in soil chemistry by acting as a pH buffer, similar to nature's innate pH regulators. It moderates soil pH fluctuations by counterbalancing soil acidity, preventing drastic changes. This buffering role mimics nature's strategy of employing buffering agents to stabilize chemical conditions. Biochar's pH buffering also extends to nutrient availability by optimizing nutrient solubility and ensuring essential elements remain accessible to plants. This adjustment reflects nature's effort to optimize nutrient accessibility for plant growth. The pH-stabilizing influence of biochar provides a microbial comfort zone, allowing microorganisms to thrive within pH ranges conducive to their growth and activity. This stabilization mirrors nature's creation of microenvironments that support diverse microbial communities. Biochar's pH-regulating impact aligns with the pH preferences of plants and microbes, creating conditions suitable for plant nutrient uptake and microbial processes. This alignment reflects nature's pH gradients that shape microbial niches.

Biochar's pH-buffering role affects cation exchange, enhancing nutrient availability to plants. Stable pH conditions foster efficient cation exchange, aligning with nature's design of ion exchange systems. Biochar's pH-dependent enzymatic activity moderates nutrient cycling and organic matter decomposition, optimizing nutrient release. It mitigates extreme pH shifts that can lead to nutrient deficiencies or toxicities, mirroring nature's mechanisms for preventing imbalances in ecosystems. Biochar's pH regulation can mitigate toxic metal availability by stabilizing pH levels, minimizing toxicological risks. pH resilience and plant tolerance are also influenced by biochar. Plants within biochar-amended soils can better withstand pH fluctuations, maintaining steady growth and development. pH also acts as a microbial modulator, shaping microbial communities and metabolic activities. Biochar's pH-buffering role fosters stable pH conditions that cater to specific microbial populations, mirroring nature's design of pH gradients that govern microbial distributions.

Biochar plays a crucial role in regulating soil pH, which is essential for maintaining the resilience and long-term balance of ecosystems. It buffers against excessive acidity and alkalinity, ensuring pH conditions conducive to plant growth and microbial activity (Gu et al., 2023). This regulation mirrors nature's mechanisms for preventing extreme pH shifts. Biochar's pH-buffering can optimize soil structure, enhancing water infiltration and root penetration. This optimization reflects nature's creation of diverse soil architectures that accommodate various ecological functions. Additionally, biochar's role in pH modulation influences biogeochemical cycles, shaping element mobility and availability, impacting nutrient cycling pathways (Gorovtsov et al., 2020). This influence aligns with nature's interconnected cycles that maintain elemental equilibrium. Overall, biochar's impact on soil pH is significant and contributes to the resilience and long-term balance of ecosystems.

Biochar's interaction with soil pH influences rhizosphere ecology, the zone around plant roots rich in microbial activity. Stable pH conditions created by biochar influence microbial diversity and plant–microbe interactions, reflecting nature's design of microenvironments. Biochar's pH-buffering impact contributes to soil resilience, supporting plant growth and microbial processes, enhancing the soil's capacity to withstand disturbances. This resilience mirrors nature's strategy of maintaining stability through dynamic processes. In the face of climate change, biochar's pH-regulating role gains significance. Stable pH conditions provide plant and microbial resilience to changing environmental factors, illustrating nature's ability to navigate shifts and maintain homeostasis. Biochar's pH harmony shapes nutrient availability, microbial communities, and ecosystem stability. In the future, we will explore metabolic adaptations, structural transformations, and functional alignments arising from the interplay of biochar and soil pH dynamics, aiming to understand the delicate balance that governs soil chemistry and its role in sustaining life.

17.12 Resilience against pathogens: unveiling biochar's immune boosting power

Biochar plays a crucial role in soil ecosystems, acting as an immune booster that enhances soil's defenses against pathogens. Its physical and chemical properties create an environment conducive to disease suppression, reflecting nature's strategies of deploying defense mechanisms to safeguard ecosystems. Biochar's porous structure acts as a barrier against soil pathogens, preventing their movement and restricting their access to plant roots. Its chemical composition contributes to its immune-boosting power, releasing compounds that inhibit pathogen growth or create unfavorable conditions for their survival. This chemical arsenal mirrors nature's deployment of secondary metabolites to deter pathogens. Biochar fosters beneficial microbial communities that act as natural antagonists to pathogens, engaging in competitive interactions, inhibiting pathogen colonization and proliferation. This alliance reflects nature's utilization of microbial competitors for disease suppression. In the complex battle between plants and soil pathogens, biochar emerges as a champion in this battle, enhancing the soil's defenses against pathogens.

Biochar has a significant impact on plant resistance against pathogens, triggering innate defense mechanisms that make plants more resistant to disease. This induced resistance mirrors nature's ability to prime plants for defense in response to external threats. Biochar contributes to disease-suppressive soils, where pathogens struggle to establish themselves, by outcompetent beneficial microbial communities. This soil defense mirrors nature's establishment of diverse microbial consortia to maintain equilibrium. Biochar's immune-boosting impact can be disease-specific, with different pathogens facing varying levels of suppression based on the interplay between biochar, plant species, and microbial communities. This specificity aligns with nature's targeted defense strategies. Biochar's immune-boosting influence enhances rhizosphere resilience, creating a microenvironment that deters pathogen colonization. Biochar also contributes to soil microbiome balance by promoting beneficial microbes and curbing pathogen growth, maintaining a diverse and functional microbial community. The immune-boosting influence of biochar fosters a plant–microbe pathogen defense partnership, with beneficial microbes working in tandem with plants to form a united front against pathogens. Biochar-mediated systemic resistance also extends to a plant's ability to fend off pathogens system-wide.

Biochar's role in disease resistance is significant, as it helps maintain ecosystem stability by creating an enduring barrier against pathogen invasion. This resilience is a result of nature's investment in building resilience over generations. Biochar's immune-boosting capacity is inspired by nature's lessons in pathogen defense, which include symbiotic relationships and chemical defenses used by plants and microbes to fend off pathogens. This aligns with resistance management strategies, which promote a diverse and resilient soil environment to mitigate the risk of pathogen resistance to chemical treatments. Biochar's impact on disease suppression aligns with principles of sustainable agriculture, as disease-resistant plants, supported by biochar-amended soils, exemplify nature's approach to building resilient and balanced agroecosystems.

Biochar plays a crucial role in ecological restoration by enhancing plant survival and disease resistance in biochar-amended soils. This mimics nature's strategies for rejuvenating degraded habitats. Biochar's immune-boosting impact reflects its adaptive response to environmental challenges, as it influences beneficial microbes to counter new threats. This aligns with nature's evolutionary arms race between pathogens and their hosts. As biochar strengthens the soil's immune system against pathogens, it aligns with nature's strategies of resistance, collaboration, and adaptation. The interplay between biochar and soil's immune mechanisms reveals metabolic dialogs, structural defenses, and functional resilience, contributing to the intricate choreography that safeguards life beneath the earth's surface.

17.13 Longevity of biochar effects: deciphering nature's timeless imprint

This chapter explores the longevity of biochar's effects on soil ecosystems, highlighting its impact on nutrient cycling, microbial communities, and plant growth. Biochar's stable carbon structure resists decomposition, allowing its effects to persist over extended periods. This resilience reflects nature's mastery in preserving essential components of ecosystems. Biochar's effects include the gradual release of nutrients, providing a sustained source of nourishment for plants and microbes. This release mirrors the natural nutrient cycles that sustain ecosystems. Biochar's enduring impact extends to microbial communities, as beneficial microbes within biochar-amended soils persist, adapting to changing conditions and influencing microbial succession. Biochar's role in stabilizing nutrient cycling also endures over time, as beneficial microbes within biochar-amended soils maintain their functionality, ensuring a consistent supply of essential nutrients. This stability reflects nature's preference for sustaining nutrient cycles through time.

Biochar's long-term impact on soil structure includes its role in shaping soil aggregation and porosity, enhancing water retention, aeration, and root penetration. It also contributes to long-term carbon sequestration, reducing atmospheric carbon dioxide concentrations and storing carbon in various reservoirs. Biochar's presence also combats soil erosion, enhancing landscape stability and retaining soil particles. Its effects extend across plant generations, providing nutrient reservoirs and favorable conditions for successive generations. Biochar-amended soils maintain enriched microbial communities and nutrient availability, buffering against disturbances and stressors. This resilience reflects nature's enduring mechanisms for maintaining soil vitality. Biochar's role in supporting biodiversity leaves a lasting legacy, as the enhanced soil conditions created by biochar persist, fostering diverse plant species and microorganisms that thrive over time. This legacy aligns with nature's design of habitats that nurture diverse life forms.

Biochar's role in ecological succession is integral to the gradual change of species and communities over time, influencing the trajectory of ecosystem development. Its enduring impact is reflected in historical soils, which show evidence of ancient carbon sequestration and enhanced fertility. Studying these soils offers insights into the enduring legacy of biochar's effects over millennia, reflecting nature's lessons in creating lasting impacts. Biochar's lasting imprint navigates the tides of time, mirroring nature's ability to withstand environmental changes and build stability and resilience within ever-changing landscapes. As we explore the longevity of biochar's effects, we uncover nature's timeless echo within soil dynamics, reflecting nature's efforts to sustain the intricate web of life. In the future, we will explore metabolic echoes, structural legacies, and functional echoes from the interplay of biochar and the passage of time, aiming to decipher the eternal symphony within the heart of the soil.

17.14 Microbial enzyme activities: unraveling biochar's enzymatic symphony

This chapter explores the role of biochar in microbial enzyme activities, highlighting its influence on soil nutrient cycling, organic matter decomposition, and ecosystem dynamics. Enzymes, the molecular catalysts of biochemical reactions, play a crucial role in the intricate ballet of microbial life within soil ecosystems. Biochar, with its physical and chemical properties, acts as a conductor, enhancing enzyme activity within the soil matrix. This enhances enzyme production, accelerating the breakdown of organic matter and nutrient cycling. Biochar's role in enhancing enzyme activities contributes to the degradation of organic matter, accelerating the breakdown of complex organic compounds into simpler forms (Xiang et al., 2022). Biochar also acts as a catalyst for nutrient cycling, expediting the transformation of nutrients, enhancing their availability to plants and microbes. This reflects nature's reliance on enzymatic reactions to drive nutrient fluxes. Overall, biochar's impact on microbial enzyme activities mirrors nature's precision in driving vital processes. Recent research on biochar application reveals its impact on soil microbial status, pH, organic matter, and nutrient elements, affecting crop yield and quality. It also explores soil enzymes (acid phosphatase, nitrate reductase, protease, and β-glucuronidase) and functional genes (*nosZ, amoA, mcrA, pmoA*, and *phoD*) (Zhao et al., 2023). A graphical illustration of the impact of biochar on soil physical properties and soil microbial ecology is shown in Fig. 17.1A and B, respectively.

Biochar, a natural substance, plays a crucial role in soil enzymatic reactions, facilitating the colonization of microbes and enzyme activity. This synergy accelerates biochemical processes, enhancing ecosystem functions. Enzymatic activities regulate soil pH through nutrient transformations, and biochar's impact on enzyme-mediated reactions contributes to pH stability. Biochar also shapes enzyme diversity within soils, encouraging the growth of diverse enzyme-producing communities. This promotes nutrient release from organic matter, allowing plants and microbes to utilize nutrients in forms that are easily absorbed by plants and microbes. Biochar's enzymatic reactions also influence the composition and structure of soil organic matter, reflecting nature's organic matter evolution. Moreover, biochar's influence on enzymatic activities contributes to disease suppression (Weller et al., 2002), as enzymes produced within biochar-amended soils inhibit pathogen growth and activate plant defense responses, reflecting nature's use of enzymes as defense mechanisms.

Biochar plays a crucial role in enzymatic reactions, facilitating plant–microbe interactions and carbon sequestration processes within soils. Enzymes produced by microbes within biochar-amended soils facilitate nutrient exchange, mirroring nature's enzyme-mediated dialogs (Gorovtsov et al., 2020). Biochar's influence on enzyme activities can enhance the stabilization of carbon compounds, reflecting nature's use of enzymatic reactions to regulate carbon dynamics. Enzymes

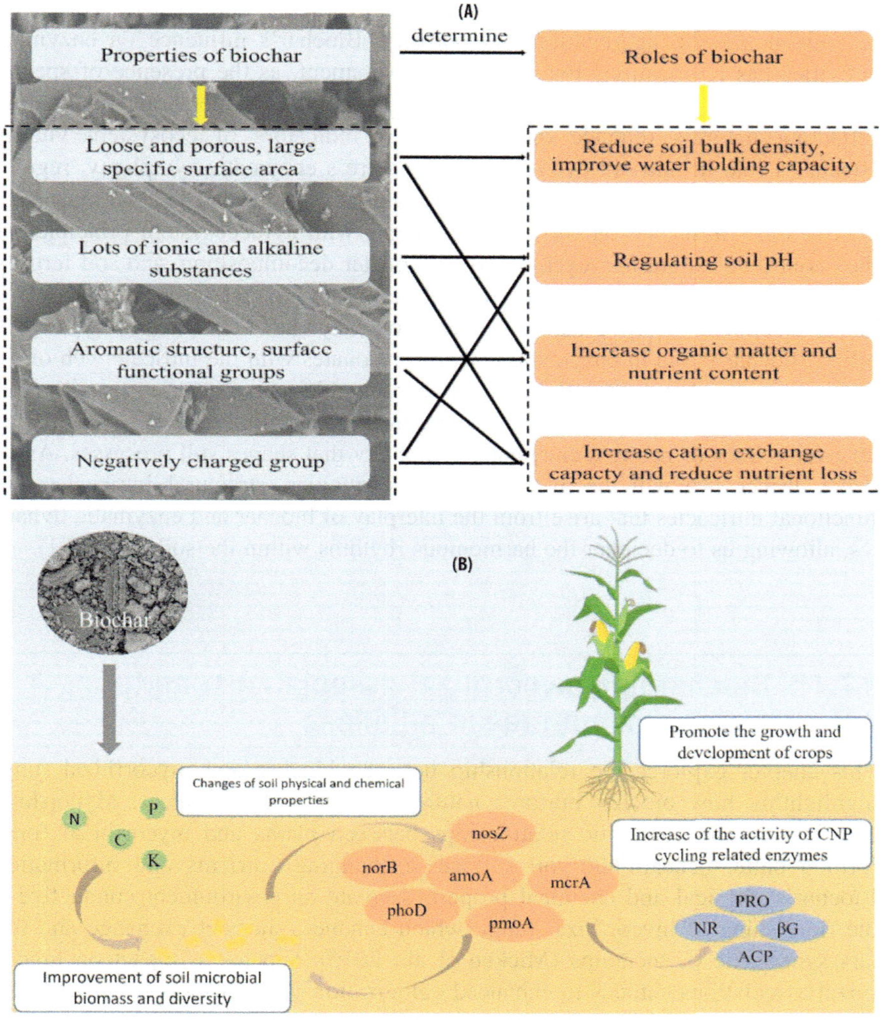

FIGURE 17.1

Biochar affecting soil enzyme activities and soil microbial ecology. (A) Impact of biochar on soil physical properties and (B) mechanisms of biochar in enzymatic actions affecting microbial ecology of soils. *Abbreviations*: *ACP*, acid phosphatase; *NR*, nitrate reductase; *PRO*, protease; *βG*, β-glucuronidase; *norB*, nitric oxide reductase; *nosZ*, N_2O reductase; *amoA*, ammonia monooxygenase; *mcrA*, methanogenic archaea; *pmoA*, methanotrophic bacteria; *phoD*, phosphorus mineralization gene.

MDPI Source: Zhao et al. (2022).

facilitated by biochar contribute to soil structure by breaking down organic matter, enhancing soil aggregation and porosity. Biochar's influence on enzymatic activities has implications for soil health assessment, as the presence of specific enzymes can indicate microbial community diversity and metabolic functioning, reflecting nature's reliance on enzymes as indicators of ecosystem vitality. Biochar's role in nutrient cycling mirrors nature's enzymatic symphony, regulating the flow of essential elements among plants, microbes, and the soil matrix.

Biochar's influence on soil enzymes aligns with agroecosystem principles, as they contribute to nutrient cycling, organic matter decomposition, and soil fertility. These enzymes regulate nutrient availability, organic matter turnover, and microbial interactions, reflecting nature's reliance on enzymatic action for ecosystem stability. Biochar's role in enzymatic reactions resonates with the intricate web of life within soils, as they bridge the gap between organic matter, nutrients, and organisms. As we explore microbial enzyme activities influenced by biochar, we uncover echoes of nature's enzymatic symphony that shapes soil processes. As we delve deeper, we will discover metabolic harmonies, structural intricacies, and functional intricacies that arise from the interplay of biochar and enzymatic dynamics, allowing us to decipher the harmonious rhythms within the soil (Fig. 17.1).

17.15 Biochar and mycorrhizal associations: nurturing nature's subterranean alliances

This chapter explores the relationship between biochar and mycorrhizal fungi, highlighting how biochar nurtures nature's subterranean alliances. Mycorrhizal associations, the symbiotic relationships between plants and mycorrhizal fungi, form a ballet beneath the soil surface, exchanging nutrients and information. Biochar's physical and chemical properties create an environment conductive to the flourishing of mycorrhizal fungi, which enhances nutrient exchange and fosters synergistic partnerships (Mickan et al., 2016). Biochar's impact on mycorrhizal associations extends to enhanced colonization of plant roots by mycorrhizal fungi. Biochar-amended soils offer a supportive habitat, encouraging mycorrhizal hyphae to colonize root surfaces. Biochar's role in mycorrhizal associations contributes to nutrient exchange facilitation, as mycorrhizal fungi extend their hyphae into soil, enhancing nutrient absorption and transferring nutrients to plant roots (Yin et al., 2023). Mycorrhizal-mediated nutrient uptake by plants is also facilitated by biochar, as mycorrhizal fungi deliver nutrients more efficiently, aligning with nature's strategy of nutrient transport through intricate underground networks. Overall, biochar's role in mycorrhizal associations is a testament to the intricate interplay between plants and fungi in nature.

Biochar enhances plant growth and health by supporting mycorrhizal-assisted nutrient uptake and stress tolerance. It also influences soil structure by enhancing soil aggregation, porosity, and water retention through mycorrhizal networks.

Biochar's role in ecosystem restoration involves supporting mycorrhizal associations, which play a pivotal role in rehabilitating degraded soils and establishing plant cover and soil stability. This aligns with biodiversity promotion, as diverse mycorrhizal communities contribute to plant species richness and ecosystem complexity. Biochar's influence on mycorrhizal associations enhances stress tolerance in plants, reflecting nature's capacity to provide resilience through interconnected partnerships. Additionally, biochar's role in mycorrhizal associations influences soil carbon sequestration, with mycorrhizal fungi contributing to carbon storage through their activities in the soil. Overall, biochar's role in plant health, biodiversity, and soil carbon sequestration is crucial for a healthy ecosystem (Alkharabsheh et al., 2021).

Biochar's impact on mycorrhizal associations contributes to disease resistance, as mycorrhizal fungi enhance plant defense responses against pathogens, pests, and stressors. This resistance mirrors nature's use of mycorrhizal interactions for ecosystem protection. Biochar's role extends to interactions with other soil biota, influencing microbial communities within biochar-amended soils. Mycorrhizal fungi play a role in nutrient acquisition, transport, and cycling, reflecting nature's reliance on them to regulate nutrient flows. Biochar fosters mutually beneficial partnerships between plants and mycorrhizal fungi, mirroring nature's symbiotic symbiosis. Mycorrhizal fungi help direct carbon resources toward below-ground structures, reflecting nature's partitioning of carbon for diverse ecosystem functions. Mycorrhizal alliances in soil restoration help stabilize soil structure, enhance nutrient availability, and promote plant establishment, reflecting nature's mechanisms for rejuvenating ecosystems. Biochar's enhancement of mycorrhizal associations aligns with agroforestry principles, as mycorrhizal fungi play a crucial role in promoting tree establishment and growth in agroforestry systems (Mickan et al., 2016).

Biochar plays a crucial role in mycorrhizal associations, which enhance nutrient cycling, soil structure, and plant health, mirroring nature's provision of services through intricate partnerships. This interplay between biochar and mycorrhizal associations reflects nature's mycorrhizal symphony, which sustains nutrient cycles, supports plant vitality, and enhances ecosystem resilience. The upcoming chapters will explore metabolic harmonies, structural intricacies, and functional resonances that arise from the interplay of biochar and mycorrhizal interactions, bringing closer to understanding the harmonious rhythms that unite the roots of life.

17.16 Soil structure enhancement: unveiling biochar's architectural resilience

This chapter explores the role of biochar in enhancing soil structure and its impact on water retention, aeration, and nutrient availability. It highlights the intricate dynamics of soil systems and how nature's soil architectures serve as blueprints for functional ecosystems. Biochar, with its porous nature, enhances soil aggregation by providing attachment points for soil particles, creating stable aggregates that resist erosion and improve water infiltration. This aggregation reflects nature's reliance on forces that bind soil particles together. Biochar also enhances pore space formation, creating pathways for air and water movement, enhancing soil aeration and drainage. This aligns with nature's strategy of creating interconnected channels for fluid transport. Biochar's role in soil structure contributes to improve water retention and drainage, as aggregates and pore spaces facilitate water storage and movement (Khan et al., 2024). Biochar's impact on soil structure also leads to the creation of nutrient reservoirs, which offer spaces for nutrient retention, minimizing nutrient leaching, and making nutrients available for plant uptake. This echoes nature's strategies for nutrient conservation.

Biochar plays a significant role in soil structure, enhancing root growth conditions and microbial habitats. Its improved aggregation and pore spaces create a favorable environment for root penetration and expansion, reflecting nature's emphasis on creating hospitable niches for plant roots. Biochar also contributes to erosion prevention by forming stable aggregates that resist the erosive forces of wind and water, reflecting nature's mechanisms for safeguarding soil integrity. It also enhances soil resilience to compaction, as well-aggregated soils with increased pore spaces resist compression, aligning with nature's creation of robust soil architectures. Biochar's impact on soil porosity is mirrored in its impact on soil structure, with improved pore size distribution facilitating air exchange, water movement, and microbial habitat development (Chang et al., 2021).

Biochar enhances soil structure, allowing it to adapt to environmental changes. Its well-structured soils show resilience to fluctuations in moisture levels and temperature, reflecting nature's ability to cope with dynamic conditions. Biochar's porous nature can alter soil texture, influencing water-holding capacity, and nutrient retention. Aggregates are designated as microbial hotspots, where microbes colonize these stable structures, engaging in nutrient cycling and organic matter decomposition. This aligns with sustainable land management practices, as improved aggregation, pore spaces, and water retention contribute to soil health and productivity (Hardy et al., 2019). Biochar also plays a role in carbon sequestration, as well-structured soils create stable habitats for carbon storage, reflecting nature's incorporation of carbon within soil matrices.

Soil architecture plays a crucial role in ecosystem stability, as well-structured soils resist erosion, support plant growth, and facilitate microbial interactions. Biochar's impact on soil structure aligns with permaculture design principles, as improved aggregation and porosity contribute to self-sustaining ecosystems. The enhancement of soil structure by biochar influences root-mycorrhizal interactions, as improved porosity and nutrient availability support mycorrhizal colonization. Well-structured soils improve water infiltration, reduce runoff, and enhance water-holding capacity, aligning with nature's regulation of hydrological processes. The architectural resonance of biochar aligns with nature's efforts to create functional soil architectures that support plant growth, microbial communities, and nutrient cycling. As we explore the enhancement of soil structure fostered by biochar, we uncover echoes of nature's architectural resonance that shape soil systems.

17.17 Nitrogen fixation and biochar: bridging nature's nitrogen-fixing prowess

This chapter explores the role of biochar in nitrogen fixation, a crucial process in ecosystems. Nitrogen fixation is a vital symphony, where nitrogen-fixing bacteria harmonize with leguminous plants to enhance soil fertility, nutrient cycling, and plant–microbe partnerships. Biochar, as a conductor, influences the dynamics of nitrogen fixation by providing a habitat conducive to nitrogen-fixing bacteria, enhancing their colonization of plant roots and the nitrogen-fixing process itself. Biochar's influence extends to enhanced microbial communities, as nitrogen-fixing bacteria within biochar-amended soils thrive in stable carbon structures, promoting efficient nitrogen fixation. This aligns with nutrient cycling, where nitrogen fixation contributes to the conversion of atmospheric nitrogen into organic forms, enriching soil nutrient pools and supporting plant nutrition. Biochar's role in nitrogen fixation supports symbiotic relationships between leguminous plants and nitrogen-fixing bacteria, as the porous nature of biochar creates niches for these bacteria, fostering their partnership with legumes (Harman et al., 2021). This support mirrors nature's design of mutualistic interactions, highlighting the importance of nitrogen fixation in ecosystems. Biochar enhances plant nitrogen uptake by enhancing the availability of nitrogen fixed by bacteria, which promotes growth and development. This supports nutrient availability and ecosystem productivity, enriching soil nutrient profiles and facilitating plant growth. Biochar's impact on nitrogen fixation aligns with soil fertility enhancement, supporting the sustainability of diverse plant communities. Nitrogen fixation also plays a role in biodiversity promotion, fostering diverse plant species and microbial communities. This aligns with nature's emphasis on nutrient cycling to maintain biodiversity. The enhancement of nitrogen fixation by biochar aligns with agroecosystem principles, benefiting plant growth in agricultural systems, demonstrating the alignment of nutrient acquisition strategies with agricultural productivity. Results show that biochar enhances soil urease, catalase, and alkaline phosphatase activity, compensates for nitrogen toxicity, and improves nitrogen utilization (Khan et al., 2022). Overall, biochar's role in nutrient management is crucial for sustainable productivity.

Biochar plays a crucial role in nitrogen fixation, contributing to nutrient cycling and soil health. It enhances nutrient availability, supporting microbial communities, organic matter decomposition, and overall soil vitality. This aligns with sustainable agriculture practices, promoting crop growth without relying on synthetic fertilizers. Biochar also plays a role in climate change mitigation, reducing the need for nitrogen-based fertilizers and minimizing greenhouse gas emissions. It also supports nutrient synchronization in ecological succession, with nitrogen-fixing bacteria contributing to nutrient availability, supporting plant establishment and ecosystem transitions. Furthermore, biochar enhances ecosystem resilience, enhancing nutrient availability during environmental fluctuations, resembling nature's reliance on nitrogen-fixing strategies for stability. Overall, biochar's impact on nitrogen fixation is significant for promoting sustainable agriculture and reducing greenhouse gas emissions. Biochar plays a crucial role in restoration ecology, as it aids in soil rehabilitation and the establishment of native plant species, promoting ecosystem renewal. It also supports nitrogen fixation in biodiversity hotspots, where nitrogen availability influences plant diversity and productivity. Nitrogen-fixing bacteria contribute to soil aggregation, reflecting nature's reliance on microbial activities to enhance soil architecture. This enhances nitrogen fixation and aligns with permaculture design principles, as nitrogen-fixing bacteria support nutrient cycling and plant health, guiding nature in creating self-sustaining ecosystems. The interplay between biochar and nitrogen fixation reveals echoes of nature's nitrogen-fixing harmony, enriching ecosystems through nutrient cycles, plant growth, and microbial interactions. As we explore the interplay, we will discover metabolic harmonies, structural intricacies, and functional resonances that arise from the interplay of biochar and nitrogen fixation, enhancing the vitality of the terrestrial realm.

17.18 Biochar and water-holding capacity: emulating nature's hydrological symphony

This chapter explores the role of biochar in enhancing water retention, infiltration, and availability in soil ecosystems. Biochar, like nature's hydrological systems, plays a crucial role in regulating rainfall, infiltration, runoff, and water availability for plants, microbes, and aquatic systems (Ndede et al., 2022). Its porous structure creates reservoirs for water storage and release, enhancing water retention. Biochar's impact on water-holding capacity extends to enhanced water retention, as its porous nature allows it to absorb and retain water, reducing water loss through evaporation and runoff. This retention enhancement aligns with nature's mechanisms for conserving water resources. Biochar's role in water dynamics contributes to improved infiltration and reduced runoff. Water absorbed by biochar-amended soils infiltrates more efficiently, minimizing surface runoff and reflecting nature's regulation of water movement. Overall, biochar plays a vital role in sustaining water balance in soil ecosystems. Biochar-amended soils supply water as needed, promoting plant growth and vitality. This enhances water dynamics, promoting nutrient movement to plant roots, reflecting nature's coupling of water and nutrient cycles. Biochar also contributes to erosion prevention by minimizing soil erosion, reflecting nature's mechanisms for safeguarding soil structure and stability. Additionally, biochar enhances soil moisture regulation in agroforestry, supporting plant growth and ecosystem functions, echoing nature's coordination of water availability in diverse landscapes. Overall, biochar's impact on water management is significant and aligns with nature's strategies for coping with water scarcity.

Biochar plays a crucial role in enhancing water dynamics, contributing to ecosystem health by supporting microbial activities, organic matter decomposition, and soil vitality. It is particularly beneficial in urban green spaces, where biochar-amended soils regulate soil moisture, supporting green infrastructure and promoting water management principles. Biochar's impact on water dynamics also extends to arid environments, where improved water retention aids plants in water-stressed regions. It also supports aquatic ecosystem resilience, with enhanced water retention in adjacent soils ensuring stable water inputs and supporting aquatic life. Biochar's role in watershed management is also evident, as improved water retention within biochar-amended soils mitigates runoff, improving downstream water quality and demonstrating nature's control over water flow.

Biochar plays a crucial role in water dynamics, impacting soil microbial activity and supporting microbial communities. Its enhanced water-holding capacity aligns with flood control strategies, reducing surface runoff and minimizing flood risks (Ndede et al., 2022). Biochar also supports root health by fostering optimal root growth conditions. Its influence on water dynamics contributes to erosion control by enhancing water retention within biochar-amended soils, stabilizing slopes and minimizing erosion risks. Biochar also supports wetland restoration efforts by improving water-holding capacity, supporting vegetation and hydrological functions (Rasool et al., 2023). The interplay between biochar and water dynamics echoes nature's hydrological harmony, which sustains ecosystems. Biochar's support of water movement, retention, and distribution aligns with nature's phenomena of water movement, retention, and distribution. As we explore the interplay between biochar and water dynamics, we will discover metabolic harmonies, structural intricacies, and functional resonances that arise from the interplay of biochar and water dynamics, allowing us to understand the harmonious rhythms that govern the flow of life-giving water in Table 17.5.

17.19 Beyond fertilization: biochar's holistic impact

This chapter explores the holistic effects of biochar on soil health, focusing on its interconnected web of influences. It highlights the interdependence of nature's elements, such as soil, water, plants, and microbes, in shaping soil structure, nutrient cycling, microbial interactions, and plant vitality. Biochar, as a linchpin in this interconnected web, influences soil structure, water dynamics, microbial

Table 17.5 Biochar's impact on water dynamics and ecosystems.

Section	Key points	Benefits/ impacts	Applications/ contexts	References
Water retention and infiltration	Biochar's porous structure creates reservoirs for water storage, reducing water loss through evaporation and runoff. Enhances infiltration and aligns with nature's water conservation mechanisms	Enhanced water retention and reduced runoff	Agriculture, landscaping, soil remediation, etc.	Vejan et al. (2021)
Drought mitigation	Biochar supports plants during dry periods, promotes plant growth, and aids nutrient movement to plant roots. Contributes to erosion prevention and soil moisture regulation in agroforestry	Mitigation of drought impacts, improved plant growth	Agriculture, reforestation, soil stabilization, etc.	Powlson et al. (2011)
Ecosystem health	Benefits soil microbial activity and organic matter decomposition. Particularly useful in urban green spaces, arid environments, and watershed management. Supports aquatic ecosystems and flood control	Enhanced soil vitality and microbial activity	Urban green infrastructure, arid regions, watershed management, etc.	Powlson et al. (2011)
Root health and erosion control	Fosters optimal root growth conditions and stabilizes slopes. Enhances water retention in biochar-amended soils, contributing to erosion control	Improved root health, reduced soil erosion	Slope stabilization, erosion-prone areas, landscaping, etc.	Jahan et al. (2020)
Wetland restoration	Improves water-holding capacity, supports vegetation and hydrological functions in wetland restoration efforts	Enhanced wetland health and ecosystem functions	Wetland restoration projects, conservation efforts, etc.	Rasool et al. (2023)

(Continued)

Table 17.5 Biochar's impact on water dynamics and ecosystems. *Continued*

Section	Key points	Benefits/ impacts	Applications/ contexts	References
Biochar and water dynamics interplay	Examining the harmonious interplay between biochar and water dynamics in sustaining ecosystems and life-giving water flow	Holistic understanding of biochar's role in water dynamics	Environmental research, ecosystem studies	Prescott et al. (2021)

communities, and nutrient cycling (Prescott et al., 2021). This reflects nature's holistic approach to ecological balance. Biochar's holistic impact on soil health aligns with the pillar of ecosystem functionality, as improved soil structure, nutrient availability, and microbial interactions promote optimal conditions for plant growth and overall ecosystem resilience. This aligns with nature's reliance on interconnected components for ecosystem health. Biochar plays a crucial role in ecosystem services, enhancing soil health, supporting nutrient cycling, water regulation, carbon sequestration, and biodiversity. It aligns with agroecological principles, mimicking nature's ecosystems by improving soil fertility, structure, and nutrient cycling. Biochar's holistic impact mirrors nature's integrated systems of nutrient flow and energy exchange. It also contributes to biodiversity promotion by improving conditions for plant growth, microbial interactions, and nutrient availability, fostering diverse ecosystems. Biochar also contributes to climate change mitigation efforts by improving carbon sequestration, nutrient cycling, and soil structure, reducing greenhouse gas emissions and promoting ecosystem resilience. These effects reflect nature's role in regulating Earth's climate.

Biochar plays a crucial role in soil health, promoting ecosystem recovery and sustainable agriculture. Its holistic effects, including enhanced fertility, structure, and nutrient cycling, align with nature's principles of optimizing soil productivity and maintaining ecological balance. Biochar's role in soil carbon management is also evident, with enhanced carbon sequestration, organic matter decomposition, and nutrient cycling within biochar-amended soils. This aligns with nature's strategies for regulating carbon dynamics. Biochar fosters microbial-ecosystem synergy, allowing microbial interactions, nutrient transformations, and organic matter decomposition within biochar-amended soils. It aligns with restoration ecology goals, supporting the rehabilitation of degraded landscapes through enhanced soil conditions, nutrient cycling, and plant establishment. Biochar's holistic impact also resonates with ecological succession, guiding ecosystem transitions through improved conditions for plant growth and microbial communities. Biochar plays a crucial role in promoting soil health, contributing to ecosystem stability through optimal nutrient cycling, water dynamics, and microbial interactions (Prescott et al., 2021). It aligns with carbon-nitrogen coupling, improving carbon sequestration, nitrogen availability, and nutrient cycling within biochar-amended soils. As we delve deeper, we will discover metabolic harmonies, structural intricacies, and functional resonances that arise from the interplay of biochar's holistic effects, helping us understand the harmonious rhythms that govern the vitality of terrestrial landscapes.

17.20 Future prospects and implications: shaping tropical cropland ecosystems with biochar

Biochar is a transformative force that holds profound implications for soil health, sustainability, biodiversity, and global food security (Wijerathna-Yapa & Pathirana, 2022). As we look ahead, a new landscape emerges, where biochar plays a central role in tropical cropland ecosystems. This landscape mirrors nature's complexity, interdependence, and adaptability, offering a glimpse of the potential biochar holds. Biochar's potential to nurture resilience and adaptation is paramount in the future of uncertainty marked by shifting climates and evolving agricultural challenges. By fostering robust soil structures, efficient nutrient cycling, and diverse microbial communities, biochar-equipped ecosystems are better equipped to weather climatic fluctuations, water scarcity, and pest pressures. Biochar stands at the crossroads of agroecology, embodying the principles of agroecology. As biochar-amended soils teem with microbial life, support nutrient cycling, and enhance water dynamics, they become microcosms of self-sustaining ecosystems. This harmony between agriculture and ecology mirrors nature's balance between productivity and conservation. Carbon sequestration, a cornerstone of climate change mitigation, takes on renewed importance in the future. Biochar's capacity to lock away carbon in soils parallels nature's carbon-storing mechanisms, and its potential as a carbon guardian resonates with nature's timeless approach to carbon balance.

Biochar is a promising solution for nutrient management, biodiversity, and food security (Wijerathna-Yapa & Pathirana, 2022). It enhances nutrient retention, availability, and cycling, aligning with nature's intricate pathways (N. Wang et al., 2022). As a catalyst for biodiversity, biochar fosters diverse microbial communities, plant–microbe partnerships, and nutrient cycling, promoting resilience in ecosystems. As global populations grow, securing food for all becomes a paramount challenge. Biochar's potential to enhance soil fertility, increase crop yields, and mitigate environmental stressors aligns with nature's strategies for sustenance provision. The future of biochar is marked by innovation and research, as researchers explore biochar's molecular interactions, metabolic pathways, and ecosystem effects. This ongoing pursuit of knowledge parallels nature's evolution, paving the way for novel applications and insights. Sustainability is emerging as a guiding star in the future, as biochar mimics nature's nutrient cycles, enhances soil health, and reduces environmental impacts. As a beacon of sustainable practices, biochar mirrors nature's blueprint for harmonious coexistence (Kwaku Armah et al., 2023).

Biochar is a promising future solution for sustainable agriculture, as it can mimic nature's nutrient cycles, improve soil health, and reduce environmental impacts. This aligns with the principles of sustainability, mirroring nature's blueprint for harmonious coexistence. Collaboration among farmers, scientists, policymakers, and communities is crucial in harnessing biochar's potential. Indigenous knowledge and traditional practices are emphasized, allowing biochar to be used for sustainable land stewardship. As biochar's potential becomes evident, policy and regulation will evolve to support its responsible use, fostering ethical use, equitable distribution, and environmental safeguards. Education and outreach will also flourish, as biochar's benefits are shared through knowledge dissemination, workshops, and public engagement. This commitment to education aligns with nature's intention to share wisdom across generations. The future of agriculture will be a landscape of collaboration, stewardship, and education, reflecting the interconnectedness of nature.

Biochar is a key player in achieving carbon-neutral agriculture, as it plays a crucial role in carbon sequestration and its multifaceted impacts. It adapts to changing landscapes due to urbanization, climate shifts, and land use changes, fostering sustainable practices in both rural and urban settings. Biochar's legacy parallels nature's intergenerational transfer of knowledge, ensuring future stewards continue to enhance soil health, foster resilience, and care for the Earth. The future prospects and implications of biochar are a resonant echo of nature's grand symphony, with its impacts on soil health, sustainability, biodiversity, and food security reverberating through time, echoing the principles that govern nature's tapestry. This reflects the intergenerational transfer of knowledge and the importance of preserving the balance between carbon sources and sinks in agriculture. Biochar, a renewable energy source, has the potential to transform tropical cropland ecosystems and contribute to a thriving future. It represents nature's blueprint for progress, as it adapts, innovates, and weaves complex relationships. As we explore biochar's role in these ecosystems, we find ourselves at the intersection of scientific inquiry and natural wonder (Salma et al., 2024). The future landscape is one where biochar stands as a testament to humanity's endeavor to align with nature's wisdom.

Biochar's impact on tropical cropland ecosystems continues to expand, with researchers, practitioners, and communities working together to refine its applications and adapt it to the changing needs of our planet. The journey of biochar's integration into agriculture and ecology is marked by ongoing discovery and innovation. The future also holds the promise of collective responsibility, as decisions made today reverberate into tomorrow. The symbiosis between biochar and tropical cropland ecosystems embodies the essence of coexistence and mutual benefit.

The exploration of biochar's impacts on soil health, microbial interactions, and nutrient cycles has revealed the profound parallels between biochar's role and nature's design (N. Wang et al., 2022). The layers of complexity underlie soil ecosystems, shedding light on the intricate mechanisms that sustain life beneath the surface. As we stand on the cusp of the future, let us remember that we are not separate from nature but intricately woven into its fabric. The story of biochar in tropical cropland ecosystems is a chapter within the larger narrative of humanity's relationship with the Earth. In its potential to transform landscapes, enhance sustainability, and emulate nature's wisdom, biochar offers a tangible pathway toward harmonizing our endeavors with the rhythms of the natural world.

17.21 Conclusion

The exploration of biochar in fertilized tropical croplands has revealed a tapestry of intricate interactions that mirror the timeless artistry of nature. Biochar's impact on soil biotic processes and microbial structures and functions has been revealing its potential to reshape agricultural landscapes in harmony with nature's design. The insights gained from this exploration hold profound implications for the future of sustainable agriculture and ecological balance. Biochar's impact on microbial abundance has been elucidated, showcasing its role in creating a favorable habitat for microbial communities to thrive. This, in turn, influences nutrient cycling, organic matter decomposition, and overall soil health. The parallel to nature's intricate web of microbial interactions highlights biochar's potential to enhance the microbial actions in the rhizosphere. The pathways through which biochar interacts with microbes have been unveiled, akin to nature's blueprints for intricate connections. These interactions influence nutrient availability, soil structure, and even carbon sequestration, reminding us of nature's ability to forge complex relationships that sustain ecosystems. The structural adaptations of microbes have been revealed, echoing nature's spatial organization strategies. Biochar fosters a dynamic environment where microbes can thrive, form biofilms, and create microhabitats, reminiscent of nature's ingenious structural adaptations. The virtuous cycle of nutrient exchange has been explored, showcasing biochar's role in enriching soil fertility. The modulation of metabolic pathways offers insights into biochar's influence on microbial metabolism, leading to enhanced nutrient transformations and ecosystem functions. The mechanisms of carbon sequestration within biochar mirror nature's own methods of locking away atmospheric carbon. Biochar fosters symbiotic relationships among beneficial microbes, echoing nature's strategies of collaboration and mutualism that underpin healthy ecosystems. Plant–microbe interplay in biochar-amended soils has demonstrated how biochar performs harmonious growth, much like nature's partnerships between plants and their microbial allies. The enhanced nutrient availability and soil structure fostered by biochar create an environment conducive to thriving plant–microbe collaborations. The influence of biochar on soil pH, as we have uncovered, reflects its role as a pH buffer, akin to nature's innate pH regulators. This buffering capacity contributes to stable pH levels that promote optimal nutrient availability and microbial functions. Looking into the future, we discern a landscape where biochar plays a transformative role. Just as nature adapts to changing circumstances, biochar adapts to evolving agricultural challenges, acting as a catalyst for innovation, sustainability, and collaborative action. Biochar stands as a symbol of humanity's quest to align with nature's wisdom, as we cocreate a future that resonates with nature's harmonies.

Acknowledgment

The authors acknowledge the global community of biochar initiatives for their global awareness for promoting biochar in agriculture and environment.

Author contributions

All authors have contributed equally in writing, revising, and finalizing this chapter besides their supervision, visualization, and conceptualization.

Ethical statement

Ethical issues do not apply to this review chapter. This chapter does not involve a study on animals.

Conflict of interest

The authors declare that there are no conflicts of interest among the authors for the publication of this chapter in the form of a book.

Funding declaration

The authors declare that they have not received any financial assistance, and they have no known competing financial interests or personal relationships that could have appeared to influence the work reported in this chapter.

Data availability statement

The authors declare that there are no specific data for preparing this review article; however, the authors declare that they will make screening data of consulted literature available whenever it is required.

Declaration on the use of AI tools

The authors declare that AI tools like Grammarly was used for improved readability and language. AI tools were not used for any graphical illustrations.

References

Afridi, M. S., Ali, S., Salam, A., César Terra, W., Hafeez, A., Sumaira., Ali, B., S. AlTami, M., Ameen, F., Ercisli, S., Marc, R. A., Medeiros, F. H. V., & Karunakaran, R. (2022). Plant microbiome engineering: Hopes or hypes. *Biology*, *11*(12). Available from https://doi.org/10.3390/biology11121782, http://www.mdpi.com/journal/biology.

Ahmad, M., Wang, X., Hilger, T. H., Luqman, M., Nazli, F., Hussain, A., Zahir, Z. A., Latif, M., Saeed, Q., Malik, H. A., & Mustafa, A. (2020). Evaluating biochar-microbe synergies for improved growth, yield of maize, and post-harvest soil characteristics in a semi-arid climate. *Agronomy*, *10*(7). Available from https://doi.org/10.3390/AGRONOMY10071055, https://www.mdpi.com/2073-4395/10/7/1055.

Ajijah, N., Fiodor, A., Pandey, A. K., Rana, A., & Pranaw, K. (2023). Plant growth-promoting bacteria (PGPB) with biofilm-forming ability: A multifaceted agent for sustainable agriculture. *Diversity*, *15*(1), 112. Available from https://doi.org/10.3390/d15010112.

Alkharabsheh, H. M., Seleiman, M. F., Battaglia, M. L., Shami, A., Jalal, R. S., Alhammad, B. A., Almutairi, K. F., & Al-Saif, A. M. (2021). Biochar and its broad impacts in soil quality and fertility, nutrient leaching and crop productivity: A review. *Agronomy*, *11*(5). Available from https://doi.org/10.3390/agronomy11050993, https://www.mdpi.com/2073-4395/11/5/993/pdf.

Almendras, K., Iannuzzi, S., Carú, M., & Orlando, J. (2022). Diversity of microbial functional genes should be considered during the interpretation of the qPCR melting curves. *Microbial Ecology*, *84*(3), 935–940. Available from https://doi.org/10.1007/s00248-021-01883-9, https://www.springer.com/journal/248.

Barone, G. D., Cernava, T., Ullmann, J., Liu, J., Lio, E., Germann, A. T., Nakielski, A., Russo, D. A., Chavkin, T., Knufmann, K., Tripodi, F., Coccetti, P., Secundo, F., Fu, P., Pfleger, B., Axmann, I. M., & Lindblad, P. (2023). Recent developments in the production and utilization of photosynthetic microorganisms for food applications. *Heliyon*, *9*(4). Available from https://doi.org/10.1016/j.heliyon.2023.e14708.

Bello, A., Liu, W., Chang, N., Erinle, K. O., Deng, L., Egbeagu, U. U., Babalola, B. J., Yue, H., Sun, Y., Wei, Z., & Xu, X. (2023). Deciphering biochar compost co-application impact on microbial communities mediating carbon and nitrogen transformation across different stages of corn development. *Environmental Research*, *219*, 115123. Available from https://doi.org/10.1016/j.envres.2022.115123.

Bengtsson-Palme, J. (2020). Microbial model communities: To understand complexity, harness the power of simplicity. *Computational and Structural Biotechnology Journal*, *18*, 3987–4001. Available from https://doi.org/10.1016/j.csbj.2020.11.043.

Bertola, M., Ferrarini, A., & Visioli, G. (2021). Improvement of soil microbial diversity through sustainable agricultural practices and its evaluation by -omics approaches: A perspective for the environment, food quality and human safety. *Microorganisms*, *9*(7). Available from https://doi.org/10.3390/microorganisms9071400.

Chang, Y., Rossi, L., Zotarelli, L., Gao, B., Shahid, M. A., & Sarkhosh, A. (2021). Biochar improves soil physical characteristics and strengthens root architecture in Muscadine grape (*Vitis rotundifolia* L.). *Chemical and Biological Technologies in Agriculture*, *8* (1). Available from https://doi.org/10.1186/s40538-020-00204-5, http://chembioagro.springeropen.com/.

Chukwuka, K.S. Akanmu, A.O. Umukoro, B.O.-J. Asemoloye, M.D. Christopher O.A., Biochar: A vital source for sustainable agriculture. (2020). Available from https://doi.org/10.13140/RG.2.2.21560.49928.

Clough, T. J., Condron, L. M., Kammann, C., & Müller, C. (2013). A review of biochar and soil nitrogen dynamics. *Agronomy*, *3*(2), 275–293. Available from https://doi.org/10.3390/agronomy3020275, https://www.mdpi.com/2073-4395/3/2/275/pdf.

Deshoux, M., Sadet-Bourgeteau, S., Gentil, S., & Prévost-Bouré, N. C. (2023). Effects of biochar on soil microbial communities: A meta-analysis. *Science of the Total Environment*, *902*, 166079. Available from https://doi.org/10.1016/j.scitotenv.2023.166079.

Dorjee, L., Nishmitha, K., Pattanayak, S., Wangmu, T., Meshram, S., Chongtham, S., & Gogoi, R. (2024). Biochar: A comprehensive review on a natural approach to plant disease management. *Journal of Pure and Applied Microbiology*, *18*(1), 29–45. Available from https://doi.org/10.22207/JPAM.18.1.58, https://microbiologyjournal.org/download/90645/.

Evangelista, S. J., Field, D. J., McBratney, A. B., Minasny, B., Ng, W., Padarian, J., Román Dobarco, M., & Wadoux, A. M. J.-C. (2023). A proposal for the assessment of soil security: Soil functions, soil services and threats to soil. *Soil Security*, *10*, 100086. Available from https://doi.org/10.1016/j.soisec.2023.100086.

Farrell, M., Kuhn, T. K., Macdonald, L. M., Maddern, T. M., Murphy, D. V., Hall, P. A., Singh, B. P., Baumann, K., Krull, E. S., & Baldock, J. A. (2013). Microbial utilisation of biochar-derived carbon. *The Science of the Total Environment*, *465*, 288–297. Available from https://doi.org/10.1016/j.scitotenv.2013.03.090, Nov,

Giagnoni, L., & Renella, G. (2022). Effects of biochar on the C use efficiency of soil microbial communities: Components and mechanisms. *Environments - MDPI*, *9*(11). Available from https://doi.org/10.3390/environments9110138, http://www.mdpi.com/journal/environments.

Gorovtsov, A. V., Minkina, T. M., Mandzhieva, S. S., Perelomov, L. V., Soja, G., Zamulina, I. V., Rajput, V. D., Sushkova, S. N., Mohan, D., & Yao, J. (2020). The mechanisms of biochar interactions with microorganisms in soil. *Federation Environmental Geochemistry and Health*, *42*(8), 2495–2518. Available from https://doi.org/10.1007/s10653-019-00412-5, http://www.wkap.nl/journalhome.htm/0269-4042.

Gu, Y. Y., Liang, X. Y., Zhang, H. Y., Fu, R., Li, M., & Chen, C. J. (2023). Effect of biochar and bioorganic fertilizer on the microbial diversity in the rhizosphere soil of *Sesbania cannabina* in saline-alkaline soil. *Frontiers in Microbiology*, *14*. Available from https://doi.org/10.3389/fmicb.2023.1190716, https://www.frontiersin.org/journals/microbiology#.

Hardy, B., Sleutel, S., Dufey, J. E., & Cornelis, J. T. (2019). The long-term effect of biochar on soil microbial abundance, activity and community structure is overwritten by land management. *Frontiers in Environmental Science*, *7*. Available from https://doi.org/10.3389/fenvs.2019.00110, journal.frontiersin.org/journal/environmental-science.

Harman, G., Khadka, R., Doni, F., & Uphoff, N. (2021). Benefits to plant health and productivity from enhancing plant microbial symbionts. *Frontiers in Plant Science*, *11*. Available from https://doi.org/10.3389/fpls.2020.610065.

Hou, J., Pugazhendhi, A., Phuong, T. N., Thanh, N. C., Brindhadevi, K., Velu, G., Lan Chi, N. T., & Yuan, D. (2022). Plant resistance to disease: Using biochar to inhibit harmful microbes and absorb nutrients. *Environmental Research*, *214*, 113883. Available from https://doi.org/10.1016/j.envres.2022.113883.

Iacomino, G., Idbella, M., Laudonia, S., Vinale, F., & Bonanomi, G. (2022). The suppressive effects of biochar on above- and belowground plant pathogens and pests: A review. *Plants*, *11*(22), 3144. Available from https://doi.org/10.3390/plants11223144.

Jahan, S., Iqbal, S., Rasul, F., & Jabeen, K. (2020). Efficacy of biochar as soil amendments for soybean (*Glycine max* L.) morphology, physiology, and yield regulation under drought. *Arabian Journal of Geosciences*, *13*(10). Available from https://doi.org/10.1007/s12517-020-05318-6.

Jatav, H. S., Rajput, V. D., Minkina, T., Singh, S. K., Chejara, S., Gorovtsov, A., Barakhov, A., Bauer, T., Sushkova, S., Mandzhieva, S., et al. (2021). Sustainable Approach and Safe Use of Biochar and Its Possible Consequences. *Sustainability*, *13*(18), 10362. Available from https://doi.org/10.3390/su131810362.

Joseph, S., Cowie, A. L., Van Zwieten, L., Bolan, N., Budai, A., Buss, W., Cayuela, M. L., Graber, E. R., Ippolito, J. A., Kuzyakov, Y., Luo, Y., Ok, Y. S., Palansooriya, K. N., Shepherd, J., Stephens, S., Weng, Z., & Lehmann, J. (2021). How biochar works, and when it doesn't: A review of mechanisms controlling soil and plant responses to biochar. *GCB Bioenergy*, *13*(11), 1731–1764. Available from https://doi.org/10.1111/gcbb.12885, http://onlinelibrary.wiley.com/journal/10.1111/(ISSN)1757-1707.

Khan, S., Irshad, S., Mehmood, K., Hasnain, Z., Nawaz, M., Rais, A., Gul, S., Wahid, M. A., Hashem, A., Abd_Allah, E. F., & Ibrar, D. (2024). Biochar production and characteristics, its impacts on soil health, crop production, and yield enhancement: A review. *Plants*, *13*(2). Available from https://doi.org/10.3390/plants13020166, http://www.mdpi.com/journal/plants.

Khan, Z., Zhang, K., Khan, M. N., Bi, J., Zhu, K., Luo, L., & Hu, L. (2022). How biochar affects nitrogen assimilation and dynamics by interacting soil and plant enzymatic activities: Quantitative assessment of 2 years potted study in a rapeseed-soil system. *Frontiers in Plant Science*, *13*. Available from https://doi.org/10.3389/fpls.2022.853449, https://www.frontiersin.org/journals/plant-science.

Kocsis, T., Ringer, M., & Biró, B. (2022). Characteristics and applications of biochar in soil–plant systems: A short review of benefits and potential drawbacks. *Applied Sciences*, *12*(8), 4051. Available from https://doi.org/10.3390/app12084051.

Kolton, M., Graber, E. R., Tsehansky, L., Elad, Y., & Cytryn, E. (2017). Biochar-stimulated plant performance is strongly linked to microbial diversity and metabolic potential in the rhizosphere. *New Phytologist*, *213*(3), 1393–1404. Available from https://doi.org/10.1111/nph.14253, http://onlinelibrary.wiley.com/journal/10.1111/(ISSN)1469-8137.

Kumar, C., Ramawat, N., & Kumar Verma, A. (2022). Organic fertigation system in saline-sodic soils: A new paradigm for the restoration of soil health. *Agronomy Journal*, *114*(1), 317–330. Available from https://doi.org/10.1002/agj2.20957.

Kwaku Armah, E., Chetty, M., Adebisi Adedeji, J., Erwin Estrice, D., Mutsvene, B., Singh, N., & Tshemese, Z. (2023). *Biochar: Production, application and the future*. IntechOpen. Available from http://doi.org/10.5772/intechopen.105070.

Küçükağa, Y., Facchin, A., Alfonsi, A., Costantini, F., Kara, S., & Torri, C. (2023). Development of a novel biochar-made porous monolith for enhanced C1 and H2 fermentation. *MethodsX*, *11*, 102296. Available from https://doi.org/10.1016/j.mex.2023.102296.

Lechón-Alonso, P., Clegg, T., Cook, J., Smith, T. P., & Pawar, S. (2021). The role of competition versus cooperation in microbial community coalescence. *PLoS Computational Biology*, *17*(11). Available from https://doi.org/10.1371/journal.pcbi.1009584, https://journals.plos.org/ploscompbiol/article/file?id = 10.1371/journal.pcbi.1009584&type = printable.

Li, S., & Tasnady, D. (2023). Biochar for soil carbon sequestration: Current knowledge, mechanisms, and future perspectives. *C*, *9*(3), 67. Available from https://doi.org/10.3390/c9030067.

Liang, Q., Pan, D., & Zhang, X. (2023). Construction and application of biochar-based composite phase change materials. *Chemical Engineering Journal*, *453*, 139441. Available from https://doi.org/10.1016/j.cej.2022.139441.

Lu, L., Yu, W., Wang, Y., Zhang, K., Zhu, X., Zhang, Y., Wu, Y., Ullah, H., Xiao, X., & Chen, B. (2020). Application of biochar-based materials in environmental remediation: from multi-level structures to specific devices. *Biochar*, *2*(1), 1−31. Available from https://doi.org/10.1007/s42773-020-00041-7.

Luo, G., Xue, C., Jiang, Q., Xiao, Y., Zhang, F., Guo, S., Shen, Q., Ling, N., Stegen, J. C., & Carbon, S. (2020). Nitrogen, and phosphorus cycling microbial populations and their resistance to global change depend on soil C:N:P stoichiometry. *mSystems*, *5*(3). Available from https://doi.org/10.1128/msystems.00162-20.

Lyu, H., Lim, J. Y., Zhang, Q., Senadheera, S. S., Zhang, C., Huang, Q., & Ok, Y. S. (2024). Conversion of organic solid waste into energy and functional materials using biochar catalyst: Bibliometric analysis, research progress, and directions. *Applied Catalysis B: Environmental*, *340*, 123223. Available from https://doi.org/10.1016/j.apcatb.2023.123223.

Mansoor, S., Kour, N., Manhas, S., Zahid, S., Wani, O. A., Sharma, V., Wijaya, L., Alyemeni, M. N., Alsahli, A. A., El-Serehy, H. A., Paray, B. A., & Ahmad, P. (2021). Biochar as a tool for effective management of drought and heavy metal toxicity. *Chemosphere*, *271*, 129458. Available from https://doi.org/10.1016/j.chemosphere.2020.129458.

Maron, P. A., Sarr, A., Kaisermann, A., Lévêque, J., Mathieu, O., Guigue, J., Karimi, B., Bernard, L., Dequiedt, S., Terrat, S., Chabbi, A., & Ranjard, L. (2018). High microbial diversity promotes soil ecosystem functioning. *Applied and Environmental Microbiology*, *84*(9). Available from https://doi.org/10.1128/AEM.02738-17, http://aem.asm.org/content/84/9/e02738-17.full.pdf#page = 1&view = FitH.

de Medeiros, E. V., Lima, N. T., de Sousa Lima, J. R., Pinto, K. M. S., da Costa, D. P., Franco Junior, C. L., Souza, R. M. S., & Hammecker, C. (2021). Biochar as a strategy to manage plant diseases caused by pathogens inhabiting the soil: a critical review. *Phytoparasitica*, *49*(4), 713−726. Available from https://doi.org/10.1007/s12600-021-00887-y.

Meki, K., Liu, Q., Wu, S., & Yuan, Y. (2022). Plant- and microbe-assisted biochar amendment technology for petroleum hydrocarbon remediation in saline-sodic soils: A review. *Pedosphere*, *32*(1), 211–221. Available from https://doi.org/10.1016/S1002-0160(21)60041-3.

Mickan, B. S., Abbott, L. K., Stefanova, K., & Solaiman, Z. M. (2016). Interactions between biochar and mycorrhizal fungi in a water-stressed agricultural soil. *Mycorrhiza*, *26*(6), 565–574. Available from https://doi.org/10.1007/s00572-016-0693-4, http://link.springer.de/link/service/journals/00572/index.htm.

Mukherjee, S., Sarkar, B., Aralappanavar, V. K., Mukhopadhyay, R., Basak, B. B., Srivastava, P., Marchut-Mikołajczyk, O., Bhatnagar, A., Semple, K. T., & Bolan, N. (2022). Biochar-microorganism interactions for organic pollutant remediation: Challenges and perspectives. *Environmental Pollution*, *308*, 119609. Available from https://doi.org/10.1016/j.envpol.2022.119609.

Murtaza, G., Ahmed, Z., Eldin, S. M., Ali, B., Bawazeer, S., Usman, M., Iqbal, R., Neupane, D., Ullah, A., Khan, A., Hassan, M. U., Ali, I., & Tariq, A. (2023). Biochar–soil–plant interactions: A cross talk for sustainable agriculture under changing climate. *Frontiers in Environmental Science*, *11*. Available from https://doi.org/10.3389/fenvs.2023.1059449, journal.frontiersin.org/journal/environmental-science.

Nakhli, S. A. A., Goy, S., Manahiloh, K. N., & Imhoff, P. T. (2021). Spatial heterogeneity of biochar (segregation) in biochar-amended media: An overlooked phenomenon, and its impact on saturated hydraulic conductivity. *Journal of Environmental Management*, *279*, 111588. Available from https://doi.org/10.1016/j.jenvman.2020.111588.

Nascimento, Í. Vd, Fregolente, L. G., Pereira, A. Pd. A., Nascimento, C. D. Vd, Mota, J. C. A., Ferreira, O. P., Sousa, H. Hd. F., Silva, D. G. Gd, Simões, L. R., Souza Filho, A. G., & Costa, M. C. G. (2023). Biochar as a carbonaceous material to enhance soil quality in drylands ecosystems: A review. *Environmental Research*, *233*. Available from https://doi.org/10.1016/j.envres.2023.116489, https://www.sciencedirect.com/journal/environmental-research.

Ndede, E. O., Kurebito, S., Idowu, O., Tokunari, T., & Jindo, K. (2022). The potential of biochar to enhance the water retention properties of sandy agricultural soils. *Agronomy*, *12*(2), 311. Available from https://doi.org/10.3390/agronomy12020311.

Nepal, J., Ahmad, W., Munsif, F., Khan, A., & Zou, Z. (2023). Advances and prospects of biochar in improving soil fertility, biochemical quality, and environmental applications. *Frontiers in Environmental Science*, *11*. Available from https://doi.org/10.3389/fenvs.2023.1114752.

Oni, B. A., Oziegbe, O., & Olawole, O. O. (2019). Significance of biochar application to the environment and economy. *Annals of Agricultural Sciences*, *64*(2), 222–236. Available from https://doi.org/10.1016/j.aoas.2019.12.006, http://www.elsevier.com/journals/annals-of-agricultural-sciences/0570-1783/.

Pang, Z., Dong, F., Liu, Q., Lin, W., Hu, C., & Yuan, Z. (2021). Soil Metagenomics Reveals Effects of Continuous Sugarcane Cropping on the Structure and Functional Pathway of Rhizospheric Microbial Community. *Frontiers in Microbiology*, *12* (March)), 1–19. Available from https://doi.org/10.3389/fmicb.2021.627569.

Powlson, D. S., Gregory, P. J., Whalley, W. R., Quinton, J. N., Hopkins, D. W., Whitmore, A. P., Hirsch, P. R., & Goulding, K. W. T. (2011). Soil management in relation to sustainable agriculture and ecosystem services. *Food Policy*, *36*(1), S72. Available from https://doi.org/10.1016/j.foodpol.2010.11.025.

Prasad, M. N. V., & Kumar, C. (2023). *Biochar applications for sustainable agriculture and environmental management agroecological approaches for sustainable soil management* (pp. 165–199). Wiley. Available from http://doi.org/10.1002/9781119911999.ch7.

Prasad, M. N. V. (2023). Biochar Applications for Sustainable Agriculture and Environmental Management. In M. N. V. Prasad, & C. Kumar (Eds.), *In: Agroecological Approaches for Sustainable Soil Management* (pp. 165–199). UK: Johns Willey & Sons Lts, ISBN 978-11-1991-196-8. Available from https://doi.org/10.1002/9781119911999.ch7.

Prescott, S. L., Wegienka, G., Kort, R., Nelson, D. H., Gabrysch, S., Hancock, T., Kozyrskyj, A., Lowry, C. A., Redvers, N., Poland, B., Robinson, J., Moubarac, J. C., Warber, S., Jansson, J., Sinkkonen, A., Penders, J., Erdman, S., Nanan, R., Van Den Bosch, M., Schneider, K., Schroeck, N. J., Sobko, T., Harvie, J., Kaplan, G. A., Moodie, R., Lengnick, L., Prilleltensky, I., Celidwen, Y., Berman, S. H., Logan, A. C., & Berman, B. (2021). Project earthrise: Proceedings of the ninth annual conference of in vivo planetary health. *International Journal of Environmental Research and Public Health, 18*(20). Available from https://doi.org/10.3390/ijerph182010654, https://www.mdpi.com/1660-4601/18/20/10654/pdf.

Rasul, M., Cho, J., Shin, H.-S., & Hur, J. (2022). Biochar-induced priming effects in soil via modifying the status of soil organic matter and microflora: A review. *Science of The Total Environment, 805*, 150304. Available from https://doi.org/10.1016/j.scitotenv.2021.150304.

Rasool, S., Rasool, T., & Gani, K. M. (2023). Unlocking the potential of wetland biomass: Treatment approaches and sustainable resource management for enhanced utilization. *Bioresource Technology Reports, 23*, 101553. Available from https://doi.org/10.1016/j.biteb.2023.101553.

Ren, H., Huang, B., Fernández-García, V., Miesel, J., Yan, L., & Lv, C. (2020). Biochar and Rhizobacteria Amendments Improve Several Soil Properties and Bacterial Diversity. *Microorganisms, 8*(4), 502. Available from https://doi.org/10.3390/microorganisms8040502.

Rombel, A., Krasucka, P., & Oleszczuk, P. (2022). Sustainable biochar-based soil fertilizers and amendments as a new trend in biochar research. *Science of the Total Environment, 816*, 151588. Available from https://doi.org/10.1016/j.scitotenv.2021.151588.

Salma, A., Fryda, L., & Djelal, H. (2024). Biochar: A key player in carbon credits and climate mitigation. *Resources, 13*(2), 31. Available from https://doi.org/10.3390/resources13020031.

Sessions, J., Smith, D., Trippe, K. M., Fried, J. S., Bailey, J. D., Petitmermet, J. H., Hollamon, W., Phillips, C. L., & Campbell, J. D. (2019). Can biochar link forest restoration with commercial agriculture? *Biomass and Bioenergy, 123*, 175–185. Available from https://doi.org/10.1016/j.biombioe.2019.02.015, http://www.journals.elsevier.com/biomass-and-bioenergy/.

Shen, J., Liang, Z., Kuzyakov, Y., Li, W., He, Y., Wang, C., Xiao, Y., Chen, K., Sun, G., & Lei, Y. (2023). Dissolved organic matter defines microbial communities during initial soil formation after deglaciation. *Science of The Total Environment, 878*, 163171. Available from https://doi.org/10.1016/j.scitotenv.2023.163171.

Smith, P., Cotrufo, M. F., Rumpel, C., Paustian, K., Kuikman, P. J., Elliott, J. A., McDowell, R., Griffiths, R. I., Asakawa, S., Bustamante, M., House, J. I., Sobocká, J., Harper, R., Pan, G., West, P. C., Gerber, J. S., Clark, J. M., Adhya, T., Scholes, R. J., & Scholes, M. C. (2015). Biogeochemical cycles and biodiversity as key drivers of ecosystem services provided by soils. *Soil*, *1*(2), 665–685. Available from https://doi.org/10.5194/soil-1-665-2015.

Suman, J., Rakshit, A., Ogireddy, S. D., Singh, S., Gupta, C., & Chandrakala, J. (2022). Microbiome as a key player in sustainable agriculture and human health. *Frontiers in Soil Science*, *2*. Available from https://doi.org/10.3389/fsoil.2022.821589.

Thakur, N., Nigam, M., Mann, N. A., Gupta, S., Hussain, C. M., Shukla, S. K., Shah, A. A., Casini, R., Elansary, H. O., & Khan, S. A. (2023). Host-mediated gene engineering and microbiome-based technology optimization for sustainable agriculture and environment. *Functional & Integrative Genomics*, *23*(1), 57. Available from https://doi.org/10.1007/s10142-023-00982-9.

Vejan, P., Khadiran, T., Abdullah, R., & Ahmad, N. (2021). Controlled release fertilizer: A review on developments, applications and potential in agriculture. *Journal of Controlled Release*, *339*, 321–334. Available from https://doi.org/10.1016/j.jconrel.2021.10.003.

Wang, H., Zhang, R., Zhao, Y., Shi, H., & Liu, G. (2022). Effect of biochar on rhizosphere soil microbial diversity and metabolism in tobacco-growing soil. *Ecologies*, *3*(4), 539–556. Available from https://doi.org/10.3390/ecologies3040040.

Wang, J., Liu, J., Chang, L., Pan, Y., Zhai, L., Shen, Z., Shi, L., & Chen, Y. (2023). The characteristic difference between non-drilosphere and drilosphere-aged biochar: Revealing that earthworms accelerate the aging of biochar. *Chemosphere*, *321*, 138141. Available from https://doi.org/10.1016/j.chemosphere.2023.138141.

Wang, M., Yu, X., Weng, X., Zeng, X., Li, M., & Sui, X. (2023). Meta-analysis of the effects of biochar application on the diversity of soil bacteria and fungi. *Microorganisms*, *11*(3). Available from https://doi.org/10.3390/microorganisms11030641, http://www.mdpi.com/journal/microorganisms.

Wang, N., Zhao, K., Li, F., Peng, H., Lu, Y., Zhang, L., Pan, J., Jiang, S., Chen, A., Yan, B., Luo, L., Huang, H., Li, H., Wu, G., & Zhang, J. (2022). Characteristics of carbon, nitrogen, phosphorus and sulfur cycling genes, microbial community metabolism and key influencing factors during composting process supplemented with biochar and biogas residue. *Bioresource Technology*, *366*, 128224. Available from https://doi.org/10.1016/j.biortech.2022.128224.

Weller, D. M., Raaijmakers, J. M., McSpadden Gardener, B. B., & Thomashow, L. S. (2002). MIcrobial populations responsible for specific soil suppressiveness to plant pathogens. *Annual Review of Phytopathology*, *40*(1), 309–348. Available from https://doi.org/10.1146/annurev.phyto.40.030402.110010.

Wijerathna-Yapa, A., & Pathirana, R. (2022). Sustainable agro-food systems for addressing climate change and food security. *Agriculture*, *12*(10), 1554. Available from https://doi.org/10.3390/agriculture12101554.

Wijitkosum, S., & Jiwnok, P. (2019). Elemental composition of biochar obtained from agricultural waste for soil amendment and carbon sequestration. *Applied Sciences*, *9*(19), 3980. Available from https://doi.org/10.3390/app9193980.

Wong, J. W. C., & Ogbonnaya, U. O. (2021). Biochar porosity: a nature-based dependent parameter to deliver microorganisms to soils for land restoration. *Environmental*

Science and Pollution Research International, *28*(34), 46894–46909. Available from https://doi.org/10.1007/s11356-021-14803-8.

Xiang, L., Harindintwali, J. D., Wang, F., Redmile-Gordon, M., Chang, S. X., Fu, Y., He, C., Muhoza, B., Brahushi, F., Bolan, N., Jiang, X., Ok, Y. S., Rinklebe, J., Schaeffer, A., Zhu, Y. G., Tiedje, J. M., & Xing, B. (2022). Integrating biochar, bacteria, and plants for sustainable remediation of soils contaminated with organic pollutants. *Environmental Science and Technology*, *56*(23), 16546–16566. Available from https://doi.org/10.1021/acs.est.2c02976, http://pubs.acs.org/journal/esthag.

Xu, W., Xu, H., Delgado-Baquerizo, M., Gundale, M. J., Zou, X., & Ruan, H. (2023). Global meta-analysis reveals positive effects of biochar on soil microbial diversity. *Geoderma*, *436*, 116528. Available from https://doi.org/10.1016/j.geoderma.2023.116528.

Yaashikaa, P. R., Senthil Kumar, P., Varjani, S., & Saravanan, A. (2020). A critical review on the biochar production techniques, characterization, stability and applications for circular bioeconomy. *Biotechnology Reports*, *28*, e00570. Available from https://doi.org/10.1016/j.btre.2020.e00570.

Yan, H., Liu, C., Yu, W., Zhu, X., & Chen, B. (2023). The aggregate distribution of Pseudomonas aeruginosa on biochar facilitates quorum sensing and biofilm formation. *Science of The Total Environment*, *856*, 159034. Available from https://doi.org/10.1016/j.scitotenv.2022.159034.

Yan, T., Xue, J., Zhou, Z., & Wu, Y. (2021). Biochar-based fertilizer amendments improve the soil microbial community structure in a karst mountainous area. *Science of The Total Environment*, *794*, 148757. Available from https://doi.org/10.1016/j.scitotenv.2021.148757, ISSN 0048-9697,

Yang, C., Liu, J., Ying, H., & Lu, S. (2022). Soil pore structure changes induced by biochar affect microbial diversity and community structure in an Ultisol. *Soil and Tillage Research*, *224*, 105505. Available from https://doi.org/10.1016/j.still.2022.105505.

Yin, D., Li, H., Wang, H., Guo, X., Wang, Z., Lv, Y., Ding, G., Jin, L., & Lan, Y. (2021). Impact of different biochars on microbial community structure in the rhizospheric soil of rice grown in albic soil. *Molecules*, *26*(16), 4783. Available from https://doi.org/10.3390/molecules26164783.

Yin, Y., Li, M., Tao, X., Yang, C., Zhang, W., Li, H., Zheng, Y., Wang, X., & Chen, R. (2023). Biochar enhanced organic matter transformation during pig manure composting: Roles of the cellulase activity and fungal community. *Journal of Environmental Management*, *333*. Available from https://doi.org/10.1016/j.jenvman.2023.117464, https://www.sciencedirect.com/journal/journal-of-environmental-management.

Zhang, J., & Shen, J. L. (2022). Effects of biochar on soil microbial diversity and community structure in clay soil. *Ann Microbiol*, *72*, 35. Available from https://doi.org/10.1186/s13213-022-01689-1.

Zhang, L., Cheng, H., Pan, D., Wu, Y., Ji, R., Li, W., Jiang, X., & Han, J. (2021). One-pot pyrolysis of a typical invasive plant into nitrogen-doped biochars for efficient sorption of phthalate esters from aqueous solution. *Chemosphere*, *280*, 130712. Available from https://doi.org/10.1016/j.chemosphere.2021.130712.

Zhang, M., Liu, Y., Wei, Q., Gu, X., Liu, L., & Gou, J. (2022). Biochar application ameliorated the nutrient content and fungal community structure in different yellow soil depths in the karst area of Southwest China. *Front Plant Sci*, *13*, 1020832. Available from https://doi.org/10.3389/fpls.2022.1020832, Oct,

Zhao, Y., Wang, X., Yao, G., Lin, Z., Xu, L., Jiang, Y., Jin, Z., Shan, S., & Ping, L. (2022). Advances in the Effects of Biochar on Microbial Ecological Function in Soil and Crop Quality. *Sustainability*, *14*, 10411. Available from https://doi.org/10.3390/su141610411.

Zhao, X. Y., Zhang, Z. Y., Huang, Y. M., & Feng, F. J. (2023). Enhancing the effect of biochar ageing on reducing cadmium accumulation in *Medicago sativa* L. *Science of the Total Environment*, *862*. Available from https://doi.org/10.1016/j.scitotenv.2022.160690, http://www.elsevier.com/locate/scitotenv.

Zhu, X., Mao, L., & Chen, B. (2019). Driving forces linking microbial community structure and functions to enhanced carbon stability in biochar-amended soil. *Environment International*, *133*(Part B), 105211. Available from https://doi.org/10.1016/j.envint.2019.105211, ISSN 0160-4120.

PART 5

Nanotechnological approach for developing biochar-fertilizer

CHAPTER

18 Biochar modification methods: property engineering for diverse value-added applications

Flavia Lega Braghiroli[1] and Gisele Amaral-Labat[2]

[1]*Research Forest Institute (Institut de recherche sur les forêts - IRF), University of Québec at Abitibi-Témiscamingue (UQAT), Rouyn-Noranda, QC, Canada*

[2]*Coordination of Applied Research and Technological Development, National Institute for Space Research (INPE), São José dos Campos- São Paulo, Brazil*

18.1 Introduction

The rising global demand for nano-based, carbon (C)-based, porous, and composite materials has sharpened interest in the production of sustainable materials prepared from bio-precursors. The aim is to (1) reduce manufacturing costs, (2) align with circular economy principles (i.e., recover and recycle locally to reduce the C footprint for transport of residual materials instead of "take, make, and throw away"), and (3) replace petroleum-based materials with natural biomass-based products. The global annual production of biomass waste from agricultural and forestry practices is in the order of 140 Gt (United Nations Environmental Programme Division of Technology, 2015). A main environmental issue in biomass waste management is that materials are either deposited in local sorting centers and/or landfills or else incinerated. This generates leachable and toxic contaminants (e.g., phenolic compounds), air toxic pollutants (e.g., formaldehyde), as well as carbon di oxide (CO_2) emissions.

In the quest for a sustainable future, addressing waste management issues is increasingly imperative, with one promising approach being the development of bio-inspired materials to help mitigate these challenges. This chapter highlights the diversity of engineered materials, especially from wastes, showcasing their vast potential for application in different fields.

A vast range of biochars and by-products can be produced thanks to the diversity of available feedstocks, their heterogeneous composition, the varied physicochemical properties and characteristics, and the different thermochemical conversion methods, reactor designs, and process parameters. These factors greatly influence the biochar's physicochemical properties and performance, which are engineered to meet the needs of targeted applications. In the last decade, these improved properties have

Biochar Ecotechnology for Sustainable Agriculture and Environment. DOI: https://doi.org/10.1016/B978-0-443-29855-4.00020-5

broadened the application potential across multiple and multidisciplinary science and engineering fields. This chapter presents the major biochar modification methods and recent developments: (1) grinding and irradiation (UV, microwave, and ultrasound); (2) amination, oxidation, and sulfonation; (3) surface modification with metals such as iron (Fe), nickel (Ni), and copper (Cu) or metal oxides; and (4) physical and chemical activation. Conclusions and prospects are then given.

18.2 Biochar production

A well-known alternative for handling biomass waste is thermochemical conversion, whereby a sustainable biorefinery process decomposes the organic matter and converts it into different bioproducts: solid (biochar), gas (syngas, containing mainly carbon monooxide [CO] and hydrogen [H_2]), and liquid (bio-oil) (Ahmad et al., 2014; Braghiroli, et al., 2020; Bridgwater, 2012; Laird et al., 2009; Lehmann & Joseph, 2015). The bioproducts are produced at different proportions depending on the thermochemical conversion processes: torrefaction; slow, intermediate, fast, and flash pyrolysis; and gasification. First, biochar is produced via different thermochemical processes at different temperatures, residence times, and heating rates (Fig. 18.1). For example, torrefaction is performed at 200°C–300°C for 10–60 minutes and the main bioproduct is torrefied biomass (up to 80%). Next, pyrolysis, a thermochemical process, is classified as slow, intermediate, fast, and flash, and is performed at 100°C–1000°C for 5–30 minutes, ~ 500°C for 10–20 seconds, 300°C–1000°C for less than 2 seconds, and 400°C–1000°C

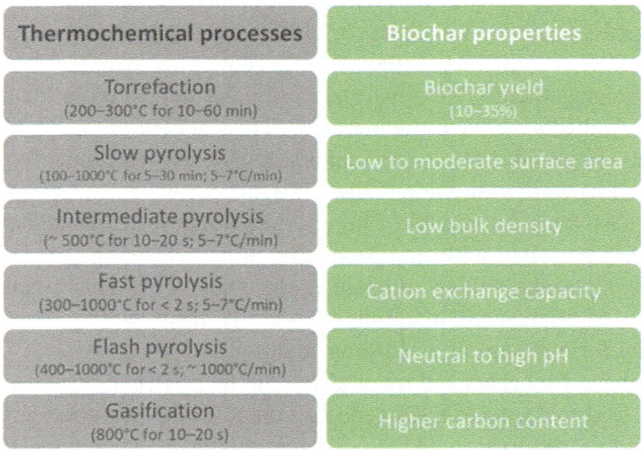

FIGURE 18.1

Different thermochemical conversion methods for biochar production and property enhancement.

for less than 2 seconds, respectively. The heating rates are 5°C–7°C/min, up to 100°C/min, 300°C–800°C/min, and ~ 1000°C/min, respectively. The main biofuels obtained are biochar (25%–35%), bio-oil (up to 50%), bio-oil (up to 75%), and bio-oil (60%–75%), respectively. Finally, gasification is performed at temperatures exceeding 800°C for 10–20 seconds, with syngas as the main biofuel produced (85%) (Ahmad et al., 2014; Braghiroli, et al., 2020; Ighalo et al., 2022; Laird et al., 2009; Lehmann & Joseph, 2015).

Biochar can be produced from a wide range of biomass sources. Of these, lignocellulosic materials are the most desirable precursors for biochar production because they contain the fibrous parts of plants (cellulose, hemicellulose, and lignin) (Chen, 2015), which provide the strength required for application in the textile, paper, construction, chemical, and renewable energy industries. Lignocellulosic compounds undergo thermochemical decomposition at different temperatures. Hemicellulose and cellulose degrade at 240–350°C and 200–260°C, respectively (Pandey et al., 2015), whereas lignin degrades at the highest temperature range, from 280°C to 500°C (Babu, 2008). This is a relatively simple method to produce highly nanoporous C. However, nonlignocellulosic biomass has a different chemical composition. For example, algal biomass comprises mainly proteins, lipids, and carbohydrates (Van Wychen et al., 2021). Therefore the thermochemical decomposition process is more complex than for lignocellulosic materials. It involves not only the degradation of carbohydrates, lipids, and protein components but also the decomposition of metal carbonates, sugars, triglycerides, and low molecular weight proteins into nitrogenated compounds present in tar (pyrrole, pyridine, and diketopiperazine) and gas (ammonium, nitrates, and carbonates) (Debiagi et al., 2017).

The biochars that are generated by thermochemical conversion of biomass or biomass wastes are C-rich materials with advantageous physicochemical properties. These value-added properties include cation exchange capacity (CEC); C sequestration potential; total and fixed C content; volatile matter content; chemical and physical recalcitrance (i.e., resistance to biodegradation); and enhanced surface area, pore volume, and average pore diameter. One of the most important biochar characteristics is the specific surface area, defined as the interstitial surface area of the voids and pores in the biochar structure per unit mass or volume of the material. This specific property indicates the biochar's adsorptive capacity and its potential applications. In a review by (Ippolito et al., 2020) of approximately 5400 peer-reviewed journal articles, mostly from the last decade, and considering over 50,800 individual data points, pyrolysis temperature was correlated to specific surface area (Fig. 18.2A) and the molar ratio of O/C and H/C (van Krevelen diagram, Fig. 18.2B) across biochars made from wood, crop wastes, manures, and biosolids. Overall, the results showed increased specific surface area of biochars with increasing (low and fast) pyrolysis temperature (Ippolito et al., 2020). In terms of material degradability, biochars with O/C ratio below 0.2 were highly stable, between 0.2 and 0.6 moderately stable, and exceeding 0.6 relatively unstable. According to Spokas (2010), the time required for half of these materials to change or decay (half-life) is a minimum of 1000 years, from

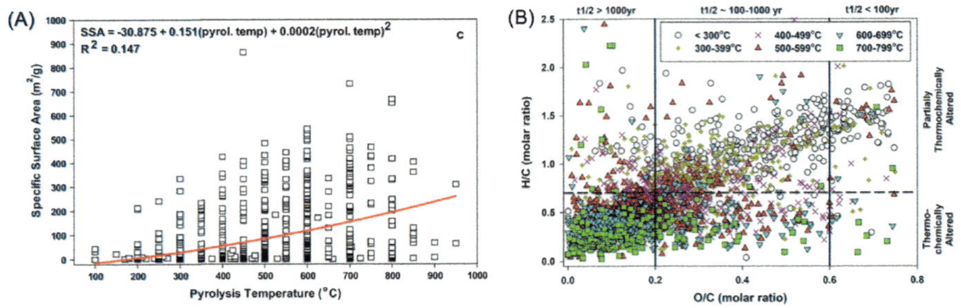

FIGURE 18.2

The relationship between: (A) pyrolysis temperature and specific surface area and (B) the molar ratio of O/C and H/C and pyrolysis temperature (van Krevelen diagram) for different biochars made from wood, crop wastes, other grasses, manures, and biosolids.

Reproduced with permission under a Creative Commons Attribution 4.0 International License from Ippolito, J. A., Cui, L., Kammann, C., Wrage-Mönnig, N., Estavillo, J. M., Fuertes-Mendizabal, T., Cayuela, M. L., Sigua, G., Novak, J., Spokas, K., & Borchard, N. (2020). Feedstock choice, pyrolysis temperature and type influence biochar characteristics: A comprehensive meta-data analysis review. Biochar, *2(4), 421–438. https://doi.org/10.1007/s42773-020-00067-x. Copyright 2020, Springer.*

100 to 1000 years, and less than 100 years, respectively. However, biochars with H/C ratio below 0.7 that were thermochemically altered had greater proportions of aromatic ring structures compared to biochars with H/C ratio exceeding 0.7 (International Biochar Initiative, 2015).

18.3 Progress of biochar applications

Conversion of biomass wastes into biochar is one of the most important waste management strategies. It prevents soil and groundwater contamination at local sorting centers and landfills and it valorizes residual organic matter (or biomass). The term "biochar" is derived from "bio" (which means "living") and "char" (which means "coal"). Biochar was first discovered in the Amazon basin in Brazil, where ancient Amerindian civilizations lived long before the Europeans arrived. Since then, researchers from around the world have studied its soil and vegetation. The soil, called *Indian Black Earth*, consists of black earth that was created under conditions similar to those used to thermochemically convert biomass into biochar (Lehmann, 2007). Soil studies have confirmed its exceptional fertility and potential for C sequestration, nutrient cycling, and improved microbial activity (Lehmann, 2015).

Currently, biochar is used mainly in low-value-added sectors such as bioenergy (e.g., energy pellets), fertilization of agricultural and horticultural soils, CO_2 sequestration from the atmosphere, and restoration of degraded sites (e.g.,

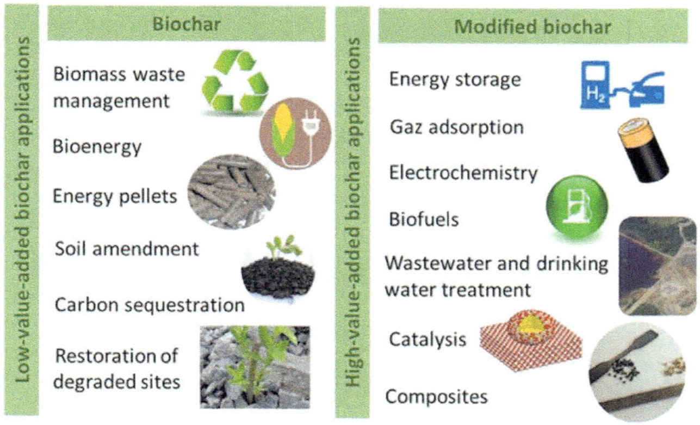

FIGURE 18.3

Illustration of low-valued and high-value-added biochar applications of biochar and modified biochar.

abandoned mine sites) (Fig. 18.3). Moreover, biochar performance is generally limited by its low to moderate porosity and surface area (from 0.1 to 500 m^2/g). However, recent modification methods have improved the textural and physicochemical properties of biochar to provide excellent performance in high-value-added sectors such as energy storage, electrochemistry, biofuels, gas adsorption, wastewater and drinking water treatment, catalysis, and composite design (e.g., biochar-polymer) (Chemerys & Baltrėnaitė, 2016; Faheem et al., 2020; Godwin et al., 2019; Liu et al., 2015; Sajjadi et al., 2019; Wang et al., 2020; Wu et al., 2020; Fig. 18.3).

18.4 Biochar modification: recent trends

Recent advances in biochar engineering have been investigated to evaluate the effects of various modification methods on biochar properties and performance in multiple fields. Biochar modifications can enhance the physical and chemical properties by developing the surface area and pore size distribution and/or by loading heteroatoms, metals, and functional groups on the biochar surface, respectively. Fig. 18.4 depicts the main physical and chemical modification methods investigated and applied in the last decade. The physical methods include mechanophysical methods such as grinding and irradiation (ultrasound, microwave, and UV) and physical activation methods in the presence of superheated steam or CO_2. The chemical methods include surface chemical methods such as amination, oxidation, and sulfonation, whereby heteroatoms or metals are introduced into the material surface, and chemical activation in the presence of inorganic bases, acids, or salts. Physical and chemical methods are conducted before,

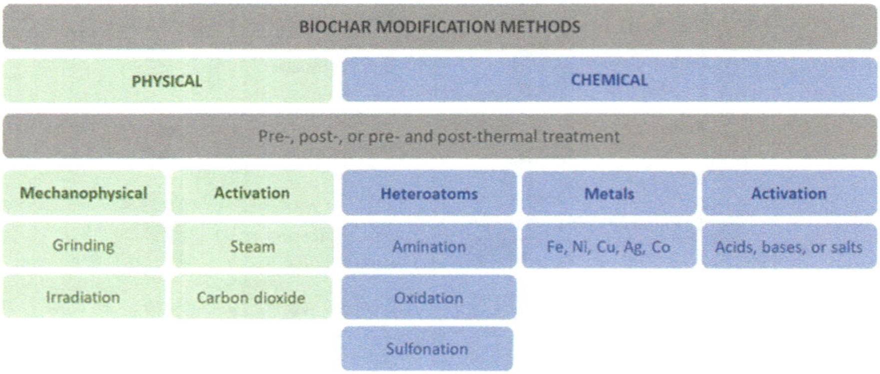

FIGURE 18.4

The main physical and chemical biochar modification methods applied before, during, or after thermal treatment.

after, or simultaneously before and after thermal treatment. The latter refers to physicochemical activation, where the material is first impregnated with chemicals, then thermally treated in the presence of steam or CO_2 (a physical agent), and then washed with water to remove excess chemicals.

18.4.1 Physical modification methods

The most commonly used physical modification methods (grinding, irradiation, and physical activation) are described in the next sections. These methods have been demonstrated the most effective for enhancing surface area, exposing functional groups, and opening closed pores to release volatile materials and expose inner pores. Their advantages and disadvantages and a comparison in terms of surface area as a function of temperature treatment are then presented.

18.4.1.1 Grinding

Two grinding methods are used: (1) manual, using a mortar and pestle and (2) mechanical, using a ball mill or an oscillating or a vibratory mill. Being more efficient, the mechanical method is used most often in industries. However, both methods are widely used, efficient, and solvent-free (i.e., eco-friendly) approaches to improve biochar physicochemical properties, including surface area, micropore volume, CEC, oxygen functional groups, and surface negative charge (Fahmi et al., 2018). Industrial ball mills generally contain multiple chambers that can process large volumes of material, while planetary ball mills or laboratory-scale mills, which rotate on a vertical or horizontal axis, are used mainly for laboratory testing.

In ball milling, the material is loaded into a cylinder filled with grinding balls. As the mill rotates along a central axis, the balls are lifted up the rising side and then dropped onto the feed (i.e., material). The mechanical impact between the cylinder walls and the milling balls depends on the rotation speed. As the balls strike the material, the impact causes continuous abrasion, or grinding. The material is ground into smaller particles (grains), causing structural changes. The final particle size is impacted by the grinding conditions: milling time, rotational speed, ball size, and ball-to-biochar ratio. The final material particles can be as small as 100 nm in diameter (Agrawal & Yi, 2020). Ball grinding improves the material's textural properties, including specific surface area and porosity, and the newly generated active surface sites enhance the adsorptive capacity (Ambika et al., 2022; Amusat et al., 2021). Ball milling can also induce mechanochemical modifications due to chemical interactions or abrasions during agitation (Wang et al., 2017). Nevertheless, the milling conditions must be well controlled: for instance, high compression can damage the crystalline properties of biochar (Barcelló, 2020).

18.4.1.2 Irradiation

Mechanophysical methods based on irradiation are also effective for developing and functionalizing biochar surfaces (Ma et al., 2021). The most recently investigated methods include UV, microwave, and ultrasound irradiation. UV irradiation is an inexpensive method in which the material is directly exposed to high-energy species or photons. The UV light penetrates the biochar surface and affects the surface properties, depending on the UV wavelength, intensity, and exposure duration (Q. Li et al., 2016). Long-wave UV (315–380 nm) irradiation is the most common and preferred industrial process due to its stronger penetration and higher power (Peng et al., 2018). Q. Li et al. (2016) found that UV irradiation results in the formation of oxygen-containing groups, mainly associated with the presence of oxygen on the biochar surface. Specifically, at 365 nm, the C = C or C–O bonds present in the biochar tend to break compared to the C–C or C–H bonds. Moreover, UV irradiation generates photoacids (i.e., molecules that become acidic under light absorption) that attach to the biochar surface. Fig. 18.5 shows the mechanism proposed by Q. Li et al. (2016), whereby benzyl is oxidized into carboxylic acid under UV light and in the presence of oxygen.

In microwave irradiation, high-frequency electromagnetic waves (900–2450 MHz range) (Pourhakkak et al., 2021) are irradiated into the material. The energy is transferred to the material via dipole rotation and ionic conduction during heating (Cai et al., 2019). Specifically, the electromagnetic radiation interacts with the molecules' dipole moment (i.e., the product of the magnitude of charges and the distance separating them) in the material. Microwave heating compares favorably with conventional heating because it provides rapid heating with less energy (Nicholas et al., 2018). Although comparisons between the two systems are inexact due to the different parameters and conditions (temperature, reactor size, and design), a few differences are worth noting: compared to conventional heating, microwaving produces less gas, yields lower biochar mass, and releases higher proportions of CH_4 and H_2 gases with increasing power or heating rate (Wu et al., 2014).

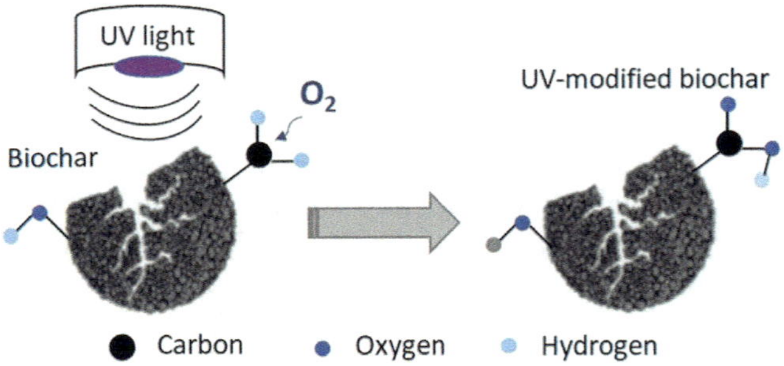

FIGURE 18.5

Schematical illustration of functional groups obtained after UV irradiation. *UV*, Ultraviolet.

Modified from Li, Q., Ding, W., Yong, Y., Zeng, X., & Gao, Y. (2016). Effects of ultraviolet modification on physicochemical property and adsorption performance of biochar. Nanoscience and Nanotechnology Letters, *8(11), 978–984. https://doi.org/10.1166/nnl.2016.2184.*

Next, in ultrasound modification, ultrasound energy is dissipated through the material via ultrasonic cavitation, or the formation and subsequent collapse of cavities or bubbles under gas or vapor pressure. The microbubbles expand under high pressure until they collapse, releasing energy to produce high pressure, high temperature, and a liquid disturbance in the form of waves (Janer et al., 2020). Ultrasound produces other effects as well, including chemical changes, sonoluminescence, emulsification, and dispersion (Wang et al., 2020). Because the released energy is strong enough to remove particles and impurities from material surfaces, ultrasound is also used for cleaning contaminated surfaces (Aghapour Aktij et al., 2020). The advantages of using ultrasound to modify a material's physicochemical properties are its selectivity, short reaction time, and no need for chemicals. Sajjadi et al. (2019) observed differences in ultrasound's effects on surface morphologies of raw and sonicated biochars made from softwood pine. Fig. 18.6A shows a scanning electron microscope (SEM) image of raw biochar, revealing a wrinkled and uneven surface texture with tiny white/gray lumps on the surface (blue circles), probably due to pyrolysis residues. In contrast, the ultrasound-modified biochar exhibits a large number of circular pores 2–3 μm in size (Fig. 18.6B). The developed porosity could be the result of cavitation, which dislodged and removed the lumps, resulting in higher surface area and increased porosity.

18.4.1.3 Physical activation

Physical or thermal activation is a thermal treatment whereby biochar is exposed to high temperatures up to 1500°C in the presence of CO_2 or steam (H_2O) in a limited or zero oxygen environment (Rodríguez-Reinoso & Molina-Sabio, 1992). The physical agents CO_2 or H_2O remove the C atoms from the biochar to produce

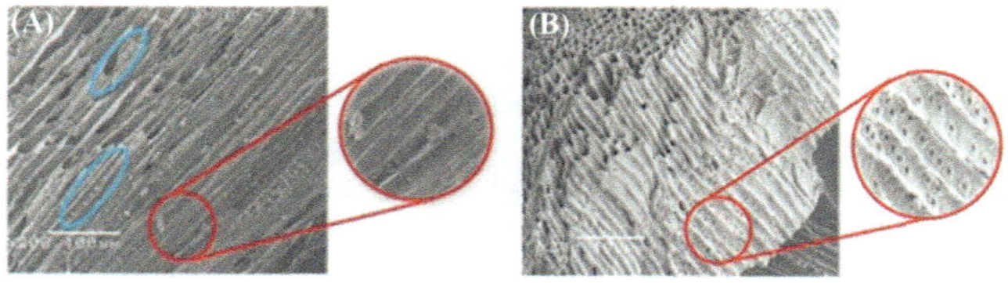

FIGURE 18.6

Scanning electron microscope images of (A) raw and (B) ultrasound-modified biochar.

Reproduced with permission from Sajjadi, B., Broome, J. W., Chen, W. Y., Mattern, D. L., Egiebor, N. O., Hammer, N., & Smith, C. L. (2019). Urea functionalization of ultrasound-treated biochar: A feasible strategy for enhancing heavy metal adsorption capacity. Ultrasonics Sonochemistry, 51*, 20–30. https://doi.org/10.1016/j.ultsonch.2018.09.015. Copyright 2019, Elsevier.*

CO (Boudouard reaction) (Calo & Perkins, 1987) or CO and H_2, according to the following equations (18.1–18.3):

$$C + CO_2 \rightarrow 2CO \tag{18.1}$$

$$C + H_2O \rightarrow CO + H_2 \tag{18.2}$$

$$CO + H_2O \rightarrow CO_2 + H_2 \tag{18.3}$$

The CO_2 penetrates the biochar's internal structure and causes the remotion of C atoms via a gasification process in which the inaccessible pores open and widen to create a porous structure. Micropores are formed during the devolatilization and the exposure of closed pores (Rodriguez-Reinos, 2002; Rodríguez-Reinoso & Molina-Sabio, 1998), whereas mesopores are formed via gasification and the disintegration of adjacent pore walls (Marsh, 2001). Thus the gases that are released during physical activation cause structural and physicochemical changes in the biochar, thereby improving the adsorptive capacity. In addition, the development of different pore sizes facilitates ion transport into micropores and the retention of larger molecules in the larger pores (mesopores) of the activated biochar (Rodriguez-Reinos, 2002). Fig. 18.7 presents SEM images of a wood residue, the derived biochar, and the activated biochar showing changes in the porous structure with thermal treatment at 450°C and 900°C for biochar and activated biochar, respectively.

Several studies have optimized pyro-gasification and activation conditions to produce materials with higher surface area and porosity, making them suitable for a wide range of applications. Fig. 18.8 shows a compilation of surface areas of activated biochars made from various feedstocks as a function of (1) pyro-gasification temperature using CO_2 or steam activation; (2) residence time, and (3) heating rate. The activated biochars made from lignocellulosic materials present higher surface area compared to other feedstocks such as burcucumber plants and broiler litter. A negative impact is seen on the porosity of activated biochars at higher pyro-gasification temperatures, longer residence times, and higher

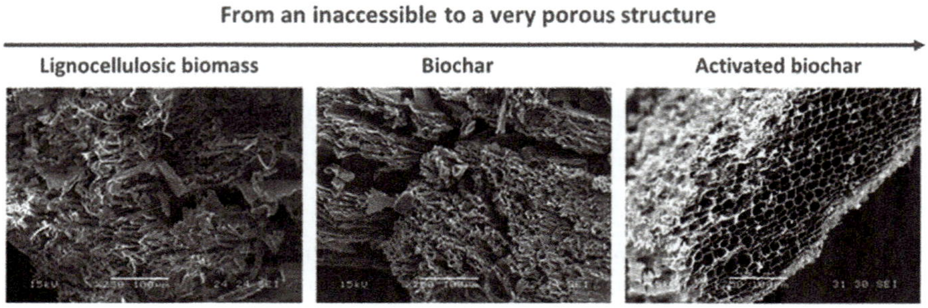

FIGURE 18.7

Scanning electron microscope images of a white birch sawmill residue, its biochar after fast pyrolysis at 450°C, and its biochar after activation in the presence of CO_2 gas at 900°C.

Reproduced with permission from Braghiroli, F. L., Bouafif, H., Neculita, C. M., & Koubaa, A. (2018). Activated biochar as an effective sorbent for organic and inorganic contaminants in water. Water, Air, and Soil Pollution, 229*(7). https://doi.org/10.1007/s11270-018-3889-8. Copyright 2020, Springer.*

heating rates. However, other studies have demonstrated that activation temperature had the greatest impact on material porosity, rather than pyro-gasification conditions (Braghiroli et al., 2018). Statistical analysis revealed that the average surface area for activated biochars made from black spruce and white birch wood residues increased to nearly 120 m^2/g just by raising the temperature from 700°C to 900°C. Işıtan et al. (2016) also found that activation temperature had the greatest impact: when temperature was increased from 800°C to 900°C, the surface area of activated biochars made from pistachio nutshells rose to nearly 300 m^2/g.

Fig. 18.8D shows a compilation of the surface areas of modified biochars made from various feedstocks (wheat stalk, wheat straw, sawdust, willow [*Salix viminalis*], waste tea, white birch and black spruce residues, green coconut pericarp, broiler litter, burcucumber plants, vine shoots, olive mill waste, pistachio nutshells, wheat straw, softwood, peach stones, and oil palm shells) via ball milling (Wei Xiang et al., 2020; W. Xiang et al., 2020), UV irradiation (Peng et al., 2019, 2018), microwave irradiation (Kołtowski et al., 2017), ultrasonication (Sajjadi et al., 2019), and activation in the presence of superheated steam (Arantes et al., 2023; Braghiroli et al., 2019; Mondal et al., 2016, 2015; Rajapaksha et al., 2015; Shen et al., 2017; Uchimiya et al., 2010; Wu et al., 2016) and CO_2 (Braghiroli et al., 2018; Kozyatnyk et al., 2021; Kołtowski et al., 2017; Lua et al., 2006; Manyà et al., 2018; Pepè Sciarria et al., 2020; Sricharoenchaikul et al., 2008) as a function of the physical treatment temperature. Although CO_2 physical activation appears to be the most efficient of the physical modification methods to develop biochar surface area and porosity, an extensive characterization must be performed to evaluate the other properties (elemental composition; porosity type: ultramicro-, micro-, and mesopores; metals composition) and match them with targeted applications. The advantage of physical over chemical

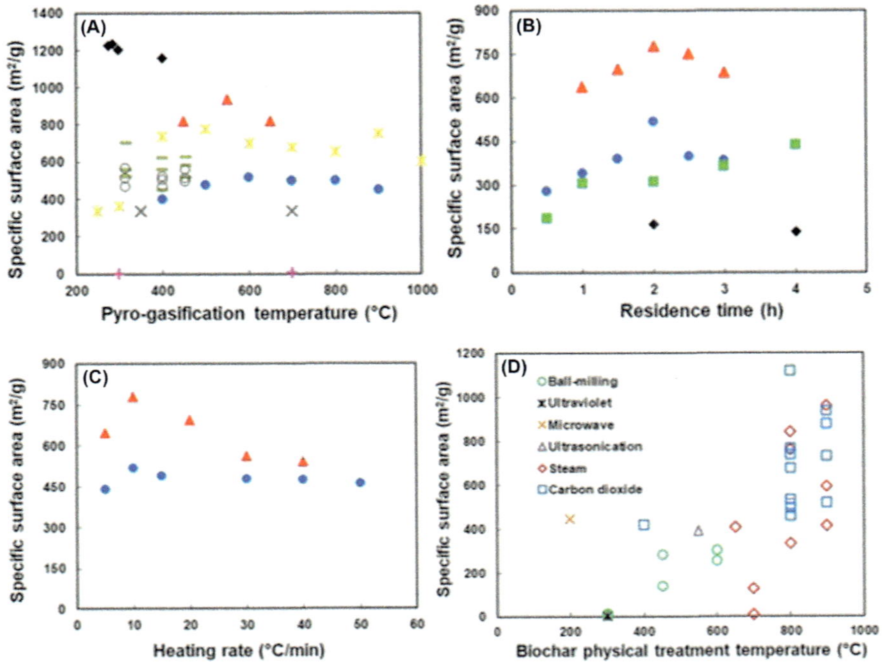

FIGURE 18.8

Surface area of activated biochars as a function of (A) pyro-gasification temperature in the presence of CO_2 obtained from pistachio nutshells ж (Lua et al., 2004) and ▲(Işıtan et al., 2016), from oil palm shells ● (Lua et al., 2006; activated at 900°C), from almond shells ♦ (Marcilla et al., 2000; activated at 780°C), and from white birch and black spruce ∘ (Braghiroli et al., 2018; activated at 700°C, 800°C, and 900°C); and steam from broiler litter x (Uchimiya et al., 2010; activated at 800°C) and from burcucumber plants (Rajapaksha et al., 2015; activated at 300°C and 700°C). (B) Residence time for biochars made from pistachio nutshells ▲ (Işıtan et al., 2016), from oil-palm shells ● (Lua et al., 2006; activated at 900°C) and ♦ (Hamza et al., 2016) activated at 800°C), and from date pits ■ ((Bouchelta et al., 2012); activated at 700°C). (C) Heating rate for biochars made with the same materials and conditions as in (B). (D) Surface areas of modified biochars as a function of physical treatment temperature according to physical modification treatment: ball milling (Wei Xiang et al., 2020; W. Xiang et al., 2020), ultraviolet irradiation (Peng et al., 2019, 2018), microwaving (Kołtowski et al., 2017), ultrasonication (Sajjadi et al., 2019), and activation in the presence of superheated steam (Arantes et al., 2023; Braghiroli et al., 2019; Mondal et al., 2016, 2015; Rajapaksha et al., 2015; Shen et al., 2017; Uchimiya et al., 2010; Wu et al., 2016) and CO_2 (Braghiroli et al., 2018; Kołtowski et al., 2017; Kozyatnyk et al., 2021; Lua et al., 2006; Manyà et al., 2018; Pepè Sciarria et al., 2020; Sricharoenchaikul et al., 2008).

(C) Reproduced with permission from Braghiroli, F. L., Bouafif, H., Neculita, C. M., & Koubaa, A. (2020). Influence of pyro-gasification and activation conditions on the porosity of activated biochars: A literature review. Waste and Biomass Valorization, 11*(9), 5079–5098. https://doi.org/10.1007/s12649-019-00797-5.*

methods is that they do not require chemicals, so the resulting biochars have high purity and do not need purification or washing before application in areas such as catalysis, drug delivery, and environmental cleanup (e.g., drinking water treatment and soil remediation). Physically modified biochars via ball milling and via UV, microwave, and ultrasound irradiation are used to adsorb volatile organic compounds (acetone, ethanol, chloroform, cyclohexane, toluene, and benzene) (Q. Li et al., 2016; W. Xiang et al., 2020) and to remove heavy metals (Peng et al., 2019, 2018; Sajjadi et al., 2019) and contaminants (Amusat et al., 2021; Sajjadi et al., 2019) from contaminated water, or as amendment to reduce the toxicity of contaminated soil (Kołtowski et al., 2017). Because CO_2 and steam-activated materials present the highest surface area with increasing activation temperature compared to other physically modified biochars, they are used in not only the above-cited applications (Braghiroli et al., 2019; Kozyatnyk et al., 2021; Mondal et al., 2016; Shen et al., 2017; Wu et al., 2016) but also in those requiring high surface area, such as catalysis (Arantes et al., 2023; Pepè Sciarria et al., 2020), energy storage (Braghiroli et al., 2020), and electromagnetic shielding (Boss et al., 2021, 2020).

18.4.2 Chemical modification methods

Surface modification methods are commonly used to alter the physicochemical properties of pristine biochar. Acids or basic functional groups are added to the bio-sourced material to enhance the rate of physical interactions and chemical reactions on the biochar surface. The methods include amination, oxidation, and sulfonation. Additionally, metals or metal oxides can be added to the biochar surface. The resulting material is generally used for environmental applications, and particularly wastewater treatment, to remove organic and inorganic compounds such as pharmaceuticals and heavy metals. The main modification processes are outlined below.

18.4.2.1 Amination

Biochar surfaces can be enriched with nitrogen-containing functional groups (N doping) to significantly enhance the biochar's performance and properties. This highly efficient method is widely used in research and industry for applications such as environmental remediation, agriculture, and energy production. Among other benefits, surface modification can substantially increase the adsorption capacity, nutrient retention, and catalytic activity of biochar. This approach holds immense potential to improve the sustainability and efficiency of biochar applications. Fig. 18.9A presents a breakdown of the various nitrogen functionalities present on biochar surfaces, categorized by specific nitrogen functionalities: graphitic, pyridinic, pyrrolic, pyridinic, pyridone, amino, and nitrogen oxidized. Adding a specific group to the biochar surface allows fine-tuning the surface properties. For example, the introduction of nitrogenous species into an amorphous C (e.g., biochar) typically yields higher electrical conductivity materials

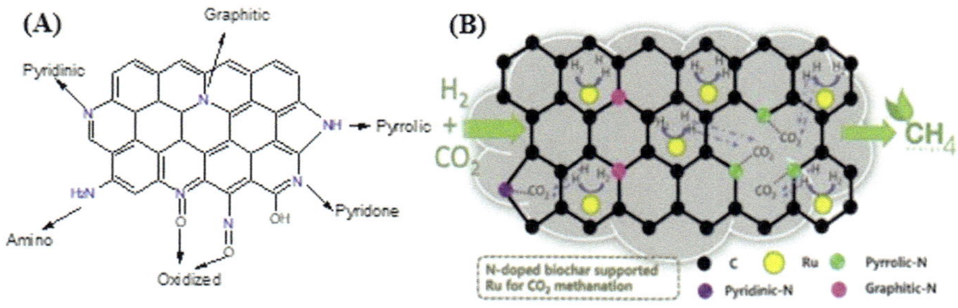

FIGURE 18.9

(A) Classification of N-functional groups: graphitic, pyridinic, pyrrolic, pyridinic, pyridone, amino, and nitrogen oxidized. (B) Schematic representation of N-doped biochar as catalyst support for CO_2 methanation.

(A) Adapted from Gao, W., Lin, Z., Chen, H., Yan, S., Huang, Y., Hu, X., & Zhang, S. (2022). A review on N-doped biochar for enhanced water treatment and emerging applications. Fuel Processing Technology, 237, *107468. https://doi.org/10.1016/j.fuproc.2022.107468. (B) From Wang, X., Liu, Y., Zhu, L., Li, Y., Wang, K., Qiu, K., Tippayawong, N., Aggarangsi, P., Reubroycharoen, P., & Wang, S. (2019). Biomass derived N-doped biochar as efficient catalyst supports for CO_2 methanation.* Journal of CO_2 Utilization, 34, *733–741. https://doi.org/10.1016/j.jcou.2019.09.003.*

(Basha et al., 2023). The material is then well suited for use as electrodes in energy storage devices (Demir et al., 2017), CO_2 capture (Rana et al., 2017), removal of organic pollutants from wastewater (Başer et al., 2021), or catalyst support for CO_2 methanation Fig. 18.9B. In the case of CO_2 methanation, the introduction of pyridinic nitrogen atoms increases the basicity of N-doped biochar, thereby improving the CO_2 adsorption, resulting in greater CH_4 production and superior catalytic performance (Wang et al., 2019).

N-containing functional groups may be loaded onto the biochar surface via reactions with nitrogen-rich precursors such as ammonia, nitric acid, ammonium salts, urea, and melamine, among others (Chatterjee et al., 2019; Gao et al., 2022; Kasera et al., 2021; Yang et al., 2021). Other materials with natural nitrogen content, such as algae or wood end-of-life panels, offer promise as precursors for producing N-doped biochar without the addition of chemicals. In practice, three main functionalization approaches are used.

1. Preoxidized biochar is exposed to ammonia at high temperatures ranging from 600°C to 900°C. This process leverages the previous oxidation of the biochar and the improved accessibility of the ammonia groups to the biochar. Thus, at high temperatures, ammonia dissociates and reacts with the oxygenated species present in the biochar. When the biochar surface is exposed to ammonia, water molecules are also released, according to Eq. (18.4) (Xiong et al., 2013):
2. Low molecular weight N-precursors, such as tetraethylenepentamine and diethanolamine (Shafawi et al., 2021), are used for amination. Here, the

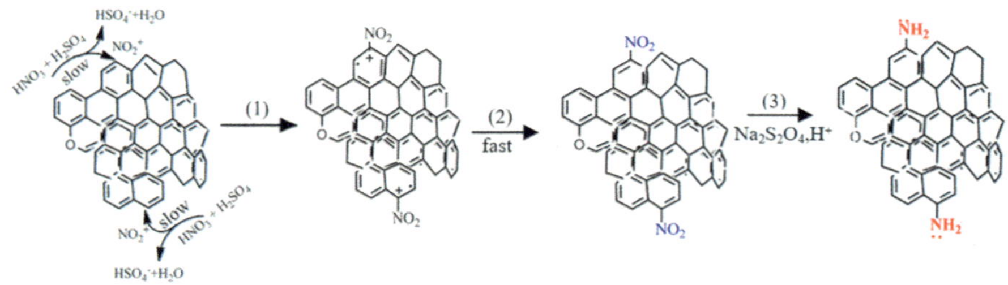

FIGURE 18.10

Schematic representation of the amination reaction of biochar.

Adapted from Yang, G. X., & Jiang, H. (2014). Amino modification of biochar for enhanced adsorption of copper ions from synthetic wastewater. Water Research, *48(1), 396–405. https://doi.org/10.1016/j.watres.2013.09.050.*

hydroxyl (−OH) groups present on the raw biochar surface play a crucial role to facilitate the introduction of N-containing functional groups, according to following equations (18.5−18.7) (Başer et al., 2021):

3. The biochar is exposed to an oxidant medium and multiple chemical reactions take place (Yang & Jiang, 2014; Fig. 18.10). In the first step, sulfuric and nitric acid react to generate nitronium ions (NO_2^+). Nitric acid then undergoes protonation and gradual dissociation to form NO_2^+ (Eqs. 18.8 and 18.9). The NO_2^+ then attacks the aromatic ring to form a complex. The final product depends on the reaction medium: for example, if $Na_2S_2O_4$ is present, the complex is reduced to aniline (Fig. 18.10).

$$\text{Biochar} - \text{COO}^- + \text{NH}_4^+ \overset{-\text{H}_2\text{O}}{\rightarrow} \text{CO} - \text{NH}_2 \tag{18.4}$$

$$\text{Biochar} - \text{C} - \text{OH} + \text{NH}_3 \rightarrow \text{Biochar} - \text{C} - \text{NH}_2 + \text{H}_2\text{O} \tag{18.5}$$

$$\text{Biochar} - \text{C} - \text{NH}_2 + \text{R} - \text{OH} \rightarrow \text{Biochar} - \text{C} - \text{N} = \text{C} - \text{Biochar} + \text{H}_2\text{O} \tag{18.6}$$

$$\text{Biochar} - \text{C} + \text{NH}_3 \rightarrow \text{Biochar} - \text{C} - \text{NH}_2 + \text{Biochar} - \text{C} - \text{NH} + \text{H}_2 \tag{18.7}$$

$$\text{HNO}_3 + \text{H}_2\text{SO}_4 \rightleftharpoons \text{H}_2\text{NO}_3^+ + \text{HSO}_4^- \tag{18.8}$$

$$\text{H}_2\text{NO}_3^+ + \text{H}_2\text{SO}_4 \rightleftharpoons \text{NO}_2^+ + \text{H}_3\text{O}^+ + \text{HSO}_4^- \tag{18.9}$$

18.4.2.2 Oxidation

In oxidation, many O-containing functional groups (−OH, COO^-, CO^-) are introduced onto the biochar surface. The oxygenated groups increase the negative charges on the material surface and enhance the biochar's ability to adsorb positively charged species via electrostatic attraction (Y. Zhu et al., 2018). These

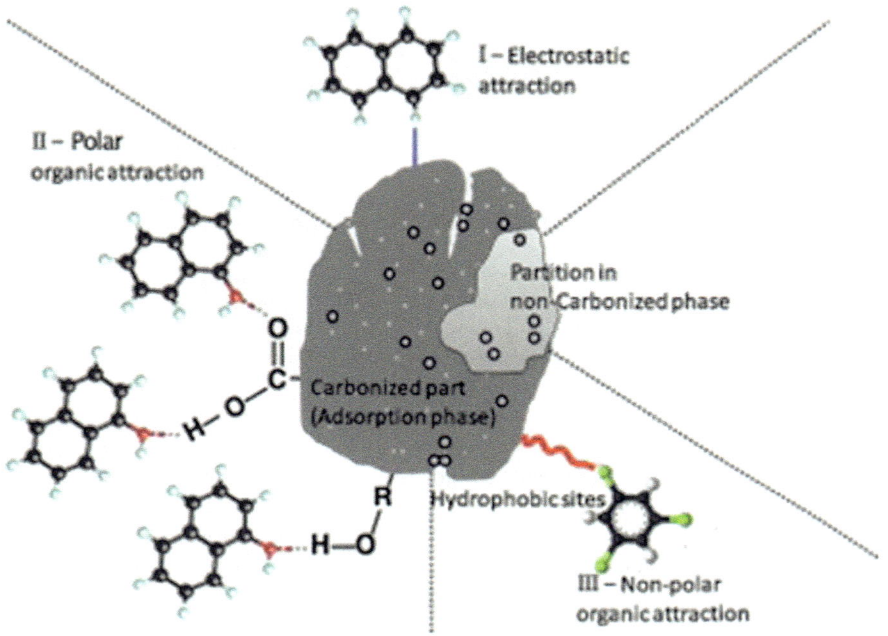

FIGURE 18.11

A diagrammatic representation of an oxygenated biochar particle demonstrating the electrostatic attraction of polar and nonpolar substances.

Reproduced with permission from Ahmad, M., Rajapaksha, A. U., Lim, J. E., Zhang, M., Bolan, N., Mohan, D., Vithanage, M., Lee, S. S., & Ok, Y. S. (2014). Biochar as a sorbent for contaminant management in soil and water: A review. Chemosphere, 99, *19–33. https://doi.org/10.1016/j.chemosphere.2013.10.071. Copyright 2014, Elsevier.*

species include metal cations such as Cu^{2+}, Cd^{2+}, Pb^{2+}, and Ni^{2+} as well as organic contaminants (Ahmad et al., 2014), as shown in Fig. 18.11. Thus the chemical surface is substantially transformed to allow more efficient interaction with the adsorbed contaminant. The most common biochar oxidation methods are thermal treatment in an air atmosphere or in a mixture of air and inert gas and strong chemical oxidation with nitric acid or ammonium persulfate.

Thermal air oxidation is a complex process that involves the reaming effect, whereby tar molecules produced during the pyrolysis of the raw material are deposited in the biochar pores. Oxidation occurs under controlled thermal conditions in the presence of air. Thus the oxygen molecules react with the tar molecules to form gases, which increases the porosity of the resulting biochar. This increased surface area increases the accessibility of the oxidizing agent into the biochar pores to obtain a highly porous and high-oxygen-content biobased material (Xiao & Pignatello, 2016), as illustrated in Fig. 18.12.

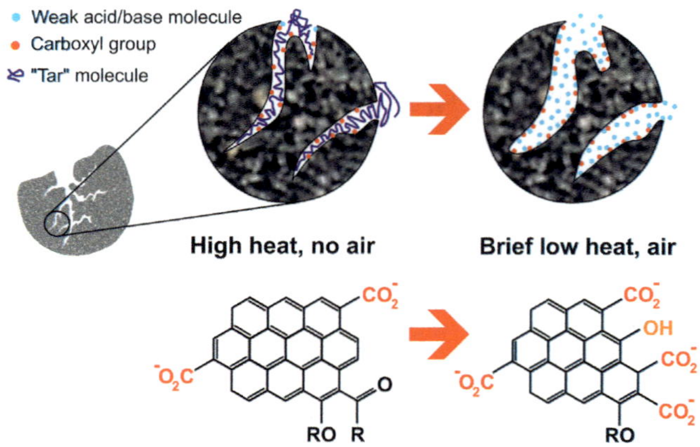

FIGURE 18.12

Schematic representation of the oxidation process on the biochar surface under air atmosphere.

Reproduced with permission under a Creative Commons Attribution 4.0 International License from Xiao, F. (2022). A review of biochar functionalized by thermal air oxidation. Environmental Functional Materials, *1(2), 187–195. https://doi.org/10.1016/j.efmat.2022.03.001, Elsevier.*

18.4.2.3 Sulfonation

Sulfonation is generally performed on incompletely carbonized material, which is essentially a C material with defects (Zhou et al., 2021). Sulfonation occurs via reactions between the biochar and sulfonated precursors under heating (Xie et al., 2020). Precursors include p-toluenesulfonic acid, chlorosulfonic acid, sulfosalicylic acid, sulfamic acid, gaseous sulfur trioxide, oleum, and concentrated sulfuric acid in a dissociated form, as shown in Fig. 18.13B. The oxygenated groups already present on the raw biochar surface can assist the sulfonation reaction, as illustrated in Fig. 18.13C and D (Xie et al., 2020). Notably, sulfonic groups can also be introduced into nitrogen-rich biochars, as illustrated in Fig. 18.13E.

Regardless of the properties of the biochar feedstock, the sulfonation process shows greater acidification efficiency when chlorosulfonic acid is used compared to sulfuric acid and p-toluenesulfonic acid (Xie et al., 2020). Fig. 18.14 depicts the chemical reactions involved in the interaction between biochar and chlorosulfonic acid. The SO_3H and oxygenated groups such as COO^- and $-OH$ on the biochar surface (Fig. 18.14A and B), combined with high porosity, contribute to the biochar's hydrophilicity (Xiong et al., 2018), resulting in improved adsorption of certain molecules and ions. The SO_3H group can also bind to nitrogen-rich biochar (Fig. 18.14C) under heating to produce active heteroatom compounds suitable for catalytic application. Sulfonated biochar may be applied to various

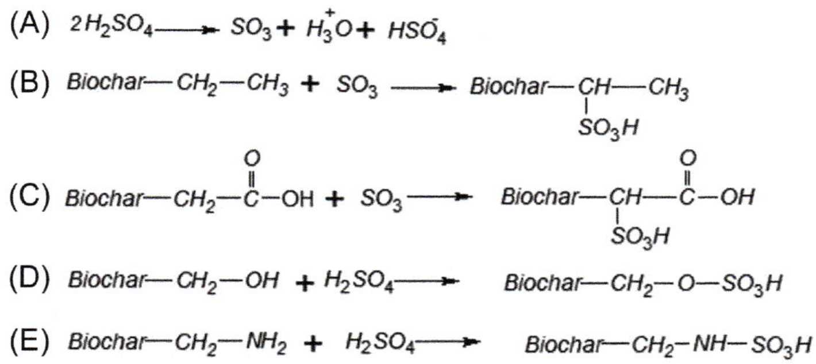

FIGURE 18.13

The main equations represent (A) sulfuric acid dissociation. Sulfonation of (B) alkyl, (C) carboxyl, (D) hydroxyl, and (E) amino groups on the biochar surface.

Adapted from Yu, W., Hu, J., Yu, Y., Ma, D., Gong, W., Qiu, H., Hu, Z., & Gao, H. (2021). Facile preparation of sulfonated biochar for highly efficient removal of toxic Pb(II) and Cd(II) from wastewater. Science of the Total Environment, 750, *141545. https://doi.org/10.1016/j.scitotenv.2020.141545.*

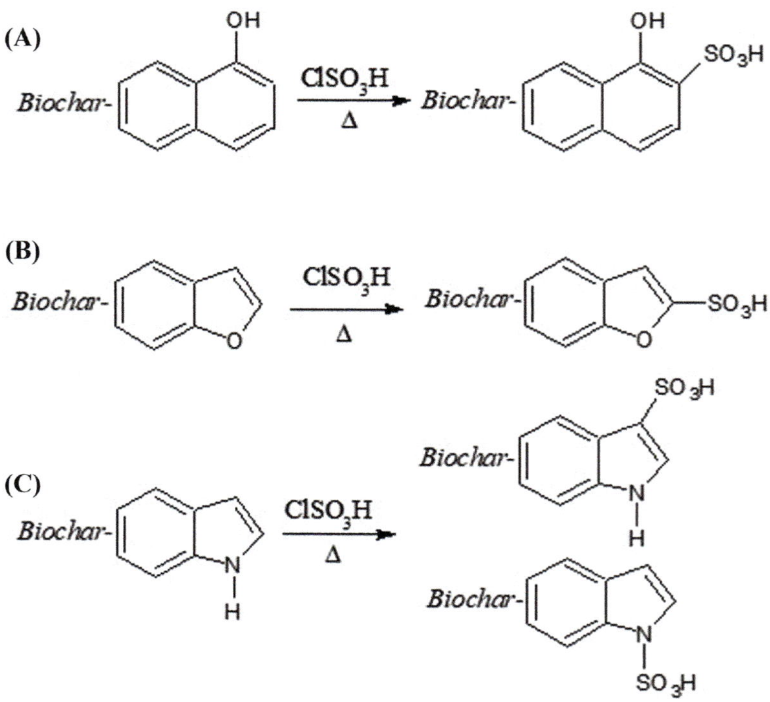

FIGURE 18.14

Sulfonation reactions of biochar rich in (A and B) oxygenated groups and (C) nitrogenated groups in the presence of chlorosulfonic acid.

Adapted from Xie, Q., Yang, X., Xu, K., Chen, Z., Sarkar, B., & Dou, X. (2020). Conversion of biochar to sulfonated solid acid catalysts for spiramycin hydrolysis: Insights into the sulfonation process. Environmental Research, 188, *109887. https://doi.org/10.1016/j.envres.2020.109887.*

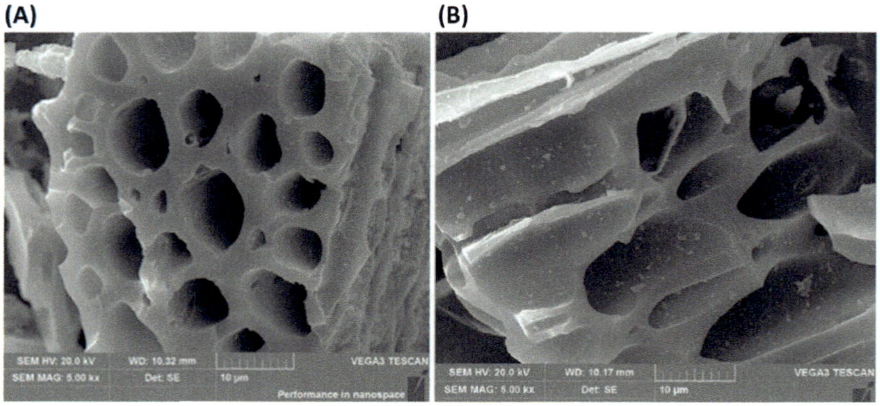

FIGURE 18.15

SEM images of (A) wood-based raw biochar and (B) its sulfonated form. *SEM*, Scanning electron microscope.

Reproduced with permission from Xiong, X., Yu, I. K. M., Chen, S. S., Tsang, D. C. W., Cao, L., Song, H., Kwon, E. E., Ok, Y. S., Zhang, S., & Poon, C. S. (2018). Sulfonated biochar as acid catalyst for sugar hydrolysis and dehydration. Catalysis Today, 314, *52–61. https://doi.org/10.1016/j.cattod.2018.02.034.*

fields, including the removal of heavy metals such as Cd^{2+} and Pb^{2+} in wastewater treatment (Yu et al., 2021; Zhou et al., 2021), as an acid catalyst for sugar hydrolysis (Kumari et al., 2022; Xiong et al., 2018), and in catalytic biodiesel production (Cao et al., 2023). The SEM images in Fig. 18.15 highlight differences in porous structure between raw and sulfonated biochar. After sulfonation, significant reductions in textural properties are observed, including average pore diameter (from 68 to 4.8 nm), pore volume (from 0.066 to 0.036 cm^3/g), and specific surface area (from 132 to 57 m^2/g). These changes suggest a collapse of mesopores into an open structure composed of large macropores. The observed decrease in the specific surface area of sulfonated biochar can be attributed to the presence of cracks on the biochar surface, indicating that the morphology and surface chemistry of sulfonated biochar play a greater role in the hydrolysis of maltose into glucose compared to the surface area. Given these findings, the properties of sulfonated biochar should be carefully considered when selecting suitable catalysts for specific reactions.

18.4.2.4 Metals and metal oxides

Ion sorption sites, and particularly for heavy metals, can be created by loading metals and metal oxides onto biochar surfaces. The biochar material serves as substrate support for the formation of composites with the deposited material having suitable synergetic properties for the final application. Adding metal to biochar is a multifaceted process involving various reactions, resulting in different forms of metals on the biochar surface, including oxides and metallic structures. A diverse range of metals can be added to modify the surface properties. Using a tailored modification approach, the unique characteristics of individual metals can provide the final biochar product with distinct advantages.

Iron is commonly added to biochar for its contribution to the desired properties of the resulting composite material. For example, iron-biochar composites are often used to remove minerals, including heavy metals (AsO_3^{-3}, Cd^{+2}, Cu^{+2}, Pb^{+2}, Cr^{+3}, Zn^{+2}), from wastewater, as well as cations (e.g., K^+, Na^+, Ca^{+2}, Mg^{+2}, Ba^{+2}, Sr^{+2}). They are also used as soil amendment and immobilizing agent for heavy metal-contaminated soil (Gao & Wan, 2023; Gueret Yadiberet Menzembere et al., 2023; Sun et al., 2019; Wang et al., 2023; Y. Zhu et al., 2018). Under certain synthesis conditions, it was noted that C reacted with certain metals through covalent bonding via doping reactions, and more specifically, in Fe_3C materials, as described below. The first step in producing iron-modified biochar is to apply iron to the biochar surface by soaking the biochar in iron salt solutions [$FeCl_3$, $Fe(NO_3)_3$] (Abdelhafez & Abbas, 2020). Structural modification occurs under inert atmosphere at high temperatures. During thermal treatment, several reactions may occur (Sun et al., 2019). First, the iron salts are initially converted to iron oxide (Eqs. 18.10 and 18.11). Then, via carbothermal reduction, the iron oxide is reduced to magnetite due to the presence of C in the biomass precursor at temperatures of 400°C–500°C (Eqs. 18.12 and 18.13). As the temperature increases to 650°C–800°C, the magnetite may be converted to protoxide (Eqs. 18.12–18.14) and iron carbide (Eq. 18.15). Lastly, iron carbide is expected to decompose at between 900°C and 1000°C (Eq. 18.16). Therefore, at the end of the reactions, multiple types of iron can be present on the final material surface.

$$2FeCl_3 + 6H_2O \rightarrow 2Fe(OH)_3 + 6HCl\uparrow \tag{18.10}$$

$$2Fe(OH)_3 \rightarrow 2FeO(OH) + 2H_2O \rightarrow Fe_2O_3 + 3H_2O \tag{18.11}$$

$$C + H_2O \rightarrow CO\uparrow + H_2\uparrow \tag{18.12}$$

$$3Fe_2O_3 + 3(H_2, CO, C) \rightarrow 2Fe_3O_4 + 3(H_2O, CO_2, CO)\uparrow \tag{18.13}$$

$$Fe_3O_4 + (H_2, CO, C) \rightarrow 3FeO + (H_2O, CO_2, CO)\uparrow \tag{18.14}$$

$$3FeO + 6C \rightarrow Fe_3C + 3CO \tag{18.15}$$

$$Fe_3C \rightarrow 3Fe + C \tag{18.16}$$

Other modified biochars include Cu-biochar, used for H_2S removal (Cui et al., 2022), and Ni-biochar, used for selective hydrogenolysis of diphenyl ether (Xie et al., 2022). Meanwhile, double-impregnated biochars have shown promising potential for various applications beyond the environmental field. The goal in these cases is to achieve synergistic effects. For example, nano α-Fe_2O_3/N-doped biochar was satisfactorily applied as electrode material for supercapacitors (Li et al., 2020). This encouraging progress in biochar technology promises to usher in more effective and sustainable energy solutions in future. A range of different metals can contribute to improve the properties of modified biochar. For example, Ag-Cu/biochar was applied to adsorb pharmaceuticals such as doxycycline and as antioxidant and antimicrobial agent against gram-negative bacteria (Hosny et al., 2022).

Furthermore, metals such as Ni, Co, and Fe can be introduced into the biochar matrix to create a magnetic bio-sourced material. In this approach, biochar grains can be recovered from aqueous phases by simple magnetic separation (Yi et al., 2020). In this way, synergic interaction is leveraged to remove various types of contaminants from soil and water. The contaminants include but are not limited to dyes, organic pollutants, oils, heavy metals, and antibiotics (Awasthi, 2022; Qu et al., 2022; Wei et al., 2024; Xiao, 2022). Not only can the overall quality of environmental resources be improved, but the removal can be easily done while preserving the positive characteristics of the biochar surface, such as pore filling and π-π interactions (Fig. 18.16; Qu et al., 2022).

In terms of biochar composition, it is imperative to consider the range of surface area values that may result from surface modification procedures. Fig. 18.17 graphically represents the potential relationships between the specific surface area

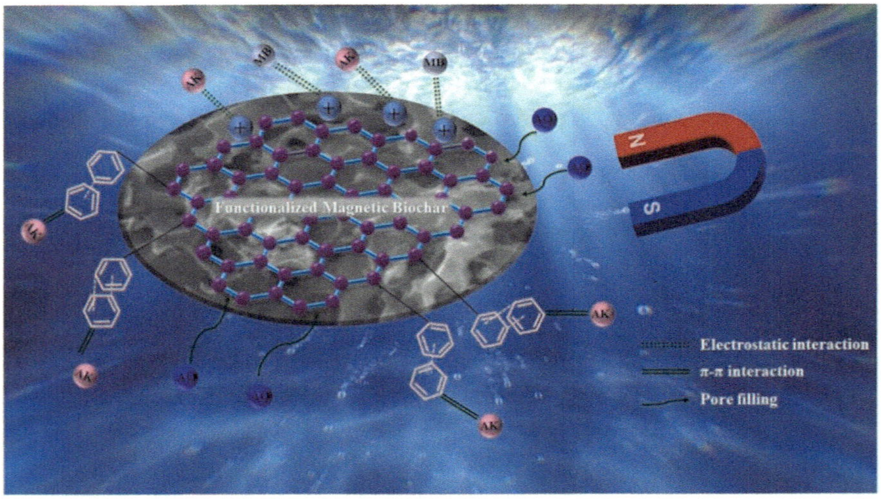

FIGURE 18.16

Schematic representation of electrostatic interactions, pore filling, and p-p interactions on the biochar surface.

Reproduced with permission from Qu, J., Shi, J., Wang, Y., Tong, H., Zhu, Y., Xu, L., Wang, Y., Zhang, B., Tao, Y., Dai, X., Zhang, H., & Zhang, Y. (2022). Applications of functionalized magnetic biochar in environmental remediation: A review. Journal of Hazardous Materials, 434, *128841. https://doi.org/10.1016/j.jhazmat.2022.128841. Copyright 2022, Elsevier.*

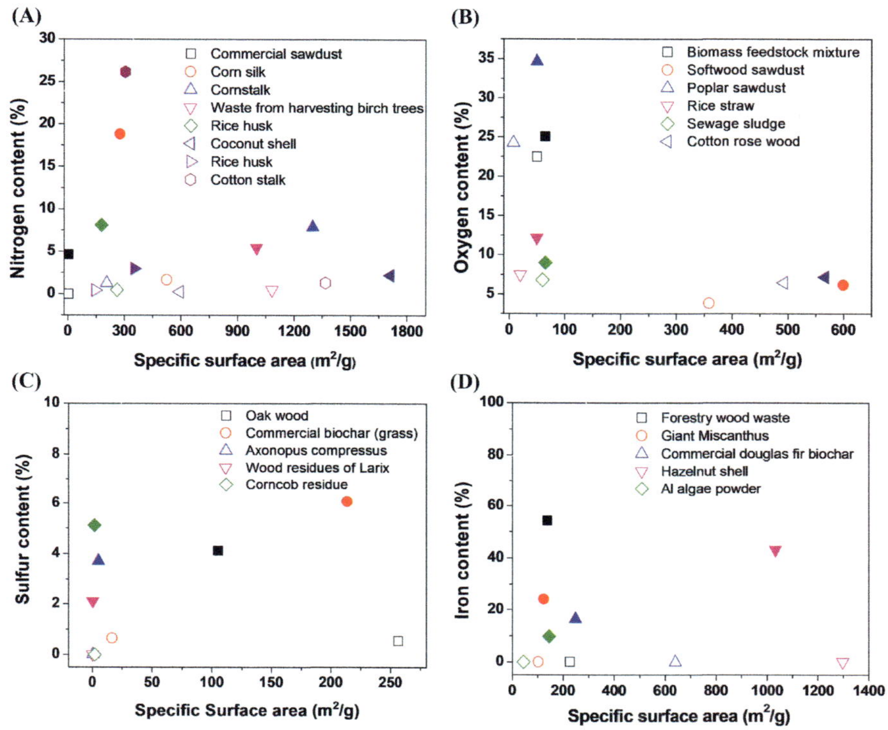

FIGURE 18.17

Surface area as a function of element content added to the biochar surface: (A) nitrogen (Chen et al., 2023; dos Reis et al., 2023; Geng et al., 2023; Liu et al., 2023; Setiawan et al., 2022; Yang & Jiang, 2014; Yang et al., 2021; Yu et al., 2023); (B) oxygen (Bardestani & Kaliaguine, 2018; Dai et al., 2020; Huang et al., 2020; Ma et al., 2021; X. Zhu et al., 2018); (C) sulfur (Cao et al., 2023; Xie et al., 2020; Yu et al., 2021; Zuo et al., 2023), and (D) iron (Ateş & Oskay, 2022; Kim et al., 2019; Samaraweera et al., 2023; Sun et al., 2019; Yu et al., 2022). Empty and full symbols indicate precursor materials before and after chemical modification, respectively.

of raw biochar (empty symbols) and surface-modified biochar (full symbols) produced by treatment with various elements: (1) nitrogen (Chen et al., 2023; Geng et al., 2023; Liu et al., 2023; Setiawan et al., 2022; Yang & Jiang, 2014; Yang et al., 2021; Yu et al., 2023); (2) oxygen (Bardestani & Kaliaguine, 2018; Dai et al., 2020; Huang et al., 2020; Ma et al., 2021; X. Zhu et al., 2018); (3) sulfur (Cao et al., 2023; Xie et al., 2020; Yu et al., 2021; Zuo et al., 2023), and (4) iron (Ateş & Oskay, 2022; Kim et al., 2019; Samaraweera et al., 2023; Sun et al., 2019; Yu et al., 2022). The specific surface areas range widely, from a few to many square meters per gram in either direction. Hence, each case must be thoroughly examined to obtain a precise understanding of the final material's properties and attributes. The synthesis conditions tend to vary within a certain range, and the loss or gain of porosity may be linked to a specific phenomenon, as illustrated in Fig. 18.15, where the surface area decreases due to pore collapse following sulfur insertion into the biochar.

Interestingly, the nitrogen content and surface area show differences between cotton stalks and coconut shells as feedstock. The nitrogen content of cotton stalk increases significantly from 1.3 to 26, resulting in a marked decrease in surface area from 1368 to 307 m^2/g. In contrast, the nitrogen content of coconut shells increases from 0.2 to 2.2, with a corresponding significant increase in surface area from 595 to 1711 m^2/g. Notably, the surface area of commercial sawdust and birch residue waste shows minimal variation with increased nitrogen content. Consequently, adding nitrogen to the biochar surface does not produce a clear trend for surface area. These findings are consistent for oxygen, sulfur, and iron (Fig. 18.17D). Hence, in surface modification, the variation in surface area according to material composition must be considered. Thus achieving greater surface area depends on multiple factors, including the elemental source, the chemical treatment temperature, and the biomass source. Furthermore, to obtain the most porous and effective biochar materials, these factors must be considered in combination.

18.4.2.5 Chemical activation

In chemical activation, biochar is exposed to chemicals (KOH, NaOH, and potassium acetate) at moderate or high temperature (depending on the chemical and the targeted physicochemical characteristics) in a limited or zero oxygen environment (Rodríguez-Reinoso & Molina-Sabio, 1992). The biochar is first impregnated with the chemical in solid or liquid form at different ratios and left to rest for several hours up to overnight. The most alkaline chemical used to date is KOH. When biochar mixed with KOH is subjected to heating, the main reactions (Eqs. 18.17–18.21) that take place are described below, depending on the temperature (Ehrburger et al., 1986; Lozano-Castelló et al., 2007). The released physical agents (CO and CO_2) during chemical activation may also act as physical agents that contribute to porosity development.

$$6KOH + 2C \rightarrow 2K + 3H_2\uparrow + 2K_2CO_3 \tag{18.17}$$

$$4KOH + CH_x \rightarrow K_2CO_3 + K_2O + \left(\frac{2+x}{2}\right)H_2\uparrow \tag{18.18}$$

$$T > 570^\circ C: \quad K_2CO_3 \rightarrow K_2O + CO_2\uparrow \tag{18.19}$$

$$T > 700^\circ C: \quad K_2CO_3 + 2C \rightarrow 2K\uparrow + 3CO\uparrow \tag{18.20}$$

$$K_2O + C \rightarrow 2K\uparrow(\text{boiling} - \text{point elevation}: 759^\circ C) + CO\uparrow \tag{18.21}$$

Fig. 18.18A shows a compilation of the surface areas of activated biochars made from various feedstocks as a function of pyro-gasification temperature in the presence of KOH or NaOH. Very high surface areas are observed, but with

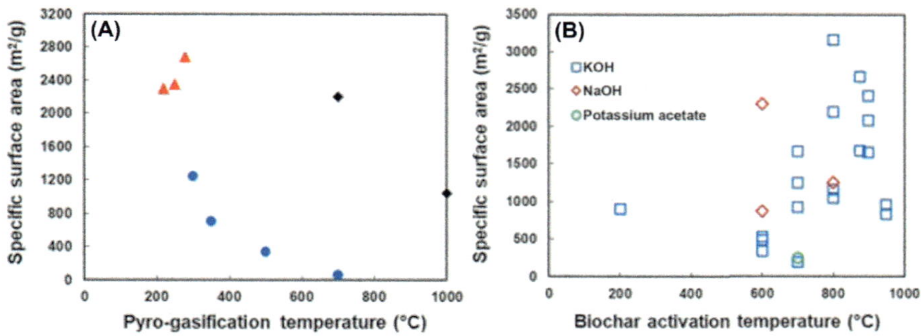

FIGURE 18.18

Surface areas of activated biochars as a function of (A) pyro-gasification temperature in the presence of KOH or NaOH from rice straw ♦ (Oh & Park, 2002), from debarked loblolly pine chips ● (Park et al., 2013) (activated at 800°C), and from rice husks ▲ (Zhang et al., 2016); activated at 825°C) and (B) activation temperature according to chemical agent type: KOH (Bhandari et al., 2014; Braghiroli et al., 2019; Dehkhoda, Ellis, & Gyenge, 2016; Dehkhoda, Gyenge, & Ellis, 2016; Gale et al., 2021; Li et al., 2016; Manyà et al., 2018; Oh & Park, 2002; Takaya et al., 2016), NaOH (Ding et al., 2016; Park et al., 2017; Zhang et al., 2016), and potassium acetate (Sakhiya et al., 2021).

a negative impact on the porosity of activated biochars at higher pyro-gasification temperatures. One explanation for this negative impact is the lower aromaticity of activated biochars made from loblolly pine chips and rice straw. This is because the smaller sized aromatic clusters and higher percentages of alkyl carbons and volatile matter act to maximize the porosity development. At higher temperatures, however, the higher aromaticity due to higher ring cluster size and consequently lower volatile matter due to the inactive C structure act to minimize the porosity development (Braghiroli et al., 2020; Brewer et al., 2009; Oh & Park, 2002; Park et al., 2017). Thus, in addition to selecting appropriate materials, the pyro-gasification and activation parameters must be optimized to target specific applications. Fig. 18.18B shows a compilation of the surface areas of chemically activated biochars in the presence of KOH, NaOH, and potassium acetate as a function of activation temperature. Chemical activation has been demonstrated effective to produce an impressively high specific surface area of up to 3500 m^2/g (Fig. 18.18B). The improved physicochemical properties are favorable for diverse adsorption mechanisms such as physisorption, chemisorption, ion exchange, hydrophobic interactions, pore filling, $\pi-\pi$ electron donor–acceptor ($\pi-\pi$ EDA) interactions, and H-bond interactions, as depicted in Fig. 18.16 (Li et al., 2022). These mechanisms are enabled by the developed pore structure and abundant oxygenated functional groups on the surface of chemically activated biochars, resulting in excellent adsorptive performance. These superior properties allow effective removal of organic, metal, and emerging contaminants (Bhandari et al., 2014; Dehkhoda, Ellis, & Gyenge, 2016; Dehkhoda, Gyenge, & Ellis, 2016; Ding et al., 2016; Gale et al., 2021; Sakhiya et al., 2021), CO_2 uptake (Li et al., 2016; Manyà et al., 2018), and energy storage (Dehkhoda, Gyenge, & Ellis, 2016).

18.5 Conclusions and prospects

Biochar can be engineered physically and chemically to improve the physicochemical properties and render them suitable for a wide range of high-value-added applications. The advantages include higher surface area, more abundant nanoporosity, more efficient electrical conductivity, upgraded aromaticity, higher graphitization, and a larger proportion of functional groups attached to the C layer (e.g., oxygenated, nitrogenated, sulfonated, metal). Recent physical methods include grinding (ball milling), irradiation (UV, microwave, ultrasound), and physical activation. Of these, physical activation appears to be the most efficient method to develop biochar surface area and porosity. However, it is important to evaluate other physicochemical properties, such as the elemental composition, porosity type (ultramicro-, micro-, and mesopores), and metals content in order to design materials with suitable characteristics for targeted applications. Unlike chemical modification methods, physical methods do not use chemicals, so physically modified biochars do not present chemical impurities. Therefore they can be used for environmentally friendly applications such as water treatment, soil remediation, catalysis, and drug delivery, where materials must be chemical-free.

Chemical methods (e.g., amination, oxidation, and sulfonation) and the addition of metals and metal oxides are also used to modify biochar surfaces. In these cases, biochar acts as substrate support to form composites with the deposited material. This enhances the biochar surface properties, including surface area, electrical conductivity, and hydrophilic or hydrophobic capacity. The various methods offer immense potential for the development of new engineered materials for a versatile range of applications across multiple fields, from environmental bioremediation to energy matrices. For instance, adding nitrogenous compounds to biochar produces materials with greater electrical conductivity. In contrast, adding oxygenated groups amplifies the negative charges on the material surface, improving the capacity to absorb positively charged species via electrostatic attraction. In addition, the combination of high porosity and the presence of sulfonated (SO_3H) and oxygenated groups ($COO-$ and $-OH$) on the biochar surface improves the hydrophilicity, which in turn improves its capacity to adsorb certain molecules and ions from wastewater or soil.

Chemical activation produces highly porous materials with surface areas up to $3500\ m^2/g$ (in the presence of KOH), greatly exceeding that for the above-mentioned physical and other chemical modification methods. The factors that significantly influence the final physicochemical properties include the feedstock type, its elemental composition, the physical or chemical treatment conditions (temperature, heating rate, residence time, and reactor design or equipment, depending on the treatment), the biochar type, and the physical or chemical activation agent. These process parameters must be optimized to improve the product's quality and efficiency and to reduce the energy or chemicals required without impacting the physicochemical characteristics of the final biochar. Optimal efficiency and economic value can be achieved by modifying the synthesis conditions. Although the handling of materials for biochar production is straightforward, control of the final properties becomes complex due to the many variables at play. Therefore producing materials with the desired properties requires fine-tuning of the process parameters. For example, controlling the thermal treatment temperature during activation can directly influence the porosity, while the presence of oxygenated groups can affect the hydrophobicity. These properties are desirable for high-value-added applications in the chemical, agricultural, and environmental fields, among others.

Some challenges remain. Among others, we may mention the high cost and lengthy time required for biochar modification and the inefficient adsorption capacity of biochars for certain contaminants. Future studies could also explore the long-term effects of biochar use. Moreover, a comprehensive economic analysis of biochar use by the Parties to the United Nations Framework Convention on Climate Change guidelines would be advisable. A key future research area where biochar can play a vital role is CO_2 sequestration. Biochar also offers a promising avenue to foster the integration of C markets. Both these directions appear essential for a sustainable future.

Acknowledgment

G. Amaral-Labat acknowledges financial support by CNPq (Conselho Nacional de Desenvolvimento Científico e Tecnológico, Brazil).

Author contribution

Flavia L. Braghiroli: conceptualization, methodology, software, writing—original draft preparation, supervision, validation, and writing—reviewing and editing. Gisele Amaral-Labat: conceptualization, methodology, software, writing—original draft preparation, validation, and writing—reviewing and editing.

Conflict of interest

The authors declare that they have no conflicts of interest to report regarding the present study.

References

A.A. Abdelhafez, M. Abbas, 2020 Applications of biochar for environmental safety. IntechOpen. https://doi.org/10.5772/intechopen.87828.

Aghapour Aktij, S., Taghipour, A., Rahimpour, A., Mollahosseini, A., & Tiraferri, A. (2020). A critical review on ultrasonic-assisted fouling control and cleaning of fouled membranes. *Ultrasonics*, *108*, 106228. Available from https://doi.org/10.1016/j.ultras.2020.106228.

Agrawal, A., & Yi, G.-C. (2020). *Chapter Two - Sample pretreatment with graphene materials. Comprehensive Analytical Chemistry* (91, pp. 21–47). Elsevier.

Ahmad, M., Rajapaksha, A. U., Lim, J. E., Zhang, M., Bolan, N., Mohan, D., Vithanage, M., Lee, S. S., & Ok, Y. S. (2014). Biochar as a sorbent for contaminant management in soil and water: A review. *Chemosphere*, *99*, 19–33. Available from https://doi.org/10.1016/j.chemosphere.2013.10.071, http://www.elsevier.com/locate/chemosphere.

Ambika, S., Kumar, M., Pisharody, L., Malhotra, M., Kumar, G., Sreedharan, V., Singh, L., Nidheesh, P. V., & Bhatnagar, A. (2022). Modified biochar as a green adsorbent for removal of hexavalent chromium from various environmental matrices: Mechanisms, methods, and prospects. *Chemical Engineering Journal*, *439*, 135716. Available from https://doi.org/10.1016/j.cej.2022.135716.

Amusat, S. O., Kebede, T. G., Dube, S., & Nindi, M. M. (2021). Ball-milling synthesis of biochar and biochar–based nanocomposites and prospects for removal of emerging contaminants: A review. *Journal of Water Process Engineering*, *41*. Available from https://doi.org/10.1016/j.jwpe.2021.101993, http://www.journals.elsevier.com/journal-of-water-process-engineering/.

Arantes, H. T. L., Machado, M. A., Santoro, M. C., Freitas, J. C. C., Ronconi, C. M., Ligiero, C. B. P., Cassini, S. T. A., Sampaio, I. C. F., & Luz, P. P. (2023). Effect of activated biochar as a low-cost catalyst on the quality of catalytic intermediate co-pyrolysis oil from waste polystyrene and green coconut pericarp. *Fuel Processing Technology*, *240*. Available from https://doi.org/10.1016/j.fuproc.2022.107539, https://www.journals.elsevier.com/fuel-processing-technology.

Ateş, A., & Oskay, K. O. (2022). Preparation and characterization of nanosized Fe_3O_4-biochar electrocatalysts with large surface area for H_2O_2 sensing. *Surfaces and Interfaces*, *29*, 101733. Available from https://doi.org/10.1016/j.surfin.2022.101733.

Awasthi, M. K. (2022). Engineered biochar: A multifunctional material for energy and environment. *Environmental Pollution*, *298*, 118831. Available from https://doi.org/10.1016/j.envpol.2022.118831.

Babu, B. V. (2008). Biomass pyrolysis: A state-of-the-art review. *Biofuels, Bioproducts and Biorefining*, *2*(5), 393–414. Available from https://doi.org/10.1002/bbb.92India, http://www3.interscience.wiley.com/cgi-bin/fulltext/121369801/PDFSTART.

Barcelló, D. (2020). *Comprehensive analytical chemistry*. Elsevier.

Bardestani, R., & Kaliaguine, S. (2018). Steam activation and mild air oxidation of vacuum pyrolysis biochar. *Biomass and Bioenergy*, *108*, 101–112. Available from https://doi.org/10.1016/j.biombioe.2017.10.011, http://www.journals.elsevier.com/biomass-and-bioenergy/.

Basha, D. B., Ahmed, S., Ahmed, A., & Gondal, M. A. (2023). Recent advances on nitrogen doped porous carbon micro-supercapacitors: New directions for wearable electronics. *Journal of Energy Storage*, *60*, 106581. Available from https://doi.org/10.1016/j.est.2022.106581.

Başer, B., Yousaf, B., Yetis, U., Abbas, Q., Kwon, E. E., Wang, S., Bolan, N. S., & Rinklebe, J. (2021). Formation of nitrogen functionalities in biochar materials and their role in the mitigation of hazardous emerging organic pollutants from wastewater. *Journal of Hazardous Materials, 416*. Available from https://doi.org/10.1016/j.jhazmat.2021.126131, http://www.elsevier.com/locate/jhazmat.

Bhandari, P. N., Kumar, A., Bellmer, D. D., & Huhnke, R. L. (2014). Synthesis and evaluation of biochar-derived catalysts for removal of toluene (model tar) from biomass-generated producer gas. *Renewable Energy, 66*, 346–353. Available from https://doi.org/10.1016/j.renene.2013.12.017.

Boss, A. F. N., Braghiroli, F. L., Amaral-Labat, G., Souza, A. A. T., Baldan, M. R., Bouafif, H., Koubaa, A., & Lenz e Silva, G. F. B. (2020). Dielectric characterization of white birch–activated biochar composites: A sustainable alternative to radar-absorbing materials. *Journal of Composite Materials, 54*(9), 1233–1244. Available from https://doi.org/10.1177/0021998319877493, http://jcm.sagepub.com/content/by/year.

Boss, A. F. N., Ferreira, H. R., Braghiroli, F. L., Amaral-Labat, G. A., Souza, A. A. T. de, Bouafif, H., Koubaa, A., Baldan, M. R., & Lenz e Silva, G. F. B. (2021). Investigation of sustainable porous carbon as radar absorbing material. *Matéria (Rio de Janeiro), 26*(2). Available from https://doi.org/10.1590/s1517-707620210002.1263.

Bouchelta, C., Medjram, M. S., Zoubida, M., Chekkat, F. A., Ramdane, N., & Bellat, J.-P. (2012). Effects of pyrolysis conditions on the porous structure development of date pits activated carbon. *Journal of Analytical and Applied Pyrolysis, 94*, 215–222. Available from https://doi.org/10.1016/j.jaap.2011.12.014.

Braghiroli, F. L., Bouafif, H., Hamza, N., Neculita, C. M., & Koubaa, A. (2018). Production, characterization, and potential of activated biochar as adsorbent for phenolic compounds from leachates in a lumber industry site. *Environmental Science and Pollution Research, 25*(26), 26562–26575. Available from https://doi.org/10.1007/s11356-018-2712-9, http://www.springerlink.com/content/0944-1344.

Braghiroli, F. L., Bouafif, H., Neculita, C. M., & Koubaa, A. (2018). Activated biochar as an effective sorbent for organic and inorganic contaminants in water. *Water, Air, and Soil Pollution, 229*(7). Available from https://doi.org/10.1007/s11270-018-3889-8, http://www.kluweronline.com/issn/0049-6979/.

Braghiroli, F. L., Bouafif, H., Neculita, C. M., & Koubaa, A. (2020). Influence of pyro-gasification and activation conditions on the porosity of activated biochars: A literature review. *Waste and Biomass Valorization, 11*(9), 5079–5098. Available from https://doi.org/10.1007/s12649-019-00797-5, http://www.springer.com/engineering/journal/12649.

Braghiroli, F. L., Cuña, A., da Silva, E. L., Amaral-Labat, G., Lenz e Silva, G. F. B., Bouafif, H., & Koubaa, A. (2020). The conversion of wood residues, using pilot-scale technologies, into porous activated biochars for supercapacitors. *Journal of Porous Materials, 27*(2), 537–548. Available from https://doi.org/10.1007/s10934-019-00823-w, https://link.springer.com/journal/10934.

Braghiroli, F. L., Bouafif, H., & Koubaa, A. (2019). Enhanced SO_2 adsorption and desorption on chemically and physically activated biochar made from wood residues. *Industrial Crops and Products, 138*, 111456. Available from https://doi.org/10.1016/j.indcrop.2019.06.019.

Brewer, C. E., Schmidt-Rohr, K., Satrio, J. A., & Brown, R. C. (2009). Characterization of biochar from fast pyrolysis and gasification systems. *Environmental Progress and Sustainable Energy, 28*(3), 386–396. Available from https://doi.org/10.1002/ep.10378, http://www3.interscience.wiley.com/cgi-bin/fulltext/122544837/PDFSTART, United States.

Bridgwater, A. V. (2012). Review of fast pyrolysis of biomass and product upgrading. *Biomass and Bioenergy*, *38*, 68–94. Available from https://doi.org/10.1016/j.biombioe.2011.01.048.

Cai, Z., Li, F., Huang, Y., Chen, X., Rong, M., Lin, L., Yao, Q., & Wang, X. (2019). *Introduction novel nanomaterials for biomedical, environmental and energy applications* (pp. 1–36). Elsevier. Available from https://www.sciencedirect.com/book/9780128144978.

Calo, J. M., & Perkins, M. T. (1987). A heterogeneous surface model for the steady-state kinetics of the boudouard reaction. *Carbon*, *25*(3), 395–407. Available from https://doi.org/10.1016/0008-6223(87)90011-X.

Cao, M., Lu, M., Yin, H., Zhu, Q., Xing, K., & Ji, J. (2023). Effect of hemicellulose extraction pretreatment on sulfonated corncob biochar for catalytic biodiesel production. *Journal of Environmental Chemical Engineering*, *11*(1), 109058. Available from https://doi.org/10.1016/j.jece.2022.109058.

Chatterjee, R., Sajjadi, B., Chen, W. Y., Mattern, D. L., Egiebor, N. O., Hammer, N., & Raman, V. (2019). Low frequency ultrasound enhanced dual amination of biochar: A nitrogen-enriched sorbent for CO_2 capture. *Energy and Fuels*, *33*(3), 2366–2380. Available from https://doi.org/10.1021/acs.energyfuels.8b03583, http://pubs.acs.org/journal/enfuem.

Chemerys, V., & Baltrėnaitė, E. (2016). Modified biochar: A review on modifications of biochar towards its enhanced adsorptive properties. In *Proceedings of the 19th conference for junior researchers "Science – Future of Lithuania"*. VGTU Technika Vilnius Gediminas Technical University

Chen, C., Sun, H., Zhang, S., & Su, X. (2023). Structure-property relationship and mechanism of peroxymonosulfate activation by nitrogen-doped biochar for organic contaminant oxidation. *Applied Surface Science*, *609*. Available from https://doi.org/10.1016/j.apsusc.2022.155294, http://www.journals.elsevier.com/applied-surface-science/.

Chen, H. (2015). *Lignocellulose biorefinery feedstock engineering* (pp. 37–86). Elsevier BV. Available from http://doi.org/10.1016/b978-0-08-100135-6.00003-x.

Cui, S., Zhao, Y., Liu, Y., & Pan, J. (2022). Preparation of copper-based biochar adsorbent with outstanding H_2S adsorption capacity and study on H_2S removal. *Journal of the Energy Institute*, *105*, 481–490. Available from https://doi.org/10.1016/j.joei.2022.11.004, http://www.journals.elsevier.com/journal-of-the-energy-institute.

Dai, L., Li, L., Zhu, W., Ma, H., Huang, H., Lu, Q., Yang, M., & Ran, Y. (2020). Post-engineering of biochar via thermal air treatment for highly efficient promotion of uranium(VI) adsorption. *Bioresource Technology*, *298*, 122576. Available from https://doi.org/10.1016/j.biortech.2019.122576.

Debiagi, P. E. A., Trinchera, M., Frassoldati, A., Faravelli, T., Vinu, R., & Ranzi, E. (2017). Algae characterization and multistep pyrolysis mechanism. *Journal of Analytical and Applied Pyrolysis*, *128*, 423–436. Available from https://doi.org/10.1016/j.jaap.2017.08.007.

Dehkhoda, A. M., Ellis, N., & Gyenge, E. (2016). Effect of activated biochar porous structure on the capacitive deionization of NaCl and $ZnCl_2$ solutions. *Microporous and Mesoporous Materials*, *224*, 217–228. Available from https://doi.org/10.1016/j.micromeso.2015.11.041, http://www.elsevier.com/inca/publications/store/6/0/0/7/6/0.

Dehkhoda, A. M., Gyenge, E., & Ellis, N. (2016). A novel method to tailor the porous structure of KOH-activated biochar and its application in capacitive deionization and energy storage. *Biomass and Bioenergy*, *87*, 107–121. Available from https://doi.org/10.1016/j.biombioe.2016.02.023, http://www.journals.elsevier.com/biomass-and-bioenergy/.

Demir, M., Saraswat, S. K., & Gupta, R. B. (2017). Hierarchical nitrogen-doped porous carbon derived from lecithin for high-performance supercapacitors. *RSC Advances*, *7* (67), 42430–42442. Available from https://doi.org/10.1039/c7ra07984b, http://pubs.rsc.org/en/journals/journalissues.

Ding, Z., Hu, X., Wan, Y., Wang, S., & Gao, B. (2016). Removal of lead, copper, cadmium, zinc, and nickel from aqueous solutions by alkali-modified biochar: Batch and column tests. *Journal of Industrial and Engineering Chemistry*, *33*, 239–245. Available from https://doi.org/10.1016/j.jiec.2015.10.007, http:http://www.sciencedirect.com/science/journal/1226086X.

dos Reis, G. S., Bergna, D., Grimm, A., Lima, E. C., Hu, T., Naushad, M., & Lassi, U. (2023). Preparation of highly porous nitrogen-doped biochar derived from birch tree wastes with superior dye removal performance. *Colloids and Surfaces A: Physicochemical and Engineering Aspects*, *669*. Available from https://doi.org/10.1016/j.colsurfa.2023.131493.

Ehrburger, P., Addoun, A., Addoun, F., & Donnet, J. B. (1986). Carbonization of coals in the presence of alkaline hydroxides and carbonates: Formation of activated carbons. *Fuel*, *65*(10), 1447–1449. Available from https://doi.org/10.1016/0016-2361(86)90121-3.

Faheem, J., Du, S. H., Kim, M. A., Hassan, S., Irshad, J., & Bao. (2020). Application of biochar in advanced oxidation processes: supportive, adsorptive, and catalytic role. *Environmental Science and Pollution Research*, *27*(30), 37286–37312. Available from https://doi.org/10.1007/s11356-020-07612-y, https://link.springer.com/journal/11356.

Fahmi, A. H., Samsuri, A. W., Jol, H., & Singh, D. (2018). Physical modification of biochar to expose the inner pores and their functional groups to enhance lead adsorption. *RSC Advances*, *8*(67), 38270–38280. Available from https://doi.org/10.1039/c8ra06867d, http://pubs.rsc.org/en/journals/journal/ra.

Gale, M., Nguyen, T., Moreno, M., & Gilliard-Abdulaziz, K. L. (2021). Physiochemical properties of biochar and activated carbon from biomass residue: Influence of process conditions to adsorbent properties. *ACS Omega*, *6*(15), 10224–10233. Available from https://doi.org/10.1021/acsomega.1c00530, http://pubs.acs.org/journal/acsodf.

Gao, A. L., & Wan, Y. (2023). Iron modified biochar enables recovery and recycling of phosphorus from wastewater through column filters and flow reactors. *Chemosphere*, *313*, 137434. Available from https://doi.org/10.1016/j.chemosphere.2022.137434.

Gao, W., Lin, Z., Chen, H., Yan, S., Huang, Y., Hu, X., & Zhang, S. (2022). A review on N-doped biochar for enhanced water treatment and emerging applications. *Fuel Processing Technology*, *237*, 107468. Available from https://doi.org/10.1016/j.fuproc.2022.107468.

Geng, Y., Du, Y., Guo, T., Zhang, Y., Bedane, A. H., & Ren, P. (2023). Isothermal and isovolumetric process of CO_2 adsorption on nitrogen-doped biochar: Equilibrium and non-equilibrium states. *Case Studies in Thermal Engineering*, *44*, 102814. Available from https://doi.org/10.1016/j.csite.2023.102814.

Godwin, P. M., Pan, Y., Xiao, H., & Afzal, M. T. (2019). Progress in preparation and application of modified biochar for improving heavy metal ion removal from wastewater. *Journal of Bioresources and Bioproducts*, *4*(1), 31–42. Available from https://doi.org/10.21967/jbb.v4i1.180, https://www.sciencedirect.com/journal/journal-of-bioresources-and-bioproducts.

Gueret Yadiberet Menzembere, E. R., He, Y., Dong, Y., Li, B., Liu, C., Lin, H., & Sambiani, L. (2023). Insight into modified biochars and their immobilizing effects on heavy metal(loid)s in contaminated soils: Mechanisms and influencing factors. *Pedosphere*, *33*(1), 23–33. Available from https://doi.org/10.1016/j.pedsph.2022.06.030, http://pedosphere.issas.ac.cn.

Hamza, U. D., Nasri, N. S., Amin, N. S., Mohammed, J., & Zain, H. M. (2016). Characteristics of oil palm shell biochar and activated carbon prepared at different carbonization times. *Desalination and Water Treatment*, *57*. Available from https://doi.org/10.1080/19443994.2015.1042068.

Hosny, M., Fawzy, M., & Eltaweil, A. S. (2022). Phytofabrication of bimetallic silver-copper/biochar nanocomposite for environmental and medical applications. *Journal of Environmental Management*, *316*, 115238. Available from https://doi.org/10.1016/j.jenvman.2022.115238.

Huang, H., Niu, Z., Shi, R., Tang, J., Lv, L., Wang, J., & Fan, Y. (2020). Thermal oxidation activation of hydrochar for tetracycline adsorption: the role of oxygen concentration and temperature. *Bioresource Technology*, *306*, 123096. Available from https://doi.org/10.1016/j.biortech.2020.123096.

Ighalo, J. O., Iwuchukwu, F. U., Eyankware, O. E., Iwuozor, K. O., Olotu, K., Bright, O. C., & Igwegbe, C. A. (2022). Flash pyrolysis of biomass: A review of recent advances. *Clean Technologies and Environmental Policy*. Available from https://doi.org/10.1007/s10098-022-02339-5, https://link.springer.com/journal/10098.

International Biochar Initiative. (2015). *Standardized product definition and product testing guidelines for biochar that is used in soil*. Author.

Ippolito, J. A., Cui, L., Kammann, C., Wrage-Mönnig, N., Estavillo, J. M., Fuertes-Mendizabal, T., Cayuela, M. L., Sigua, G., Novak, J., Spokas, K., & Borchard, N. (2020). Feedstock choice, pyrolysis temperature and type influence biochar characteristics: A comprehensive meta-data analysis review. *Biochar*, *2*(4), 421–438. Available from https://doi.org/10.1007/s42773-020-00067-x, http://springer.com/journal/42773.

Işıtan, S., Ceylan, S., Topcu, Y., Hintz, C., Tefft, J., Chellappa, T., Guo, J., & Goldfarb, J. L. (2016). Product quality optimization in an integrated biorefinery: Conversion of pistachio nutshell biomass to biofuels and activated biochars via pyrolysis. *Energy Conversion and Management*, *127*, 576–588. Available from https://doi.org/10.1016/j.enconman.2016.09.031.

Janer, M., Plantà, X., & Riera, D. (2020). Ultrasonic moulding: Current state of the technology. *Ultrasonics*, *102*, 106038. Available from https://doi.org/10.1016/j.ultras.2019.106038.

Kasera, N., Hall, S., & Kolar, P. (2021). Effect of surface modification by nitrogen-containing chemicals on morphology and surface characteristics of N-doped pine bark biochars. *Journal of Environmental Chemical Engineering*, *9*(2), 105161. Available from https://doi.org/10.1016/j.jece.2021.105161.

Kim, J., Song, J., Lee, S. M., & Jung, J. (2019). Application of iron-modified biochar for arsenite removal and toxicity reduction. *Journal of Industrial and Engineering Chemistry*, *80*, 17–22. Available from https://doi.org/10.1016/j.jiec.2019.07.026, http: http://www.sciencedirect.com/science/journal/1226086X.

Kozyatnyk, I., Oesterle, P., Wurzer, C., Mašek, O., & Jansson, S. (2021). Removal of contaminants of emerging concern from multicomponent systems using carbon dioxide activated biochar from lignocellulosic feedstocks. *Bioresource Technology*, *340*, 125561. Available from https://doi.org/10.1016/j.biortech.2021.125561.

Kołtowski, M., Charmas, B., Skubiszewska-Zięba, J., & Oleszczuk, P. (2017). Effect of biochar activation by different methods on toxicity of soil contaminated by industrial activity. *Ecotoxicology and Environmental Safety*, *136*, 119–125. Available from https://doi.org/10.1016/j.ecoenv.2016.10.033, http://www.elsevier.com/inca/publications/store/6/2/2/8/1/9/index.htt.

Kumari, N., Chhabra, T., Kumar, S., & Krishnan, V. (2022). Nanoarchitectonics of sulfonated biochar from pine needles as catalyst for conversion of biomass derived chemicals to value added products. *Catalysis Communications*, *168*, 106467. Available from https://doi.org/10.1016/j.catcom.2022.106467.

Laird, D. A., Brown, R. C., Amonette, J. E., & Lehmann, J. (2009). Review of the pyrolysis platform for coproducing bio-oil and biochar. *Biofuels, Bioproducts and Biorefining*, *3* (5), 547–562. Available from https://doi.org/10.1002/bbb.169, http://www3.interscience.wiley.com/cgi-bin/fulltext/122589511/PDFSTART, United States.

Lehmann, J. (2007). Bio-energy in the black. *Frontiers in Ecology and the Environment*, *5*(7), 381–387. Available from https://doi.org/10.1890/1540-9295(2007)5[381:BITB]2.0.CO;2.

Lehmann, J. (2015). *Biochar for environmental management science, technology and implementation*. Taylor and Francis.

Lehmann, J., & Joseph, S. Biochar for environmental management: Science, technology and implementation. Routledge (2015). Available from https://doi.org/10.4324/9780203762264.

Li, P., Xie, H., Wang, X., Xie, Y., Wang, Y., & Zhang, Y. (2020). Sustainable production of nano α-Fe_2O_3/N-doped biochar hybrid nanosheets for supercapacitors. *Sustainable Energy & Fuels*, *4*(9), 4522–4530. Available from https://doi.org/10.1039/d0se00627k.

Li, Q., Ding, W., Yong, Y., Zeng, X., & Gao, Y. (2016). Effects of ultraviolet modification on physicochemical property and adsorption performance of biochar. *Nanoscience and Nanotechnology Letters*, *8*(11), 978–984. Available from https://doi.org/10.1166/nnl.2016.2184, http://www.ingentaconnect.com/contentone/asp/nnl/2016/00000008/00000011/art00012.

Li, Y., Ruan, G., Jalilov, A. S., Tarkunde, Y. R., Fei, H., & Tour, J. M. (2016). Biochar as a renewable source for high-performance CO_2 sorbent. *Carbon*, *107*, 344–351. Available from https://doi.org/10.1016/j.carbon.2016.06.010, http://www.journals.elsevier.com/carbon/.

Li, Y., Shang, H., Cao, Y., Yang, C., Feng, Y., & Yu, Y. (2022). High performance removal of sulfamethoxazole using large specific area of biochar derived from corncob xylose residue. *Biochar*, *4*(1). Available from https://doi.org/10.1007/s42773-021-00128-9, http://springer.com/journal/42773.

Liu, S., Feng, X., Liu, Z., Zhao, W., Li, Y., & Zhang, J. (2023). Nitrogen doping in porous biochar from cotton stalk with H_3PO_4 activation for reduction of NOx with NH_3-SCR at low temperatures: Characteristics and catalytic activity analysis. *Fuel*, *332*, 126256. Available from https://doi.org/10.1016/j.fuel.2022.126256.

Liu, W. J., Jiang, H., & Yu, H. Q. (2015). Development of biochar-based functional materials: Toward a sustainable platform carbon material. *Chemical Reviews*, *115* (22), 12251–12285. Available from https://doi.org/10.1021/acs.chemrev.5b00195, http://pubs.acs.org/journal/chreay.

Lozano-Castelló, D., Calo, J. M., Cazorla-Amorós, D., & Linares-Solano, A. (2007). Carbon activation with KOH as explored by temperature programmed techniques, and the effects of hydrogen. *Carbon*, *45*(13), 2529–2536. Available from https://doi.org/10.1016/j.carbon.2007.08.021.

Lua, A. C., Lau, F. Y., & Guo, J. (2006). Influence of pyrolysis conditions on pore development of oil-palm-shell activated carbons. *Journal of Analytical and Applied Pyrolysis*, *76* (1-2), 96–102. Available from https://doi.org/10.1016/j.jaap.2005.08.001.

Lua, A.C., Yang, T., Guo, J. (2004). Effects of pyrolysis conditions on the properties of activated carbons prepared from pistachio-nut shells. *Journal of Analytical and Applied Pyrolysis, 72*, 279–287. Available from https://doi.org/10.1016/j.jaap.2004.08.001.

Ma, Y., Liang, H., Yin, J., Yao, D., Xia, Y., Zuo, K., & Zeng, Y. P. (2021). Air activation of charcoal monoliths for capacitive energy storage. *RSC Advances*, *11*(25), 15118–15130. Available from https://doi.org/10.1039/d1ra02192c, http://pubs.rsc.org/en/journals/journal/ra.

Manyà, J. J., González, B., Azuara, M., & Arner, G. (2018). Ultra-microporous adsorbents prepared from vine shoots-derived biochar with high CO_2 uptake and CO_2/N_2 selectivity. *Chemical Engineering Journal*, *345*, 631–639. Available from https://doi.org/10.1016/j.cej.2018.01.092, http://www.elsevier.com/inca/publications/store/6/0/1/2/7/3/index.htt.

Marcilla, A., García-García, S., Asensio, M., & Conesa, J. A. (2000). Influence of thermal treatment regime on the density and reactivity of activated carbons from almond shells. *Carbon, 38*, 429–440. Available from https://doi.org/10.1016/S0008-6223(99)00123-2.

Marsh, H. (2001). *Activated carbon compendium: A collection of papers from the journal Carbon*. Elsevier.

Mondal, S., Bobde, K., Aikat, K., & Halder, G. (2016). Biosorptive uptake of ibuprofen by steam activated biochar derived from mung bean husk: Equilibrium, kinetics, thermodynamics, modeling and eco-toxicological studies. *Journal of Environmental Management*, *182*, 581–594. Available from https://doi.org/10.1016/j.jenvman.2016.08.018, https://www.sciencedirect.com/journal/journal-of-environmental-management.

Mondal, S., Sinha, K., Aikat, K., & Halder, G. (2015). Adsorption thermodynamics and kinetics of ranitidine hydrochloride onto superheated steam activated carbon derived from mung bean husk. *Journal of Environmental Chemical Engineering*, *3*(1), 187–195. Available from https://doi.org/10.1016/j.jece.2014.11.021, http://www.journals.elsevier.com/journal-of-environmental-chemical-engineering/.

Nicholas, A. F., Hussein, M. Z., Zainal, Z., & Khadiran, T. (2018). *Activated carbon for shape-stabilized phase change material. Synthesis, technology and applications of carbon nanomaterials* (pp. 279–308). Elsevier. Available from http://www.sciencedirect.com/science/book/9780128157572.

Oh, G. H., & Park, C. R. (2002). Preparation and characteristics of rice-straw-based porous carbons with high adsorption capacity. *Fuel*, *81*(3), 327–336. Available from https://doi.org/10.1016/S0016-2361(01)00171-5.

Pandey, A., Bhaskar, T., Stöcker, M., & Sukumaran, R. K. (2015). *Recent advances in thermochemical conversion of biomass* (pp. 1–491). Elsevier Inc. Available from http://www.sciencedirect.com/science/book/9780444632890.

Park, C. M., Han, J., Chu, K. H., Al-Hamadani, Y. A. J., Her, N., Heo, J., & Yoon, Y. (2017). Influence of solution pH, ionic strength, and humic acid on cadmium adsorption onto activated biochar: Experiment and modeling. *Journal of Industrial and Engineering Chemistry*, *48*, 186–193. Available from https://doi.org/10.1016/j.jiec.2016.12.038, http://www.sciencedirect.com/science/journal/1226086X.

Park, J., Hung, I., Gan, Z., Rojas, O. J., Lim, K. H., & Park, S. (2013). Activated carbon from biochar: Influence of its physicochemical properties on the sorption characteristics of phenanthrene. *Bioresource Technology*, *149*, 383–389. Available from https://doi.org/10.1016/j.biortech.2013.09.085.

Peng, Z., Liu, X., Chen, H., Liu, Q., & Tang, J. (2019). Characterization of ultraviolet-modified biochar from different feedstocks for enhanced removal of hexavalent chromium from water. *Water Science and Technology*, *79*(9), 1705–1716. Available from https://doi.org/10.2166/wst.2019.170, https://watermark.silverchair.com/wst079091705.pdf.

Peng, Z., Zhao, H., Lyu, H., Wang, L., Huang, H., Nan, Q., & Tang, J. (2018). UV modification of biochar for enhanced hexavalent chromium removal from aqueous solution. *Environmental Science and Pollution Research*, *25*(11), 10808–10819. Available from https://doi.org/10.1007/s11356-018-1353-3, http://www.springerlink.com/content/0944-1344.

Pepè Sciarria, T., de Oliveira, M. A. C., Mecheri, B., D'Epifanio, A., Goldfarb, J. L., & Adani, F. (2020). Metal-free activated biochar as an oxygen reduction reaction catalyst in single chamber microbial fuel cells. *Journal of Power Sources*, *462*. Available from https://doi.org/10.1016/j.jpowsour.2020.228183, https://www.journals.elsevier.com/journal-of-power-sources.

Pourhakkak, P., Taghizadeh, M., Taghizadeh, A., & Ghaedi, M. (2021). Adsorbent. *Interface Science and Technology*, *33*, 71–210. Available from https://doi.org/10.1016/B978-0-12-818805-7.00009-6, http://www.elsevier.com/wps/find/bookdescription.cws_home/BS_3710/description#description.

Qu, J., Shi, J., Wang, Y., Tong, H., Zhu, Y., Xu, L., Wang, Y., Zhang, B., Tao, Y., Dai, X., Zhang, H., & Zhang, Y. (2022). Applications of functionalized magnetic biochar in environmental remediation: A review. *Journal of Hazardous Materials*, *434*, 128841. Available from https://doi.org/10.1016/j.jhazmat.2022.128841.

Rajapaksha, A. U., Vithanage, M., Ahmad, M., Seo, D. C., Cho, J. S., Lee, S. E., Lee, S. S., & Ok, Y. S. (2015). Enhanced sulfamethazine removal by steam-activated invasive plant-derived biochar. *Journal of Hazardous Materials*, *290*, 43–50. Available from https://doi.org/10.1016/j.jhazmat.2015.02.046, http://www.elsevier.com/locate/jhazmat.

Rana, M., Subramani, K., Sathish, M., & Gautam, U. K. (2017). Soya derived heteroatom doped carbon as a promising platform for oxygen reduction, supercapacitor and CO_2 capture. *Carbon*, *114*, 679–689. Available from https://doi.org/10.1016/j.carbon.2016.12.059, http://www.journals.elsevier.com/carbon/.

Rodriguez-Reinos, F. (2002). *Handbook of porous solids Production and applications of activated carbons* (pp. 1766–1827). Wiley-VCH Verlag GmbH. Available from https://doi.org/10.1002/9783527618286.ch24a.

Rodríguez-Reinoso, F., & Molina-Sabio, M. (1992). Activated carbons from lignocellulosic materials by chemical and/or physical activation: An overview. *Carbon*, *30*(7), 1111–1118. Available from https://doi.org/10.1016/0008-6223(92)90143-K.

Rodríguez-Reinoso, F., & Molina-Sabio, M. (1998). Textural and chemical characterization of microporous carbons. *Advances in Colloid and Interface Science*, *76-77*, 271–294. Available from https://doi.org/10.1016/s0001-8686(98)00049-9.

Sajjadi, B., Broome, J. W., Chen, W. Y., Mattern, D. L., Egiebor, N. O., Hammer, N., & Smith, C. L. (2019). Urea functionalization of ultrasound-treated biochar: A feasible strategy for enhancing heavy metal adsorption capacity. *Ultrasonics Sonochemistry*, *51*, 20–30. Available from https://doi.org/10.1016/j.ultsonch.2018.09.015, http://www.elsevier.com/inca/publications/store/5/2/5/4/5/1.

Sajjadi, B., Chen, W. Y., & Egiebor, N. O. (2019). A comprehensive review on physical activation of biochar for energy and environmental applications. *Reviews in Chemical Engineering*, *35*(6), 735–776. Available from https://doi.org/10.1515/revce-2017-0113, http://www.reference-global.com/loi/revce.

Sakhiya, A. K., Baghel, P., Anand, A., Vijay, V. K., & Kaushal, P. (2021). A comparative study of physical and chemical activation of rice straw derived biochar to enhance Zn + 2 adsorption. *Bioresource Technology Reports*, *15*. Available from https://doi.org/10.1016/j.biteb.2021.100774, https://www.journals.elsevier.com/bioresource-technology-reports.

Samaraweera, H., Rivera, A., Carter, K., Felder, T., Nawalage, S., Chui, I., Perez, F., Khan, A. H., & Mlsna, T. (2023). Green iron oxide-modified biochar for methylene blue removal from aqueous solutions. *Groundwater for Sustainable Development*, *21*, 100945. Available from https://doi.org/10.1016/j.gsd.2023.100945.

Setiawan, H., Sakamoto, M., Fujisaki, T., Lyth, S. M., & Shiratori, Y. (2022). Development of a sustainable nitrogen-doped biochar desulfurizer for solid oxide fuel cell systems. *Biomass and Bioenergy*, *167*, 106631. Available from https://doi.org/10.1016/j.biombioe.2022.106631.

Shafawi, A. N., Mohamed, A. R., Lahijani, P., & Mohammadi, M. (2021). Recent advances in developing engineered biochar for CO_2 capture: An insight into the biochar modification approaches. *Journal of Environmental Chemical Engineering*, *9*(6). Available from https://doi.org/10.1016/j.jece.2021.106869, http://www.journals.elsevier.com/journal-of-environmental-chemical-engineering/.

Shen, B., Tian, L., Li, F., Zhang, X., Xu, H., & Singh, S. (2017). Elemental mercury removal by the modified bio-char from waste tea. *Fuel*, *187*, 189–196. Available from https://doi.org/10.1016/j.fuel.2016.09.059, http://www.journals.elsevier.com/fuel/.

Spokas, K. A. (2010). Review of the stability of biochar in soils: predictability of O:C molar ratios. *Carbon Management*, 289–303. Available from https://doi.org/10.4155/cmt.10.32.

Sricharoenchaikul, V., Pechyen, C., Aht-Ong, D., & Atong, D. (2008). Preparation and characterization of activated carbon from the pyrolysis of physic nut (*Jatropha curcas* L.) waste. *Energy and Fuels*, *22*(1), 31–37. Available from https://doi.org/10.1021/ef700285u.

Sun, Y., Yu, I. K. M., Tsang, D. C. W., Cao, X., Lin, D., Wang, L., Graham, N. J. D., Alessi, D. S., Komárek, M., Ok, Y. S., Feng, Y., & Li, X. D. (2019). Multifunctional iron-biochar composites for the removal of potentially toxic elements, inherent cations, and hetero-chloride from hydraulic fracturing wastewater. *Environment International*, *124*, 521–532. Available from https://doi.org/10.1016/j.envint.2019.01.047, http://www.elsevier.com/locate/envint.

Takaya, C. A., Fletcher, L. A., Singh, S., Anyikude, K. U., & Ross, A. B. (2016). Phosphate and ammonium sorption capacity of biochar and hydrochar from different wastes. *Chemosphere*, *145*, 518–527. Available from https://doi.org/10.1016/j.chemosphere.2015.11.052.

Uchimiya, M., Wartelle, L. H., Lima, I. M., & Klasson, K. T. (2010). Sorption of deisopropylatrazine on broiler litter biochars. *Journal of Agricultural and Food Chemistry*, *58* (23), 12350–12356. Available from https://doi.org/10.1021/jf102152q.

United Nations Environmental Programme Division of Technology. (2015). *Converting waste agricultural biomass into a resource – compendium of technologies*. Japan: Industry and Economics International Environmental Technology Centre Osaka/Shiga. Available from https://www.build-a-biogas-plant.com/PDF/WasteAgriculturalBiomassEST_Compendium.pdf.

Wang, B., Gao, B., & Fang, J. (2017). Recent advances in engineered biochar productions and applications. *Critical Reviews in Environmental Science and Technology*, *47*(22), 2158–2207. Available from https://doi.org/10.1080/10643389.2017.1418580, http://www.tandf.co.uk/journals/titles/10643389.asp.

Wang, J., Riaz, M., Babar, S., Xia, H., Li, Y., Xia, X., Wang, X., & Jiang, C. (2023). Iron-modified biochar reduces nitrogen loss and improves nitrogen retention in Luvisols by adsorption and microbial regulation. *Science of the Total Environment*, *879*, 163196. Available from https://doi.org/10.1016/j.scitotenv.2023.163196.

Wang, L., Ok, Y. S., Tsang, D. C. W., Alessi, D. S., Rinklebe, J., Wang, H., Mašek, O., Hou, R., O'Connor, D., & Hou, D. (2020). New trends in biochar pyrolysis and modification strategies: Feedstock, pyrolysis conditions, sustainability concerns and implications for soil amendment. *Soil Use and Management*, *36*(3), 358–386. Available from https://doi.org/10.1111/sum.12592, http://onlinelibrary.wiley.com/journal/10.1111/(ISSN)1475-2743.

Wang, X., Liu, Y., Zhu, L., Li, Y., Wang, K., Qiu, K., Tippayawong, N., Aggarangsi, P., Reubroycharoen, P., & Wang, S. (2019). Biomass derived N-doped biochar as efficient catalyst supports for CO_2 methanation. *Journal of CO_2 Utilization*, *34*, 733–741. Available from https://doi.org/10.1016/j.jcou.2019.09.003, http://www.journals.elsevier.com/journal-of-co2-utilization/.

Wang, Z., Fang, R., & Guo, H. (2020). Advances in ultrasonic production units for enhanced oil recovery in China. *Ultrasonics Sonochemistry*, *60*, 104791. Available from https://doi.org/10.1016/j.ultsonch.2019.104791.

Wei, Z., Wei, Y., Liu, Y., Niu, S., Xu, Y., Park, J. H., & Wang, J. J. (2024). Biochar-based materials as remediation strategy in petroleum hydrocarbon-contaminated soil and water: Performances, mechanisms, and environmental impact. *Journal of Environmental Sciences (China)*, *138*, 350–372. Available from https://doi.org/10.1016/j.jes.2023.04.008, http://www.journals.elsevier.com/journal-of-environmental-sciences/.

Wu, B., Cheng, G., Jiao, K., Shi, W., Wang, C., & Xu, H. (2016). Mycoextraction by *Clitocybe maxima* combined with metal immobilization by biochar and activated carbon in an aged soil. *Science of the Total Environment*, *562*, 732–739. Available from https://doi.org/10.1016/j.scitotenv.2016.04.079, http://www.elsevier.com/locate/scitotenv.

Wu, C., Budarin, V. L., Gronnow, M. J., De Bruyn, M., Onwudili, J. A., Clark, J. H., & Williams, P. T. (2014). Conventional and microwave-assisted pyrolysis of biomass under different heating rates. *Journal of Analytical and Applied Pyrolysis*, *107*, 276–283. Available from https://doi.org/10.1016/j.jaap.2014.03.012.

Wu, P., Wang, Z., Wang, H., Bolan, N. S., Wang, Y., & Chen, W. (2020). Visualizing the emerging trends of biochar research and applications in 2019: A scientometric analysis and review. *Biochar*, *2*(2), 135–150. Available from https://doi.org/10.1007/s42773-020-00055-1, http://springer.com/journal/42773.

Van Wychen, S., Rowland, S. M., Lesco, K. C., Shanta, P. V., Dong, T., & Laurens, L. M. L. (2021). Advanced mass balance characterization and fractionation of algal biomass composition. *Journal of Applied Phycology*, *33*(5), 2695–2708. Available from https://doi.org/10.1007/s10811-021-02508-x, http://www.wkap.nl/journalhome.htm/0921-8971.

Xiang, W., Zhang, X., Chen, K., Fang, J., He, F., Hu, X., Tsang, D. C. W., Ok, Y. S., & Gao, B. (2020). Enhanced adsorption performance and governing mechanisms of ball-milled biochar for the removal of volatile organic compounds (VOCs). *Chemical*

Engineering Journal, *385*. Available from https://doi.org/10.1016/j.cej.2019.123842, http://www.elsevier.com/inca/publications/store/6/0/1/2/7/3/index.htt.

Xiang, W., Wan, Y., Zhang, X., Tan, Z., Xia, T., Zheng, Y., & Gao, B. (2020). Adsorption of tetracycline hydrochloride onto ball-milled biochar: Governing factors and mechanisms. *Chemosphere*, *255*, 127057. Available from https://doi.org/10.1016/j.chemosphere.2020.127057.

Xiao, F., & Pignatello, J. J. (2016). Effects of post-pyrolysis air oxidation of biomass chars on adsorption of neutral and ionizable compounds. *Environmental Science and Technology*, *50*(12), 6276–6283. Available from https://doi.org/10.1021/acs.est.6b00362, http://pubs.acs.org/journal/esthag.

Xiao, F. (2022). A review of biochar functionalized by thermal air oxidation. *Environmental Functional Materials*, *1*(2), 187–195. Available from https://doi.org/10.1016/j.efmat.2022.03.001.

Xie, J. X., Cao, J. P., Jiang, W., Zhao, X. Y., Zhao, L., Zhang, C., & Bai, H. C. (2022). Nickel loaded on biochar prepared from different carbon sources for selective hydrogenolysis of diphenyl ether. *Fuel Processing Technology*, *231*. Available from https://doi.org/10.1016/j.fuproc.2022.107219, https://www.journals.elsevier.com/fuel-processing-technology.

Xie, Q., Yang, X., Xu, K., Chen, Z., Sarkar, B., & Dou, X. (2020). Conversion of biochar to sulfonated solid acid catalysts for spiramycin hydrolysis: Insights into the sulfonation process. *Environmental Research*, *188*, 109887. Available from https://doi.org/10.1016/j.envres.2020.109887.

Xiong, X., Yu, I. K. M., Chen, S. S., Tsang, D. C. W., Cao, L., Song, H., Kwon, E. E., Ok, Y. S., Zhang, S., & Poon, C. S. (2018). Sulfonated biochar as acid catalyst for sugar hydrolysis and dehydration. *Catalysis Today*, *314*, 52–61. Available from https://doi.org/10.1016/j.cattod.2018.02.034, http://www.sciencedirect.com/science/journal/09205861.

Xiong, Z., Shihong, Z., Haiping, Y., Tao, S., Yingquan, C., & Hanping, C. (2013). Influence of NH_3/CO_2 modification on the characteristic of biochar and the CO_2 capture. *Bioenergy Research*, *6*(4), 1147–1153. Available from https://doi.org/10.1007/s12155-013-9304-9.

Yang, G. X., & Jiang, H. (2014). Amino modification of biochar for enhanced adsorption of copper ions from synthetic wastewater. *Water Research*, *48*(1), 396–405. Available from https://doi.org/10.1016/j.watres.2013.09.050, http://www.elsevier.com/locate/watres.

Yang, X., Zheng, X., Yan, Z., Huang, Z., Yao, Y., Li, H., Kuang, Y., & Zhou, H. (2021). Construction and preparation of nitrogen-doped porous carbon material based on waste biomass for lithium-ion batteries. *International Journal of Hydrogen Energy*, *46*(33), 17267–17281. Available from https://doi.org/10.1016/j.ijhydene.2021.02.131, http://www.journals.elsevier.com/international-journal-of-hydrogen-energy/.

Yu, B., Man, Y., Wang, P., Wu, C., Xie, J., Wang, W., Jiang, H., Zhang, L., Zhang, Y., Mao, L., Zhu, L., Zheng, Y., & Liu, X. (2023). Catalytic degradation of dimethomorph by nitrogen-doped rice husk biochar. *Ecotoxicology and Environmental Safety*, *257*, 114908. Available from https://doi.org/10.1016/j.ecoenv.2023.114908.

Yu, C., Tang, J., Su, H., Huang, J., Liu, F., Wang, L., & Sun, H. (2022). Development of a novel biochar/iron oxide composite from green algae for bisphenol-A removal: Adsorption and Fenton-like reaction. *Environmental Technology & Innovation*, *28*, 102647. Available from https://doi.org/10.1016/j.eti.2022.102647.

Yi, Y., Huang, Z., Lu, B., Xian, J., Tsang, E. P., Cheng, W., Fang, J., & Fang, Z. (2020). Magnetic biochar for environmental remediation: A review. *Bioresource Technology*, *298*.

Yu, W., Hu, J., Yu, Y., Ma, D., Gong, W., Qiu, H., Hu, Z., & Hong-wen, G. (2021). Facile preparation of sulfonated biochar for highly efficient removal of toxic Pb(II) and Cd(II) from wastewater. *Science of The Total Environment*, *750*, 141545. Available from https://doi.org/10.1016/j.scitotenv.2020.141545.

Zhang, S., Hu, B., Zhang, L., & Xiong, Y. (2016). Effects of torrefaction on yield and quality of pyrolysis char and its application on preparation of activated carbon. *Journal of Analytical and Applied Pyrolysis*, *119*, 217–233. Available from https://doi.org/10.1016/j.jaap.2016.03.002.

Zhou, Z., Yao, D., Li, S., Xu, F., Liu, Y., Liu, R., & Chen, Z. (2021). Sustainable production of value-added sulfonated biochar by sulfuric acid carbonization reduction of rice husks. *Environmental Technology & Innovation*, *24*, 102025. Available from https://doi.org/10.1016/j.eti.2021.102025.

Zhu, X., Li, C., Li, J., Xie, B., Lü, J., & Li, Y. (2018). Thermal treatment of biochar in the air/nitrogen atmosphere for developed mesoporosity and enhanced adsorption to tetracycline. *Bioresource Technology*, *263*, 475–482. Available from https://doi.org/10.1016/j.biortech.2018.05.041, http://www.elsevier.com/locate/biortech.

Zhu, Y., Li, H., Zhang, G., Meng, F., Li, L., & Wu, S. (2018). Removal of hexavalent chromium from aqueous solution by different surface-modified biochars: Acid washing, nanoscale zero-valent iron and ferric iron loading. *Bioresource Technology*, *261*, 142–150. Available from https://doi.org/10.1016/j.biortech.2018.04.004, http://www.elsevier.com/locate/biortech.

Zuo, M., Niu, Q., Zhu, Y., Li, S., Jia, W., Zhou, Z., Zeng, X., & Lin, L. (2023). Biochar catalysts for efficiently 5-hydroxymethylfurfural (HMF) synthesis in aqueous natural deep eutectic solvent (A-NADES). *Industrial Crops and Products*, *192*, 115953. Available from https://doi.org/10.1016/j.indcrop.2022.115953.

CHAPTER

19 Biochar-based smart fertilizers for sustainable agriculture and environment health

Riti Thapar Kapoor[1] and Mohd Rafatullah[2]

[1]*Centre for Plant and Environmental Biotechnology, Amity Institute of Biotechnology, Amity University Uttar Pradesh, Noida, Uttar Pradesh, India*

[2]*Division of Environmental Technology, School of Industrial Technology, Universiti Sains Malaysia, Penang, Malaysia*

19.1 Introduction

The global population is predicted to rise to 9.6 billion by the year 2050, which will increase food demand. The continuous rise in global population and reduction in arable land pose tremendous pressure on agriculture sector as it is main channel to supply food. Soil is a natural source of nutrients for plants and soil fertilization is one of the strategy to enhance crop productivity. Agricultural yield and food security can be improved with efficient use of fertilizers (Abiola et al., 2023). Farmers are frequently using synthetic fertilizers in an uncontrolled manner to meet the food demand at global level (Czekała et al., 2019). The chemical fertilizers are soluble that can easily release nutrients into the soil. However, a more than 50% increase in food production has been reported in the past decades with utilization of chemical fertilizers (Chen et al., 2018). Excessive use of chemical fertilizers alters microbial composition of the soil by lowering soil pH and negatively affect crop productivity and causes serious environmental problems as well as economic burden (Sun et al., 2023). The ideal N:P:K fertilizer ratio is 4:2:1; however, farmers are using very high ratio, that is, 10:2.7:1. Production and consumption of fertilizers is continuously increasing worldwide. Rombel et al. (2022) reported worldwide mineral fertilizers consumption was 185 million tons in 2017 and it was six times more as compared with the consumption reported in the year 1960. In recent years, approximately 34 times more consumption of fertilizers have been reported in developing nations. Nitrogen (N), phosphorus (P), and potassium (K) are indispensable elements and needed for growth and development of plants. Piwowar (2022) reported that N fertilizer consumption accounts for 57% among the total amount of fertilizers used. Requirement of N and

Biochar Ecotechnology for Sustainable Agriculture and Environment. DOI: https://doi.org/10.1016/B978-0-443-29855-4.00021-7

phosphorus fertilizers is predicted to be more than 112 and 49 million tons in coming years (Wang et al., 2022).

The need of nutrients by plants depends on their age as less amount is required in vegetative stage. Plants cannot absorb excessive nutrients, as most of them are lost through leaching into the soil and water. Plants use only 40%–60% N, 15%–20% P, and 50%–60% K after supplementation of chemical fertilizers (Beig et al., 2022). The rest are discharged into environment and show adverse impacts on soil and water eutrophication (Khan et al., 2021). Approximately 40%–70% N, 50%–70% K, and 80%–90% P are lost into environment (Duhan et al., 2017). Nitrogen-containing fertilizers undergo volatilization of NH_3, leaching of nitrate, and nitrous oxide emission, which have adverse effects on natural sources and pose risks to human health (Li & Chen, 2020). The inorganic nutrients present in conventional fertilizers generally become fixed in the soil in insoluble forms and are leached through rainfall or irrigation water, contaminating natural resources (Alshaal & El-Ramady, 2017). The excessive availability of phosphorus promotes eutrophication of water, which results in proliferation of algal blooms and reduces water quality parameters. Some of the phosphorus get attached in soil with minerals such as calcium, magnesium, iron, and zinc and make them unavailable for crop growth (Zhang et al., 2021).

Thus slow and steady delivering of nutrients to the plants is needed without any loss in the environment. In addition to food security and environmental quality, human health and economic benefits are also needed to achieve sustainable development goals (Seleiman et al., 2023). To address the drawbacks of traditional fertilizers, application of smart fertilizers have been recommended. Smart fertilizers are also known as slow- or controlled-release fertilizers and eco-friendly or improved efficiency fertilizers etc (Chen et al., 2018; Liu et al., 2019). The slow-release fertilizers have less rate of releasing nutrients in comparison with traditional fertilizers. They allow plants to absorb them for long duration, make nutrients available according to crop requirement, and reduce nutrient loss into the environment (Venugopalan et al., 2022). This chapter describes different types of biochar, properties of biochar and function, synthesis of biochar-based fertilizers, and future prospects.

19.2 Development of smart fertilizers

Smart fertilizers have been prepared from nanoparticles, polymers, and zeolites. Nanofertilizers play significant function in biochemical processes of plants by enhancing nutrients availability. Foliar spray of N nanofertilizer enhanced 165% peppermint leaf dry weight as compared to control. Zinc nanofertilizer foliar application enhanced length of root and shoot, as well as the biomass of pearl millet (Tarafdar et al., 2014). The yield of cotton was enhanced by foliar application of nano-zinc under stress conditions (Hussein & Abou-Baker, 2018). Despite of several advantages, nanofertilizers release toxic substances which may be responsible for environmental pollution. The foliar application of nanofertilizers can close stomata pore, generate toxic layer onto stigmatic surface, inhibit penetration of pollen tube, and inhibit water and nutrients transportation. Polysulfone, polyurethane, and polystyrene are mainly used for the development of polymers-based smart fertilizers. High price, burst effect, and soil acidity are major limitations for application of polymers-based smart fertilizers. Zeolites are crystalline alumino silicates that promote slow release of nutrients in environment. They also enhance water use efficiency apart from carriers of nutrients. However, zeolites are sensitive to deactivation by adsorption or blockage by secondary products and their micropores cannot be used for synthesis of bulky molecules. Biochar-based fertilizers, also known as slow- or controlled-released fertilizers, can be generated from natural sources and waste substances (Zhang et al., 2020). Biochar amendment to soil enhances soil fertility, water-holding capacity, and crop yield; helps control plant diseases; improves soil microbial community; and aids in the immobilization of heavy metals and organic pollutants (He et al., 2020). Xiao and Meng (2020) reported significant decline in nutrient loss in soil after biochar amendment. Hence, the application of biochar-based fertilizers can promote sustainable development by boosting circular economy (Lateef et al., 2019).

19.3 Biochar as smart fertilizer

The unused agrowastes/crop residues are mostly burnt or dumped into landfills, which generate different environmental problems and loss of nutrients. For effective management of agricultural wastes and their transformation into biochar via thermochemical procedure is gaining importance. Calabi-Floody et al. (2018) suggested that smart fertilizers can be produced from biochar through waste recycling. The closed-loop economy promotes change in economy from linear to circular. In circular economy, waste generated in one process can be used as raw materials for another process; in this way, it helps in closing the loop, and avoid loss of nutrients and generation of pollutants. Waste produced from crop residues, food processing units, slaughterhouse, and sewage sludge can be applied as raw material for production of fertilizers. European Fertilizer Manufacturers Organization emphasized on the implementation of circular economy concept in fertilizer sector (Chojnacka et al., 2020). Hossain et al. (2020) observed that biochar increases potassium and N availability by reducing its leaching from soil. Biochar-based fertilizers promote carbon sequestration and greenhouse gases (GHGs) emission reduction (Xiang et al., 2020). The difference between conventional and smart fertilizers is given in Table 19.1.

Nitrogen recovery can be obtained via reutilization of sewage sludge and cattle manure (Grigatti et al., 2019). Different animal body parts, such as bones, horns, hooves, and claws, can be used to increase phosphate minerals (Kataki

Table 19.1 Difference between conventional and smart fertilizers.

Properties	Conventional fertilizers	Smart fertilizers
Solubility and dispersion of mineral micronutrients	Due to large particle size and less solubility, plants uptake is less	High bioavailability to plants due to improved solubility, less soil absorption, and fixation
Nutrient uptake efficiency	Bulk components are not available for roots and less efficiency	High nutrient uptake and use efficiency
Mode of release	Excessive release of fertilizers disrupt ecological balance	Release rate and pattern of fertilizers is precisely controlled
Duration of nutrient release	Plants can use at delivery time and rest amount is transformed into insoluble salts and remain in soil	Nutrients can be used by plants for long duration
Fertilizers loss rate	High loss rate by leaching or leakage in soil, rain, and wind	Decreased loss rate by less leaching or leakage in soil

et al., 2017). The ash and cattle wastes are also rich source of potassium and can replace potassium salts applications such as potassium nitrate and potassium sulfate (Maeda et al., 2018). The materials mentioned above are rich in organic carbon and other nutrients such as magnesium, calcium, manganese, zinc, sulfur, copper, and iron, which are essential for soil fertility (Chojnacka et al., 2020). The change in climatic conditions can be mitigated by fixing carbon in soil and mitigation of greenhouse gases while improving soil properties.

19.4 Benefits of slow- or controlled-release fertilizers

The application of slow- or controlled-release fertilizers has several benefits such as:

1. Enhanced nutrient efficiency: These fertilizers slowly release nutrients into the soil, improving the ability of crops to uptake and utilize nutrients, by increasing effectiveness and reducing nutrient waste.
2. Reduced loss of nutrients: Traditional fertilizers can discharge nutrients rapidly, which can be lost in soil due to runoff and leaching and pollute water. Slow- or controlled-release fertilizers minimize the losses and reduce environmental contamination.
3. Long-term effect: Slow- or controlled-release fertilizers release nutrients for the long duration, that is, few weeks to months. Hence, there is no need of frequent application of fertilizers and it saves time and money.

4. Steady crop growth: Slow- or controlled-release fertilizers can be used in different types of soil, which encourage consistent plant growth due to continuous nutrients supply.
5. Enhanced quality of crop: If nutrients are absorbed by plants in controlled manner, they can enhance quality of crops in terms of flavor, color, nutritional constituents and so on.
6. Environmental protection: The application of slow- or controlled-release fertilizers reduces nutrients leakage and runoff, therby protecting terrestrial and aquatic ecosystems.

19.5 Biochar production methods

Biochar, carbon negative material, rich in recalcitrant carbon can be produced by thermochemical decomposition of organic matter under oxygen-limited conditions. Due to the presence of meso and micropores, biochar reflected high surface area as compared with activated carbon (Qu et al., 2021). Biochar is dark in color, which alters soil thermal dynamics and facilitates crop yield, enhances water-holding capacity of soil, decreases crop disease incidence, restricts bioavailability of heavy metals, and decreases emission of nitrous oxide from soil and nutrient loss and fertilizer requirements for soil. Hence, biochar-based fertilizers are considered as smart fertilizers. Biochar is sustainable product as it holds carbon in the soil for hundred of years (Zhang et al., 2016). Biochar contains high carbon content, a low H:C ratio, and reflects high stability, porosity, and specific surface area (Kapoor et al., 2023).

Biomass waste is cheap, widely available at zero cost, renewable source of energy can be used to produce biochar. Biochar can be generated from various feedstocks such as crop residues, wood waste, food and paper waste, sewage sludge, and animal manure (Novotny et al., 2023). Bones are natural biocomposites, rich in proteins and lipids can be used for biochar preparation (Ren et al., 2021). Plant biomass is useful because of less N and sulfur (S) contents and less emission of carbon dioxide (Biswas et al., 2017). However, it has high water content and requires grinding before processing (Yousaf et al., 2021). Wastes generated from slaughterhouse such as animal bones (Ren et al., 2021) and wastes generated from food industries such as dairy, fruit, vegetable, and brewery can also be used for biochar production. Biochar can be produced by different methods such as slow and fast pyrolysis, gasification, hydrothermal carbonization, and microwave radiation (Lin et al., 2022). When biomass materials are exposed to high temperatures, they generate biochar, bio-oil, and syngas (Sakhiya et al., 2020). Fast pyrolysis is conducted at high temperature $>500°C$, and at high heating rate ($>200°C/s$) with short residence time. The slow pyrolysis is conducted at temperature from 300°C to 600°C at low heating rate (5–15°C/min) with long residence time up to hours.

In microwave pyrolysis, biomass decomposition takes place under anaerobic conditions by using microwave radiation at temperature between 300°C and 800°C. Gasification is used for production of syngas and hydrogen can be generated via this process. The end product generated through hydrothermal process shows less moisture content and hydrophobicity. Carbonization is conducted at low temperature 180°C–280°C in the absence of oxygen (Vlaskin & Vladimirov, 2018). Torrefaction also takes place at low temperature 200°C–300°C under atmospheric pressure. Physicochemical properties and quality of biochar depend on its production process (Yaashikaa et al., 2020). It is essential to analyze the rate of nutrient release and carbon stability when biochar is used to enhance fertility of soil or carbon sequestration (Saletnik et al., 2019). According to International Biochar Initiative (IBI), biochar-based materials should contain at least 10% organic carbon and H/C ratio <0.7, which reflect stability of biochar. The European Biochar Certificate (EBC) classifies biochar as a soil conditioner and fertilizer. It also provides guidelines for the sustainable production of biochar, outlines quality parameters, and specifies the dosage required for agricultural applications. The assessment of quality control parameters of biochar inhibits its misuse, protects the environment, and meets consumer demand. Different procedures can be applied to characterize biochar such as inductively coupled plasma mass spectrometry, Fourier transform infrared spectrometer, Brunauer–Emmett–Teller, X-ray diffraction, thermogravimetric analysis, and scanning electron microscopy (Yaashikaa et al., 2020).

19.6 Applications of biochar

19.6.1 Biochar enhances soil fertility

Supplementation with biochar can improve physicochemical and biological characteristics of soil, positively affect soil quality and fertility by improving soil texture, aeration, and water-retention capacity (Simanský et al., 2022). Fig. 19.1 shows various functions of biochar. Biochar can act as carrier of microbes, which remediate contaminated soil and decrease the negative effects of change in climate conditions (Ayaz et al., 2021; Bolan et al., 2022). The alkaline nature and ion exchange ability of biochar make it an effective support material for microbial cells proliferation (Pant & Rai, 2021).

Biochar can be used for various purposes such as soil conditioner, carbon sequestration, adsorption, catalysis, energy storage, and waste management because of its extraordinary properties (Liu et al., 2019; O'Connor et al., 2021). Biochar-based fertilizers reduce economic burden of traditional fertilizers by promoting controlled release of nutrients. Chen et al. (2021) exhibited that application of 0.1% fertilizer prepared from biochar enhanced corn productivity and soil fertility, while also reducing cadmium content in corn grains. Yan et al. (2021) reported positive effects of biochar-based fertilizers on tea quality, N uptake, and its utilization efficiency by vegetables. Utilization of biochar fertilizers increases

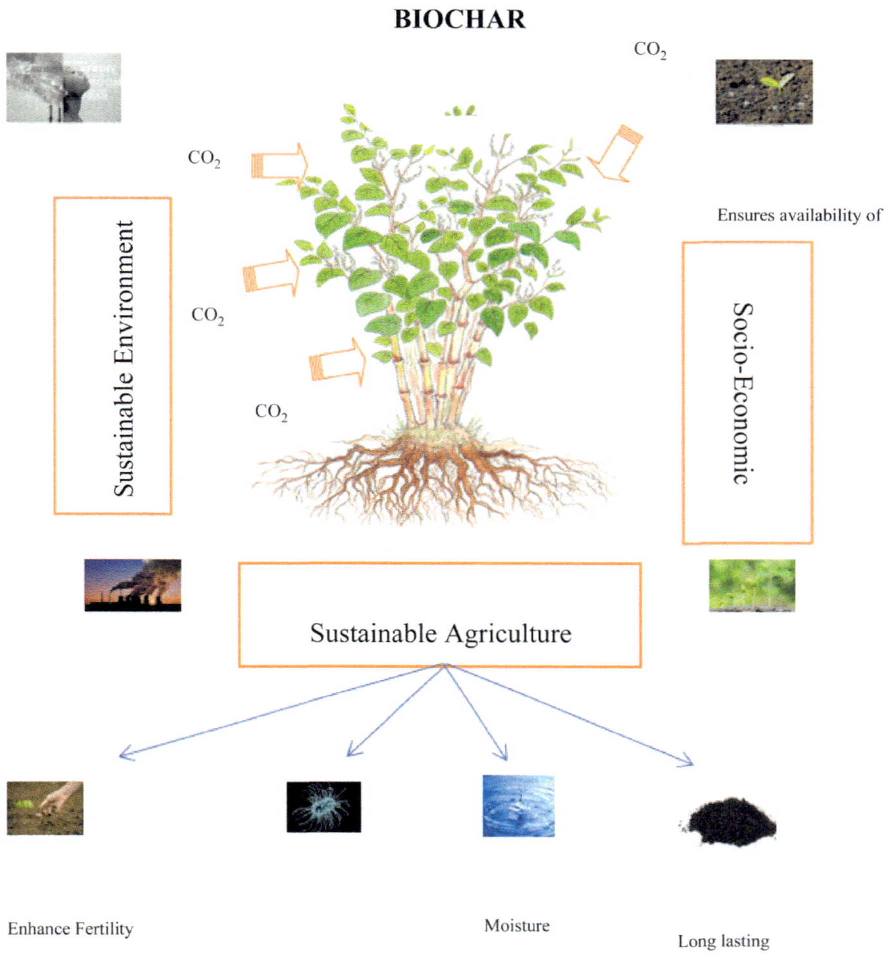

FIGURE 19.1

Different applications of biochar.

Reprinted from Ayaz, M., Feizienė, D., Tilvikienė, V., Akhtar, K., Stulpinaitė, U., & Iqbal, R. (2021). Biochar role in the sustainability of agriculture and environment. Sustainability, *13(3), 1330. https://doi.org/10.3390/su13031330.*

soil porosity, water-retention capacity, and aggregate stability, and enhances resistance against wind erosion and desertification (Fu et al., 2021). The development of pores improves the movement of air and water in soil layers, promoting the growth of symbiotic microbes, and functionality and fertility of soil (Shukla et al., 2021). Fang et al. (2020) reported that biochar supplementation in the soil changes composition and structure of microbial communities by enhancing its diversity up to 25%.

19.6.2 Reduction of greenhouse gas emissions

Rapid increase in industrial and anthropogenic activities enhanced emission of GHGs, which has become one of the major environmental problem. The sequestration of carbon in the soil can assist in reduction of GHGs (Bolan et al., 2022). Biochar amendment can capture and store carbon in the soil (Bolan et al., 2022). The unstable form of carbon can be transformed into its resistant form during pyrolysis, which reduces CO_2 concentration and promotes carbon sequestration. Van der Waals forces promote sorption of CO_2 at the biochar surface. Gwenzi et al. (2021) reported that biochar surface modification can significantly promote binding of carbon dioxide. Biochar addition can stimulate nitrifying and denitrifying bacteria and decrease N_2O emission form soil up to 50% by decreasing mineral N availability through sorption and N immobilization by microbes or reduction in nitrous oxide production (Gillingham et al., 2022; Lee et al., 2021). The nitrous oxide emission was inhibited from acidic soil of tea fields by addition of biochar, which improves soil pH and population of ammonia oxidizing bacteria by reducing fungal communities (Ji et al., 2022). Dong et al. (2021) observed biochar-based fertilizers amendment decreased methane emission by 33% in rice fields and it was due to availability of N and K in the biochar fertilizers.

19.6.3 Soil remediation

Biochar can significantly adsorb complex inorganic and organic compounds (Kamali et al., 2022). The development of pores and the presence of functional sites on the surface of biochar enhance pollutants sorption and can be used for soil remediation (Zhu et al., 2022). The presence of carboxyl, hydroxyl, carbonyl, phenolic groups, high pH, and cation exchange ability of biochar can reduce mobility of heavy metals or immobilization of heavy metals takes place via cation interaction, surface precipitation, and complexation mechanism (Kamali et al., 2022). Biochar amendment enhances pH and decreases metals mobility (Mailakeba & Rajashekhar Rao, 2021). Researches reflected that biochar application decreases lead, cadmium, and nickle toxicity by decreasing their solubility and their immobilization in stable form which cannot be absorbed by the plants (Mailakeba & Rajashekhar Rao, 2021).

Polycyclic aromatic hydrocarbons can be adsorbed by biochar and its content reduces in the soil (Kong et al., 2021). Biochar assists in sorption of pesticides and reduces their toxicity to soil microbes by restricting leakage (Ogura et al., 2021). Biochar can effectively remediate organochlorine and organophosphorus pesticides contaminated soil (Zhen et al., 2018). Difenoconazole, metalaxyl, and thiophanate methyl contaminated soil can be treated with biochar application, which reduces their availability for plants (You et al., 2021). Zhen et al. (2018) reported concentration of pesticides decreased after biochar amendment because of enhanced activity of microbes.

19.7 Biochar-based smart fertilizers

Biochar can be fabricated with plant- or animal-derived biomass feedstocks and biochar reflects excellent properties, such as large specific surface area, porosity, graphitic structure, and various binding sites. Because of their diverse role, biochar-based fertilizers are eco-friendly and promote circular economy. Due to large surface area, biochar has been found suitable for slow release of nutrients because of more sorption (Sha et al., 2019). Various methods have been applied for biochar-based slow release fertilizers production such as in situ pyrolysis, copyrolysis, impregnation, encapsulation, and granulation (Wang et al., 2022). These procedures can be grouped into three main categories (Abiola et al., 2023).

19.7.1 Direct treatment

It involves the application of nutrient-rich feedstock to generate biochar. The volatilization of nutrients can be decreased via slow pyrolysis at low temperature. Different nutrient-rich feedstocks, such as cattle manure, poultry litter, and human feces, can be utilized for biochar production, which may contain 2%–5% N, 4%–8% P, and 2%–6% K.

19.7.2 Pretreatment

In this method, feedstock is treated with nutrients such as mineral fertilizers or nutrient-rich materials before pyrolysis. Pretreatment can be done by impregnation or copyrolysis, which results in the production of biochar-based fertilizers with high nutrient content.

19.7.3 Posttreatment

Biochar is processed with nutrient-rich materials such as manure and mineral fertilizer under controlled environmental conditions. Posttreatment can be done with encapsulation, granulation, coating, and blending procedures. The posttreatment procedure is simple and can be used by farmers for biochar-based smart fertilizers production.

19.8 Synthesis of biochar-based fertilizers

19.8.1 In situ pyrolysis

Biomass wastes rich in nutrients such as banana peduncle and seaweeds, mushroom compost can be used in preparation of potassium-rich biochar. Phosphorus-rich biochar can be produced from cattle and human manure, sewage sludge, poultry litter, and bone waste (Wang et al., 2022). Karim et al. (2019) reported production of biochar potassium-rich banana peduncle and phosphorus-containing waste sludge after copyrolysis or plasma treatment. Biochar produced via pyrolysis showed high productivity and total carbon content in comparison with plasma technique; however, more nutrients were present in biochar generated through plasma method. Biochar generated from plasma and slow pyrolysis procedures showed 4500–6000 and 5600 mg/kg bioavailable phosphorus, respectively. Kim et al. (2018) generated biochar from fermentation industry waste in a muffle furnace at 600°C in the presence of N. The resulting biochar was rich in phosphorus (85 mg/g) and did not produce any toxic by-products.

19.8.2 Copyrolysis

Biochar composite was prepared by copyrolysis of montmorillonite and bamboo biochar (Chen et al., 2017). In the composite, biochar and montmorillonite provide sorption sites for nutrients. Release of ammonium and phosphate from biochar composite was 0.30%–4.92% and 2.6%–5%, reflected slow-release of N and P. Chew et al. (2020) reported biomass (200 g) was infused with demineralized water (100 g), 15 g each of urea, bentonite clay, rock phosphate, and 5 g iron sulfate. After 1 day of mixing, biomass was dried at 110°C for 3 hours. Pyrolysis was carried out at 400°C for half an hour at heating rate of 5°C/min. Biochar-based fertilizer was composed with 3% N, 1% P, 2% K, and 43% C. The produced fertilizer enhanced rice biomass by up to 67% and increased the absorption of N by 40%, P by 46%, and K by 26% compared to the control (Chew et al., 2020). The phosphoric acid and magnesium oxide (1:1) were added in mixture of poultry litter and coffee husks (1:0.5). After drying mixture was kept for pyrolysis at 500°C for 2 hours at 10°C/min heating rate. The application of new fertilizer stimulated the growth of maize and bean (Carneiro et al., 2021). Farrar et al. (2019) mixed manure, bamboo biochar, iron oxide, sulfate, and fertilizers, and conducted postpyrolysis at 180°C for 24 hours for the preparation of bamboo biochar-based fertilizer. New fertilizer was tested on ginger with two doses such as 3 and 7.5 t/ha and compared with traditional fertilizer. Biochar-based fertilizer increased ginger stem height 22% in comparison with commercial fertilizer whereas 7.5 t/ha rate enhanced ginger commercial price up to 36%. Phosphorus-containing biochar–based fertilizer was prepared with copyrolysis of cotton straw with phosphoric acid at 400°C. The new biochar-based fertilizer showed high adsorption capacity 56 mg/g for cyhalothrin pesticide (An et al., 2021).

19.8.3 Impregnation

Pristine biochar impregnation in nutrient solution significantly enhances content of nutrients. The presence of pores and a large surface area provide sites for nutrients storage and provide nutrients to plant for long duration. Date palm leaf waste was pyrolyzed at 600°C and enriched with phosphorus. Application of a 3% biomass-based fertilizer increased shoot and root fresh and dry weight by up to

66% and 60%, respectively, and improved shoot and root length by 73% and 76% compared to the control (Ahmad et al., 2018). The pyrolysis of biomass was conducted at 450°C–550°C for 50 minutes. Raw materials (mineral:biochar:urea) ratio was 9:5:10 and bentonite and sepiolite were utilized as minerals. The 31% less leaching of N and root and shoot biomass of maize was increased up to 25% and 14%, respectively, whereas N use efficiency was 24% enhanced as compared with application of urea alone (Shi et al., 2020). Impregnation corncob biochar with macronutrients such as N, P, K, calcium, and magnesium, as well as micronutrients such as sodium, zinc, and iron, improved water retention and enhanced discharge of nutrients (Lateef et al., 2019). Palm kernel shell biochar impregnation with ammonium nitrate and potassium dihydrogen phosphate reflected nutrients retention such as 53% nitrate, 77% ammonia, and 55% phosphate (Dominguez et al., 2020).

19.8.4 Encapsulation

Slow-release fertilizers can be fabricated via encapsulation, which coats the fertilizers onto inorganic or organic substrates to decrease nutrient loss and enhance nutrient use efficacy. Various coating materials, such as wax, polyolefin, paraffin, and polyvinylidene chloride, have been used for the preparation of SRFs but these are nondegradable and expensive (Lawrencia et al., 2021). However, release of nutrients, coating materials retain in soil for long duration because of less degradation and accumulate continuously. The accumulation of of coating polymers can degrade soil texture, inhibit nutrient cycling and microbiol activity, and induce environmental threat (Dong et al., 2020). Thus naturally biodegradable or green coating materials, such as sodium alginate, chitosan, cellulose, starch, and maltodextrin, have been applied for preparation of biochar-based fertilizers (An et al., 2021). Biochar-coated urea or biochar polymer–coated urea have been prepared with encapsulation procedure (Puga et al., 2020). Khajavi-Shojaei et al. (2020) stated the use of polyvinyl alcohol and corn starch in preparation of granular $MgCl_2$ modified biochar. Jia et al. (2020) prepared biochar-coated urea by oxidized starch. Urea-coated biochar reflected controlled release of nitrogen fertilizer and reduced nitrogen loss and improved efficiency of nitrogen application.

19.8.5 Granulation

Formation of granules or pellet is mostly used in fertilizers generation. Because of less density and small particle size, biochar storage and transport is tedious task. Due to wind and rain, biochar can be lost and fine biochar particles induce eye and skin irritation, as well as disorders of respiratory system (Ye et al., 2019). Granulation process protects biochar from loss and save handling cost for application in soil. Biochar and nutrients mixture granulation effectively reduces release of nutrients by enhancing diffusion resistance. Diammonium phosphate, potassium chloride, and bentonite were added in wheat straw biochar drum granulation device with steam vapor as a binder was used for granulation (Zheng et al., 2017). The granulated biochar fertilizer contained 37% nutrients and field experiments reflected that fertilizer enhanced 11% yield of grains and production of maize up to 46%. Rice-husk biochar was infused with urea hydrogen peroxide, and pelletized by using kaolin as a binder. The release of nitrogen from fertilizer was 17% and its application enhanced growth of cabbage.

19.8.6 Mixing or blending with fertilizers

The mixing or blending of biochar with fertilizers and other soil additives such as manure, compost, or lime enhanced fertilizing properties of soil and crop yield. The production of uncoated slow-release fertilizer was proposed by El Sharkawi et al. (2018). In this method, rice-husk biochar was blended with phosphoric acid, ammonia gas, and granulated in stainless-steel drum. Discharge of the nitrogen from the granular fertilizers was slow in comparison with ammonium phosphate. High nitrogen content enhanced plant growth after granular biochar–ammonium phosphate amendment. Positive effects of addition of biochar-based fertilizers in soil are given in Table 19.2. The application of biochar-based fertilizers to plants and their entry into plant cells is given in Fig. 19.2 (Akhtar et al., 2022).

Orchard pruning residue biochar (2 and 10 t/ha) was used with pig manure and ammonium sulfate (500 kg N/ha) and pyrolysis was conducted at 500°C–600°C. Japanese pear trees mixed with biochar, pig manure, and ammonium sulfate showed a significant reduction in nitrous oxide emissions (Oo et al., 2018). The mixture of NPK fertilizer and biochar enhanced rice yield from 2 to 3 t/ha (Hirst, 2019). Rice husk, rice straw, and rubber twigs (5:3:2) with 20% starch, 30% water, and NPK fertilizers were applied for preparation of biochar-based fertilizer. In pot experiments, biochar-based fertilizer enhanced pigment content, root weight, and plant nutrient use efficiency such as 17% N, 5% P, and 46% K for sweet corn (Lee et al., 2021).

Liao et al. (2020) prepared a fertilizer by mixing biochar with urea. This fertilizer enhanced 17% productivity and 59% nitrogen use efficiency of *Brassica napus* via slow release of nitrogen, enhanced nitrification, and reduced denitrification in comparison with fertilization with urea. The combination of biochar with minerals enhanced nutrient absorption and enhancement in pumpkin, cabbage, tea, maize, chili, tomato, potato, onion, and cauliflower as compared to control (Schmidt et al., 2017). Biochar and compost mixture reflected 123% increase in productivity in comparison with traditional organic fertilizers.

19.9 Disadvantages of biochar-based fertilizers

The application of biochar-based fertilizers is environmentally benign (Rombel et al., 2022), improves soil characteristics (Maikol et al., 2021), and provides an

Table 19.2 Effect of addition of biochar-based fertilizers in soil.

Biochar feedstock	Modification approach	Effect	References
Corncob	Nutrients uptake	Enhancement in soil fertility, slow release of nutrients, decrease in nutrient leaching and volatilization	Lateef et al. (2019)
Sugarcane residue	Soaking in KOH and H_3PO_4 solution, then granulation with bentonite clay	Increase potassium and phosphorus contents and yield, high phosphorus efficiency use by plants	Borges et al. (2020)
Pig manure, poultry litter, sewage sludge	Enrichment with Mg^{2+}	Improvement in soil properties	Nardis et al. (2021)
Peanut shell, sugarcane bagasse	Soaking in $MgCl_2$ and $CaCl_2$ solution	Enhanced nutrients content	Fang et al. (2020)
Cotton straw	Coating with starch, cellulose, and chitosan	Enhanced plant growth and root length, reduction in phosphorus release rate up to 50%	An et al. (2021)
Wheat straw	Coating with bio-oil	Slow nutrients release, better nutrients use efficacy by plants	Ye et al. (2019)
Wood chips	Mixing and coating withtriple superphosphate	Slow release of phosphorus in soil	Pogorzelski et al. (2020)
Rice husk	Mixing with urea	Slow release of nitrogen in soil	Petrus et al. (2020)
Sewage sludge	Mixing with cattle manure compost	High carbon content in soil	Goldan et al. (2019)
Pea	Composting with manure	Increased height, biomass of root, shoot, grain, husk	Bashir et al. (2020)
Residues of oak	Mixing with plants and sheep manure	High yield of crops	Oldfield et al. (2018)
Palm empty fruit bunch	Blending with palm waste compost	Enhanced bioavailability of calcium, magnesium, nitrogen, phosphorus, and potassium and soil pH	Radin et al. (2018)
Forest shrub	Composting with green waste and cattle manure	Increased bioavailability of nitrogen, phosphorus, calcium, magnesium, and potassium in soil	Pandit et al. (2020)
Rice straw	Blending with maize straw and cattle manure	Increased contents of nitrogen and potassium and enhanced fungal diversity in soil	Bello et al. (2021)
Cow manure	Mixing with cow manure compost	Enhancement in maize plant height, biomass, yield, and harvest index	Zahra et al. (2022)

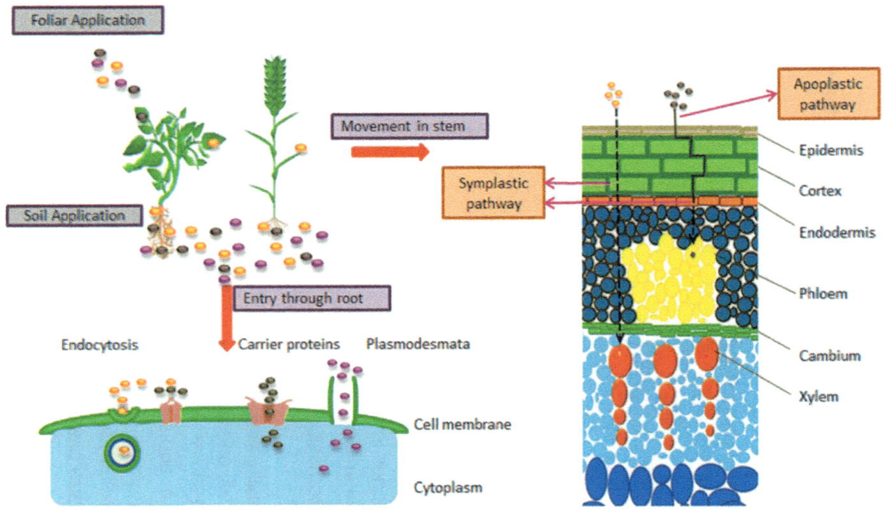

FIGURE 19.2

Different modes of application of biochar-based fertilizers on plants and their entry in plant cells.

Reprinted from Akhtar, N., Ilyas, N., Meraj, T. A., Pour-Aboughadareh, A., Sayyed, R. Z., Mashwani, Z. U. R., & Poczai, P. (2022). Improvement of plant responses by nanobiofertilizer: A step towards sustainable agriculture. Nanomaterials, 12*(6). https://doi.org/10.3390/nano12060965.*

effective solution for recycling of wastes (Czekała et al., 2019) and remediation of soil because its sorption properties (Rombel et al., 2022). Although various biochar-based fertilizers are available on the market, their prices are very high as compared with traditional fertilizers (Table 19.3).

Therefore economic feasibility is needed for large-scale application of biochar-based fertilizers. Utilization of locally available cattle manure or agro-wastes biochar easily available waste biomass or crop residues can give a favorable economic feasibility. Oldfield et al. (2018) reported that requirement of mineral fertilizers can be replaced by release of nutrients from biochar with compost or organic wastes mixture rich in micro- and macronutrients. The application of biochar-based fertilizers is reported in few geographical regions. However, there is need to study impact of biochar fertilizers in different regions under varied climate conditions. Some reports revealed that due to the presence of heavy metal, polyaromatic hydrocarbons, and dioxins, biochar may increase soil pollution. Kuppusamy et al. (2016) reported high doses of biochar can negatively affect survival of earthworms. Further investigations are needed to assess the efficacy of biochar-based fertilizers under real field conditions to confirm its impact on agricultural productivity.

Table 19.3 Commercial available biochar-based fertilizers.

Product	Producer	Components (%)	Application	Price (per kg)
Biochar fertilizer	Carbon Gold	N:P:K, 5:5:5 (each 5%)	Vegetables, fruit, tree, plants	GB£8.49
Biogranules	CharGrow	Microbes and microbial food	Field crop	US$14.1
Bamboo biochar with bio-organic fertilizer	Seek	NPK and amino acid >5, organic matter > 45, humic acid > 10, microbes > 20 million/g	Field crops, vegetables, fruits	US$0.6
CarbonVital plus special	Biohelp	N:P:K, 6:3:6; Ca, 2; Mg, 0.1; S, 2.5 Biochar, lime, mineral, microbes	Vegetables, fruits	GB£8.9
OceanMagic 2-2-0.3	Hendrikus organics	N:P:K, 2:2:0.3; phosphoric acid; Fish and seafood protein hydrolysate	Tree, shrub	US$8.9
Biochar fertilizer 18-5-12	Lewis bamboo	N:P:K, 18:5:12; biochar 30	Bamboo	US$21.9

19.10 Conclusion

The production of biochar-based slow- or controlled-release smart fertilizers can enhance crop yield, soil fertility, and less loss of nutrient in comparison with traditional fertilizers. Biochar-based fertilizers can partially or fully replace the need of traditional fertilizers. The application of biochar-based slow-release fertilizers is feasible method for sustainable agriculture; however, its actual use is still in infancy. Several challenges have been reported in the formulation of effective biochar-based fertilizers such as nutrient release and bioavailability evaluation, long-term impact of fertilizers on physicochemical characteristics of soil, changes in the structure and fertilizers properties after their retention in soil, and analysis of the environmental risks related with biochar generated from sewage sludge and animal wastes. Utilization of biochar-based fertilizers is not only a "win–win," but a "multiwin" solution as it fulfills objectives of sustainability and circular economy. There is a need to develope a database to raise awareness about the application of biochar-based fertilizers, including details on plant species, types of waste materials used, and experimental conditions. This information can promote development and application of biochar-based fertilizers.

Biochar generation from wastes ensures recycling of wastes and assists in achievement of sustainable developmental goals. However, feedstock selection, process parameters, and additives not only determine functional features of biochar but also effect the economic feasibility of the fertilizers. Future research should continue to explore composition, manufacture, and agronomic and environmental performance of different smart fertilizers. Biochar-based fertilizers have simple fabrication process, excellent slow-release performance for nutrients, and high bioavailability for crop growth. Therefore the development of strategy for synthesis of biochar-based fertilizers with all the above-mentioned requirements is imperative for future studies. The long-term assessment of impact of biochar-based fertilizers on reduction in GHG emission should be further examined.

19.11 Recommendations and future perspectives of biochar-based smart fertilizers

1. There is a need of production of integrated products by mixing with biochar-based fertilizers, micro- and macronutrients. This type of integrated fertilizers with reduce application of nutrients to the crop plants and save money and time.
2. There is a need of end of life assessment of slow- or controlled-release fertilizer by using life cycle assessment tools.
3. Researchers should focus on cradle-to-grave analysis by using life cycle assessment and sustainability analysis tools for both positive and negative impact assessment of slow-release fertilizers.
4. The manufacturing of biochar-based smart fertilizers may release emission and loss of materials during synthesis and processing. The process parameters and design can be modified by exploring different strategies; hence, this area should be further explored.
5. Earlier studies have shown that presence of nitrification and urease inhibitors delay in the conversion of ammonium to nitrate or plant-available nitrogen.
6. Detailed investigations are needed after combination of slow-release fertilizers and inhibitors on nutrient use efficiency and crop yield.
7. Most of the researchers have been conducted in laboratory under controlled environmental conditions. The field trial for a specific crop under different seasons is needed for validation of application of slow-release fertilizers for large scale.
8. The engineered materials, metal-organic frameworks, graphene, and layered double hydroxide can be mixed in biochar-based fertilizers, which can improve physical and chemical properties and enhance nutrients retention capacity in soil.

Acknowledgments

The authors extend their appreciation to Amity University Uttar Pradesh, Noida, India and Universiti Sains Malaysia, Malaysia for providing all the necessary facilities and valuable support.

Author contributions

Conceptualization, R.T.K., M.R.; writing—original draft preparation, R.T.K., M. R.; and writing—review and editing, R.T.K., M.R. Authors have read and agreed to the published version of this chapter.

Conflicts of interest

The authors declare no conflict of interest.

Statement on the use of AI tools for writing

We declare that we have not used any of the AI tools for writing or editing this chapter.

References

Abiola, W. A., Diogo, R. V. C., Tovihoudji, P. G., Mien, A. K., & Schalla, A. (2023). Research trends on biochar-based smart fertilizers as an option for the sustainable agricultural land management: Bibliometric analysis and review. *Frontiers in Soil Science*, *3*. Available from https://doi.org/10.3389/fsoil.2023.1136327, http://www.frontiersin.org/journals/soil-science.

Ahmad, M., Usman, A. R. A., Al-Faraj, A. S., Ahmad, M., Sallam, A., & Al-Wabel, M. I. (2018). Phosphorus-loaded biochar changes soil heavy metals availability and uptake potential of maize (*Zea mays* L.) plants. *Chemosphere*, *194*, 327–339. Available from https://doi.org/10.1016/j.chemosphere.2017.11.156, http://www.elsevier.com/locate/chemosphere.

Akhtar, N., Ilyas, N., Meraj, T. A., Pour-Aboughadareh, A., Sayyed, R. Z., Mashwani, Z. U. R., & Poczai, P. (2022). Improvement of plant responses by nanobiofertilizer: A step towards sustainable agriculture. *Nanomaterials*, *12*(6). Available from https://doi.org/10.3390/nano12060965, https://www.mdpi.com/2079-4991/12/6/965/pdf.

Alshaal, T., & El-Ramady, H. (2017). Foliar application: From plant nutrition to biofortification. *Environment, Biodiversity and Soil Security*, *1*(2017), 71–83. Available from https://doi.org/10.21608/jenvbs.2017.1089.1006.

An, X., Wu, Z., Shi, W., Qi, H., Zhang, L., Xu, X., & Yu, B. (2021). Biochar for simultaneously enhancing the slow-release performance of fertilizers and minimizing the pollution of pesticides. *Journal of Hazardous Materials*, *407*, 124865. Available from https://doi.org/10.1016/j.jhazmat.2020.124865.

Ayaz, M., Feizienė, D., Tilvikienė, V., Akhtar, K., Stulpinaitė, U., & Iqbal, R. (2021). Biochar role in the sustainability of agriculture and environment. *Sustainability*, *13*(3), 1330. Available from http://doi.org/10.3390/su13031330.

Bashir, A., Rizwan, M., Zia ur Rehman, M., Zubair, M., Riaz, M., Qayyum, M. F., Alharby, H. F., Bamagoos, A. A., & Ali, S. (2020). Application of co-composted farm manure and biochar increased the wheat growth and decreased cadmium accumulation in plants under different water regimes. *Chemosphere*, *246*. Available from https://doi.org/10.1016/j.chemosphere.2019.125809, http://www.elsevier.com/locate/chemosphere.

Beig, B., Niazi, M. B. K., Sher, F., Jahan, Z., Malik, U. S., Khan, M. D., Américo-Pinheiro, J. H. P., & Vo, D. V. N. (2022). Nanotechnology-based controlled release of sustainable fertilizers. A review. *Environmental Chemistry Letters*, *20*(4), 2709–2726. Available from https://doi.org/10.1007/s10311-022-01409-w, http://springerlink.metapress.com/app/home/journal.asp?wasp = d86tgdwvtg0yvw9gvkwp&referrer = parent&backto = browsepublicationsresults,140,541.

Bello, A., Wang, B., Zhao, Y., Yang, W., Ogundeji, A., Deng, L., Egbeagu, U. U., Yu, S., Zhao, L., Li, D., & Xu, X. (2021). Composted biochar affects structural dynamics, function and co-occurrence network patterns of fungi community. *Science of the Total Environment*, *775*, 145672. Available from https://doi.org/10.1016/j.scitotenv.2021.145672.

Bolan, N., Hoang, S. A., Beiyuan, J., Gupta, S., Hou, D., Karakoti, A., Joseph, S., Jung, S., Kim, K. H., Kirkham, M. B., Kua, H. W., Kumar, M., Kwon, E. E., Ok, Y. S., Perera, V., Rinklebe, J., Shaheen, S. M., Sarkar, B., Sarmah, A. K., ... Van Zwieten, L. (2022). Multifunctional applications of biochar beyond carbon storage. *International Materials Reviews*, *67*(2), 150–200. Available from https://doi.org/10.1080/09506608.2021.1922047, http://www.tandfonline.com/loi/yimr20#.VwHbh01f1Qs.

Borges, B. M. M. N., Strauss, M., Camelo, P. A., Sohi, S. P., & Franco, H. C. J. (2020). Re-use of sugarcane residue as a novel biochar fertiliser – increased phosphorus use efficiency and plant yield. *Journal of Cleaner Production*, *262*. Available from https://doi.org/10.1016/j.jclepro.2020.121406, https://www.journals.elsevier.com/journal-of-cleaner-production.

Calabi-Floody, M., Medina, J., Rumpel, C., Condron, L. M., Hernandez, M., Dumont, M., & Mora, MdlL. (2018). Smart fertilizers as a strategy for sustainable agriculture. *Advances in Agronomy*, *147*, 119–157. Available from https://doi.org/10.1016/bs.agron.2017.10.003, http://www.elsevier.com/wps/find/bookdescription.cws_home/675816/description#description.

Carneiro, J. Sd. S., Ribeiro, I. C. A., Nardis, B. O., Barbosa, C. F., Lustosa Filho, J. F., & Melo, L. C. A. (2021). Long-term effect of biochar-based fertilizers application in tropical soil: Agronomic efficiency and phosphorus availability. *Science of the Total Environment*, *760*. Available from https://doi.org/10.1016/j.scitotenv.2020.143955, http://www.elsevier.com/locate/scitotenv.

Chen, J., Lü, S., Zhang, Z., Zhao, X., Li, X., Ning, P., & Liu, M. (2018). Environmentally friendly fertilizers: A review of materials used and their effects on the environment. *Science of the Total Environment*, *613–614*, 829–839. Available from https://doi.org/10.1016/j.scitotenv.2017.09.186, http://www.elsevier.com/locate/scitotenv.

Chen, L., Chen, X. L., Zhou, C. H., Yang, H. M., Ji, S. F., Tong, D. S., Zhong, Z. K., Yu, W. H., & Chu, M. Q. (2017). Environmental-friendly montmorillonite-biochar composites: Facile production and tunable adsorption-release of ammonium and phosphate. *Journal of Cleaner Production*, *156*, 648–659. Available from https://doi.org/10.1016/j.jclepro.2017.04.050.

Chen, Z., Pei, J., Wei, Z., Ruan, X., Hua, Y., Xu, W., Zhang, C., Liu, T., & Guo, Y. (2021). A novel maize biochar-based compound fertilizer for immobilizing cadmium and improving soil quality and maize growth. *Environmental Pollution*, *277*, 116455. Available from https://doi.org/10.1016/j.envpol.2021.116455.

Chew, J., Zhu, L., Nielsen, S., Graber, E., Mitchell, D. R. G., Horvat, J., Mohammed, M., Liu, M., van Zwieten, L., Donne, S., Munroe, P., Taherymoosavi, S., Pace, B., Rawal, A., Hook, J., Marjo, C., Thomas, D. S., Pan, G., Li, L., . . . Fan, X. (2020). Biochar-based fertilizer: Supercharging root membrane potential and biomass yield of rice. *Science of the Total Environment*, *713*, 136431. Available from https://doi.org/10.1016/j.scitotenv.2019.136431.

Chojnacka, K., Moustakas, K., & Witek-Krowiak, A. (2020). Bio-based fertilizers: A practical approach towards circular economy. *Bioresource Technology*, *295*, 122223. Available from https://doi.org/10.1016/j.biortech.2019.122223.

Czekała, W., Jeżowska, A., & Chełkowski, D. (2019). The use of biochar for the production of organic fertilizers. *Journal of Ecological Engineering*, *20*(1), 1−8. Available from https://doi.org/10.12911/22998993/93869.

Dominguez, E. L., Uttran, A., Loh, S. K., Manero, M. H., Upperton, R., Tanimu, M. I., & Thomas Bachmann, R. T. (2020). Characterisation of industrially produced oil palm kernel shell biochar and its potential as slow release nitrogen-phosphate fertilizer and carbon sink. *Materials Today: Proceedings*, *31*, 221−227. Available from https://doi.org/10.1016/j.matpr.2020.05.143, https://www.sciencedirect.com/journal/materials-today-proceedings.

Dong, D., Wang, C., Van Zwieten, L., Wang, H., Jiang, P., Zhou, M., & Wu, W. (2020). An effective biochar-based slow-release fertilizer for reducing nitrogen loss in paddy fields. *Journal of Soils and Sediments*, *20*(8), 3027−3040. Available from https://doi.org/10.1007/s11368-019-02401-8, http://www.springerlink.com/content/1439-0108.

Dong, D., Li, J., Ying, S., Wu, J., Han, X., Teng, Y., Zhou, M., Ren, Y., & Jiang, P. (2021). Mitigation of methane emission in a rice paddy field amended with biochar-based slow-release fertilizer. *The Science of the Total Environment*, *792*, 148460. Available from https://doi.org/10.1016/j.scitotenv.2021.148460.

Duhan, J. S., Kumar, R., Kumar, N., Kaur, P., Nehra, K., & Duhan, S. (2017). Nanotechnology: The new perspective in precision agriculture. *Biotechnology Reports*, *15*, 11−23. Available from https://doi.org/10.1016/j.btre.2017.03.002, http://www.journals.elsevier.com/biotechnology-reports/.

El Sharkawi, H. M., Tojo, S., Chosa, T., Malhat, F. M., & Youssef, A. M. (2018). Biochar-ammonium phosphate as an uncoated-slow release fertilizer in sandy soil. *Biomass and Bioenergy*, *117*, 154−160. Available from https://doi.org/10.1016/j.biombioe.2018.07.007, http://www.journals.elsevier.com/biomass-and-bioenergy/.

Fang, W., Song, Z., Tao, S., Zhang, D., Huang, B., Ren, L., Cheng, H., Yan, D., Li, Y., Cao, A., & Wang, Q. (2020). Biochar mitigates the negative effect of chloropicrin fumigation on beneficial soil microorganisms. *Science of the Total Environment*, *738*, 139880. Available from https://doi.org/10.1016/j.scitotenv.2020.139880.

Farrar, M. B., Wallace, H. M., Xu, C. Y., Nguyen, T. T. N., Tavakkoli, E., Joseph, S., & Bai, S. H. (2019). Short-term effects of organo-mineral enriched biochar fertiliser on ginger yield and nutrient cycling. *Journal of Soils and Sediments*, *19*(2), 668−682. Available from https://doi.org/10.1007/s11368-018-2061-9, http://www.springerlink.com/content/1439-0108.

Fu, B., Chen, L., Huang, H., Qu, P., & Wei, Z. (2021). Impacts of crop residues on soil health: A review. *Environmental Pollutants and Bioavailability*, *33*(1), 164−173. Available from https://doi.org/10.1080/26395940.2021.1948354, https://www.tandfonline.com/loi/tcsb21.

Gillingham, M. D., Gomes, R. L., Ferrari, R., & West, H. M. (2022). Sorption, separation and recycling of ammonium in agricultural soils: A viable application for magnetic biochar? *Science of the Total Environment*, *812*, 151440. Available from https://doi.org/10.1016/j.scitotenv.2021.151440.

Goldan, E., Nedeff, V., Sandu, I., Barsan, N., Mosnegutu, E., & Panainte, M. (2019). The use of biochar and compost mixtures as potential organic fertilizers. *Revista de Chimie*, *70*(6), 2192–2197. Available from https://doi.org/10.37358/rc.19.6.7303, http://www.revistadechimie.ro/pdf/63%20GOLDAN%206%2019.pdf.

Grigatti, M., Cavani, L., di Biase, G., & Ciavatta, C. (2019). Current and residual phosphorous availability from compost in a ryegrass pot test. *Science of the Total Environment*, *677*, 250–262. Available from https://doi.org/10.1016/j.scitotenv.2019.04.349, http://www.elsevier.com/locate/scitotenv.

Gwenzi, W., Chaukura, N., Wenga, T., & Mtisi, M. (2021). Biochars as media for air pollution control systems: Contaminant removal, applications and future research directions. *Science of The Total Environment*, *753*, 142249. Available from https://doi.org/10.1016/j.scitotenv.2020.142249.

He, K., He, G., Wang, C., Zhang, H., Xu, Y., Wang, S., Kong, Y., Zhou, G., & Hu, R. (2020). Biochar amendment ameliorates soil properties and promotes Miscanthus growth in a coastal saline-alkali soil. *Applied Soil Ecology*, *155*, 103674. Available from https://doi.org/10.1016/j.apsoil.2020.103674.

Hossain, M. Z., Bahar, M. M., Sarkar, B., Donne, S. W., Ok, Y. S., Palansooriya, K. N., Kirkham, M. B., Chowdhury, S., & Bolan, N. (2020). Biochar and its importance on nutrient dynamics in soil and plant. *Biochar*, *2*(4), 379–420. Available from https://doi.org/10.1007/s42773-020-00065-z, http://springer.com/journal/42773.

Ji, C., Han, Z., Zheng, F., Wu, S., Wang, J., Wang, J., Zhang, H., Zhang, Y., Liu, S., Li, S., & Zou, J. (2022). Biochar reduced soil nitrous oxide emissions through suppressing fungal denitrification and affecting fungal community assembly in a subtropical tea plantation. *Agriculture, Ecosystems & Environment*, *326*, 107784. Available from https://doi.org/10.1016/j.agee.2021.107784.

Jia, Y., Hu, Z., Mu, J., Zhang, W., Xie, Z., & Wang, G. (2020). Preparation of biochar as a coating material for biochar-coated urea. *Science of the Total Environment*, *731*, 139063. Available from https://doi.org/10.1016/j.scitotenv.2020.139063.

Kamali, M., Sweygers, N., Al-Salem, S., Appels, L., Aminabhavi, T. M., & Dewil, R. (2022). Biochar for soil applications-sustainability aspects, challenges and future prospects. *Chemical Engineering Journal*, *428*, 131189. Available from https://doi.org/10.1016/j.cej.2021.131189.

Kapoor, R. T., Rafatullah, M., Tajarudin, H. A., Siddiqui, M. R., & Alam, M. (2023). Treatment of direct red 28 dye through *Phoenix dactylifera* L. fruit seed biochar: Equilibrium, kinetics, thermodynamics, and phytotoxicity studies. *Sustainability*, *15*(21), 15266. Available from https://doi.org/10.3390/su152115266.

Karim, A. A., Kumar, M., Mohapatra, S., & Singh, S. K. (2019). Nutrient rich biomass and effluent sludge wastes co-utilization for production of biochar fertilizer through different thermal treatments. *ournal of Cleaner Production*, *228*, 570–579. Available from https://doi.org/10.1016/j.jclepro.2019.04.330.

Kataki, S., Hazarika, S., & Baruah, D. C. (2017). Assessment of by-products of bioenergy systems (anaerobic digestion and gasification) as potential crop nutrient. *Waste Management*, *59*, 102–117. Available from https://doi.org/10.1016/j.wasman.2016.10.018, http://www.elsevier.com/locate/wasman.

Khajavi-Shojaei, S., Moezzi, A., Masir, M. N., & Taghavi, M. (2020). Synthesis modified biochar-based slow-release nitrogen fertilizer increases nitrogen use efficiency and corn growth. *Biomass Conversion and Biorefinery.*, *13*, 593–601.

Khan, H. A., Naqvi, S. R., Mehran, M. T., Khoja, A. H., Khan Niazi, M. B., Juchelková, D., & Atabani, A. (2021). A performance evaluation study of nano-biochar as a potential slow-release nano-fertilizer from wheat straw residue for sustainable agriculture. *Chemosphere*, *285*. Available from https://doi.org/10.1016/j.chemosphere.2021.131382, http://www.elsevier.com/locate/chemosphere.

Kim, J. A., Vijayaraghavan, K., Reddy, D. H. K., & Yun, Y. S. (2018). A phosphorus-enriched biochar fertilizer from bio-fermentation waste: A potential alternative source for phosphorus fertilizers. *Journal of Cleaner Production*, *196*, 163–171. Available from https://doi.org/10.1016/j.jclepro.2018.06.004, https://www.journals.elsevier.com/journal-of-cleaner-production.

Kong, L., Song, B., Zhang, T., Gao, K., & Liu, J. (2021). Effects of soil organic matter on biochar application in developing the biodegradation potentials of polycyclic aromatic hydrocarbons (PAHs). *Applied Soil Ecology*, *167*, 104046. Available from https://doi.org/10.1016/j.apsoil.2021.104046.

Kuppusamy, S., Thavamani, P., Megharaj, M., Venkateswarlu, K., & Naidu, R. (2016). Agronomic and remedial benefits and risks of applying biochar to soil: Current knowledge and future research directions. *Environment International*, *87*, 1–12. Available from https://doi.org/10.1016/j.envint.2015.10.018, http://www.elsevier.com/locate/envint.

Lateef, A., Nazir, R., Jamil, N., Alam, S., Shah, R., Khan, M. N., Saleem, M., & Rehman, S.-ur (2019). Synthesis and characterization of environmental friendly corncob biochar based nano-composite – a potential slow release nano-fertilizer for sustainable agriculture. *Environmental Nanotechnology, Monitoring & Management*, *11*, 100212. Available from https://doi.org/10.1016/j.enmm.2019.100212.

Lawrencia, D., Wong, S. K., Low, D. Y. S., Goh, B. H., Goh, J. K., Ruktanonchai, U. R., Soottitantawat, A., Lee, L. H., & Tang, S. Y. (2021). Controlled release fertilizers: A review on coating materials and mechanism of release. *Plants*, *10*(2), 1–26. Available from https://doi.org/10.3390/plants10020238, https://www.mdpi.com/2223-7747/10/2/238/pdf.

Lee, Y. L., Ahmed, O. H., Wahid, S. A., & AB Aziz, Z. F. (2021). Biochar tablets with and without embedded fertilizer on the soil chemical characteristics and nutrient use efficiency of *Zea mays*. *Sustainability*, *13*(9), 4878. Available from https://doi.org/10.3390/su13094878.

Li, S., & Chen, G. (2020). Contemporary strategies for enhancing nitrogen retention and mitigating nitrous oxide emission in agricultural soils: Present and future. *Environment, Development and Sustainability*, *22*(4), 2703–2741. Available from https://doi.org/10.1007/s10668-019-00327-2, http://www.wkap.nl/journalhome.htm/1387-585X.

Lin, J., Sun, S., Xu, D., Cui, C., Ma, R., Luo, J., Fang, L., & Li, H. (2022). Microwave directional pyrolysis and heat transfer mechanisms based on multiphysics field stimulation: Design porous biochar structure via controlling hotspots formation. *Chemical Engineering Journal*, *429*, 132195. Available from https://doi.org/10.1016/j.cej.2021.132195.

Liu, Y., Tang, H., Muhammad, A., & Huang, G. (2019). Emission mechanism and reduction countermeasures of agricultural greenhouse gases – a review. *Greenhouse Gases: Science and Technology*, *9*(2), 160–174. Available from https://doi.org/10.1002/ghg.1848, https://onlinelibrary.wiley.com/page/journal/21523878/homepage/forauthors.html.

Maeda, N., Fukasawa, T., Katakura, T., Ito, M., Ishigami, T., Huang, A. N., & Fukui, K. (2018). Existence form of potassium components in woody biomass combustion ashes and estimation method of its enrichment degree. *Energy and Fuels*, *32*(1), 517–524. Available from https://doi.org/10.1021/acs.energyfuels.7b03090, http://pubs.acs.org/journal/enfuem.

Maikol, N., Haruna, A. O., Maru, A., Asap, A., & Medin, S. (2021). Utilization of urea and chicken litter biochar to improve rice production. *Scientific Reports*, *11*(1). Available from https://doi.org/10.1038/s41598-021-89332-y, http://www.nature.com/srep/index.html.

Mailakeba, C. D., & Rajashekhar Rao, B. K. (2021). Biochar application alters soil Ni fractions and phytotoxicity of Ni to pakchoi (*Brassica rapa* L. ssp. *chinensis* L.) plants: Effects of Ni and biochar on Ni fractions. *New Guinea Environmental Technology and Innovation*, *23*. Available from https://doi.org/10.1016/j.eti.2021.101751, http://www.journals.elsevier.com/environmental-technology-and-innovation/.

Nardis, B. O., Santana Da Silva Carneiro, J., Souza, I. M. G. D., Barros, R. G. D., & Azevedo Melo, L. C. (2021). Phosphorus recovery using magnesium-enriched biochar and its potential use as fertilizer. *Archives of Agronomy and Soil Science*, *67*(8), 1017–1033. Available from https://doi.org/10.1080/03650340.2020.1771699, http://www.tandf.co.uk/journals/titles/03650340.asp.

Novotny, M., Markovic, M., Raček, J., Sipka, M., Chorazy, T., Tošić, I., & Hlavínek, P. (2023). The use of biochar made from biomass and biosolids as a substrate for green infrastructure: A review. *Sustainable Chemistry and Pharmacy.*, *32*.

O'Connor, J., Hoang, S. A., Bradney, L., Dutta, S., Xiong, X., Tsang, D. C. W., Ramadass, K., Vinu, A., Kirkham, M. B., & Bolan, N. S. (2021). A review on the valorisation of food waste as a nutrient source and soil amendment. *Environmental Pollution*, *272*. Available from https://doi.org/10.1016/j.envpol.2020.115985, https://www.journals.elsevier.com/environmental-pollution.

Ogura, A. P., Lima, J. Z., Marques, J. P., Massaro Sousa, L., Rodrigues, V. G. S., & Espíndola, E. L. G. (2021). A review of pesticides sorption in biochar from maize, rice, and wheat residues: Current status and challenges for soil application. *Journal of Environmental Management*, *300*. Available from https://doi.org/10.1016/j.jenvman.2021.113753, https://www.sciencedirect.com/journal/journal-of-environmental-management.

Oldfield, T. L., Sikirica, N., Mondini, C., López, G., Kuikman, P. J., & Holden, N. M. (2018). Biochar, compost and biochar-compost blend as options to recover nutrients and sequester carbon. *Journal of Environmental Management*, *218*, 465–476. Available from https://doi.org/10.1016/j.jenvman.2018.04.061, https://www.sciencedirect.com/journal/journal-of-environmental-management.

Oo, A. Z., Gonai, T., Sudo, S., Win, K. T., & Shibata, A. (2018). Surface application of fertilizers and residue biochar on N_2O emission from japanese pear orchard soil. *Plant, Soil and Environment*, *64*(12), 597–604. Available from https://doi.org/10.17221/114/2018-PSE, https://www.agriculturejournals.cz/publicFiles/114_2018-PSE.pdf.

Pandit, N. R., Schmidt, H. P., Mulder, J., Hale, S. E., Husson, O., & Cornelissen, G. (2020). Nutrient effect of various composting methods with and without biochar on soil fertility and maize growth. *Archives of Agronomy and Soil Science*, *66*(2), 250–265. Available from https://doi.org/10.1080/03650340.2019.1610168, http://www.tandf.co.uk/journals/titles/03650340.asp.

Pant, A., & Rai, J. P. N. (2021). Application of biochar on methane production through organic solid waste and ammonia inhibition. *Environmental Challenges*, *5*, 100262. Available from https://doi.org/10.1016/j.envc.2021.100262.

Petrus, H. T. B. M., Putera, A. D. P., Wangi, I. P., Ramadhian, M. A., Setiawan, H., & Prasetya, A. (2020). Characterization of nitrogen release in modified controlled-release-fertilizer using rice husk biochar. *International Journal of Technology*, *11*(4), 774. Available from https://doi.org/10.14716/ijtech.v11i4.3520.

Piwowar, A. (2022). Consumption of mineral fertilizers in the polish agriculture – trends and directions of changes. *Agricultural Research*, *11*(3), 477–487. Available from https://doi.org/10.1007/s40003-021-00591-7, http://www.springer.com/life + sciences/cell + biology/journal/40003.

Pogorzelski, D., Filho, J. F. L., Matias, P. C., Santos, W. O., Vergütz, L., & Melo, L. C. A. (2020). Biochar as composite of phosphate fertilizer: Characterization and agronomic effectiveness. *Science of the Total Environment*, *743*. Available from https://doi.org/10.1016/j.scitotenv.2020.140604, http://www.elsevier.com/locate/scitotenv.

Puga, A. P., Grutzmacher, P., Cerri, C. E. P., Ribeirinho, V. S., & Andrade, C. Ad (2020). Biochar-based nitrogen fertilizers: Greenhouse gas emissions, use efficiency, and maize yield in tropical soils. *Science of the Total Environment*, *704*. Available from https://doi.org/10.1016/j.scitotenv.2019.135375, http://www.elsevier.com/locate/scitotenv.

Qu, J., Wang, Y., Tian, X., Jiang, Z., Deng, F., Tao, Y., Jiang, Q., Wang, L., & Zhang, Y. (2021). KOH-activated porous biochar with high specific surface area for adsorptive removal of chromium (VI) and naphthalene from water: Affecting factors, mechanisms and reusability exploration. *Journal of Hazardous Materials*, *401*, 123292. Available from https://doi.org/10.1016/j.jhazmat.2020.123292.

Radin, R., Abu Bakar, R., Ishak, C. F., Ahmad, S. H., & Tsong, L. C. (2018). Biochar-compost mixture as amendment for improvement of polybag-growing media and oil palm seedlings at main nursery stage. *International Journal of Recycling of Organic Waste in Agriculture*, *7*(1), 11–23. Available from https://doi.org/10.1007/s40093-017-0185-3, http://www.springer.com/environment/pollution + and + remediation/journal/40093.

Ren, X., Wang, J., Yu, J., Song, B., Feng, H., Shen, M., Zhang, H., Zou, J., Zeng, G., Tang, L., & Wang, J. (2021). Waste valorization: Transforming the fishbone biowaste into biochar as an efficient persulfate catalyst for degradation of organic pollutant. *Journal of Cleaner Production*, *291*, 125225. Available from https://doi.org/10.1016/j.jclepro.2020.125225.

Rombel, A., Krasucka, P., & Oleszczuk, P. (2022). Sustainable biochar-based soil fertilizers and amendments as a new trend in biochar research. *Science of the Total Environment*, *816*, 151588. Available from https://doi.org/10.1016/j.scitotenv.2021. 151588.

Sakhiya, A. K., Anand, A., & Kaushal, P. (2020). Production, activation, and applications of biochar in recent times. *Biochar*, *2*(3), 253–285. Available from https://doi.org/10.1007/s42773-020-00047-1, http://springer.com/journal/42773.

Saletnik, B., Zaguła, G., Bajcar, M., Tarapatskyy, M., Bobula, G., & Puchalski, C. (2019). Biochar as a multifunctional component of the environment—a review. *Applied Sciences*, *9*(6), 1139. Available from https://doi.org/10.3390/app9061139.

Schmidt, H. P., Pandit, B. H., Cornelissen, G., & Kammann, C. I. (2017). Biochar-based fertilization with liquid nutrient enrichment: 21 field trials covering 13 crop species in Nepal. *Land Degradation and Development*, *28*(8), 2324–2342. Available from https://doi.org/10.1002/ldr.2761, http://onlinelibrary.wiley.com/journal/10.1002/(ISSN)1099-145X.

Seleiman, M. F., Al-Selwey, W. A., Ibrahim, A. A., Shady, M., & Alsadon, A. A. (2023). Foliar applications of ZnO and SiO_2 nanoparticles mitigate water deficit and enhance potato yield and quality traits. *Agronomy*, *13*(2). Available from https://doi.org/10.3390/agronomy13020466, http://www.mdpi.com/journal/agronomy/.

Sha, Z., Li, Q., Lv, T., Misselbrook, T., & Liu, X. (2019). Response of ammonia volatilization to biochar addition: A meta-analysis. *Science of the Total Environment*, *655*, 1387−1396. Available from https://doi.org/10.1016/j.scitotenv.2018.11.316, http://www.elsevier.com/locate/scitotenv.

Shi, W., Ju, Y., Bian, R., Li, L., Joseph, S., Mitchell, D. R. G., Munroe, P., Taherymoosavi, S., & Pan, G. (2020). Biochar bound urea boosts plant growth and reduces nitrogen leaching. *Science of the Total Environment*, *701*, 134424. Available from https://doi.org/10.1016/j.scitotenv.2019.134424.

Shukla, P., Giri, B. S., Mishra, R. K., Pandey, A., & Chaturvedi, P. (2021). Lignocellulosic biomass-based engineered biochar composites: A facile strategy for abatement of emerging pollutants and utilization in industrial applications. *Renewable and Sustainable Energy Reviews*, *152*, 111643. Available from https://doi.org/10.1016/j.rser.2021.111643.

Simanský, V., Horak, J., & Bordoloi, S. (2022). Improving the soil physical properties and relationships between soil properties in arable soils of contrasting texture enhancement using biochar substrates: case study in Slovakia. *Geoderma Regional*, *28*.

Sun, Y., Xian, J., Chen, X., Cao, D., Yao, R., Luo, Y., & Zhang, X. (2023). Chemical fertilizer reduction potential estimation and fertilization optimization strategy based on a 10-year application summary and status questionnaires in a typical Yellow River irrigated area. *Agronomy*, *13*(8), 2047. Available from https://doi.org/10.3390/agronomy13082047.

Tarafdar, J. C., Raliya, R., Mahawar, H., & Rathore, I. (2014). Development of zinc nanofertilizer to enhance crop production in pearl millet (*Pennisetum americanum*). *Agricultural Research*, *3*(3), 257−262. Available from https://doi.org/10.1007/s40003-014-0113-y, http://www.springer.com/life + sciences/cell + biology/journal/40003.

Venugopalan, V. K., Nath, R., & Sarath Chandran, M. A. (2022). Smart fertilizers − a way ahead for sustainable agriculture. *Journal of Plant Nutrition*, *45*(13), 2068−2076. Available from https://doi.org/10.1080/01904167.2022.2046065.

Vlaskin, M. S., & Vladimirov, G. N. (2018). Hydrothermal carbonization of organic components from municipal solid waste. *Theoretical Foundations of Chemical Engineering*, *52*(6), 996−1003. Available from https://doi.org/10.1134/S0040579518050421, https://rd.springer.com/journal/11236.

Wang, C., Luo, D., Zhang, X., Huang, R., Cao, Y., Liu, G., Zhang, Y., & Wang, H. (2022). Biochar-based slow-release of fertilizers for sustainable agriculture: A mini review. *Environmental Science and Ecotechnology*, *10*, 100167. Available from https://doi.org/10.1016/j.ese.2022.100167.

Xiang, A., Qi, R., Wang, M., Zhang, K., Jiang, E., Ren, Y., & Hu, Z. (2020). Study on the infiltration mechanism of molten urea and biochar for a novel fertilizer preparation. *Industrial Crops and Products*, *153*. Available from https://doi.org/10.1016/j.indcrop.2020.112558, http://www.elsevier.com/inca/publications/store/5/2/2/8/2/5.

Xiao, L., & Meng, F. (2020). Evaluating the effect of biochar on salt leaching and nutrient retention of Yellow River Delta soil. *Soil Use and Management*, *36*(4), 740−750. Available from https://doi.org/10.1111/sum.12638, http://onlinelibrary.wiley.com/journal/10.1111/(ISSN)1475-2743.

Yaashikaa, P. R., Senthil Kumar, P., Varjani, S., & Saravanan, A. (2020). A critical review on the biochar production techniques, characterization, stability and applications for circular bioeconomy. *Biotechnology Reports*, *28*, e00570. Available from https://doi.org/10.1016/j.btre.2020.e00570.

Yan, P., Shen, C., Zou, Z., Fu, J., Li, X., Zhang, L., Zhang, L., Han, W., & Fan, L. (2021). Biochar stimulates tea growth by improving nutrients in acidic soil. *Scientia Horticulturae*, *283*, 110078. Available from https://doi.org/10.1016/j.scienta.2021.110078.

Ye, Z., Zhang, L., Huang, Q., & Tan, Z. (2019). Development of a carbon-based slow release fertilizer treated by bio-oil coating and study on its feedback effect on farmland application. *Journal of Cleaner Production*, *239*, 118085. Available from https://doi.org/10.1016/j.jclepro.2019.118085.

You, X., Suo, F., Yin, S., Wang, X., Zheng, H., Fang, S., Zhang, C., Li, F., & Li, Y. (2021). Biochar decreased enantioselective uptake of chiral pesticide metalaxyl by lettuce and shifted bacterial community in agricultural soil. *Journal of Hazardous Materials*, *417*, 126047. Available from https://doi.org/10.1016/j.jhazmat.2021.126047.

Yousaf, B., Liu, G., Ali, U., Abbas, M., Liu, Q., Ullah, Y., & Imtiyaz Cheema, A. (2021). Decisive role of vacuum-assisted carbonization in valorization of lignin-enriched lignin-enriched (Juglans regia-shell) biowaste. *Bioresource Technology*, *323*.

Zahra, Z., Habib, Z., Hyun, H., & Muhammad Aamir Shahzad, H. (2022). Overview on recent developments in the design, application, and impacts of nanofertilizers in agriculture. *Sustainability*, *14*(15), 9397. Available from https://doi.org/10.3390/su14159397.

Zhang, D., Pan, G., Wu, G., Kibue, G. W., Li, L., Zhang, X., Zheng, J., Zheng, J., Cheng, K., Joseph, S., & Liu, X. (2016). Biochar helps enhance maize productivity and reduce greenhouse gas emissions under balanced fertilization in a rainfed low fertility inceptisol. *Chemosphere*, *142*, 106–113. Available from https://doi.org/10.1016/j.chemosphere.2015.04.088, http://www.elsevier.com/locate/chemosphere.

Zhang, W., Tan, X., Gu, Y., Liu, S., Liu, Y., Hu, X., Li, J., Zhou, Y., Liu, S., & He, Y. (2020). Rice waste biochars produced at different pyrolysis temperatures for arsenic and cadmium abatement and detoxification in sediment. *Chemosphere*, *250*, 126268. Available from https://doi.org/10.1016/j.chemosphere.2020.126268.

Zhang, Y., Luo, J., Peng, F., Xiao, Y., & Du, A. (2021). Application of bag-controlled release fertilizer facilitated new root formation, delayed leaf, and root senescence in peach trees and improved nitrogen utilization efficiency. *Frontiers in Plant Science*, *12*. Available from https://doi.org/10.3389/fpls.2021.627313, https://www.frontiersin.org/journals/plant-science.

Zhen, M., Song, B., Liu, X., Chandankere, R., & Tang, J. (2018). Biochar-mediated regulation of greenhouse gas emission and toxicity reduction in bioremediation of organophosphorus pesticide-contaminated soils. *Chinese Journal of Chemical Engineering*, *26* (12), 2592–2600. Available from https://doi.org/10.1016/j.cjche.2018.01.028.

Zheng, J., Han, J., Liu, Z., Xia, W., Zhang, X., Li, L., Liu, X., Bian, R., Cheng, K., Zheng, J., & Pan, G. (2017). Biochar compound fertilizer increases nitrogen productivity and economic benefits but decreases carbon emission of maize production. *Agriculture, Ecosystems and Environment*, *241*, 70–78. Available from https://doi.org/10.1016/j.agee.2017.02.034, http://www.elsevier.com/inca/publications/store/5/0/3/2/9/8.

Zhu, J., Song, Y., Wang, L., Zhang, Z., Gao, J., Tsang, D. C. W., Ok, Y. S., & Hou, D. (2022). Green remediation of benzene contaminated groundwater using persulfate activated by biochar composite loaded with iron sulfide minerals. *Chemical Engineering Journal*, *429*. Available from https://doi.org/10.1016/j.cej.2021.132292, http://www.elsevier.com/inca/publications/store/6/0/1/2/7/3/index.htt.

CHAPTER

20 Biochar-based slow-release fertilizers for sustainable agriculture: a mechanistic overview on process development

Assima Firdoos[1], Mannat Parbhakar[2], Praveen Guleria[3] and Vineet Kumar[2]

[1]*Department of Botany, School of Bioengineering and Biosciences, Lovely Professional University, Phagwara, Punjab, India*

[2]*Department of Biotechnology, School of Bioengineering and Biosciences, Lovely Professional University, Phagwara, Punjab, India*

[3]*Department of Biotechnology, Faculty of Life Sciences, DAV University, Jalandhar, Punjab, India*

20.1 Introduction

The primary inputs that are administered in large quantities to boost food production for the growing population are chemical fertilizers. However, their widespread use has led to soil salinization, decreased fertilizer efficiency, and serious degradation of agricultural areas, all of which pose serious environmental risks (Khan et al., 2021). For instance, nitrogen (N) and phosphorus (P) fertilizers have been extensively used for agricultural practices as nutrients for promoting plant growth and development. Of the N fertilizer supplied only 20%–30% is consumed by plants (Naz & Sulaiman, 2016), and the rest undergoes various transformations in the environment such as ammonia volatilization, nitrous oxide release, and leaching as nitrate ions that impose serious environmental issues (Wen et al., 2017). Similarly, fertilizers containing P are discharged to the environment through runoff, erosion, and leaching (Li et al., 2020) and are a primary cause of eutrophication (Chen, Qin, et al., 2018). Moreover, the phosphate reserves (rocks) will be depleted in the coming 50–400 years, making P avalibility insufficient for plant growth applications (Van Dijk et al., 2016). Therefore, increasing crop yields without worsening ecological repercussions is a challenge as the world's demand for food production rises. This problem is addressed by incorporating slow-release fertilizers (SRFs) into agriculture system which uses a carbon-rich substance produced by the thermochemical transformation of biomass feedstocks to enhance soil health and lessen the environmental impact of traditional

Biochar Ecotechnology for Sustainable Agriculture and Environment. DOI: https://doi.org/10.1016/B978-0-443-29855-4.00036-9

fertilization (Wang et al., 2022; Zhang, Gang, et al., 2021). Moreover, it represents an exciting new way of managing nutrients in line with resource sustainability and environmental stewardship. Compared to conventional fertilizers, SRFs are fertilizers that following application tardily make their nutrients available for plant absorption thus extending their availability for plant uptake (Kim et al., 2017) and thereby minimizing environmental hazard (Hossain et al., 2020; Li et al., 2016). The inherent characteristics of biochar, such as its large surface area, porous structure, and diverse surface functional groups, allow it to efficiently absorb and retain nutrients (Wang et al., 2021). However, its properties are dependent on the feedstock used (Finalis et al., 2020), and synthesis conditions (Guo, 2015). Owing to its distinct characteristics, biochar has been studied for a wide range of uses, including carbon sequestration (Feng et al., 2020), nutrient delivery and yield enhancement (Melo et al., 2019), strengthening soil texture and assets (Yu et al., 2017), lowering greenhouse gas (GHG) emissions (Dong et al., 2021), immobilizing heavy metals/ organic pollutants (Qian et al., 2021), and garbage disposal (Wang et al., 2020). Furthermore, SRFs are far more efficient at using fertilizers than ordinary fertilizers, as they require in smaller amount (Zhong et al., 2013). These boosts crop nutrient recovery (Nelson et al., 2009). To prevent fertilizer loss into the environment, they are also made to release nutrients in accordance with the needs of the plants. The number of studies on biochar-based SRFs has steadily been rising, indicating a growing interest in the field.

Biochar-based SRFs work through a multifaceted mechanism, all of which hinge on the physicochemical relationships that exist between biochar, nutrients, and soil. The production atmosphere for biochar, soil microbial activity, pH, moisture content, and other variables can all affect these associations, which control the adsorption, retention, and eventual release of nutrients. In addition to helping crops develop steadily, these slow-release properties reduce nutrient runoff and leaching, which lessens the environmental problems associated with conventional fertilizers (Das & Ghosh, 2023). This overview will cover the process development to produce biochar-based SRFs, their significance in promoting sustainable agriculture, and an in-depth examination of the complex mechanisms that underlie their operation. Understanding these mechanisms will help us to fully comprehend the potential of biochar-based SRFs to promote future farming practices that are more ecologically friendly and productive.

20.2 Biochar production

Biochar, a carbon-rich material, is produced by using various types of waste materials, such as sludge, animal manure, food waste, green waste, and agricultural waste under anaerobic conditions (Zhang, Huang, et al., 2021). It can be produced by several traditional and modern methods including pyrolysis, gasification, torrefaction, and hydrothermal carbonization. In the process of pyrolysis, biomass undergoes thermal degradation in an anaerobic or a low-oxygen environment resulting in biochar and other byproducts. The optimal temperature range for pyrolysis is 300°C–700°C (Suárez-Abelenda et al., 2017) based on variable pyrolysis temperatures of the constituents (silica, cellulose, hemicelluloses, and lignin) present in biomass (Czernik & Bridgwater, 2004). Pyrolysis is further subdivided into slow pyrolysis, fast pyrolysis, flash pyrolysis, and vacuum pyrolysis. In slow pyrolysis, different feedstocks can be used to produce biochar with a heating range of 5–7°C/minute and at a temperature of 300°C–600°C (Lai et al., 2013) in an auger/screw pyrolizer reactor (Verma et al., 2012). In addition, fast pyrolysis is carried out with a heating rate of 300°C/minute and at a temperature above 500°C (Dai et al., 2017) by using a fixed-bed quartz reactor or auger pyrolysis reactor (Dong et al., 2015; Li et al., 2016). The third kind of pyrolysis is flash pyrolysis which is a modification of fast pyrolysis, where biomass is subjected to degradation at a higher heating rate of more than 1000°C/second and a higher temperature range of 900°C–1200°C (Li et al., 2013) by using a twin-screw mixing reactor and fluidized bed reactor. The last type of pyrolysis is vacuum pyrolysis which involves the thermal breakdown of biomass at a low pressure and temperature range of 0.05–0.20 MPa and 450°C–600°C, respectively, by using a vacuum (Tripathi et al., 2016). However, different types of pyrolysis yield different end products with variable amounts of biochar. Among this vacuum, pyrolysis generates high-quality biochar, with greater porosity (Uras et al., 2012). Besides pyrolysis, gasification is another technique for biochar production, this process involves the thermochemical conversion of biomass into biochar at a temperature greater than 700°C (Guan et al., 2016) in an environment with a controlled supply of oxygen and steam (Turner et al., 2008). Furthermore, another method for producing biochar is called torrefaction, which entails altering biomass by treating it at 230°C–300°C temperatures (Pentananunt et al., 1990) in the absence of oxygen under atmospheric pressure in a fixed bed reactor. This improves the biomass's performance for gasification and combustion (Tumuluru et al., 2011). Additionally, hydrothermal carbonization is another technique for biochar production using moisture-enriched feedstocks such as compost, sewage sludge, and animal waste (Titirici et al., 2007). These organic matters are subjected to heating at a temperature varying from 220°C to 240°C in the presence of 2–10 MPa pressure in rotary drums, stoves, and kilns.

Biochar exhibits a complex pore structure. Its chemical and physical properties are influenced by factors such as pyrolysis temperature, feedstock type, and retention time (Finalis et al., 2020; Guo, 2015). In recent years, biochar has gained a wide range of applications in the treatment of wastewater and contaminated soils. It has the ability to enhance water and nutrient holding capacity, mineralization, and cation exchange capacity of the soil. It serves as a soil amendment, positively impacting soil health and fertility status. However, there are not enough nutrients in the biochar produced by direct thermochemical treatment of biomass feedstocks, so to increase the fertilizer efficiency, biochar must be altered.

Incorporating different nutrients with biochar as a carrier has been found to increase nutrient use efficiency and crop growth while minimizing losses as compared to conventional fertilizers (Lateef et al., 2019). To alter biochar to convert them into SRFs, various methods have been used such as copyrolysis, impregnation, encapsulation, and granulation. Thus, the creation of SRFs offers a chance to lower the negative environmental effects of conventional fertilizers while also increasing their usage efficiency (Ye et al., 2019).

20.2.1 Copyrolysis

SRFs formation via copyrolysis is a new and efficient method for the formation of fertilizer-loaded biochar (Gao et al., 2018). It relies on the properties of biochar such as its carbon content, and porous nature with variable functional groups for the loading of fertilizer. The nutrients are bonded to the surface of biochar through adsorption, electrostatic attraction, and by the development of chemical bonds (An et al., 2020a). Specifically, copyrolyzing biomass with P sources offers a viable method for creating biochar-based SRFs (Fig. 20.1). This process allows for the incorporation of P into biochar via adsorption and the formation of chemical bonds. Advantages of copyrolyzing include straightforward production, enhanced P content, and extended nutrient release periods (An et al., 2020a). Biochar can be prepared from varied biological sources such as coffee husk, poultry litter, biogas residues, pine tree sawdust, switchgrass, etc. In a study, to

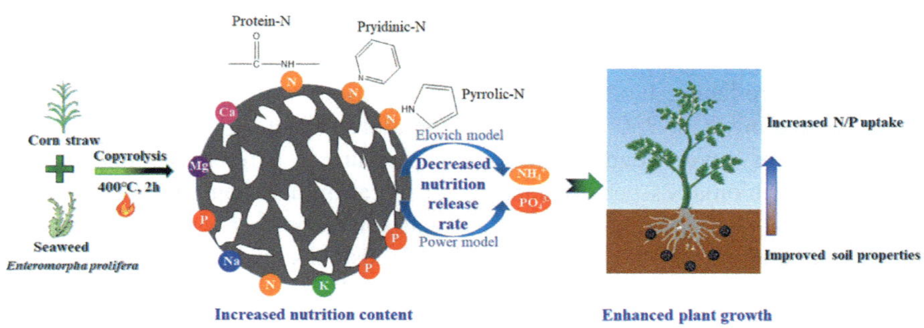

FIGURE 20.1 Copyrolysis of mixed biomass.

Biochar synthesized from lignocellulose-rich corn straw (CPECs) and *Enteromorpha prolifera* by copyrolysis. The CPECs possess higher N/P content, and surface area while it exhibited slow nutrient release rate. The modified biochar exhibited an enhancement in cherry tomato plant growth by altering the soil properties and enriching plant nutrient uptake. Biochar produced from lignocellulose-rich biomass through copyrolysis (Suo *et al.*, 2021).

From Suo, F., You, X., Yin, S., Wu, H., Zhang, C., Yu, X., Sun, R., & Li, Y. (2021). Preparation and characterization of biochar derived from co-pyrolysis of Enteromorpha prolifera *and corn straw and its potential as a soil amendment.* Science of the Total Environment, 798, *149167. https://doi.org/10.1016/j.scitotenv.2021.149167.*

synthesize biochar-concentrated phosphoric acid, magnesium oxide, and coffee husk were combined with poultry litter. Thereafter it was put to rest for several hours after adding water. The mixture was dried at 60°C, and subjected to pyrolysis for 2 hours at 500°C. When tested it was reported that total P emitted from them in an hour was around 6%–9% (Carneiro et al., 2021). Likewise, in another study, residues from biogas synthesizing fermenters were collected, air-dried, and then subjected to pyrolysis for 2 hours at 500°C under N-enriched conditions. The biochar synthesized was combined with commercial NPK fertilizers and converted to pallet form. SRF in pellet form possessed a very regulated release of nutrients (Yu et al., 2021). Similarly, Zhao et al. (2016) have reported that copyrolyzed biomass (pine sawdust or switchgrass) with phosphate fertilizers (triple superphosphate or bone meal) to synthesize SRFs that exhibit stable chemical structures such as C-O-PO_3 or C-P bonds with reduced P-release. Likewise, wheat straw copyrolysis with various P sources, such as phosphoric acid (H_3PO_4), phosphate rock, or triple superphosphate resulted in a decrease in carbon loss during pyrolysis. The stability of the biochar structure was improved due to less microbial breakdown of the biochar (Zhao et al., 2014). According to Karim et al. (2019), copyrolysis of P-rich effluent sludge with banana peduncles is a viable technique to produce SRFs that can be used on nonfood crops. Furthermore, SRF developed by copyrolyzing biomass, nutrients, and bentonite possess excellent slow-release performance of P and K compared to SRF lacking bentonite (An et al., 2020b). In some cases, mixing biomass with an acid phosphate source may lead to an acidic biochar product that is less suitable for soil application due to low pH levels. To mitigate this issue and enhance the overall nutrient profile of the resulting biochar, substances like MgO can be introduced before pyrolysis to adjust the pH level favorably while adding beneficial phosphates and magnesium (Mg). According to Lustosa Filho et al. (2017), impregnating poultry litter with triple superphosphate and phosphoric acid, while including or excluding MgO in the mix, resulted in SRFs that effectively reduced P release rates and augmented soil pH levels. These modified SRFs facilitated better P absorption and stimulated maize development as revealed by subsequent studies (Lustosa Filho et al., 2019). Moreover, when compared to traditional triple superphosphate applications, these tailored biochar-based SRFs yielded superior. Even though it has been observed that SRFs produced by copyrolyzing biomass along with other feedstocks comprising nutrients have outstanding slow-release nutrient performance, there is still a great need to further improve their efficiency by combining it with some other synthesis techniques.

20.2.2 Impregnation

The enhancement of nutrient content in biochar can be achieved through its impregnation in a nutrient solution. The innate porous nature and substantial surface area of biochar create reserves and binding sites for the retention of nutrients, thus extending their availability to plants within the biochar matrix. This impregnation technique is noted for its simplicity and cost-effectiveness. According to studies, the N–P–K nutrient solution can be prepared by dissolving urea, diammonium phosphate, and muriate of potash, or ammonium nitrate (NH_4NO_3) potassium dibasic phosphate (KH_2PO_4), and single super phosphate in deionized water (Das & Ghosh, 2023; Gwenzi et al., 2018). The synthesized biochar (pyrolysis) was then mixed with the nutrient solution to facilitate nutrient adsorption. Thereafter a starch–polyvinyl alcohol (PVA) (1:1) binder was mixed with N–P–K nutrient-impregnated biochar to obtain an even coating (Das & Ghosh, 2023). Following this process, the biochar is separated by filtration before being dried. The method has shown promising results in the creation of SRFs, with the aim of lessening nutrient depletion and promoting plant growth.

Biochar through impregnation can be synthesized using variable sources such as corn stover, cotton straw, corn straw, oil palm kernel shell, etc. Corn stover has been used to synthesize biochar by subjecting it to pyrolysis at 500°C for 30 minutes under a low oxygen environment, and is then immersed in a solution containing KH_2PO_4. The mixture is dried at 60°C. Furthermore, this fertilizer exhibited excellent slow-release performance by releasing only 40% of the bioavailable P after 5 days (Sepúlveda-Cadavid et al., 2021). Likewise, biochar can be synthesized using cotton straw by mixing cotton straw with bentonite and kept in deionized water for 45 minutes. They were then dried and subjected to pyrolysis for 2 hours at 600°C under a N environment. Biochar thus prepared was mixed with KH_2PO_4 aqueous solution for 24 hours at 120 rpm. Biochar synthesized this way exhibited an excellent slow-release performance of P with just 72.6% release for 15 days (An et al., 2020a).

The methodology of biochar impregnation is notably constrained due to the pristine biochar's minimal nutrient adsorption capability. To meet the demands of real-world application, refinements to biochar with various substances like MgO, K_2O, bentonite, and coal gangue have been instituted, augmenting its adsorption aptitude (An et al., 2020a; Chandra et al., 2020; Wu et al., 2019). Biochar cultivated from Mg-rich tomato residues has demonstrated a robust ability to recover phosphate from liquid solutions, achieving an uppermost adsorption rate surpassing 100 mg/g (Yao et al., 2013). Besides this, enhancements through MgO in biochar considerably boosted the uptake of phosphate and ammonium from effluent water (Li et al., 2017). Furthermore, Mg-augmented biochar produced from corn straw has registered superior results in recapturing nutrients from the outflow of biogas systems (Luo et al., 2021). Impressively, the cumulative release rate of N and P from this Mg-fortified biochar was sevenfold and sixfold less compared to synthetic fertilizers reflecting an efficient release of nutrients from these SRFs, respectively (Fig. 20.2). Therefore its application has remarkably fortified nutrient absorption and has been instrumental in fostering better crop development, showcasing its high potential for agricultural uses.

Urea-impregnated biochar (UIB) has been synthesized using peppercorn waste comprising urea:biochar (0.6:1) ratio, at a pyrolysis temperature of 400°C for 2 hours. The tapioca starch and palm oil were used to coat UIB to synthesize

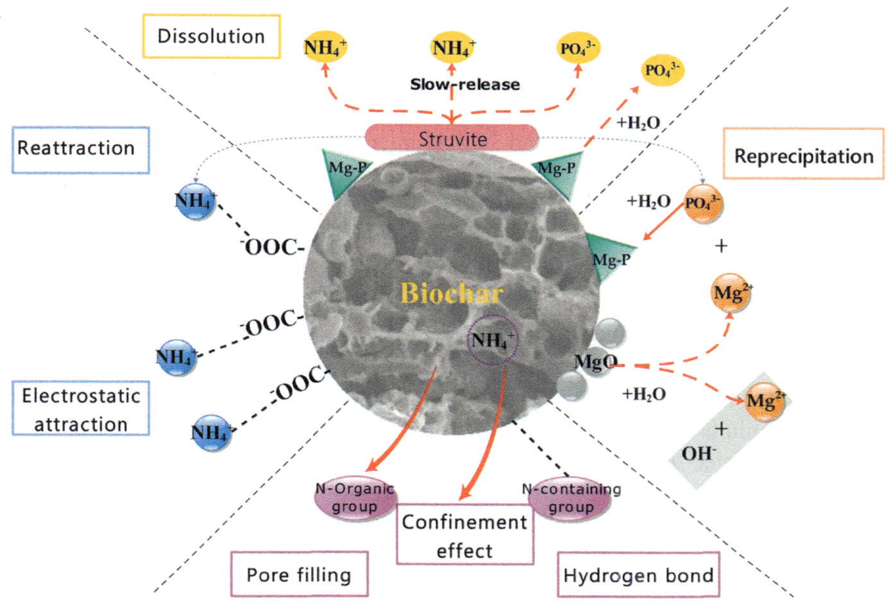

FIGURE 20.2 Mechanism involved in the slow release of nutrient from Mg-enriched biochar fertilizer.

Schematic representation of mechanism involved in the slow release of nutrient from Mg-enriched biochar fertilizer. Structural characteristics of biochar involved in the slow release of nutrients (Luo *et al.*, 2021)

Reproduced with permission from Luo, W., Qian, L., Liu, W., Zhang, X., Wang, Q., Jiang, H., Cheng, B., Ma, H., & Wu, Z. (2021). A potential Mg-enriched biochar fertilizer: Excellent slow-release performance and release mechanism of nutrients. Science of the Total Environment, 768, *Article 144454. https://doi.org/10.1016/j.scitotenv.2020.144454.*

encapsulated urea-impregnated biochar with an enhanced potential for a sustained and controlled release of fertilizers for agriculture.

20.2.3 Encapsulation

The encapsulating technique has recently been developed to overcome the drawback associated with other methods to achieve the persistent release of nutrients and to enhance nutrient use efficiency as it employs coating materials as diffusion barriers to regulate nutrient release. Encapsulation is a commonly employed technique for fabricating SRFs, which involves coating fertilizers onto either inorganic or organic substrates (Lawrencia et al., 2021). The efficiency of biochar-coated fertilizers varies depending on the properties of the biochar, the coating materials used, and the encapsulation processes employed. Biochar with a large surface area is advantageous for controlling nutrient release due to its strong adsorption capabilities (Sha et al., 2019). Polymers such as polyolefin, copolymers, polyvinylidene chloride, wax, and paraffin, can be used for coating; however, they are costly, nondegradable, and can negatively impact soil structure, suppress soil nutrient cycling, and inhibit microbial activity (Dong et al., 2020). Therefore the use of naturally biodegradable materials, such as waste palm oil, starch, and a blend of chitosan and PVA, is preferred for coating SRFs (Li et al., 2017; Tian et al., 2019; Vo et al., 2021). Several studies have explored the use of biochar-based SRFs prepared through encapsulation, such as biochar-biodegradable polymers-coated urea, biochar-waterborne copolymers-coated urea, and biochar-coated urea (Chen, Qin, et al., 2018; González et al., 2015; Puga et al., 2020). The biochar was prepared from cotton stalks heated at 600°C for 1 hour under N_2. Subsequently, sodium alginate (NaAlg)-g-poly, bentonite, biochar, ammonium chloride (NH_4Cl), and potassium phosphate (K_3PO_4) were added to distilled water and stirred at 40°C for 45 minutes. Then acrylic acid (AA), polyvinylpyrrolidone (PVP), *N*-nitro-*N*-methyl-4-aminobutyric acid (N-MBA), and ammonium persulfate (APS) solutions were added to the above mixture. They were then subjected to polymerization by microwave irradiation at 300 W for 4.5 minutes under N_2 as described in Fig. 20.3. The release ratios of N, P, and K within 30 days were less than 80.0% which can then be used as SRF (An et al., 2020c). Similarly Ye et al. (2019) used a hydrated spray system having a gas speed of 1.0–3.0 m/second and an atomizing air pressure of 0.1–0.5 MPa to

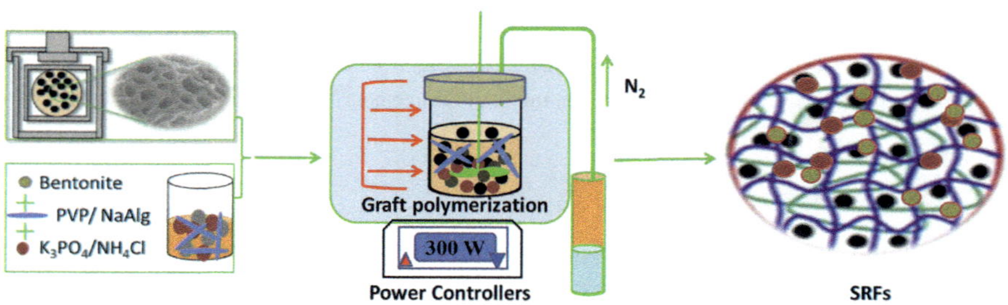

FIGURE 20.3 Biochar encapsulation with mineral nutrients.

The biochar was prepared from cotton stalks heated at 600°C for 1 h under N_2 thereafter sodium alginates (NaAlg)-g-poly, bentonite, biochar, ammonium chloride (NH_4Cl), and potassium phosphate (K_3PO_4) were added to distilled water and stirred at 40°C for 45 min. Then acrylic acid (AA), polyvinylpyrrolidone (PVP), N-nitro-N-methyl-4-aminobutyric acid (N-MBA), and ammonium persulfate (APS) solutions were added to the above mixture. The reaction mixture was then subjected to polymerization by microwave irradiation at 300 W for 4.5 min **under** N_2. Reproduced with permission from (An et al.,2020c).

An, X., Yu, J., Yu, J., Tahmasebi, A., Wu, Z., Liu, X. and Yu, B., 2020c. Incorporation of biochar into semi-interpenetrating polymer networks through graft co-polymerization for the synthesis of new slow-release fertilizers. Journal of cleaner production, 272, *122731.*

develop a bio-oil encapsulated biochar-based fertilizer. This fertilizer has improved the sustained release of nutrients in comparison to pristine biochar fertilizer. A combination of biochar and feedstocks of poultry manures and green waste was composted together, resulting in an organic coating enriched in nutrients, particularly N and potassium (K), formed on the surface and inner pores of biochar (Hagemann et al., 2017). By combining bentonite, PVA, and urea-loaded biochar, (Liu et al., 2019) created an encapsulated biochar-based fertilizer with higher water retention capacity and a lower rate of nutrient release. In another study, a SRFs was synthesized by graft-copolymerizing rape seed meal biochar and sodium alginate, with a significantly enhanced water retention capability, N content, and slow release of the graft copolymers (Baki & Abedi-Koupai, 2018). Another approach involved the preparation of biochar-hydrogel composites using a microwave-assisted method, which resulted in improved water absorbency and N release (Wu, Brickler, et al., 2021). Jia and colleagues developed biochar-coated urea using oxidized starch, which proved to be an effective slow-release N fertilizer, reducing N loss, and increasing N utilization efficiency (Jia et al., 2020). The SRF not only releases nutrients slowly but also helps in stabilizing soil characteristics and immobilizing hazardous heavy metals from the soil once the covering material has broken down. Additionally, the leftover biochar can be utilized as a material for carbon sequestration, lowering carbon dioxide emissions, and halting global warming. As a result of this, biochar-based and biochar-encapsulated fertilizers are superior to the majority of other SRFs.

20.2.4 Granulation

Granulation or palletization is a commonly used technique for fertilizer production (Fig. 20.4). This method involves the granulation of a mixture consisting of biochar, chemical fertilizer, and a binder by granulator or pelletizer (Sim et al., 2021). The storage and transportation of biochar for soil application pose significant challenges due to its brittleness, low density, and small particle size. Wind-blown or surface runoff during intense rain events results in a high loss of biochar, and the fine particles of biochar can irritate the human eyes, skin, and respiratory system (Chen, Qin, et al., 2018; Ye et al., 2019). To address these issues, granulation is an effective strategy to reduce biochar loss and handling costs for soil application, allowing for practical large-scale implementation in agricultural practices. Various binders, including molasses, ammonium nitrate (Briens & Bowden-Green, 2020), clay, starch, modified lignin, sepiolite, and bentonite (Xiang et al., 2020), have been utilized to support the palletization and granulation processes in biochar because of its hydrophobic nature (Zhong et al., 2018). In a study the biochar was prepared through pyrolysis using wheat straw; the material was grounded to pass through a 0.3 mm sieve then the biochar was mixed with diammonium phosphate, potassium chloride, and bentonite. The granulation was carried out using a revolving drum granulation device with steam vapor as a binder (Zheng et al., 2017). The resulting granulated biochar fertilizer

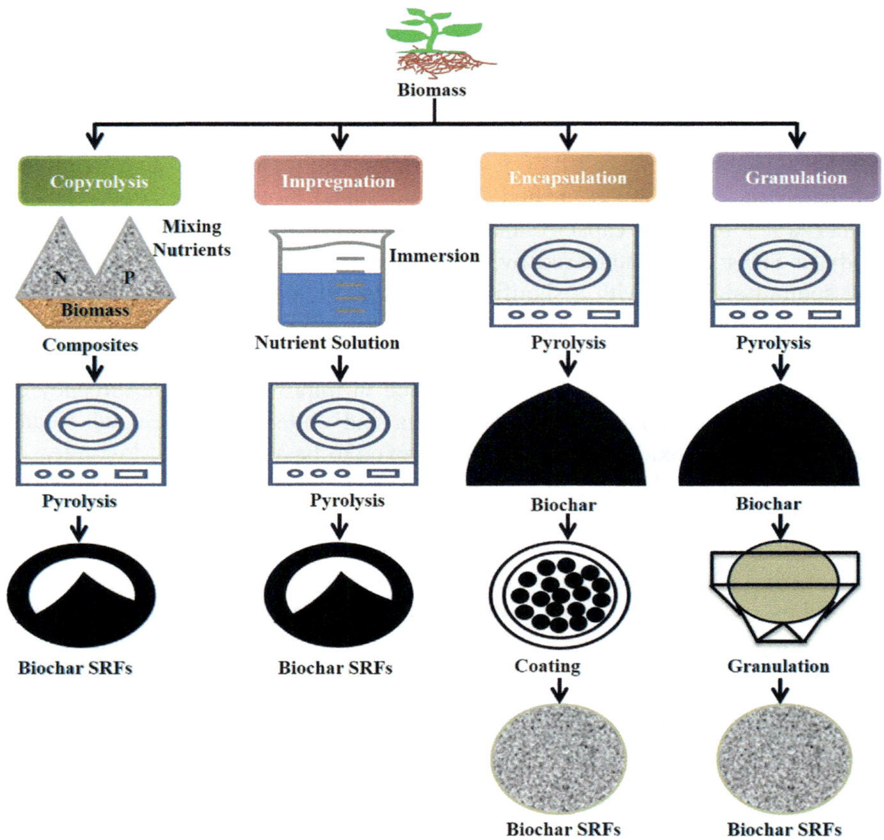

FIGURE 20.4

Different processes utilized for the synthesis of biochar-based slow-release fertilizer from biomass. Representing different processes utilized for the synthesis of biochar-based slow-release fertilizer. Synthesis of biochar-based slow-release fertilizer.

contained a total of 37% nutrients (18% N, 9% P_2O_5, and 10% K_2O). A field experiment demonstrated that the application of granular fertilizer increased grain yield by 10.7% and maize production by 46.2%. These beneficial effects were attributed to increased phosphorus supply, slow nutrient release, and improved moisture regime for maize growth. Granulation of biochar and nutrient mixtures effectively reduces nutrient release by increasing diffusion resistance. There are two ways to perform granulation for biochar-based SRFs: (1) carbonization of pelletized biomass containing nutrients and (2) pelletization of biochar/fertilizer mixtures. In the former method, additional nutrients need to be impregnated into the pelletized biochar due to the low nutrient content derived from biomass. For example, poultry litter and switchgrass were blended, pelletized, and pyrolyzed at different temperatures to produce biochar pellets (Cantrell & Martin, 2012). In the latter method, after a thorough mixing of biochar and fertilizer particles with uniform size, granulation is carried out using disckgranulation or mechanical extrusion, often employing binders in the process. Lignin and liquid fertilizer were mixed with biochar derived from switchgrass, and the mixtures were granulated using a pellet mill (Kim et al., 2014). Furthermore, biochar pellets produced by mixing biochar and manure compost proved effective in ammonium adsorption, resulting in improved plant growth (Yu et al., 2021). In one study, wheat straw biochar was mixed with diammonium phosphate, potassium chloride, and bentonite, and granulation was carried out using a revolving drum granulation device with steam vapor as a binder (Zheng et al., 2017). The resulting granulated biochar fertilizer contained a total of 37% nutrients (18% N, 9% P_2O_5, and 10% K_2O). A field experiment demonstrated that the application of granular fertilizer increased grain yield by 10.7% and maize production by 46.2%. These beneficial effects were attributed to increased P supply, slow nutrient release, and improved moisture regime for maize growth. The pellet properties depend on the characteristics of the biomass and the operating conditions. It has been reported that drying temperature has a positive impact on the mechanical properties and nutrient release of granular fertilizer (Kim et al., 2014). To synthesize the binder-free granular fertilizer, biogas residue biochar was mixed with commercial fertilizer and then compressed using a compression mold (Yu et al., 2021). The granulation process was optimized using response surface methodology. The physical properties of the pellets were found to be optimal at a diameter of 7 mm, moisture content of 7.84%, and compression speed of 49.54 mm/minute. The granular fertilizer demonstrated excellent sustained nutrient release, with a slow-release effect showing a preference for $P > N > K$. Although water or steam can also be used for granulation, the mechanical performance of binder-free granular biochar for application is poor. To enhance the mechanical strength of biochar-based SRFs, binders are employed. However, a lot of organic binders are hazardous to the environment because they degrade poorly and obstruct the pores in the biochar, which stops the biochar from benefiting the soil environment when fertilizer is released (Shi et al., 2020). Therefore further research is required to develop granular fertilizers to improve their efficiency and widespread application to achieve sustainability in food production.

20.3 Nutrient loading and release mechanism

The physical structure and varied chemical features of biochar, such as its porosity and range of functional groups, allow it to hold different nutrients through both chemical and physical means of adsorption (Lateef et al., 2019). It serves as an effective carrier for essential nutrients like N, P, and K because of its high capacity to adsorb nitrate, ammonium, and phosphate ions. Various techniques, such as impregnation, encapsulation, granulation, and copyrolysis, have been developed to prepare biochar-SRFs. These methods differ in terms of the mechanism for nutrient loading. Impregnation heavily relies on the characteristics of biochar, such as its large surface area, porous structure, and abundant surface groups. These properties provide favorable locations for nutrient attachment by adsorption. Granulation, in contrast, aims to increase resistance to diffusion and decrease nutrient leaching. Encapsulation involves embedding nutrients primarily within the internal sites of SRFs through diverse interactions. Thus, gaining a deeper understanding of the essential characteristics and nutrient-loading mechanisms of biochar-based SRFs can significantly enhance their overall performance. Biochar exhibits complex interactions with different macronutrients, such as N (urea, NH_4^+, NO_3^-), K^+, and P (PO_4^{3-}), as well as micronutrients, such as zinc (Zn^{2+}), manganese (Mn^{2+}), and copper (Cu^{2+}). According to Liu et al. (2019), examples of polar and nonpolar chemisorption processes that aid in the sorption of nutrients into biochar include feeble H-bonds, $\pi-\pi$ electron donor−acceptor interactions, and hydrophobic bonding. Mg^{2+}, Ca^{2+}, Al^{3+}, and Fe^{3+} are the various positive ions found on the biochar's surface that interact with phosphorus ions by forming bridge bonds, thereby retaining phosphorus on the biochar surface (Ahmad et al., 2018). In a separate study, it was observed that phosphorus binds to biochar during pyrolysis through chemical bonding with the carbon and oxygen atoms of biochar (Zhao et al., 2016). The adsorption mechanism for reclaiming phosphorus from water is primarily driven by the precipitation and deposition of phosphorus on Mg particles in biochar (Yao et al., 2013). According to Li et al. (2016), the processes of precipitation, electrostatic attraction, ligand exchange, and metallic complexation between surface groups and metals were thought to be responsible for P adsorption onto MgO-modified magnetic biochar. Nevertheless, Chen et al. (2017) discovered that surface precipitation mostly regulated the phosphorus adsorption onto cow dung biochar. Meanwhile, surface precipitation, ligand exchange, and electrostatic adsorption were found to be the primary processes for phosphate adsorption onto coal gangue-modified biochar (Wang et al., 2021). When Zhang, Gang, et al. (2021) used a goethite/biochar composite to recover phosphate from synthetic urine, they found that phosphate adsorption occurred through the formation of inner-sphere complexes (Fe−O−P) and Fe−OH bonds. So, different scientists have found different perspectives for phosphorus adsorption on biochar surfaces and it needs further exploration. Likewise for N, its variable sources were reported to bind differently. Urea interacts with oxygen-containing groups in biochar through the creation of hydrogen bonds through $-NH_2$ and $C=O$ groups (Zhang et al., 2015). Whereas, Wen et al. (2017) showed that ammonium interacts with oxygen groups in biochar via both electrostatic interaction and hydrogen bonding. Regarding K, it has been documented to engage in interactions with a variety of function groups on the surface of biochar, such as $-COO^-$, COOH, R-O^-, SiO_4, and OH via means of electrostatic attraction, coprecipitation, surface complexation, and inner sphere complexation (Ye et al., 2019). Whereas, PO_4^{3-} and NO_3^- adsorbs through electrostatic interaction and hydrogen bond onto biochar surface (Ye et al., 2019). According to Chen et al. (2017), the absorption of essential nutrients onto biochar/montmorillonite composites revealed that phosphate (PO_4^{3-}) adhesion was primarily mediated by electrostatic attraction or ionic bonding within PO_4^{3-} and metal cations, whereas ammonium adsorption was mainly determined by surface adsorption linked to ion exchange. In the case of micronutrients, Mn^{2+} was bound to the biochar surface through physical adsorption, precipitation, and cation exchange, whereas Cu^{2+} and Zn^{2+} bound via physical adsorption, cation−π bond, cation exchange, and precipitation (Deng et al., 2020). Furthermore, Zhao et al. (2016) anticipated a copyrolysis approach that uses biomass and P fertilizers to produce biochar-based SRFs.

Lower phosphorus release was ascribed to the establishment of stable C−O−P or C−P bonds. An et al., 2020a synthesized modified biochar by copyrolyzing bamboo and bentonite. The formation of a porous structure, a decrease in surface negative charge, and precipitation brought about by the addition of calcium and magnesium were all considered to be responsible for the improved phosphorus adsorption and discharge efficiency of the bentonite-modified biochar. Moreover, the phosphorus release kinetics was influenced by the bentonite quantity in a comparable manner. Coated SRFs employ barrier materials on the surface of fertilizers to hinder nutrient diffusion (An et al., 2020a). The process of releasing nutrients from these coated SRFs usually entails three steps: introducing water into the fertilizer's interior, dissolving nutrients, and releasing nutrients through the coating (Mikula et al., 2020). Co-composted biochar has been shown to develop an outer and inner layer of nutrient-rich organic coating that improves mesoporosity, hydrophilicity, and redox-active moieties to retain more water and nutrients. When nutrient release from blended and coated SRFs was evaluated, the blended biochar-based SRFs displayed comparatively poor P sluggish release in contrast to coat, which was constrained by swelling/relaxation and diffusion (Pogorzelski et al., 2020). The porosity and surface functional groups of the biochar varied according to the type of feedstock and synthesis circumstances. Thus, understanding the nutrient bonding mechanism may help in tailoring the properties of the biochar to optimize the nutrient retention capacity when synthesizing biochar-based SRFs.

The usual methods for evaluating the release behavior of nutrients include column leaching in soil or sand, water, or solution media (An et al., 2020b; Khan et al., 2021; Luo et al., 2021). It is difficult to compare the findings published by many studies because of the variations in techniques, procedures, and equipment used. The transformation of nutrients at the elemental level after applying them to soil remains inadequately investigated. The impact of biochar and its special qualities on the biochar-based fertilizer are frequently mentioned by researchers (Dong et al., 2020; Jia et al., 2020; Lateef et al., 2019; Petrus et al., 2020), but they seldom delve into the scientific explanation. Moreover, there is little information available regarding how soil conditions affect nutrient release in biochar-based SRFs. The role of biochar in the controlled release of nutrients has been the subject of some investigation (Jia et al., 2020; Luo et al., 2021). Kinetic modeling of nutrient release has also been modeled using a variety of mathematical models, including the Korsmeyer-Peppas model, Hixson-Crowell model, and the Baker-Lansdale model (Dias et al., 2018; Sim et al., 2021). In comparison to unmodified fertilizer, Jia et al. (2020) investigated the delayed release of nitrates from biochar-coated urea fertilizer that revealed much less leaching. The authors elucidated that the vast porous structure of biochar, along with the existence of functional groups on its surface, expedites N binding via adsorption. N release into the soil is impeded by these physiochemical interactions as well as the physical obstacle the porous structure creates. Wu, Yan, et al. (2021) explained that the porous structure of biochar and, more precisely, the internal forces that stabilize the functional groups holding N and oxygen (primarily phenolic groups), are responsible for the gradual nutrient release from biochar-based fertilizer. Wang et al. (2021) further investigated the process of the gradual release of N from urea-biochar fertilizer and formulated three different phases for the release of N. Labile forms of N are released from urea that has been spread throughout the surface of the biochar in the first stage. N bounded by intermolecular interactions and hydrogen bonds with biochar were freed in the second stage. Whereas the N bound by strong chemical bonds is released in the last phase. Furthermore, the diffusion of nutrient molecules along a concentration gradient and their solubility in the soil were reported to be the main mechanisms responsible for the release of nutrients from the biochar-based fertilizers (An et al., 2020c, 2021; Sim et al., 2021).

In addition, an in-depth study on the mechanisms was done by Luo et al. (2021), who also put out some significant recommendations. They found that the poor solubility of Mg-phosphate precipitates generated on the surface of the biochar was found to correlate with the sluggish release of P from the Mg-enriched biochar fertilizer. However, the confinement effect and the electrostatic interaction between ammonium ions and the negatively charged surface of the biochar controlled the process of the delayed release of N. Moreover, N compounds might form hydrogen bonds with functional groups on the surface of the biochar, which could potentially delay N desorption. In light of this, using biochar-based SRFs is an innovative method of fertilizing soil, and therefore, it is crucial to thoroughly investigate the mechanisms underlying their efficacy, particularly in the context of environmental conditions, and release kinetics under varying conditions.

20.4 Soil and plant health improvement

Many biochar-based SRFs have been made using biochar as an adsorber as it is a capable support material for their synthesis (Das et al., 2015). Biochar-based SRFs alter the soil properties to make it suitable for the growth of different crops (Fig. 20.5). According to Chan et al. (2007), biochar is known to raise soil pH,

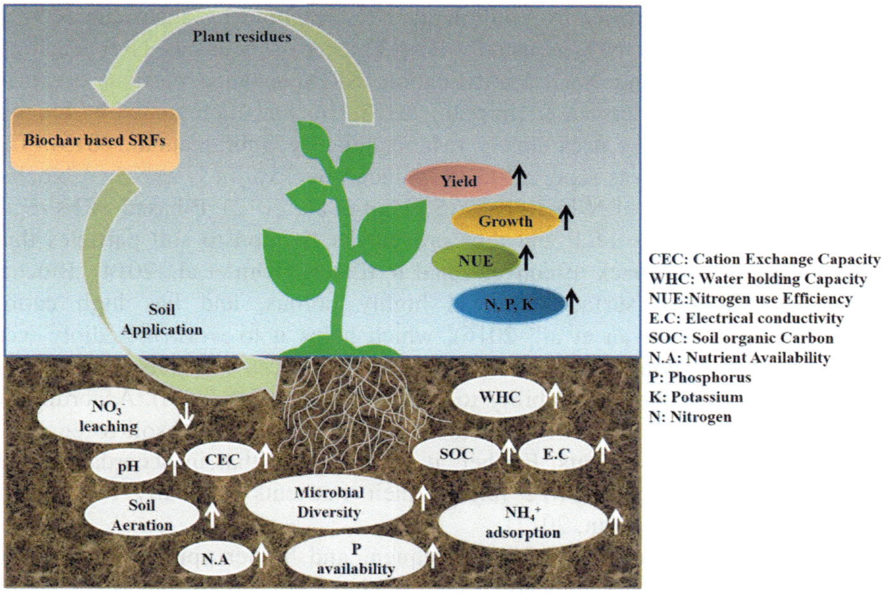

FIGURE 20.5

Mechanism of action illustrating the impact of biochar-based slow-release fertilizers on soil and plant health.

which improves the availability of nutrients for crop planting in acidic soils. Soil pH is a significant element in influencing the bioavailability of nutrients. However, the use of biochar in alkaline soil is quite challenging because biochar is alkaline; however, according to Ippolito et al. (2012), biochar produced at low photolysis temperatures may have a low pH, which could enhance environmental quality by lowering nutrient losses in calcareous soils. A biochar composite has been formulated to reduce soil pH and boot nutrients bioavailability (Ippolito et al., 2015). So, biochar as a nutrient carrier could provide an environmentally friendly way to address the issues associated with alkaline soils.

SRFS reduces N_2O emission in contrast to conventional N-fertilizers (Trinh et al., 2017). A reduction in the production of N_2O by 21%–56% has been reported in rice fields by biochar supplementation (Zhang et al., 2016). Likewise, UIB enriched soil C-content and minimized the production of NH_3 and N_2O (Dimin et al., 2014). Grutzmacher et al. (2018) reported that coapplication of NH_4NO_3 and biochar reduced N_2O emission by 69%–96%. Furthermore, SRFs also reduce NH_3 volatilization, around 43% less NH_3 was volatilized when urea was encapsulated into eucalyptus biochar (Paiva et al., 2012). This could be attributed to the direct adsorption by biochar or by promoting the soil microbial activity that in turn immobilizes N (Taghizadeh-Toosi et al., 2012). Biochar supplementation further assists in carbon sequestration and enriching soil fertility by minimizing NO_3 leaching by adsorbing NO_3 as soil's ability to retain NO_3 is usually less. Thus, by preventing NO_3 leaching, NH_3 volatilization, and N_2O denitrification, SRFs enhance nutrient use efficiency by plants (Biederman & Harpole, 2013). Biochar impregnated by biochar was reported to release urea slowly (Manikandan & Subramanian, 2015). The nitrification process was reported to be reduced by 75% in N-enriched biochar thus delaying the rate of N-hydrolysis (Sarkhot et al., 2012). P-based SRFs regulate the bioavailability of P and thus prevent its fixation to soil particles thus overcoming the drawback of conventional fertilizers (Kim et al., 2014). Biochar possesses extensive surface area, is highly porous, and has high cation exchangeability (Rizwan et al., 2016), which helps it to work incredibly well as an adsorber (Thomas et al., 2013) hence it enhances soil nutrient retention (N, P, and K) and their availability to plants (Major et al., 2010). According to Rop et al. (2019), SRFs limit the solubility of nutrients in the soil to ensure a continuous supply for plants. Further, it was reported that in accordance with the crop's requirements, SRFs release their nutrients gradually and slowly (Lubkowski & Pomeranian, 2017).

Furthermore, by using SRFs the frequent and higher application of conventional fertilizers was reduced (Buckthought et al., 2015), thus SRFs lessen the ill effects of traditional fertilizers such as GHGs emission, nutrient overflow, and leaching (Das and Ghosh, 2023; Zhang et al., 2019). Further, its porous structure enhances soil aeration and water retention (Soliman et al., 2023), which contributes to the improvement of soil fertility and enhanced root growth. Like during temperature stress, when there is a shortage of water, it helps to retain water resulting in improved hydration status of plants (Karhu et al., 2011). Das and Ghosh (2023) developed NPK SRFs and it reported to enrich soil pH, bioavailability of nutrients, and carbon content, slowly releasing the impregnated nutrients in the soil and thus improving nutrient use efficiency and minimizing environmental risk by reducing leaching of essential nutrients.

In addition to the impact of biochar-based SRFs on soil health, they have been reported to regulate the growth and development of many crops. *Brassica napus* yield was promoted by the supplementation of biochar-based compound fertilizer by modulating its nutrient use efficiency, activity of N-assimilation enzymes, and root activity (Liao et al., 2020). NPK SRFs improved the yield of maize and black gram (Das & Ghosh, 2023). The grain and stover yield of maize was promoted by nutrient-enriched biochar by modulating the nutrient retention of soil (Eazhilkrishna, 2017). Similarly, rice yield, height, grain quality, and chlorophyll content were improved by biochar-based fertilizers (Huang et al., 2019). Cotton biomass was also reported to be promoted by biochar-based SRF (Wen et al., 2017). Other studies related to the impact of biochar-based SRFs on soil and plant health are listed in Table 20.1. Thus biochar-based SRFs can be used efficiently to improve the soil quality and plant growth without harming the environment.

Table 20.1 Represents feedstock and synthesis methods used for preparing different biochar-derived slow-release fertilizers, and their impact on soil and plant health.

Feedstock used	Synthesis technique used for fertilizer	Nutrients released	Nutrient source	Outcome	References
Maize stalk, black gram, pine needle, and *Lantana camara*	Impregnation granulation	N, P, and K	N–P–K nutrient solution (urea, diammonium phosphate, and muriate of potash)	Reduced nutrient release, and leaching into soil. Improved soil quality by increasing available nutrients, oxidizable carbon, and altering soil pH. Improved nutrient uptake efficiency, and overall crop yield in maize and black gram.	Das and Ghosh (2023)
Corncob	Impregnation	Macro- (N, P, K, Ca, Mg) and micronutrients (Fe, Na, Zn)	Macro- and micronutrients	Releases nutrients gradually, perk up soil fertility, halt the volatilization, and leaching of nutrients	Lateef et al. (2019)
Corn stover	Impregnation	P	KH_2PO_4	Enhanced soil P bioavailability, along with increased protein, ascorbic acid content, P, N, and K levels, as well as yield	Sepúlveda-Cadavid et al. (2021)
Corn stover	Impregnation, granulation, encapsulation	N	Ammonium sulfate	Enhanced rate and duration of N release in soil	Cen et al. (2021)
Corn stalk	Copyrolysis Impregnation	N and P	Biogas effluent and KH_2PO_4	Reduce the N and P release as compared to chemical fertilizer. Promoted nutrient uptake efficiency and the corn growth	Luo et al. (2021)

(Continued)

Table 20.1 Represents feedstock and synthesis methods used for preparing different biochar-derived slow-release fertilizers, and their impact on soil and plant health. *Continued*

Feedstock used	Synthesis technique used for fertilizer	Nutrients released	Nutrient source	Outcome	References
Cotton straw	Copyrolysis	P	$Mg_3(PO_4)_2$	Reduce the release of nutrient. Promoted water retention, nutrient uptake efficiency, and the growth of pepper seedlings	An et al. (2020a)
Cotton stalk	Impregnation encapsulation	P	KH_2PO_4	Promoted the early growth of maize	An et al. (2020b)
Cotton straw	Impregnation, granulation, encapsulation	N	Ammonium nitrate	Reduce the nutrient release (nitrate and ammonium). Promoted water retention, and N content in soil and plant. Enhanced plant growth, leaf area, pigment content, and N use efficiency.	Khajavi-Shojaei et al. (2023)
Palm kernel shell	Impregnation	N and P	NH_4NO_3 and KH_2PO_4	Exhibited maximum nutrient release, and carbon sequestration potential	Dominguez et al. (2020)
By-product of straw and wheat shell used for bio-oil synthesis	Impregnation	N and P	Magnesium ammonium orthophosphate hexahydrate ($MgNH_4PO_4{\cdot}6H_2O$)	Exhibited great potential for slow release of N, P, and Mg for a longer period	Hu et al. (2019)
Pine biomass	Encapsulation, granulation	N	Urea	Regulates urea release, minimized N release, and NH_3-N volatilization and thus improved N-use efficiency	Bakshi et al. (2021)

Wood chips	Granulation, encapsulation	P	Triple superphosphate fertilizer TSP-Ca $(H_2PO_4)_2$, commercial source	Promoted P use efficiency in high P-fixing tropical soils, P recovery, and uptake when compared with triple superphosphate	Pogorzelski et al. (2020)
Wheat straw	Granulation encapsulation	N, P, and K	KH_2PO_4 and KNO_3	Reduced the nutrients release rate in water, and leaching rate in soil	Ye et al. (2019)
Maize straw, rice straw, and forest litter	Encapsulation	N	Urea	Improved the slow-release property of coated urea and nutrient leaching	Chen, Yang, et al. (2018)
Cow dung	Encapsulation	P	Potassium phosphate monobasic (KH_2PO_4)	Enhanced soil fertility, pH, organic carbon, accessible P, and moisture content. Whereas, enhances the germination of seeds, growth, yield, and P concentration in lettuce.	Chen, Qin, et al. (2018)
Coffee husk and poultry litter	Copyrolysis	P	Triple superphosphate	Mombasa grass, maize, and common beans, demonstrated greater or comparable biomass yields, relative agronomic effectiveness, and P uptake	Carneiro et al. (2021)
Biogas residue-based biochar	Granulation	N, P, K	Compound fertilizer ($N:P_2O_5:K_2O = 4:3:3$)	Promoted N, P, and K sustained release properties	Yu et al. (2021)
Pine tree sawdust and switchgrass	Copyrolysis	P	Triple superphosphate and bone meal	Decreased P release, enhanced carbon retention, and biochar capability for stabilizing metals in soil such as lead, copper, and cadmium by immobilizing them through precipitation	Zhao et al. (2016)

(*Continued*)

Table 20.1 Represents feedstock and synthesis methods used for preparing different biochar-derived slow-release fertilizers, and their impact on soil and plant health. *Continued*

Feedstock used	Synthesis technique used for fertilizer	Nutrients released	Nutrient source	Outcome	References
Wood shavings, medicinal plant, and willow	Granulation	N	Urea superphosphate granules	Compared to fertilizer without BC layer, the wheat yield increased by 35%, and compared to commercial NPK fertilizer, it improved by 55%	Mikos-Szymańska et al. (2019)
Rice husks	Encapsulation	N	Ureic fertilizer	Slower N release, enhance the ability of fertilizer to hold N	Petrus et al. (2020)
Rice straw, chicken manure, *Phyllostachys pubescens*, vinasse, *Arundo donax*, and sugarcane bagasse	Encapsulation	N	Urea	Decrease in N elution and release	Jia et al. (2020)
Rice straw	Encapsulation	N	Compound fertilizer (15% N, 15% P_2O_5, 15% K_2O) and N as $(NH_4)_2HPO_4$ and NH_4NO_3	Reduction of nitrogen leaching, elevation of rice stems height, and improvement in SPAD value	Dong et al. (2020)
Poultry litter	Pyrolysis	P	Triple superphosphate and phosphoric acid (H_3PO_4)	Enriches soil P content by its slow-release capability, and by restricting its fixation in soil thus promoted better P use efficiency by plants leading to their improved growth	Lustosa Filho et al. (2020)
Sugarcane residue	Impregnation, granulation	P	H_3PO_4	Sugarcane seedlings exhibit increased K and P content, plant yields, and P use efficiency	Borges et al. (2020)

N, nitrogen; P, phosphorus; K, potassium; Ca, calcium; Fe, iron; Na, sodium; Zn, zinc; Mg, magnesium.

20.5 Conclusion

SRFs based on biochar have the potential to enhance soil health by boosting soil structure, microbial activity, nutrient retention, and carbon sequestration while reducing environmental hazards including GHG emissions and nutrient leaching. Sustainable waste management is also aided by the production of fertilizers based on biochar from diverse waste materials. However, to optimize their advantage and minimize any potential disadvantage, a comprehensive evaluation of variables including biochar type, application rate, and soil condition is required. There have been very few field studies to gauge biochar's potential as SRF for the crop improvement, the mechanism for effective utilization of nutrients, and nutrient adsorption and desorption mechanism. Therefore, additional studies are required to fully comprehend the mechanisms of biochar's nutrient sorption and desorption, and its effect on plant health in field conditions, which will ultimately increase the effectiveness of SRFs.

Acknowledgments

AF, MP, and VK are thankful to Lovely Professional University for providing an encouraging environment. PG would like to thank DAV University, Jalandhar for their encouragement to carry out research.

Author contribution

AF—writing-original draft, literature survey. MP—writing-original draft, literature survey. VK—conceptualization, and editing, supervision. PG—conceptualization, editing, supervision.

Conflict of interest

The authors declare that they have no conflict of interest.

Use of AI tools

Not applicable.

References

Ahmad, M., Usman, A. R. A., Al-Faraj, A. S., Ahmad, M., Sallam, A., & Al-Wabel, M. I. (2018). Phosphorus-loaded biochar changes soil heavy metals availability and uptake potential of maize (*Zea mays* L.) plants. *Chemosphere*, *194*, 327–339. Available from https://doi.org/10.1016/j.chemosphere.2017.11.156.

An, X., Wu, Z., Yu, J., Ge, L., Li, T., Liu, X., & Yu, B. (2020a). High-efficiency reclaiming phosphate from an aqueous solution by bentonite modified biochars: a slow release fertilizer with a precise rate regulation. *ACS Sustainable Chemistry & Engineering*, *8* (15), 6090–6099.

An, X., Wu, Z., Yu, J., Cravotto, G., Liu, X., Li, Q., & Yu, B. (2020b). Copyrolysis of biomass, bentonite, and nutrients as a new strategy for the synthesis of improved biochar-based slow-release fertilizers. *ACS Sustainable Chemistry & Engineering*, *8*(8), 3181–3190.

An, X., Yu, J., Yu, J., Tahmasebi, A., Wu, Z., Liu, X., & Yu, B. (2020c). Incorporation of biochar into semi-interpenetrating polymer networks through graft co-polymerization for the synthesis of new slow-release fertilizers. *Journal of cleaner production*, *272*, 122731.

An, X., Wu, Z., Shi, W., Qi, H., Zhang, L., Xu, X., & Yu, B. (2021). Biochar for simultaneously enhancing the slow-release performance of fertilizers and minimizing the pollution of pesticides. *Journal of Hazardous Materials*, *407*124865. Available from https://doi.org/10.1016/j.jhazmat.2020.124865.

Baki, M., & Abedi-Koupai, J. (2018). Preparation and characterization of a superabsorbent slow-release fertilizer with sodium alginate and biochar. *Journal of Applied Polymer Science*, *135*(10). Available from https://doi.org/10.1002/app.45966.

Bakshi, S., Banik, C., Laird, D. A., Smith, R., & Brown, R. C. (2021). Enhancing biochar as scaffolding for slow release of nitrogen fertilizer. *ACS Sustainable Chemistry and Engineering*, *9*(24), 8222–8231. Available from https://doi.org/10.1021/acssuschemeng.1c02267.

Biederman, L. A., & Harpole, W. S. (2013). Biochar and its effects on plant productivity and nutrient cycling: A meta-analysis. *GCB Bioenergy*, *5*(2), 202–214. Available from https://doi.org/10.1111/gcbb.12037.

Borges, B. M. M. N., Strauss, M., Camelo, P. A., Sohi, S. P., & Franco, H. C. J. (2020). Re-use of sugarcane residue as a novel biochar fertiliser - Increased phosphorus use efficiency and plant yield. *Journal of Cleaner Production*, *262*. Available from https://doi.org/10.1016/j.jclepro.2020.121406.

Briens, L., & Bowden-Green, B. (2020). A comparison of liquid binders for drum granulation of biochar powder. *Powder Technology*, *367*, 487–496. Available from https://doi.org/10.1016/j.powtec.2020.03.065.

Buckthought, L. E., Clough, T. J., Cameron, K. C., Di, H. J., & Shepherd, M. A. (2015). Fertiliser and seasonal urine effects on N2O emissions from the urine-fertiliser interface of a grazed pasture. *New Zealand Journal of Agricultural Research*, *58*(3), 311–324. Available from https://doi.org/10.1080/00288233.2015.1031405.

Cantrell, K. B., & Martin, J. H. (2012). *Poultry litter and switchgrass blending and pelletizing characteristics for biochar production*, *American society of agricultural and biological engineers annual international meeting 2012* (3, pp. 2047–2059). ASABE. Available from 10.13031/2013.42276.

Carneiro, J. S. D. S., Ribeiro, I. C. A., Nardis, B. O., Barbosa, C. F., Lustosa Filho, J. F., & Melo, L. C. A. (2021). Long-term effect of biochar-based fertilizers application in tropical soil: Agronomic efficiency and phosphorus availability. *Science of the Total Environment*, *760*. Available from https://doi.org/10.1016/j.scitotenv.2020.143955.

Cen, Z., Wei, L., Muthukumarappan, K., Sobhan, A., & McDaniel, R. (2021). Assessment of a biochar-based controlled release nitrogen fertilizer coated with polylactic acid. *Journal of Soil Science and Plant Nutrition*, *21*(3), 2007–2019. Available from https://doi.org/10.1007/s42729-021-00497-x.

Chan, K. Y., Van Zwieten, L., Meszaros, I., Downie, A., & Joseph, S. (2007). Agronomic values of greenwaste biochar as a soil amendment. *Soil Research*, *45*(8), 629. Available from https://doi.org/10.1071/sr07109.

Chandra, S., Medha, I., & Bhattacharya, J. (2020). Potassium-iron rice straw biochar composite for sorption of nitrate, phosphate, and ammonium ions in soil for timely and controlled release. *Science of the Total Environment*, *712*136337. Available from https://doi.org/10.1016/j.scitotenv.2019.136337.

Chen, L., Chen, X. L., Zhou, C. H., Yang, H. M., Ji, S. F., Tong, D. S., Zhong, Z. K., Yu, W. H., & Chu, M. Q. (2017). Environmental-friendly montmorillonite-biochar composites: Facile production and tunable adsorption-release of ammonium and phosphate. *Journal of Cleaner Production*, *156*, 648–659. Available from https://doi.org/10.1016/j.jclepro.2017.04.050.

Chen, Q., Qin, J., Sun, P., Cheng, Z., & Shen, G. (2018). Cow dung-derived engineered biochar for reclaiming phosphate from aqueous solution and its validation as slow-release fertilizer in soil-crop system. *Journal of Cleaner Production*, *172*, 2009–2018. Available from https://doi.org/10.1016/j.jclepro.2017.11.224.

Chen, S., Yang, M., Ba, C., Yu, S., Jiang, Y., Zou, H., & Zhang, Y. (2018). Preparation and characterization of slow-release fertilizer encapsulated by biochar-based waterborne copolymers. *Science of the Total Environment*, *615*, 431–437. Available from https://doi.org/10.1016/j.scitotenv.2017.09.209.

Czernik, S., & Bridgwater, A. V. (2004). Overview of applications of biomass fast pyrolysis oil. *Energy and Fuels*, *18*(2), 590–598. Available from https://doi.org/10.1021/ef034067u.

Dai, L., Fan, L., Liu, Y., Ruan, R., Wang, Y., Zhou, Y., Zhao, Y., & Yu, Z. (2017). Production of bio-oil and biochar from soapstock via microwave-assisted co-catalytic fast pyrolysis. *Bioresource Technology*, *225*, 1–8. Available from https://doi.org/10.1016/j.biortech.2016.11.017.

Das, S. K., & Ghosh, G. K. (2023). Developing biochar-based slow-release N-P-K fertilizer for controlled nutrient release and its impact on soil health and yield. *Biomass Conversion and Biorefinery*, *13*(14), 13051–13063. Available from https://doi.org/10.1007/s13399-021-02069-6.

Das, S. K., Mukherjee, I., & Kumar, A. (2015). Effect of soil type and organic manure on adsorption–desorption of flubendiamide. *Environmental Monitoring and Assessment*, *187*(7). Available from https://doi.org/10.1007/s10661-015-4623-2.

Deng, H., Li, Q., Huang, M., Li, A., Zhang, J., Li, Y., Li, S., Kang, C., & Mo, W. (2020). Removal of Zn(II), Mn(II) and Cu(II) by adsorption onto banana stalk biochar: Adsorption process and mechanisms. *Water Science and Technology*, *82*(12), 2962–2974. Available from https://doi.org/10.2166/wst.2020.543, https://watermark.silverchair.com/wst082122962.pdf?.

Dias, D. S., Crespi, M. S., Torquato, L. D. M., Kobelnik, M., & Ribeiro, C. A. (2018). Torrefied banana tree fiber pellets having embedded urea for agricultural use. *Journal of Thermal Analysis and Calorimetry*, *131*(1), 705–712. Available from https://doi.org/10.1007/s10973-016-6049-7.

Dimin, M. F., Se, S. M., Shaaban, A., & Hashim, M. M. (2014). Urea impregnated biochar to minimize nutrients loss in paddy soils. *International Journal of Automotive and Mechanical Engineering*, *10*(1), 2016–2024. Available from https://doi.org/10.15282/ijame.10.2014.18.0169, http://ijame.ump.edu.my/images/Volume_10/18%20Dimin%20et%20al.pdf.

Dominguez, E. L., Uttran, A., Loh, S. K., Manero, M. H., Upperton, R., Tanimu, M. I., & Thomas Bachmann, R. T. (2020). Characterisation of industrially produced oil palm kernel shell biochar and its potential as slow release nitrogen-phosphate fertilizer and carbon sink. *Materials Today: Proceedings*, *31*, 221–227. Available from https://doi.org/10.1016/j.matpr.2020.05.143.

Dong, D., Wang, C., Van Zwieten, L., Wang, H., Jiang, P., Zhou, M., & Wu, W. (2020). An effective biochar-based slow-release fertilizer for reducing nitrogen loss in paddy fields. *Journal of Soils and Sediments*, *20*(8), 3027–3040. Available from https://doi.org/10.1007/s11368-019-02401-8.

Dong, D., Li, J., Ying, S., Wu, J., Han, X., Teng, Y., Zhou, M., Ren, Y., & Jiang, P. (2021). Mitigation of methane emission in a rice paddy field amended with biochar-based slow-release fertilizer. *Science of the Total Environment*, *792*148460. Available from https://doi.org/10.1016/j.scitotenv.2021.148460.

Dong, T., Gao, D., Miao, C., Yu, X., Degan, C., Garcia-Pérez, M., Rasco, B., Sablani, S. S., & Chen, S. (2015). Two-step microalgal biodiesel production using acidic catalyst generated from pyrolysis-derived bio-char. *Energy Conversion and Management*, *105*, 1389–1396. Available from https://doi.org/10.1016/j.enconman.2015.06.072.

Eazhilkrishna, N. (2017). Eco friendly utilization of nutrient enriched biochar from sugar industry wastes and its effect on yield and NPK uptake by hybrid maize grown in vertisol. *International Journal of Agriculture Sciences*, 975–3710.

Feng, D., Wang, S., Zhang, Y., Zhao, Y., Sun, S., Chang, G., Lai, X., Tan, H., & Qin, Y. (2020). Review of carbon fixation evaluation and emission reduction effectiveness for biochar in China. *Energy and Fuels*, *34*(9), 10583–10606. Available from https://doi.org/10.1021/acs.energyfuels.0c02396.

Finalis, E. R., Djangkung, S. M. S., Arfiana., Noor, I., Suratno, H., Rosyadi, E., Saputra, H., & Noda, R. (2020). Development of carbon based NPK slow release fertilizer using biochar from oil palm empty fruits bunch. *Indonesian Journal of Energy*, *3*(1), 19–24. Available from https://doi.org/10.33116/ije.v3i1.45.

Gao, R., Wang, Q., Liu, Y., Zhu, J., Deng, Y., Fu, Q., & Hu, H. (2018). Co-pyrolysis biochar derived from rape straw and phosphate rock: Carbon retention, aromaticity, and Pb removal capacity. *Energy & Fuels*, *33*(1), 413–419.

González, M. E., Cea, M., Medina, J., González, A., Diez, M. C., Cartes, P., Monreal, C., & Navia, R. (2015). Evaluation of biodegradable polymers as encapsulating agents for the development of a urea controlled-release fertilizer using biochar as support material. *Science of the Total Environment*, *505*, 446–453. Available from https://doi.org/10.1016/j.scitotenv.2014.10.014.

Grutzmacher, P., Puga, A. P., Bibar, M. P. S., Coscione, A. R., Packer, A. P., & de Andrade, C. A. (2018). Carbon stability and mitigation of fertilizer induced N2O emissions in soil amended with biochar. *Science of the Total Environment*, *625*, 1459–1466. Available from https://doi.org/10.1016/j.scitotenv.2017.12.196.

Guan, G., Kaewpanha, M., Hao, X., & Abudula, A. (2016). Catalytic steam reforming of biomass tar: Prospects and challenges. *Renewable and Sustainable Energy Reviews*, *58*, 450–461. Available from https://doi.org/10.1016/j.rser.2015.12.316.

Guo, M. (2015). *Application of biochar for soil physical improvement. Agricultural and environmental applications of biochar: Advances and barriers* (pp. 101−122). Wiley. Available from http://onlinelibrary.wiley.com/book/9780891189671, 10.2136/sssaspecpub63.2014.0039.5.

Gwenzi, W., Nyambishi, T. J., Chaukura, N., & Mapope, N. (2018). Synthesis and nutrient release patterns of a biochar-based N−P−K slow-release fertilizer. *International Journal of Environmental Science and Technology*, *15*(2), 405−414. Available from https://doi.org/10.1007/s13762-017-1399-7.

Hagemann, N., Joseph, S., Schmidt, H. P., Kammann, C. I., Harter, J., Borch, T., Young, R. B., Varga, K., Taherymoosavi, S., Elliott, K. W., McKenna, A., Albu, M., Mayrhofer, C., Obst, M., Conte, P., Dieguez-Alonso, A., Orsetti, S., Subdiaga, E., Behrens, S., & Kappler, A. (2017). Organic coating on biochar explains its nutrient retention and stimulation of soil fertility. *Nature Communications*, *8*(1). Available from https://doi.org/10.1038/s41467-017-01123-0.

Hossain, M. Z., Bahar, M. M., Sarkar, B., Donne, S. W., Ok, Y. S., Palansooriya, K. N., Kirkham, M. B., Chowdhury, S., & Bolan, N. (2020). Biochar and its importance on nutrient dynamics in soil and plant. *Biochar*, *2*(4), 379−420. Available from https://doi.org/10.1007/s42773-020-00065-z.

Hu, P., Zhang, Y., Liu, L., Wang, X., Luan, X., Ma, X., Chu, P. K., Zhou, J., & Zhao, P. (2019). Biochar/struvite composite as a novel potential material for slow release of N and P. *Environmental Science and Pollution Research*, *26*(17), 17152−17162. Available from https://doi.org/10.1007/s11356-019-04458-x.

Huang, M., Fan, L., Jiang, L. G., Yang, S. Y., Zou, Y. B., & Uphoff, N. (2019). Continuous applications of biochar to rice: Effects on grain yield and yield attributes. *Journal of Integrative Agriculture*, *18*(3), 563−570. Available from https://doi.org/10.1016/S2095-3119(18)61993-8.

Ippolito, J. A., Novak, J. M., Busscher, W. J., Ahmedna, M., Rehrah, D., & Watts, D. W. (2012). Switchgrass biochar affects two aridisols. *Journal of Environmental Quality*, *41*(4), 1123−1130. Available from https://doi.org/10.2134/jeq2011.0100.

Ippolito, J. A., Spokas, K. A., Novak, J. M., Lentz, R. D., & Cantrell, K. B. (2015). Biochar elemental composition and factors influencing nutrient retention. In J. Lehmann, & S. Joseph (Eds.), *Biochar for environmental management: Science, technology and implementation* (2nd edition, pp. 137−161). Routledge.

Jia, Y., Hu, Z., Mu, J., Zhang, W., Xie, Z., & Wang, G. (2020). Preparation of biochar as a coating material for biochar-coated urea. *Science of the Total Environment*, *731*139063. Available from https://doi.org/10.1016/j.scitotenv.2020.139063.

Karhu, K., Mattila, T., Bergström, I., & Regina, K. (2011). Biochar addition to agricultural soil increased CH4 uptake and water holding capacity - Results from a short-term pilot field study. *Agriculture, Ecosystems and Environment*, *140*(1-2), 309−313. Available from https://doi.org/10.1016/j.agee.2010.12.005.

Karim, A. A., Kumar, M., Mohapatra, S., & Singh, S. K. (2019). Nutrient rich biomass and effluent sludge wastes co-utilization for production of biochar fertilizer through different thermal treatments. *Journal of Cleaner Production*, *228*, 570−579. Available from https://doi.org/10.1016/j.jclepro.2019.04.330.

Khajavi-Shojaei, S., Moezzi, A., Norouzi Masir, M., & Taghavi, M. (2023). Synthesis modified biochar-based slow-release nitrogen fertilizer increases nitrogen use efficiency and corn (*Zea mays* L.) growth. *Biomass Conversion and Biorefinery*, *13*(2), 593−601. Available from https://doi.org/10.1007/s13399-020-01137-7.

Khan, H. A., Naqvi, S. R., Mehran, M. T., Khoja, A. H., Khan Niazi, M. B., Juchelková, D., & Atabani, A. (2021). A performance evaluation study of nano-biochar as a potential slow-release nano-fertilizer from wheat straw residue for sustainable agriculture. *Chemosphere*, *285*. Available from https://doi.org/10.1016/j.chemosphere.2021.131382.

Kim, J., Yoo, G., Kim, D., Ding, W., & Kang, H. (2017). Combined application of biochar and slow-release fertilizer reduces methane emission but enhances rice yield by different mechanisms. *Applied Soil Ecology*, *117-118*, 57–62. Available from https://doi.org/10.1016/j.apsoil.2017.05.006.

Kim, P., Hensley, D., & Labbé, N. (2014). Nutrient release from switchgrass-derived biochar pellets embedded with fertilizers. *Geoderma*, *232-234*, 341–351. Available from https://doi.org/10.1016/j.geoderma.2014.05.017.

Lai, W. Y., Lai, C. M., Ke, G. R., Chung, R. S., Chen, C. T., Cheng, C. H., Pai, C. W., Chen, S. Y., & Chen, C. C. (2013). The effects of woodchip biochar application on crop yield, carbon sequestration and greenhouse gas emissions from soils planted with rice or leaf beet. *Journal of the Taiwan Institute of Chemical Engineers*, *44*(6), 1039–1044. Available from https://doi.org/10.1016/j.jtice.2013.06.028.

Lateef, A., Nazir, R., Jamil, N., Alam, S., Shah, R., Khan, M. N., Saleem, M., & Rehman, S.-ur (2019). Synthesis and characterization of environmental friendly corncob biochar based nano-composite – a potential slow release nano-fertilizer for sustainable agriculture. *Environmental Nanotechnology, Monitoring & Management*, *11*100212. Available from https://doi.org/10.1016/j.enmm.2019.100212.

Lawrencia, D., Wong, S. K., Low, D. Y. S., Goh, B. H., Goh, J. K., Ruktanonchai, U. R., Soottitantawat, A., Lee, L. H., & Tang, S. Y. (2021). Controlled release fertilizers: A review on coating materials and mechanism of release. *Plants*, *10*(2), 1–26. Available from https://doi.org/10.3390/plants10020238, https://www.mdpi.com/2223-7747/10/2/238/pdf.

Li, H., Li, Y., Xu, Y., & Lu, X. (2020). Biochar phosphorus fertilizer effects on soil phosphorus availability. *Chemosphere*, *244*125471. Available from https://doi.org/10.1016/j.chemosphere.2019.125471.

Li, L., Rowbotham, J. S., Christopher Greenwell, H., & Dyer, P. W. (2013). An introduction to pyrolysis and catalytic pyrolysis: Versatile techniques for biomass conversion. In S. L. Suib (Ed.), *New and future developments in catalysis: Catalytic biomass conversion* (pp. 173–208). Elsevier B.V. Available from http://www.sciencedirect.com/science/book/9780444538789, 10.1016/B978-0-444-53878-9.00009-6.

Li, R., Wang, J. J., Zhou, B., Zhang, Z., Liu, S., Lei, S., & Xiao, R. (2017). Simultaneous capture removal of phosphate, ammonium and organic substances by MgO impregnated biochar and its potential use in swine wastewater treatment. *Journal of Cleaner Production*, *147*, 96–107. Available from https://doi.org/10.1016/j.jclepro.2017.01.069.

Li, X., Li, Q., Xu, X., Su, Y., Yue, Q., & Gao, B. (2016). Characterization, swelling and slow-release properties of a new controlled release fertilizer based on wheat straw cellulose hydrogel. *Journal of the Taiwan Institute of Chemical Engineers*, *60*, 564–572. Available from https://doi.org/10.1016/j.jtice.2015.10.027.

Liao, J., Liu, X., Hu, A., Song, H., Chen, X., & Zhang, Z. (2020). Effects of biochar-based controlled release nitrogen fertilizer on nitrogen-use efficiency of oilseed rape (*Brassica napus* L.). *Scientific Reports*, *10*(1). Available from https://doi.org/10.1038/s41598-020-67528-y.

Liu, X., Liao, J., Song, H., Yang, Y., Guan, C., & Zhang, Z. (2019). A biochar-based route for environmentally friendly controlled release of nitrogen: Urea-loaded biochar and bentonite composite. *Scientific Reports*, *9*(1). Available from https://doi.org/10.1038/s41598-019-46065-3.

Lubkowski, K., & Pomeranian, W. (2017). Fertilizers with a delayed nutrient release. *Advanced Mater Technol Environ Appl*, *1*, 38–47.

Luo, W., Qian, L., Liu, W., Zhang, X., Wang, Q., Jiang, H., Cheng, B., Ma, H., & Wu, Z. (2021). A potential Mg-enriched biochar fertilizer: Excellent slow-release performance and release mechanism of nutrients. *Science of the Total Environment*, *768*144454. Available from https://doi.org/10.1016/j.scitotenv.2020.144454.

Lustosa Filho, J. F., Barbosa, C. F., Carneiro, J. S. D. S., & Melo, L. C. A. (2019). Diffusion and phosphorus solubility of biochar-based fertilizer: Visualization, chemical assessment and availability to plants. *Soil and Tillage Research*, *194*. Available from https://doi.org/10.1016/j.still.2019.104298.

Lustosa Filho, J. F., Carneiro, J. S. D. S., Barbosa, C. F., de Lima, K. P., Leite, A. D. A., & Melo, L. C. A. (2020). Aging of biochar-based fertilizers in soil: Effects on phosphorus pools and availability to *Urochloa brizantha* grass. *Science of the Total Environment*, *709*. Available from https://doi.org/10.1016/j.scitotenv.2019.136028.

Lustosa Filho, J. F., Penido, E. S., Castro, P. P., Silva, C. A., & Melo, L. C. A. (2017). Co-pyrolysis of poultry litter and phosphate and magnesium generates alternative slow-release fertilizer suitable for tropical soils. *ACS Sustainable Chemistry and Engineering*, *5*(10), 9043–9052. Available from https://doi.org/10.1021/acssuschemeng.7b01935.

Major, J., Lehmann, J., Rondon, M., & Goodale, C. (2010). Fate of soil-applied black carbon: Downward migration, leaching and soil respiration. *Global Change Biology*, *16* (4), 1366–1379. Available from https://doi.org/10.1111/j.1365-2486.2009.02044.x.

Manikandan, A., & Subramanian, K. S. (2015). Ability of urea impregnated biochar fertilizers for securing the slow release of nitrogen in soils—Preliminary study. *International Journal of Agriculture Sciences*, 975–3710.

Melo, T. M., Bottlinger, M., Schulz, E., Leandro, W. M., Botelho de Oliveira, S., Menezes de Aguiar Filho, A., El-Naggar, A., Bolan, N., Wang, H., Ok, Y. S., & Rinklebe, J. (2019). Management of biosolids-derived hydrochar (Sewchar): Effect on plant germination, and farmers' acceptance. *Journal of Environmental Management*, *237*, 200–214. Available from https://doi.org/10.1016/j.jenvman.2019.02.042.

Mikos-Szymańska, M., Schab, S., Rusek, P., Borowik, K., Bogusz, P., & Wyzińska, M. (2019). Preliminary study of a method for obtaining brown coal and biochar based granular compound fertilizer. *Waste and Biomass Valorization*, *10*(12), 3673–3685. Available from https://doi.org/10.1007/s12649-019-00655-4.

Mikula, K., Izydorczyk, G., Skrzypczak, D., Mironiuk, M., Moustakas, K., Witek-Krowiak, A., & Chojnacka, K. (2020). Controlled release micronutrient fertilizers for precision agriculture – A review. *Science of the Total Environment*, *712*136365. Available from https://doi.org/10.1016/j.scitotenv.2019.136365.

Naz, M. Y., & Sulaiman, S. A. (2016). Slow release coating remedy for nitrogen loss from conventional urea: A review. *Journal of Controlled Release*, *225*, 109–120. Available from https://doi.org/10.1016/j.jconrel.2016.01.037.

Nelson, K. A., Paniagua, S. M., & Motavalli, P. P. (2009). Effect of polymer coated urea, irrigation, and drainage on nitrogen utilization and yield of corn in a claypan soil.

Agronomy Journal, *101*(3), 681–687. Available from https://doi.org/10.2134/agronj2008.0201.

Paiva, D. M. D., Cantarutti, R. B., Guimarães, G. G. F., & Silva, I. R. D. (2012). Urea coated with oxidized charcoal reduces ammonia volatilization. *Revista Brasileira de Ciência do Solo*, *36*(4), 1221–1230. Available from https://doi.org/10.1590/S0100-06832012000400016.

Pentananunt, R., Rahman, A. N. M. M., & Bhattacharya, S. C. (1990). Upgrading of biomass by means of torrefaction. *Energy*, *15*(12), 1175–1179. Available from https://doi.org/10.1016/0360-5442(90)90109-f.

Petrus, H. T. B. M., Putera, A. D. P., Wangi, I. P., Ramadhian, M. A., Setiawan, H., & Prasetya, A. (2020). Characterization of nitrogen release in modified controlled-release-fertilizer using rice husk biochar. *International Journal of Technology*, *11*(4), 774. Available from https://doi.org/10.14716/ijtech.v11i4.3520.

Pogorzelski, D., Filho, J. F. L., Matias, P. C., Santos, W. O., Vergütz, L., & Melo, L. C. A. (2020). Biochar as composite of phosphate fertilizer: Characterization and agronomic effectiveness. *Science of the Total Environment*, *743*. Available from https://doi.org/10.1016/j.scitotenv.2020.140604.

Puga, A. P., Queiroz, M. C. D. A., Ligo, M. A. V., Carvalho, C. S., Pires, A. M. M., Marcatto, J. D. O. S., & Andrade, C. A. D. (2020). Nitrogen availability and ammonia volatilization in biochar-based fertilizers. *Archives of Agronomy and Soil Science*, *66*(7), 992–1004. Available from https://doi.org/10.1080/03650340.2019.1650916.

Qian, Z. Y., Xue, S. G., Cui, M. Q., Wu, C., & Li, W. C. (2021). Arsenic availability and transportation in soil-rice system affected by iron-modified biochar. *Journal of Central South University*, *28*(6), 1901–1918. Available from https://doi.org/10.1007/s11771-021-4738-2.

Rizwan, M., Ali, S., Qayyum, M. F., Ibrahim, M., Zia-ur-Rehman, M., Abbas, T., & Ok, Y. S. (2016). Mechanisms of biochar-mediated alleviation of toxicity of trace elements in plants: A critical review. *Environmental Science and Pollution Research*, *23*(3), 2230–2248. Available from https://doi.org/10.1007/s11356-015-5697-7.

Rop, K., Karuku, G. N., Mbui, D., Njomo, N., & Michira, I. (2019). Evaluating the effects of formulated nano-NPK slow release fertilizer composite on the performance and yield of maize, kale and capsicum. *Annals of Agricultural Sciences*, *64*(1), 9–19. Available from https://doi.org/10.1016/j.aoas.2019.05.010.

Sarkhot, D. V., Berhe, A. A., & Ghezzehei, T. A. (2012). Impact of biochar enriched with dairy manure effluent on carbon and nitrogen dynamics. *Journal of Environmental Quality*, *41*(4), 1107–1114.

Sepúlveda-Cadavid, C., Romero, J. H., Torres, M., Becerra-Agudelo, E., & López, J. E. (2021). Evaluation of a biochar-based slow-release P fertilizer to improve *Spinacia oleracea* P use, yield, and nutritional quality. *Journal of Soil Science and Plant Nutrition*, *21*(4), 2980–2992. Available from https://doi.org/10.1007/s42729-021-00583-0.

Sha, Z., Li, Q., Lv, T., Misselbrook, T., & Liu, X. (2019). Response of ammonia volatilization to biochar addition: A meta-analysis. *Science of the Total Environment*, *655*, 1387–1396. Available from https://doi.org/10.1016/j.scitotenv.2018.11.316.

Shi, W., Ju, Y., Bian, R., Li, L., Joseph, S., Mitchell, D. R. G., Munroe, P., Taherymoosavi, S., & Pan, G. (2020). Biochar bound urea boosts plant growth and reduces nitrogen leaching. *Science of the Total Environment*, *701*134424. Available from https://doi.org/10.1016/j.scitotenv.2019.134424.

Sim, D. H. H., Tan, I. A. W., Lim, L. L. P., & Hameed, B. H. (2021). Encapsulated biochar-based sustained release fertilizer for precision agriculture: A review. *Journal of Cleaner Production*, *303*127018. Available from https://doi.org/10.1016/j.jclepro.2021.127018.

Soliman, M. H., Alnusairi, G. S. H., Khan, A. A., Alnusaire, T. S., Fakhr, M. A., Abdulmajeed, A. M., Aldesuquy, H. S., Yahya, M., & Najeeb, U. (2023). Biochar and selenium nanoparticles induce water transporter genes for sustaining carbon assimilation and grain production in salt-stressed wheat. *Journal of Plant Growth Regulation*, *42*(3), 1522–1543. Available from https://doi.org/10.1007/s00344-022-10636-y.

Suárez-Abelenda, M., Kaal, J., & McBeath, A. V. (2017). Translating analytical pyrolysis fingerprints to thermal stability indices (TSI) to improve biochar characterization by pyrolysis-GC-MS. *Biomass and Bioenergy*, *98*, 306–320. Available from https://doi.org/10.1016/j.biombioe.2017.01.021.

Suo, F., You, X., Yin, S., Wu, H., Zhang, C., Yu, X., Sun, R., & Li, Y. (2021). Preparation and characterization of biochar derived from co-pyrolysis of *Enteromorpha prolifera* and corn straw and its potential as a soil amendment. *Science of the Total Environment*, *798*149167. Available from https://doi.org/10.1016/j.scitotenv.2021.149167.

Taghizadeh-Toosi, A., Clough, T. J., Sherlock, R. R., & Condron, L. M. (2012). Biochar adsorbed ammonia is bioavailable. *Plant and Soil*, *350*(1-2), 57–69. Available from https://doi.org/10.1007/s11104-011-0870-3.

Thomas, S. C., Frye, S., Gale, N., Garmon, M., Launchbury, R., Machado, N., Melamed, S., Murray, J., Petroff, A., & Winsborough, C. (2013). Biochar mitigates negative effects of salt additions on two herbaceous plant species. *Journal of Environmental Management*, *129*, 62–68. Available from https://doi.org/10.1016/j.jenvman.2013.05.057.

Tian, H., Liu, Z., Zhang, M., Guo, Y., Zheng, L., & Li, Y. C. (2019). Biobased polyurethane, epoxy resin, and polyolefin wax composite coating for controlled-release fertilizer. *ACS Applied Materials and Interfaces*, *11*(5), 5380–5392. Available from https://doi.org/10.1021/acsami.8b16030.

Titirici, M. M., Thomas, A., & Antonietti, M. (2007). Back in the black: Hydrothermal carbonization of plant material as an efficient chemical process to treat the CO2 problem? *New Journal of Chemistry*, *31*(6), 787–789. Available from https://doi.org/10.1039/b616045j.

Trinh, M. V., Tesfai, M., Borrell, A., Nagothu, U. S., Bui, T. P. L., Quynh, V. D., & Thanh, L. Q. (2017). Effect of organic, inorganic and slow-release urea fertilisers on CH4 and N2O emissions from rice paddy fields. *Paddy and Water Environment*, *15*(2), 317–330. Available from https://doi.org/10.1007/s10333-016-0551-1.

Tripathi, M., Sahu, J. N., & Ganesan, P. (2016). Effect of process parameters on production of biochar from biomass waste through pyrolysis: A review. *Renewable and Sustainable Energy Reviews*, *55*, 467–481. Available from https://doi.org/10.1016/j.rser.2015.10.122.

Tumuluru, J. S., Wright, C. T., Hess, J. R., & Kenney, K. L. (2011). A review of biomass densification systems to develop uniform feedstock commodities for bioenergy application. *Biofuels, Bioproducts and Biorefining*, *5*(6), 683–707. Available from https://doi.org/10.1002/bbb.324.

Turner, J., Sverdrup, G., Mann, M. K., Maness, P. C., Kroposki, B., Ghirardi, M., Evans, R. J., & Blake, D. (2008). Renewable hydrogen production. *International Journal of Energy Research*, *32*(5), 379–407. Available from https://doi.org/10.1002/er.1372.

Uras, U., Carrier, M., Hardie, A. G., & Knoetze, J. H. (2012). Physico-chemical characterization of biochars from vacuum pyrolysis of South African agricultural wastes for application as soil amendments. *Journal of Analytical and Applied Pyrolysis*, *98*, 207–213. Available from https://doi.org/10.1016/j.jaap.2012.08.007.

Van Dijk, K. C., Lesschen, J. P., & Oenema, O. (2016). Phosphorus flows and balances of the European Union Member States. *Science of the Total Environment*, *542*, 1078–1093.

Verma, M., Godbout, S., Brar, S. K., Solomatnikova, O., Lemay, S. P., & Larouche, J. P. (2012). Biofuels production from biomass by thermochemical conversion technologies. *International Journal of Chemical Engineering*, *2012*, 1–18. Available from https://doi.org/10.1155/2012/542426.

Vo, P. T., Nguyen, H. T., Trinh, H. T., Nguyen, V. M., Le, A. T., Tran, H. Q., & Nguyen, T. T. T. (2021). The nitrogen slow-release fertilizer based on urea incorporating chitosan and poly(vinyl alcohol) blend. *Environmental Technology and Innovation*, *22*. Available from https://doi.org/10.1016/j.eti.2021.101528.

Wang, B., Ma, Y., Lee, X., Wu, P., Liu, F., Zhang, X., Li, L., & Chen, M. (2021). Environmental-friendly coal gangue-biochar composites reclaiming phosphate from water as a slow-release fertilizer. *Science of the Total Environment*, *758*143664. Available from https://doi.org/10.1016/j.scitotenv.2020.143664.

Wang, C., Luo, D., Zhang, X., Huang, R., Cao, Y., Liu, G., Zhang, Y., & Wang, H. (2022). Biochar-based slow-release of fertilizers for sustainable agriculture: A mini review. *Environmental Science and Ecotechnology*, *10*100167. Available from https://doi.org/10.1016/j.ese.2022.100167.

Wang, C., Zhang, X., Sun, R., & Cao, Y. (2020). Neutralization of red mud using bio-acid generated by hydrothermal carbonization of waste biomass for potential soil application. *Journal of Cleaner Production*, *271*122525. Available from https://doi.org/10.1016/j.jclepro.2020.122525.

Wen, P., Wu, Z., Han, Y., Cravotto, G., Wang, J., & Ye, B. C. (2017). Microwave-assisted synthesis of a novel biochar-based slow-release nitrogen fertilizer with enhanced water-retention capacity. *ACS Sustainable Chemistry and Engineering*, *5*(8), 7374–7382. Available from https://doi.org/10.1021/acssuschemeng.7b01721.

Wu, L., Wei, C., Zhang, S., Wang, Y., Kuzyakov, Y., & Ding, X. (2019). MgO-modified biochar increases phosphate retention and rice yields in saline-alkaline soil. *Journal of Cleaner Production*, *235*, 901–909. Available from https://doi.org/10.1016/j.jclepro.2019.07.043.

Wu, W., Yan, B., Sun, Y., Zhong, L., Lu, W., & Chen, G. (2021). Potential of yak dung-derived hydrochar as fertilizer: Mechanism and model of controlled release of nitrogen. *Science of the Total Environment*, *781*146665. Available from https://doi.org/10.1016/j.scitotenv.2021.146665.

Wu, Y., Brickler, C., Li, S., & Chen, G. (2021). Synthesis of microwave-mediated biochar-hydrogel composites for enhanced water absorbency and nitrogen release. *Polymer*

Testing, *93*106996. Available from https://doi.org/10.1016/j.polymertesting.2020.106996.

Xiang, A., Qi, R., Wang, M., Zhang, K., Jiang, E., Ren, Y., & Hu, Z. (2020). Study on the infiltration mechanism of molten urea and biochar for a novel fertilizer preparation. *Industrial Crops and Products*, *153*112558. Available from https://doi.org/10.1016/j.indcrop.2020.112558.

Yao, Y., Gao, B., Chen, J., & Yang, L. (2013). Engineered biochar reclaiming phosphate from aqueous solutions: Mechanisms and potential application as a slow-release fertilizer. *Environmental Science and Technology*, *47*(15), 8700–8708. Available from https://doi.org/10.1021/es4012977.

Ye, Z., Zhang, L., Huang, Q., & Tan, Z. (2019). Development of a carbon-based slow release fertilizer treated by bio-oil coating and study on its feedback effect on farmland application. *Journal of Cleaner Production*, *239*118085. Available from https://doi.org/10.1016/j.jclepro.2019.118085.

Yu, L., Lu, X., He, Y., Brookes, P. C., Liao, H., & Xu, J. (2017). Combined biochar and nitrogen fertilizer reduces soil acidity and promotes nutrient use efficiency by soybean crop. *Journal of Soils and Sediments*, *17*(3), 599–610. Available from https://doi.org/10.1007/s11368-016-1447-9.

Yu, Z., Zhao, J., Hua, Y., Li, X., Chen, Q., & Shen, G. (2021). Optimization of granulation process for binder-free biochar-based fertilizer from digestate and its slow-release performance. *Sustainability*, *13*(15), 8573. Available from https://doi.org/10.3390/su13158573.

Zhang, C., Zeng, G., Huang, D., Lai, C., Chen, M., Cheng, M., Tang, W., Tang, L., Dong, H., Huang, B., Tan, X., & Wang, R. (2019). Biochar for environmental management: Mitigating greenhouse gas emissions, contaminant treatment, and potential negative impacts. *Chemical Engineering Journal*, *373*, 902–922. Available from https://doi.org/10.1016/j.cej.2019.05.139.

Zhang, S., Yang, Y., Gao, B., Wan, Y., Li, Y. C., & Zhao, C. (2016). Bio-based interpenetrating network polymer composites from locust sawdust as coating material for environmentally friendly controlled-release urea fertilizers. *Journal of Agricultural and Food Chemistry*, *64*(28), 5692–5700. Available from https://doi.org/10.1021/acs.jafc.6b01688.

Zhang, X., Gang, D. D., Sun, P., Lian, Q., & Yao, H. (2021). Goethite dispersed corn straw-derived biochar for phosphate recovery from synthetic urine and its potential as a slow-release fertilizer. *Chemosphere*, *262*127861. Available from https://doi.org/10.1016/j.chemosphere.2020.127861.

Zhang, X., Huang, R., Cao, Y., & Wang, C. (2021). Rapid conversion of red mud into soil matrix by co-hydrothermal carbonization with biomass wastes. *Journal of Environmental Chemical Engineering*, *9*(5)106039. Available from https://doi.org/10.1016/j.jece.2021.106039.

Zhang, Y., Li, Z., & Mahmood, I. B. (2015). Effects of corn cob produced biochars on urea recovery from human urine and their application as soil conditioners. *Clean – Soil, Air, Water*, *43*(8), 1167–1173. Available from https://doi.org/10.1002/clen.201400489.

Zhao, L., Cao, X., Zheng, W., Kan, Y., & Zhou, Z. (2014). Phosphorus-assisted biomass thermal conversion: Reducing carbon loss and improving biochar stability. *PLoS One*, *9*(12), e115373. Available from https://doi.org/10.1371/journal.pone.0115373.

Zhao, L., Cao, X., Zheng, W., Scott, J. W., Sharma, B. K., & Chen, X. (2016). Copyrolysis of biomass with phosphate fertilizers to improve biochar carbon retention, slow nutrient release, and stabilize heavy metals in soil. *ACS Sustainable Chemistry and Engineering*, *4*(3), 1630–1636. Available from https://doi.org/10.1021/acssuschemeng.5b01570.

Zheng, J., Han, J., Liu, Z., Xia, W., Zhang, X., Li, L., Liu, X., Bian, R., Cheng, K., Zheng, J., & Pan, G. (2017). Biochar compound fertilizer increases nitrogen productivity and economic benefits but decreases carbon emission of maize production. *Agriculture, Ecosystems and Environment*, *241*. Available from https://doi.org/10.1016/j.agee.2017.02.034.

Zhong, K., Lin, Z. T., Zheng, X. L., Jiang, G. B., Fang, Y. S., Mao, X. Y., & Liao, Z. W. (2013). Starch derivative-based superabsorbent with integration of water-retaining and controlled-release fertilizers. *Carbohydrate Polymers*, *92*(2), 1367–1376. Available from https://doi.org/10.1016/j.carbpol.2012.10.030.

Zhong, X., Wang, M., & Jiang, E. (2018). Preparation of an environment-friendly biochar fertilizer. *IOP Conference Series: Materials Science and Engineering*, *301*(1), 012157. Available from https://doi.org/10.1088/1757-899x/301/1/012157.

CHAPTER

21 Mineral-enriched biochar nutritional nanocomposites for enhanced soil properties, biogeochemical cycle, and sustainable crop production: a mechanistic overview

Vishnu D. Rajput[1], Swati Priya[2], Sagar Mondal[2], Jutishna Bora[2], Priyadarshani Rajput[1], Abhishek Singh[3], Karen Ghazaryan[3], Sumira Malik[2,4], Saglara Mandzhieva[1] and Tatiana Minkina[1]

[1]*Academy of Biology and Biotechnology, Southern Federal University, Rostov-on-Don, Russia*
[2]*Amity Institute of Biotechnology, Amity University, Ranchi, Jharkhand, India*
[3]*Laboratory of Applied Biology and Ecology, Faulty of Biology, Yerevan State University, Yerevan, Armenia*
[4]*Department of Biotechnology, University Center for Research & Development (UCRD) Chandigarh University, Mohali, Punjab, India*

21.1 Introduction

Biochar, commonly known as bio-charcoal, is produced through the pyrolysis of organic material, including feedstock and waste collected from forests and farmland. This phenomenon takes place in conditions where there is limited availability of oxygen, leading to the formation of carbonaceous materials that bear a resemblance to charcoal. Studies have demonstrated that biochar efficiently increases the organic carbon levels in soil and reduces the harmful effects of salt and drought stress on vegetation (Hammer et al., 2015). Biochar effectively stabilizes and immobilizes heavy metals in soil, particularly in agricultural fields with significant use of chemical fertilizers and in garbage disposal areas with high concentrations of heavy metals in the soil. Studies on the effects of biochar on soil microbiota have shown that biochar can enhance the development of soil microbial communities. The discovery implies the possibility of developing economical approaches to enhance the quality of soil (Rajput et al., 2022). Biochar can influence and modifies the physical and chemical attributes of the soil, as well as the

Biochar Ecotechnology for Sustainable Agriculture and Environment. DOI: https://doi.org/10.1016/B978-0-443-29855-4.00023-0

organisms that inhabit there (Zhang et al., 2017). Biochar has proven to be highly effective in purifying liquid solutions, including wastewater, in addition to its use in soil remediation (Pi et al., 2015). Before implementing it, it is crucial to evaluate and establish the physicochemical characteristics of biochar to attain best results in terms of soil health and wastewater treatment. This chapter offers a comprehensive investigation of the effects of biochar application on the preservation of soil quality and the remediation of soil pollutants.

Traditional chemical vas fertilizers rapidly release a significant portion of their nutrients upon application to the soil. A substantial fraction of the nutrients supplied is not effectively utilized by crops, resulting in the ,eventual loss of the unused portion. As a result, the nitrogen utilization efficiency (NUE) of these inorganic nutrients is notably low, which means that a large amount of fertilizer needs to be applied, resulting in increased agricultural production costs (Verma et al., 2023; Yadav et al., 2023). Moreover, the leftover waste generated by these unnecessary fertilizers has substantial environmental repercussions, including eutrophication (Fig. 21.1) water quality degradation, and nitrous oxide emission (Akinnawo, 2023). Furthermore, the stocks of minerals used in the manufacturing of chemical fertilizers are diminishing rapidly, as evidenced by Vaccari and Strigul (2011) (Table 21.1).

Given these circumstances, it is imperative to have alternative sources of nutrients to sustain agricultural productivity. Recycling agricultural waste rich in nutrients in an efficient manner is a practical option, as many agricultural waste materials, such as manures, contain substantial amounts of nutrients. However, the prompt application of manure might lead to rapid mineralization and release of excess nutrients that plants cannot effectively utilize (Liu et al., 2020). Therefore the development of a slow-release fertilizer made from nutrient-rich

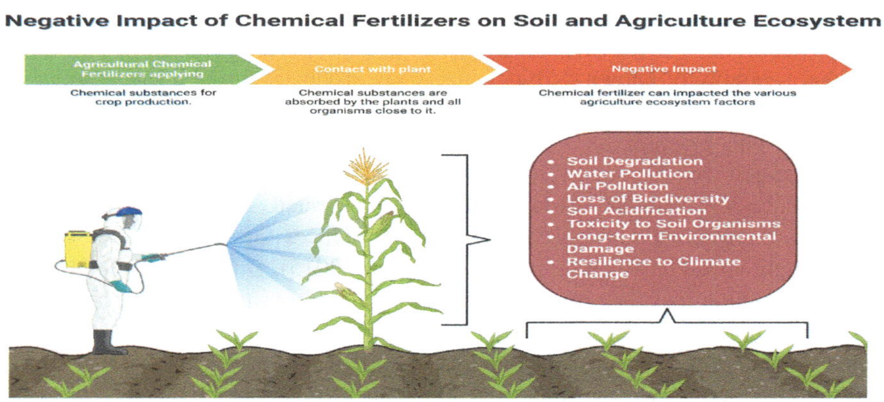

FIGURE 21.1 Adverse effect of chemical fertilizers in soil and agriculture ecosystem.

Negative impacts of chemical fertlizers on soil health and plant growth. Huge input chemical fertlizers affecting human health via accumulating edible crops.

Table 21.1 Application of nanobiochar for improvement of plant response against different abiotic stresses.

Plant species	Scientific name	Type of nanobiochar/ biochar-based nanomaterials	Applied concentration	Mode of action	References
Black cumin	*Nigella sativa* L.	Sugarcane press mud nano-BC	100 mg/L	Lower concentrations of oxidative stress indicators (H_2O_2) and Malondialdehyde and higher concentrations of chlorophyll and osmo-protectants were observed in the plants	Ramzan et al. (2023)
Tobbaco	*Nicotiana benthamiana* L.	Corn straw nano-BC	400 mg/L	Enhancement of the plant's defense system through the induction of systematic resistance by salicylic acid	Kong et al. (2022)
Rice	*Oryza sativa* L.	Nano-BC	5.0 mg/L	Enhancing the levels of antioxidant enzymes superoxide dismutase, peroxidase, and catalase reduced Cd-phytotoxicity while simultaneously increasing biomass and chlorophyll content	Yue et al. (2019)
Rice	*O. sativa* L.	Nano-BC	5–10 mg/L	Improved root growth and development by nutrient uptake resulting in increased plant growth	Akhtar et al. (2022)
Shallot	*Allium cepa* var. *Bima*	Nano-BC	0.5 mg/kg	Increased the phytoremediation process of lead	Purbalisa et al. (2021)

BC, *Biochar.*

organic agricultural waste has the capacity to function as a sustainable alternative to chemical fertilizers, providing enhanced NUE and reduced ecological impacts.

Biochar is a carbon-rich material produced by subjecting agricultural or organic waste to thermochemical decomposition in an oxygen-limited atmosphere (Lehmann & Joseph, 2015). Due to its highly porous structure, diverse functional groups, and extensive surface area, biochar exhibits significant potential as a stable medium for environmentally friendly slow-release fertilizers (Shi et al., 2020). Researchers have been doing extensive studies on the development of slow-release fertilizers by combining chemical fertilizers with biochar (Chaudhary et al., 2023). This strategy is favored because biochar alone is believed to enhance NUE without supplying extra nutrients (Ndoung et al., 2021). According to recent research by Guo et al. (2020), biochar has been found to contain certain nutrients that plants can easily absorb. However, it should be noted that these nutrients may not always be sufficient.

In contrast, nano biochar has emerged as a remedy to overcome the constraints of biochar by augmenting its physical, chemical, and structural attributes. For example, the small size of the material results in a greater surface area, and the existence of many functional groups improves its mechanical and thermal stability. These attributes render it unique and advantageous, as recently stated. Moreover, increased negative zeta potentials, reduced hydrodynamic radius, and elevated concentration of functional groups with oxygen and carbon defects might result in the generation of reactive organic species. Consequently, this improves the material's ability to adsorb substances when compared to its larger biochar equivalents (Sani et al., 2023). Carbon nanomaterials frequently contain carbon defects, specifically topological or edge defects, that greatly influence their physiochemical properties.

Their production yields a material characterized by a high concentration of unpaired electrons, resulting in a reduced energy barrier for their formation and an accelerated rate of electron transmission. The catalytic activity during the oxygen reduction process was notably affected by carbon defects. At present, nano biochar is regarded as a practical choice for strengthening soil quality, boosting enzyme activity, and functioning as an effective adsorbent (Jin et al., 2022). This is because of its distinctive physical and chemical characteristics and its porous structure. The nano biochar particles possess an increased specific surface area, microporosity, and surface hydrophobicity. Consequently, they possess a heightened capacity to assimilate diverse contaminants, including pesticides, heavy metals, polychlorinated biphenyls, and polycyclic aromatic hydrocarbons (PAHs). Furthermore, they possess the ability to counteract soil acidity as a result of their elevated alkalinity and pH levels (Wang et al., 2013). Nano biochar is currently being used to produce nanocomposites for different purposes, including wastewater treatment, energy generation, and the elimination of organic dyes and other pollutants (Inyang et al., 2014; Tan et al., 2015, 2016; Yao et al., 2011, 2013). Moreover, the material's porous structure, ample surface functional groups, and heightened surface-to-volume ratio made it a highly efficient support for immobilizing enzymes (Naghdi et al., 2018). Nanocomposites frequently exhibit substantial improvements in diverse attributes, including pore properties and functional groups, leading to enhanced catalytic degradation capability and more active sites. In addition, they offer convenient and swift separation (Ramadan et al., 2020; Tan et al., 2015).

In 2014 Inyang et al. conducted a study that found the carbon nanotube (CNT) Carb-biochar hybridized nanocomposite to be a cost-effective and efficient material for the removal of organic pollutants and dyes. The findings suggest that these nanocomposites possess the capacity to function as a new and efficient adsorbent for the cost-effective and proficient elimination of heavy metals, particularly Cr(VI), from water-based systems (Ramadan et al., 2020). Nanobiochar, despite its benefits, has the potential to cause groundwater contamination due to the migration of biochar nanoparticles through drainage, surface runoff, and irrigation. The researchers have expressed concerns about this matter. The absorption of heavy metals by nanobiochar-heavy-metal composites has the capacity to penetrate the soil easily (Kuang et al., 2012). Consequently, these metallic elements with high atomic weights can be liberated through desorption and disseminated to different areas, contaminating soil and water (Rajput et al., 2024). Therefore, with its unique features and specified applications, nano biochar, as shown in Fig. 21.2, offers new prospects for tackling existing agricultural and environmental issues in a sustainable, economical, and promising manner. Prior research mostly focused on the utilization of nano biochar and biochar nanocomposite materials in the domains of agriculture and environment, along with the methodologies applied for their production. This chapter seeks to offer a thorough account of both novel and traditional production methods of nano biochar. The text discusses the diverse uses of nano biochar in agriculture and the environment,

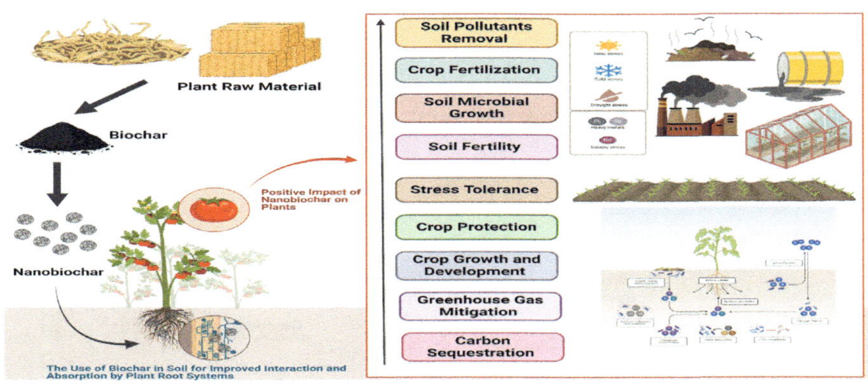

FIGURE 21.2 Application nanobiochar in plants and their various positive impact on plants.

Emerging role of biochar in soil system to improve growth of crop plants. The various types of biochar positively improving soil health and soil microorganism which help crop production.

encompassing recent progress in energy, catalysis, biofuel, and material science. Furthermore, it examines the prospective advancements of nano biochar.

21.2 Synthesis of nano biochar

Biochar is commonly produced through the pyrolysis method, which involves subjecting the feedstock to high temperatures between 350°C and 700°C in an oxygen-deprived environment (<1% oxygen) (Oleszczuk et al., 2016). Using different thermochemical methods, biochar is derived from lignocellulosic biomass, including crop residues, timber biomass, stabilized waste products, and animal litter (Fig. 21.3). The processes encompass rapid pyrolysis, slow combustion, torrefaction, carbonization (such as hydrothermal treatment or flash), and gasification techniques (Meyer et al., 2011).

In general, slow pyrolysis is used in an environment without oxygen to produce biochar, which results in around 35% of dry biomass (Tomczyk et al., 2020). In addition, a little amount of bio-oil is also generated (Mohan et al., 2014; Ramadan et al., 2020). Fast pyrolysis is commonly used for biofuel production, while gasification is generally used to produce syngas, which may then be used to generate heat and energy (Cheah et al., 2016). In addition, Ashiq et al. (2019) discovered that lignocellulosic material yielded a higher biochar production in comparison with municipal

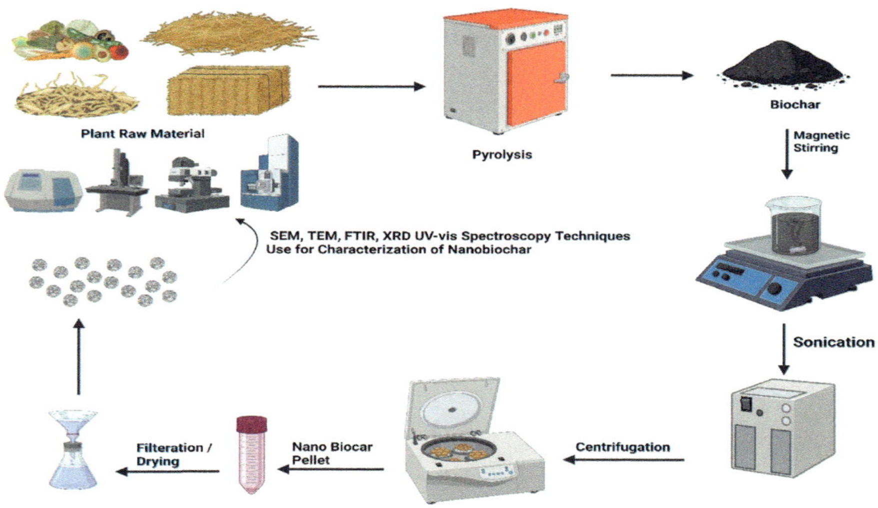

FIGURE 21.3 A method and process for formation of nanobiochar from raw material of plants.

Several methods to produce nano size biochar. Small size of biochars are highly effective to improve soil health and crop growth.

solid waste. Furthermore, several chemical and physical modification procedures are utilized to improve the effectiveness of biochar for specific purposes. Chemical modification treatments encompass several approaches, including chemical oxidation, CO_2 induction, vapor wrapping, acid/base administration, and penetration with natural and synthetic nanoparticles. Conversely, the physical alteration method referred to as ball milling has not gained widespread popularity. Nevertheless, the scant study carried out in this field has produced encouraging outcomes, justifying the need for more investigation (Lyu et al., 2018; Ramanayaka et al., 2020).

Currently, nano biochar is synthesis by employing various ways as illustrated in Fig. 21.1, starting from regular biochar. Typically, nanosized particle production involves the use of mechanical grinding. Flash heating during direct nanobiochar manufacture yielded graphitic nanosheets, as reported by Ramanayaka et al. (2020). Oleszczuk et al. (2016) used an ultrasonic vibrator to mechanically disperse the biochars, followed by the application of sonication to produce nanoparticles of nanoscale dimensions. Based on the literature, the ball milling process has been determined as the most appropriate and advised technique for manufacturing biochar (Fan et al., 2016). Ball milling is a process where particles collide, causing them to come together and increasing particle size (Bayram & Öner, 2007; Karinkanta et al., 2018). Double-disk milling is another option for producing nano biochar, although its use is restricted due to its high operational costs. Studies have demonstrated that vibrating disk milling surpasses the ball mill method in terms of generating a larger amount of nano biochar with enhanced consistency in size and shape. This is accomplished by employing attrition and shear stress (Bayram & Öner, 2007; Karinkanta et al., 2018), where Guo et al. (2020) presented an additional method for manufacturing nano biochar.

This was accomplished by a hydrothermal technique that made use of agricultural remnants and by-products. Bulk biochar was synthesized using soybean straw and animal dung as source materials. Afterwards, the material was digested using concentrated sulfuric acid and nitric acid in a hydrothermal reactor under high pressure circumstances. The digestion suspension was filtered using a 0.22-μm membrane and then dried in an oven at 65°C. The synthesis of biochar nanodots measuring 5 and 4 nm in size was achieved through the utilization of this approach. Different methodologies have been devised to produce nano biochar, each possessing its own advantages and disadvantages (Liu et al., 2018). There is significant interest in biochar-based nanocomposites due to their unique features and potential uses. These qualities and applications will be further described in Section 1.1.

21.2.1 Classification of biochar nanocomposites and their fabrication methods

The manufacturing techniques used for various nanomaterials based on biochar are illustrated in Fig. 21.3. The nanomaterials utilized in nano biochar can be categorized into three types: oxide/hydroxide biochar nanocomposites, magnetic biochar nanocomposites, and biochar coated with functional nanoparticles such as graphene, graphene oxide, chitosan, carbon nanotubes, ZnS nanocrystals, and graphitic C_3N_4. These composites are manufactured utilizing a variety of synthetic processes. According to Tan et al. (2016), both the preparation and treatment of biomass are customary steps in all synthesis techniques.

21.2.1.1 Biochar nanocomposites containing oxides and hydroxides

Oxide/hydroxide biochar nanocomposites are predominantly synthesized using three methods: (1) augmenting the concentration of the required element via bioaccumulation, (2) subjecting the biomass to metal salt treatment prior to the pyrolysis process, and (3) injecting nanometal oxide after the pyrolysis stage. The first two methods consisted of saturating biomass with metal before subjecting it to pyrolysis, while the third method entailed directly incorporating nanometal oxide/hydroxide into the pyrolyzed biochar (Ramadan et al., 2020; Tan et al., 2016).

21.2.1.2 Biochar coated with functional nanoparticles

Nanomaterial composites, created by applying a layer of functional nanoparticles such as graphene and its oxides, chitosan, carbon nanotubes, and stacked double hydroxides onto charcoal, can effectively remove various contaminants remove various contaminants.

21.2.1.3 Magnetic biochar composites

Two approaches can produce biochar-based magnetic composites: impregnating biomass with iron ions or chemically depositing iron oxides onto biochar. These procedures are used because of the difficulty in isolating the composites from the aqueous solution. The following techniques can be used to deposit nanoscale magnetic iron oxide particles, including $CoFe_2O_4$ (Harikishore Kumar Reddy & Lee, 2014), Fe_3O_4 (Baig et al., 2014), and c-Fe_2O_3 (Wang et al., 2015), onto the biochar surface. This method ensures the induction of magnetism and the existence of active sites for iron oxide, facilitating the elimination of contaminants. The magnetic property of biochar allowed its easy extraction from the solution, and also enhanced its capacity to attract and retain substances (Tan et al., 2016).

Composites with improved properties, such as specific surface area, presence of active sites, ability to catalytically degrade, distribution of pore sizes, volume of pores, and ease of separation, have shown great efficiency in eliminating pollutants. Multiple studies have shown that these composites possess improved adsorption capability and can effectively adsorb various pollutants from water-based solutions (Yan et al., 2015; Yao et al., 2011; Zhang et al., 2012). Moreover, these nanocomposites, which consist of several reductive/oxidative and catalytic nanoparticles such graphitic C_3N_4 (full name) and nanoscale zero-valent iron, have the capability to both adsorb and decompose multiple harmful chemicals (Tan et al., 2016). Hence, biochar-based nanocomposites exhibit significant promise for the remediation of diverse waste materials and can be utilized in a broad spectrum of environmental scenarios.

21.3 Utilization of nano biochar and biochar nanocomposites in the field of agriculture

Biochar is an organic material that has various applications in agriculture, including improving plant growth, disease control, pesticide removal, nutrient provision, soil enhancement, and promotion of microorganism growth (Helal et al., 2019; Wang et al., 2017). Moreover, research has demonstrated that biochar can enhance productivity (Alburquerque et al., 2013), soil efficiency (Liu et al., 2013), and plant biomass (Soja et al., 2013). Utilizing biochar in agricultural soil has demonstrated a reduction in the need for chemical fertilizers, specifically N fertilizers. The leaching of N into groundwater is diminished as a consequence.

21.3.1 Stimulation of plant growth and improvement of soil quality

Human activities such as industrialization, deforestation, overgrazing, and others have significantly reduced the nutritional quality of soil. The decline in nutrient availability resulted in diminished agricultural yield and impeded plant development, adversely affecting food security (Ramadan et al., 2020). Excessive utilization of chemical fertilizers has been used to safeguard soil nutrients. Conversely, leaching and the release of nutrients decreased their presence in the soil and promoted the reliance on chemical fertilizers. Bah et al. (2014) documented the cause of soil contamination and its harmful impact on soil biota in different places. The application of biochar resulted in positive effects on soil characteristics, including improved water retention, permeability, and fertility. Furthermore, the high charge density of the substance enabled it to effectively retain substantial amounts of nutrients, resulting in a notable enhancement in crop yield (Helal et al., 2019). P is a vital component for plant growth. Plants obtain P from the soil solution through phosphate anions ($H_2PO_4^-$). The process is affected by the adsorption and desorption of P and the conversion between soil particles and the solution (Debicka et al., 2016; Helal et al., 2019). Hence, the adsorption and desorption of P in soil are pivotal in defining the accessibility of P, thereby impacting the utilization and absorption of P by plants.

Biochar, renowned for its capacity to impact soil characteristics, can be used as a soil amendment to enhance the accessibility of P and N (Zeng et al., 2013). In addition, it has been shown by Dong et al. (2017) and Pan et al. (2017) that it can lower phosphorus (P) fixation, limit nitrogen (N) leaching, and decrease N_2O emission. The user's input is devoid of any content. The salinity exerted a notable influence on the effective assimilation of P, and it was demonstrated that the assimilation was augmented with a rise in soil salinity. After being treated with 0.20% nano biochar, the Freundlich constant increased and reached a value of 497.60 mg/kg, compared to the original value of 64.60 mg/kg in the saline soil sample. According to some studies, it was confirmed that P's maximum adsorption capacity (q_{max}) rose in both nonsaline and saline soils, reaching values of 9.43 and 63.29 mg/kg, respectively. Therefore biochar improves plant growth and plays a vital function in soil nutrient management to promote agricultural yield. In addition, the incorporation of biochar into the soil influenced several soil properties, such as texture, porosity, particle size distribution, and density (Mangrich et al., 2015). The organic composition of the soil causes an increase in soil pH, cation exchange capacity (CEC), electrical conductivity, and nutrient levels such as carbon (C), N, and P (Liang et al., 2006). As a result, this improves the efficiency of metal absorption. Moreover, it substantially impacted the concentrations of air and oxygen, the soil's water retention capacity, the microbial composition, and the nutritional status of the soil in the rhizosphere area. Furthermore, Chan et al. (2008) showed that the use of biochar reduced the probability of soil compaction by weakening its tensile strength. Several research initiatives have indicated that combining biochar with advanced technologies and engineered nanomaterials could produce superior outcomes. Bopp et al. (2016) found that the application of alkaline biochar resulted in a notable improvement in both root development and root biomass, in contrast to acidic biochar. The use of biochar mixed with nonpyrogenic organic amendments in agriculture and vegetation is limited due to the changing behaviors exhibited by these materials, despite their ability to enhance plant growth and development (Bonanomi et al., 2017). Furthermore, a 2019 study conducted by Yue et al. examined the application of nanobiochar in mitigating the detrimental impacts of metal toxicity, notably Cd^{2+}, and oxidative stress on rice seedlings. The researchers noted a substantial decrease in both oxidative stress and toxicity caused by Cd^{2+}.

21.4 Methods for creating biochar-based solid recovered fuels

The biochar produced through the direct thermochemical conversion of biomass feedstocks generally lacks essential nutrients, such as fertilizer. Altering biochar is essential for optimizing the effectiveness of fertilizers. The remarkable attributes of biochar provide a strong basis for developing biochar-based sustainable release fertilizers (SRFs). To enhance the efficiency of biochar-based SRFs, many methods have been proposed, such as impregnation, encapsulation, granulation, copyrolysis, and more approaches.

21.4.1 In situ pyrolysis

The rapid depletion of nonrenewable resources such as phosphate rocks and sylvinite minerals requires the extraction of essential nutrients from waste items.

Biomass wastes with a high nutritional content are highly attractive for manufacturing biochar-based fertilizers. Biomass feedstocks can be easily converted into fertilizers using biochar when they have sufficient nutrients and specific conditions are met. Biochar rich in potassium (K) can be produced from banana peduncle, mushroom compost, and seaweed. Multiple studies have investigated biochar with high levels of P, which has been produced from different sources including chicken litter, cattle carcasses, human manure, sewage sludge, bone debris, and bio-fermentation waste. The P concentration in biochar produced from different biomass sources varies between 16.8 and 84.7 mg/g. The existence of an elevated P concentration in biochar suggests its possible use as a fertilizer. The copyrolysis or plasma treatment of banana peduncle, which is high in K, and waste sludge, which is rich in P, to make biochar. Pyrolysis resulted in a larger quantity of biochar and exhibited a higher overall carbon content than plasma processing. Nevertheless, plasma processing exhibited a greater nutritious composition. The biochar produced using heat plasma and slow pyrolysis has bioavailable P contents of 4500–6000 and 5600 mg/kg, respectively.

A substantial amount of P-rich bacterial waste is produced in the fermentation industry by *Escherichia coli*. The pyrolysis process was conducted in a muffle furnace under a N atmosphere at a temperature of 600°C. This procedure aimed to generate biochar that is enhanced with P (Khodakovskaya et al., 2013). The biochar exhibited a substantial phosphorus content of 84.7 mg/g, which remained within the acceptable thresholds for potentially harmful constituents. Pyrolysis involves the removal of water from biomass, as well as the vaporization of organic substances and the formation of char. The Fourier-transform infrared spectroscopy (FTIR) spectra indicated the existence of surface groups containing N–H, C=O, C–O, and PO_4^{3-}. Furthermore, scanning electron microscope (SEM) and energy-dispersive X-ray spectroscopy (EDS) measurements revealed that P was evenly dispersed across the macropore biochar. The P-enriched biochar was evaluated as a potential P fertilizer and was observed to improve the initial growth of *Lactuca sativa*. The thermal conversion of bacterial wastes into P-enriched biochar offers a feasible method for recycling biomass wastes derived from the fermentation industry. P-enriched biochar produced from feedstocks such as sewage sludge and animal manures has certain drawbacks as a potential fertilizer. These factors encompass insufficient provision of nutrients to the soil and crops, along with the existence of perilous compounds.

21.4.2 Copyrolysis

The simultaneous thermal decomposition of biomass with P sources provides an alternative approach for generating SRFs based on biochar. Copyrolysis is a method that enables the capture of P in biochar by both adsorption and chemical bonding. Biochar-based SRFs produced via copyrolysis offer notable advantages, such as simple synthesis, high P content, and exceptional slow-release properties. Some researchers conducted a study where they carried out copyrolysis of wheat straw utilizing several sources of P (H_3PO_4, phosphate rock, and triple superphosphate). They witnessed a decline in carbon loss during the process of pyrolysis, a decrease in microbial mineralization, and an improvement in the stability of biochar. They witnessed a decline in carbon loss during the process of pyrolysis, a decrease in microbial mineralization, and an improvement in the stability of biochar. C-O-PO_3 or C−P molecules formed from the chemical reaction between Ca $(H_2PO_4)_2$ and carbon obtained from biomass, occurring during the pyrolysis process. This interaction resulted in a decrease in the rate of P release.

Premixing biomass with an acid phosphate source leads to the biochar having a low pH, hence restricting its suitability for soil applications. Before undergoing pyrolysis, the addition of alkaline sources such as MgO can efficiently regulate the pH of biochar and increase the concentrations of phosphate and magnesium nutrients. Used triple superphosphate and phosphoric acid to poultry litter, with or without the inclusion of MgO, in order to produce SRFs based on biochar. The application of SRFs led to a reduction in P release, an elevation in soil pH, and a promotion of both P absorption and maize growth. SRFs made from biochar improved the efficiency of P utilization in tropical soil and promoted the growth of *Marandu grass*, when compared with triple superphosphate.

The montmorillonite catalyst facilitated cellulose conversion in bamboo biomass to biochar under hydrothermal conditions. Multiple research works have recorded the utilization of biochar/clay mineral composites, such as bentonite/biochar and calcite/biochar, biochar/bentonite (Ashiq et al., 2019), montmorillonite/biochar, and attapulgite/biochar. To develop a biochar composite, researchers used a copyrolysis procedure where montmorillonite and bamboo powder were simultaneously decomposed (Chen et al., 2017). Montmorillonite functions as a catalyst during the pyrolysis of bamboo, reducing the temperature needed to produce biochar. The composite consisted of biochar and montmorillonite, which offered a substantial surface area and adsorption sites for nutrients. The NH_4^+ adsorption capacity was 12.52 mg/g, but the PO_4^{3-} adsorption capacity was 105.28 mg/g. The biochar composite, which contained a high amount of nutrients, demonstrated a beneficial slow-release impact for N and P. The release rates for NH_4^+ and PO_4^{3-} ranged from 0.30% to 4.92% and 2.63%−5.09%, respectively. An innovative approach to producing slow-release fertilizer was suggested by An et al. (2020), using microwave-assisted copyrolysis of biomass, nutrient, and bentonite. The procedure entailed the amalgamation of cotton straw with K_3PO_4 and bentonite in a solution, and thereafter, it was exposed to pyrolysis under microwave irradiation in an oxygen-limited environment. The use of bentonite and K_3PO_4 enhanced the thermal properties. The SEM image demonstrated the biochar's remarkably porous structure, while the energy-dispersive X-ray spectroscopy (EDX) test confirmed the presence of P and K adsorbed onto the biochar. The X-ray photoelectron spectroscopy (XPS) spectra investigation of P 2p verified the presence of HPO_4^{2-}, PO_4^{3-}, and H_2PO^{4-}. The FTIR study demonstrated that the surface groups underwent substantial attenuation due to the fast conversion of biomass components into aromatic carbon, enhanced by the presence of

K_3PO_4 or bentonite. The X-ray diffraction (XRD) analysis identified the presence of $Mg_3(PO_4)_2$ and $Ca(H_2PO_4)_2$. The augmented long-lasting releasing effect of P was verified by the observed enhancements in utilization efficiency and decrease in leaching loss. The growth of pepper seedlings was stimulated in pot experiments, presumably as a result of the slow release of nutrients.

21.4.3 Impregnation

The nutrient content is effectively enhanced by saturating pure biochar in the nutrient solution owing to its low nutrient concentration. The biochar's porous structure and significant surface area offer vessels and sites for nutrient retention, hence prolonging the period of nutrient accessibility for plants. Its straightforwardness and economical nature distinguish the impregnation process. The use of biochar in conjunction with ammonium salts, phosphates, and nitrates has been scientifically recorded as a technique for producing slow-release fertilizer, thereby mitigating nutrient loss and stimulating accelerated crop development. Biochar was produced from wheat using traditional slow pyrolysis, carried out at either 300°C or 350°C. The nutrient-enriched biochar was created by mixing biochar with saline solutions containing vital components (P, K, calcium, and N). The mixture was agitated for a duration of 3–4 hours to facilitate the absorption of nutrients, subsequently filtered and desiccated. The biochar was subjected to multiple characterizations. The biochar's microstructure predominantly derived from wheat biomass, and the presence of nutrients after impregnation was verified by EDX. Surface groups and nutrient incorporation on the biochar surface were identified using FTIR analysis. The thermogravimetric analysis study verified the intrinsic stability of the biochar. The impregnation of nutrients with biochar convincingly indicated the improvement of salt index, water retention, and slow-release nutrition.

In addition, the corncob biochar was enriched with vital macronutrients (N, calcium, P, K, and magnesium) and micronutrients (sodium, zinc, and iron). Some studies demonstrated that this led to improved water retention and prolonged release of nutrients. A study examined the effects of unmodified and P-enriched biochar (saturated with KH_2PO_4) on the accessibility and dispersion of P in soil. A study examined the effects of unmodified and P-enriched biochar (saturated with KH_2PO_4) on the accessibility and dispersion of P in soil. The introduction of pure biochar decreased the accessibility of P in the soil, while biochar that was enhanced with P maintained a consistently high level of P accessibility throughout the incubation period. The study by biochar containing P can effectively improve the utilization of P in soil, making it a valuable factor for soil rehabilitation. A study was conducted to infuse charcoal produced from oil palm kernel shells with NH_4NO_3 and KH_2PO_4. The impregnated biochar showed a high capacity for retaining nutrients, with NO_3^- at 52.9%, NH_3 at 77.4%, and PO_4^{3-} at 55.2%. A comparative analysis assessed the N release capabilities of urea-impregnated biochar, hydro-char, zeolite, and hydroxyapatite. The study identified a significant discrepancy in the N release, and the progressive release of N likely reduced N loss and improved the efficiency of fertilizer application. The cow dung was subjected to pyrolysis to produce biochar, which was then used to adsorb P from a liquid solution. The greatest P adsorption capacity was 345 mg/g. The application of XRD and XPS analyses verified that the adsorption of P took place via a precipitation mechanism. Moreover, adding charcoal enriched with P led to improved seed germination and lettuce growth.

The impregnation process is significantly hindered by the restricted adsorption ability of unaltered biochar for nutrients. Biochar has been modified to enhance adsorption capacity using various minerals such as MgO, K_2O, bentonite, and coal gangue (Wu et al., 2019). The biochar produced from tomato tissues, which were enhanced with magnesium, effectively reclaimed phosphate from a liquid solution, with a maximum adsorption capacity exceeding 100 mg/g (Yao et al., 2013). The incorporation of MgO into biochar significantly enhanced its capacity to adsorb phosphate and ammonium from wastewater. The application of Mg-enriched biochar, derived from maize straw, shown remarkable effectiveness in extracting nutrients from biogas effluent. The leaching ratio of N and P from magnesium (Mg)-enriched biochar was significantly reduced by a factor of 7 and 6, respectively, compared to that of chemical fertilizer. The treatment substantially improved nutrient absorption and promoted maize growth, demonstrating its intriguing potential for agricultural applications.

The α-FeOOH-biochar composites, produced through coprecipitation, were utilized for the adsorption of phosphate from urine. Phosphate adsorption was controlled using chemisorption and monolayer adsorption. The phosphate exhibited a peak adsorption capacity of 57.4 mg/g and an extractable phosphate concentration of 967.5 mg/kg. This indicates that it possesses favorable bioavailability and exhibits promise as a very efficient slow-release fertilizer. The production of biochar composites was achieved by directly subjecting feedstocks and coal gangue to pyrolysis. The maximum level of P adsorption measured was 7.9 mg/g, which is 4.6 times higher than the original biochar. The P-enriched charcoal was utilized as a fertilizer with a regulated release rate to augment the germination and growth of bean seeds. The study examined the application of bentonite modified biochar, which was created by copyrolyzing cotton straw and bentonite, to adsorb phosphate from a liquid solution. The transmission electron microscopy images exhibited distinct lattice fringes that corresponded to the presence of magnesium oxide (MgO) and calcium oxide (CaO) in the biochar. CaO and MgO were verified using XRD analysis and XPS inspection.

At times, wastewater from different sources may contain substantial nutrients, such as N and P. After biochar has absorbed and concentrated nutrients from wastewater, it can be used as a soil amendment or fertilizer. The application of biochar produced from wastewater for N retrieval encourages the adoption of biochar-based soil remediation fertilizers, thus mitigating environmental concerns. Previous research has shown that biochar may effectively adsorb N, phosphorous (P), and K from several sources, including urine, anaerobic digestate, landfill leachates, and livestock effluent (Kizito et al., 2015; Takaya et al., 2016). The use of biochar supplemented with digestate was discovered to improve soil quality by substantially increasing levels of macronutrients (110%–230%) and organic matter content (232%–514%).

The researchers in Hu et al.'s study (2019) produced biochar/struvite composites using a combination of crystallization and adsorption techniques. The biochar was mixed with a solution containing magnesium ions (Mg^{2+}), phosphate ions (PO_4^{3-}), and ammonium ions (NH^{4+}) in equal molar proportions (1:1:1), leading to the formation of biochar/struvite. The aqueous solution, which consisted of 10% biochar/struvite, had an impressive N recovery rate of 99.02%, P recovery rate of 97.23%, and magnesium (Mg) recovery rate of 95.22%. The XRD analysis indicated the existence of a distinct crystalline arrangement corresponding to struvite ($MgNH_4PO_4{\cdot}6H_2O$). The SEM images demonstrated that the biochar had a surface characterized by porosity and roughness, but the struvite adhered to the charcoal displayed a smooth and impermeable surface. The EDS examination has verified the presence of struvite. The compound of biochar and struvite has demonstrated exceptional efficacy as a slow-release fertilizer, surpassing the nutrient release capabilities of traditional chemical fertilizers. Wastewater containing elevated levels of N and P can be effectively utilized as a suitable resource for producing struvite. Some examples of wastewater include swine wastewater, calf dung wastewater, leather tanning wastewater, sewage sludge, dairy wastewater, and semiconductor effluent.

Impregnation has clear advantages in terms of simplicity and cost-effectiveness. However, the adsorption capacity of charcoal limits the fertilizer efficiency of SRFs. It is feasible to improve the adsorption capacity of biochar for nutrients by making modifications; however, it is crucial to consider the expenses connected with the modification procedure. The idea of reclaiming nutrients from wastewater using biochar impregnation is quite attractive. This method not only reduces the negative impact on the environment caused by wastewater, such as the excessive growth of plants due to an abundance of nutrients, but also optimizes the use of the nutrients that are released. An evaluation of the limited assimilation of nutrients in biochar and the existence of perilous chemicals is imperative for practical application.

21.4.4 Encapsulation

Encapsulation is a frequently used method in the manufacturing of SRFs. The process entails the application of a safeguarding coating of fertilizers onto either inorganic or organic surfaces. This process aims to reduce nutrition loss and optimize nutrient utilization. Multiple studies have documented the process of creating SRFs based on biochar through the use of encapsulating methods. Notable instances include of urea coated with biochar-biodegradable polymers, urea coated with biochar-waterborne copolymers, and biochar-coated urea. The combination of poultry manures and green waste was subjected to cocomposting with biochar, leading to the development of an organic coating that is more abundant in N and K nutrients on both the surface and inner pores of the biochar. A slow-release fertilizer was manufactured using the graft-copolymerization method, which involved the integration of rapeseed meal biochar and sodium alginate. As a result, there were significant improvements in the ability to retain water, the amount of N present, and the progressive release of the graft copolymers. The granular SRFs were produced by utilizing modified starch derived from cassava or maize as coating agents. The utilization of granular biochar that was treated with $MgCl_2$, followed by coating it with polyvinyl alcohol and maize starch. Compared to uncoated $MgCl_2$-modified biochar and chemical fertilizer, the granular coated biochar exhibited the slowest N release, the highest N concentration in both the soil and maize plants, and the greatest NUE. The scientists used a microwave-assisted method to fabricate composites consisting of biochar and hydrogel. Microwave irradiation accelerated the alteration of the structure and promoted the development of cross-links between molecules. The composite demonstrated improved water absorption and N release capacities. Jia and her colleagues synthesized biochar-coated urea by using oxidized starch as a precursor . Urea-coated biochar was successful in reducing N loss and improving NUE.

The variability in the efficacy of biochar-coated fertilizer can be ascribed to the attributes of biochar, the constituents employed for coating, and the encapsulating techniques. The utilization of biochar with a significant surface area proved beneficial in controlling the release of nutrients due to its strong adsorption capabilities. The existence of alkaline groups and inorganic salts in biochar increases the emission of NH_3 by volatilization. The existence of surface functional groups on biochar significantly influences the development of granular fertilizer. Various coating materials, such as organic polymers and inorganic minerals, are used to manufacture SRFs. Polymers such as polyolefin, copolymers, polyvinylidene chloride, wax, and paraffin are costly and do not possess the ability to biodegrade. Furthermore, the accumulation of coated polymers deteriorates the arrangement of soil, hampers the process of soil nutrient circulation, and limits microbiological functioning. The coating SRFs commonly employ environmentally degradable materials such as discarded palm oil, starch, and a blend of chitosan and poly (vinyl alcohol). Researchers have utilized a range of eco-friendly coating materials, such as cotton stalk-based polyol, wood vinegar, sodium alginate, cellulose, starch, chitosan, and maltodextrin, to create sustainable release fertilizers (SRFs) based on biochar. A water-based copolymer was utilized as a covering substance for the fabrication of a biochar-based SRF. The incorporation of biochar decreased the water absorbency of the copolymer, while simultaneously improving its degradability. As a consequence, there was a decelerated discharge of urea that had been coated. The researchers used bio-oil derived from biomass pyrolysis to apply a layer on biochar that was enhanced with N, P, and K nutrients. Using bio-oil coating greatly improved the capacity of nutrients to be gradually released over time.

Encapsulation is a vital method utilized in manufacturing SRFs, in which coating materials are used as a barrier to control the diffusion and release of nutrients. The encapsulation process is typically complex, involving the infusion of biochar with nutrients, the blending of different components (biochar, minerals, glue, binders, etc.), and the production of pellets. The complex and expensive process of encapsulation greatly limits the use of coated SRFs in agriculture. Coating materials have inherent attributes in terms of agronomy, environment, and economy.

After the nutrients are released, the coating materials remain in the soil because they do not degrade easily. This causes the materials to accumulate, which can potentially create environmental hazards gradually. Coating materials can potentially introduce dangerous organic chemicals, reduce the ability of soil to allow substances to pass through, and hinder the activity of microorganisms. The future trajectory entails the application of eco-friendly, biodegradable, and cost-efficient coating materials.

21.5 Biochar-based fertilizers

Conventional fertilizers pose a significant hazard to the environment through various means, including leaching nutrients, contribution to global warming, and volatilization from fertilized croplands. They developed a composite material called granular biochar mineral urea composite (Bio-MUC), which demonstrated superior N retention compared to urea fertilizer (Shi et al., 2020). The Bio-MUC facilitated plant growth by increasing shoot growth by 14% and root growth by 25%. In their study, they demonstrated that biochar compound fertilizers derived from various sources (such as pig manure compost, maize straw, peanut husk, and municipal waste) were significantly more effective in enhancing NUE and decreasing greenhouse gas emissions in rice cultivation compared with conventional chemical fertilizers. Biochar-based fertilizers have been utilized to investigate the interactions between biochar and the rhizosphere.

In a study conducted by Chew et al. (2020), the introduction of biochar to agricultural fields led to notable alterations in the electron status of the rhizospheric soils. This resulted in a 65 mV increase in the potential difference between the rhizospheric soil and the root membrane. Consequently, there was a substantial 67% enhancement in both plant nutrient content and biomass. This occurred because there was a decreased need for free energy to accumulate nutrients. Therefore it can be inferred that biochar-based fertilizers offer many agronomic advantages, such as enhancing the absorption of nutrients (N and P) and promoting plant biomass growth.

21.6 Biochar stimulates soil microbial response to heavy metal contaminations

Various factors can disturb ecosystems, and heavy metals are a substantial environmental hazard. They have the potential to cause life-threatening disorders in people, such as Minamata disease and itai–itai disease (Taty-Costodes et al., 2003). Cadmium, a heavy metal, has notable characteristics including its high solubility, mobility, and ability to accumulate in living organisms, which have negative effects on the quality of soil and water (Belhalfaoui et al., 2009). The listed contaminants, such as those present in nickel–cadmium batteries, metal coatings, soil waste incineration, and agricultural effluents, have a tendency to collect in particular environmental areas. The accumulation of pollutants increases their concentration, subsequently causing contamination of groundwater, surface water, and other aquatic systems.

The current scenario has sparked interest in researching the utilization of biochar derived from different pyrolyzed biomass to enhance soil quality and hinder the dispersion of heavy metals. An abundant and varied microbial community is necessary to effectively convert and move organic carbon in soil. Nevertheless, the presence of toxic heavy metals, such as lead (Pb), arsenic (As), cadmium (Cd), zinc (Zn), copper (Cu), and others, impedes the proliferation of these indigenous bacteria, resulting in soil degradation. As a result of this suppression, the amount of phospholipid fatty acids in the soil decreases, leading to a decrease in microbial carbon-use efficiency. The physicochemical features of biochar make it a valuable technique for soil remediation. The carbonaceous material possesses the capacity to adsorb heavy metal pollutants onto its surface, thus diminishing the accessibility of these metals to plants and soil microbes (Mackie et al., 2015). The utilization of biochar derived from rice straw and rice hull resulted in a substantial reduction in the bioavailability of cadmium. This decrease significantly contributed to the increase in soil microbial biomass, soil N, and organic matter. The references for Bashir et al.'s works in 2018 are labeled as a, b, and c. The utilization of different raw materials in the synthesis of biochar has specific effects on the mobility of metals, leading to a variety of routes for ion adsorption based on the inherent properties of the resulting pyrolyzed product. The study found that animal- and plant-derived manures with distinct surface functional groups like –OH and –COOH aided in the chelation of Cd^{2+} ions. The interaction between sewage sludge and animal-derived biochars results in the formation of solid particles due to the reaction of Cd^{2+} ions with minerals present in the biochar, namely phosphate ions (PO_4^{3-}) and carbonate ions (CO_3^{2-}) (Chen et al., 2015; Xu et al., 2013). Plant-based biochars, such as those derived from wheat straw, can immobilize and adsorb Cd^{2+} cations. This is achieved by substituting them with other cations, such as K^+, Ca^{2+}, Mg^{2+}, and Na^+, which are present in the biochar matrix. Arsenic (As), a distinct metallic element, is known for its capacity to cause diverse health effects in humans. Arsenic is subject to methylation and volatilization processes, which are promoted by various soil microorganisms, such as fungi, bacteria, and yeast. These actions lead to the conversion of arsenic into less harmful forms.

The phenomenon in question has undergone thorough examination and recording by Zeng et al. (2013). However, it is essential to use biostimulation tactics to enhance the metabolic activity of soil microbial communities to optimize arsenic methylation and volatilization. A recent study discovered that the use of biochar, along with genetically modified Pseudomonas putida, led to an augmentation in the process of methylation and volatilization of arsenic, causing it to be released into the atmosphere. Moreover, the nature of the soil also exerts a substantial influence on this restoration process (Chen et al., 2017). Therefore biochar, due to its capacity to enhance microbial activity, has great potential as a technique to combat metal pollution in many areas efficiently.

21.7 Nanomaterial-based modified biochar and their application

The chemical makeup of biochar is heavily influenced by the type of material used as the raw feedstock and the specific circumstances of the pyrolysis process used for its synthesis. Although biochar possesses a highly favorable nutritional content, the pyrolysis process may result in the loss of certain essential food elements. Thus, by employing sophisticated technologies such as nanotechnology, biochar can be strengthened to improve its nutrient contents and other crucial characteristics. The addition of nanomaterials to biochar leads to a significant improvement in its physical and chemical properties. This includes increased pores, enhanced ability to exchange cations, immobilization of heavy metals, larger specific surface area, and improved nutrient composition (Tan et al., 2016). This enhancement in physicochemical qualities improves the effectiveness of biochar in promoting plant development, soil fertility, and the availability of nutrients in the soil. Biochars that have been altered with nanoparticles, including magnetic iron oxide particles, graphene, chitosan, carbon nanotubes, nanoscale zero-valent iron, and nanometal oxides/hydroxides, have been widely used in various environmental remediation approaches (Tan et al., 2016). The use of biochar treated with nanocomposites consisting of magnesium (Mg) and manganese (Mn) metal oxides had a notable effect on reducing the impact of salinity stress in Safflower (*Carthamus tinctorius*) plants. Additionally, this treatment enhanced the accessibility of essential nutrients like Mg and Mn, improving the plants' physiological and biochemical processes. Biochar-based nanocomposites have been increasingly popular in wastewater treatment due to their superior ability to adsorb pollutants from polluted water, surpassing that of raw biochar (Pi et al., 2015). Nano biochar, which consists of biochar combined with catalytic nanoparticles, shows potential as a technique for effectively eliminating organic contaminants, including methylene blue, phenol, and sulfamethoxazole, from water that has been contaminated. Biochars have desirable qualities, including electrical solid conductivity, distinctive surface properties with surface functional groups, and chemical stability. These properties make them beneficial for blending with catalytic nanoparticles, such as biochar loaded with TiO_2. Biochar boosts the stability, adsorption, and photocatalytic degradation capacity of blended nanoparticles when compared with nanoparticles alone. This improvement is observed in organic pollutant removal from wastewater (Zhang et al., 2017). Studies have also examined the efficacy of biochar-based nanomaterials in soil repair and carbon sequestration.

Liu et al. (2020) demonstrated that using nano biochar positively reduced the harmful impacts of soil heavy metal (Cd^{2+}) contamination and enhanced the soil microbial biomass. The application of Ni/Fe biochar in the soil–plant system has been found to be effective in adsorbing polybrominated diphenyl esters, reducing their phytotoxicity, and minimizing their accumulation and translocation. Creamer et al. (2018) utilized biochar with stabilized MgO nanoparticles to capture CO_2. The researchers discovered that the ratio between biomass and nanoparticles substantially impacted the nanocomposite's ability to capture carbon dioxide. Hence, using nanomaterial in conjunction with biochar remediation technology can emerge as a proficient, dependable, and cost-effective approach for environmental remediation.

21.8 Future perspective and conclusions

Strategic methodologies and innovative methods are crucial for addressing the increasing presence of environmental pollutants. These contaminants have disastrous effects on all living organisms, including bacteria, plants, animals, and humans. The presence of factors that pose a threat to the existence of plants raises serious concerns about the ability of species that rely on plants for sustenance to survive. Considering these conditions, the use of biochar in environmental management has garnered considerable interest due to its physicochemical and biological characteristics. Biochar is an affordable and sustainable technology that involves pyrolyzing undesired organic waste into a carbon-rich substance. Additionally, biochar enhances soil fertility by providing abundant carbon content and remaining in the soil for an extended period, allowing for the sequestration of carbon and the enrichment of mineral composition. Moreover, the use of biochar has been shown to be useful in alleviating various biological and environmental pressures on plants. While there are drawbacks associated with biochar usage, its application rate is a major concern, which might lead to a decrease in herbicide availability in agricultural fields due to surface adsorption. However, sophisticated and innovative methods can be used to analyze the properties of biochar and uncover distinct characteristics unique to the material used as feedstock. The application possibilities and limitations of biochar extend to various disciplines, such as gas adsorption, supercapacitors, energy storage, fuel cell systems, pharmaceuticals, and the removal of PAH contaminants. The recent progress in nanotechnology for enhancing the physicochemical characteristics of biochar has garnered the interest of researchers due to its superior abilities in soil and wastewater purification when compared with regular biochar. Thorough analysis is necessary to assess the potential of biochar as a technology-driven environmental management tool, considering its advantages and disadvantages. An inquiry should be conducted to analyze the properties of the lignocellulosic components in the feedstock material, understand the pyrolysis mechanisms, including the process parameters, and gather data through extensive field studies.

Therefore biochar has emerged as a highly promising instrument for waste management. Modified biochars have been widely utilized for many environmental purposes, particularly as soil fertilizers, providing a novel approach to improving soil quality. To maximize the benefits, it is necessary to conduct further research on novel techniques for preparing and modifying biochar, and to investigate their long-term impact on soil properties. To mitigate the harmful consequences of utilizing these fertilizers in agricultural fields, it is imperative to conduct a thorough examination of the manufacture of biochar from contaminated waste materials.

To optimize the utilization of biochar, it is advisable to establish a comprehensive database that encompasses detailed information regarding the source materials used in the production of biochar-based fertilizers, the characteristics of biochar, and its effects on the soil ecosystem. Such a database would facilitate the effective application of biochar-based fertilizers across a wide range of crops. Biochar enriched with naturally occurring minerals can be used as a substitute for chemical fertilizers, reducing their usage in agriculture. The enhanced characteristics of biochar, such as specific surface area, pore volume, CEC, and nutrient composition, have a significant impact on preserving and improving soil quality, ensuring long-term crop production. Additionally, nano biochar has several applications in enhancing plant growth and yield, recycling nutrients, remediating pesticides, applying fertilizers, amending soil, and eliminating pathogenic effects. Despite its numerous benefits, further research is needed to understand the impact of nano additives produced from biochar on plant growth, microbial communities, and the environment. Biochar can also function as a nanocarrier to transport pesticides, nucleic acid, and fertilizers directly to crops and plants, making it a viable option for enhancing soil fertility and supporting agricultural output.

Acknowledgments

VDR and TM were supported by the Strategic Academic Leadership Program of Southern Federal University, "Priority 2030," and the Ministry of Science and Higher Education of the Russian Federation (grant number: FENW-2023-0008).

Credit author statement

Conceptualization: Vishnu D. Rajput; methodology: Vishnu D. Rajput; software: Karen Ghazaryan and Sumira Malik; validation: Karen Ghazaryan and Sumira Malik; formal analysis: Vishnu D. Rajput; investigation: Sagar Mondal, Jutishna Bora, Priyadarshani Rajput, and Abhishek Singh; resourcesm data curation: Sagar Mondal, Jutishna Bora, Priyadarshani Rajput, and Abhishek Singh; writing—original draft and writing—review and editing: Swati Priya, Sagar Mondal, Jutishna Bora, and Priyadarshani Rajput; visualization and supervision: Tatiana Minkina; project administration: Tatiana Minkina; and funding acquisition: Vishnu D. Rajput.

Conflict of interest

The authors declare that they have no known competing financial interests or personal relationships that could have appeared to influence the work reported in this work.

AI and AI-assisted technologies

AI and AI-assisted technologies were not used to prepare this work.

References

Akhtar, N., Khan, S., Rehman, S. U., Rha, E. S., & Jamil, M. (2022). Combined effect of zinc oxide nanoparticles and bacteria on osmolytes and antioxidative parameters of rice (*Oryza sativa* L.) plant grown in heavy metal-contaminated water. *Adsorption Science and Technology*, *2022*. Available from https://doi.org/10.1155/2022/4148765, https://www.hindawi.com/journals/ast/.

Akinnawo, S. O. (2023). Eutrophication: Causes, consequences, physical, chemical and biological techniques for mitigation strategies. *Environmental Challenges*, *12*. Available from https://doi.org/10.1016/j.envc.2023.100733, https://www.journals.elsevier.com/environmental-challenges.

Alburquerque, J. A., Salazar, P., Barrón, V., Torrent, J., Del Campillo, M. D. C., Gallardo, A., & Villar, R. (2013). Enhanced wheat yield by biochar addition under different mineral fertilization levels. *Agronomy for Sustainable Development*, *33*(3), 475–484. Available from https://doi.org/10.1007/s13593-012-0128-3.

Ashiq, A., Adassooriya, N. M., Sarkar, B., Rajapaksha, A. U., Ok, Y. S., & Vithanage, M. (2019). Municipal solid waste biochar-bentonite composite for the removal of antibiotic ciprofloxacin from aqueous media. *Journal of Environmental Management*, *236*, 428–435. Available from https://doi.org/10.1016/j.jenvman.2019.02.006, https://www.sciencedirect.com/journal/journal-of-environmental-management.

Bah, A., Husni, M. H. A., Teh, C. B. S., Rafii, M. Y., Syed Omar, S. R., & Ahmed, O. H. (2014). Reducing runoff loss of applied nutrients in oil palm cultivation using controlled-release fertilizers. *Advances in Agriculture*, *2014*, 1–9. Available from https://doi.org/10.1155/2014/285387.

Baig, S. A., Zhu, J., Muhammad, N., Sheng, T., & Xu, X. (2014). Effect of synthesis methods on magnetic Kans grass biochar for enhanced As(III, V) adsorption from aqueous solutions. *Biomass and Bioenergy*, *71*, 299–310. Available from https://doi.org/10.1016/j.biombioe.2014.09.027, http://www.elsevier.com/inca/publications/store/9/8/6.

Bayram, M., & Öner, M. D. (2007). Bulgur milling using roller, double disc and vertical disc mills. *Journal of Food Engineering*, *79*(1), 181–187. Available from https://doi.org/10.1016/j.jfoodeng.2006.01.042.

Belhalfaoui, B., Aziz, A., Elandaloussi, E. H., Ouali, M. S., & De Ménorval, L. C. (2009). Succinate-bonded cellulose: A regenerable and powerful sorbent for cadmium-removal from spiked high-hardness groundwater. *Journal of Hazardous Materials*, *169*(1-3), 831–837. Available from https://doi.org/10.1016/j.jhazmat.2009.04.021.

Bonanomi, G., Ippolito, F., Cesarano, G., Nanni, B., Lombardi, N., Rita, A., Saracino, A., & Scala, F. (2017). Biochar as plant growth promoter: Better off alone or mixed with organic amendments? *Frontiers in Plant Science*, *8*. Available from https://doi.org/10.3389/fpls.2017.01570.

Bopp, C., Christl, I., Schulin, R., & Evangelou, M. W. H. (2016). Biochar as possible long-term soil amendment for phytostabilisation of TE-contaminated soils. *Environmental Science and Pollution Research*, *23*(17), 17449–17458. Available from https://doi.org/10.1007/s11356-016-6935-3.

Chan, K. Y., Van Zwieten, L., Meszaros, I., Downie, A., & Joseph, S. (2008). Using poultry litter biochars as soil amendments. *Australian Journal of Soil Research*, *46*(5), 437–444. Available from https://doi.org/10.1071/SR08036.

Chaudhary, P., Rajput, V. D., Chen, S., & Jaiswal, D. K. (2023). *Bioinoculants with nano-compounds to improve soil health: A step toward sustainable*. *Agriculture frontiers research topics*. Frontiers Media SA. Available from https://doi.org/10.3389/978-2-8325-3287-4.

Cheah, S., Jablonski, W. S., Olstad, J. L., Carpenter, D. L., Barthelemy, K. D., Robichaud, D. J., Andrews, J. C., Black, S. K., Oddo, M. D., & Westover, T. L. (2016). Effects of thermal pretreatment and catalyst on biomass gasification efficiency and syngas composition. *Green Chemistry*, *18*(23), 6291–6304. Available from https://doi.org/10.1039/c6gc01661h, http://pubs.rsc.org/en/journals/journal/gc.

Chen, P., Li, J., Wang, H. Y., Zheng, R. L., & Sun, G. X. (2017). Evaluation of bioaugmentation and biostimulation on arsenic remediation in soil through biovolatilization. *Environmental Science and Pollution Research*, *24*(27), 21739–21749. Available from https://doi.org/10.1007/s11356-017-9816-5, http://www.springerlink.com/content/0944-1344.

Creamer, A. E., Gao, B., Zimmerman, A., & Harris, W. (2018). Biomass-facilitated production of activated magnesium oxide nanoparticles with extraordinary CO_2 capture capacity. *Chemical Engineering Journal*, *334*, 81–88. Available from https://doi.org/10.1016/j.cej.2017.10.035, http://www.elsevier.com/inca/publications/store/6/0/1/2/7/3/index.htt.

Debicka, M., Kocowicz, A., Weber, J., & Jamroz, E. (2016). Organic matter effects on phosphorus sorption in sandy soils. *Archives of Agronomy and Soil Science*, *62*(6), 840–855. Available from https://doi.org/10.1080/03650340.2015.1083981, http://www.tandf.co.uk/journals/titles/03650340.asp.

Dong, X., Li, G., Lin, Q., & Zhao, X. (2017). Quantity and quality changes of biochar aged for 5 years in soil under field conditions. *Catena*, *159*, 136–143. Available from https://doi.org/10.1016/j.catena.2017.08.008, http://www.elsevier.com/inca/publications/store/5/2/4/6/0/9.

Fan, X., Chang, D. W., Chen, X., Baek, J. B., & Dai, L. (2016). Functionalized graphene nanoplatelets from ball milling for energy applications. *Current Opinion in Chemical Engineering*, *11*, 52–58. Available from https://doi.org/10.1016/j.coche.2016.01.003, http://www.elsevier.com/wps/find/journaldescription.cws_home/725837/description#description.

Guo, F., Bao, L., Wang, H., Larson, S. L., Ballard, J. H., Knotek-Smith, H. M., Zhang, Q., Su, Y., Wang, X., & Han, F. (2020). A simple method for the synthesis of biochar nanodots using hydrothermal reactor. *MethodsX*, *7*. Available from https://doi.org/10.1016/j.mex.2020.101022, http://www.journals.elsevier.com/methodsx/.

Hammer, E. C., Forstreuter, M., Rillig, M. C., & Kohler, J. (2015). Biochar increases arbuscular mycorrhizal plant growth enhancement and ameliorates salinity stress. *Applied Soil Ecology*, *96*, 114–121. Available from https://doi.org/10.1016/j.apsoil.2015.07.014, http://www.elsevier.com/inca/publications/store/5/2/4/5/1/8/index.htt.

Harikishore Kumar Reddy, D., & Lee, S. M. (2014). Magnetic biochar composite: Facile synthesis, characterization, and application for heavy metal removal. *Colloids and Surfaces A: Physicochemical and Engineering Aspects*, *454*(1), 96–103. Available from https://doi.org/10.1016/j.colsurfa.2014.03.105, http://www.elsevier.com/locate/colsurfa.

Helal, M. I., Husein, M. E., Walaa, G., & Mostafa, E. D. (2019). Characterization of agricultural residues-based nano biochar and its efficiency in adsorption/desorption of nutrients. *International Journal of Environment*, *8*(2), 130–141.

Inyang, M., Gao, B., Zimmerman, A., Zhang, M., & Chen, H. (2014). Synthesis, characterization, and dye sorption ability of carbon nanotube–biochar nanocomposites. *Chemical Engineering Journal*, *236*, 39–46. Available from https://doi.org/10.1016/j.cej.2013.09.074.

Jin, Z., Xiao, S., Dong, H., Xiao, J., Tian, R., Chen, J., Li, Y., & Li, L. (2022). Adsorption and catalytic degradation of organic contaminants by biochar: Overlooked role of biochar's particle size. *Journal of Hazardous Materials*, *422*, 126928. Available from https://doi.org/10.1016/j.jhazmat.2021.126928.

Karinkanta, P., Ämmälä, A., Illikainen, M., & Niinimäki, J. (2018). Fine grinding of wood – overview from wood breakage to applications. *Biomass and Bioenergy*, *113*, 31–44. Available from https://doi.org/10.1016/j.biombioe.2018.03.007.

Khodakovskaya, M. V., Kim, B. S., Kim, J. N., Alimohammadi, M., Dervishi, E., Mustafa, T., & Cernigla, C. E. (2013). Carbon nanotubes as plant growth regulators: Effects on tomato growth, reproductive system, and soil microbial community. *Small (Weinheim an der Bergstrasse, Germany)*, *9*(1), 115–123. Available from https://doi.org/10.1002/smll.201201225.

Kong, M., Liang, J., White, J. C., Elmer, W. H., Wang, Y., Xu, H., He, W., Shen, Y., & Gao, X. (2022). Biochar nanoparticle-induced plant immunity and its application with the elicitor methoxyindole in Nicotiana benthamiana. *Environmental Science: Nano*, *9*(9), 3514–3524. Available from https://doi.org/10.1039/d1en00953b, http://pubs.rsc.org/en/journals/journal/en.

Kuang, R., Gui, D., Wu, L., Zeng, G., Si, D., & Liu, J. (2012). Preparation of thermal interface material filled with micro-nano-composite particles into the polymer. In *Proceedings—2012 13th international conference on electronic packaging technology and high density packaging (ICEPT-HDP 2012)* (pp. 695–699), China. Available from https://doi.org/10.1109/ICEPT-HDP.2012.6474713.

Lehmann, J., & Joseph, S. (2015). *Biochar for environmental management: an introduction* (pp. 33–46). Informa UK Limited. Available from https://doi.org/10.4324/9780203762264-8.

Liang, B., Lehmann, J., Solomon, D., Kinyangi, J., Grossman, J., O'Neill, B., Skjemstad, J. O., Thies, J., Luizão, F. J., Petersen, J., & Neves, E. G. (2006). Black carbon increases cation exchange capacity in soils. *Soil Science Society of America Journal*, *70*(5), 1719–1730. Available from https://doi.org/10.2136/sssaj2005.0383.

Liu, G., Zheng, H., Jiang, Z., Zhao, J., Wang, Z., Pan, B., & Xing, B. (2018). Formation and physicochemical characteristics of nano biochar: Insight into chemical and colloidal stability. *Environmental Science & Technology*, *52*(18), 10369–10379. Available from https://doi.org/10.1021/acs.est.8b01481.

Liu, W., Li, Y., Feng, Y., Qiao, J., Zhao, H., Xie, J., Fang, Y., Shen, S., & Liang, S. (2020). The effectiveness of nanobiochar for reducing phytotoxicity and improving soil remediation in cadmium-contaminated soil. *Scientific Reports*, *10*(1). Available from https://doi.org/10.1038/s41598-020-57954-3.

Liu, X., Zhang, A., Ji, C., Joseph, S., Bian, R., Li, L., Pan, G., & Paz-Ferreiro, J. (2013). Biochar's effect on crop productivity and the dependence on experimental conditions—a meta-analysis of literature data. *Plant and Soil*, *373*(1-2), 583–594. Available from https://doi.org/10.1007/s11104-013-1806-x.

Lyu, H., Gao, B., He, F., Zimmerman, A. R., Ding, C., Huang, H., & Tang, J. (2018). Effects of ball milling on the physicochemical and sorptive properties of biochar: Experimental observations and governing mechanisms. *Environmental Pollution*, *233*, 54–63. Available from https://doi.org/10.1016/j.envpol.2017.10.037, http://www.elsevier.com/inca/publications/store/4/0/5/8/5/6.

Mackie, K. A., Marhan, S., Ditterich, F., Schmidt, H. P., & Kandeler, E. (2015). The effects of biochar and compost amendments on copper immobilization and soil microorganisms in a temperate vineyard. *Agriculture, ecosystems & environment*, *201*, 58–69. Available from https://doi.org/10.1016/j.agee.2014.12.001.

Mangrich, A. S., Cardoso, E. M. C., Doumer, M. E., Romão, L. P. C., Vidal, M., Rigol, A., & Novotny, E. H. (2015). Improving the water holding capacity of soils of Northeast Brazil by biochar augmentation. *ACS Symposium Series*, *1206*, 339–354. Available from https://doi.org/10.1021/bk-2015-1206.ch016, http://global.oup.com/academic/content/series/a/acs-symposium-series-acsss/?cc = in&lang = en.

Meyer, S., Glaser, B., & Quicker, P. (2011). Technical, economical, and climate-related aspects of biochar production technologies: A literature review. *Environmental Science & Technology*, *45*(22), 9473–9483. Available from https://doi.org/10.1021/es201792c.

Mohan, D., Kumar, S., & Srivastava, A. (2014). Fluoride removal from ground water using magnetic and nonmagnetic corn stover biochars. *Ecological Engineering*, *73*, 798–808. Available from https://doi.org/10.1016/j.ecoleng.2014.08.017.

Naghdi, M., Taheran, M., Brar, S. K., Kermanshahi-pour, A., Verma, M., & Surampalli, R. Y. (2018). Pinewood nanobiochar: A unique carrier for the immobilization of crude laccase by covalent bonding. *International Journal of Biological Macromolecules*, *115*, 563–571. Available from https://doi.org/10.1016/j.ijbiomac.2018.04.105, http://www.elsevier.com/locate/ijbiomac.

Ndoung, O. C. N., Figueiredo, C. Cd, & Ramos, M. L. G. (2021). A scoping review on biochar-based fertilizers: enrichment techniques and agro-environmental application. *Heliyon*, *7*(12). Available from https://doi.org/10.1016/j.heliyon.2021.e08473, http://www.journals.elsevier.com/heliyon/.

Oleszczuk, P., Ćwikła-Bundyra, W., Bogusz, A., Skwarek, E., & Ok, Y. S. (2016). Characterization of nanoparticles of biochars from different biomass. *Journal of Analytical and Applied Pyrolysis*, *121*, 165–172. Available from https://doi.org/10.1016/j.jaap.2016.07.017.

Pan, F., Chapman, S. J., Li, Y., & Yao, H. (2017). Straw amendment to paddy soil stimulates denitrification but biochar amendment promotes anaerobic ammonia oxidation.

Journal of Soils and Sediments, *17*(10), 2428–2437. Available from https://doi.org/10.1007/s11368-017-1694-4, http://www.springerlink.com/content/1439-0108.

Pi, L., Jiang, R., Zhou, W., Zhu, H., Xiao, W., Wang, D., & Mao, X. (2015). g-C_3N_4 modified biochar as an adsorptive and photocatalytic material for decontamination of aqueous organic pollutants. *Applied Surface Science*, *358*, 231–239. Available from https://doi.org/10.1016/j.apsusc.2015.08.176.

Purbalisa, W., Zulaehah, I., & Paputri, D. M. W. (2021). Application of remediation treatment to reduce lead in the soil of shallot cropping. *IOP Conference Series: Earth and Environmental Science*, *648*(1), 012063. Available from https://doi.org/10.1088/1755-1315/648/1/012063.

Rajput, P., Kumar, P., Priya, A. K., Kumari, S., Shiade, S. R. G., Rajput, V. D., Fathi, A., Pradhan, A., Sarfraz, R., Sushkova, S., Mandzhieva, S., Minkina, T., Soldatov, A., Wong, M. H., & Rensing, C. (2024). Nanomaterials and biochar mediated remediation of emerging contaminants. *Science of the Total Environment*, *916*. Available from https://doi.org/10.1016/j.scitotenv.2024.170064, https://www.sciencedirect.com/science/journal/00489697.

Rajput, V. D., Minkina, T., Upadhyay, S. K., Kumari, A., Ranjan, A., Mandzhieva, S., Sushkova, S., Singh, R. K., & Verma, K. K. (2022). Nanotechnology in the restoration of polluted soil. *Nanomaterials*, *12*(5). Available from https://doi.org/10.3390/nano12050769, https://www.mdpi.com/2079-4991/12/5/769/pdf.

Ramadan, M. M., Asran-Amal., & Abd-Elsalam, K. A. (2020). *Micro/nano biochar for sustainable plant health: Present status and future prospects* (pp. 323–357). Elsevier BV. Available from https://doi.org/10.1016/b978-0-12-819786-8.00016-5.

Ramanayaka, S., Vithanage, M., Alessi, D. S., Liu, W. J., Jayasundera, A. C. A., & Ok, Y. S. (2020). Nanobiochar: Production, properties, and multifunctional applications. *Environmental Science: Nano*, *7*(11), 3279–3302. Available from https://doi.org/10.1039/d0en00486c, http://pubs.rsc.org/en/journals/journal/en.

Ramzan, M., Zia, A., Naz, G., Shahid, M., Shah, A. A., & Farid, G. (2023). Effect of nanobiochar (nBC) on morpho-physio-biochemical responses of black cumin (*Nigella sativa* L.) in Cr-spiked soil. *Plant Physiology and Biochemistry*, *196*, 859–867. Available from https://doi.org/10.1016/j.plaphy.2023.02.037, http://www.journals.elsevier.com/plant-physiology-and-biochemistry/.

Sani, M. N. H., Amin, M., Siddique, A. B., Nasif, S. O., Ghaley, B. B., Ge, L., Wang, F., & Yong, J. W. H. (2023). Waste-derived nanobiochar: A new avenue towards sustainable agriculture, environment, and circular bioeconomy. *Science of the Total Environment*, *905*. Available from https://doi.org/10.1016/j.scitotenv.2023.166881, http://www.elsevier.com/locate/scitotenv.

Shi, W., Ju, Y., Bian, R., Li, L., Joseph, S., Mitchell, D. R. G., Munroe, P., Taherymoosavi, S., & Pan, G. (2020). Biochar bound urea boosts plant growth and reduces nitrogen leaching. *Science of the Total Environment*, *701*. Available from https://doi.org/10.1016/j.scitotenv.2019.134424, http://www.elsevier.com/locate/scitotenv.

Soja, G., Kloß, S., Karer, J., Kitzler, B., Wimmer, B., & Zehetner, F. (2013). Biochar as yield promoter for agricultural crops? *Conjectures and observations 1st FOREBIOM workshop*. ÖAW.

Tan, X., Liu, Y., Zeng, G., Wang, X., Hu, X., Gu, Y., & Yang, Z. (2015). Application of biochar for the removal of pollutants from aqueous solutions. *Chemosphere*, *125*,

70–85. Available from https://doi.org/10.1016/j.chemosphere.2014.12.058, http://www.elsevier.com/locate/chemosphere.

Tan, X. F., Liu, Y. G., Gu, Y. L., Xu, Y., Zeng, G. M., Hu, X. J., Liu, S. B., Wang, X., Liu, S. M., & Li, J. (2016). Biochar-based nano-composites for the decontamination of wastewater: A review. *Bioresource Technology*, *212*, 318–333. Available from https://doi.org/10.1016/j.biortech.2016.04.093, http://www.elsevier.com/locate/biortech.

Taty-Costodes, V. C., Fauduet, H., Porte, C., & Delacroix, A. (2003). Removal of Cd(II) and Pb(II) ions, from aqueous solutions, by adsorption onto sawdust of Pinus sylvestris. *Journal of Hazardous Materials*, *105*(1-3), 121–142. Available from https://doi.org/10.1016/j.jhazmat.2003.07.009, http://www.elsevier.com/locate/jhazmat.

Tomczyk, A., Sokołowska, Z., & Boguta, P. (2020). Biochar physicochemical properties: Pyrolysis temperature and feedstock kind effects. *Reviews in Environmental Science and Bio/Technology*, *19*(1), 191–215. Available from https://doi.org/10.1007/s11157-020-09523-3.

Vaccari, D. A., & Strigul, N. (2011). Extrapolating phosphorus production to estimate resource reserves. *Chemosphere*, *84*(6), 792–797. Available from https://doi.org/10.1016/j.chemosphere.2011.01.052, http://www.elsevier.com/locate/chemosphere.

Verma, K. K., Song, X. P., Degu, H. D., Guo, D. J., Joshi, A., Huang, H. R., Xu, L., Singh, M., Huang, D. L., Rajput, V. D., & Li, Y. R. (2023). Recent advances in nitrogen and nano-nitrogen fertilizers for sustainable crop production: A mini-review. *Chemical and Biological Technologies in Agriculture*, *10*(1). Available from https://doi.org/10.1186/s40538-023-00488-3, https://chembioagro.springeropen.com.

Wang, D., Zhang, W., Hao, X., & Zhou, D. (2013). Transport of biochar particles in saturated granular media: Effects of pyrolysis temperature and particle size. *Environmental Science and Technology*, *47*(2), 821–828. Available from https://doi.org/10.1021/es303794d.

Wang, S., Gao, B., Zimmerman, A. R., Li, Y., Ma, L., Harris, W. G., & Migliaccio, K. W. (2015). Removal of arsenic by magnetic biochar prepared from pinewood and natural hematite. *Bioresource Technology*, *175*, 391–395. Available from https://doi.org/10.1016/j.biortech.2014.10.104, http://www.elsevier.com/locate/biortech.

Wang, S., Dai, G., Yang, H., & Luo, Z. (2017). Lignocellulosic biomass pyrolysis mechanism: A state-of-the-art review. *Progress in Energy and Combustion Science*, *62*, 33–86. Available from https://doi.org/10.1016/j.pecs.2017.05.004.

Wu, L., Wei, C., Zhang, S., Wang, Y., Kuzyakov, Y., & Ding, X. (2019). MgO-modified biochar increases phosphate retention and rice yields in saline-alkaline soil. *Journal of Cleaner Production*, *235*, 901–909. Available from https://doi.org/10.1016/j.jclepro.2019.07.043.

Yadav, M. R., Kumar, S., Lal, M. K., Kumar, D., Kumar, R., Yadav, R. K., Kumar, S., Nanda, G., Singh, J., Udawat, P., Meena, N. K., Jha, P. K., Minkina, T., Glinushkin, A. P., Kalinitchenko, V. P., & Rajput, V. D. (2023). Mechanistic understanding of leakage and consequences and recent technological advances in improving nitrogen use efficiency in cereals. *Agronomy*, *13*(2). Available from https://doi.org/10.3390/agronomy13020527, http://www.mdpi.com/journal/agronomy/.

Yan, L., Kong, L., Qu, Z., Li, L., & Shen, G. (2015). Magnetic biochar decorated with ZnS nanocrytals for Pb (II) removal. *ACS Sustainable Chemistry & Engineering*, *3*(1), 125–132. Available from https://doi.org/10.1021/sc500619r.

Yao, Y., Gao, B., Chen, J., & Yang, L. (2013). Engineered biochar reclaiming phosphate from aqueous solutions: Mechanisms and potential application as a slow-release fertilizer. *Environmental Science and Technology*, *47*(15), 8700–8708. Available from https://doi.org/10.1021/es4012977.

Yao, Y., Gao, B., Inyang, M., Zimmerman, A. R., Cao, X., Pullammanappallil, P., & Yang, L. (2011). Biochar derived from anaerobically digested sugar beet tailings: Characterization and phosphate removal potential. *Bioresource Technology*, *102*(10), 6273–6278. Available from https://doi.org/10.1016/j.biortech.2011.03.006.

Yue, L., Lian, F., Han, Y., Bao, Q., Wang, Z., & Xing, B. (2019). The effect of biochar nanoparticles on rice plant growth and the uptake of heavy metals: Implications for agronomic benefits and potential risk. *Science of the Total Environment*, *656*, 9–18. Available from https://doi.org/10.1016/j.scitotenv.2018.11.364, http://www.elsevier.com/locate/scitotenv.

Zeng, Z., Zhang, S. D., Li, T. Q., Zhao, F. L., He, Z. L., Zhao, H. P., Yang, X. E., Wang, H. L., Zhao, J., & Rafiq, M. T. (2013). Sorption of ammonium and phosphate from aqueous solution by biochar derived from phytoremediation plants. *Journal of Zhejiang University: Science B*, *14*(12), 1152–1161. Available from https://doi.org/10.1631/jzus.B1300102.

Zhang, H., Wang, Z., Li, R., Guo, J., Li, Y., Zhu, J., & Xie, X. (2017). TiO_2 supported on reed straw biochar as an adsorptive and photocatalytic composite for the efficient degradation of sulfamethoxazole in aqueous matrices. *Chemosphere*, *185*, 351–360. Available from https://doi.org/10.1016/j.chemosphere.2017.07.025.

Zhang, M., Gao, B., Yao, Y., Xue, Y., & Inyang, M. (2012). Synthesis, characterization, and environmental implications of graphene-coated biochar. *Science of the Total Environment*, *435-436*, 567–572. Available from https://doi.org/10.1016/j.scitotenv.2012.07.038.

CHAPTER

22 Nanotechnological aspects in producing biochar-mineral fertilizer for sustainable and intensive agricultural practices: pre- and postpyrolysis biochar modification

Pallavi Kumari[1], Smitha John Kundukulam[2], Manoj Kumar Solanki[3], Naveen Kumar Singh[4], Majeti Narasimha Vara Prasad[5] and Abhay Kumar[6]

[1]*Centre of Polymer and Carbon Materials, Polish Academy of Sciences, Zabrze, Poland*

[2]*Department of Soil Science & Agricultural Chemistry, College of Agriculture, Kerala Agricultural University, Vellanikkara, Kerala, India*

[3]*Department of Life Sciences and Biological Sciences, IES University, Bhopal, Madhya Pradesh, India*

[4]*Department of Life Science, Central University of South Bihar, Gaya, Bihar, India*

[5]*Retired Professor, School of Life Sciences, University of Hyderabad, Hyderabad, Telangana, India*

[6]*Department for Innovation in Biological, Agro-food and Forest Systems, University of Tuscia, Viterbo, Italy*

22.1 Introduction

The "Green Revolution" refers to substantial agricultural breakthroughs that began in the mid-20th century, including the adoption of high-yield crops, increased use of synthetic chemical nitrogen (N) fertilizer (CF), and improved irrigation (Mahmud et al., 2021). While pivotal for global food production, it unintentionally led to over-reliance on CF, where farmers extensively used CF such as urea and ammonium nitrate in the belief that adding more N would increase yields (Xin et al., 2019). However, crops frequently used applied N inefficiently, resulting in a low "N use efficiency" (NUE) (Lea & Azevedo, 2006). Excessive fertilizer application to soil or low plant NUE has a direct impact on the N budget, resulting in N losses through leaching, runoff, and the release of reactive N compounds (Nr) into the environment (Sharma & Bali, 2018). Therefore, implementing a fertilization management system based on natural

Biochar Ecotechnology for Sustainable Agriculture and Environment. DOI: https://doi.org/10.1016/B978-0-443-29855-4.00037-0

resources is crucial to boost food productivity for ever growing populations, while also preserving soil health and ensuring environmental sustainability, particularly in developing countries, where the majority of the global population resides.

Among various traditional soil management strategies currently available to address these issues, applying biochar to the soil emerges as an economically viable strategy compared to adopting modern agricultural technology for intensification and sustainability (Idbella et al., 2024). Biochar, produced through the pyrolysis of biomass under limited oxygen condition, serves as a valuable tool in agroecotechnology promoting physical and chemical soil characteristics, offering advantages such as improved water infiltration, reduced greenhouse gas (GHG) emissions, and enhanced plant productivity (Joseph et al., 2021). Its integration into farming practices aligns with the holistic approach of agroecology, fostering synergetic relation between agricultural productivity and ecological well-being. Integrating ecological principles into agricultural practices to promote sustainable and resilient systems and brings several benefits to agricultural ecosystems (Lehmann & Joseph, 2024). Biochar persists in the soil for extended periods, positively impacting soil properties at both physical and chemical levels. Unlike other organic amendments, its resistance to decomposition makes it cost-effective for small-scale farming (Keske et al., 2020). Beyond its soil-related benefits, biochar has applications in environmental remediation, as it can adsorb pollutants and contaminants (Kumar, Joseph, et al., 2018). Overall, biochar is a versatile material with potential applications in agriculture, environmental management, and climate change mitigation.

To elicit an effective response, a significant amount of biochar is required for soil augmentation, often ranging from 10 to 40 metric tons/hectare, depending on soil conditions and the features inherent in biochar itself (Bandara et al., 2017). However, adopting higher application rates may impose a financial burden on producers and farmers, limiting its general adoption. Furthermore, constraints like as increased production costs, excessive consumption, and the use of feedstocks contaminated with both organic and inorganic contaminants limit the widespread use of biochar. Given the importance of these multidimensional issues, there is a need for a new and improved approach to fertilizer management that not only promotes sustainability but also increases plant productivity while reducing dependency on chemical fertilizers. As a result, incorporating biochar into fertilizer formulations at lower rates, as indicated by Taherymoosavi et al. (2016), or using it as a coamendment alongside conventional fertilizers, as advocated by Saha et al. (2019), appears to be a viable option. Such technique promises not just increased yield but also economic benefits (Zheng et al., 2017). This integrated approach necessitates a thorough rethinking of traditional techniques, spanning from resource management to economic and environmental implications (Blenis et al., 2023). Therefore the ultimate goal of this chapter is to discuss and develop a mechanistic understanding on the optimization of pyrolysis feedstock management and the impact of mineral-enriched biochar-fertilizer formulations on improving soil performance in the rhizosphere and facilitating improvements in plant/crop vitality.

22.2 Methodology

The primary aim of this systematic review is to assess the collective impact of biochar-mineral fertilizer applications on both plant growth and soil fertility across a spectrum of scientific investigations. A thorough exploration of pertinent literature was conducted through the evaluation of electronic databases, notably PubMed, Web of Science, Scopus, and Google Scholar. The search strategy employed a meticulous amalgamation of keywords pertinent to biochar-based mineral fertilizer (BMF), encompassing variants such as biochar fertilizer, biochar-compost fertilizer, biochar nanocompost, and biochar slow-release fertilizer, among others, with a specific focus on their influence on soil fertility, soil biological augmentation, as well as plant growth and productivity dynamics. Inclusion criteria stipulated studies published in peer-reviewed journals, available in the English language, and explicitly addressing the impacts of biochar-mineral fertilizer on both plant growth and soil fertility within the timeframe spanning 2014 to 2024. Conversely, studies not directly appraising the combined effects of biochar and mineral fertilizer on these parameters were systematically excluded. Data extraction was executed utilizing a standardized template encompassing key facets such as study design, biochar typology and application dosage, mineral fertilizer characteristics and application rates, plant species under investigation, as well as diverse growth metrics (e.g., biomass, stature, yield) and soil fertility indices (e.g., nutrient concentrations, pH levels, microbial activity), alongside the temporal span of the experiments. Synthesis of findings was predominantly qualitative, encapsulating observed impacts of BMF applications on plant growth and soil fertility, and subsequently summarized through the presentation of synthesized tables and figures to facilitate comprehension and interpretation.

22.3 Biochar production

Biochar is produced through pyrolysis, a thermochemical processes that converts organic materials into carbon (C)-rich solid residue (char) at temperatures between 350°C and 800°C (Joseph et al., 2021). Various feedstocks can be used, including forestry residues, agricultural byproducts, wood waste, crop residue, poultry waste, dairy and cattle manure, and organic components of municipal solid waste (refer chapter 1 for details) (Ghodake et al., 2021; Liu et al., 2023; Wang et al., 2020). Generally, feedstocks with high lignin concentration produce higher biochar quantity (Yaashikaa et al., 2019). The pyrolysis process itself is dependent to various variables, such as heating rate, final temperature (highest treatment temperature [HTT]), and total processing duration, which directly and indirectly affect the quality and quantity of the resultant biochar (Chatterjee et al., 2020). Based on the pyrolysis temperature, the thermochemical processes are categorized into different groups such as fast pyrolysis (high HTT and short residence time), slow pyrolysis (low HTT and long residence time) (Kizito et al., 2015; Russell et al., 2017), hydrothermal carbonization, and torrefaction (Zhang et al., 2019). These processes can also yield liquid (bio-oil) and gaseous product rich in CO, CH_4, CO_2, and H_2, which can be used as energy sources (Wang et al., 2013).

22.4 Factors affecting biochar properties and potential

Biochar is widely used in agriculture environmental remediation, with research highlighting its effectiveness in removing toxic trace metals from contaminated soil due to its high specific surface area (SSA), porosity, and adsorption capacity (Kumar et al., 2016). Nonetheless, the physicochemical profile of biochar, including parameters such as pH, electrical conductivity, SSA, pore size distribution, and surface functional groups, plays a pivotal role in regulating its efficacy for soil and water applications (Foong et al., 2022; Papageorgiou et al., 2021). Beyond intrinsic biochar properties, various extrinsic factors including feedstock composition, pyrolysis temperature and conditions, and residence time further modulate its contaminant removal potential (Cao et al., 2022; Schreiter et al., 2018). An extensive array of organic materials or biomass residues, encompassing agricultural byproducts, forestry waste, food scraps, and crop residues, serves as feedstocks for biochar production (Kizito et al., 2015; Liu et al., 2017; Zhang et al., 2019).

Slow pyrolysis processes, typically conducted within the temperature range of 350°C–600°C, characterized by low heating rates (5–7°C/min), and prolonged residence times (> 1 hour), yield relatively high biochar output alongside minor quantities of bio-oil and gases (Al Arni, 2018). Conversely, fast pyrolysis entails significantly higher heating rates (> 200°C/min) and markedly shorter residence times (≤10 seconds) (Codignole Luz et al., 2018). Elevated pyrolysis temperatures generally lead to reduced biochar yield, typically ranging between 10% and 30% of the initial dry biomass, concomitant with heightened SSA, porosity, aromaticity, ash content, and fixed C content (Cao et al., 2022; Chatterjee et al., 2020). Additionally, pyrolysis temperature exerts a pronounced influence on the abundance of oxygenated functional groups present on biochar surfaces, with lower temperature pyrolysis yielding biochars characterized by a higher concentration of such groups compared to those generated at higher temperatures. Lower pyrolysis temperatures tend to engender a greater proportion of volatile C within the resultant biochar, while higher temperatures yield biochars endowed with enhanced stability in soil environments (Roberts & de Nys, 2016). However, due to multidimensional nature of pyrolysis-driven biochar synthesis, which is defined by several relevant variables, biochar has a wide range of additional properties rather than a single structure or chemical composition.

A comprehensive understanding of the underlying mechanisms governing physical transformations and chemical reactions during biochar production is imperative for process control and customization of the resulting biochar's physicochemical properties (Liu et al., 2023). These properties exhibit variability depending on the feedstock utilized and the applied HTT. The pH of biochar is contingent upon feedstock composition and pyrolysis conditions, typically spanning a range of 5–12 (Meier et al., 2017; Yang et al., 2019). Feedstocks rich in mineral elements tend to yield biochar with elevated pH and ash content, as evidenced by biochar derived from cattle manure exhibiting high pH and mineral concentrations, while grain husk-derived biochar demonstrates comparatively lower pH and ash contents (Kumar, Tsechansky, et al., 2018).

Elevation in pyrolysis temperature augments the pH of resultant biochar, attributed to increased concentrations of alkali mineral compounds and alkaline substances on its surface, concurrently fostering an augmentation in the abundance of negatively charged organic functional groups (Kumar et al., 2021). This augmentation enhances the cation exchange capacity (CEC) and active adsorption sites of the biochar, enhancing its capacity to sequester cations (Kumar, Joseph, et al., 2018). High pyrolysis temperatures foster higher levels of fixed C, indicative of condensed and thermally stable components, as well as increased ash or mineral contents within the biochar, compared to biochar produced at lower temperatures (Yang et al., 2019). However, escalating pyrolysis temperatures generally result in reduced biochar yields, potentially attributable to heightened thermal decomposition of lignin and cellulose constituents within the feedstock (Yang et al., 2019). Feedstock compositional variations significantly influence pyrolysis outcomes, with feedstocks high in lignin content typically yielding greater biochar quantities (Duku et al., 2011). Given the multifactorial nature of biochar properties, targeted customization can be achieved by modulating pyrolysis parameters or employing specific feedstock combinations during the prepyrolysis stage (Taherymoosavi et al., 2016). Additionally, postpyrolysis modification techniques are increasingly employed to tailor biochar properties to desired specifications (Rafael et al., 2019; Zhu et al., 2019).

22.5 Biochar-mineral fertilizer concepts

The development of BMF attempts to provide targeted increases in soil fertility and crop yield via a number of key conceptual approaches (Agarwal et al., 2022). To begin, it aims to optimize biochar application rates in order to minimize the environmental impact while maximizing benefits (Greenberg et al., 2019). Second, BMF aims to reduce and replace dependency on the current synthetic CF with more sustainable alternatives (Banik et al., 2023). Third, it seeks to reduce the decomposition of applied organic resources such as manure or compost, hence lowering GHG emissions (Puga et al., 2020). Biochar, owing to its intricate pore structures and diverse surface functional groups, has the unique ability to sequester and retain nutrients, making it a perfect carrier for organic matter, mineral elements, and CF as demonstrated in Table 22.1. BMFs are prepared and used to take advantage of these qualities. Notably, the oxygen-containing functional groups found on the biochar's surface provide superior slow-release characteristics to BMF formulations (Marcińczyk & Oleszczuk, 2022), prolonging nutrient availability to plants while minimizing leaching (Liu et al., 2023). Furthermore, biochar is an efficient substrate carrier for slow-release nutrients, providing a low-cost approach for increasing soil fertility and crop development. This multifaceted approach underscores the significant potential of biochar in enhancing agricultural sustainability and productivity, independently contributing to soil health and crop yield optimization (Brtnicky et al., 2023).

22.5.1 Process development

BMF can be synthesized through two primary processes: prepyrolysis biochar modification and postpyrolysis biochar modification (Samoraj et al., 2022). Prepyrolysis modification involves incorporating minerals, synthetic CF, or combinations of different feedstock sources with the biomass prior to pyrolysis (Fig. 22.1; An et al., 2020; Marcińczyk & Oleszczuk, 2022). This process, which may include copyrolysis or optimized feedstock management strategies, aims to produce biochar with tailored properties and enhanced nutrient retention capabilities. Using prepyrolysis feedstock management, (Kumar et al., 2021) produced two types of biochars using prepyrolysis feedstock management: (1) pine wood: clay:sand (PCS-BC; 70:15:15, w/w) and (2) wheat straw:bird (chicken) manure (WB-BC; 50:50, w/w). Advanced spectroscopic and microscopic techniques revealed distinct physicochemical and nanostructural properties (Fig. 22.2). PCS-BC consisted of woody biochar encapsulated with sand and clay, whereas WB-BC exhibited complex surface structures with higher organic and inorganic mineral content. X-ray spectroscopy analysis identified elemental compositions of C, O, Si, Ca, K, Al, Mg, and Na, with lower concentrations of P, S, and Cl. Notably, WB-BC featured Ca/P-rich nanoparticles from bird manure mixed with Si, C, Al, Mg, and Na on wheat straw surfaces, and aluminosilicate clusters combined with K, Mg, Na, and P within the inner layers (Fig. 22.1; Kumar et al., 2021). Postpyrolysis modification, on the other hand, involves treating the biochar produced with CF or mineral nutrients using procedures such as impregnation or granulation (Fig. 22.3; Peter et al., 2021). The impregnation approach infuses nutrient solutions into biochar (Dominguez et al., 2020), whereas the granulation process blends biochar and solid CF, grinds them into a powder, and pelletizes them impregnation. These postpyrolysis techniques, as demonstrated by Shin et al. (2019), result in the creation of BMF pellets with regulated nutrient release

Table 22.1 Production of various biochar-based mineral fertilizers (BMFs) from diverse feedstocks, combined with numerous inorganic or organic materials, and the specification of the final product along with the associated responses.

Feedstock	Pyrolysis conditions		Mineral added	Biochar: mineral ratio	Modification methods	Product type	Specific remark	References
	Temperature (°C)	Residence time (min)						
Rice straw	500	240	Na_2SeO_3 or/and hydroxyapatite	Na_2SeO_3: bicoahr, 1:1477 and hydroxyapatite: biochar, 1:5	Chemical mixing	Selenium and P comodified biochar	↑ Surface area, pore volume, and Hg adsorption capacity of biochar due to selenium and P co-modification and ↓ mercury forms, transforming soil mercury into less bioavailable states	Qin et al. (2024)
Apple tree branches	500	120	Urea and oxidized starch as a binder to granulate	BC:urea, 1:1 and 2:1	Water-soluble impregnation	Biochar-based slow-release nitrogen fertilizer	BMF to be easily-synthesized and low-cost, representing slow-release N performance and increasing crop yield	Zhang et al. (2023)

(Continued)

Table 22.1 Production of various biochar-based mineral fertilizers (BMFs) from diverse feedstocks, combined with numerous inorganic or organic materials, and the specification of the final product along with the associated responses. *Continued*

Feedstock	Pyrolysis conditions		Mineral added	Biochar: mineral ratio	Modification methods	Product type	Specific remark	References
	Temperature (°C)	Residence time (min)						
Banana peels	600	360	NH_2CONH_2; $NaH_2PO_4.2H_2O$; and K_2SO_4	5% solution of each	Simple impregnation approach	Biochar nanofertilizer	BMF demonstrated noticeably better water absorbance ratio (68%), equilibrium water content (79%), and swelling ratio (3.6 g/g) than biochar, the salt index values (0.087 mmhos/cc) of the created nanocomposite performed better when compared to the conventional fertilizers due to their safer application	Tarar et al. (2023)

Rice straw	500	120	Urea 28.6%, Ca $(H_2PO_4)_2$ (32.4%), and KCl (26%) with biochar (13%)	H_2O_2-modified starch (HS) 30% added to BMF 70%	Simple impregnation and blending approach	Starch/biochar-based compound fertilizer	↑ 10% soil water retention by adding starch into BMF. After 30 days, N (50%), P (13%), and K (87%) leaching amounts of HS-BMF in soil followed diffusion and relaxation-controlled mechanism. Hydrotalcite could bind with P in HS-BMF, contributing to the enhanced durability of P in HS-BMF	Lu et al. (2023)
Cotton stalk	550	180	g-C_3N_4, $MgCl_2$	BC-g-C_3N_4: MgO at 90:10, 80:20, 70:30, 60:40, and 50:50	Thermal polymerization	Bi-functional biochar-g-C_3N_4-MgO composites	MgO in BC-g-C_3N_4-MgO composite ↑ P adsorption capacity to 111 mg/g and ↑ visible light utilization efficiency. Biochar in the composite acts as a charge transporter with good conductivity, ↑ efficient transfer of photo-generated charge carriers.	An et al. (2023)

(*Continued*)

Table 22.1 Production of various biochar-based mineral fertilizers (BMFs) from diverse feedstocks, combined with numerous inorganic or organic materials, and the specification of the final product along with the associated responses. *Continued*

Feedstock	Pyrolysis conditions		Mineral added	Biochar: mineral ratio	Modification methods	Product type	Specific remark	References
	Temperature (°C)	Residence time (min)						
Chicken manure	300	60	Ca-bentonite	Ca-bentonite at 10% or 40%	Copyrolysis	Biochar-based fertilizers	Ca-bentonite ↓ soluble orthophosphates by forming less-soluble Ca/Mg-phosphates. P and K release from BMFs in coarse sand was 38% and 24% higher, respectively, than in clay loam.	Piash et al. (2022)
Citrus trees	600	–	KNO_3, C-nanoparticles	–	Chemical activation	Eco-friendly nanocomposite fertilizer	Spectroscopy analysis indicated silicon, C, P, Ca aluminum, manganese, sodium, and sulfur distribution on the BMF	Salama et al. (2021)
Apple branches	450	480	$CuSO_4$	–	Chemical precipitation method	Biochar-based Cu_2O nanoparticles	The Cu_2O/biochar composites have a larger specific surface area compared to pure Cu_2O nanoparticles. Cu_2O/biochar composites exhibit superior antibacterial performance compared to pure Cu_2O nanoparticles	Yang et al. (2019)

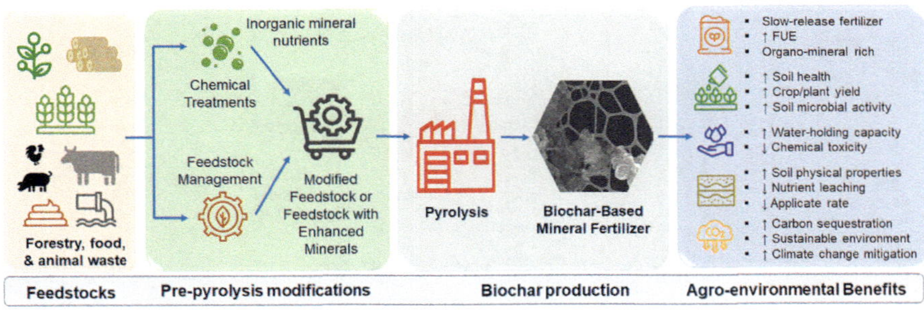

FIGURE 22.1 Prepyrolysis modification for biochar-based mineral fertilizer production.

Schematic representation of biochar-based mineral fertilizer production, illustrating the prepyrolysis modification process. This includes the incorporation of mineral nutrients into the biomass feedstock before pyrolysis, resulting in an enriched biochar product. The figure also describing the agronomic and environmental benefits associated with the production and soil application of biochar-based mineral fertilizer.

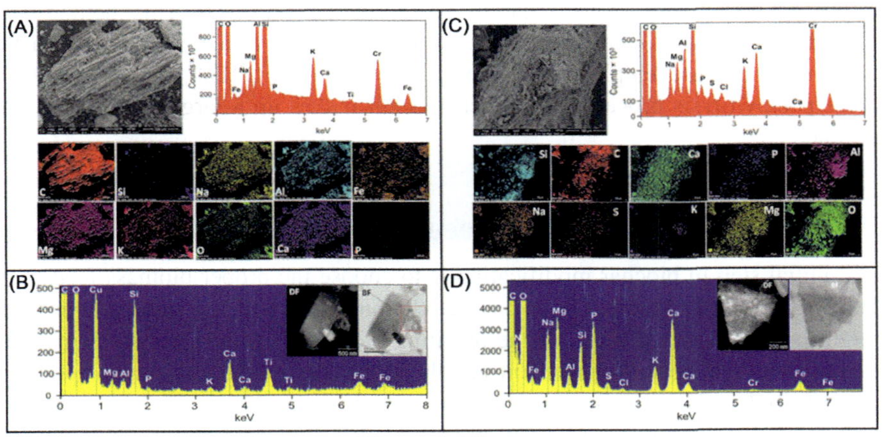

FIGURE 22.2 Microscopy and spectroscopic examination of pine wood:clay:sand (PCS-BC; 70:15:15) and wheat straw:bird manure (WB-BC; 50:50) biochar. (A) Secondary electron image of a biochar particle with an energy dispersive X-ray spectroscopy (EDS) spectrum and X-ray elemental map of the total area; (B) scanning transmission electron microscopy (STEM) dark field (DF) and bright field (BF) images along with an EDS spectrum; (C) secondary electron image of biochar particle with an EDS spectrum and X-ray elemental map of the total area; and (D) STEM DF and BF images along with an EDS spectrum.

Microscopy and spectroscopic examination of two biochars, that is, PCS-BC and WB-BC produced using prepyrolysis feedstock management.

From Kumar, A., Joseph, S., Graber, E. R., Taherysoosavi, S., Mitchell, D. R. G., Munroe, P., Tsechansky, L., Lerdahl, O., Aker, W., & Sæbø, M. (2021). Fertilizing behavior of extract of organomineral-activated biochar: Low-dose foliar application for promoting lettuce growth. Chemical and Biological Technologies in Agriculture, 8*(1). https://doi.org/10.1186/s40538-021-00222-x.*

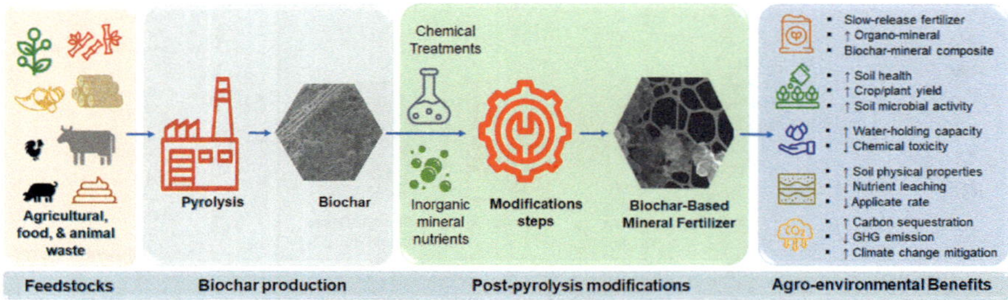

FIGURE 22.3 Postpyrolysis modification for biochar-based mineral fertilizer production.

Schematic representation of biochar-based mineral fertilizer production, illustrating the postpyrolysis modification process. This includes the incorporation of mineral nutrients into the biochar after pyrolysis, resulting in an mineral-enriched biochar product. The figure also describing the agronomic and environmental benefits associated with the production and soil application of biochar-based mineral fertilizer.

features. Furthermore, cocomposting biochar and animal manure is another type of postpyrolysis modification (Marcińczyk & Oleszczuk, 2022). This method typically incorporates biochar into the composting process at a lower ratio than the amount of manure employed (Agyarko-Mintah et al., 2017). This varies from other postpyrolysis processes in which biochar makes up a larger part of the end product.

22.5.2 Biochar modification: feedstock management

The initial step in biochar modification is to choose the proper biomass feedstock. Different forms of biomass have distinct chemical and physical compositions that might influence the qualities of the resulting biochar. Furthermore, biochar modification via feedstock management entails intentional modification and optimization of feedstock constituents, structural configuration, and process variables in order to tailor the physicochemical properties, functional moiety distribution, and elemental composition of the resulting biochar (Ghorbani et al., 2024; Wu et al., 2021; Zhou, Qu, et al., 2021). For example, two different biochars produced through feedstock management have shown different physicochemical properties and are detailed in Table 22.2. This multifaceted approach combines biomass analytical methodologies, pretreatment protocols, and enhanced pyrolysis optimization techniques to create biochars with unique properties tailored to specific usage requirements.

22.5.2.1 Biomass feedstock characterization

Biomass feedstock characterization investigates the chemical, physical, and structural qualities required for pyrolysis bioconversion. This integrative approach, which draws on analytical chemistry, materials science, spectroscopy, and

Table 22.2 Physicochemical characteristics of biochar produced from two different combinations of feedstock: pine wood:clay:sand (PCS-BC; 70:15:15, w/w) and wheat straw:bird manure (WB-BC; 50:50, w/w).

Chemical properties	PCS biochar	WB biochar
Ca (mg/kg)	42.0	38.3
K (mg/kg)	168	7980
Mg (mg/kg)	18.5	4.7
Na (mg/kg)	69.9	4497
P (mg/kg)	4.5	262
S (mg/kg)	12.1	1758
Si (mg/kg)	74.6	1746
As (mg/kg)	18.4	316
Al (mg/kg)	12,400	367
B (mg/kg)	840	32,400
Cd (mg/kg)	0.4	0.3
Co (mg/kg)	0.3	0.4
Cr (mg/kg)	1019	19.7
Cu (mg/kg)	2.5	4.6
Fe (mg/kg)	241	150
Mn (mg/kg)	174	38
Mo (mg/kg)	721	545
Ni (mg/kg)	5	4.2
Pb (mg/kg)	0.5	0.3
Ti (mg/kg)	7	51
Zn (mg/kg)	15	11.4
Surface area—Dubinin-Radushkevich (m^2/g)	123.7	186.7
Micropore volume (cm^2/kg)	45	71
Average particle size (nm)	173.3 ± 36.9	228.6 ± 41.7
Zeta potential (mV)	−26.1 ± 6.11	−38.1 ± 7.09
Total dissolved C (mg/kg)	197	1680
Total dissolved inorganic-C (mg/kg)	89	1233
Total dissolved organic-C (mg/kg)	108	449
Total dissolved N (mg/kg)	8.3	18.0
Dissolved organic carbon (mg/kg)	200	4700
Chromatographable organic carbon (mg/kg)	200	4600
Bio-polymers (mg/kg)	ND	40
Humics (mg/kg)	60	1300
Building blocks (mg/kg)	40	1300
LMW neutrals (mg/kg)	100	400
LMW acids (mg/kg)	ND	1500
Aromaticity (L/(mg m))	0.11	4.10

Source: *From Kumar, A., Joseph, S., Graber, E. R., Taherysoosavi, S., Mitchell, D. R. G., Munroe, P., Tsechansky, L., Lerdahl, O., Aker, W., & Sæbø, M. (2021). Fertilizing behavior of extract of organomineral-activated biochar: Llow-dose foliar application for promoting lettuce growth.* Chemical and Biological Technologies in Agriculture, 8*(1). https://doi.org/10.1186/s40538-021-00222-x.*

thermodynamics, is driving progress in bioenergy, biorefining, and sustainable bioproducts (Kim et al., 2018). It uses techniques such as X-ray fluorescence and inductively coupled plasma emission spectrometry to determine elemental composition, nutritional content, ash, and pollutants (Kenda et al., 2024). Thermogravimetric analysis evaluates energy density, emission levels, and thermal stability. Advanced technologies including Fourier transform infrared, nuclear magnetic resonance, and chromatography provide information about cellulose, hemicellulose, and lignin, indicating structural integrity and carbohydrate content (Barbosa et al., 2022; Ramesh & Raghavan, 2024). These investigations provide a thorough understanding of biomass chemical reactivity, molecular complexity, and catalyst interactions (Lin et al., 2021). Morphological investigations employ electron microscopy to analyze biomass surface characteristics, pore distribution, and particle morphology, thereby elucidating physical properties and mass transport mechanisms (Lu et al., 2017; Mahdi et al., 2019).

22.5.2.2 Biomass pretreatment

Pretreatment technologies for biomass modification in biochar synthesis help improve the physicochemical qualities and reactivity of the feedstock prior to pyrolysis (Peter et al., 2021). Pretreatment methods, such as drying, grinding, and sieving, can be employed to enhance the uniformity and consistency of the feedstock (Liu et al., 2018). Grinding or milling the biomass feedstock can improve particle size and enhance surface area, hence improving thermal conductivity (Shukla et al., 2023; Smith et al., 2024). Furthermore, thermal treatment like drying, torrefaction, and steam explosion improve biomass thermal stability and pyrolytic susceptibility, resulting in higher biochar output and quality (He et al., 2021). Chemical treatments use acids, bases, and oxidizing agents to change the composition of biomass, eliminating lignin and hemicellulose while increasing cellulose depolymerization, resulting in biochars with higher porosity and durability (Mahdi et al., 2019). Additionally, conditioning processes using biological substances, like composting and microbial inoculation, can also be used to enrich the feedstock with beneficial microorganisms or nutrients, which can contribute to the biochar's or BMF nutrient retention and soil amendment capabilities (Conversa et al., 2015). Biological pretreatments use microbial consortia and enzymes to break down complex polymers, releasing fermentable sugars and increasing C precursor bioavailability.

22.5.2.3 Posttreatment and functionalization

Advanced postproduction treatments of biomass feedstock including chemical activation, oxidative modification, surface functionalization, and copyrolysis with additives (e.g., metals, metal oxides, organic polymers) (Hu et al., 2024; Tarar et al., 2023) and combination of feedstocks obtained from different biomass sources (Kumar et al., 2021). This approach involves applying nanomaterials onto the biochar surface or encapsulating nanoparticles within the biochar matrix postpyrolysis (Agilandeswari et al., 2024). Utilizing nanotechnology, it aims to customize and enhance biochar properties to address specific agricultural challenges, optimizing soil fertility and overall sustainability (Verma & Singh, 2023). These strategies are employed to tailor the biochar or BMF surface chemistry, enhance SSA, and introduce functional groups (e.g., oxygen-containing moieties, N-containing groups, and sulfur-containing species) to optimize its adsorption capacity, catalytic activity, and environmental compatibility (Lalmalsawmdawngliani et al., 2024).

22.5.3 Biochar modification: biochar impregnation

Biochar impregnation is an advanced approach for improving the physical, chemical, and biological properties of biochar (Mishra et al., 2023). This process includes infusing biochar with certain compounds and embedding them inside its matrix, so improving its performance for specific purposes (Peter et al., 2021). The process starts with the selection of high-quality biochar derived from wood chips, agricultural wastes, or green waste and serves as the base material for modification. Following that, impregnating agents are carefully selected based on their intended use. These agents can include nutrients like N, phosphate (P), and potassium (K), as well as helpful microorganisms, organic polymers, and inorganic substances like zeolite and clay minerals (Dominguez et al., 2020; Lustosa Filho et al., 2019). During the impregnation process, the selected agents are coupled with biochar using methods such as soaking, spraying, or tumbling to guarantee thorough penetration and equitable dispersion inside the porous structure of the biochar. The modified biochar is then dried and cured under controlled circumstances to stabilize the infused chemicals and improve their connection with the carbonaceous matrix (Kenda et al., 2024). The advantages of this modified biochar include superior nutrient retention capacities (Romero et al., 2021), making it an excellent soil amendment and fertilizer transporter (Marcińczyk & Oleszczuk, 2022). Furthermore, its increased water-holding capacity aids effective soil moisture management and plant growth. The impregnation procedure also enables customized functioning by tailoring biochar qualities to specific soil types, crop needs, or environmental cleanup. In terms of uses, modified biochar is extremely useful in soil enhancement, where it enhances soil structure, fertility, and microbial activity, increasing crop yield and sustainability (Chen et al., 2020; Ren et al., 2023). It is also a valuable aid in environmental rehabilitation efforts, acting as an effective adsorbent for contaminants in soil and water (Shaikh et al., 2022; Tran et al., 2022). Furthermore, tailored biochar-based formulations release nutrients gradually, ensuring that plants have a steady supply of vital elements for optimal growth and development.

22.5.4 Biochar-compost application

Biochar-compost application combines the use of biochar and compost to enhance soil quality, fertility, and crop productivity in agricultural and gardening settings.

Compost, produced from organic waste decomposition, enriches soil with nutrients, boosts microbial activity, and enhances organic matter content (Seehausen et al., 2017). This combined application proves beneficial across agriculture, gardening, and land remediation, supporting soil fertility, water management, and plant health while aiding in soil remediation and pollutant reduction (Bashir et al., 2020; Nadeem et al., 2017; Qayyum et al., 2017; Schulz et al., 2013). Application methods include layering and mixing to ensure effective distribution and integration into various soil types. For example, a BMF by impregnating biochar from quinoa, maize, rice, and sugarcane residues with biogas slurry from anaerobic digestion of municipal solid waste were developed (Villada et al., 2024). Analysis showed that developed BMF positively impacted cucumber plant nutrition by increasing N, P, and K concentrations by 61%, 32%, and 19%, respectively in comparison to control. This study significantly demonstrated the potential of BMFs for controlled nutrient release and improved plant nutrition. Thus, integrating biochar with biogas slurry presents a promising and sustainable approach for NPK recovery and fertilizer production in agriculture (Villada et al., 2024).

Research on soil amendments highlights significant improvements in crop performance and soil quality. In a Mediterranean environment, the residual effects of compost and biochar on soil properties and durum wheat were evaluated by comparing mineral fertilizer (100 kg N/ha), compost (25 Mg/ha), and biochar (10 and 30 Mg/ha) (Leogrande et al., 2024). Results indicated that compost and biochar significantly increased soil total organic carbon (OC). Compost showed the highest soil water extractable OC (76.9 mg/kg) and improved grain yield (3.36 Mg/ha), while high-rate biochar had the lowest yield (2.18 Mg/ha) (Leogrande et al., 2024). For Chinese cabbage, a study examined the effects of biochar-amended chicken manure and wood chips compost (BMFs) in varying proportions (1%–7%) on soil fertility, enzyme activity, microbial communities, and cabbage growth (Chen et al., 2024). Increased BMF levels enhanced cabbage growth and soil fertility, especially at 3%–5% BMF. However, higher BMF levels reduced soil bacterial diversity, suggesting the need for a reevaluation of BMF application strategies (Chen et al., 2024).

Another study focused on the short-term effects of biochar, compost, and their combination on soil biochemical properties related to C and N cycling, microbial communities, and lettuce yield (Massaccesi et al., 2024). All amendments increased lettuce biomass, with compost having a more substantial impact on enzymatic activities and microbial communities than biochar. Biochar amendments indicated C immobilization and N cycling enhancement. Combined treatments showed potential long-term soil quality benefits but no immediate synergistic effects (Massaccesi et al., 2024). These findings underscore the crucial role of soil amendments of co-composted fertilizer in enhancing crop performance and maintaining soil health. Overall, biochar-compost application provides a comprehensive, sustainable solution for soil improvement and environmental enhancement by leveraging the complementary strengths of biochar and compost.

22.6 Multifaceted impact of biochar-based mineral fertilizer

The utilization of BMFs in agriculture presents a multifaceted approach essential for advancing sustainable agriculture and environment management. By harmoniously integrating biochar's soil-conditioning properties with the nutrient-rich characteristics of mineral fertilizers, these advanced formulations promote soil health, amplify agricultural productivity, reduce environmental footprint, and proactively contribute to climate resilience, positioning them as indispensable resources in modern sustainable agriculture (Agarwal et al., 2022; Ahmadian et al., 2021; Ramesh & Raghavan, 2024). Since biochar is renowned for its C sequestration capabilities, its incorporation into organomineral fertilizer formulations bolsters climate resilience by reducing the C footprint associated with synthetic fertilizer production and application. BMF represent a transformative innovation in agricultural nutrient and soil management strategies. These innovative fertilizers synergistically amalgamate the soil-conditioning attributes of biochar's C-rich matrix with the nutrient profiles inherent in mineral fertilizers, thereby addressing both soil enhancement and nutrient supplementation needs comprehensively (Rombel et al., 2022). Biochar's porous structure serves as a nutrient reservoir, facilitating the gradual and sustained release of essential elements such as N, P, and K (El Sharkawi et al., 2018; Rafael et al., 2019). This controlled-release mechanism not only enhances plant nutrient uptake but also mitigates fertilizer leaching and runoff (Brtnicky et al., 2023), thereby improving nutrient use efficiency (Khajavi-Shojaei et al., 2023). Additionally, the integration of biochar into mineral fertilizers augments soil aggregation, porosity, and water retention capabilities, leading to enhanced soil aeration, chemical properties, and fertility (Zhao et al., 2021), and plant root development (Tarar et al., 2023) that are pivotal for sustainable agricultural practices.

22.6.1 Impact on soil physicochemical properties and fertility

The application of BMFs significantly influences soil physicochemical properties and fertility, acting as a cornerstone for enhancing soil health, structure, and productivity. BMF enhances soil aggregation, creating a stable framework that binds soil particles and fosters the formation of robust soil aggregates (Chen et al., 2020). BMFs porous nature boosts soil structure and porosity, facilitating improved aeration, root penetration, and water-holding capacity, which mitigates water runoff and augments drought resistance (Oladele et al., 2019). Furthermore, incorporating BMF enhances the soil's organic C content, promoting long-term soil fertility and C sequestration, that are essential for nutrient cycling and overall soil health (Wali et al., 2020). The increase in organic matter also supports microbial diversity and activity, crucial for organic matter decomposition, disease suppression, and overall soil vitality. Furthermore, it raises the soil pH and CEC

(Rafael et al., 2019), creating a nutrient reservoir that adsorbs and retains key cationic nutrients such as ammonium, K, and Mg.

For example, in a 70-day pot experiment on acid sandy soil (Arenosol), biochars from baby corn peel, mango tree branches, and rice husk were evaluated with and without half or full NPK fertilizers (Rafael et al., 2019). Biochars combined with NPK significantly increased soil pH and CEC, while enhancing available P and enzymatic activity related to P cycling, resulting in higher cowpea yields. Both full and half NPK doses with biochars demonstrated similar soil enhancements and yield advantages, highlighting biochars' potential to improve soil conditions and economize fertilizer use in acidic soils (Rafael et al., 2019). Similarly, in a five-year field trial, different biochar rates (5–40 t/ha) combined with N fertilizer (60, 90, 120 kg/ha) were tested for their effects on soil pH and available P (Jin et al., 2019). The study found that higher biochar rates (20 and 40 t/ha) with N significantly increased both soil pH and P availability, though the effects decreased over time, suggesting biochar re-application every three years. The optimal biochar rate for maximizing crop yields and economic benefits was identified as 11.4 t/ha, demonstrating the potential of BMF to enhance productivity in upland red soil (Jin et al., 2019).

From a fertility perspective, the porous structure of biochar functions as a nutrient reservoir, facilitating the gradual and sustained release of essential nutrients over time. This controlled-release mechanism boosts plant nutrient uptake, reduces fertilizer leaching, and minimizes environmental nutrient losses (Tarar et al., 2023). For example, nutrient release from pristine and modified chicken manure biochars combined with quartz sand and clay loam soil, compared to prescribed sweet corn fertilizer doses, revealed that Ca-bentonite reduced orthophosphate solubility, resulting in less-soluble Ca/Mg-P (Piash et al., 2022). Biochar with 25% Ca-bentonite released 22% K slower due to high adsorption during copyrolysis. P and K releases from BMFs in coarse sand were 38% and 24% greater, respectively, than in clay loam, whereas N release was 49% higher in clay loam due to microbial activity (Piash et al., 2022).

22.6.2 Impact on soil biological functioning

The integration of BMFs into soil significantly enhances biological attributes, promoting a vibrant and diverse soil microbiome essential for nutrient cycling, OM decomposition, and overall soil health (Massaccesi et al., 2024; Table 22.3). The high OM content, primarily C, synergizes with mineral fertilizers to create an optimal environment for soil microorganisms (Ren et al., 2023). Pyrolyzed C serves as both habitat and C reservoir, stimulating microbial structure, diversity, activity, and biomass (Yan et al., 2021b; Zhao et al., 2021). This fosters the growth of beneficial microorganisms crucial for OM breakdown, disease suppression, and soil structure improvement. The heightened microbial activity facilitated by BMF improves nutrient transformation, mineralization, and availability, leading to enhanced soil fertility and productivity (Rashid et al., 2023).

Table 22.3 Impacts of various biochar-based mineral fertilizers (BMFs) at different application rates on soil microbial responses and other specific parameters.

BMF types and compositions	Application rate (%)	Specific responses	Microbial response	Associated parameters	References
Nanobiochar	2, 5, 10	Influence on NH_4 emission, N, P, K mineralization and the aforementioned nutrients uptake by maize	↑ Microbial activity and microbial biomass C and N	↓ NH_4 emission by 43% from vermicompost. ↑ Soil mineral N, P, and K by 71, 120, 95% compared to vermicompost	Rashid et al. (2023)
Ca-bentonite and chicken manure derived biochar	10% and 25% Ca-bentonite	Slow-release performance of BMFs	↑ Ammonia-oxidizing bacteria > ammonia-oxidizing archaea	↑ P (38%) and K (24%) release from BMF in coarse sand than clay loam. Ca-bentonite-addition provided BMF a slow-release properties and satisfy plant nutritional need	Piash et al. (2022)
Biochar-based fertilizer	1333 kg/ha	Soil N_2O emissions in a subtropical Moso bamboo plantation	↑ Microbial biomass C	↓ N_2O emissions under BMF over CF and CF with biochar and was corelated with ↓ soil water soluble organic N, urease, and protease activities	Zhou, Ying, et al. (2021)
Biochar-based fertilizer	750 kg/ha	Impact on brown earth soil bacterial communities	↑ Acidobacteria and ↓ actinobacteria	↑ Genes related to energy production, transport and metabolism of amino acid, nucleotide transport, and carbohydrate in BMF than CF	Gao et al. (2021)
	–				

(*Continued*)

Table 22.3 Impacts of various biochar-based mineral fertilizers (BMFs) at different application rates on soil microbial responses and other specific parameters. *Continued*

BMF types and compositions	Application rate (%)	Specific responses	Microbial response	Associated parameters	References
Biochar-based fertilizer		Impact on bacterial community network complexity	↑ Microbial biomass C, N, and P, ↑ gammaproteo-bacteria, ↑ acidobacteria, and ↑ correlation network complexity	↑ Bacterial diversity (Simpson and Shannon), microbial biomass and soil nutrient content in biochar-based fertilizer compared to control or CF	Zhou et al. (2019)
Coapplication of biochar and CF	Biochar: 5 t/ha & NPK, 60:20:40 kg/ha	Yield, quality of *Andrographis paniculata* and soil health	↑ Soil microbial biomass C, fluorescein diacetate hydrolytic, dehydrogenase and alkaline phosphatase activity	↓ Stomatal conductance and ↑ water use efficiency indicates it's potential to mitigate water deficit stress	Saha et al. (2019)
Coapplication of biochar and CF	Biochar and CF: 2.5 t/ha	Soil C dynamics	↑ Microbial biomass C	↑ Total OC, particulate OC, humic acid C, and fulvic acid C. ↑ Plant growth and physiology	Sarma et al. (2018)

Moreover, biochar from this organomineral fertilizers also supports the formation of symbiotic relationships between plants and beneficial microbes like mycorrhizal fungi, enhancing nutrient uptake efficiency, particularly for P and micronutrients (Yang et al., 2021), by expanding the root system's reach and nutrient absorption capabilities. This leads to increased crop yields, superior quality, and greater resilience against environmental stresses such as drought, nutrient imbalances, and diseases. BMFs also promotes disease control by encouraging antagonistic microorganisms that suppress soil-borne pathogens, reducing the need for chemical pesticides and supporting sustainable agricultural practices. Additionally, OC from BMF contributes to soil OM accumulation, providing a consistent food source and energy reserve for soil microbes. This sustains microbial growth and activity, ensuring a resilient and adaptable soil microbiome capable of withstanding environmental fluctuations and preserving long-term soil health. A high-throughput sequencing approach evaluated the impact of BMF on microbial communities in field trials with and without compost and CF (Yan et al., 2021a). Study showed that biochar combined with manure and/or CF increased soil C, N, P, and K levels more than other treatments. These amendments significantly boosted soil microbial abundance, diversity, and the relative abundance of dominant microbes, altering microbial community structure. Network analysis revealed that BMF enhanced microbial cooccurrence complexity, promoting keystone species in soil C management and nutrient cycling, benefiting the restoration of degraded soils (Yan et al., 2021a).

The impact of BMF on soil fertility and microbial diversity was assessed relative to no fertilizer, biochar, and CF (Yang et al., 2021). After one season, BMF significantly enhanced the relative abundance of 10 key bacterial genera and 13 fungal genera, including beneficial microbes such as *Rhodanobacter*, *Mizugakiibacter*, *Pedobacter*, *Sphingomonas*, *Devosia*, *Rhodosporidiobolus*, *Chloridium*, *Amylocorticium*, *Clavulina*, *Inocybe*, and *Mycofalcella*. These microbes were positively correlated with improved plant yield and quality. Redundancy analysis revealed that soil pH, total N, and available K were primary factors shaping bacterial communities, while dissolved OC, exchangeable Ca, pH, and total N influenced fungal communities (Yang et al., 2021).

22.6.3 Impact on crop growth and performances

The incorporation of BMFs to soil significantly enhance plant/crop growth and physiology, creating an environment primed for enhanced plant development, resilience, and yield potential (Agarwal et al., 2022) and demonstrated in details in Table 22.4. The controlled-release properties inherent to BMFs ensure a consistent and balanced provision of crucial nutrients, including N, P, and K, throughout the plant's growth phases (Tarar et al., 2023). This sustained nutrient delivery supports uninterrupted growth, mitigates nutrient-associated stress, and fortifies plant resilience against a gamut of environmental challenges, encompassing drought, salinity, and temperature oscillations (Bashir et al., 2020; Ghassemi-

Table 22.4 Examining the production of biochar-based mineral fertilizers (BMFs) from diverse feedstock materials, combined with various mineral or organic elements, and their specific impacts on plant growth responses. This study also investigates how BMF regulates nitrogen (N) mineralization in soil and reduces leaching loss.

Feedstock for biochar production	Minerals used for enrichment	BMF production method	Soil type	Experimental type	Crop	BMF-induced NUE and nutrient release	BMF-induced Plant growth	References
Rice straw	NPK fertilizer	Postpyrolysis	–	Pot/soil column leaching	*Tomato*	Leaching of N, P, and K were 44.5%, 13.3%, and 87.4%	↑ Tomato seedlings growth	Lu et al. (2023)
Southern yellow pine biochar	N-urea pellets	Postpyrolysis	Mesic typic/aquic hapludolls	Pot	Maize	↑ N-recovery efficiency by 17%–50%	↑ Shoot and root biomass by 34% and 23%	Banik et al. (2023)
Distillers grain	Urea, monoammonium phosphate, potassium sulfate, and solid binder	Postpyrolysis	Yellow zonal soils	Field	Eggplant	↑ Nutrient recovery efficiency	↑ Fresh eggplant yield by 23%–49% compared to CA. ↑ soluble sugar and Vitamin C	Zhang et al. (2022)
Vinasse	Urea	Postpyrolysis	Red soil	Pot	Rape seed	↑ NUE & ↓ N loss by 30% compared with urea	↑ 10% oilseed rape biomass with BMF compared to urea alone	Jia et al. (2021)
Wheat straw	CF	Postpyrolysis	Ferralsol	Field	Moso bamboo plantation	↓ N_2O emission with BMF over CF	–	Zhou, Ying, et al. (2021)

Spent mushroom, cottonseed shell, sawdust, lime, and bran	Urea (22%), diammonium phosphate (10%), K-sulfate	Postpyrolysis	Acidic red soil	Field	Tea	—	↑ Yield (39.2%), sprout weight (26.6%), and sprout density (10.7%) under BMF over control	Yang et al. (2021)
Lemongrass	NPK, 60:20:40 (kg/ha)	Postpyrolysis	Fluventic, Ustochrept and sandy loam	Pot	Andrographis paniculata (kalmegh) herb	—	↑ Root length (102%) and herbage yield (29%–95%) with BMF compared control	Saha et al. (2019)
Maize straw	Mineral fertilizer and bentonite clay	Postpyrolysis	—	Field	Water spinach	↑ Plant N uptake by 116% and ↑ N recovery efficiency 27.6%–40.8%	Higher yield increase of 217% compared to control treatment	Zhou, Qu, et al. (2021)
Corn Straw	$MgCl_2$ solution	Prepyrolysis	Sandy loam soil	Pot	Maize	↑ NUE by 56%	↑ Chlorophyll (15.7%), root dry weight (95%), and shoot dry weight (79%) over control	Khajavi-Shojaei et al. (2023)
Eucalyptus wood	CF	Postpyrolysis	Oxisol	Field	Maize	↑ NUE by 12%	↑ Crop yield by 26%	Puga et al. (2020)
Rape seed straw	Urea and benthonite	Postpyrolysis (hydrothermal synthesis)	Lateritic soil	Pot	Rape seed (*Brassica napus*)	↑ NUE by 58.8% than the urea treatment	↑ Grain yield by 16.6% compared to urea	Liao et al. (2020)

(*Continued*)

Table 22.4 Examining the production of biochar-based mineral fertilizers (BMFs) from diverse feedstock materials, combined with various mineral or organic elements, and their specific impacts on plant growth responses. This study also investigates how BMF regulates nitrogen (N) mineralization in soil and reduces leaching loss. *Continued*

Feedstock for biochar production	Minerals used for enrichment	BMF production method	Soil type	Experimental type	Crop	BMF-induced NUE and nutrient release	BMF-induced Plant growth	References
Baby corn peel and mango branches	NPK fertilizer	Postpyrolysis (combined)	Acid sandy soil	Pot	Cowpea	↑ NUE compared to CF	↑ Total dry biomass (76%), plant height, stem diameter, and number of leaves	Rafael et al. (2019)
Wheat straw	Chicken manure	Postpyrolysis	Alfisol	Field	Tea	↓ N_2O emission compared to CF	↑ 10% yield compared to CF	He et al. (2019)
Rice husk biochar and oilpalm biochar	compost	Prepyrolysis	Typic Kanhapludults	Field	Maize	–	↑ 291% dry weight of grain	Nurida and Jubaedah (2019)
E. coli fermentation waste	Phosphorus	Prepyrolysis	–	Hydroponics	*Lactuca sativa*		↑ Early growth, root length, and larger leaf size under BMF	Kim et al. (2018)
Rice husk	Ammonium phosphate	Postpyrolysis	Sandy soil	Pot	Japanese mustard spinach	↑ NUE by 28% compared to ammonium phosphate	↑ Plant yield	El Sharkawi et al. (2018)
Cotton stalk	N load	Postpyrolysis	–	Pot	Cotton	↑ NUE by 64% as compared to NH_4Cl	↑ Plant growth and yield	Wen et al. (2017)

Golezani & Rahimzadeh, 2023). The enhanced crop yields or productivity associated with BMFs can be attributed to several factors: (1) the enrichment of nutrient content in the final product due to nutrient doping in biochar (Han et al., 2018), (2) the gradual release of nutrients throughout the crop's growth period, reducing nutrient leaching and improving fertilizer efficiency (Zhang et al., 2023), and (3) the synergistic effect of biochar and CF acting as a supportive matrix for plant growth (Ibrahim et al., 2022; Marcińczyk & Oleszczuk, 2022). The bioactive constituents inherent to biochar activate innate plant defense mechanisms, reinforcing resistance against pests, and diseases adversities (Kolton et al., 2017; Zhang et al., 2019). Additionally, biochar's beneficial impact on soil microbiota stimulates the production and secretion of growth-promoting substances by soil microbes, further supporting plant vigor, developmental balance, and overall physiological homeostasis. Through adopting of BMFs, farmers can capitalize on the combined benefits of both biochar and fertilizer (Oladele et al., 2019).

Numerous studies have demonstrated the positive effects of combining biochar with fertilizer on plant yield and growth (Pogorzelski et al., 2020; Rafael et al., 2019; Yang et al., 2021). For instance, a meta-analysis showing that the application of BMFs increased yields by up to 10% at rate of 0.9 t/ha (Melo et al., 2022). Banana peel biochar and its nanocomposite were investigated for their long-term mineral availability to plants (Tarar et al., 2023). The biochar-fertilizer increased cucumber, broccoli, and okra germination parameters (index, final percentage, and mean time, respectively). In addition, it increased shoot and root lengths as compared to traditional fertilizers, where the best dose for germination and growth promotion was 37.5 kg/ha through improved soil quality by introducing OM and slow-release nutrients (Tarar et al., 2023). C nanoparticles were immersed in potassium nitrate to synthesize a nanocomposite fertilizer for common beans (*Phaseolus vulgaris* L.), which was administered as a foliar spray at concentrations of 10–40 mg/L, 25 days postsowing (Salama et al., 2021). During two growing seasons, application of the ≥ 20 mg/L nanocomposite significantly enhanced seed yields and improved growth parameters, including plant height, leaf count, flower count, and fresh weight (Salama et al., 2021).

From a physiological perspective, BMFs regulate plant hormone homeostasis, enhance photosynthetic efficiency, and stimulate secondary metabolite biosynthesis. Engineered biochar is a superior carrier for high-quality slow-release fertilizers. Soils treated with $MgCl_2$-modified BMF retained more water and released NO_3^- and NH_4^+ 2.5 and 1.5 times slower than ammonium nitrate (Khajavi-Shojaei et al., 2023). $MgCl_2$-modified biochar significantly increased plant height, shoot and root dry weight, chlorophyll content, and leaf area, resulting in the highest soil and plant N content and NUE. Thus, $MgCl_2$-modified biochar is promising for synthesizing slow-release fertilizers to improve corn growth and NUE (Khajavi-Shojaei et al., 2023). Application of these fertilizers markedly elevated net photosynthetic rate and stomatal conductance during early and mid-transplanting phases, along with increased transpiration rate and intercellular CO_2 concentration throughout the growth period (Ren et al., 2023). Additionally, Ren et al. (2023) have demonstrated that even with a 50% reduction in N, BMFs combined with drip irrigation significantly augmented soil microbial biomass C, oxidizable C, and soluble C, thereby enhancing photosynthetic activity and dry matter accumulation in tobacco plants, facilitating improved crop cultivation.

BMF alter root physiology by enhancing root system architecture and function. Biochar's porous structure and C composition improve soil aggregation, porosity, and moisture retention, facilitating deeper root growth and proliferation (Kumar et al., 2020). These enhancements increase water and nutrient accessibility, as well as improve plant anchorage, stability, and resilience to environmental stresses (Abideen et al., 2020). Moreover, biochar or C from BMS participate in modulating root exudation patterns and establishing symbiotic relationships with beneficial soil microbes (Conversa et al., 2015). These synergistic associations faciliate nutrient uptake efficiency by optimizing the root system's spatial distribution and nutrient absorption kinetics, particularly for P and micronutrients (An et al., 2021). This enhanced nutrient assimilation fosters improved root biomass and physiological functions (Banik et al., 2023). The addition of activated wheat straw BMF has demonstrated significant enhancements in crop yields and modifications to rhizosphere soil properties such as plant biomass (67%), herbage N (40%), and P uptake (46%) in rice plants compared to CF (Chew et al., 2020). Micron and submicron-sized biochar particles embedded in the root plaque layer enhanced soil Eh by 85 mV and increased the potential difference between rhizosphere soil and root membranes by 65 mV, thereby lowering nutrient uptake energy. Moreover, biochar fostered plant-growth-promoting bacteria and fungi, underscoring its redox properties as crucial for agronomic benefits (Chew et al., 2020). Yang et al. (2021) reported a remarkable 39% increase in tea leaf yield using biochar-based NPK fertilizer. Similarly, Jia et al. (2012) demonstrated that, in compared to urea alone, incorporating biochar with urea and manure at rates of 20 or 40 Mg/ha increased vegetable production by 32% and 48%, respectively, while simultaneously reducing N_2O-N emissions by 77%–86%.

22.7 Conclusion and future perspective

In conclusion, the strategic use of mineral-enriched biochar-fertilizer or BMFs holds great promise for enhancing agricultural productivity by improving rhizospheric soil function and enhancing plant/crop performance. Integrating biochar with mineral nutrients offers substantial benefits, including enhanced soil structure, water retention, and nutrient availability, crucial for optimizing root system development and enhancing overall plant vigor and physiological functionality. This approach underscores biochar-fertilizers' multifaceted roles in agriculture fostering beneficial microbial activities in the rhizosphere, and promoting a conducive environment for plant growth. By mitigating nutrient leaching, enhancing soil fertility, and reducing GHG emissions, biochar-fertilizers support sustainable farming practices. Future research directions should focus on exploring the long-term impacts of BMFs applications across diverse soil types and climates. Investigating the mechanisms underlying biochar's interactions with plant roots and soil microbiota will deepen our understanding of its role in bolstering crop resilience to environmental stresses such as drought and salinity. Moreover, advancing biochar production methods through innovative technologies and sustainable biomass sources could further enhance its effectiveness as a soil amendment. Effective progress in biochar-fertilizer technologies requires collaborative efforts across multidisciplinary research teams, encompassing agronomists, soil scientists, microbiologists, and environmental engineers. Such collaborations are pivotal for translating experimental findings into practical applications that benefit global food security and foster sustainable agricultural practices. Overall, the continued development and deployment of BMFs offer immense potential for addressing global food security challenges while promoting sustainable agricultural practices that benefit both agriculture and the environment.

Acknowledgments

P.K. thankfully acknowledges financial support by the European Union's Horizon 2020 research and Innovation Programme under the Marie Skłodowska-Curie grant agreement No 847639 and from the Ministry of Education and Science. A.K. is thankful to University of Tuscia, Viterbo, Italy, for providing researcher grant from the EXCELLENCE Landscape 4.0.

CRediT author statement

P.K.: conceptualization, methodology, investigation, resources, visualization, writing—original draft, writing—review and editing; S.J.K.: methodology, resources, writing-original draft; M.N.V.P.: conceptualization, writing—review and editing, supervision; M.K.S.: conceptualization, methodology, writing—review and editing; N.K.S.: conceptualization, writing—review and editing; A.K.: conceptualization, formal analysis, investigation, methodology, resources, visualization, writing—original draft, writing—review and editing, supervision.

Conflicts of interest

The authors declare that they have no known competing financial interests or personal relationships that could have appeared to influence the work reported in this paper.

References

Abideen, Z., Koyro, H. -W., Huchzermeyer, B., Ansari, R., Zulfiqar, F., Gul, B., & Leegood, R. (2020). Ameliorating effects of biochar on photosynthetic efficiency and antioxidant defence of Phragmites karka under drought stress. *Plant Biology*, *22*(2), 259–266. Available from https://doi.org/10.1111/plb.13054.

Agarwal, H., Kashyap, V. H., Mishra, A., Bordoloi, S., Singh, P. K., & Joshi, N. C. (2022). Biochar-based fertilizers and their applications in plant growth promotion and protection. *3 Biotech*, *12*(6). Available from https://doi.org/10.1007/s13205-022-03195-2, https://www.springer.com/journal/13205.

Agilandeswari, P., Venkateshbabu, S., Sarojini, G., & Rajasimman, M. (2024). Sustainable development and analysis of a novel bio-derived (biochar) nanocomposite for the remediation of carbamazepine from aqueous solution. *Chemosphere*, *347*.

Agyarko-Mintah, E., Cowie, A., Singh, B. P., Joseph, S., Van Zwieten, L., Cowie, A., Harden, S., & Smillie, R. (2017). Biochar increases nitrogen retention and lowers greenhouse gas emissions when added to composting poultry litter. *Waste Management*, *61*, 138–149. Available from https://doi.org/10.1016/j.wasman.2016.11.027, http://www.elsevier.com/locate/wasman.

Ahmadian, K., Jalilian, J., & Pirzad, A. (2021). Nano-fertilizers improved drought tolerance in wheat under deficit irrigation. *Agricultural Water Management*, *244*, 106544. Available from https://doi.org/10.1016/j.agwat.2020.106544.

Al Arni, S. (2018). Comparison of slow and fast pyrolysis for converting biomass into fuel. *Renewable Energy*, *124*, 197–201. Available from https://doi.org/10.1016/j.renene.2017.04.060, http://www.journals.elsevier.com/renewable-and-sustainable-energy-reviews/.

An, T., Cheng, H., Qin, Y., Su, W., Deng, H., Wu, J., Liu, Z., & Guo, X. (2021). The dual mechanisms of composite biochar and biofilm towards sustainable nutrient release control of phosphate fertilizer: Effect on phosphorus utilization and crop growth. *Journal of Cleaner Production*, *311*, 127329.

An, X., Wu, Z., Yu, J., Cravotto, G., Liu, X., Li, Q., & Yu, B. (2020). Copyrolysis of biomass, bentonite, and nutrients as a new strategy for the synthesis of improved biochar-based slow-release fertilizers. *ACS Sustainable Chemistry and Engineering*, *8*(8), 3181–3190. Available from https://doi.org/10.1021/acssuschemeng.9b06483, http://pubs.acs.org/journal/ascecg.

An, X., Xu, X., Guo, W., Chen, Z., Miao, Z., Yuan, J., & Wu, Z. (2023). Bi-functional biochar-g-C_3N_4-MgO composites for simultaneously minimizing pollution: Photocatalytic degradation of pesticide and phosphorus recovery as slow-release fertilizer. *Journal of Environmental Management*, *344*, 118489. Available from https://doi.org/10.1016/j.jenvman.2023.118489.

Bandara, T., Herath, I., Kumarathilaka, P., Hseu, Z. Y., Ok, Y. S., & Vithanage, M. (2017). Efficacy of woody biomass and biochar for alleviating heavy metal bioavailability in serpentine soil. *Environmental geochemistry and health*, *39*(2), 391–401. Available from https://doi.org/10.1007/s10653-016-9842-0, http://www.wkap.nl/journalhome.htm/0269-4042.

Banik, C., Bakshi, S., Laird, D. A., Smith, R. G., & Brown, R. C. (2023). Impact of biochar-based slow-release N-fertilizers on maize growth and nitrogen recovery efficiency. *Journal of environmental quality*, *52*(3), 630–640. Available from https://doi.org/10.1002/jeq2.20468, https://acsess.onlinelibrary.wiley.com/loi/15372537.

Barbosa, B. M., Vaz, S., Colodette, J. L., de Aguiar, A. R., Cabral, C. P. T., & de Freitas Homem de Faria, B. (2022). Structural and chemical characterization of lignin and hemicellulose isolated from corn fibers toward agroindustrial residue valorization. *Cellulose*, *29*(15), 8117–8132. Available from https://doi.org/10.1007/s10570-022-04769-2.

Bashir, A., Rizwan, M., Zia ur Rehman, M., Zubair, M., Riaz, M., Qayyum, M. F., Alharby, H. F., Bamagoos, A. A., & Ali, S. (2020). Application of co-composted farm manure and biochar increased the wheat growth and decreased cadmium accumulation in plants under different water regimes. *Chemosphere*, *246*, 125809. Available from https://doi.org/10.1016/j.chemosphere.2019.125809.

Blenis, N., Hue, N., Maaz, T. M. C., & Kantar, M. (2023). Biochar production, modification, and its uses in soil remediation: A review. *Sustainability*, *15*(4), 3442. Available from https://doi.org/10.3390/su15043442.

Brtnicky, M., Mustafa, A., Hammerschmiedt, T., Kintl, A., Trakal, L., Beesley, L., Ryant, P., Omara-Ojungu, C., Baltazar, T., & Holatko, J. (2023). Pre-activated biochar by fertilizers mitigates nutrient leaching and stimulates soil microbial activity. *Chemical and Biological Technologies in Agriculture*, *10*(1). Available from https://doi.org/10.1186/s40538-023-00430-7, https://chembioagro.springeropen.com.

Cao, Q., An, T., Xie, J., Liu, Y., Xing, L., Ling, X., & Chen, C. (2022). Insight to the physiochemical properties and DOM of biochar under different pyrolysis temperature and modification conditions. *Journal of Analytical and Applied Pyrolysis*, *166*, 105590. Available from https://doi.org/10.1016/j.jaap.2022.105590.

Chatterjee, R., Sajjadi, B., Chen, W. Y., Mattern, D. L., Hammer, N., Raman, V., & Dorris, A. (2020). Effect of pyrolysis temperature on physicochemical properties and acoustic-based amination of biochar for efficient CO_2 adsorption. *Frontiers in Energy Research*, *8*. Available from https://doi.org/10.3389/fenrg.2020.00085, http://www.frontiersin.org/Energy_Research/.

Chen, K., Peng, J., Li, J., Yang, Q., Zhan, X., Liu, N., & Han, X. (2020). Stabilization of soil aggregate and organic matter under the application of three organic resources and biochar-based compound fertilizer. *Journal of Soils and Sediments*, *20*(10), 3633–3643. Available from https://doi.org/10.1007/s11368-020-02693-1, http://www.springerlink.com/content/1439-0108.

Chen, L., Zhang, H., Jia, X., Fang, Y., & Lin, C. (2024). Soil fertility and bacterial community composition in response to the composting of biochar-amended chicken manure. *Agronomy*, *14*(5), 886. Available from https://doi.org/10.3390/agronomy14050886.

Chew, J., Zhu, L., Nielsen, S., Graber, E., Mitchell, D. R. G., Horvat, J., Mohammed, M., Liu, M., van Zwieten, L., Donne, S., Munroe, P., Taherymoosavi, S., Pace, B., Rawal, A., Hook, J., Marjo, C., Thomas, D. S., Pan, G., Li, L., … Fan, X. (2020). Biochar-based fertilizer: Supercharging root membrane potential and biomass yield of rice. *Science of the Total Environment*, *713*. Available from https://doi.org/10.1016/j.scitotenv.2019.136431, http://www.elsevier.com/locate/scitotenv.

Codignole Luz, F., Cordiner, S., Manni, A., Mulone, V., & Rocco, V. (2018). Biomass fast pyrolysis in a shaftless screw reactor: A 1-D numerical model. *Energy*, *157*, 792–805.

Available from https://doi.org/10.1016/j.energy.2018.05.166, http://www.elsevier.com/inca/publications/store/4/8/3/.

Conversa, G., Bonasia, A., Lazzizera, C., & Elia, A. (2015). Influence of biochar, mycorrhizal inoculation, and fertilizer rate on growth and flowering of pelargonium (*Pelargonium zonale* L.) plants. *Frontiers in Plant Science*, *6*(June). Available from https://doi.org/10.3389/fpls.2015.00429, http://journal.frontiersin.org/article/10.3389/fpls.2015.00429/pdf.

Dominguez, E. L., Uttran, A., Loh, S. K., Manero, M. H., Upperton, R., Tanimu, M. I., & Thomas Bachmann, R. T. (2020). Characterisation of industrially produced oil palm kernel shell biochar and its potential as slow release nitrogen-phosphate fertilizer and carbon sink. *Materials Today: Proceedings*, *31*, 221–227. Available from https://doi.org/10.1016/j.matpr.2020.05.143, https://www.sciencedirect.com/journal/materials-today-proceedings.

Duku, M. H., Gu, S., & Hagan, E. B. (2011). Biochar production potential in Ghana – a review. *Renewable and Sustainable Energy Reviews*, *15*(8), 3539–3551. Available from https://doi.org/10.1016/j.rser.2011.05.010.

El Sharkawi, H. M., Tojo, S., Chosa, T., Malhat, F. M., & Youssef, A. M. (2018). Biochar-ammonium phosphate as an uncoated-slow release fertilizer in sandy soil. *Biomass and Bioenergy*, *117*, 154–160. Available from https://doi.org/10.1016/j.biombioe.2018.07.007, http://www.journals.elsevier.com/biomass-and-bioenergy/.

Foong, S. Y., Chan, Y. H., Chin, B. L. F., Lock, S. S. M., Yee, C. Y., Yiin, C. L., Peng, W., & Lam, S. S. (2022). Production of biochar from rice straw and its application for wastewater remediation – an overview. *Bioresource Technology*, *360*. Available from https://doi.org/10.1016/j.biortech.2022.127588, http://www.elsevier.com/locate/biortech.

Gao, M., Yang, J., Liu, C., Gu, B., Han, M., Li, J., Li, N., Liu, N., An, N., Dai, J., Liu, X., & Han, X. (2021). Effects of long-term biochar and biochar-based fertilizer application on brown earth soil bacterial communities. *Agriculture, Ecosystems & Environment*, *309*, 107285.

Ghassemi-Golezani, K., & Rahimzadeh, S. (2023). Biochar-based nutritional nanocomposites: a superior treatment for alleviating salt toxicity and improving physiological performance of dill (*Anethum graveolens*). *Environmental Geochemistry and Health*, *45*(6), 3089–3111. Available from https://doi.org/10.1007/s10653-022-01397-4, https://www.springer.com/journal/10653.

Ghodake, G. S., Shinde, S. K., Kadam, A. A., Saratale, R. G., Saratale, G. D., Kumar, M., Palem, R. R., AL-Shwaiman, H. A., Elgorban, A. M., Syed, A., & Kim, D. Y. (2021). Review on biomass feedstocks, pyrolysis mechanism and physicochemical properties of biochar: State-of-the-art framework to speed up vision of circular bioeconomy. *Journal of Cleaner Production*, *297*. Available from https://doi.org/10.1016/j.jclepro.2021.126645, https://www.journals.elsevier.com/journal-of-cleaner-production.

Ghorbani, M., Konvalina, P., Neugschwandtner, R. W., Soja, G., Bárta, J., Chen, W. H., & Amirahmadi, E. (2024). How do different feedstocks and pyrolysis conditions effectively change biochar modification scenarios? A critical analysis of engineered biochars under H_2O_2 oxidation. *Energy Conversion and Management*, *300*. Available from https://doi.org/10.1016/j.enconman.2023.117924, https://www.sciencedirect.com/science/journal/01968904.

Greenberg, I., Kaiser, M., Gunina, A., Ledesma, P., Polifka, S., Wiedner, K., Mueller, C. W., Glaser, B., & Ludwig, B. (2019). Substitution of mineral fertilizers with biogas digestate plus biochar increases physically stabilized soil carbon but not crop biomass in a field trial. *Science of the Total Environment*, *680*, 181–189. Available from https://doi.org/10.1016/j.scitotenv.2019.05.051, http://www.elsevier.com/locate/scitotenv.

Han, J., Chen, X., Xi, B., Mao, H., Feng, J., & Xiong, S. (2018). High-surface-area nitrogen/phosphorus dual-doped hierarchical porous carbon derived from biochar for sulfur holder. *ChemistrySelect*, *3*(36), 10175–10181. Available from https://doi.org/10.1002/slct.201802296, http://onlinelibrary.wiley.com/journal/10.1002/(ISSN)2365-6549.

He, Q., Raheem, A., Ding, L., Xu, J., Cheng, C., & Yu, G. (2021). Combining wet torrefaction and pyrolysis for woody biochar upgradation and structural modification. *Energy Conversion and Management*, *243*, 114383. Available from https://doi.org/10.1016/j.enconman.2021.114383.

He, T., Yuan, J., Luo, J., Wang, W., Fan, J., Liu, D., & Ding, W. (2019). Organic fertilizers have divergent effects on soil N_2O emissions. *Biology and Fertility of Soils*, *55*(7), 685–699. Available from https://doi.org/10.1007/s00374-019-01385-4, http://link.springer.de/link/service/journals/00374/index.htm.

Hu, Y., Cao, Y., Ma, C., & Yan, W. (2024). Nano-biochar as a potential amendment for metal(loid) remediation: Implications for soil quality improvement and stress alleviation. *Journal of Environmental Management*, *351*. Available from https://doi.org/10.1016/j.jenvman.2023.119658, https://www.sciencedirect.com/science/journal/03014797.

Ibrahim, M. M., Guo, L., Wu, F., Liu, D., Zhang, H., Zou, S., Xing, S., & Mao, Y. (2022). Field-applied biochar-based MgO and sepiolite composites possess CO_2 capture potential and alter organic C mineralization and C-cycling bacterial structure in fertilized soils. *Science of the Total Environment*, *813*, 152495. Available from https://doi.org/10.1016/j.scitotenv.2021.152495.

Idbella, M., Baronti, S., Giagnoni, L., Renella, G., Becagli, M., Cardelli, R., Maienza, A., Vaccari, F. P., & Bonanomi, G. (2024). Long-term effects of biochar on soil chemistry, biochemistry, and microbiota: Results from a 10-year field vineyard experiment. *Applied Soil Ecology*, *195*, 105217.

Igalavithana, A. D., Kwon, E. E., Vithanage, M., Rinklebe, J., Moon, D. H., Meers, E., Tsang, D. C. W., & Ok, Y. S. (2019). Soil lead immobilization by biochars in short-term laboratory incubation studies. *Environment International*, *127*, 190–198. Available from https://doi.org/10.1016/j.envint.2019.03.031, http://www.elsevier.com/locate/envint.

Jia, J., Li, B., Chen, Z., Xie, Z., & Xiong, Z. (2012). Effects of biochar application on vegetable production and emissions of N_2O and CH_4. *Soil Science and Plant Nutrition*, *58*(4), 503–509. Available from https://doi.org/10.1080/00380768.2012.686436.

Jia, Y., Hu, Z., Ba, Y., & Qi, W. (2021). Application of biochar-coated urea controlled loss of fertilizer nitrogen and increased nitrogen use efficiency. *Chemical and Biological Technologies in Agriculture*, *8*(1). Available from https://doi.org/10.1186/s40538-020-00205-4, http://chembioagro.springeropen.com/.

Jin, Z., Chen, C., Chen, X., Jiang, F., Hopkins, I., Zhang, X., Han, Z., Billy, G., & Benavides, J. (2019). Soil acidity, available phosphorus content, and optimal biochar and nitrogen fertilizer application rates: A five-year field trial in upland red soil, China. *Field Crops Research*, *232*, 77–87.

Joseph, S., Cowie, A. L., Van Zwieten, L., Bolan, N., Budai, A., Buss, W., Cayuela, M. L., Graber, E. R., Ippolito, J. A., Kuzyakov, Y., Luo, Y., Ok, Y. S., Palansooriya, K. N., Shepherd, J., Stephens, S., Weng, Z., & Lehmann, J. (2021). How biochar works, and when it doesn't: A review of mechanisms controlling soil and plant responses to biochar. *GCB Bioenergy*, *13*(11), 1731–1764. Available from https://doi.org/10.1111/gcbb.12885, http://onlinelibrary.wiley.com/journal/10.1111/(ISSN)1757-1707.

Kenda, G. T., Kouteu, P. A. N., Tchuifon, D. R. T., Fotsop, C. G., Bopda, A., Kuete, H.-I. T., Ndifor-Angwafor, N. G., & Anagho, S. G. (2024). Green synthesis of magnetic biochars derived from biobased orange peel materials as sustainable heterogeneous catalytic supports for the Fenton process. *Arabian Journal of Chemistry*, *17*, 105502.

Keske, C., Godfrey, T., Hoag, D. L. K., & Abedin, J. (2020). Economic feasibility of biochar and agriculture coproduction from Canadian black spruce forest. *Food and Energy Security*, *9*(1). Available from https://doi.org/10.1002/fes3.188, http://onlinelibrary.wiley.com/journal/10.1002/(ISSN)2048-3694.

Khajavi-Shojaei, S., Moezzi, A., Norouzi Masir, M., & Taghavi, M. (2023). Synthesis modified biochar-based slow-release nitrogen fertilizer increases nitrogen use efficiency and corn (*Zea mays* L.) growth. *Biomass Conversion and Biorefinery*, *13*(2), 593–601. Available from https://doi.org/10.1007/s13399-020-01137-7, https://www.springer.com/journal/13399.

Kim, J. A., Vijayaraghavan, K., Reddy, D. H. K., & Yun, Y. S. (2018). A phosphorus-enriched biochar fertilizer from bio-fermentation waste: A potential alternative source for phosphorus fertilizers. *Journal of Cleaner Production*, *196*, 163–171. Available from https://doi.org/10.1016/j.jclepro.2018.06.004, https://www.journals.elsevier.com/journal-of-cleaner-production.

Kizito, S., Wu, S., Kipkemoi Kirui, W., Lei, M., Lu, Q., Bah, H., & Dong, R. (2015). Evaluation of slow pyrolyzed wood and rice husks biochar for adsorption of ammonium nitrogen from piggery manure anaerobic digestate slurry. *Science of the Total Environment*, *505*, 102–112. Available from https://doi.org/10.1016/j.scitotenv.2014.09.096, http://www.elsevier.com/locate/scitotenv.

Kolton, M., Graber, E. R., Tsehansky, L., Elad, Y., & Cytryn, E. (2017). Biochar-stimulated plant performance is strongly linked to microbial diversity and metabolic potential in the rhizosphere. *New Phytologist*, *213*(3), 1393–1404. Available from https://doi.org/10.1111/nph.14253, http://onlinelibrary.wiley.com/journal/10.1111/(ISSN)1469-8137.

Kumar, A., Schreiter, I. J., Wefer-Roehl, A., Tsechansky, L., Schüth, C., & Graber, E. R. (2016). *Production and utilization of biochar from organic wastes for pollutant control on contaminated sites. Environmental Materials and Waste: Resource Recovery and Pollution Prevention* (pp. 91–116). Available from http://www.sciencedirect.com/science/book/9780128038376, 10.1016/B978-0-12-803837-6.00005-6.

Kumar, A., Joseph, S., Tsechansky, L., Privat, K., Schreiter, I. J., Schüth, C., & Graber, E. R. (2018). Biochar aging in contaminated soil promotes Zn immobilization due to changes in biochar surface structural and chemical properties. *Science of the Total Environment*, *626*, 953–961. Available from https://doi.org/10.1016/j.scitotenv.2018.01.157, http://www.elsevier.com/locate/scitotenv.

Kumar, A., Tsechansky, L., Lew, B., Raveh, E., Frenkel, O., & Graber, E. R. (2018). Biochar alleviates phytotoxicity in Ficus elastica grown in Zn-contaminated soil.

Science of the Total Environment, *618*, 188–198. Available from https://doi.org/10.1016/j.scitotenv.2017.11.013, http://www.elsevier.com/locate/scitotenv.

Kumar, A., Joseph, S., Graber, E. R., Taherysoosavi, S., Mitchell, D. R. G., Munroe, P., Tsechansky, L., Lerdahl, O., Aker, W., & Sæbø, M. (2021). Fertilizing behavior of extract of organomineral-activated biochar: Low-dose foliar application for promoting lettuce growth. *Chemical and Biological Technologies in Agriculture*, *8*(1). Available from https://doi.org/10.1186/s40538-021-00222-x, http://chembioagro.springeropen.com/.

Kumar, A., Joseph, S., Tsechansky, L., Schreiter, I. J., Schüth, C., Taherysoosavi, S., Mitchell, D. R. G., & Graber, E. R. (2020). Mechanistic evaluation of biochar potential for plant growth promotion and alleviation of chromium-induced phytotoxicity in Ficus elastica. *Chemosphere*, *243*, 125332. Available from https://doi.org/10.1016/j.chemosphere.2019.125332.

Lalmalsawmdawngliani., Lalhriatpuia, C., & Tiwari, D. (2024). Biochar-derived nanocomposites for environmental remediation: The insights and future perspectives. *Journal of Environmental Chemical Engineering*, *12*(1), 111840. Available from https://doi.org/10.1016/j.jece.2023.111840.

Lea, P. J., & Azevedo, R. A. (2006). Nitrogen use efficiency. 1. Uptake of nitrogen from the soil. *Annals of Applied Biology*, *149*(3), 243–247. Available from https://doi.org/10.1111/j.1744-7348.2006.00101.x.

Lehmann, J., & Joseph, S. (2024). *Biochar for environmental management: Science, technology and implementation*. Routledge. Available from 10.4324/9781003297673.

Leogrande, R., Vitti, C., Castellini, M., Garofalo, P., Samarelli, I., Lacolla, G., Montesano, F. F., Spagnuolo, M., Mastrangelo, M., & Stellacci, A. M. (2024). Residual effect of compost and biochar amendment on soil chemical, biological, and physical properties and durum wheat response. *Agronomy*, *14*(4), 749. Available from https://doi.org/10.3390/agronomy14040749.

Liao, J., Liu, X., Hu, A., Song, H., Chen, X., & Zhang, Z. (2020). Effects of biochar-based controlled release nitrogen fertilizer on nitrogen-use efficiency of oilseed rape (*Brassica napus* L.). *Scientific Reports*, *10*(1). Available from https://doi.org/10.1038/s41598-020-67528-y, http://www.nature.com/srep/index.html.

Lin, L., Han, X., Han, B., & Yang, S. (2021). Emerging heterogeneous catalysts for biomass conversion: Studies of the reaction mechanism. *Chemical Society Reviews*, *50*(20), 11270–11292. Available from https://doi.org/10.1039/d1cs00039j, http://pubs.rsc.org/en/journals/journal/cs.

Liu, G., Zheng, H., Jiang, Z., Zhao, J., Wang, Z., Pan, B., & Xing, B. (2018). Formation and physicochemical characteristics of nano biochar: Insight into chemical and colloidal stability. *Environmental Science and Technology*, *52*(18), 10369–10379. Available from https://doi.org/10.1021/acs.est.8b01481, http://pubs.acs.org/journal/esthag.

Liu, N., Zhou, J., Han, L., & Huang, G. (2017). Characterization of lignocellulosic compositions' degradation during chicken manure composting with added biochar by phospholipid fatty acid (PLFA) and correlation analysis. *Science of the Total Environment*, *586*, 1003–1011. Available from https://doi.org/10.1016/j.scitotenv.2017.02.081, http://www.elsevier.com/locate/scitotenv.

Liu, Z., Jia, M., Li, Q., Lu, S., Zhou, D., Feng, L., Hou, Z., & Yu, J. (2023). Comparative analysis of the properties of biochars produced from different pecan feedstocks and pyrolysis temperatures. *Industrial Crops and Products*, *197*, 116638. Available from https://doi.org/10.1016/j.indcrop.2023.116638.

Lu, J., Li, Y., Cai, Y., Jiang, P., & Yu, B. (2023). Co-incorporation of hydrotalcite and starch into biochar-based fertilizers for the synthesis of slow-release fertilizers with improved water retention. *Biochar*, *5*(1). Available from https://doi.org/10.1007/s42773-023-00242-w, https://www.springer.com/journal/42773.

Lu, Z., Hu, X., & Lu, Y. (2017). Particle morphology analysis of biomass material based on improved image processing method. *International Journal of Analytical Chemistry*, *2017*. Available from https://doi.org/10.1155/2017/5840690, http://www.hindawi.com/journals/ijac/.

Lustosa Filho, J. F., Barbosa, C. F., Carneiro, J. S. d. S., & Melo, L. C. A. (2019). Diffusion and phosphorus solubility of biochar-based fertilizer: Visualization, chemical assessment and availability to plants. *Soil and Tillage Research*, *194*, 104298.

Mahdi, Z., El Hanandeh, A., & Yu, Q. J. (2019). Preparation, characterization and application of surface modified biochar from date seed for improved lead, copper, and nickel removal from aqueous solutions. *Journal of Environmental Chemical Engineering*, *7* (5), 103379. Available from https://doi.org/10.1016/j.jece.2019.103379.

Mahmud, K., Panday, D., Mergoum, A., & Missaoui, A. (2021). Nitrogen losses and potential mitigation strategies for a sustainable agroecosystem. *Sustainability*, *13*(4), 2400. Available from https://doi.org/10.3390/su13042400.

Marcińczyk, M., & Oleszczuk, P. (2022). Biochar and engineered biochar as slow- and controlled-release fertilizers. *Journal of Cleaner Production*, *339*, 130685. Available from https://doi.org/10.1016/j.jclepro.2022.130685.

Massaccesi, L., Nogués, I., Mazzurco Miritana, V., Passatore, L., Zacchini, M., Pietrini, F., Carloni, S., Marabottini, R., Moscatelli, M. C., & Marinari, S. (2024). Short-term effects of biochar and compost on soil microbial community, C and N cycling, and lettuce (*Lactuca sativa* L.) yield in a Mediterranean environment. *Applied Soil Ecology*, *199*, 105411. Available from https://doi.org/10.1016/j.apsoil.2024.105411.

Meier, S., Curaqueo, G., Khan, N., Bolan, N., Cea, M., Eugenia, G. M., Cornejo, P., Ok, Y. S., & Borie, F. (2017). Chicken-manure-derived biochar reduced bioavailability of copper in a contaminated soil. *Journal of Soils and Sediments*, *17*(3), 741–750. Available from https://doi.org/10.1007/s11368-015-1256-6, http://www.springerlink.com/content/1439-0108.

Melo, L. C. A., Lehmann, J., Carneiro, J. Sd. S., & Camps-Arbestain, M. (2022). Biochar-based fertilizer effects on crop productivity: A meta-analysis. *Plant and Soil*, *472*(1-2), 45–58. Available from https://doi.org/10.1007/s11104-021-05276-2, http://www.wkap.nl/journalhome.htm/0032-079X.

Mishra, D., Jain, S., Khare, P., & Singh, R. (2023). Comparative Investigation of Biochar-Based Nanocomposites Over Pristine Biochar: An Overview. In D. Mishra, R. Singh, & P. Khare (Eds.), *Biochar-Based Nanocomposites for Contaminant Management: Synthesis, Contaminants Removal, and Environmental Sustainability* (pp. 57–68). Cham: Springer International Publishing.

Nadeem, S. M., Imran, M., Naveed, M., Khan, M. Y., Ahmad, M., Zahir, Z. A., & Crowley, D. E. (2017). Synergistic use of biochar, compost and plant growth-promoting rhizobacteria for enhancing cucumber growth under water deficit conditions. *Journal of the Science of Food and Agriculture*, *97*(15), 5139–5145. Available from https://doi.org/10.1002/jsfa.8393, http://onlinelibrary.wiley.com/journal/10.1002/(ISSN)1097-0010.

Nurida, N. L., & Jubaedah. (2019). Biochar based soil amendment formulation for improvement upland acidic soil in east lampung: Soil properties and maize (*Zea mays*) yield. *Journal of Tropical Soils*, *24*(1), 33. Available from https://doi.org/10.5400/jts.2019.v24i1.33-41.

Oladele, S., Adeyemo, A., Awodun, M., Ajayi, A., & Fasina, A. (2019). Effects of biochar and nitrogen fertilizer on soil physicochemical properties, nitrogen use efficiency and upland rice (*Oryza sativa*) yield grown on an Alfisol in Southwestern Nigeria. *International Journal of Recycling of Organic Waste in Agriculture*, *8*(3), 295–308. Available from https://doi.org/10.1007/s40093-019-0251-0, http://www.springer.com/environment/pollution + and + remediation/journal/40093.

Papageorgiou, A., Azzi, E. S., Enell, A., & Sundberg, C. (2021). Biochar produced from wood waste for soil remediation in Sweden: Carbon sequestration and other environmental impacts. *Science of the Total Environment*, *776*, 145953. Available from https://doi.org/10.1016/j.scitotenv.2021.145953.

Peter, A., Chabot, B., & Loranger, E. (2021). Pre- and post-pyrolysis effects on iron impregnation of ultrasound pre-treated softwood biochar for potential catalysis applications. *SN Applied Sciences*, *3*(6). Available from https://doi.org/10.1007/s42452-021-04636-y, http://springer.com/snas.

Piash, M. I., Iwabuchi, K., & Itoh, T. (2022). Synthesizing biochar-based fertilizer with sustained phosphorus and potassium release: Co-pyrolysis of nutrient-rich chicken manure and Ca-bentonite. *Science of the Total Environment*, *822*, 153509. Available from https://doi.org/10.1016/j.scitotenv.2022.153509.

Pogorzelski, D., Filho, J. F. L., Matias, P. C., Santos, W. O., Vergütz, L., & Melo, L. C. A. (2020). Biochar as composite of phosphate fertilizer: Characterization and agronomic effectiveness. *Science of the Total Environment*, *743*. Available from https://doi.org/10.1016/j.scitotenv.2020.140604, http://www.elsevier.com/locate/scitotenv.

Puga, A. P., Grutzmacher, P., Cerri, C. E. P., Ribeirinho, V. S., & Andrade, C. Ad (2020). Biochar-based nitrogen fertilizers: Greenhouse gas emissions, use efficiency, and maize yield in tropical soils. *Science of the Total Environment*, *704*. Available from https://doi.org/10.1016/j.scitotenv.2019.135375, http://www.elsevier.com/locate/scitotenv.

Qayyum, M. F., Liaquat, F., Rehman, R. A., Gul, M., ul Hye, M. Z., Rizwan, M., & Rehaman, M. Z. U. (2017). Effects of co-composting of farm manure and biochar on plant growth and carbon mineralization in an alkaline soil. *Environmental Science and Pollution Research*, *24*(33), 26060–26068. Available from https://doi.org/10.1007/s11356-017-0227-4, http://www.springerlink.com/content/0944-1344.

Qin, D., Luo, G., Qin, A., He, T., Wu, P., & Yin, D. (2024). Selenium–phosphorus modified biochar reduces mercury methylation and bioavailability in agricultural soil. *Environmental pollution (Barking, Essex: 1987)*.

Rafael, R. B. A., Fernández-Marcos, M. L., Cocco, S., Ruello, M. L., Fornasier, F., & Corti, G. (2019). Benefits of biochars and NPK fertilizers for soil quality and growth of cowpea (*Vigna unguiculata* L. Walp.) in an acid arenosol. *Pedosphere*, *29*(3), 311–333. Available from https://doi.org/10.1016/S1002-0160(19)60805-2, http://pedosphere.issas.ac.cn.

Ramesh, K., & Raghavan, V. (2024). Agricultural waste-derived biochar-based nitrogenous fertilizer for slow-release applications. *ACS Omega*, *9*(4), 4377–4385. Available from https://doi.org/10.1021/acsomega.3c06687, http://pubs.acs.org/journal/acsodf.

Rashid, M. I., Shah, G. A., Iqbal, Z., Shahzad, K., Ali, N., Rehan, M., Alhakamy, N. A. A., & Klemeš, J. J. (2023). Nanobiochar reduces ammonia emission, increases nutrient mineralization from vermicompost, and improves maize productivity. *Journal of Cleaner Production*, *414*. Available from https://doi.org/10.1016/j.jclepro.2023.137694, https://www.journals.elsevier.com/journal-of-cleaner-production.

Ren, T., Fan, P., Zuo, W., Liao, Z., Wang, F., Wei, Y., Cai, X., & Liu, G. (2023). Biochar-based fertilizer under drip irrigation: More conducive to improving soil carbon pool and promoting nitrogen utilization. *Ecological Indicators*, *154*, 110583. Available from https://doi.org/10.1016/j.ecolind.2023.110583.

Roberts, D. A., & de Nys, R. (2016). The effects of feedstock pre-treatment and pyrolysis temperature on the production of biochar from the green seaweed Ulva. *Journal of Environmental Management*, *169*, 253–260. Available from https://doi.org/10.1016/j.jenvman.2015.12.023, https://www.sciencedirect.com/journal/journal-of-environmental-management.

Rombel, A., Krasucka, P., & Oleszczuk, P. (2022). Sustainable biochar-based soil fertilizers and amendments as a new trend in biochar research. *Science of the Total Environment*, *816*, 151588. Available from https://doi.org/10.1016/j.scitotenv.2021.151588.

Romero, C. M., Hao, X., Li, C., Owens, J., Schwinghamer, T., McAllister, T. A., & Okine, E. (2021). Nutrient retention, availability and greenhouse gas emissions from biochar-fertilized Chernozems. *CATENA*, *198*, 105046.

Russell, S. H., Turrion-Gomez, J. L., Meredith, W., Langston, P., & Snape, C. E. (2017). Increased charcoal yield and production of lighter oils from the slow pyrolysis of biomass. *Journal of Analytical and Applied Pyrolysis*, *124*, 536–541. Available from https://doi.org/10.1016/j.jaap.2017.01.028.

Saha, A., Basak, B. B., Gajbhiye, N. A., Kalariya, K. A., & Manivel, P. (2019). Sustainable fertilization through co-application of biochar and chemical fertilizers improves yield, quality of *Andrographis paniculata* and soil health. *Industrial Crops and Products*, *140*, 111607. Available from https://doi.org/10.1016/j.indcrop.2019.111607.

Salama, D. M., Abd El-Aziz, M. E., El-Naggar, M. E., Shaaban, E. A., & Abd El-Wahed, M. S. (2021). Synthesis of an eco-friendly nanocomposite fertilizer for common bean based on carbon nanoparticles from agricultural waste biochar. *Pedosphere*, *31*(6), 923–933. Available from https://doi.org/10.1016/S1002-0160(21)60024-3, http://pedosphere.issas.ac.cn.

Samoraj, M., Mironiuk, M., Witek-Krowiak, A., Izydorczyk, G., Skrzypczak, D., Mikula, K., Baśladyńska, S., Moustakas, K., & Chojnacka, K. (2022). Biochar in environmental friendly fertilizers – prospects of development products and technologies. *Chemosphere*, *296*, 133975. Available from https://doi.org/10.1016/j.chemosphere.2022.133975.

Sarma, B., Farooq, M., Gogoi, N., Borkotoki, B., Kataki, R., & Garg, A. (2018). Soil organic carbon dynamics in wheat - Green gram crop rotation amended with vermicompost and biochar in combination with inorganic fertilizers: A comparative study. *Journal of Cleaner Production*, *201*, 471–480.

Schreiter, I. J., Schmidt, W., & Schüth, C. (2018). Sorption mechanisms of chlorinated hydrocarbons on biochar produced from different feedstocks: Conclusions from single- and bi-solute experiments. *Chemosphere*, *203*, 34–43. Available from https://doi.org/10.1016/j.chemosphere.2018.03.173, http://www.elsevier.com/locate/chemosphere.

Schulz, H., Dunst, G., & Glaser, B. (2013). Positive effects of composted biochar on plant growth and soil fertility. *Agronomy for Sustainable Development*, *33*(4), 817–827. Available from https://doi.org/10.1007/s13593-013-0150-0.

Seehausen, M. L., Gale, N. V., Dranga, S., Hudson, V., Liu, N., Michener, J., Thurston, E., Williams, C., Smith, S. M., & Thomas, S. C. (2017). Is there a positive synergistic effect of biochar and compost soil amendments on plant growth and physiological performance? *Agronomy*, *7*(1). Available from https://doi.org/10.3390/agronomy7010013, http://www.mdpi.com/2073-4395/7/1/13/pdf.

Shaikh, W. A., Kumar, A., Chakraborty, S., Islam, R. U., Bhattacharya, T., & Biswas, J. K. (2022). Biochar-based nanocomposite from waste tea leaf for toxic dye removal: From facile fabrication to functional fitness. *Chemosphere*, *291*, 132788.

Sharma, L., & Bali, S. (2018). A review of methods to improve nitrogen use efficiency in agriculture. *Sustainability*, *10*(1), 51. Available from https://doi.org/10.3390/su10010051.

Shin, J. D., Park, S. W., & Lee, S. I. (2019). Optimum method uploaded nutrient solution for blended biochar pellet with application of nutrient releasing model as slow release fertilizer. *Applied Sciences*, *9*(9), 1899. Available from https://doi.org/10.3390/app9091899.

Shukla, A., Kumar, D., Girdhar, M., Kumar, A., Goyal, A., Malik, T., & Mohan, A. (2023). Strategies of pretreatment of feedstocks for optimized bioethanol production: distinct and integrated approaches. *Biotechnology for Biofuels and Bioproducts*, *16*(1). Available from https://doi.org/10.1186/s13068-023-02295-2, https://biotechnologyforbiofuels.biomedcentral.com/.

Smith, Z., Isaac, B., Tumuluru, J. S., & Yancey, N. (2024). Grinding and pelleting characteristics of municipal solid waste fractions. *Energies*, *17*(1). Available from https://doi.org/10.3390/en17010029, http://www.mdpi.com/journal/energies/.

Taherymoosavi, S., Joseph, S., & Munroe, P. (2016). Characterization of organic compounds in a mixed feedstock biochar generated from Australian agricultural residues. *Journal of Analytical and Applied Pyrolysis*, *120*, 441–449. Available from https://doi.org/10.1016/j.jaap.2016.06.017.

Tarar, O. F., Asghar, A., Qayyum, S. A., Kanwal, H., Lateef, A., Nazir, R., Imam Abidi, S. H., Naeem, M. K., & Shahid, B. (2023). Synthesis and surface morphology of banana biochar-based nano-fertilizer and its effect on first stages of growth parameters of cucumber, broccoli, and red okra. *Journal of the Saudi Society of Agricultural Sciences*, *22*(8), 535–545. Available from https://doi.org/10.1016/j.jssas.2023.06.002, https://www.sciencedirect.com/science/journal/1658077X.

Tran, D.-T., Pham, T.-D., Dang, V.-C., Pham, T.-D., Nguyen, M.-V., Dang, N.-M., Ha, M.-N., Nguyen, V.-N., & Nghiem, L. D. (2022). A facile technique to prepare MgO-biochar nanocomposites for cationic and anionic nutrient removal. *Journal of Water Process Engineering*, *47*, 102702.

Verma, L., & Singh, J. (2023). Production of biochar-based nanocomposites from chemical and biological methods. *Advances in Science, Technology and Innovation*, *741*, 47–56.

Available from https://doi.org/10.1007/978-3-031-28873-9_4, https://www.springer.com/series/15883.

Villada, E., Velasquez, M., Gómez, A. M., Correa, J. D., Saldarriaga, J. F., López, J. E., & Tamayo, A. (2024). Combining anaerobic digestion slurry and different biochars to develop a biochar-based slow-release NPK fertilizer. *Science of the Total Environment*, *927*. Available from https://doi.org/10.1016/j.scitotenv.2024.171982, https://www.sciencedirect.com/science/journal/00489697.

Wali, F., Naveed, M., Bashir, M. A., Asif, M., Ahmad, Z., Alkahtani, J., Alwahibi, M. S., & Elshikh, M. S. (2020). Formulation of biochar-based phosphorus fertilizer and its impact on both soil properties and chickpea growth performance. *Sustainability*, *12* (22), 9528. Available from https://doi.org/10.3390/su12229528.

Wang, K., Brown, R. C., Homsy, S., Martinez, L., & Sidhu, S. S. (2013). Fast pyrolysis of microalgae remnants in a fluidized bed reactor for bio-oil and biochar production. *Bioresource Technology*, *127*, 494–499. Available from https://doi.org/10.1016/j.biortech.2012.08.016, http://www.elsevier.com/locate/biortech.

Wang, L., Ok, Y. S., Tsang, D. C. W., Alessi, D. S., Rinklebe, J., Wang, H., Mašek, O., Hou, R., & Hou, D. O. 'C. , D. (2020). New trends in biochar pyrolysis and modification strategies: feedstock, pyrolysis conditions, sustainability concerns and implications for soil amendment. *Soil Use and Management*, *36*(3), 358–386. Available from https://doi.org/10.1111/sum.12592, http://onlinelibrary.wiley.com/journal/10.1111/(ISSN)1475-2743.

Wen, P., Wu, Z., Han, Y., Cravotto, G., Wang, J., & Ye, B. C. (2017). Microwave-assisted synthesis of a novel biochar-based slow-release nitrogen fertilizer with enhanced water-retention capacity. *ACS Sustainable Chemistry and Engineering*, *5*(8), 7374–7382. Available from https://doi.org/10.1021/acssuschemeng.7b01721, http://pubs.acs.org/journal/ascecg.

Wu, Y., Brickler, C., Li, S., & Chen, G. (2021). Synthesis of microwave-mediated biochar-hydrogel composites for enhanced water absorbency and nitrogen release. *Polymer Testing*, *93*, 106996. Available from https://doi.org/10.1016/j.polymertesting.2020.106996.

Xin, J., Liu, Y., Chen, F., Duan, Y., Wei, G., Zheng, X., & Li, M. (2019). The missing nitrogen pieces: A critical review on the distribution, transformation, and budget of nitrogen in the vadose zone-groundwater system. *Water Research*, *165*, 114977. Available from https://doi.org/10.1016/j.watres.2019.114977.

Yaashikaa, P. R., Senthil Kumar, P., Varjani, S. J., & Saravanan, A. (2019). Advances in production and application of biochar from lignocellulosic feedstocks for remediation of environmental pollutants. *Bioresource Technology*, *292*, 122030. Available from https://doi.org/10.1016/j.biortech.2019.122030.

Yan, T., Xue, J., Zhou, Z., & Wu, Y. (2021a). Biochar-based fertilizer amendments improve the soil microbial community structure in a karst mountainous area. *Science of the Total Environment*, *794*, 148757. Available from https://doi.org/10.1016/j.scitotenv.2021.148757.

Yan, T., Xue, J., Zhou, Z., & Wu, Y. (2021b). Effects of biochar-based fertilizer on soil bacterial network structure in a karst mountainous area. *Catena*, *206*, 105535. Available from https://doi.org/10.1016/j.catena.2021.105535.

Yang, B., Zhang, M., Wu, M., Zhang, H., Song, Q., & Yu, S. (2019). Synthesis of biochar-based Cu_2O nanoparticles and their antibacterial activity against *Escherichia coli*.

Inorganic and Nano-Metal Chemistry, *49*(1), 12–16. Available from https://doi.org/10.1080/24701556.2019.1571512, http://www.tandfonline.com/loi/lsrt21.

Yang, W., Li, C., Wang, S., Zhou, B., Mao, Y., Rensing, C., & Xing, S. (2021). Influence of biochar and biochar-based fertilizer on yield, quality of tea and microbial community in an acid tea orchard soil. *Applied Soil Ecology*, *166*, 104005. Available from https://doi.org/10.1016/j.apsoil.2021.104005.

Yang, X., Ng, W., Wong, B. S. E., Baeg, G. H., Wang, C. H., & Ok, Y. S. (2019). Characterization and ecotoxicological investigation of biochar produced via slow pyrolysis: Effect of feedstock composition and pyrolysis conditions. *Journal of Hazardous Materials*, *365*, 178–185. Available from https://doi.org/10.1016/j.jhazmat.2018.10.047, http://www.elsevier.com/locate/jhazmat.

Zhang, H. W., Xing, L. B., Liang, H. X., Liu, S. Z., Ding, W., Zhang, J. G., & Xu, C. Y. (2023). Preparation and characterization of biochar-based slow-release nitrogen fertilizer and its effect on maize growth. *Industrial Crops and Products*, *203*. Available from https://doi.org/10.1016/j.indcrop.2023.117227, http://www.elsevier.com/inca/publications/store/5/2/2/8/2/5.

Zhang, M., Muhammad, R., Zhang, L., Xia, H., Cong, M., & Jiang, C. (2019). Investigating the effect of biochar and fertilizer on the composition and function of bacteria in red soil. *Applied Soil Ecology*, *139*, 107–116. Available from https://doi.org/10.1016/j.apsoil.2019.03.021, http://www.elsevier.com/inca/publications/store/5/2/4/5/1/8/index.htt.

Zhang, M., Liu, Y., Wei, Q., Liu, L., Gu, X., & Gou, J. (2022). Biochar-based fertilizer enhances the production capacity and economic benefit of open-field eggplant in the Karst Region of Southwest China. *Agriculture*, *12*(9), 1388. Available from https://doi.org/10.3390/agriculture12091388.

Zhang, Z., Zhu, Z., Shen, B., & Liu, L. (2019). Insights into biochar and hydrochar production and applications: A review. *Energy*, *171*, 581–598. Available from https://doi.org/10.1016/j.energy.2019.01.035, http://www.elsevier.com/inca/publications/store/4/8/3/.

Zhao, Z. Z., Wang, X. L., Li, H. B., Ren, S. P., Chen, J., & Wang, L. L. (2021). Slow-release property and soil remediation mechanism of biochar-based fertilizers. *Journal of Plant Nutrition and Fertilizers*, *27*(5), 886–897. Available from https://doi.org/10.11674/zwyf.20472, http://www.plantnutrifert.org/en/article/doi/10.11674/zwyf.20472.

Zheng, J., Han, J., Liu, Z., Xia, W., Zhang, X., Li, L., Liu, X., Bian, R., Cheng, K., Zheng, J., & Pan, G. (2017). Biochar compound fertilizer increases nitrogen productivity and economic benefits but decreases carbon emission of maize production. *Agriculture, Ecosystems and Environment*, *241*, 70–78. Available from https://doi.org/10.1016/j.agee.2017.02.034, http://www.elsevier.com/inca/publications/store/5/0/3/2/9/8.

Zhou, Z., Gao, T., Zhu, Q., Yan, T., Li, D., Xue, J., & Wu, Y. (2019). Increases in bacterial community network complexity induced by biochar-based fertilizer amendments to karst calcareous soil. *Geoderma*, *337*, 691–700.

Zhou, J., Qu, T., Li, Y., Van Zwieten, L., Wang, H., Chen, J., Song, X., Lin, Z., Zhang, X., Luo, Y., Cai, Y., & Zhong, Z. (2021). Biochar-based fertilizer decreased while chemical fertilizer increased soil N_2O emissions in a subtropical Moso bamboo plantation. *Catena*, *202*, 105257. Available from https://doi.org/10.1016/j.catena.2021.105257.

Zhou, M., Ying, S., Chen, J., Jiang, P., & Teng, Y. (2021). Effects of biochar-based fertilizer on nitrogen use efficiency and nitrogen losses via leaching and ammonia volatilization from an open vegetable field. *Environmental Science and Pollution Research*, *28* (46), 65188–65199. Available from https://doi.org/10.1007/s11356-021-15210-9, https://link.springer.com/journal/11356.

Zhu, Y., Li, H., Wu, Y., Yin, X. A., & Zhang, G. (2019). Effects of surface-modified biochars and activated carbon on the transformation of soil inorganic nitrogen and growth of maize under chromium stress. *Chemosphere*, *227*, 124–132. Available from https://doi.org/10.1016/j.chemosphere.2019.04.042, http://www.elsevier.com/locate/chemosphere.

Index

Note: Page numbers followed by "*f*" and "*t*" refer to figures and tables, respectively.

A

C

D

E

F

M

N

O

P

Q

R

S

9780443298554